From Catastrophe to Recovery

Stories of Fishery Management Success

From Catastrophe to Recovery

Stories of Fishery Management Success

Edited by

Charles C. Krueger, William W. Taylor, and So-Jung Youn
Center for Systems Integration and Sustainability
Department of Fisheries and Wildlife
Michigan State University
East Lansing, Michigan 48823, USA

American Fisheries Society
Bethesda, Maryland, USA
2019

Suggested citation formats follow.

Entire book

Krueger, C. C., W. W. Taylor, and S. Youn, editors. 2019. From catastrophe to recovery: stories of fishery management success. American Fisheries Society, Bethesda, Maryland.

Chapter in book

Hansen, M. J., and C. R. Bronte. 2019. Restoration of Lake Trout in Lake Superior through Interagency Cooperative. Pages 385–409 *in* C. C. Krueger, W. W. Taylor, and S. Youn, editors. From catastrophe to recovery: stories of fishery management success. American Fisheries Society, Bethesda, Maryland.

Cover images: top-right photograph by Donald J. Stewart, left photograph by Captain John McMurray, and bottom-right photograph by Eileen Randall.

Printed in the United States of America on acid-free paper.

Library of Congress Control Number 2019943041
ISBN 978-1-934874-55-4

American Fisheries Society Web site: www.fisheries.org

American Fisheries Society
425 Barlow Place, Suite 110
Bethesda, Maryland 20814
USA

Contents

Section V: Epilogue

Preface

The primary purpose of this book is to identify, describe, and highlight the successes of fishery professionals in restoration of fish populations that had moved from vibrant fisheries to the brink of extinction. The decline of populations was due to a variety of reasons, principally linked to overfishing and habitat destruction caused by anthropogenic changes to land and waterscapes.

We have spent much of our lives thinking about fish and fisheries, trying to better understand the variables that influence their production dynamics and yields. Our contemplation was done not only to serve the profession, but also was generated by our love of these animals, recognizing their value to ecosystem dynamics, and their influence on food security, recreation, livelihoods, and societal well-being. Given our fascination with these magnificent animals, we chose to dedicate our professional lives toward working to provide ways and means for ensuring their survival. In the process, we became more and more convinced that a fish's success or failure was directly or indirectly related to the activities associated with a growing human population. From our childhood days onward, we witnessed species after species move from being healthy and productive to a state of collapse of populations, and sometimes even to the brink of extinction.

The continued demise of our fisheries and the ramification that their loss meant to healthy ecosystems and to human communities led us to choose a career in fisheries. Our goal was to stem the flow of fisheries and fish populations headed toward various states of demise. As youth, we were often witnesses of these types of changes in the world around us, especially for the fish and water we loved so greatly, and it saddened us as to why we could not do better.

Being historians of fisheries demises was important, as it provided for us and others lessons to learn, such as warning signs of populations in trouble and actions to avoid. However, case studies of successes in fishery management seemed far and few between. Many of known successes seemed to hinge solely on the reduction of exploitation in combination with hatchery-based stock supplementation. Unfortunately, these methods worked only on a few stocks and only at a few places. As we became fisheries professionals working in fishery research, management, or administration, we vowed we would make a difference. We would not send the fish we valued so much into a trajectory that resulted in their collapse and ultimately extinction. If not us, who will be the safe keepers and voice for fish to survive so that they can continue to be a key component of our aquatic ecosystems and of human food security around the world?

As time passed, populations continued to fail and we asked why the battle was being lost. We thought we had provided what was good, science-based information on the structure and dynamics of fishery ecosystems and their importance. Sometimes we were not heard because fish and fisheries were seen as less valuable or easily substituted for other uses that society needed (e.g., water for agriculture, power generation). Such trade-offs impinged on the quality and quantity of fisheries production. We thus began to wonder if we caused any successes as fisheries professionals in stemming the loss of fish productivity and maintaining the immense and ecologically important diversity of this animal group that we knew was crucial to human survival in the long run. We were convinced that our efforts and those before us must have made a difference and there must be some good-news stories out there that could be learned from and must be told. We were tired of the gloom and doom heard at many of our professional meetings—there just had to be success stories waiting to be told and from which lessons could be learned that would enable future generations of fishery professionals to succeed.

This book was designed to document the successes of good-news fisheries, to demonstrate the successes we have had and highlight challenges that future generations will face. Bad news travels quickly while good news rarely is heralded. While the case studies highlighted in this book tend to showcase extremes of fish population collapse and recovery, in reality fish populations and fisheries exist on a continuum that ranges from being extinct to being abundant. In this book, our hope is to show that success and recovery can occur even in cases where there seems to be little hope for recovery. Many of the strategies and advice given in these case studies can be implemented as soon as declines in fish populations are noted. After all, keeping fisheries from ever reaching the brink of collapse is the best news of all! We hope the case studies, provided from a variety of ecosystems around the world, will facilitate the process needed to enhance our fishery profession's successes and fulfilment. Fishery resources are crucial to the viability and health of human communities and the aquatic ecosystems on which they depend. The book will hopefully encourage fishery professionals to stay the course, to conserve and enhance fish populations. This is a GOOD-NEWS FISHERIES book, one meant to inspire, teach, and provide case examples of successful fishery science in action!

William W. Taylor
So-Jung Youn
Charles C. Krueger

Acknowledgments

The editors greatly appreciate the insights and recommendations provided by the American Fisheries Society past presidents and current leadership with regards to appropriate fisheries to evaluate and to the knowledgeable authors who could write these chapters. Insightful discussions occurred during the early stages of this project regarding the concept of this book with Michael Hansen, Keith Jefferts, Danial Josephson, Jim Martin, Stan Moberly, Andrew Muir, Donald Pereira, John Robertson, George Spangler, and Paul (Jack) Wingate. This volume would not have been possible without the administrative and financial support of the Center for Systems Integration and Sustainability, the Department of Fisheries and Wildlife within Michigan State University, and the Great Lakes Fishery Commission. Jennifer Carducci cheerfully aided with managing a variety of details. Aaron Lerner, director of publications for the American Fisheries Society, provided excellent, helpful advice that ensured the highest quality product would be produced by this multiauthored book. Careful editing was provided by Deborah Lehman, book production editor, American Fisheries Society, and much improved all the chapters. A plethora of peer reviewers volunteered their precious time to carefully review the chapters and are gratefully acknowledged.

Peer Reviewers of Chapters

Barry P. Baldigo
U.S. Geological Survey

Casey Baldwin
Colville Tribes

David A. Beauchamp
U.S. Geological Survey

Barry Berejikian
NOAA Fisheries, Northwest Fisheries Science Center

Frank Borsuk
U.S. Environmental Protection Agency, Region III

Mike Bradford
Fisheries and Oceans Canada

Charles R. Bronte
U.S. Fish and Wildlife Service

David "Bo" Bunnell
U.S. Geological Survey

Stephen M. Coghlan, Jr.
University of Maine

Mark Dimech
Food and Agriculture Organization of the United Nations

Jaakko Erkinaro
Natural Resources Institute Finland

Karim Erzini
University of Algarve

Randy Eshenroder
Great Lakes Fishery Commission

David G. Fielder
Michigan Department of Natural Resources

Thomas A. Friesen
Oregon Department of Fish and Wildlife

Michael Fris
U.S. Fish and Wildlife Service

Simon Funge-Smith
Food and Agriculture Organization of the United Nations, Fisheries and Aquaculture Department

Rosemary Furfey
NOAA Fisheries, West Coast Region

Michael Goulding
Amazon Conservation Association; and Wildlife Conservation Society

Brian D. S. Graeb
South Dakota State University

Jim Habera
Tennessee Wildlife Resource Agency

Fred G. Henson
New York State Department of Environmental Conservation

Robert M. Hughes
Amnis Opes Institute; and Oregon State University

David W. Hulse
University of Oregon

John Janssen
University of Wisconsin–Milwaukee

Nicholas Johnson
U.S. Geological Survey

Daniel Josephson
Cornell University

Arthur (Chuck) Knutson, Jr.
California Department of Fish and Wildlife (retired)

Matthew Kulp
Great Smoky Mountains National Park

Ted Lawrence
Great Lakes Fishery Commission

David Lentz
California Department of Fish and Wildlife

Christopher N. Lorentz
Thomas More University

Charles Madenjian
U.S. Geological Survey

Nadir Mamilov
Al-Farbi Kazakh National University

David G. McGrath
Earth Innovation Institute

Mike Mulvey
Oregon Department of Environmental Quality

John Navarro
Ohio Department of Natural Resources

Fiona Nunan
University of Birmingham

Julian Olden
University of Washington

Donald Pereira
Minnesota Department of Natural Resources (retired)

Edwin (Phil) Pister
California Department of Fish and Wildlife (retired)

Thomas P. Quinn
University of Washington

Edward F. Roseman
U.S. Geological Survey

Lars Rudstam
Cornell University

Alexei Sharov
Maryland Department of Natural Resources

Gary Shepherd
NOAA Fisheries, Northeast Fisheries Science Center

Dan Schill
Idaho Department of Fish and Game

Michael R. Schwemm
U.S. Fish and Wildlife Service

Shawn Sitar
Michigan Department of Natural Resources

Christopher Vandergoot
U.S. Geological Survey

Vassiliki Celia Vassilopoulou
Hellenic Centre for Marine Research

Mark Vinson
U.S. Geological Survey

Martinus VanDerKnaap
Food and Agriculture Oranization of the United Nations, Regional Office for Africa

Paul (Jack) Wingate
Minnesota Department of Natural Resources (retired)

Troy Zorn
Michigan Department of Natural Resources

Symbols and Abbreviations

The following symbols and abbreviations may be found in this book without definition. Also undefined are standard mathematical and statistical symbols given in most dictionaries.

A	ampere
AC	alternating current
Bq	becquerel
C	coulomb
°C	degrees Celsius
cal	calorie
cd	candela
cm	centimeter
Co.	Company
Corp.	Corporation
cov	covariance
DC	direct current; District of Columbia
D	dextro (as a prefix)
d	day
d	dextrorotatory
df	degrees of freedom
dL	deciliter
E	east
E	expected value
e	base of natural logarithm (2.71828…)
e.g.	(exempli gratia) for example
eq	equivalent
et al.	(et alii) and others
etc.	et cetera
eV	electron volt
F	filial generation; Farad
°F	degrees Fahrenheit
fc	footcandle (0.0929 lx)
ft	foot (30.5 cm)
ft^3/s	cubic feet per second (0.0283 m^3/s)
g	gram
G	giga (10^9, as a prefix)
gal	gallon (3.79 L)
Gy	gray
h	hour
ha	hectare (2.47 acres)
hp	horsepower (746 W)
Hz	hertz
in	inch (2.54 cm)
Inc.	Incorporated
i.e.	(id est) that is
IU	international unit
J	joule
K	Kelvin (degrees above absolute zero)
k	kilo (10^3, as a prefix)
kg	kilogram
km	kilometer
l	levorotatory
L	levo (as a prefix)
L	liter (0.264 gal, 1.06 qt)
lb	pound (0.454 kg, 454g)
lm	lumen
log	logarithm
Ltd.	Limited
M	mega (10^6, as a prefix); molar (as a suffix or by itself)
m	meter (as a suffix or by itself); milli (10^{-3}, as a prefix)
mi	mile (1.61 km)
min	minute
mol	mole
N	normal (for chemistry); north (for geography); newton
N	sample size
NS	not significant
n	ploidy; nanno (10^{-9}, as a prefix)
o	ortho (as a chemical prefix)
oz	ounce (28.4 g)
P	probability
p	para (as a chemical prefix)
p	pico (10^{-12}, as a prefix)
Pa	pascal
pH	negative log of hydrogen ion activity
ppm	parts per million
qt	quart (0.946 L)
R	multiple correlation or regression coefficient
r	simple correlation or regression coefficient

rad radian
S siemens (for electrical conductance); south (for geography)
SD standard deviation
SE standard error
s second
T tesla
tris tris(hydroxymethyl)-aminomethane (a buffer)
UK United Kingdom
U.S. United States (adjective)
USA United States of America (noun)
V volt
V, Var variance (population)
var variance (sample)
W watt (for power); west (for geography)
Wb weber
yd yard (0.914 m, 91.4 cm)
α probability of type I error (false rejection of null hypothesis)
β probability of type II error (false acceptance of null hypothesis)
Ω ohm
μ micro (10^{-6}, as a prefix)
′ minute (angular)
″ second (angular)
° degree (temperature as a prefix, angular as a suffix)
% per cent (per hundred)
‰ per mille (per thousand)

Fishery Management Success: Action, Collaboration, Communication, and Commitment

Charles C. Krueger*, William W. Taylor, and So-Jung Youn
Center for Systems Integration and Sustainability
Department of Fisheries and Wildlife, Michigan State University
East Lansing, Michigan 48823, USA

Abstract.—Success achieving fishery management goals is possible but often requires concurrent strategies addressing ecology, politics, and public communication combined with some level of good fortune. As an introduction to this book, we identify several themes consistently highlighted among the fish management stories that follow, regardless of species, their life history, habitat needs, or type of waters they live in—streams, lakes, or ocean. In almost every case, success of management relied first and foremost on the abilities of professionals to restore the quality and quantity of a fish's habitat. The success of these efforts varied in magnitude but was accomplished by a combination of effective environmental regulation, substantial public and private investment, and direct habitat manipulation—whether in Lake Erie (Canada and USA), the Vindeln River in northern Sweden, an Adirondack Mountain lake of New York (USA), or Sea Lamprey *Petromyzon marinus* along the Atlantic coast (USA). Fish need acceptable water quality and habitat for living: simply stated and obvious—fish need water! When water and fish habitat are restored, fish populations can naturally recover through colonization from remnant populations, as was experienced in the Scioto River, Ohio. In some cases, populations were restored by stocking fish, using careful genetic considerations, such as told for Snake River Sockeye Salmon *Oncorhynchus nerka*. Public engagement was a common theme among case studies presented in this text. Public support for management yielded the political will to provide funding, regulation, and enforcement. Public involvement was a critical component of stories told about Great Smoky Mountains Brook Trout *Salvelinus fontinalis*, Pacific salmon in British Columbia and Idaho, and Tonle Sap fisheries of Cambodia. Consistently, management success came when goals were clearly articulated and combined with an effective consensus-built management plan that had the long-term commitment of personnel and support of their agencies. These attributes yielded programs where actions were taken and long-term monitoring and assessment were implemented to gauge success. Assessment information allowed programs to be adaptive over time to changes in the ecological system and society and thereby helped address new, as well as ongoing, challenges the fish and fishery were experiencing. The stories in this text provide incontrovertible evidence that good things can happen with the development and implementation of effective fish management programs, demonstrating the value of our profession and providing clear evidence that suc-

* Corresponding author: krueger62@msu.edu

cess is not an impossible allusion but rather an achievable event. These success stories of restored fish and fisheries throughout the world should be celebrated within fishery science.

INTRODUCTION

Fishery management is the process of decision making that implements actions to achieve goals and objectives for the conservation and restoration of fish and aquatic populations and to provide sustainable fishery harvests. All types of information, ecological, economic, political, and sociological, are used in decisions that result in needed actions (e.g., regulations, stocking, and enforcement) to achieve goals that society has set for aquatic resources (Krueger and Decker 1999). Essential to the success of this process is the assessment and monitoring of fish populations and their fisheries, which provide feedback information so that management plans and actions can be adaptive to changes in the ecological and social systems. Documented examples of successful fishery management programs exist, but they seem relatively rare and often are not heralded in the profession or news media, which could lead one to question the value of natural resources management and its practices, not to mention the competence of its professionals.

Management case histories of restoration and recovery of North American wildlife species are well documented and, as land-based animals, are often visible to the public. For example, successful species recovery from the brink of extinction has occurred for the Florida panther *Puma concolor coryi* (Kautz et al. 2006; Thatcher et al. 2009; Johnson et al. 2010) and Kirtland's warbler *Setophaga kirtlandii* (Corace et al. 2010; Probst et al. 2003). A remarkable recovery of wild turkey *Meleagris gallopavo* has occurred across eastern North America to levels of abundance, now permitting relatively high levels of sustainable harvest through hunting activities (Dickson 1992; Latham 2017).

In contrast to the wildlife management case histories listed above, collapses of fisheries while being managed by governments seem to be better known than the many fishery management success stories that have occurred throughout the world, unseen and unknown by many due to their habitats not being visible to most people. For instance, strong evidence exists that the collapse of Atlantic Cod *Gadus morhua* in the North Atlantic was due, in large measure, to unsustainable, and hence inadequately controlled, fishing mortality (e.g., Hutchings and Myers 1994; Hutchings 1996). Changes in habitat, specifically the interruption of migratory corridors by hydroelectric dams, characterize the collapse and struggle to recover Pacific salmon fisheries of the western United States and Canada (e.g., Lichatowich 1999) and elsewhere. Documented stories of successful fish management seem to be much less common in contrast to accounts of fishery collapses and their causes. The perception of failure can lead to the questioning of the value of the profession by society and, perhaps worse, deterring the best and brightest from joining in the grand cause of saving fish.

The present text stands in contrast to these examples of unfortunate fishery catastrophes and highlights the complexity of fishery systems, the breadth of the fishery profession, and the value that wise management provides to society at local, regional, national, and global levels. Fish need to go from being a hidden member of the world's biota to being recognized as a valuable member essential to human and ecosystem health and well-being!

Contained within this volume are 20 stories about fishery management in action that led to success in the restoration of fish and fisheries. The chapters detail the authors' journeys, including their trials and tribulations, toward management victory. Success was variably defined in the chapters and included a gradient from preventing extinction of a species to restoration of a population that now supports a sustainable fishery. In every case, populations in question were imperiled by a variety of threats, often working together, including poor water quality, blockage of migratory corridors, stream channelization, invasive nonnative species, and excessive unsustainable fishing mortality. Yet, the obstacles to recovery were overcome by well-developed, scientifically based fishery management actions. These stories inspired us to delve into the question as to whether common patterns or themes exist in actions undertaken by managers that were crucial to success. Identifying these themes prior to reading the text may help the reader connect the case studies presented and consider the common approaches used by authors that led to success.

As an introductory chapter to the volume, we identify themes common among the chapters. We conclude with a few observations inferred from our interactions with the authors and their chapters and identify common patterns to the fish management success stories.

COMMON THEMES OF MANAGEMENT SUCCESS STORIES

Physical Habitat Restoration

Sometimes fish populations are in trouble because of human-caused changes to the physical structure of their aquatic habitats. In rivers, these alterations often have taken the form of dams (discussed in the section that follows), irrigation diversions, bank stabilization (rip rap), removal of woody debris, and channelization to serve purposes of hydroelectric generation, navigation, and flood control. These changes alter the natural seasonal flow dynamics and temperature cycles of rivers. As a result, the natural meander of streams can be lost, including side channels, backwaters, and the regular distribution of pools and riffles. These habitat features can serve as important spawning grounds and nursery areas, and for feeding and growth. In addition, the riparian vegetation in the stream corridors can be altered through activities such as timber cutting and livestock grazing.

Several chapters recount actions to reverse effects of anthropogenically caused changes to streams as part of efforts to bring about population recovery. Channelization to combat effects of floods or to enhance floating logs (timber industry) downstream was a problem that affected streams in Oregon, USA (Hughes et al. 2019, this volume); British Columbia, Canada (Hyatt and Stockwell 2019, this volume); and Sweden (Hellström et al. 2019, this volume). For example, over a 150-year period, channelization of the Vindel River (Sweden) to improve log floating combined with the removal of massive amounts of boulders caused losses of side-channel habitats and spawning grounds for Atlantic Salmon *Salmo salar* and Brown Trout *S. trutta* (Hellström et al. 2019). With careful planning, restoration of stream habitats involved placing boulders and woody debris back into the river, opening closed-off side channels, developing spawning areas by depositing gravel and cobble, and removing splash dams blocking access to productive spawning areas in tributaries. As a result, thousands of hectares of habitat were restored. By 2014, more than 40 yearlings/100 m^2 were reported from the annual monitoring program, a 40,000% increase since the early 1990s in juvenile abundance, making the Vindel River the third most important river for wild Atlantic Salmon production in the Baltic Sea.

The riparian vegetation in the watershed surrounding stream channels can affect fish habitat, such as through bank stability, interception and runoff of precipitation, shading of water, and quantity and quality of allochthonous energy inputs. For example, within the Golden Trout Wilderness Area in California (USA), grazing by livestock over more than 145 years in mountain meadows along the Little Kern River, South Fork Kern River, and Golden Trout Creek altered species composition and abundance of the vegetation and caused a variety of changes to the streams, including loss of pool habitat, sedimentation, reduced instream cover, loss of undercut streambanks, and widening of stream channels (Stephens et al. 2019, this volume). Now, fencing to restrict access of livestock to the streams and carefully controlled grazing allotments have reduced these effects, but vegetation recovery within the stream corridor may take decades to recover.

Passage Past Dams, and Dams to Block Fish Passage

Dams constructed as flood control and water retention structures for the irrigation of agricultural fields and to generate hydroelectric power served as barriers to the migration of juveniles and adults and often were obstacles preventing fish population restoration. The problems that dams cause for fish movements are well known, but solutions have proven technologically difficult and expensive (e.g., Coutant and Whitney 2006).

Fish passage both for adults migrating upstream from the ocean and for juveniles migrating downstream was an important problem preventing population recovery of Sockeye Salmon *Oncorhynchus nerka* in Redfish Lake (Idaho, USA; Kline et al. 2019, this volume) and the Okanagan River basin (British Columbia, Canada; Hyatt and Stockwell 2019, this volume), Atlantic Salmon in the Vindel River (Sweden; Hellström et al. 2019), and Sea Lamprey *Petromyzon marinus* (Gephard 2019, this volume). Opening migratory corridors was essential for anadromous fish recovery. Dams not only create fish barriers, but also change the natural patterns of daily and seasonal discharge that can affect recruitment dynamics of fish spawning downstream. Improved reproduction and survival of larval Striped Bass *Morone saxatilis* occurred when the daily hydropower-influenced flow fluctuations in the Roanoke River were regulated to resemble more natural flow conditions (Essig et al. 2019, this volume)

Dams within the Connecticut River watershed blocked Atlantic Salmon and American Shad *Alosa sapidissima* from their spawning grounds and juvenile rearing areas. Fish passage facilities at these dams were constructed specifically to restore these two species. However, these dams had also blocked native Sea Lamprey (Gephard 2019) not considered in the original restoration efforts. Unintentionally, the new passage facilities also benefited Sea Lamprey by allowing them to move past these barriers to upstream spawning and rearing grounds. Restoration of anadromous Sea Lamprey to the Connecticut River has been an ongoing success story as defined by increasing numbers of fish to the watershed and expansion of the restored geographic range of the species in the watershed. Nest surveys confirm that once adult Sea Lamprey pass through fishways, they move upstream and distribute throughout the watershed, find suitable spawning habitat, build nests, and successfully spawn. The larvae that result easily pass downstream.

In contrast, dams can benefit fish restoration, instead of posing a problem. Dams can block movement of nonnative fishes from downstream areas and thereby prevent their colonization upstream of the dam. For example, in the Laurentian Great Lakes, nonnative Sea Lamprey prey on native fishes such as Lake Trout *Salvelinus namaycush*. Dams on tributary streams block their access upstream to spawning grounds and help reduce their reproductive potential (Hansen and Bronte 2019, this volume). Dams can also prevent nonnative fishes from colonizing and becoming resident in upstream areas. If the invasive species successfully colonize, they can genetically and ecologically interact with vulnerable resident nonmigratory native species. As an example, barriers to fish movement were constructed on the South Fork Kern River to prevent the upstream movement of fish and to protect California Golden Trout *Oncorhynchus mykiss aguabonita* from hybridization with downstream stocked nonnative Rainbow Trout *O. mykiss* (Stephens et al. 2019). These structures did not impound water, but created a formidable set of barriers to block fish passage in downstream areas from gaining access to upstream areas. Barrier construction in combination with removals of nonnative species upstream of these barriers on the upper South Fork Kern River helped prevent local extirpation and extinction of California Golden Trout in this river. Thus, dams or barriers to fish passage can be both good and bad for fishes, a finding that we found to also be true for stocking and transplantation of fish, the next topic to be discussed.

Stocking and Transplantation

Fish for stocking can come from either hatchery propagation or from the capture and movement of wild fish from other waters (transplantation). Hatchery propagation typically involves securing gametes, either from captive broodstocks or from wild sources, and the subsequent incubation and rearing of juvenile fish to a suitable size for stocking. Hatchery propagation and stocking has a long history of use by fish managers (Benson 1970; Stroud 1986; Nickum et al. 2004) and was, in part, the motivation for the formation in late 1870 of the American Fisheries Society, initially called the American Fish Culturalists' Association. Over the past century, stocking often was chosen as a management action for promoting recovery of populations through reintroduction or by adding to remnant populations (supplementation).

Stocking fish was identified in these chapters as an important management action that assisted with the recovery of fish populations in rivers, lakes, and oceans. For example, Hyatt and Stockwell (2019) note that hatchery-origin Sockeye Salmon fry introduced into Skaha Lake in the Okanagan River watershed (British Columbia, Canada) contributed to the naturalization and range expansion of the species. In Saginaw Bay, consistent stocking in Lake Huron (USA) of Walleye *Sander vitreus* maintained the presence of the species and helped ensure that feral broodfish were available for spawning when environmental conditions for natural recruitment improved. Water quality did improve and nonnative species declined, which allowed the hatchery-origin Walleye to reproduce to such an extent that stocking was discontinued in 2006 (Fielder and Baker 2019, this volume). Similarly, Walleye fry stocking was also effective in the Red Lakes, Minnesota (USA; Barnard et al. 2019, this volume). In Lake Superior (Canada and USA), early and intensive lakewide stocking of hatchery-reared Lake Trout was used to supplement failing stocks and, when combined with Sea Lamprey control and restrictive fishing regulations, permitted recovery of populations lakewide (Hansen and Bronte 2019). Likewise, in the Caspian Sea, stocking in combination with other management actions such as regulation of fishing in Iran, Azerbaijan, and Dagestan (Russia) allowed the recovery of Caspian Kutum *Rutilus kutum* stocks to 1990s levels, which allowed for resumption of fishing (Rabazanov et al. 2019, this volume).

Use of stocking in recent years has been controversial among fish managers, primarily over concerns regarding potential genetic alterations caused by propagation and rearing of fish in the artificial environments of hatcheries and by wild, interbreeding populations with genetically different hatchery-reared fish (e.g., Fraser 2008; Araki and Schmid 2010; Christie et al. 2012; Le Luyer et al. 2017). As a result, hatchery practices have evolved over time, with many propagation and stocking programs now using ancestry guidance in breeding procedures and propagation protocols to reduce these concerns. Kline et al. (2019) recount the use of stocking Sockeye Salmon to successfully recover the population in Redfish Lake, Idaho (USA) by using carefully designed captive broodstock gene-rescue technologies.

The success stories also note times when stocking was found to be ineffective or even detrimental to the recovering populations. In the Vindel River, stocking was ineffective in increasing returning anadromous Atlantic Salmon (Hellström et al. 2019). Stocking of nonnative genetic sources of Rainbow Trout posed a threat through genetic introgression to recovering native populations of Rainbow Trout in the upper Middle Fork Willamette River (Oregon, USA) (Hughes et al. 2019) and to recovering California Golden Trout in the South Fork Kern River (California, USA; Stephens et al. 2019). Last, large gamete collections for propagation made from a wild Brook Trout *Salvelinus fontinalis* population recovering in Honnedaga Lake (New York, USA) were found to suppress wild natural reproduction and, as a result, were suspended after 2 years (Kraft 2019, this volume). Successful restoration of this lake later occurred through natural reproduction of the remnant population.

Transplantation, involving the capture and movement of wild fish from other waters to waters to be restored, has the advantage of avoiding the hatchery propagation stage and the potential for domestication effects. Transplantation from wild sources was used for Sea Lamprey restoration in the Atlantic Ocean coastal streams of Connecticut (USA; Gephard 2019) and Brook Trout restoration in Lynn Camp Prong within the Great Smoky Mountain National Park (Tennessee, USA; Moore and Kulp 2019, this volume).

Recovery from Refuges

Population recolonization of restored habitat can occur not only via stocking and transplantation, as discussed above, but also by straying of fish from populations that occur in adjacent connected waters or from remnant populations that have survived in refuges. The Tippecanoe Darter *Etheostoma tippecanoe* in the main-stem Scioto River in Ohio (USA) and tributaries was virtually extirpated as a result of heavy pollution during the 19th and 20th centuries except for a few small, refuge populations within two tributaries, Big Darby and Deer creeks. With pollution abatement, recolonization occurred, most likely from fish from in these two tributaries, and by 2016, the species had demonstrated a near-complete return to its known historical range (Yoder et al. 2019, this volume). Similarly, in Honnedaga Lake (New York, USA), Brook Trout recolonized the lake from fish that had persisted for years in groundwater-sourced tributary refuges while the lake had suffered toxic conditions from atmospheric acidic deposition. Once suitable lake water chemistry conditions returned, fish produced from these groundwater areas provided recruitment to the lake (Kraft 2019).

Aquatic Protected Areas

Aquatic protected areas (marine protected areas) are zones where fishing may not occur during certain seasons or year-round, and regulations to establish such areas have been recently implemented in both freshwater (e.g., Suski and Cooke 2007; Hedges et al. 2010) and marine waters (e.g., Johnson and Sandell 2014). Aquatic protected areas have played an important role in the recovery of depressed populations. As a freshwater example, community fish refuges were established in Tonle Sap Lake by the Cambodian government beginning in 2009 (Fiorella et al. 2019, this volume). These areas are managed by local communities and are protected from fishing during the November–April dry season and designed to increase fish production and protect biodiversity. These refuges provided fish and other aquatic animals with protection so that during the subsequent wet season, they could migrate from refuges to surrounding waterways for growth and reproduction, thereby providing sustainable fisheries. In the Mediterranean Sea at Gökova Bay (Turkey), six no-fishing zones (similar to marine protected areas) were established in July 2010 to support small-scale fisheries. When combined with improved enforcement by marine rangers with fast boats, and working in close cooperation with Turkey's Coast Guard, these actions and many others (e.g., support for the marketing of invasive species) led to the conservation of fish stocks in the bay, as well as an increase in fishers' incomes (Unal and Kizilkaya 2019, this volume). Importantly, without enforcement of fishery regulations, no-fishing zones did not work.

Fishery Regulations

Control of fishing mortality while populations recover is a logical, obvious requirement for programs intending to develop naturally reproducing self-sustaining fish populations and fisheries. Struggling populations must be able to survive to maturity and subsequently spawn if they are to become more abundant. Establishment of regulations was especially difficult for the protection of arapaima (genus *Arapaima*) fisheries in Brazil and Guyana and, first, required residents to understand the need for regulations and, then, required their involvement in enforcement (Gurdak et al. 2019, this volume). The same issue was true for management of Gökova Bay in the Mediterranean Sea (Turkey; Unal and Kizilkaya 2019). In the Great Lakes, population recoveries suffered and may have been slowed because of mortality from ongoing fisheries for Lake Trout in Lake Superior and Walleye in Lake Huron and Lake Erie (Fielder and Baker 2019; Hansen and Bronte 2019; Vandergoot et al. 2019, this volume). In all three cases, fishing during the recovery period was allowed. However, regulations were in place with the intent to control fishing mortality at levels believed to not foreclose achievement of the goal of population recovery. Whereas complete closures of the fisheries likely would have speeded recovery, the advantage of allowing some fishing could be that local community interest in the species was maintained and, hence, societal value for the species, and thus their recovery programs continued. Some anadromous species, such as Sockeye Salmon, Atlantic Salmon, and Striped Bass, ranged widely, sometimes outside of jurisdictional authority of the agencies implementing recovery programs, creating substantial challenges to the effective control of fishing mortality (e.g., Essig et al. 2019; Hellström et al. 2019; Hyatt and Stockwell 2019; Kline et al. 2019).

Environmental Regulation

Improvements in water quality, air quality, and endangered species protection through government regulations were key elements contributing to successes described in many chapters. The importance of this aspect to success, especially water quality regulation, was impressive. Fish need high quality water and adequate quantities of it! What is different about these regulations in contrast to other management actions, such as stocking, is that water quality regulations were enacted, not specifically to benefit a particular species or fish community or a specific body of water, but instead for general purposes, such as having safe drinking water, by the federal government in the national interest, typically far away from actual sites and fish populations targeted for restoration. The importance of water and air regulations to the recovery of fish populations was so great that without them, chapters could not have been written about the Willamette River (Oregon, USA; Hughes et al. 2019), Ohio River (USA; Thomas et al. 2019, this volume), Scioto River (Ohio, USA; Yoder et al. 2019, this volume), Connecticut River and other streams (Connecticut, USA; Gephard 2019), Saginaw Bay Lake Huron (Fielder and Baker 2019), Lake Erie (Vandergoot et al. 2019, this volume), and Honnedaga Lake (New York, USA; Kraft 2019). The Clean Water Act (Copeland 2016), Clean Air Act (Waxman 1991), and Great Lakes Water Quality Agreement between Canada and the United States (e.g., International Joint Commission 1994; Hartig 1997) were essential requirements to these stories.

Federally legislated protections for endangered or threatened species were also cited as being important elements leading to the restoration of fish populations, as recounted in several chapters in this book. Similar to legislation for water and air quality, enabling legislation for endangered species regulations was not enacted specifically for the sites and species recounted in the chapters, but instead based on a perceived broad national interest for the perpetuation of a species. For example, the Little Kern Golden Trout of the Sierra Nevada (California, USA) was listed as threatened in 1978 (Stephens et al. 2019), and in 1991, the Snake River Sockeye Salmon *Oncorhynchus nerka* (Idaho, USA) of Redfish Lake (Kline et al. 2019) was listed as endangered under the U.S. Endangered Species Act (NMFS 1991). As a result, species at these locations were afforded special considerations in their management and recovery. The listing of Redfish Lake Sockeye Salmon significantly constrained harvest of all Sockeye Salmon in traditional fishing areas of the lower Columbia River and inadvertently but importantly assisted in the Sockeye Salmon recovery further upstream in the Okanagan River watershed (British Columbia, Canada; Hyatt and Stockwell 2019).

Importance of Monitoring Information

Assessment (monitoring) information conducted to generate time-series data was essential to most stories of restoration success. Without such information, responses of populations to management actions, to determine whether they were effective or not, would not have been known, nor whether success had been achieved. Assessment information provided essential feedback to managers, guiding them to continue an action or to revise and try a different tactic (e.g., Kraft 2019). Recovery of Walleye in the Red Lakes (Minnesota, USA) required assessment information to ensure management of spawning stock densities so that wild fry production occurred within the optimal range of a density dependent stock–recruitment relationship and to maximize the likelihood of producing strong year-classes to support harvest levels commensurate with the potential productivity of the system (Barnard et al. 2019). In the case of arapaima management in Brazil and Guyana (Gurdak et al. 2019), the methods for assessment did not exist and their development were an essential aspect contributing to population recovery. Based on knowledge of local fishers, a replicable approach for estimating arapaima populations was developed, taking advantage of the arapaima's physiology as an obligate air breather to count the number of arapaima coming up to breathe during a set interval of time. This technique for estimating arapaima populations proved effective. By involving local fishers in the data collection, residents of the region gained ownership and trust in data collected.

Interjurisdictional Collaboration

When fish cross political boundaries of management jurisdictions, if success is to be realized, authorities must work in concert for common goals of recovery and restoration. Without such collaboration, management actions implemented by different agencies can be at cross purposes and failure can result. Successful interjurisdictional collaboration was true for stories about fish species that were as wide-ranging as anadromous Pacific salmon (Kline et al. 2019; Hyatt and Stockwell 2019) and Striped Bass (Essig et al. 2019), as well as for fish that simply used lakes governed by multiple jurisdictions (e.g., Barnard et al. 2019). In the Laurentian Great Lakes, interjurisdictional collaboration is fostered and coordinated by the joint Canada–U.S. Great Lakes Fishery Commission. The commission has established management and technical committees on each of the five Great Lakes, with federal, provincial, state, and tribal authorities serving as members (Gaden and Krueger 2018). Often, scientists and other experts also serve on technical committees, providing advice to the management (lake) committees. The restoration of Lake Trout and Walleye benefited by these institutional arrangements that promoted coordinated management actions and shared assessment information (Fielder and Baker 2019; Hansen and Bronte 2019; Vandergoot et al. 2019).

Public Engagement: A Participatory Process in Governance

The public, and especially stakeholders benefiting from the resource, must understand the goals and actions of management so as to provide local and political support for restoration programs. Political support generated through public engagement can encourage governments and nongovernmental organizations to provide funds and the long-term commitment required for programs to be successful. Without local support, achievement of management goals can be thwarted, as described by Moore and Kulp (2019). In this case, native Brook Trout restoration was the goal and required that nonnative Rainbow Trout be removed from Lynn Camp Prong in the Great Smoky Mountain National Park. After Rainbow Trout were removed with chemical piscicides, a few local, unsupportive residents released Rainbow Trout back into the stream, re-establishing this nonnative species. In response, managers held several public hearings around the park to educate local residents about the Lynn Camp Prong restoration and to gain their support. The stream then was treated again to remove Rainbow Trout, and Brook Trout were reintroduced successfully. By 2015, Brook Trout abundance exceeded that of Rainbow Trout prior to the restoration project. Important stories of the role of public engagement also were described for Atlantic Salmon restoration in the Vindel River (Sweden; Hellström et al. 2019) and Sea Lamprey in Connecticut (USA; Gephard 2019).

Taking Action in the Face of Uncertainty

Although reaching scientific certainty and consensus before taking action may be ideal, resource management almost always occurs with some degree of uncertainty. Many success stories began with taking action in the face of many unknowns, such as not knowing population sizes of arapaima in Brazil and Guyana (Gurdak et al. 2019) and not knowing the engineering technology that would be required to meet water quality standards for the Scioto River in Ohio (USA; Yoder et al. 2019). Nevertheless, these challenges were met over time and programs went forward and were successful. Unknowns were sometimes met by searching widely for the information required. For example, processes at work to cause the collapse of the Brook Trout population in Honnedaga Lake, New York

(USA) were imperfectly known but were discovered in part by accessing Swedish information about atmospheric acidification (Kraft 2019). Similarly, in Sweden, the best engineering facilities for passing Atlantic Salmon around the Stornorrfors Dam on the Vindel River were sought from fish passage experts of the Columbia River, Washington (USA; Hellström et al. 2019). As these examples clearly demonstrate, global access to information can assist with understanding the obstacles that stand in the way of success and can help with choosing the correct management action to solve a problem.

Chance: Fortuitous Environmental Events

We must acknowledge that chance environmental circumstances favorable for fish growth and survival can work in combination with the effects of good management programs and together can yield remarkable restoration progress in a relatively short time period. As an example, chance events occurred in Saginaw Bay Lake Huron (USA) when a series of cascading food-web changes took place resulting in the sudden collapse of nonnative Alewife *Alosa pseudoharengus* (Fielder and Baker 2019). Alewives were believed to have been a major predator of juvenile Walleyes and a block to natural reproduction. Because Walleyes had been stocked previously and had survived and matured, these adults were able to immediately take advantage of the favorable environmental circumstances and spawning began to provide significant natural recruitment. Walleye natural reproduction surged beginning in 2003 after Alewives collapsed. Walleye stocking was discontinued in 2006 and recovery targets were first achieved in 2009. Similarly, restoration of a nearly extirpated species occurred for Deepwater Sculpin *Myoxecephalus thompsonii* in Lake Ontario. Population recovery in this case history was a consequence of no specific management actions directed at the species but instead as a result of stocking salmonid predators that controlled nonnative Alewives and Rainbow Smelt *Osmerus mordax*, in combination with a shift in the lake from mesotrophic to oligotrophic conditions as a result of the U.S. Clean Water Act and the Great Lakes Water Quality Agreement (Weidel et al. 2019, this volume).

CONCLUSION: UNSPOKEN THEMES

Role of Skeptics

Rarely mentioned in the chapters was the role of people, either as politicians, agency leaders, or the public, who questioned, sometime vehemently, the management goals or the validity of the science supporting the restoration programs. In our interactions with the authors and reading between the lines of chapter text, we sensed considerable emotion and frustration over the degree and magnitude of the skepticism that fisheries professionals had to endure. The authors and their colleagues clearly had a vision of the future and were able to act on that vision. They gained needed resources, overcame social-political and ecological obstacles, and executed actions over the long term to achieve restoration success, providing examples of genuine leadership. Moore and Kulp (2019) addressed the skepticism in the local community by holding public meetings to present the Brook Trout restoration program and its goals and to answer questions. Yoder et al. (2019) noted that water quality standards were able to be established for the Scioto River, even though the technology to treat the pollution was not yet available. Kraft (2019) detailed how some public and government officials did not believe that atmospheric acidic deposition had substantial effects on fish. In this case, the skepticism was substantial and became political; however, the ensuing arguments challenging science were possibly what drove the U.S. Congress to support the needed long-term research funding for this topic during the 1980s and 1990s. Similarly, the 60+ year Lake Trout restoration program of Canada and the United States in the Great Lakes costing millions of dollars has consistently had critics arguing its ecological feasibility and value to society (Kernen 1995; Muir et al. 2012; Hansen and Bronte 2019). As a result, this scrutiny likely encouraged managers to carefully consider the feasibility of Lake Trout restoration and the scientific basis for management actions taken. We suspect that most restoration successes faced critics, especially given the long-term nature needed for bringing a species back from the brink of extinction.

Are skeptics or critics good or bad? They certainly can drain the resources of many projects and sap the emotional energy of project leaders. However, questioning current findings of science is part of the process of science. Conclusions regarding ecological processes are constantly being refined, honed, and improved as new information and interpretation becomes available. Skepticism, from whatever source, provides motivation to continue the process of science, to continue to improve our understanding, and to strive for excellence.

Long-Term Commitment of Managers and their Agencies

Every success story involved managers and scientists, and sometimes the public and politicians, all committed to a common goal of restoration. No story contained within this volume recounts a story that was accomplished in 2 or 3 years (except Pister 2019, this volume). Successful fishery management programs often required decades of effort characterized by commitment, action, assessment, and adaptation. Lake Trout restoration in Lake Superior required three generations of fishery managers and scientists from the 1950s to 2000s and beyond and the intergenerational transfer of the restoration goal to new, upcoming professionals (Hansen and Bronte 2019).

A Common Process used for Success

Above, we have identified some common elements or themes that occurred in the chapters that led to fishery management success; however, we also note that virtually every chapter followed a similar process over time that led to success (Figure 1). Programs of fishery management that were successful had a clear vision of what the future could be. Achievement of that goal provided the endpoint of when success could be claimed (Figure 1). The goal might be saving a species such as the Owens Pupfish *Cyprinodon radiosus* (Pister 2019) or the establishment of Walleye and Striped Bass populations that can provide sustainable fisheries (e.g., Barnard et al. 2019; Essig et al. 2019). These goals were then combined with the long-term personal and emotional commitments of the management programs' leaders (Figure 1). We believe that this commitment transcended the normal level; it often became a passion that they lived with, day in and day out. As such, this personal and professional commitment was successfully transferred to fishery management agencies, eventually becoming a part of their organizational cultures, or as some professionals like to say, it became part of the organization's DNA. The commitment by agencies then became translated into long-term management programs, sometimes spanning decades, where problems or obstacles that stood in the way of success were identified and actions were taken to solve those problems (Figure 1). Monitoring and assessment occurred over time to determine whether actions implemented were successful in meeting challenges that stood in the way of future success. Monitoring and assessment allowed programs to be adaptive. When actions did not yield the desired effect, programs could then adjust to changes in ecology, economics,

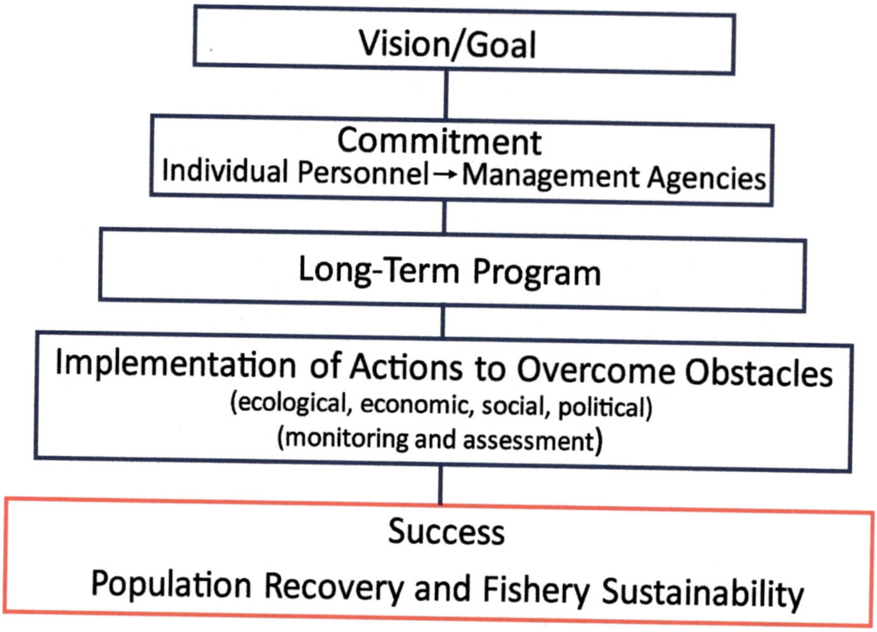

FIGURE 1. Pattern of management success described in most chapters that led to population restoration and fishery sustainability.

culture, and politics. Importantly, monitoring and assessment provided information to track the progress towards success, to communicate that progress, and to be able to declare success when achieved (Figure 1).

Can fishery management programs be successful? Yes! We intentionally chose stories such that readers would be faced, at their conclusion, with unequivocal success. Most often, these stories were of imperiled populations brought back from the brink, sometimes to abundance that could now sustain fisheries. Our desire is that these stories will give the next generation of fishery professionals the vision that success is possible and worth the struggle, that our profession can yield personal, and thatprofessional satisfaction with one's career and make a difference for the fish, their ecosystems, and all humankind!

ACKNOWLEDGMENTS

We thank chapter authors of this volume for their patience with our editorial nitpicking. Dr. Shawn Riley (Michigan State University) provided the details and helpful citations regarding successful wildlife management programs.

REFERENCES

Araki, H., and C. Schmid. 2010. Is hatchery stocking a help or harm? Evidence, limitations and future directions in ecological and genetic surveys. Aquaculture 308:S2–S11.

Barnard, G. C., A. J. Kennedy, W. P. Brown, and D. L. Pereira. 2019. Walleye management in the Red Lakes, Minnesota: collapse, recovery, and cooperative management. Pages 323–352 *in* C. C. Krueger, W. W. Taylor, and S. Youn, editors. From catastrophe to recovery: stories of fishery management success. American Fisheries Society, Bethesda, Maryland.

Benson, N. G. 1970. A century of fisheries in North America. American Fisheries Society, Special Publication 7, Bethesda, Maryland.

Christie, M. R., M. L. Marine, R. A. French, and M. S. Blouin. 2012. Genetic adaptation to captivity can occur in a single generation. Proceedings of the National Academy of Sciences of the United States of America 109:238–242.

Copeland, C. 2016. Clean Water Act: a summary of the law. Prepared for Members and Committees of Congress, Congressional Research Service, Report 7-5700. RL30030. Available: www.crs.gov and https://fas.org/sgp/crs/misc/RL30030.pdf (August 2018).

Corace, R. G., III, P. C. Goebel, and D. L. McCormick. 2010. Kirtland's warbler habitat management and multi-species bird conservation: considerations for planning and management across jack pine (*Pinus banksiana* Lamb.) habitat types. Natural Areas Journal 30:174–190.

Coutant, C. C., and R. R. Whitney. 2006. Hydroelectric system development: effects on juvenile and adult migration. Pages 249–324 *in* R. N. Williams, editor. Return to the river: restoring salmon to the Columbia River. Elsevier Academic Press, Burlington, Massachusetts.

Dickson, J. G., editor. 1992. The wild turkey: biology and management. Stackpole Books, Mechanicsburg, Pennsylvania.

Essig, R. J., R. W. Laney, M. H. Appelman, F. A. Harris, R. A. Rulifson, and K. L. Nelson. 2019. Atlantic Striped Bass: an interjurisdictional fisheries management success story. Pages 533–563 *in* C. C. Krueger, W. W. Taylor, and S. Youn, editors. From catastrophe to recovery: stories of fishery management success. American Fisheries Society, Bethesda, Maryland.

Fielder, D. G., and J. P. Baker. 2019. Recovery of Saginaw Bay Walleye, Lake Huron. Pages 411–430 *in* C. C. Krueger, W. W. Taylor, and S. Youn, editors. From catastrophe to recovery: stories of fishery management success. American Fisheries Society, Bethesda, Maryland.

Fraser, D. J. 2008. How well can captive breeding programs conserve biodiversity? A review of salmonids. Evolutionary Applications 1:535–586.

Gaden, M., and C. C. Krueger. 2018. Multijurisdictional governance of the shared Great Lakes fishery: can a non-binding agreement work? Chapter 4 *in* S. Brooks and A. Olive, editors. Transboundary environmental governance across the world's longest border. Michigan State University Press, Lansing.

Gephard, S. R. 2019. Restoration of anadromous Sea Lamprey to the Connecticut River and adjacent watersheds. Pages 185–209 *in* C. C. Krueger, W. W. Taylor, and S. Youn, editors. From catastrophe to recovery: stories of fishery management success. American Fisheries Society, Bethesda, Maryland.

Gurdak, D. J., C. C. Arantes, L. Castello, D. J. Stewart, and L. C. Watson. 2019. Evidence of recoveries from tropical floodplain fisheries: three examples of management gains for South American Giant Arapaima. Pages 267–295 *in* C. C. Krueger, W. W. Taylor, and S. Youn, editors. From catastrophe to recovery: stories of fishery management success. American Fisheries Society, Bethesda, Maryland.

Hansen, M. J., and C. R. Bronte. 2019. Restoration of Lake Trout in Lake Superior through interagency cooperative managewment. Pages 385–409 *in* C. C. Krueger, W. W. Taylor, and S. Youn, editors. From catastrophe to recovery: stories of fishery management success. American Fisheries Society, Bethesda, Maryland.

International Joint Commission. 1994. Great Lakes Water Quality Agreement 1978, revised. Available: www.ijc.org/files/tinymce/uploaded/GLWQA_e.pdf. (August 2018).

Hartig, J. H. 1997. Great Lakes remedial action plans: fostering adaptive ecosystem-based management processes. The American Review of Canadian Studies 27:437–458.

Hedges, K. J., N. E. Mandrak, M. A. Koops, and O. E. Johannsson. 2010. Use of aquatic protected areas in the management of large lakes. Aquatic Ecosystem Health and Management 13:135–142.

Hellström, G., C. Nilsson, S. Westberg, D. Palm, K. Leonardsson, D. Holmqvist, S. Ågren, J. Gardeström, A. Alanärä, and H. Lundqvist. 2019. Restoring the Vindel River, Sweden for Atlantic Salmon and Brown Trout. Pages 125–156 *in* C. C. Krueger, W. W. Taylor, and S. Youn, editors. From catastrophe to recovery: stories of fishery management success. American Fisheries Society, Bethesda, Maryland.

Hughes, R. M, B. L. Bangs, S. V. Gregory, P. D. Scheerer, R. C. Wildman, and J. S. Ziller. 2019. Recovery of Willamette River (Oregon, USA) fish assemblages: successes and remaining threats. Pages 157–184 *in* C. C. Krueger, W. W. Taylor, and S. Youn, editors. From catastrophe to recovery: stories of fishery management success. American Fisheries Society, Bethesda, Maryland.

Hutchings, J. A. 1996. Spatial and temporal variation in the density of Northern Cod and a review of hypotheses for the stock's collapse. Canadian Journal of Fisheries and Aquatic Sciences 53:943–962.

Hutchings, J. A., and R. A. Myers. 1994. What can be learned from the collapse of a renewable resource? Atlantic Cod, *Gadus morhua*, of Newfoundland and Labrador. Canadian Journal of Fisheries and Aquatic Sciences 51:2126–2146.

Hyatt, K. D., and M. M. Stockwell. 2019. Chasing an illusion? Successful restoration of Okanagan River Sockeye Salmon in a sea of uncertainty. Pages 65–100 *in* C. C. Krueger, W. W. Taylor, and S. Youn, editors. From catastrophe to recovery: stories of fishery management success. American Fisheries Society, Bethesda, Maryland.

Johnson, M. L., and J. Sandell, editors. 2014. Marine managed areas and fisheries. Advances in Marine Biology 69:2–416.

Johnson, W. E., D. P. Onorato, M. E. Roelke, E. D. Land, M. Cunningham, R. C. Belden, R. McBride, D. Jansen, M. Lotz, D. Shindle, and J. Howard. 2010. Genetic restoration of the Florida panther. Science 329:1641–1645.

Kautz, R., R. Kawula, T. Hoctor, J. Comiskey, D. Jansen, D. Jennings, J. Kasbohm, F. Mazzotti, R. McBride, L. Richardson, and K. Root. 2006. How much is enough? Landscape-scale conservation for the Florida panther. Biological conservation 130:118–133.

Kernen, L. T. 1995. Lake trout: native fauna or just grease? Journal of Great Lakes Research 21(Supplement 1):198–201.

Kim, M., E. R. Bageant, S. H. Thilsted, and K. J. Fiorella. 2019. Community management of fish refuges in Cambodian rice-field fisheries. Pages 353–372 *in* C. C. Krueger, W. W. Taylor, and S. Youn, editors. From catastrophe to recovery: stories of fishery management success. American Fisheries Society, Bethesda, Maryland.

Kline, P. A., T. A. Flagg, C. C. Kozfkay, D. J. Baker, D. J. Maynard, and W. C. McAuley. 2019. Snake River Sockeye Salmon—a perspective on twenty-five years of effort to recover a population in central Idaho, USA. Pages 101–123 *in* C. C. Krueger, W. W. Taylor, and S. Youn, editors. From catastrophe to recovery: stories of fishery management success. American Fisheries Society, Bethesda, Maryland.

Kraft, C. 2019. Adirondack Brook Trout and acid rain: environmental legislation fosters successful restoration. Pages 299–322 *in* C. C. Krueger, W. W. Taylor, and S. Youn, editors. From catastrophe to recovery: stories of fishery management success. American Fisheries Society, Bethesda, Maryland.

Krueger, C. C., and D. J. Decker. 1999. The process of fisheries management. Pages 31–59 *in* C. C. Kohler and W. A. Hubert, editors. Inland fisheries management in North America, 2nd edition. American Fisheries Society, Bethesda, Maryland.

Latham, R. M. 2017. Complete book of the wild turkey. Stackpole Books, Mechanicsburg, Pennsylvania.

Le Luyer, J., M. Laporte, T. D. Beacham, K. H. Kaukinen, R. E. Withler, J. S. Leong, E. B.

Rondeau, B. F. Koop, and L. Bernatchez. 2017. Parallel epigenetic modifications induced by hatchery rearing in a Pacific salmon. Proceedings of the National Academy of Sciences of the United States of America 114:12964–12969.

Lichatowich, J. 1999. Salmon without rivers: a history of the Pacific salmon crisis. Island Press, Washington, D.C.

Moore, S. E., and M. A. Kulp. 2019. Overcoming challenges to native fish restoration in a national park. Pages 45–64 *in* C. C. Krueger, W. W. Taylor, and S. Youn, editors. From catastrophe to recovery: stories of fishery management success. American Fisheries Society, Bethesda, Maryland.

Muir, A. M., C. C. Krueger, and M. J. Hansen. 2012. Re-establishing Lake Trout in the Laurentian Great Lakes: the past, present, and future. Pages 533–588 *in* W. W. Taylor and C. P. Ferreri, editors. Great Lakes fishery policy and management: a binational perspective. Michigan State University Press, East Lansing.

Nickum, M. J., P. M Mazik, J. G. Nickum, and D. D. MacKinlay, editors. 2004. Propagated fish in resource management. American Fisheries Society, Symposium 44, Bethesda, Maryland.

NMFS (National Marine Fisheries Service). 1991. Endangered and threatened species; endangered status for Snake River Sockeye Salmon. Federal Register 56:224(20 November 1991):58619–58624.

Pister, E. P. 2019. Species salvation: saving the Owens Pupfish from extinction. Pages 373–381 in C. C. Krueger, W. W. Taylor, and S. Youn, editors. From catastrophe to recovery: stories of fishery management success. American Fisheries Society, Bethesda, Maryland.

Probst, J. R., D. M. Donner, C. I. Bocetti, and S. Sjogren. 2003. Population increase in Kirtland's warbler and summer range expansion to Wisconsin and Michigan's Upper Peninsula, USA. Oryx 37:365–373.

Rabazanov, N. I., A. M. Orlov, A. S. Abdusamadov, R. M. Barkhalov, and K. M. Akhemedkhanov. 2019. Caspian Kutum *Rutilus kutum*: a story of exploitation, survival, and revival. Pages 485–508 *in* C. C. Krueger, W. W. Taylor, and S. Youn, editors. From catastrophe to recovery: stories of fishery management success. American Fisheries Society, Bethesda, Maryland.

Stephens, M. R., S. J. Stephens, and C. C. Krueger. 2019. Lessons learned from more than one-hundred years of golden trout management and recovery in California. Pages 19–43 *in* C. C. Krueger, W. W. Taylor, and S. Youn, editors. From catastrophe to recovery: stories of fishery management success. American Fisheries Society, Bethesda, Maryland.

Stroud, R. H., editor. 1986. Fish culture in fisheries management. American Fisheries Society, Fish Culture Section and Fisheries Management Section, Bethesda, Maryland.

Suski, C. D., and S. J. Cooke. 2007. Conservation of aquatic resources through the use of freshwater protected areas: opportunities and challenges. Biodiversity and Conservation 16:2015–2029.

Thatcher, C. A., F. T. van Manen, and J. D. Clark. 2009. A habitat assessment for Florida panther population expansion into central Florida. Journal of Mammalogy 90:918–925.

Thomas, J. A., J. G. Schulte, P. A. Tennant, and D. R. Argo. 2019. Recovery of a great river fishery: the story of the Ohio River. Pages 211–231 *in* C. C. Krueger, W. W. Taylor, and S. Youn, editors. From catastrophe to recovery: stories of fishery management success. American Fisheries Society, Bethesda, Maryland.

Ünal, V., and Z. Kizilkaya. 2019. A long and participatory process towards successful fishery management of Gökova Bay, Turkey. Pages 509–532 *in* C. C. Krueger, W. W. Taylor, and S. Youn, editors. From catastrophe to recovery: stories of fishery management success. American Fisheries Society, Bethesda, Maryland.

Vandergoot, C. S., M. D. Faust, J. T. Francis, D. W. Einhouse, R. Drouin, C. Murray, and R. L. Knight. 2019. Back from the brink: sustainable management of the Lake Erie Walleye fishery. Pages 431–466 *in* C. C. Krueger, W. W. Taylor, and S. Youn, editors. From catastrophe to recovery: stories of fishery management success. American Fisheries Society, Bethesda, Maryland.

Waxman, H. A. 1991. An overview of the Clean Air Act amendments of 1990. Environmental Law 21:1721–1816.

Weidel, B. C., M. J. Connerton, M. G. Walsh, J. P. Holden, K. T. Holeck, and B. F. Lantry. 2019. Lake Ontario Deepwater Sculpin recovery: an unexpected outcome of ecosystem change. Pages 467–482 *in* C. C. Krueger, W. W. Taylor, and S. Youn, editors. From catastrophe to recovery: stories of fishery management success. American Fisheries Society, Bethesda, Maryland.

Yoder, C. O., E. T. Rankin, V. L. Gordon, L. E. Hersha, and C. E. Boucher. 2019. Degradation and recovery of the Scioto River (Ohio, USA) fish assemblages from presettlement to present-day conditions. Pages 233–265 *in* C. C. Krueger, W. W. Taylor, and S. Youn, editors. From catastrophe to recovery: stories of fishery management success. American Fisheries Society, Bethesda, Maryland.

Section I

Rivers

Lessons Learned from More than One-Hundred Years of Golden Trout Management and Recovery in California

Molly R. Stephens*

School of Natural Sciences, University of California, Merced
5200 North Lake Road, Merced, California 95343, USA

Stanley J. Stephens[1]

California Department of Fish and Wildlife
1234 East Shaw Avenue, Fresno, California 93710, USA

Charles C. Krueger

Center for Systems Integration and Sustainability
Department of Fisheries and Wildlife, Michigan State University
East Lansing, Michigan 48823, USA

Abstract.—Native golden trout of California's upper Kern River basin have inspired anglers and scientists alike with their beauty, ecology, and evolutionary history. Three Rainbow Trout *Oncorhynchus mykiss* subspecies comprise the golden trout complex: California Golden Trout *O. m. aguabonita*, Little Kern Golden Trout *O. m. whitei*, and Kern River Rainbow Trout *O. m. gilberti*. This chapter focuses on restoration and management of the first two subspecies, California Golden Trout and Little Kern Golden Trout. Agency biologists, other scientists, and citizens have all worked for more than 100 years to protect and save these subspecies from a range of threats, some of fishery managers' own making and some threats evolving over time. Major problems have included overharvest by anglers, overgrazing of habitat by livestock, competition and predation by nonnative Brown Trout *Salmo trutta* and Brook Trout *Salvelinus fontinalis*, and hybridization with introduced nonnative forms of Rainbow Trout subspecies; these factors drove both species perilously close to extinction by the late 1960s. Little Kern Golden Trout were listed as threatened in 1978 under the U.S. Federal Endangered Species Act. That same year, the federal government designated large portions of California Golden Trout and Little Kern Golden Trout native watersheds as the Golden Trout Wilderness. Management actions to prevent extinction and conserve these subspecies have included angling regulations, livestock grazing restrictions, barrier construction, chemical treatments to remove nonnative trout, and research to identify and quantify genetic introgression of nonnative Rainbow Trout genes into native golden trout populations. Restoration efforts have, thus far, averted extinction, allowed populations to rebound, and provided several important lessons on genetic management of closely related subspecies, including pitfalls of a zero-introgression target for conservation, the potential need to continue management indefinitely, being responsive to emerging threats, recognizing that barriers to up-

* Corresponding author: valleymolly@gmail.com

[1] Retired.

stream fish movement can be useful, the caveats of using hatcheries for conservation, the potential role of native trout donor populations to facilitate restoration, and the need to harness public and stakeholder understanding and support for species conservation.

INTRODUCTION

Three beautiful and unique forms of inland Rainbow Trout *Oncorhynchus mykiss* are endemic to the upper Kern River basin of California's southern Sierra Nevada (Figure 1). These three subspecies comprise the golden trout complex: California Golden Trout *O. m. aguabonita*, Little Kern Golden Trout *O. m. whitei*, and Kern River Rainbow Trout *O. m. gilberti*. Their limited evolutionary separation from other forms of Rainbow Trout (reviewed by Behnke 2002) is a result of their recent post-Pleistocene isolation and radiation, and episodic hydrographic connectivity (and likely secondary contact) with their lower-elevation coastal Rainbow Trout ancestors and with one another. The receding of hydrologic connections at the end of the Pleistocene, combined with regional volcanic activity that created formidable isolating fish barriers, eventually put each subspecies onto its own evolutionary trajectory. Both morphological and genetic characteristics readily separate California Golden Trout, Little Kern Golden Trout, and Kern River Rainbow Trout from other Rainbow Trout and from one another (Schreck and Behnke 1971; Gold and Gall 1975; Gold 1977; Bagley and Gall 1998). Genetic analyses show significant nuclear and mitochondrial DNA variation differentiating all three golden trout subspecies (Bagley and Gall 1998; Erickson 2013; Stephens 2007). This chapter focuses on management and restoration of two subspecies: California Golden Trout and Little Kern Golden Trout (Figures 2 and 3). Kern River Rainbow Trout have been excluded for the sake of simplicity, as their story and conservation status is less straightforward.

Golden Trout Range Expansion

Golden trout were recognized as unique and beautiful as early as the late 1800s by early European-American explorers and settlers and had also been an important food source for the Tubatulabal and other native American tribes in the Kern River basin (Voegelin 1938; Figure 4). Consequently, upon their discovery by early westerners, golden trout were almost immediately transplanted outside their native range into naturally fishless Sierra Nevada lakes and streams, mostly to the north. In an unpublished manuscript, Ardis Manley Walker, a Kern River valley resident and local conservationist, describes the gold fever of those who first encountered these alluring fish:

> Many of the pioneer visitors to golden trout waters reacted with a desire that was almost compulsive; they shared a common missionary urge to spread the golden beauty and life of this native habitat to the barren waters of more elevated and more easterly and northerly lakes and streams. These men included early sheep herders, cow men, and prospectors after timber and gold.... Although we do not have complete records of their transplants; legend tells us that these men launched many golden trout on coffee-pot and frying-pan migrations.

Subsequently, in the late 19th and early 20th centuries, the California Fish and Game Commission began to propagate and distribute California Golden Trout to fishless waters throughout the region (Evermann 1906; Ellis and Bryant 1920; Figure 4). California Golden Trout were also transplanted to waters outside of California, such as in Wyoming and Idaho (Pister 1991). The early introductions into the Sierra Nevada region used a more systematic

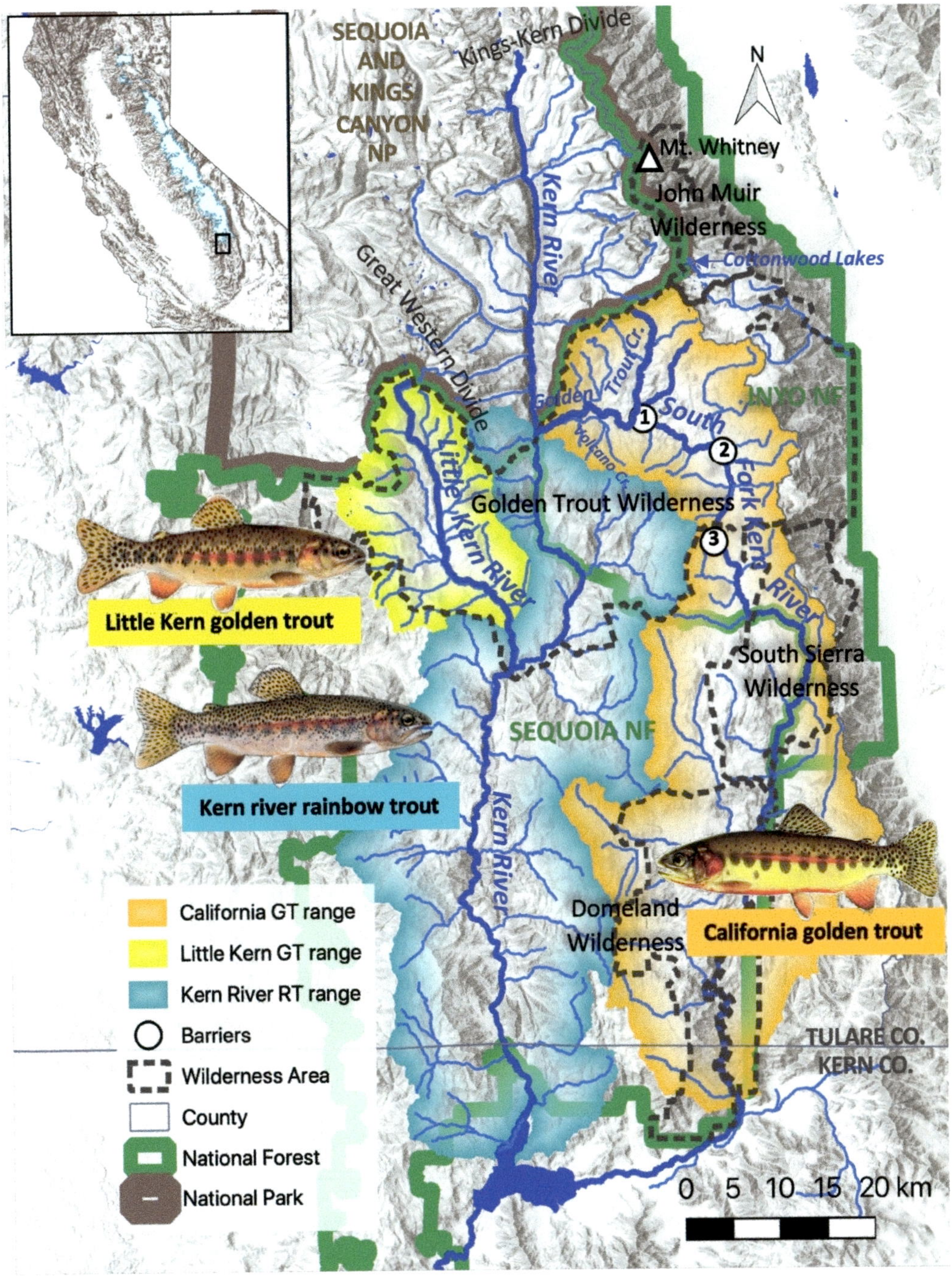

FIGURE 1. Map of California (inset) shows the approximate maximum extent of Pleistocene (Tahoe) glaciation in the Sierra Nevada (light blue polygon) and box of magnified map area. Map depicts the Kern Plateau, a 2,450-km^2 tableland bordered by the Great Western Divide and Kings-Kern Divide, Mount Whitney, and the Sierra Nevada Crest and encompassing the Kern River valley and entire Kern River basin. Historic native ranges for each Golden Trout subspecies are depicted and major barriers—Ramshaw, Templeton, and Schaeffer—denoted. Trout subspecies illustrations copyright by Joseph Tomelleri, used with permission.

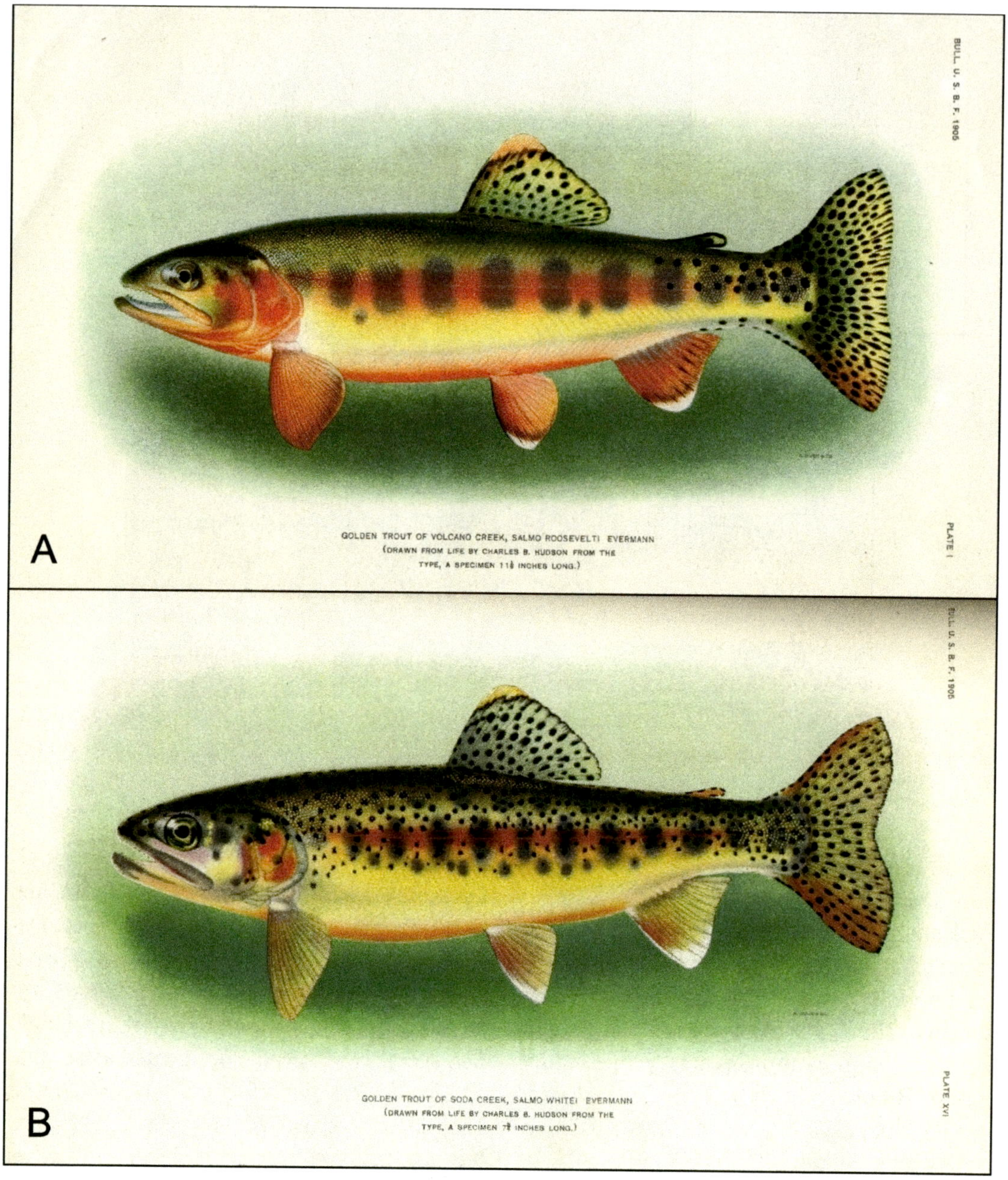

FIGURE 2. Plates from Evermann (1906) depicting drawings by Charles Bradford Hudson. **(A)** California Golden Trout from Volcano Creek (former name of Golden Trout Creek; not to be confused with the current-day Volcano Creek), denoted as *Salmo roosevelti.* **(B)** Little Kern Golden Trout, denoted as *Salmo whitei.* Both taxa would undergo subsequent naming revisions as a consequence of increased genetic understanding regarding their evolutionary placement within the Pacific salmon genus *Oncorhynchus* and their relationship to other trout within the genus.

FIGURE 3. **(A)** Little Kern Golden Trout from Upper Fish Creek, tributary to Little Kern River (2011). **(B)** California Golden Trout from South Fork Kern River above Templeton Barrier (2004). Photo credit, Molly Stephens.

and advanced approach than those early coffee pots and frying pan plants of the late 1800s. Pack animals carried fish in milk cans wrapped in wet burlap to keep them cool, transporting fish to southern Sierra Nevada lakes and streams. The commission introduced golden trout into "high summit streams" while putting eastern Brook Trout *Salvelinus fontinalis* and Brown Trout *Salmo trutta* into other tributary waters with the goal "to reach and plant with desirable varieties of trout all of the barren waters of the mountainous sections of the Fresno Division." They rationalized that new railroads and highways would "enable many thousands of people," both local and statewide, "to reap the benefits of the Commission's efforts toward making this magnificent public 'outing' ground doubly attractive" (Anonymous 1913). Around 1914, the commission delegated implementation of trout management activities to the California Department of Fish and Game (which later became the California Department of Fish and Wildlife [CDFW]) but retained many of its regulatory powers. California Golden Trout were California's iconic fish species and became designated as its state fish in 1947; at that time, the California State Legislature recognized the need to preserve the golden trout "due to its beauty, scarcity, and unique quality of being a truly California trout" (California State Assembly 1947).

Change in Management Direction

Despite a long era of Johnny Appleseed-style stewardship that scattered golden trout to waters throughout California, the nation, and even the world, the status of these subspecies within their native ranges became precarious

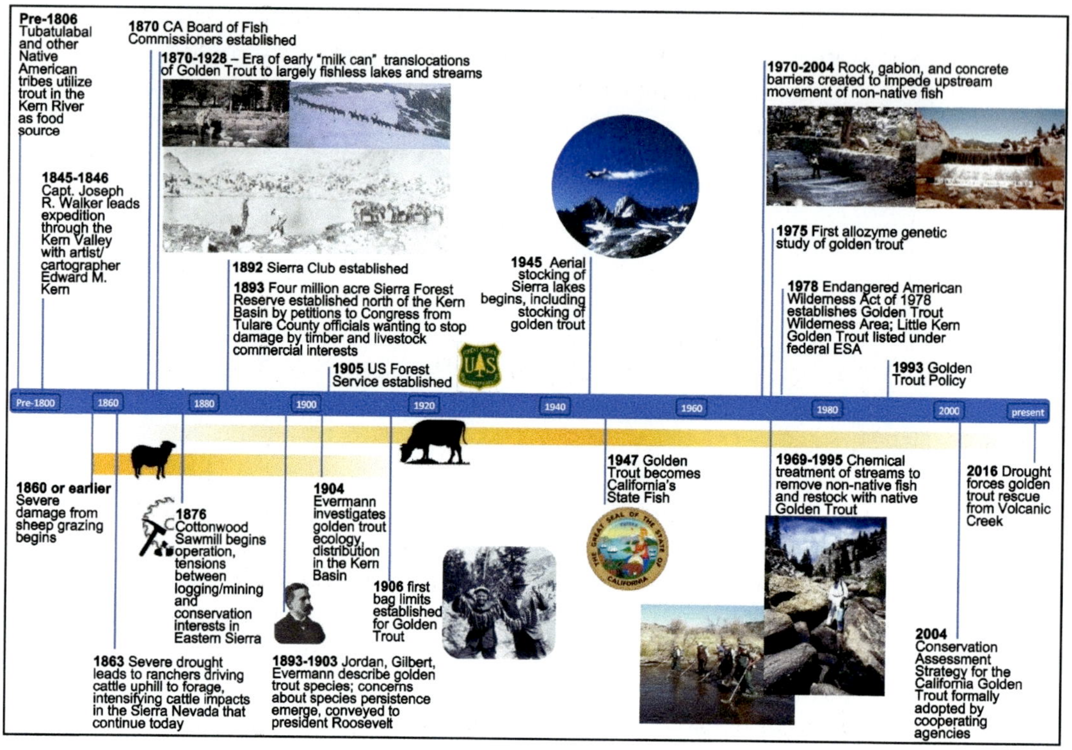

FIGURE 4. Selected threats, policies, and management strategies applied to golden trout over time.

as the 20th century progressed. Overharvest by anglers, habitat degradation mainly due to sheep grazing, competition and predation by introduced nonnative Brook Trout and Brown Trout, and hybridization with introduced hatchery Rainbow Trout (Coastal Rainbow Trout *Oncorhynchus mykiss irideus* and *O. mykiss* ssp.) threatened to drive native golden trout to extinction. As attitudes toward fish stocking and fishery management evolved (Pister 2010), environmental legislation and litigation began to influence California's fishery management programs (Lentz and Clifford 2014). Several key protections were put into place to stem the decline of golden trout. The federal government established the Golden Trout Wilderness as part of the Endangered American Wilderness Act (1978). The California State Legislature passed the Trout and Steelhead Conservation and Management Planning Act of 1979, which changed the California Fish and Game Code and amended it many times over subsequent years with additional legislation, increasingly emphasizing native and wild trout management over hatchery fish and limiting where hatchery fish could be planted (California Fish and Game Code 1979). Last, the commission adopted policies that included a Golden Trout Policy, Trout Policy, and Designated Wild Trout Waters Policy (California Fish and Game Commission, various dates), which bolstered protections for native species and wild trout populations and emphasized native and wild trout conservation and catch-and-release angling opportunities.

Subspecies Conservation Status

The Little Kern Golden Trout was listed as threatened in 1978 by the U.S. Fish and Wildlife Service (USFWS) under the U.S. Federal Endangered Species Act (ESA; USFWS 1978). At the time of listing, the Little Kern Golden Trout had been reduced to an estimated 10% of its historic range (from 160 river kilometers, originally, down to 16 km) and contained fewer than 5,000 individuals in reaches of five tributary streams (Christenson 1984; Moyle 2002). Though perhaps less charismatic than its sister subspecies, California Golden Trout, Little Kern Golden Trout's new status greatly influenced the establishment of the Golden Trout Wilderness. Despite being listed for 40 years, no federal recovery plan has been developed for Little Kern Golden Trout. The existing state management plan is 34 years old and lacks necessary updated information to guide future conservation.

In 2000, the nongovernmental organization (NGO) Trout Unlimited petitioned the USFWS to list California Golden Trout as endangered under the ESA. The USFWS deemed listing not warranted in its final rule in 2011, asserting that major threats to the species lacked "sufficient imminence, intensity, or magnitude" qualifying as endangerment (USFWS 2002, 2011). The decision noted the collaborative conservation plans formalized in the "Conservation Assessment and Strategy for the California Golden Trout" (Stephens et al. 2004). This strategy detailed ongoing and planned management activities necessary to stem decline and promote conservation of the species through cooperative interagency efforts by the CDFW, USFWS, and U.S. Forest Service (USFS), along with support from several NGOs (Table 1).

Jurisdictional authority for management of golden trout is a complicated mixture of state and federal authorities based on federal land management and state fishery management laws, combined with a variety of special legislative designations. Kern River watersheds are found within the Sequoia and Inyo national forests and Sequoia National Park (Figure 1). Within these federal lands, South Sierra, Domeland, and Golden Trout Wilderness areas have been designated, which provide special protections to maintain and perpetuate their wilderness character. The Golden Trout Wilderness includes the upper portion of the South Fork Kern River and all of Golden Trout Creek, and management is directed towards recovery and maintenance of California Golden Trout habitat. In addition, the South Fork Kern River, described above, is a federally designated wild and scenic river, which prevents instream activities that would harm the river's free-flowing condition, water quality, and outstanding resource values (Wild and Scenic Rivers Act 1968). Nested within this federal land and endangered species management, the state of California retains authority to regulate angling for golden trout, has designated California Golden Trout and Kern River Rainbow Trout as species of special concern, and identified all three subspecies as species of greatest conservation need (CDFW 2015).

At the time of this writing, many decades of active fishery management and resources have been invested in golden trout conservation; on the one hand, the preservation of these two subspecies constitutes a major fishery management success, preventing local extirpation and extinction. On the other hand, golden trout subspecies are still at risk. This chapter describes the management history of California Golden Trout and Little Kern Golden Trout. It focuses on the various threats to the California Golden Trout and Little Kern Golden Trout, responses to these threats with management actions, and concludes with lessons learned during the saving of golden trout.

DISTRIBUTION AND ECOLOGY

California Golden Trout are native to the South Fork Kern River and Golden Trout Creek (Figure 1). Golden Trout Creek en-

TABLE 1. Select examples of management activities and partnerships that have built conservation support for golden trout. NGO refers to nongovernmental organizations, whose individual descriptions are given in the footnotes below.

Activity	Effect/desired outcome
Participation by Trout Unlimited[a], The Nature Conservancy[b], and other volunteers in annual fencing off areas to seasonally exclude grazing cattle from sensitive watersheds in the California Golden Trout native range, plant vegetation restore streambanks, and collect tissue samples for genetic research.	Provides needed labor for managers and researchers, promotes public perception and stewardship of native species by publicizing high-profile restoration opportunities via social media.
Agency biologists work with grazing permittees and conduct vegetation use monitoring.	Minimizes grazing impacts to sensitive streams and riparian vegetation.
Communication with backcountry user groups about nonnative fish removal efforts in high mountain lakes and alternative fishing opportunities for native species.	Educates backcountry user groups regarding the rationale for fish removal and work to mitigate real or perceived negative impacts of fish removal to their business interests.
Increased trailhead signage and outreach.	Educates the public about the value of native species fisheries; advocates reasonable use levels on public land; discourages illegal translocation of nonnative or hybridized fish above barriers or into fishless waters.
"California Heritage Trout Challenge" angler recognition program; outreach to and partnerships with NGO angling and conservation organizations (Trout Unlimited[a], CalTrout[c], Federation of Fly Fishers, Kern River Flyfishers, and Aguabonita Flyfishers[d]).	Promotes awareness of and angling for California native trout species; enhances public appreciation for native species; offers alternative experience for anglers, different from anglers wangint to catch only large hatchery fish.
CalTrout commissioned reports documenting the economic valuation of angling in the Golden Trout Wilderness (Alkire 2003) and state of the salmonids (Moyle et al. 2008, 2017.)	Documents the tangible benefits with clear monetary value to the local community and businesses; highlights the threats to and conservation status of native trout species.
Research by geneticists and ecologists.	Informs management of habitat and populations to prevent decline.
Regular communication between NGO, state and federal biologists, and academic researchers (e.g., through conservation assessment and strategy meetings, scientific meetings, etc.).	Coordinates conservation efforts and informs adaptive species management; maintains a pulse on species status and pressure on government agencies to prevent Endangered Species Act listing (California Golden Trout) or encourage species recovery (Little Kern Golden Trout).

TABLE 1. Continued.

Activity	Effect/desired outcome
Funding, including federal and state species conservation funds for species of concern and unique private–public partnerships such as the Upper Kern Basin Fishery Resource Enhancement Trust[e].	Provides research and mangement funds for Little Kern Golden Trout and select California Golden Trout (populations in tributaries to the upper Kern River). An important source of matching funds for state and federal grants.
Petitions to list species, uphold environmental laws, and protect critical habitat.	Legal action (or threat of legal action) keeps pressure on management agencies to prevent endangerment and extinction of the species and promote recovery.

[a] Trout Unlimited is a national nonprofit organization whose mission is to conserve, protect, and restore North America's coldwater fisheries and their watersheds.
[b] The Nature Conservancy is a national nonprofit organization whose mission is to conserve the lands and waters on which all life depends.
[c] California Trout, or CalTrout, is a nonprofit organization whose mission is to protect and restore wild trout, steelhead, salmon, and their waters throughout California.
[d] Federation of Fly Fishers, Aguabonita Flyfishers, and Kern River Flyfishers: local flyfishing organizations with interests in both angling and trout conservation.
[e] The Upper Kern Basin Fishery Resource Enhancement Trust was established in 1996 by the Federal Energy Regulatory Commission as part of the relicensing the Southern California Edison Company (SCE) Kern River No. 3 (KR3) Hydroelectric Project, located on the upper Kern River near Kernville. Fund is administered by the Kern Community Foundation and grants are reviewed by a committee made up state and federal resource agency managers and Southern California Edison.

ters the upper Kern River from the east. Historically, it drained into the South Fork Kern River, but past volcanic activity redirected its flow into the upper Kern River (Figure 1; Webb 1946). The headwaters of both streams begin near the base of Mount Whitney at an elevation of approximately 3,200 m above mean sea level, flowing through high-elevation meadows to reaches with narrow, rocky, steep-walled canyons. In the South Fork Kern River, California Golden Trout likely occurred from near the headwaters just south of maximum extent of the last Pleistocene glaciation, downstream to as far as the lower end of the Domeland Wilderness (Walker, no date; Figure 1); it has been speculated that they may have episodically extended downstream of Lake Isabella (Stephens et al. 2004). In the Golden Trout Creek drainage, they likely occurred in waters below glaciated headwater lakes, downstream to a series of bedrock falls at the mouth of Golden Trout Creek, including Agua-Bonita Falls (18–24 m). These natural falls notably protected the watershed from invasion by native (and, later, nonnative) species from the main-stem Kern River (Evermann 1906).

Little Kern Golden Trout are endemic to the Little Kern River. This drainage enters the upper Kern River from the west, just downstream from the entry point of Golden Trout Creek (Figure 1). Similar to California Golden Trout waters, the majority of this drainage is located on USFS land, with a small portion located in Sequoia National Park. Most of the Little Kern River drainage occurs within the Golden Trout Wilderness.

Golden trout of both subspecies are found in clear, cold (<20°C), low-productivity streams flowing in lush, wet meadows with dense grasses (Gramineae), sedges (Cyperaceae), and other plants. In their native stream habitats, these fish do not attain particularly large sizes, reaching only 200–300 mm in length. The only other fish species originally in these streams was the Sacramento Sucker *Catostomus occidentalis*, a nonpredatory species. Thus, golden trout, having little evolutionary background for competition with, or predation by, other fish species, were vulnerable to nonnative fish invasions and ultimately dominated, preyed upon, or displaced, by Brown Trout or Brook Trout in many locations (Christenson 1984; Stephens et al. 2004). Golden trout mature at ages 3 or 4 (>95 mm), spawn in late May or June at average water temperatures between 7°C and 10°C after peak snowmelt discharge, and build redds in fine gravel and sand typically in less than 25-mm-diameter substrate (Pister 1991; Knapp and Vredenburg 1996). Golden trout easily hybridize with other closely related subspecies of Rainbow Trout, forming fertile offspring (Cordes et al. 2006).

THREATS TO GOLDEN TROUT SUBSPECIES AND MANAGEMENT ACTIONS

Many threats exist for golden trout and their habitat, some of which date back to the early 1900s. However, perhaps the greatest current threat is genetic introgression with nonnative forms of introduced Rainbow Trout; therefore, management of this threat is a major focus.

Overfishing

The California Golden Trout and Little Kern Golden Trout have historically suffered a number of common threats and have, at times, hovered on the brink of extinction. Overfishing in the early days of California's European-origin colonization was of concern to Stewart White (1904), who brought to President Theodore Roosevelt's attention the ease with which California Golden Trout could be caught and potentially brought to extinction (Evermann 1906; Figure 4). President Roosevelt directed Fisheries Commissioner George M. Bowers to send Barton Warren Evermann of the U.S. Bureau of Fisheries to investigate the matter, which culminated in the publication of *The Golden Trout of the Southern High Sierras* (Evermann 1906). Here, the vulnerability of the species to angling and the need for protection was described: "[Golden trout] are easily captured, as they are so voracious and rise to the lure so readily. Two years ago, the members of the Sierra Club and others accompanying them on their annual outing to Mount Whitney are said to have taken 600 or 700 trout from Volcano Creek [prior name for Golden Trout Creek] in one day." (Evermann 1906:28–29). "Although now abundant the Golden Trout cannot long remain so unless afforded some protection," Evermann warned. To combat the overharvest, Evermann (1906) recommended a minimum size restriction, a limit on the number that may be caught and kept, and a prohibition of all fishing during the spawning season; these actions were implemented. In addition, he recommended stocking nearby barren waters (naturally fishless lakes and streams) with California Golden Trout to expand their range. This action had already begun as early as 1876. California Golden Trout were transplanted from Mulkey Creek to Cottonwood Creek in 1879 and then to Cottonwood Lakes in 1891 (Pister 1991). The CDFW then used Cottonwood Lakes as a source of California Golden Trout broodstock. They established an egg collecting station at Cottonwood Lakes; eggs were then transported to Mount Whitney Hatchery to be hatched (Leitritz 1970). Fingerling-sized trout were stocked in other Sierra Nevada lakes and streams and elsewhere in the United States.

Current angling regulations in the Golden Trout Wilderness include a bag limit of five trout and a fishing season that extends from the last Saturday in April through November 15. Also, only artificial lures with barbless hooks may be used (7.50[b][71]; www.eregulations.com/california/fishing/freshwater/waters-special-regulations-f-m/). Overall, overharvest no longer appears to be a threat to golden trout survival.

Habitat Degradation

Although off-road vehicle activity, road construction, logging, stream pollution from mineral extraction, hiking, and horseback riding and packstock use have all had adverse, or potentially adverse, impacts on golden trout habitat on the Kern Plateau (Figure 1), livestock grazing has generated the largest and most lasting impacts on habitat quality (USFWS 1978; Stephens et al. 2004). Livestock grazing of mountain meadows along the Little Kern River, South Fork Kern River, and Golden Trout Creek has occurred for more than 145 years, initially and most heavily by sheep *Ovis aries* beginning in the mid- to late 1800s and later predominantly by cattle *Bos taurus* (Stephens et al. 2004; USFWS 2011; see Kattelmann and Embury 1996 and Kinney 1996 for comprehensive reviews of grazing in the Sierra Nevada). Thousands of sheep and cattle grazed the Kern Plateau each season from the 1860s until the 1920s. Clarence King (1902) noted a before-and-after observation of the effects of sheep grazing within his book *Mountaineering in the Sierra Nevada*:

> The Kern Plateau, so green and lovely in my former visit in 1864, was now a gray sea of rolling granite ridges, darkened at intervals by forest, but no longer velveted with meadows and upland grasses. The indefatigable shepherds have camped every-where, leaving hardly a spear of grass behind them.

Overgrazing caused loss of pool habitat, sedimentation, reduced instream cover, riparian cover loss, and loss of undercut stream banks. Stream channels became wide and shallow, developed an inability to moderate air temperature extremes in both summer and winter, lost spawning habitat, and suffered reductions of invertebrate food production for the trout (Armour et al. 1994; Stephens et al. 2004). In addition, species composition of plants in the meadows changed, with liverworts *Riccia* sp. and willows *Salix* sp. decreasing and sagebrush *Artemisia* sp. and sedges (Cyperaceae) increasing after approximately 1900 (Dull 1999).

Cattle grazing continues on the Kern Plateau based on annual allotments provided to cattle permittees by the USFS but at much reduced levels (Stephens et al. 2004). From 2001 to 2010, the Inyo National Forest established a 10-year period of rest and recovery from grazing on two (Templeton and Whitney allotments; https://data.ecosystem-management.org/nepaweb/nepa_project_exp.php?project = 34380) of four grazing allotments on the Kern Plateau. The Inyo National Forest has initiated analysis and review of these four grazing allotments in preparation for a National Environmental Policy Act review, which is not complete at the time of this writing. Some restoration efforts, such as fencing and active vegetation planting, have been made to recover meadow areas (Matthews 1996; Stephens et al. 2004; Table 1).

Over recent decades, several studies have investigated the effects of grazing on stream habitat and golden trout populations. One study found that overall trends in meadow plant species richness, diversity, and frequency of soil-stabilizing species were not significantly different between the limited-grazing versus rested study sites over a 10-year period (Freitas et al. 2014). Several other studies, however, found significant differences between grazed and ungrazed areas. Ungrazed areas protected by fencing consistently had more canopy shading, deeper stream depths, and greater bank-full heights and smaller stream widths than grazed areas (Knapp and Matthews 1996). California Golden Trout were observed to use pools, runs, and selected habitat features such as undercut banks that were present in ungrazed stream portions and avoided bare and collapsed banks caused by grazing (Matthews 1996). Interestingly, California Golden Trout abundance was higher in grazed sections than in ungrazed sections but showed low growth rates (Knapp et al. 1998). Cattle grazing widened the studied stream sections, resulting in more spawning habitat and recruitment, population overabundance, and subsequent negative density dependent growth of individual fish. In a study of meadows within the South Fork Kern River, where cattle exclusions had been placed in 1991, riverbank vegetation more than 20 years later was larger and denser and maximum water temperatures cooler than outside the exclosure, where cattle were present (Nusslé et al. 2015). Vegetation recovery from grazing could take several decades because some "rested" areas had yet to recover to full vegetation height, even after 25 years (Nusslé et al. 2017). Clearly, recovery of the landscapes and streams supporting golden trout is an ongoing process that will likely continue for decades and require management oversight to determine whether current grazing levels and regimes are compatible with ongoing trout management.

Competition and Predation by Invasive Species

Stocking of nonnative trout into the Little Kern River basin and streams of the Kern Plateau may have begun as early as the late 19th century and included Brown Trout, Brook Trout, and nonnative forms of Rainbow Trout (Christenson 1984; Stephens et al. 2004). Brown Trout are not now present in Golden Trout Creek or the Little Kern River but do still occur in the South Fork Kern River downstream of Templeton Barrier (Figure 1). When investigated in 1969, Brown Trout outnumbered golden trout by nearly 100 to 1 (Pister 2008, 2010). Brown Trout preyed on golden trout, competed for food and space, and successfully displaced California Golden Trout in some stream reaches, especially in deep pools with overhead cover. Brook Trout were stocked in the Little Kern River drainage from 1930 to 1941, including into many of the tributaries and headwater lakes (Christenson 1984; USFWS 2011). Brook Trout successfully competed with Little Kern Golden Trout, especially in lakes, where they have been known to eliminate golden trout populations. Nonnative forms of Rainbow Trout, mostly of hatchery origins, were derived from a complex variety of Rainbow Trout and steelhead (anadromous Rainbow Trout) sources (described in Busack and Gall 1980; Needham and Behnke 1962) and stocked prolifically into native golden trout waters.

Management actions to combat the problems (competition, predation, and interbreeding) caused by nonnative trout introductions began in 1969 and continue to the present. These included the construction of barriers to prevent upstream invasion, physical removal of fish by electrofishing, and chemical treatments of stream segments using piscicides such as rotenone and antimycin (Christenson 1984; Pister 2008, 2010). These actions helped address the invasion of nonnative trout forms, including Rainbow Trout and their hybridization with golden trout (see below). The sequence of steps used to remove nonnative trout was to first construct barriers, remove golden trout by electrofishing and hold them alive off-site, then use piscicides to kill all fish from the selected stream reach or lake, and, last, to restock the treated stream section or lake with the trout that were held off site. For example, after Brown Trout were detected in the upper South Fork Kern River in the late 1960s, Ramshaw Barrier was constructed in the early 1970s and was followed by the downstream construction of two additional barriers, Templeton and Schaeffer, followed by several chemical treatments (Figure 1; Pister 2008). More than 160 km of stream were chemically treated to remove nonnative trout (Pister 2010).

Barrier construction on the Little Kern River began in 1970 and continued into the 1980s, mostly on tributaries and with one barrier being placed on the main river. All barrier construction in this drainage consisted of modifying naturally occurring water falls. Chemical treatments on the Little Kern River occurred between 1974 and 1995 (USFWS 2011; Stephens 2014). The only notable difference in methodology from that used for the South Fork Kern River was the eventual use of a hatchery (Kern River Planting Base—also a production hatchery) to propagate Little Kern Golden Trout offspring for reintroduction into the Little Kern River drainage after the chemical treatments. This decision, made under political pressure from backcountry user groups to speed up the treatment-reintroduction process, resulted in the inadvertent reintroduction of nonnative Rainbow Trout genes from the hatchery, as discussed below. The use of barriers combined with chemical treatments have effectively eliminated Brook Trout from the Little Kern River watershed and Brown Trout from the upper portions of the South Fork Kern River above Templeton Barrier; however, both Brown Trout and introgressed Rainbow Trout and golden trout are currently present between Templeton and Schaffer barriers and downstream.

Genetic Introgression from Nonnative Forms of Rainbow Trout

The most serious ongoing threat common to the two subspecies is the genetic introgression with naturalized nonnative forms of Rainbow Trout and stocking with already introgressed golden trout. Introgression denotes that hybrids are fertile and have crossed beyond just a single generation. Not only do these fish potentially compete with and prey on native golden trout, but they also breed with them, thus creating a hybrid form potentially altering the golden trout genome forever.

Stocking of golden trout watersheds occurred for many years beginning in the 1930s, and these stocked fish were a source for the problem of introgression (Stephens et al. 2004; Pister 2010). For example, a catchable-size Rainbow Trout stocking program was implemented in the lower South Fork Kern River to satisfy anglers (Pister 2008). In this case, Rainbow Trout migrated upstream into the upper South Fork Kern River and resulted in introgression with the native California Golden Trout. Beginning in 2004, triploid (sterile) Rainbow Trout were used for stocking these waters, and this practice continued until 2009 when trout stocking in the lower South Fork Kern River was terminated (Stephens et al. 2004, 2013). However, the genetic damage from decades of upstream movement of introgressed fish had already occurred.

California Golden Trout were also stocked in Sierra Nevada lakes as part of the CDFW fishery management goal to support recreational angling. These stocked fish were presumed by most to be pure California Golden Trout (but see Calhoun 1961) but were later found to possess hybridized genomes introgressed with Rainbow Trout (Cordes et al. 2006). A population of presumably genetically pure California Golden Trout had been established sometime prior to 1900 just outside of their native range in the previously fishless Cottonwood Lakes, which drain into the Owens River, east of the Sierra Nevada. California Golden Trout from these lakes were used as a broodstock source for hatchery propagation and stocking throughout California and elsewhere. Prior to the 1930s, fish produced by this program were most likely nonhybridized and represented a pure source of California Golden Trout. However, sometime during the 1930s, or possibly earlier, the Cottonwood Lakes broodstock became introgressed with Rainbow Trout (Cordes et al. 2006; R. F. Leary, University of Montana, personal communication). One probable source was fingerling Rainbow Trout inadvertently included with California Golden Trout from Mount Whitney Hatchery, which were planted in Cottonwood Lakes to maintain the golden trout broodstock (CDFW, unpublished records). Likely, these introgressed fish, when stocked into native waters, further promoted the genetic link between Rainbow Trout and California Golden Trout. For example, stocking of these introgressed fish occurred in the headwater lakes of tributaries to Golden Trout Creek (Cordes et al. 2006); these fish were later removed through extensive gill-netting efforts over several years, after they had already genetically contaminated trout in downstream creeks, based on genetic analyses. The Cottonwood Lakes broodstock continues to be used as a California Golden Trout source for propagation and stocking outside the native golden trout range.

Like California Golden Trout, Little Kern Golden Trout inadvertently introgressed with Rainbow Trout were also directly introduced into watersheds during restoration efforts after chemical treatment to eradicate nonnative fish in Little Kern River and tributaries. Initially, genetically pure fish were marked and held in nearby streams during chemical treatment to remove Brook Trout and Rainbow Trout and then reintroduced. The CDFW was pressured by a backcountry user group to speed up the recovery process and thus began using the Kern River Planting Base to propagate Little Kern Golden Trout offspring for planting back into the Little Kern River watershed. Because this facility was a production rather than a conservation hatchery, inadvertent incorporation of Rainbow Trout at the hatchery likely occurred, resulting in the introduction of either Rainbow Trout, introgressed Little Kern Golden Trout, or both back into some waters of the Little Kern River drainage (Stephens 2007).

Regardless of the source of either Rainbow Trout or hybridized golden trout, introgression is a stealthy and insidious threat because it is not often obvious or visibly detectable. Sometimes a hybrid fish in the hand can be identified based on intermediate characters between forms, such as spotting patterns or lateral band coloration that looks more Rainbow Trout-like than golden trout-like. However, a seemingly pure golden trout in the hand of a biologist actually may have a small portion of its genome compromised with Rainbow Trout genes. Without genetic analysis, the genetic threat cannot be easily detected. This problem is hard to eliminate but can be managed and reduced.

Actions used to manage the threat of introgression have been the same as those used to manage nonnative species: barriers and chemical eradication (or mechanical removal through gill netting and electrofishing in small waters, when feasible). However, monitoring and assessment of the effectiveness of these actions requires genetic analysis of the restored populations. Because the chance of introgression is always present, genetic monitoring over time should be an important activity identified within restoration plans and undertaken as a part of ongoing management assessment (Stephens et al. 2013; Stephens 2014). For example, headwater lakes (Johnson, Rocky Basin, and Chicken Spring) that are connected to Golden Trout Creek tributaries were unwittingly stocked with introgressed California Golden Trout propagated from the Cottonwood Lakes broodstock between 1955 and 1993 (Stephens et al. 2004). Six microsatellite loci and a minisatellite DNA marker were used to estimate levels of introgression in approximately 700 California Golden Trout samples from 23 locations in the Golden Trout Creek drainage (Cordes et al. 2006). Low to moderate introgression levels (1–17% nonnative Rainbow Trout) were found in all but one of the locations, with the highest levels observed within transplanted lake populations, their respective downstream outlets, and Cottonwood Lakes populations. Efforts to halt further introgression of Rainbow Trout alleles into the Golden Trout Creek system were implemented and included gill netting to remove all fish from Johnson, Rocky Basin, and Chicken Spring lakes, electrofishing of stream sections, and discontinuation of fish stocking into the Golden Trout Creek drainage. Multiple gill nets were set in late fall and left in the lakes over the winter period. This type of netting was conducted for three winters and proved to be effective, as all lakes became fishless (Stephens et al. 2004); thus, the threat of introgression from this source has been successfully reduced in the Golden Trout Creek drainage, and genetic monitoring continues over a decade later.

Genetic assessment and monitoring of nonnative Rainbow Trout introgression in California Golden Trout and Little Kern Golden Trout has been ongoing within the native ranges of both subspecies over the past several decades (e.g., Gall et al. 1976; Cordes et al. 2006; Stephens 2007). For California Golden Trout, the lowest introgression levels were found within the Golden Trout Creek drainage and included several populations that showed less than 5% introgression (Stephens and May 2013). However, introgression levels were generally higher within the South Fork Kern River than in Golden Trout Creek, though assignment of exact percentages is complicated by the observation of more variable introgression estimates in the 2013 microsatellite analysis of South Fork Kern River populations than previous studies. Introgression levels were highest in the lower reaches of the South Fork Kern River and its tributaries, decreasing incrementally to zero or nearly zero introgression in headwater tributaries upstream (Stephens 2007; Stephens et al. 2013). This result was consistent with the stocking history of the lower reaches and upstream movement of nonnative or introgressed trout, likely happening prior to the construction of barriers. For Little Kern Golden Trout, low levels of introgression were found in the Little Kern River drainage, with 85% of locations showing less than 5% introgression, suggesting that barriers, chemical treatments, and reintroduction efforts over the past decades were largely successful. However, some localities exist that contain estimates of 8–20% introgression that are of concern: Mountaineer Creek in the southwest portion of the drainage, Shotgun Creek in the northeast, and the main-stem Little Kern River. These locations will require ongoing genetic monitoring and possibly management action, though determining the necessary actions based on introgression percentages is a perilous process (see Lesson 1 below). In addition, the legacy of low genetic diversity (likely a consequence of low population sizes, postchemical-treatment reintroductions, and consequent genetic bottlenecks) remains a threat to Little Kern Golden Trout persistence, as well as to some bottlenecked California Golden Trout populations. It is encouraging that the majority of populations in the Little Kern River and Golden Trout Creek watersheds show low (0–3%) levels of introgression, as do some headwater populations in the South Fork Kern River. Both California Golden Trout and Little Kern

Golden Trout will benefit from updated genetic analyses that use more comprehensive genomic sequencing data, which can better inform questions regarding evolutionary ancestry, genomic, and adaptive consequences of hybridization and provide more specific direction for nonnative trout eradications.

Fire, Drought, and Climate Change

Even though the majority of the golden trout's native habitat falls on federal lands with certain protections, its populations cannot elude the realities that climate change could intensify the effects of drought and wildfires (Diffenbaugh et al. 2015; Abatzoglou and Williams 2016). Forest fires threaten stream habitats and trout populations as vegetation is lost (removing habitat and shade), water temperature increases (resulting in fish kills and physiological stress), and water quality decreases (caused by erosion, siltation, and decreased dissolved oxygen content) during and after fire events. For example, the 2011 Lion Fire burned more than 20,000 acres (8,094 ha) in the Little Kern River basin (Stephens 2014; Hogan and Buchanan 2015) and threatened several important native Little Kern Golden Trout populations. Since 2011, several wildfires (Fish in 2013, Soda in 2014, Cabin in 2015, Jacobsen in 2016, Lion in 2017, and Mountaineer in 2018) have subsequently occurred in the golden trout native range. Altered precipitation patterns due to climate change could also affect the most critical habitat characteristic essential for golden trout survival: the quality and abundance of water (Hogan and Buchanan 2015; Nusslé et al. 2015; Swain et al. 2018). The unprecedented 2012–2016 California drought necessitated a fish rescue in 2016, after population and habitat monitoring identified precipitous declines in fish numbers and habitat in Volcano Creek, tributary to Golden Trout Creek, which harbors a valuable nonintrogressed lineage of California Golden Trout. The California Department of Fish and Wildlife temporarily relocated 52 fish, collected from three reaches of the creek, to the American River Hatchery and eventually returned them in 2017, once habitat conditions improved. If the drought had not ended and conditions continued to decline, these actions would have preserved a valuable lineage.

SUMMARY

New and continuing challenges remain for managing California Golden Trout and Little Kern Golden Trout. Continuing efforts to restore habitat damage caused by more than a century of overgrazing are essential and may require many decades for riparian vegetation and stream channels to fully recover. Intensifying effects of fire, drought, and climate change are already beginning to require additional management planning and actions. The threat of introgression of the golden trout genome by nonnative Rainbow Trout persists as most problematic. Genetic monitoring over time will be essential, and when new introgression is detected, management actions such as chemical eradication, barrier construction and maintenance, and reintroduction may be needed.

Management actions implemented in the past to address the challenges of overharvest by anglers; habitat destruction by overgrazing, competition, and predation by nonnative trout; and genetic dilution by introgression often generated amazing success. Golden trout could possibly be extinct now if not for the early intervention by fishery managers, beginning with controlling overharvest in the early 1900s and then, in 1969, with barrier construction and habitat reclamation through chemical treatment to stem the flood of nonnative fish. Fortunately, golden trout are not extinct! California Golden Trout populations in the South Fork Kern River were estimated to be surprisingly abundant (Knapp and Matthews 1996), though likely a consequence of increased spawning habitat created by grazing-induced bank widening). Brown Trout and most introgressed golden trout have been removed from the upper South Fork Kern River (Pister 2008, 2010). Nonnative Brook Trout are now absent from the Little Kern River drainage (USFWS 2011). Low to near-zero levels of introgression are now found in the majority of the populations of the Little Kern River drainage and in Golden Trout Creek indicating successful preservation of the genetic heritage of these populations (Stephens 2007, 2014; Stephens et al. 2013). Collaboration and commitment among state and federal agencies, along with many NGO partners, have been a requirement for this amazing achievement.

LESSONS LEARNED FROM OVER 100 YEARS OF GOLDEN TROUT MANAGEMENT

Lesson 1: Genetic Purity of a Species Is a Vague, Relative, and Sometimes Unattainable Goal

Defining a species can be problematic.—Scientific debate continues about how to precisely delimit species and subspecies boundaries (or distinct population segments under the federal ESA). The hierarchical, bifurcating nature of Linnaean systematics performs poorly in many salmonid groups, particularly with species or subspecies that have encountered natural secondary contact in their recent evolutionary histories. Defining species "purity" in these groups, therefore, can be extremely problematic, given the difficulty of defining the groups themselves. Establishing specific cutoff levels for what constitutes a pure or pure-enough population for conservation purposes is complicated. When faced with this dilemma, engaging population biologists and geneticists early and often in the process will help development of management options, monitor outcomes, and apply adaptive management as needed.

Genetic purity (0% detectable introgression) is perhaps unattainable in some cases, yet conservation should proceed.—Choosing to eliminate hybridized individuals from a population may result in sacrificing diversity to obtain genetic purity or throwing the proverbial "baby out with the bathwater." This approach is not recommended! In the case of golden trout, the protection of a portfolio of nonintrogressed populations that contain no apparent influence from introduced Rainbow Trout is possible; however, it is important to recognize that reduced genetic diversity exists at the individual population level. Some slightly introgressed populations may still have conservation value, despite having low to moderate levels of introgression, due to the unique genetic diversity that they may contribute to the overall species portfolio. Careful consideration is required to identify those populations that have genetic characteristics that are the highest priority for conservation.

Interpretation of measures of introgression should not strictly be informed by a percentage measure based on a limited subset of genetic loci, but rather should be concerned with the answers to many important questions, such as those below:

- Does introgression threaten evolutionary processes, ecological function, or population persistence for a species?
- Have populations been monitored to evaluate changes in population structure and genetic changes over time?
- Do introgression levels show changes over time, and if so, do these changes correspond to changes in with environmental or climatic factors (e.g., stream conditions, drought) or adaptive genetic variation?
- What proportion of the species' range is under threat?

- What is the geographic and genetic diversity distribution of extant populations, and what, if any, nonintrogressed populations exist from which to draw upon as potential donor source populations (e.g., if the potential to obtain nonintrogressed populations or sources is poor, managers might relax their acceptable levels of introgression or use selective breeding or rearing in the hatchery to recover populations)?
- What are the available resources and the desired outcomes for management actions?
- What are the consequences of inaction?

Lesson 2: Management Actions for Restoration May Require Monitoring and Maintenance Indefinitely if Success Is to Continue

Perpetual active management may be a fact of life for an increasing number of "conservation-reliant" species (Goble et al. 2012) and difficult-to-recover species for which ESA listing may never end (Doremus and Pagel 2001). For example, Lake Trout Salvelinus namaycush in Lake Superior required a 40+ year effort by a variety of fishery agencies before restoration was realized (Hansen and Bronte 2019, this volume) and continues to depend on ongoing Sea Lamprey Petromyzon marinus control. Large portions of the native ranges of California Golden Trout (mostly South Fork Kern River populations) and some Little Kern Golden Trout still contain fish with low to moderate levels of introgression, despite past management efforts. Continued removal of some introgressed and nonnative fish is needed for as long as these fish pose a threat to native species' existence. Although gill-net use in small lakes has been an effective tool (Knapp and Matthews 1998), chemical treatment methods of fish removal in select golden trout waters will likely be necessary. Despite facing public and regulatory scrutiny of potential environmental impacts, chemical treatments remain the only tool available for most locations where physical removal methods (gill nets and electroshockers) are ineffective or inefficient, such as in remote streams in difficult terrain or in waters that are too shallow, too vegetated, or too complex in habitat features; judicious use of best management practices can minimize impacts to nontarget species (Finlayson et al. 2010). The political challenges and negative public perceptions associated with using chemicals in California have delayed progress in furthering golden trout conservation and will need to be addressed in order to proceed with high-priority restoration actions to eliminate nonnative fish threatening native populations in key locations.

Lesson 3: Managing Species Means Dealing with Evolving Threats

Restoration can feel like a game of "whack-a-mole," where threats are initially addressed only to reappear later, emerge in a different form, or yield and reveal other pressing threats to a species that were previously unknown. For example, hybridization of Little Kern Golden Trout was initially a primary threat to species persistence; although this threat has been ameliorated in much of the species' native range, managers are now left to deal not only with remaining small pockets of low-to-moderate hybridization, but also with low genetic diversity (bottlenecks) and genetically isolated populations (Stephens 2014; Lusardi et al. 2015). Having the courage to respond immediately and adaptively to threats as they emerge is a key to dealing with ongoing threats to population and species persistence.

Lesson 4: Physical Barriers: Good and Bad for Conservation

Physical barriers, both natural and constructed, receive considerable attention from fishery professionals for their detrimental effects—of blocking upstream migratory corridors and causing genetic isolation and bottlenecks. Despite their potential drawbacks, barriers should also be recognized for their conservation benefits—for protecting upstream populations from invasion by downstream nonnative species (Rahel 2013). Natural and artificial barriers have helped to isolate nonintrogressed populations and protect native California Golden Trout and Little Kern Golden Trout populations from the spread of both introgressed and nonnative fish from downstream waters. Artificial barriers can be expensive (millions of dollars of state funds invested for Templeton and Schaeffer barriers constructed in the Golden Trout Wilderness), require maintenance to prevent deterioration over time, and require monitoring to evaluate their functioning under various environmental scenarios such as high-water years. Striking a balance between these positive and negative aspects of physical barriers will benefit native species management in the future.

Lesson 5: Hatcheries: Good and Bad for Conservation

Hatcheries have garnered notoriety for their negative impacts on wild fish species (e.g., Fraser 2008; Araki and Schmid 2010; Christie et al. 2012; Le Luyer et al. 2017). However, recent decades have seen the emergence of conservation hatchery approaches that are increasingly improving their operations to minimize domestication selection, increase the representation of wild fish, and improve their conservation applications (e.g., Kline et al. 2019, this volume). The goal of a conservation hatchery is to put yourself out of business by stocking native fish that will survive and successfully reproduce. Some related lessons regarding the use of hatcheries have emerged from their use for golden trout conservation.

- Use dedicated facilities (i.e., whole hatcheries or physically isolated operations within a hatchery dedicated to a single native species) to avoid unintentional introgression of native stocks that are used in restoration, as likely occurred during the 1930s for the California Golden Trout broodstock held in Cottonwood Lakes.
- The originating fish used to establish a broodstock must not be introgressed. Past efforts to ensure the genetic integrity of native California Golden Trout and Little Kern Golden Trout broodstock source populations were developed, in hindsight, used inadequate information, used false assumptions, and ignored warning signs of nonnative Rainbow Trout influence. Early introgression of the Cottonwood Lakes population resulted in the hatchery propagation of introgressed California Golden Trout that were subsequently stocked by the thousands into waters both within and outside the California Golden Trout's native range. Contamination of broodstock also likely occurred in Little Kern Golden Trout at the Kern River Hatchery and Planting Base, resulting in Rainbow Trout genes being introduced into the Little Kern River (Stephens 2007).
- Use best hatchery practices for broodstock propagation. Management should consider whether additional measures such as careful outcrossing, which involves the breeding of individuals from genetically different populations (Krueger et al. 1981) could potentially improve the overall genetic diversity of native golden trout. However, outcrossed populations should be monitored in both hatchery and wild environments, as the potential negative effects of outcrossing remain poorly understood. Best practices for managing hatcheries for restoration

purposes have been described by several authors (e.g., Miller and Kapuscinski 2003; George et al. 2009; Kostow 2012) and include using a dedicated conservation hatchery, following a hatchery genetic management plan, and conducting regular genetic monitoring of broodstock and progeny.

- Invoke the use of hatchery stocking sparingly and strategically. Relying on regular hatchery propagation and stocking rather than on natural reproduction for the purposes of maintaining a species can lead to a status quo with no real progress toward restoration. Alternatives to hatchery propagation are the use of intrabasin transfers of genetically pure fish to waters that have been chemically reclaimed or the rescue of populations in peril, or otherwise at risk, and transferring them to fishless lakes or streams where nonnative fish have been removed and habitat is suitable (or temporarily to a hatchery). Hatchery-produced fish should not be introduced on top of donor/source or sensitive fish populations unless intentional outcrossing is a carefully planned part of an overall genetic management strategy.

Lesson 6: Having an Insurance Policy Is a Good Idea, Even if You Hope Never to Use It

Out-of-basin transplanted populations represent insurance policies—gene banks that should be considered as a backup measure should a population catastrophe occur. Transplanted populations should not be needed to save a species, but if needed, their availability for propagation and stocking or for transplantation may be critical. Populations that have been transplanted outside their native range are subject to a variety of additional stressors, such as potential for exposure to different natural selection pressures in different environments and different diseases and microbial communities that can affect their value for reintroduction into the native range. However, sometimes transplanted populations may be the best available option for species restoration, such as was done for Lake Trout (Krueger et al. 1983; Kincaid et al. 1993). A handful of early California Golden Trout transplants to formerly fishless waters in the Kern River basin have been shown to have created small reservoirs of California Golden Trout genetic diversity outside the native range; despite having reduced genetic diversity, these nonintrogressed populations could be useful in the future, should native-range populations decline or experience local extirpations (Stephens et al. 2013). Low to moderately introgressed populations should also be carefully considered as potential insurance policies for the same reasons; they may also represent otherwise extinct unique genetic lineages or contain important adaptive diversity characteristics worth preserving, despite having some limited fraction of a Rainbow Trout genome. Even the Cottonwood Lakes, if restored with nonintrogressed California Golden Trout, may represent an opportunity to act as a reserve for species conservation. Genetic management plans (Stephens et al. 2013; Stephens 2014) provide a useful framework in which to carefully evaluate the genetic status of extant native-range populations and undertake careful risk-benefit analyses of both potential donor and introgressed populations (Lesson 1 above).

Lesson 7: Harnessing Public and Stakeholder Support Benefits Native Species, Managers, and the Public

Involving stakeholders—private individuals, grazing permittees, property owners, nongovernmental conservation organizations, and angling and conservation groups—in golden trout restoration and recovery produced proven benefits for all parties involved, as demonstrated by past and ongoing efforts for California Golden Trout and Little Kern Golden Trout (Table 1). Stakeholder involvement can include a wide range of activities, from providing labor for implementation of species management projects for accomplishing conservation goals to initiating legal action to pressure government agencies to follow through on their legal obligations to prevent species extinction.

Involvement in species management in the field, as well as developing conservation assessments and strategies, has provided opportunities to dispel misconceptions, forge partnerships and unlikely alliances, implement management solutions, diffuse conflict, promote environmental stewardship, and allow people to gain a sense of ownership in a species' future (see Moore and Kulp 2019, this volume). One example where future stakeholder involvement could be valuable for California Golden Trout is in Chicken Springs Lake and Rocky Basin Lakes in the Golden Trout Creek headwaters, where CDFW has successfully removed introgressed fish. Ironically, these empty lakes are now vulnerable to the threat of introgression through the potential illegal restocking of nonnative or introgressed fish by users who desire angling opportunities. Management alternatives exist for these now-fishless lakes: stocking them with native California Golden Trout only, or leaving lakes in their fishless in their natural state to provide habitat for other vulnerable species, such as the ESA-listed mountain yellow-legged frog *Rana muscosa*. Educating the public, highlighting existing angling opportunities, and involving stakeholders in the process of evaluating alternatives could reduce conflicts and minimize the likelihood of illegal translocations (Table 1).

People learn as they participate and can become ardent supporters with knowledge and motivation. Participation builds constituencies that support native species and counteract small but often vocal and politically influential groups from dictating management policies and activities. Building public support creates necessary political pressure to stay the course to achieve restoration. These supporters were the very forces that helped establish the Golden Trout Wilderness and brought attention to ESA-listing needs at the state and federal levels. Golden trout enjoy a huge amount of support from a range of groups—anglers who enjoy catching and observing them and people who care about native fish, are invested in their conservation, and are dedicated to preserving them indefinitely as a part of their responsibility to future generations.

ACKNOWLEDGMENTS

Bernie May, Graham Gall, and the Genomic Variation Laboratory at University of California, Davis have provided decades of golden trout research oversight and insight. Edwin Philip "Phil" Pister, CDFW (retired), has devoted decades of his career to managing golden trout and inspiring generations of fisheries biologists. Numerous agency biologists and personnel have contributed to golden trout management, including many CDFW and U.S. Department of Agriculture and National Park Service biologists and staff and members of the Threatened Trout Committee. The CDFW Heritage and Wild Trout Program and crews have provided decades of field sampling for genetic analyses. Nongovernmental organization partners (Table 1) have been essential participants in the conservation strategy, educational outreach, tissue collection for genetic analysis, and species conservation activities.

REFERENCES

Abatzoglou, J. T., and A. P. Williams. 2016. Impact of anthropogenic climate change on wildfire across western US forests. Proceedings of the National Academy of Sciences of the United States of America 113:11770–11775.

Alkire, C. 2003. Economic value of golden trout fishing in the Golden Trout Wilderness, California. California Trout, San Francisco.

Anonymous. 1913. Golden trout. California Fish and Game Commission Fish Bulletin 1:45–48.

Araki, H., and C. Schmid. 2010. Is hatchery stocking a help or harm? Evidence, limitations and future directions in ecological and genetic surveys. Aquaculture 308:S2–S11.

Armour, C., D. Duff, and W. Elmore. 1994. The effects of livestock grazing on western riparian and stream ecosystem. Fisheries 19:9–12.

Bagley, M., and G. Gall. 1998. Mitochondrial and nuclear DNA sequence variability among populations of Rainbow Trout (*Oncorhynchus mykiss*). Molecular Ecology 7:945–961.

Behnke, R. J. 2002. Trout and salmon of North America. The Free Press, New York.

Busack, C., A., and G. A. E. Gall. 1980. Ancestry of artificially propagated California Rainbow Trout strains. California Fish and Game 66:17–24.

Calhoun, A, 1961. Genetic purity of golden trout in Cottonwood Lakes. California Department of Fish and Game, Bishop, California.

California Fish and Game Commission, various dates. Golden Trout Policy, Trout Policy, and Designated Wild Trout Waters Policy. Available: www.fgc.ca.gov/policy/p2fish.aspx. (October 2018).

California Fish and Game Code. 1979. Trout and Steelhead Conservation and Management Planning Act of 1979, §1725–1730, 1979. Available: http://leginfo.legislature.ca.gov/faces/codes_displayText.xhtml?lawCode=FGC&division=2.&title=&part=&chapter=7.2.& article. (October 2018).

California State Assembly. 1947. Assembly concurrent resolution no. 52—relative to declaring the golden trout the official state fish of California. Session of the Legislature, regular session. Statutes of California (Resolution). State of California. Ch. 90 p. 3512.

CDFW (California Department of Fish and Wildlife). 2015. California State Wildlife action plan, 2015 update: a conservation legacy for Californians. A. G. Gonzales and J. Hoshi, editors. Prepared by CDFW, Sacramento, with assistance from Ascent Environmental, Inc., Sacramento, California.

Christenson, D. P. 1984. The revised fishery management plan for Little Kern Golden Trout. California Department of Fish and Game, Fresno.

Christie, M. R., M. L. Marine, R. A. French, and M. S. Blouin. 2012. Genetic adaptation to captivity can occur in a single generation. Proceedings of the National Academy of Sciences of the United States of America 109:238–242.

Cordes, J. R., M. R. Stephens, M.A. Blumberg, and B. May. 2006. Identifying introgressive hybridization in native populations of California Golden Trout based on molecular markers. Transactions of the American Fisheries Society 135:110–128.

Diffenbaugh, N. S., D. L. Swain, and D. Touma. 2015. Anthropogenic warming has increased drought risk in California. Proceedings of the National Academy of Sciences of the United States of America 112:3931–3936.

Doremus, H., and J. E. Pagel. 2001. Why listing may be forever: perspectives on delisting under the US Endangered Species Act. Conservation Biology 15:1258–1268.

Dull, R. A. 1999. Palynological evidence for 19th century grazing induced vegetation change in the southern Sierra Nevada, California, USA. Journal of Biogeography 26:899–912.

Endangered American Wilderness Act. 1978. U.S. Code, volume 16, sections 1131–1136.

Ellis, S. L. N., and H. C. Bryant. 1920. Distribution of the golden trout in California. California Fish and Game 6:141–152.

Erickson, P. B. 2013. The search for the Kern River Rainbow: conservation genetics of a native trout. University of California Davis, Davis.

Evermann, B. W. 1906. The golden trout of the southern high Sierras. U.S. Bureau of Fisheries Bulletin 25:1–51.

Finlayson, B., R. Schnick, D. Skaar, J. Anderson, L. Demong, D. Duffield, W. Horton, and J. Steinkjer. 2010. Planning and standard operating procedures for the use of rotenone in fish management: rotenone SOP manual. American Fisheries Society, Bethesda, Maryland.

Fraser, D. J. 2008. How well can captive breeding programs conserve biodiversity? A review of salmonids. Evolutionary Applications 1:535–586.

Freitas, M. R., L. M. Roche, D. Weixelman, and K. W. Tate. 2014. Montane meadow plant community response to livestock grazing. Environmental Management 54:301–308.

Gall, G. A. E., C. A. Busack, R. C. Smith, J. R. Gold, and B. J. Kornblatt. 1976. Biochemical genetic variation in populations of Golden Trout, *Salmo aguabonita*. The Journal of Heredity 67:330–335.

George, A. L., B. R. Kuhajda, J. D. Williams, M. A. Cantrell, P. L. Rakes, and J. R. Shute. 2009. Guidelines for propagation and translocation for freshwater fish conservation. Fisheries 34:529–545.

Goble, D. D., J. A. Wiens, J. M. Scott, T. D. Male, and J. A. Hall. 2012. Conservation-reliant species. Bioscience 62:869–873.

Gold, J. R. 1977. Systematics of western North American trout (*Salmo*), with notes on the redband trout of Sheepheaven Creek, California. Canadian Journal of Zoology 55:1858–1873.

Gold, J. R., and G. A. E. Gall. 1975. The taxonomic structure of six Golden Trout (*Salmo aguabonita*) populations from the Sierra Nevada, California (Pisces: Salmonidae). Proceedings of the California Academy of Sciences 40:243–263.

Hansen, M. J., and C. R. Bronte. 2019. Restoration of Lake Trout in Lake Superior through interagency cooperative management. Pages 385–409 *in* C. C. Krueger, W. W. Taylor, and S. Youn, editors. From catastrophe to recovery: stories of fishery management success. American Fisheries Society, Bethesda, Maryland.

Hogan, S., and C. Buchanan. 2015. Drought assessments of Little Kern Golden Trout (*Oncorhynchus mykiss whitei*). California Department of Fish and Wildlife, Heritage and Wild Trout Program, Rancho Cordova.

Kattelmann, R., and M. Embury. 1996. Riparian areas and wetlands. Chapter 5 in Sierra Nevada ecosystem project: final report to Congress, volume 3. Assessment and scientific basis for management options. Centers for Water and Wildland Resources, University of California, Davis.

King, C. 1902. Mountaineering in the Sierra Nevada. Charles Scribner and Son, New York.

Kincaid, H. L. C. C. Krueger, and B. May. 1993. Preservation of genetic variation in the Green Lake strain Lake Trout derived from remnant domestic and feral populations. North American Journal of Fisheries Management 13:318–325.

Kinney, W. C. 1996. Conditions of rangelands before 1905. Chapter 3 *in* Sierra Nevada ecosystem project: final report to Congress, volume 2. Assessment and scientific basis for management options. Centers for Water and Wildland Resources, University of California, Davis.

Kline, P. A., T. A. Flagg, C. C. Kozfkay, D. J. Baker, D. J. Maynard, and W. C. McAuley. 2019. Snake River Sockeye Salmon—a perspective on twenty-five years of effort to recover a population in central Idaho, USA. Pages 101–123 *in* C. C. Krueger, W. W. Taylor, and S. Youn, editors. From catastrophe to recovery: stories of fishery management success. American Fisheries Society, Bethesda, Maryland.

Knapp, R. A., and K. R. Matthews. 1998. Eradication of nonnative fish by gill netting from a small mountain lake in California. Restoration Ecology 6:207–213.

Knapp, R. A., and K. R. Matthews. 1996. Livestock grazing, golden trout, and streams in the Golden Trout Wilderness, California: impacts and management implications. North American Journal of Fisheries Management 16:805–820.

Knapp, R. A., and V. T. Vredenburg. 1996. Spawning by California Golden Trout: characteristics of spawning fish, seasonal and daily timing, redd characteristics, and microhabitat preferences. Transactions of the American Fisheries Society 125:519–531.

Knapp, R. A., V. T. Vredenburg, and K. R. Matthews. 1998. Effects of stream channel morphology on golden trout spawning habitat and recruitment. Ecological Applications 8:1104–117.

Kostow, K. 2012. Strategies for reducing the ecological risks of hatchery programs: case studies from the Pacific Northwest. Environmental Biology of Fishes 94:285–310.

Krueger, C. C., A. J. Gharrett, T. R. Dehring, and F. W. Allendorf. 1981. Genetic aspects of fisheries rehabilitation programs. Canadian Journal of Fisheries and Aquatic Sciences 38:1877–1881.

Krueger, C. C., R. M. Horrall, and J. Gruenthal. 1983. Strategy for the use of Lake Trout strains in Lake Michigan. Wisconsin Department of Natural Resources, Administrative Report Number 17, Madison.

Le Luyer, J., M. Laporte, T. D. Beacham, K. H. Kaukinen, R. E. Withler, J. S. Leong, E. B. Rondeau, B. F. Koop, and L. Bernatchez. 2017. Parallel epigenetic modifications induced by hatchery rearing in a Pacific salmon. Proceedings of the National Academy of Sciences of the United States of America 114:12964–12969.

Leitritz, E. 1970. A history of California's fish hatcheries, 1870–1960. Department of Fish and Game Bulletin 150.

Lentz, D. C., and M. A. Clifford. 2014. A synopsis of recent history of California's inland trout management programs: litigation and legislation. California Fish and Game 100:727–739.

Lusardi, R., M. R. Stephens, P. B. Moyle, C. McGuire, and J. Hull. 2015. Threat evolution: negative feedbacks between management action and species recovery in threatened trout (Salmonidae). Reviews in Fish Biology and Fisheries 25:521–535.

Matthews, K. R. 1996. Habitat selection and movement patterns of California Golden Trout in degraded and recovering stream sections in the Golden Trout Wilderness, California. North American Journal of Fisheries Management 16:579–590.

Miller, L. M., and A. R. Kapuscinski. 2003. Genetic guidelines for hatchery supplementation programs. Pages 329–355 *in* E. M. Hallerman, editor. Population genetics: principles and applications for fisheries scientists. American Fisheries Society, Bethesda, Maryland.

Moore, S. E., and M. A. Kulp. 2019. Overcoming challenges to native fish restoration in a national park. Pages 45–64 *in* C. C. Krueger, W. W. Taylor, and S. Youn, editors. From catastrophe to recovery: stories of fishery management success. American Fisheries Society, Bethesda, Maryland.

Moyle, P. B. 2002. Inland fishes of California. University of California Press, Berkeley.

Moyle, P. B., R. A. Lusardi, P. J. Samuel, and J. V. E. Katz. 2017. State of the salmonids: status of California's emblematic fishes 2017. California Trout, San Francisco.

Moyle, P. B., J. A. Israel, and S. E. Purdy. 2008. Salmon, steelhead, and trout in California: status of an emblematic fauna. California Trout, San Francisco.

Needham, P. R., and R. J. Behnke. 1962. The origin of hatchery Rainbow Trout. The Progressive Fish-Culturist 24:156–158.

Nusslé, S., K. R. Matthews, and S. M. Carlson. 2015. Mediating water temperature increases due to livestock and global change in high elevation meadow streams of the Golden Trout Wilderness. PLOS (Public Library of Science) One [online serial] 10(11) e0142426.

Nusslé, S., K. R. Matthews, and S. M. Carlson. 2017. Patterns and dynamics of vegetation recovery following grazing cessation in the

California Golden Trout habitat. Ecosphere [online serial] 8(7):e01880.

Pister, E. P. 1991. Golden Trout (*Onchorhynchus aguabonita*). Pages 280–285 *in* J. Stolz and J. Schnell, editors. Trout. Stackpole, Harrisburg, Pennsylvania.

Pister, E. P. 2008. Restoration of the California Golden Trout in the South Fork Kern River, Kern Plateau, Tulare County, California, 1966–2004, with reference to Golden Trout Creek. California Department of Fish and Game, Administrative Report Number 2008-1, Sacramento.

Pister, E. P. 2010. California Golden Trout: perspectives on restoration and management. Fisheries 35:550–553.

Rahel, F. J. 2013. Intentional fragmentation as a management strategy in aquatic systems. Bioscience 63:362–372.

Schreck, C. B., and R. J. Behnke. 1971. Trouts of the upper Kern River basin, California, with reference to systematics and evolution of western North American *Salmo*. Journal Fisheries Research Board of Canada 28:987–998.

Stephens, M. R. 2007. Systematics, genetics and conservation of golden trout. Doctoral dissertation. University of California, Davis.

Stephens, M. R., and B. P. May. 2013. Microsatellite analysis of California Golden Trout populations. Report to U.S. Fish and Wildlife Service and California Department of Fish and Wildlife (Agreement #F09AC00468 and #P1181006).

Stephens, M. R. 2014. Genetic management plan for Little Kern Golden Trout, *Oncorhynchus mykiss whitei*. Genomic Variation Laboratory, University of California, Davis.

Stephens, M. R., B. Erickson, A. Schreier, K. Tomalty, M. Baerwald, B. May, and M. Meek. 2013. Genetic management plan for California Golden Trout. Genomic Variation Laboratory, University of California, Davis.

Stephens, S. J., C. McGuire, and L. Sims. 2004 Conservation assessment and strategy for the California Golden Trout (*Oncorhynchus mykiss aguabonita*) Tulare County, California. California Fish and Game, Sacramento. Available: https://nrm.dfg.ca.gov/FileHandler.ashx?DocumentID=132635. (February 2019).

Swain, D. L., B. Langenbrunner, J. D. Neelin, and A. Hall. 2018. Increasing precipitation volatility in twenty-first-century California. Nature Climate Change 8:427–433.

USFWS (U.S. Fish and Wildlife Service). 1978. Endangered and threatened wildlife and plants; listing of the Little Kern Golden Trout as a threatened species with critical habitat. Federal Register 43:72(13 April 1978):15427–15429.

USFWS (U.S. Fish and Wildlife Service). 2001. Endangered and threatened wildlife and plants; 90-day finding on a petition to list the California Golden Trout as endangered. Federal Register 67:183(20 September 2002):59241–59243.

USFWS (U.S. Fish and Wildlife Service). 2011. Endangered and threatened wildlife and plants; 12-month finding for a petition to list the California Golden Trout as endangered. Federal Register 76:186(11 October 2011):63094–63115.

USFWS (U.S. Fish and Wildlife Service). 2011. Little Kern Golden Trout (*Oncorhynchus mykiss whitei*). 5-year review: summary and evaluation. U.S. Fish and Wildlife Office, Sacramento, California.

Voegelin, E. W. 1938. Tübatulabal ethnography. Anthropological records, volume 2, no. 1. University of California Press, Berkeley.

Walker, A. No date. The home of the golden trout—Kern Plateau. The Kern Plateau Association, Kernville, California.

Webb, R. W. 1946 Geomorphology of the middle Kern River basin, southern Sierra Nevada, California. Bulletin of the Geological Society of America. 57:367–382.

White, S. E. 1904. The mountains. McClure, Phillips & Company, New York.

Wild and Scenic Rivers Act. 1968. U.S. Code, volume 16, sections 1271–1287.

Overcoming Challenges to Native Fish Restoration in a National Park

Stephen E. Moore*[1,2]
Great Smoky Mountains National Park
107 Park Headquarters Road, Gatlinburg, Tennessee 37738, USA

Matt A. Kulp
Great Smoky Mountains National Park
107 Park Headquarters Road, Gatlinburg, Tennessee 37738, USA

Abstract.—The primary mission of the U.S. National Park Service is to protect and preserve native species. Control of nonnative species is also a principal management objective. Historic land management and stocking of nonnative Rainbow Trout *Oncorhynchus mykiss* in Great Smoky Mountains National Park in the eastern United States resulted in native Brook Trout *Salvelinus fontinalis* losing approximately 75% of their historic range. Consequently, the park initiated a program in 1976 to evaluate electrofishing for removal of Rainbow Trout in six small streams and found it to be successful upstream of impassable barriers. Later (mid-1990s), park biologists effectively used Fintrol (antimycin) to remove nonnative Rainbow Trout from low-elevation streams with impassable barriers. The National Environmental Policy Act and the Environmental Assessment process required public meetings be held in communities around the park prior to the use of Fintrol to explain the need for and purpose of these projects. In September 2008, Fintrol was used to renovate a large system—12.8 km (8 mi) of upper Lynn Camp Prong. Native Brook Trout were reintroduced in 2009, but monitoring efforts in 2010 revealed the presence of illegally stocked Rainbow Trout in upstream sections of the project area. Approximately 4.6 km (3.0 mi) of the project area was successfully retreated in 2011 to remove the reintroduced Rainbow Trout. Several public hearings were held successfully around the park to educate local residents about the Lynn Camp Prong project. Monitoring efforts during 2012–2015 showed that the Brook Trout population increased steadily, with abundance ultimately exceeding that of Rainbow Trout prior to restoration. Lessons learned are that (1) public education, buy-in, and involvement are crucial to success; (2) partnerships with state and federal agencies, local conservation groups, and the local community are essential; (3) fisheries professionals must be steadfastly committed to success and adaptable to changing conditions; and (4) restoration of native species can be controversial. If sabotage happens, reach out to the public, provide them with up-to-date information, and enlist their help.

---

* Corresponding author: smoore4fish@comcast.net
[1] Retired.
[2] Current address: 3523 Lawson Lane, Sevierville, Tennessee 3782, USA

INTRODUCTION

Throughout history, anthropogenic activities have significantly altered the world's terrestrial and aquatic habitats, resulting in habitat fragmentation and destruction and the alteration of natural hydrological processes (Simon and Lyons 1995). Ecosystems worldwide have been—and continue to be—threatened by rapid human population growth and uncontrolled consumption of natural resources (Groom 2006). Additionally, settlers from many European countries brought many of their favorite plants, animals, and fish species to their new country, which in some cases displaced native species. During the industrial revolution (approximately 1760–1840), many people in the United States became concerned about the uncontrolled destruction of natural resources such as timber, wildlife, fishes, and water quality (Meffe et al. 2006), and the concept of conservation of these resources arose within the United States. Gifford Pinchot, John Muir, Aldo Leopold, Theodore Roosevelt, and others promoted the protection of wilderness for the benefit of all creatures.

Today, many ideological facets of conservation and preservation have converged to provide federal protective status to many vast unspoiled wildernesses across the United States. National Parks and wilderness areas provide outstanding examples of preserving large land areas in their natural state. Many lands administered by the U.S. Fish and Wildlife Service, U.S. Forest Service, Bureau of Land Management, and National Park Service provide excellent examples of multiple land-use activities that have allowed sustainable harvest of fish and wildlife for many decades.

President Woodrow Wilson in 1916 created the National Park Service (NPS), a federal bureau in the U.S. Department of the Interior (DOI). The mission of this new agency was to unify management of established and future national parks to "conserve the scenery and the natural and historic objects and the wild life therein and to provide for the enjoyment of the same in such manner and by such means as will leave them unimpaired for the enjoyment of future generations" (NPS 2009). Since inception, 356 park system areas, including parks, monuments, preserves, reserves, lakeshores, seashores, rivers, wild and scenic river ways, scenic trails, historic sites, memorials, recreation areas, and parkways, have been added to the 35 original parks and monuments under NPS supervision (NPS 2009). Early in the history of the NPS, Great Smoky Mountains National Park (GRSM) was created in the eastern United States.

The legislation establishing GRSM was signed in Congress on May 22, 1926. The mission of GRSM is "to preserve exceptionally diverse resources and to provide for public benefit from and enjoyment of those resources in ways which will leave them basically unaltered by modern human influences" (Ditmanson et al. 2008). Great Smoky Mountains National Park, located in the southern Appalachian Mountains along the Tennessee and North Carolina border, is more biologically diverse than any other temperate region of comparable size on Earth. The park's extensive richness of flora and fauna has resulted in its designation as an international biosphere reserve as well as a World Heritage Site (Ditmanson et al. 2008). Given the location of GRSM, the park is within a day's automobile drive of more than half of the United States' population and thus has become the most visited park within the National Park System, exceeding 9 million visitors annually (Ditmanson et al. 2008).

The history of land use and settlement prior to the creation of GRSM provides a localized picture of the destruction of natural resources nationally. According to Lambert (1958), extensive logging activities occurred from the late 1800s until about 1936 in steep gradient mountainsides and high-elevation valleys. Pyle (1988) found that large valleys such as Cades Cove and Cataloochee were subsequently converted to cattle pasture and row crop agriculture. Logging activities practically denuded the mountainsides, and the agricultural activities left no buffer zones along stream corridors. King (1937) found that these activities caused increased stream sedimentation from surface runoff and bank erosion, as well as increased water temperature due to canopy cover loss. King (1939), Jones (1959), and Vitousek et al. (1997) provided evidence that introductions of nonnative species such as feral hogs *Sus scrofa*, Rainbow Trout *Oncorhynchus mykiss*, and many plant species within GRSM and the surrounding area were intended to improve wildlife and fisheries resources. However, many of these species deleteriously affected native flora and fauna—sometimes eliminating them altogether from segments of their historic ranges (King 1939). The majority of these introductions occurred before park establishment, and managers recently have attempted to eradicate many of these nonnative species within portions the historic range of impacted native species (Gibbs 2013).

This chapter chronicles the struggles and corresponding lessons learned while working to remove invasive Rainbow Trout and restore native Brook Trout *Salvelinus fontinalis* in some stream segments within the GRSM, with a special focus on Lynn Camp Prong. Critical to the success of these projects has been a long-term commitment from a consortium of federal and state agencies, local conservation groups, and the local community.

CHALLENGES BROOK TROUT FACED WITHIN THE PARK

Reestablishment of displaced native species within park boundaries through implementation of techniques that minimize effects on nontarget native species has been essential to fulfill native species preservation as outlined by the Park Service's mission (Ditmanson et al. 2008). Brook Trout are among the native fauna contributing to the rich natural heritage of GRSM. Brook Trout is the only native salmonid in Tennessee, and the southern Appalachian Mountains represent the southern extent of the species' distribution (Etnier and Starnes 1993). Historically, Brook Trout range extended north along the Appalachian Mountains through the northeast Atlantic coast to Hudson Bay and west across the Great Lakes to eastern Minnesota and northeastern Iowa. Initial genetic studies by Stoneking et al. (1981), McCracken et al. (1993), and Hayes et al. (1996) indicated that Brook Trout in GRSM were genetically different from northern counterparts. However, more recent research has provided evidence that populations throughout native Brook Trout range do not warrant subspecific status but emphasizes the need to protect pure populations not affected by past stocking efforts. Protection of those populations within GRSM is paramount for preservation of unique "Southern Appalachian" Brook Trout genotypes (Habera and Moore 2005; Richards et al. 2008).

Prior to settlement and logging activities, Brook Trout probably inhabited 680 km (425 mi) of 1,173 km (733 mi) of fishable streams in GRSM (Kelly et al. 1980). These authors also stated that Brook Trout had lost approximately 75% of their range within GRSM since the turn of the century. Prior to the establishment of the park, initial range loss (about 50%) was caused by logging and resultant water-quality degradation and virtually eliminated Brook Trout in streams below about 914 m (3,000 ft) in elevation (King 1937). Around 1910, to meet the demand for angling at the time, logging companies began stocking the area's streams with nonnative Rainbow Trout and hatchery-produced Brook

Trout. This activity continued until the park was established in 1934, and the park subsequently stocked both species for recreational purposes until 1974. Brook Trout distribution studies by Kelly et al. (1980) demonstrated that Brook Trout had lost 45% of the range they exclusively occupied in the mid-1930s. Larson and Moore (1985) demonstrated that the decline in stream miles exclusively occupied by Brook Trout was the direct result of Rainbow Trout encroachment into previously unstocked Brook Trout streams. Native Brook Trout had become restricted to marginal headwater streams above 1,067 m (3,500 ft), characterized by steep gradients (>8%) and pH that was naturally slightly acidic (pH 6.2–6.9). According to Kelly et al. (1980), the only places where Brook Trout were not displaced were streams above waterfalls that Rainbow Trout could not ascend and where they had not been introduced. The mechanism by which Rainbow Trout displaced Brook Trout in GRSM streams remains unknown but is thought to be competitive in nature (Larson and Moore 1985). Prior to 1974, Park staff saw no harm in stocking Rainbow Trout as this activity seemed to meet the NPS mandate for "use and enjoyment" of national parks.

Parkwide water quality monitoring was initiated in 1992 as part of an inventory and monitoring program. Data from this program clearly showed that stream acidity increased with increasing elevation as a result of increased atmospheric acidic deposition (Robinson et al. 2001). Models indicated that if these pH trends continued, stream segments above 1,067 m (3,500 ft) would have a pH of 6.0 in 15 years and that this would decline to 5.5 in 26 years, thus eliminating Brook Trout and other aquatic species. Brook Trout distribution surveys in the 1990s documented the loss of all Brook Trout upstream of a barrier waterfall in one stream and the loss of upstream most segments of Brook Trout in five other streams. Monitoring surveys since 2010 have documented headwater range loss in an additional stream. All losses have been in streams at or above 1,067 m, and field pH measurements indicated that streams were too acidic (<6.0) to support fish life (Cai et al. 2010; Neff et al. 2009). These data increased the urgency of restoring, as rapidly as possible, mid- to low-elevation stream segments with native Brook Trout identified in the GRSM fishery management plan (Moore 1993).

PARTNERSHIP DEVELOPMENT FOR BROOK TROUT RESTORATION

Given that wild trout fisheries in Tennessee and North Carolina have characteristics similar to those in GRSM and that state and federal agencies managing those resources have similar information needs, a partnership program was developed in the late 1980s. This partnership is particularly beneficial for sampling monitoring stations on large (≥10 m wide) wild trout streams, as well as for planning and conducting native fish restorations. Staff from agencies in both states and GRSM also reached out to local universities with fisheries programs and offered their students a chance to participate in these cooperative efforts. This multi-faceted partnership continues today, yielding scientific data needed to assess annual abundance variation, restore important native fish populations (not limited to Brook Trout), and guide various management decisions.

A new partner joined the cooperative program in 1993 when the Little River Chapter of the nongovernmental organization (NGO) Trout Unlimited (TU) was founded. The primary purposes of this chapter, located close to GRSM, were to provide the park with volunteer labor for fishery sampling activities and to assist with procuring funds for Brook Trout restoration and other fishery research needs. Project funds were secured from sources including TU's Embrace-a-Stream grants, the Trout and Salmon Foundation, and local fundraising events. Chapter members immediately went to work procuring funds for a streambank stabilization project in Cades Cove to repair damage caused by cattle. Upon completion of this initial project, stream sampling workdays were scheduled and Chapter members came out in force. Little River Chapter of Trout Unlimited members also leveraged assistance from members of other area TU chapters by informing them about these events and inviting them to participate.

Concurrently during the early 1990s, water quality monitoring data obtained through the University of Tennessee provided evidence that pH in stream segments above 3,500 ft was decreasing because of atmospheric deposition of sulfates (SO_4^{-2}) and nitrates (NO_3-). This deposition was acidifying watershed soils and leaching cations from them, resulting in lowered stream pH levels that were potentially dangerous to native Brook Trout populations. When the University of Tennessee proposed a more intensive sampling program than available staff could conduct, the Little River Chapter of Trout Unlimited helped collect the additional water samples.

Subsequently, after GRSM fishery staff identified several Brook Trout restoration projects that could not be internally funded because of budget shortfalls, members of the Little River Chapter of Trout Unlimited applied for Embrace-A-Stream grants through TU. These grant applications were successful and provided project funds almost every year for more than a decade. Additionally, thanks to information received from TU members, GRSM fishery staff applied for and received funds from the Trout and Salmon Foundation, The Orvis Company, and Friends of the Smokies. Park staff also applied for and received funds from the Eastern Brook Trout Joint Venture, a group of East Coast fishery managers and biologist dedicated to the preservation of Brook Trout throughout its native range (Eastern Brook Trout Joint Venture 2011). The funds received from these sources were used to hire seasonal staff necessary to complete three large restoration projects with Fintrol (antimycin, a piscicide used to remove Rainbow Trout). Little River Chapter of Trout Unlimited members also provided valuable onsite help during each project as well as with the collection and reintroduction of native Brook Trout into the restored stream sections.

GUIDELINES FOR RESTORATION

The goal of restoring or enhancing portions of historic native Brook Trout range within GRSM presented managers with a challenge—how to remove nonnative Rainbow Trout without harming other native species. The enabling legislation of the National Park System mandates that park resources were to be managed in such manner and by such means that would leave them unimpaired for the enjoyment of future generations (NPS 2006). National Park Service management policies further state that management of exotic (nonnative) species, up to and including eradication, will be undertaken whenever such species threaten park resources or public health and when control is prudent and feasible.

Re-establishment of historically displaced native species within park boundaries required the implementation of techniques that minimized effects to nontarget native species consistent with the native species preservation principle as outlined by the NPS' mission (Ditmanson et al. 2008). Thus, when GRSM fishery managers began to evaluate techniques for the removal of nonnative Rainbow Trout, they had to ensure that there would be no detrimental long-term effects on aquatic macroinvertebrates and other nontarget organisms. A technique that minimized effort and cost (Moore et al. 2008) was also desirable.

When Brook Trout restoration effort began in 1976, restoration efforts were limited to backpack electrofishing techniques. The goal during this time was to determine if this technique could be used successfully to restore segments of small headwater streams back to allopatric Brook Trout. Lessons learned from this work and additional fishery management needs led to the development of a fishery management plan in 1993 (Moore 1993) This plan had section that defined a program for monitoring Brook Trout distribution, research needs, and a proposed restoration strategy.

HISTORY OF ELECTROFISHING RESTORATION EFFORTS IN THE PARK

Great Smoky Mountains National Park funded a study in 1976 to evaluate the effectiveness of electrofishing as a tool for eliminating nonnative Rainbow Trout from six small headwater streams. Nonnative Rainbow Trout were removed from the streams annually using one backpack electrofishing unit and a crew of one or two assistants. Moore et al. (1981) indicated that the annual efforts had significantly reduced (≥95%) but not eliminated the Rainbow Trout population in each stream after 4 years. A review of the data from this study clearly showed that annual removals collected a significant portion (60–70%) of adult Rainbow Trout and a small segment (≤30%) of the age-0 fish each year. The adults that remained spawned the next spring, thus perpetuating the population and making eradication by this technique a difficult goal to achieve, even after several years.

Moore and Larson (1989) and West et al. (1990) later evaluated 7 of the 10 streams in which Brook Trout restoration had been attempted by electrofishing in GRSM. They determined that native Brook Trout had been successfully restored in two streams (Silers Creek and Taywa Creek). Apparently, the last removal effort in 1981 either eliminated Rainbow Trout or reduced their density to the point where reproduction ceased. Additionally, each stream had a barrier to prevent recolonization and a mean width of ≤4.0 m, making them ideal for restoration with one backpack electrofishing unit. The authors concluded that the variation in removal success using one backpack electrofishing unit was due to stream width, stream channel complexity, presence or absence of a barrier, and the amount of effort expended. Similarly, Habera et al. (1992) found that complexity of stream channel morphology significantly reduced electrofishing efficiency for nonnative Rainbow Trout. Specifically, streams with a mean width of ≥6 m and ≥4 pools/km greater than 1 m deep permitted a residual population of Rainbow Trout not vulnerable to electrofishing to remain in the deep pools. Habera et al. (1992) also stated that the use of piscicides (fish toxins) was the only restoration tool available to ensure complete removal of Rainbow Trout in streams of this size.

EVALUATION OF ALTERNATIVE METHODS TO ELECTROFISHING

Experiences using electrofishing led to the conclusion that if Brook Trout were to be successfully restored in large streams (≥6 m mean width) with numerous deep (>1.0 m deep) pools at elevations lower than 1,000 m, an alternative technique would be needed. The only other method that has proven successful for eliminating nonnative or unwanted fish species has been the application of piscicides. Because of increased concerns related to atmospheric acid deposition and headwater Brook Trout range loss, park fishery staff initiated efforts in 1996 to evaluate the use of the piscicides antimycin and rotenone for restoration efforts in selected streams. Extensive literature surveys, staff training at Crater Lake National Park (Oregon) in connection with a native Bull Trout *Salvelinus confluentus* restoration project, and research in GRSM (Walker 2003) indicated that antimycin would have no long-term impacts on the environment. Based on this information, GRSM managers concluded that antimycin could be safely used to restore native Brook Trout in stream segments upstream of natural barriers.

Any native fish restoration project in GRSM involving piscicides could have major environmental consequences and thus would be governed by the National Environmental Policy Act (NEPA). The initial step in the NEPA compliance procedure was to complete an Environmental Assessment (EA) for the project. The EA process included (1) an evaluation of alternatives for the removal of nonnative trout, (2) a description of the affected environment, (3) an assessment of the potential impact of each alternative, (4) public review and comment related to the proposed project (including public meetings in communities around GRSM), and (5) selection of the preferred alternative. An EA advisory team was formed to assist GRSM staff with each step of the EA. Biologists from the states of Tennessee and North Carolina, the U.S. Forest Service, and TU representatives served on the team. The EA identified six streams in which antimycin could be used to restore native Brook Trout populations.

The EA for the experimental use of antimycin was approved in May 2000, but because of public and internal concerns related to potential long-term impacts to nontarget organisms, the initial approval included only Sams Creek. Restoration of native Brook Trout populations in the other five streams using antimycin could proceed only if subsequent evaluations of aquatic macroinvertebrates and other nontarget organisms indicated no adverse long-term effects. Research did provide conclusive evidence that antimycin had no long-term impacts on nontarget organisms in Sam's Creek or on the environment (Walker 2003). Therefore, GRSM managers determined that antimycin could be safely used in the other five streams identified in the EA.

A RESTORTION CASE STUDY—LYNN CAMP PRONG

The Lynn Camp Prong subwatershed was identified as a stream that could be restored for native Brook Trout in the "Environmental Assessment for Using a Piscicide for Brook Trout Restoration, Great Smoky Mountains National Park" (GRSM 2000). Primary reasons for identifying this stream were that (1) it was a natural barrier to prevent reinvasion of nonnative salmonids from downstream; (2) the Brook Trout population in upper Lynn Camp Prong was possibly the most threatened Brook Trout population in GRSM as it was open to invasion by Rainbow Trout from downstream sources and threatened by acidic deposition in headwater reaches; (3) the project would reconnect three isolated headwater populations; (4) the project would restore 12.8 km of stream below 1,067 m, which is less susceptible to acidic deposition and pH effects on trout; and (5) the completion of this project helped GRSM meet restoration goals for native Brook Trout identified in the park's fishery management plan. An additional benefit of the project is that the successful completion of the project would provide park anglers with the opportunity to fish for native Brook Trout in a large, low-elevation stream. Roughly 12 public meetings were held in the 2 years prior to treatment in order to brief the public and respond to public questions and comments. No significant opposition to the project was voiced during any of these public meetings.

The Lynn Camp Prong subwatershed is located in Sevier and Blount counties

(Tennessee) in the southwestern interior of GRSM (Figure 1). Lynn Camp Prong is a fourth-order stream with a large cascade (30 m high) located roughly 400 m upstream of its confluence with Thunderhead Prong. The cascade on Lynn Camp Prong serves as a barrier to upstream fish movement and marks the downstream boundary of the 12.8 km of stream that was to be restored for native Brook Trout. Five tributaries enter Lynn Camp Prong within the restoration zone: Marks Creek (third-order), Panther Creek (second-order), Indian Flats Prong (third-order), Buckeye Cove (first-order), and an unnamed tributary (first-order) just downstream of Buckeye Cove Creek. Rainbow Trout occupied Lynn Camp Prong from the barrier up to the last 400-m reach, which comprised a short segment of sympatric Brook Trout and Rainbow Trout (200 m) followed by one of allopatric Southern Appalachian Brook Trout (200 m). A small population of Blacknose Dace *Rhinichthys atratulus* is present in the downstream-most 1.8 km of Lynn Camp Prong. Prior to treatment, backpack electrofishing techniques were used to collect as many Blacknose Dace as possible. These fish were placed in holding cages in Thunderhead Prong until treatment efforts were completed and then returned to Lynn Camp Prong. Rainbow Trout were present on Indian Flats up to Indian Flats Falls, above which only nonnative, hatchery origin Brook Trout were present. Rainbow Trout were also present in

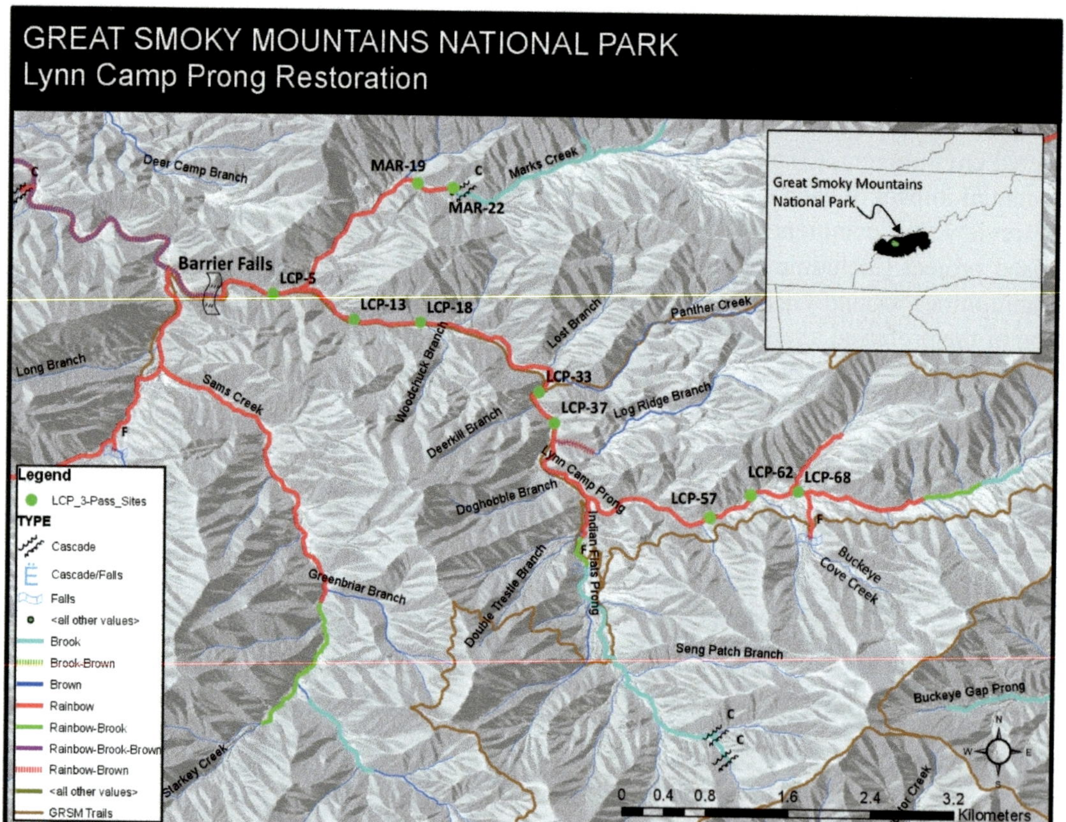

FIGURE 1. Map of Lynn Camp Prong treatment area within the Middle Prong Little River watershed, Great Smoky Mountains National Park. Tags were placed every 100 m starting at the barrier falls (LCP-1) upstream 8.6 km to the end of the treatment zone at tag LCP-86.

the first 60 m of Panther Creek and occupied 2.2 km of Marks Creek up to a barrier falls, above which only native Brook Trout were present.

Lynn Camp Prong and its main tributaries (Marks Creek, Indian Flats Prong, and Buckeye Cove) have physical and chemical characteristics (Table 1) similar to those of other GRSM streams (Fakhraei et al. 2016). Mean wetted widths of Lynn Camp Prong ranged from 2.2 to 10.8 m, stream gradient ranged from 3% to 12%, discharge ranged from 0.20 to 0.34 m^3/s, mean conductivity was 10.46 (SE = 0.37) μS/cm, and mean stream pH was 6.56 (SE = 0.07).

INITIAL TREATMENT OF LYNN CAMP PRONG

Great Smoky Mountains National Park partnered with numerous federal, state, and nongovernmental organizations in September 2008 to restore 10.2 km (6.4 mi) of Lynn Camp Prong and 2.2 km (1.4 mi) of Marks Creek (a tributary) for Brook Trout using antimycin to remove nonnative Rainbow Trout (Figure 2). The 2-week removal effort involved 31 people from four federal agencies, three state agencies, and one university.

Mainstem Lynn Camp Prong was divided into numbered 100-m increments or sites starting with site 1 at the downstream barrier and extending to site 86 at the upper end of the Rainbow Trout distribution on Lynn Camp Prong. Sites in tributaries within the treatment area were designated similarly, beginning at their confluence with Lynn Camp Prong.

Holding cages with sentinel Rainbow Trout (Figure 3) were initially used to verify piscicide toxicity (Moore et al. 2008); however, problems occurred when black bears *Ursus americanus* learned to associate these cages with food. Crews placed sentinel cages every 200 m throughout the upper section of Lynn Camp Prong on the day prior to treatment, but a bear found one, tore it open, and ate or released 10 live Rainbow Trout. Over the next 5 d, the bear(s) destroyed nearly 30 cages, likely releasing some Rainbow Trout back into the stream. Accordingly, a decision was made to stop using sentinel cages and move piscicide application stations closer together to ensure effective treatment of each segment.

Although running blind without the use of sentinel fish to ensure adequate antimy-

TABLE 1. Physical and chemical characteristics of Lynn Camp Prong and major tributaries treated during the Brook Trout restoration project. All ranges represent those observed during base flow treatment conditions. Tributary 1 was located just downstream of campsite 28. N/A = no data.

Stream segment	Treated length (km)	Stream width (range m)	Stream gradient (range %)	Discharge (range m^3/s)	Mean conductivity (μS/cm) (SE)	Mean pH (SE)
Lynn Camp Prong	8.4	2.2–10.8	0–28	0.020–0.340	10.46 (0.37)	6.56 (0.07)
Marks Creek	2	3.6–5.5	0–24	0.010–0.166	14.06 (0.45)	N/A
Indian Flats Prong	3.4	3.1–6.5	0–49	0.014–0.102	12.08 (0.59)	N/A
Tributary 1 below campsite 28	0.4	1.8–3.1	0–35	0.026–0.043	N/A	N/A
Buckeye Cove	0.6	1.4–1.6	0–35	0.009–0.011	N/A	N/A

FIGURE 2. Fluorescein dye imparts a greenish tint to the water as it moves downstream, marking the progress of antimycin during a treatment of Lynn Camp Prong to remove nonnative Rainbow Trout. National Park Service photo.

cin toxicity, managers were confident that the 200-m separation and ≤15-m elevation change between treatment stations would be effective. Data from previous antimycin treatments in GRSM (Moore et al. 2008) indicated that an elevation change of more than 15 m between treatment stations resulted in loss of efficacy near the lower station as oxidization permitted some fish to survive 8 h exposure at 8 parts per billion. Daily visual inspections during treatment indicated that Rainbow Trout were being killed within each segment.

The Lynn Camp Prong project area was surveyed in early June 2009 to determine if the September 2008 treatment had successfully eliminated Rainbow Trout from the stream. The original restoration plan called for conducting electrofishing surveys on a subset of the treatment sites in the main stem and tributaries to make this determination. Accordingly, project managers selected eight assessment sites across the entire treatment area of Lynn Camp Prong, two sites on Marks Creek, and the upper 200 m of two tributary streams (Figure 1). Sites with habitat characteristics that hindered complete eradication of nonnative Rainbow Trout (i.e., major woody debris piles, low-flow and high-organic-load headwater sections, and large deep pools) were selected for monitoring as eradication of Rainbow Trout in these areas would infer eradication throughout the treatment area. If Rainbow Trout were found in a site, the 100 m

FIGURE 3. A coauthor (M. A. Kulp) preparing to add antimycin to a treater on Lynn Camp Prong. Note the cage in the background used to hold sentinel Rainbow Trout during treatment. National Park Service photo.

sites upstream and downstream of the affected site were also monitored. These assessment efforts produced more than 30 Rainbow Trout. Two adults were collected in the upper reach of Buckeye Cove, two adults and one age-0 fish were captured in the unnamed tributary at site 62, and three adults (two males and one female) were collected in the main stem at site 57. Additionally, 24 Rainbow Trout (including 17 age-0 fish) were collected in Marks Creek at sites 19, 20, 21, and 22. Based upon these results, project managers determined that the remaining 1.8 km of Marks Creek should be sampled to determine if Rainbow Trout were present there. This additional effort resulted in the collection of three additional adult male Rainbow Trout.

The reason Rainbow Trout remained in each of these areas was not immediately clear but appeared to be related to two possible issues encountered during treatment. The first was relatively slow streamflow velocity (≥30 min/100 m) in headwater sections during treatment, along with the potentially excessive vertical elevation drop (~12.2–15.2 m) in the main-stem reach containing station 57. In addition, stream gradients in sites 55–60 (9–12%) exceeded the preferred treatment range (<8%) to minimize antimycin neutralization due to hydrolysis (Moore et al. 2008). Second, excessive organic material (e.g., leaves) in the headwater section may have affected the antimycin toxicity. Apparently, slow flow in these headwater reaches (≥30 min/100 m) allowed the antimycin to bind with the organics in the stream, thus decreasing the effective travel distance of the chemical and rendering it ineffective for the removal of nonnative trout. Similar results have been observed in previous GRSM Rainbow Trout removal projects where water travel times exceeded 30 min/100 m.

Interference by black bears was another issue that potentially reduced the effectiveness of the 2008 treatment. Bears persistently tore into most of the sentinel fish cages during treatment, thus hindering or eliminating the best means for determining the effectiveness and travel distance of each day's treatment. Many times, sentinel fish that died after limited exposure were replaced with fresh Rainbow Trout and the section was retreated with antimycin to ensure that the treatment was effective. Project managers also suspected that some of the Rainbow Trout found in 2009 were these replacement sentinel fish that escaped from cages destroyed by bears and simply did not have enough antimycin exposure time to kill them. All sections with residual Rainbow Trout were retreated with antimycin in late July 2009. Follow-up electrofishing surveys of these and other stream sections 7 d after retreatment produced no Rainbow Trout.

Approximately 340 native Southern Appalachian Brook Trout were collected from three local streams and released into the upper 1.6 km of the project area. Source stocks were selected from populations with relatively robust genetic fitness (i.e., effective population size, allelic richness, and expected heterozygosity) located in streams within the same or next closest major watershed based upon the guidelines suggested by the Trout Committee of the Southern Division of the American Fisheries Society (Habera and Moore 2005). Additional releases were delayed until late September and early October 2009 because of increasing water temperatures (>20°C). During that time frame, approximately 1,500 native Brook Trout were collected from seven source streams (Two of three original streams and five additional streams) and released throughout the project area using the same selection guidelines described above (Habera and Moore 2005). Translocated Brook Trout were comprised of all age-classes (ages 0–4) to ensure recruitment in the newly restore stream. Most of those fish (809) were released in the uppermost 4.3 km of the main stem, with about half released in sites 46–53 (800 m) and half in sites 66–70 (500 m), yielding a stocking density of about 50–80 Brook Trout/100 m of stream in those areas. No Brook Trout were introduced into the more remote stream sections, which were to rely on dispersal. Based upon results of previous restoration efforts, the Lynn Camp Prong native Brook Trout population was predicted to reach pretreatment Rainbow Trout densities within 3 years (Kanno et al. 2016).

INTERFERENCE WITH RESTORATION EFFORTS AND MITIGATION

Before those 3 years elapsed, a new challenge to the success of the project arose. Surveys conducted in June 2010 to evaluate the survival and reproductive success of the reintroduced Brook Trout revealed that Rainbow Trout had been illegally stocked into a ~200-m portion of sites 32–34 centered on the Panther Creek horse trail crossing (Figure 1). The nature of this introduction was verified by the presence of several Rainbow Trout with features indicating that they were of hatchery origin. Distinguishing feature included unusually large size (302–332 mm) and pectoral fins that were either clipped or eroded—as often occurs from contact with concrete hatchery raceways. Additionally, a local informant verified that hatchery-raised

Rainbow Trout were obtained from a stocked stream outside the park, loaded into coolers, and transported by horses 3.2 km up Lynn Camp Prong where they were released. When asked about the motive for the stocking, the informant stated that seeing the supervisory biologist on TV "bragging" about the project's success motivated the perpetrators to "stick it to the federal government" by reintroducing Rainbow Trout. Fishery staff electrofished the remainder of the project area in 2010 and collected an additional 29 adult and 33 age-0 Rainbow Trout, providing evidence that the fish had been stocked prior to the spring spawning season. Managers decided at that point to continue electrofishing efforts with the goal of eliminating adult Rainbow Trout from the stream to prevent the possibility of spawning in the spring of 2011. During this process, project managers also realized that a 60-m section of a first-order tributary stream containing fewer than 10 Rainbow Trout had not been treated in 2008, and steps were immediately taken to eliminate this source of Rainbow Trout using electrofishing. After three passes through the 60-m section (two passes on one trip and one pass on a second trip netting zero Rainbow Trout), the small tributary was considered devoid of Rainbow Trout. By July 2011, 232 age-0 and 59 adult Rainbow Trout were removed from 2.9 km (1.8 mi) of the original project area (Figure 4).

Local residents, partner groups like TU, and other NGOs were furious that someone had illegally reintroduced Rainbow Trout into the Lynn Camp Prong restoration area. Some wanted game cameras placed on the trails to monitor activity, volunteers offered to patrol the stream to thwart further illegal stocking, and requests were made to investigate and prosecute those responsible

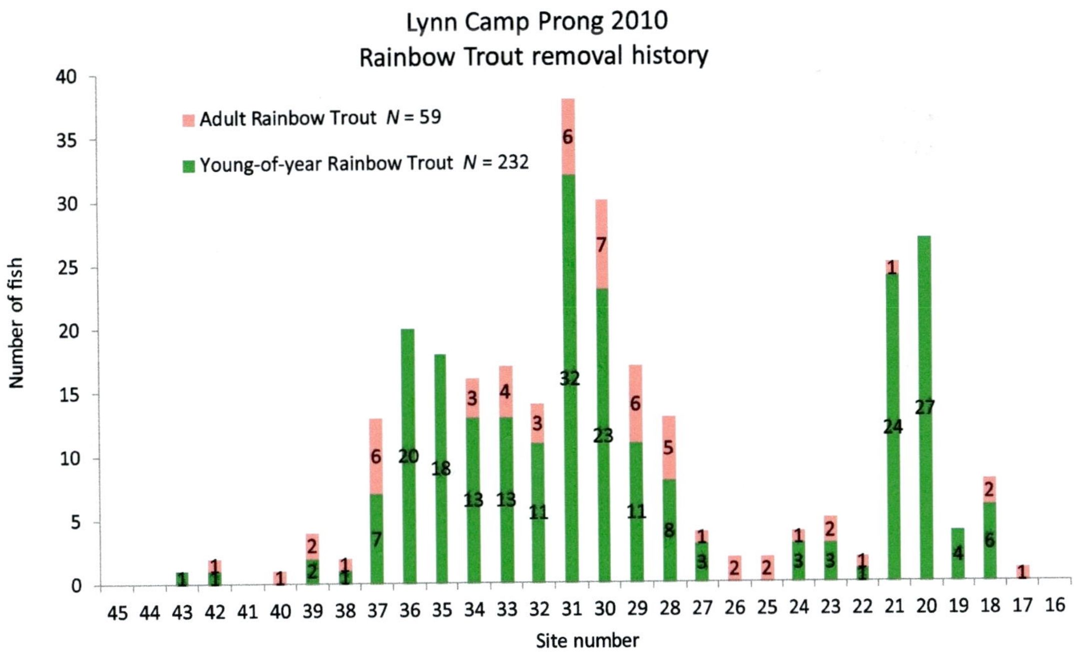

FIGURE 4. Distribution of Rainbow Trout removed from a 2-km section of Lynn Camp Prong using backpack electrofishing gear during the summer of 2010. Sites were numbered in 100-m increments starting with site 1 at the downstream barrier and ending with site 86 at the upper end of the original Rainbow Trout removal zone. Rainbow Trout were found in sites 17 through 43.

for illegally reintroducing Rainbow Trout. The NPS made it known that if the violators were caught and sabotage proven, they would be prosecuted in federal court, fined, and charged with the initial cost of the project (US$327,999) as well as the costs to retreat the stream ($250,295). After a short but thorough investigation, no charges were filed due to insufficient evidence necessary to prosecute the perpetrators.

Based upon the presence of numerous age-0 and adult Rainbow Trout in the lower portion of the Lynn Camp Prong treatment area, the decision was made to retreat the initial 4.6 km with antimycin to eliminate the Rainbow Trout. However, some issues needed to be addressed before the treatment could proceed. First, GRSM did not have enough antimycin available for the complete treatment. Also, Brook Trout had already been reintroduced to the lower 4.6 km segment so they would have to be removed, temporarily held offsite during treatment, then returned to the treatment zone. Large pools in local streams (within 6.4 km) were identified to serve as holding areas for Brook Trout removed from Lynn Camp Prong. Knowing that not all Brook Trout could be recovered from the holding pools, additional Brook Trout transplants were scheduled to also be stocked after the retreatment was completed. Finally, two questions remained. "Why would someone purposely sabotage a native Brook Trout restoration project?" And, more importantly, "Would it be attempted again?" Incomplete Rainbow Trout mortality in a few areas could be mitigated by retreating, but changing public sentiment about the federal government or overcoming opposition to a particular management project would be much more difficult.

After much consideration, a strategy was developed that focused upon engaging the public and reaffirming the primary reasons the NPS seeks to restore native species across the country: NPS policy and preservation of the cultural and natural heritage of the region. The NPS is unique in that one of its foremost resource management priorities is the restoration of native species where feasible—which may include nonnative species removal (NPS 2006). Although native species restoration is often a resource management goal of other fish and wildlife management agencies, they (and their constituents) can be reluctant to remove a popular nonnative recreational resource (e.g., Rainbow Trout or Brown Trout *Salmo trutta* fishery). The preservation of local and regional heritage, as defined in the enabling legislation of all National Parks, is a cause that most people can relate to and will also fight to preserve. Consequently, park biologists believed that local residents would rally around the Lynn Camp Prong project if this link to the area's natural heritage were promoted. Therefore, eight public meetings were held in and around GRSM during spring 2011 to discuss native fish restoration goals, with the major focus on relating these goals to the preservation of local heritage. The meetings were held in places familiar and welcoming to local residents, such as fire halls, public libraries, and other local civic meeting places. Meetings were well attended by people of all ages and response was very positive.

Planning for retreatment of the lower 4.6 km of Lynn Camp Prong began in fall 2011. Antimycin left over from the 2008 and 2009 treatments, along with some additional product acquired from Rocky Mountain National Park, was enough to complete the project. Prior to retreatment in September 2011, 725 Brook Trout were removed by electrofishing and transported to nearby refuge streams outside of the treatment area. No Rainbow Trout were found during follow-up electrofishing surveys throughout the treatment area. Afterward, 472 (65%) Brook Trout were recaptured from the holding areas in refuge streams and returned to Lynn Camp Prong. An additional 759 Brook Trout were collected from three of the original GRSM source stock streams used previously and translocated to Lynn Camp Prong.

RE-ESTABLISHMENT OF NATIVE BROOK TROUT

Park fisheries staff and volunteers from TU annually surveyed the five Lynn Camp monitoring sites (sites 5, 13, 17, 33, and 37) from 2012 to 2016 to (1) determine if Rainbow Trout had been eliminated during the 2011 retreatment, and (2) document the survival and reproductive success of reintroduced Brook Trout. Surveys in 2012 indicated that Brook Trout had successfully spawned and that population density was about 34–90% of pretreatment Rainbow Trout densities in locations where Brook Trout were released. Data from previous restoration projects indicated that adult Brook Trout density and biomass 2–5 years after reintroduction were not statistically different from prerestoration Rainbow Trout density and biomass levels (Kanno et al. 2016). Monitoring data from the five sites in the lower 5.3 km of Lynn Camp Prong indicated that Brook Trout survived but reproductive success was poor in 2012 and 2013, likely as a result of above-average total precipitation during January–May of those years (Kanno et al. 2017). Consequently, limited recruitment to age 1 and age 2 occurred in subsequent years (Kanno et al. 2017). Data from five sites in 2013 indicated that Brook Trout biomass had declined to 18–45% of maximum pretreatment Rainbow Trout biomass (34 kg/ha; Figure 5). The above-average precipitation in 2012 and 2013 appeared to delay the level of population establishment expected in 3 years (Kanno et al. 2017). Electrofishing surveys in 2015 on Lynn Camp Prong (Figure 5) and Marks Creek indicated that Brook Trout populations were within (14–34 kg/ha) or above (>34 kg/ha) the pretreatment Rainbow Trout biomass range at all sites. Biomass exceeding 35 kg/ha would be considered exceptional in terms of trout abundance for any GRSM stream. Both the 2014 and 2015 cohorts were relatively abundant, with age-0 fish comprising 53% of the total catch across all sites. Additionally, 23% of adults collected at all sites were greater than 178 mm (7 in) and the abundance of fish greater than 203 mm (8 in) was above average for GRSM streams. Given these attributes, native southern Brook Trout were considered to be successfully established in Lynn Camp Prong.

SUCCESS AND LESSONS LEARNED

The Lynn Camp Prong project produced a wide variety of natural, technical, and cultural challenges during its course. However, each of these challenges was overcome and native Brook Trout were successfully restored to the upper Lynn Camp Prong system. Accordingly, Lynn Camp Prong was reopened to fishing and harvest under general park regulations on March 6, 2015. This success was not only a milestone for the park's Brook Trout restoration program, but was also historic in that it marked the first time since the park was established in 1934 that every stream was open to fishing and harvest. Nearly all park streams had been closed to Brook Trout fishing and harvest from 1975 through 2008. One of the goals of the park's Brook Trout restoration program was to reopen streams to fishing once monitoring indicated that Brook Trout biomass had reached pretreatment Rainbow Trout biomass range (14–34 kg/ha). Recent angler catch rates have been reported (Kulp, personal observation) in the 20–40 fish/h range, with the highest catch rates typically occurring farther away from the trailhead and from access points. Given Lynn Camp Prong's history, re-establishment of native Brook Trout fishery restores a significant cultural and natural resource that can

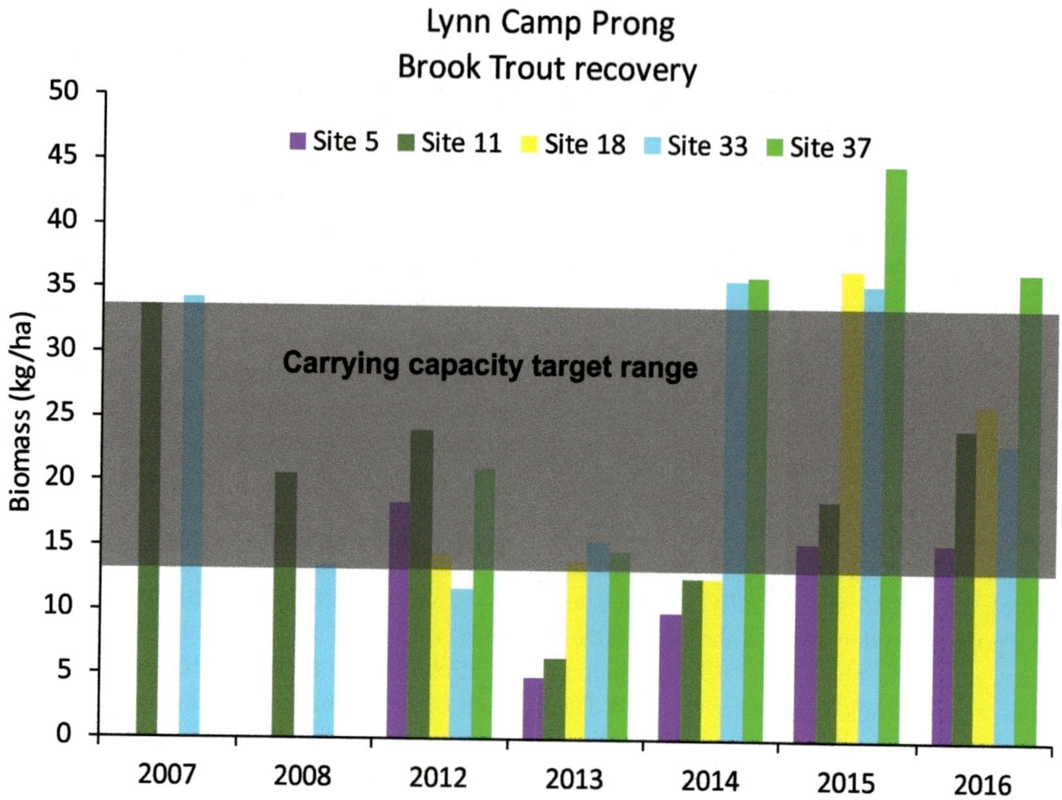

FIGURE 5. Biomass of pretreatment Rainbow Trout (gray bar) and posttreatment Brook Trout (vertical bars) after restoration of 12.8 km (8.5 mi) of Lynn Camp Prong, Great Smoky Mountains National Park using the fish piscicide antimycin. The pretreatment Rainbow Trout biomass range (14–34 kg/ha) was used as a recovery target for the newly established Brook Trout population. Each site number represents 100-m increments from the barrier falls.

now be enjoyed by current generations and those to come.

A number of lessons were learned while dealing with the challenges that arose during the Lynn Camp Prong that we believe would be useful to other fisheries professionals.

1. Public education, buy-in, and involvement are crucial to success. Educating the public prior to project implementation informs them of the project objectives, helps them understand underlying policies and justifications, and dispels any unfounded or mistruths surrounding a project. Early engagement with the public also provides an opportunity to obtain feedback and encourage involvement. Public involvement is critical for gaining support through a sense of inclusion in, and co-ownership of, the project. So, hold meetings and afford the public opportunities to find out what is being done and to ask questions. Be sure to hand out literature that can be taken home, and use social media to provide information as well.
2. Most successful restoration projects do not occur in a vacuum; rather, they typically require a large number (e.g., 12–20) of technically trained staff to complete successfully. State and federal agency personnel are essential sources for technical assistance during project

implementation, but can also serve in an advisory role during the development and planning phases. Local user groups and other NGOs were also helpful in project setup, as well as community outreach and education. All of these partnerships also had key roles in follow-up monitoring and evaluation of a project's short- and long-term success.

3. Fisheries professionals must be steadfastly committed to success and adaptable to changing conditions. Every native fish restoration project can provide unforeseen challenges that fisheries professionals must be prepared to recognize, adapt to, and mitigate. Examples include gear/technique failure or ineffectiveness, public interference, safety issues, personnel issues, and changing environmental conditions. All such issues must be effectively dealt with if the project is to be safely and successfully completed. If project staff are not prepared or willing to adapt to these challenges, failure is likely.
4. Restoration of native species can be controversial. Whether its reintroducing predators such as gray wolves *Canis lupus* (Wilson 1997), altering river flows to restore native salmon (Poff et al. 2003), or using controversial techniques like prescribed fire (Ryan et al. 2013) or biological controls (Louda et al. 2003), native species restoration can be very controversial with the public. Be clear and honest about restoration project objectives and provide policies that support them. Be prepared for controversial public meetings, allow opposing opinions to be heard, and consider their merit. Agree to disagree if necessary, but always let the public know their voices have been heard and that their opinions are valued, even if some are disagreeable.

Native species restoration is one of the most meaningful and rewarding resource management activities a fisheries professional can undertake. Re-establishing native species in a portion of their historic range not only helps restore more natural ecosystem functioning, but also restores or helps maintain a legacy for future generations to enjoy. Fishery managers should seek opportunities to restore native species where possible with the knowledge that the rewards—both ecological and societal—are worth the challenges that will undoubtedly have to be faced. Careful planning and diligent implementation are critical to successful native fish restoration efforts. If these measures are taken and public support is cultivated, success is achievable.

ACKNOWLEDGMENTS

The authors would like to thank the Eastern Brook Trout Joint Venture for project funding. We would also like to thank the dedicated and professional staff from the Tennessee Wildlife Resources Agency, North Carolina Wildlife Resources Commission, Georgia Department of Natural Resources, Tennessee Valley Authority, Tennessee Technological University, University of Tennessee, Trout Unlimited, Orvis, Federation of Flyfishers, and Little River Outfitters, as well as the many other volunteers who assisted with the Lynn Camp Prong restoration project and follow-up monitoring.

REFERENCES

Cai, M., J. S. Schwartz, R. B. Robinson, S. E. Moore and M. A. Kulp. 2010. Long-term effects of acidic deposition on water quality in a high-elevation Great Smoky Mountains National Park Watershed: use of an ion input–output budget. Water, Air, and Soil Pollution 209:143–156.

Ditmanson, D., K. Fitzgerald, D. Losson, A. Sumeriski, C. Cook, N. Finley, W. Wright, L. Jewell, J. Click, I. Wegwerth, J. Ossoff, K. Johnson, J. Renfro, J. Rock, E. Kreusch,

M. Kulp, A. Mayor, W. Stiver, G. Minnigh, M. Taylor, and T. Wales. 2008. Strategic plan for Great Smoky Mountains National Park: October 1, 2008–September 30, 2012. National Park Service, Great Smoky Mountains National Park, Gatlinburg, Tennessee. Available: http://npshistory.com/publications/grsm/strategic-plan-2005-2008.pdf. (July 2018).

Eastern Brook Trout Joint Venture. 2011. Conserving the Eastern Brook Trout: action strategies. Conservation Strategy/Habitat Work Group Eastern Brook Trout Joint Venture. Available: www.fishhabitat.org/files/uploads/EBTJV_Conservation_Strategy-Nov2011.pdf. (July 2018).

Etnier, D. A., and W. C. Starnes. 1993. The fishes of Tennessee. University of Tennessee Press, Knoxville.

Fakhraei, H., C. T. Driscoll, J. R. Renfro, M. A. Kulp, T. F. Blett, P. F. Brewer, and J. S. Schwartz. 2016. Critical loads and exceedances for nitrogen and sulfur atmospheric deposition in Great Smoky Mountains National Park, United States. Ecosphere [online serial] 7(10):e01466.

Gibbs, W. K. 2013. Effects of Fintrol (Antimycin A) application on benthic macroinvertebrate communities in the Lynn Camp Prong Watershed, Great Smoky Mountains National Park. Doctoral dissertation. Tennessee Tech University, Cookeville.

Groom, M. J. 2006. Threats to biodiversity. Pages 63–110 *in* M. J. Groom, G. K. Meffe, and C. R. Carroll, editors. Principles of conservation biology. Sinauer Associates, Sunderland, Massachusetts.

GRSM (Great Smoky Mountains National Park). 2000. Environmental assessment for using a piscicide for Brook Trout restoration, Great Smoky Mountains National Park. National Park Service, Great Smoky Mountains National Park, Gatlinburg, Tennessee.

Habera, J. W., R. J. Strange, and S. E. Moore. 1992. Stream morphology affects capture efficiency of an AC backpack electrofisher. Journal of the Tennessee Academy of Science 67(3):55–58.

Habera, J., and S. E. Moore. 2005. Managing Southern Appalachian Brook Trout: a position statement. Fisheries 30(7):10–20.

Hayes, J. P., S. Z. Guffey, F. J. Kriegler, G. F. McCracken, and C. R. Parker. 1996. The genetic diversity of native, stocked, and hybrid populations of Brook Trout in the southern Appalachians. Conservation Biology 10:1403–1412.

Jones, P. 1959. The European wild boar in North Carolina. North Carolina Wildlife Resources Commission, Game Division, Raleigh.

Kanno, Y., M. A. Kulp and S. E. Moore. 2016. Recovery of native Brook Trout populations following the eradication of nonnative Rainbow Trout in southern Appalachian Mountain streams. North American Journal of Fisheries Management 36:1325–1335.

Kanno, Y., M. A. Kulp, S. E. Moore, and G. D. Grossman. 2017. Native Brook Trout and invasive Rainbow Trout respond differently to seasonal weather variation: spawning timing matters. Freshwater Biology 62:868–879.

Kelly, G. A., J. S. Griffith, and R. D. Jones. 1980. Changes in the distribution of trout in Great Smoky Mountains National Park, 1900–1977. U.S. Fish and Wildlife Service, Technical Paper 102, Washington, D.C.

King, W. 1937. Notes on the distribution of native Speckled and Rainbow trout in the streams of Great Smoky Mountains National Park. Journal of the Tennessee Academy of Science 12:351–361.

King, W. 1939. A program for the management of fish resources in Great Smoky Mountains National Park. Transactions of the American Fisheries Society 68(1):86–95.

Lambert, R. S. 1958. Logging in the Great Smoky Mountains National Park: a report to the superintendent. National Park Service, Smoky Mountains National Park, Gatlinburg, Tennessee.

Larson, G. L., and S. E. Moore. 1985. Encroachment of Rainbow Trout in the Southern Appalachian Mountains. Transactions of the American Fisheries Society 114:195–203.

Louda, S. M., R. W. Pemberton, M. T. Johnson, and P. Follett. 2003. Nontarget effects—the Achilles' heel of biological control? Retrospective analyses to reduce risk associated with biocontrol introductions. Annual Review of Entomology 48:365–396.

McCracken, G. F., C. R. Parker, and S. Z. Guffey. 1993. Genetic differentiation and hybridization between stocked hatchery and native Brook Trout in Great Smoky Mountains National Park. Transactions of the American Fisheries Society 122:533–542.

Meffe, G. K., C. R. Carroll, and M. J. Groom. 2006. What is conservation biology? Pages 3–26 *in* M. J. Groom, G. K. Meffe, and C. R. Carroll, editors. Principles of conservation biology. Sinauer Associates, Sunderland, Massachusetts.

Moore, S. E. 1993. Fishery management plan, Great Smoky Mountains National Park. National Park Service, Washington, D.C. Available: https://irma.nps.gov/DataStore/Reference/Profile/2224431. (July 2018).

Moore, S. E., M. A . Kulp, B. Rosenlund, J. Brooks, and D. Propst. 2008. A field manual for the use of Antimycin A for restoration of native fish populations. National Park Service, Natural Resource Program Center, Natural Resource Report NPS/NRPC/NRR-2008/033, Washington, D.C. Available: https://irma.nps.gov/DataStore/DownloadFile/152709. (July 2018).

Moore, S. E., and G. L. Larson. 1989. Native Brook Trout Restoration Program in Great Smoky Mountains National Park. Pages 62–66 *in* F. Richardson and R. H. Hamre, editors. Wild Trout IV: proceedings of the symposium. Wild Trout Symposium, Bozeman, Montana. Available: www.wildtroutsymposium.com/proceedings-4.pdf. (July 2018).

Moore, S. E., B. L. Ridley, and G. L. Larson. 1981. Changes in standing crop of Brook Trout concurrent with removal of exotic trout species, Great Smoky Mountains National Park. Report prepared for the National Park Service, NPS-SER Research/Resources Management Report No. 37, Washington, D.C.

Neff, K. J., J. S. Schwartz, T. B. Henry, R. B. Robinson, S. E. Moore, and M. A. Kulp. 2009. Physiological stress in native southern Brook Trout during episodic stream acidification in the Great Smoky Mountains National Park. Archives of Environmental Contamination and Toxicology 57:366–376.

NPS (National Park Service). 2006. Management policies 2006: the guide to managing the National Park System. Available: www.nps.gov/policy/mp/policies.html#_Toc157232681. (July 2018).

NPS (National Park Service) 2009. The national parks: index 2009–2011. Government Printing Office, Washington, D.C. Available: www.nps.gov/parkhistory/online_books/nps/nps/part1.htm. (July 2018).

Poff, N. L., J. D. Allan, M. A. Palmer, D. D. Hart, B. D. Richter, A. H. Arthington, K. H. Rogers, J. L. Meyer, and J. A. Stanford. 2003. River flows and water wars: emerging science for environmental decision making. Frontiers in Ecology and the Environment 1(6):298–306.

Pyle, C. 1988. The type and extent of anthropogenic vegetation disturbance in the Great Smoky Mountains before National Park Service acquisition. Castanea 53(3):183–196.

Richards, A. L., T. L. King, B. A. Lubinski, S. E. Moore, M. A. Kulp, and L. S. Webb. 2008. Characterization of the genetic structure among Brook Trout in LeConte Creek, Tennessee. Proceedings of the Southeastern Association of Fish and Wildlife Agencies 62:195–202.

Robinson, R. B., J. Smoot, G. Harwell, and J. Shubdza. 2001. Water quality characteristics, temporal trends, and influencing factors for selected streams in Great Smoky Mountains National Park. Report prepared for the National Park Service, Washington, D.C. Available: https://irma.nps.gov/DataStore/Reference/Profile/2195018. (July 2018).

Ryan, K. C., E. E. Knapp, and J. M. Varner. 2013. Prescribed fire in North American forests and woodlands: history, current practice, and challenges. Frontiers in Ecology and the Environment 11(Online Issue 1):e15–e24.

Simon, T. P., and J. Lyons. 1995. Application of the index of biotic integrity to evaluate water resource integrity in freshwater ecosystems. Pages 245–262 *in* W. S. Davis and T. P. Simon editors. Biological assessment and criteria: tools for water resource planning and decision making. CRC Press, Inc., Boca Raton, Florida.

Stoneking, M., D. J. Wagner, and A. C. Hildebrand. 1981. Genetic evidence suggesting subspecific differences between northern and southern populations of Brook Trout (*Salvelinus fontinalis*). Copeia 1981:810–819.

Vitousek, P. M., C. M. D'antonio, L. L. Loope, M. Rejmanek, and R. Westbrooks. 1997. Introduced species: a significant component of human-caused global change. New Zealand Journal of Ecology 21:1–16.

Walker, C. A. 2003. Effects of antimycin treatment on benthic macroinvertebrates in Sam's Creek and Starkey Creek, Great Smoky Mountains National Park, Blount/Sevier counties, Tennessee. Master's thesis. University of Tennessee. Knoxville.

West, J. L., S. E. Moore, and M. R. Turner. 1990. Evaluation of electrofishing as a management technique for restoring Brook Trout in Great Smoky Mountains National Park. National Park Service, Southeast Regional Office, Research/Resources Management Report SER-90/01, Atlanta, Georgia.

Wilson, M. A. 1997. The wolf in Yellowstone: science, symbol, or politics? Deconstructing the conflict between environmentalism and wise use. Society and Natural Resources 10:453–468.

Chasing an Illusion? Successful Restoration of Okanagan River Sockeye Salmon in a Sea of Uncertainty

Kim D. Hyatt* and Margot M. Stockwell

Fisheries and Oceans Canada, Science Branch, Pacific Region
Pacific Biological Station
3190 Hammond Bay Road, Nanaimo, British Columbia V9T 6N7, Canada

Abstract.—Since the mid-1800s, human activities have increasingly dominated ecosystems within the Okanagan River basin, which spans the Canada–United States border between British Columbia and Washington State. Over the past 50 years, fisheries for anadromous salmon in the Okanagan River basin virtually disappeared as once abundant stocks, such as Sockeye Salmon *Oncorhynchus nerka*, declined to fewer than 10,000 adults returning annually (on average) in the 1990s. Threat assessments suggested degradation of freshwater habitat in the Columbia River basin as the general cause for the decline. However, recent record returns (2008–2016 average >200,000 adults) indicated surprising resilience and recovery. Review of recent stock management and restoration efforts focused on Okanagan Sockeye Salmon indicated that management actions and fortuitous events facilitated the restoration of salmon to levels exceeding recorded, historic maxima. Actions and events identified include (1) assessment to determine whether large increases in escapement provided evidence of historic underuse of spawning (Okanagan River) and rearing environment (Osoyoos Lake) capacities; (2) development of a decision support system to facilitate fish-friendly water management, which reduced losses of eggs or fry to density-independent events (Okanagan River and Lake); (3) a small contribution (<10% of total production) of hatchery-origin fish; and (4) a coincidental return to favorable marine conditions for Okanagan Sockeye Salmon. Recovery success also involved development of an ecosystem-based sustainability strategy incorporating a shared vision for dealing with human and natural system impacts on salmon from local (Okanagan River basin) to global (North Pacific Ocean) scales. Key elements that characterized efforts to restore Okanagan Sockeye Salmon were the development of ecosystem-based management (including elevated levels of engagement, cooperation, and collaboration among responsible parties to support a common cause); the creation of new knowledge of complex cause-and-effect ecological, economic, and cultural associations; and the creation of new resource management tools (e.g., models and decision support systems). Science-based collaboration to restore aquatic ecosystems and Okanagan salmon is an example of positive outcomes resulting from implementation of Canada's 2005 Wild Salmon Policy.

* Corresponding author: kim.hyatt@dfo-mpo.gc.ca

INTRODUCTION

Anadromous salmon *Oncorhynchus* spp. using the Okanagan River basin are part of a multi-species complex that occupies hundreds of freshwater lakes and rivers scattered throughout the eastern and western rims of the North Pacific Ocean (Figure 1A; Groot and Margolis 1991). Anadromous salmon populations have supported important subsistence and commercial fisheries in fresh and marine waters throughout their geographic range (Augerot 2005). However, by the mid- to late 1990s, salmon returning through the Columbia River system and its Okanagan subbasin (Figure 1B) exhibited a time-weighted trend of declining numbers, decade over decade for more than 50 years, similar to many other salmon populations at the southern end of their range (Nehlsen et al. 1991; Slaney et al. 1996; Hyatt and Rankin 1999). Most of the declines have been attributed to overexploitation and human-induced, habitat degradation (Lichatowich 1999; Lackey 2000).

Lackey (2000) provided a sobering analysis of the state of wild salmon populations in the Pacific Northwest, including the Columbia River basin. He concluded that many, and perhaps most, stocks of wild salmon in the Pacific Northwest are likely to remain at their current relatively low levels or continue to decline despite heroic, expensive, and socially turbulent attempts at restoration. Lackey also concluded that society is currently "chasing an illusion" that wild salmon restoration is possible in the Pacific Northwest. Lackey's grim prognosis for the future is supported by numerous independent assessments of historic stock status (Nehlsen et al. 1991; Slaney et al. 1996; Augerot 2005) and the failure of massive, salmon-restoration efforts (see details in Williams et al. 1999; National Research Council 1996) to lift aggregate stock size of Columbia River salmon above even 5% of historic levels in the recent decade (Table 1). However, restoration projects initiated in the early 2000s in the Columbia River's Okanagan subbasin have been associated with recent returns (2008–2017) of Sockeye Salmon *Oncorhynchus nerka* exceeding 100% of maximum historic run sizes observed over the previous 50+ years (1950s to 2007) and suggest that successful restoration is not always just an illusion.

The general objective of this chapter is to explore the history of the Okanagan anadromous salmon stocks, human-induced changes to their habitat, and their apparent responses to various types of restoration work undertaken in the Okanagan River subbasin since the early 2000s that have contributed to this salmon recovery success. Lessons from this system may provide some reasons for optimism and guidance to facilitate salmon restoration success.

RESTORATION PLANNING

Conceptually, restoration planning follows a linear, stepwise process (Figure 2). However, the reality is much less tidy because successful restoration planning and implementation must frequently satisfy complex requirements. These include (1a) a clear, common vision or goal and (1b) the will to set aside competing, mutually exclusive objectives to support attainment of the goal; (2) attention to frameworks of enabling legislation and policy that constrain actions; (3) engagement of multiple regulatory authorities and interest groups; (4a) assembly and analysis of historic observations of status and trends exhibited by a target species and their associated ecosystems in order to (4b) explain the underlying biophysical processes accounting for the status and trends shown by a species of interest; (5) identifying threats likely to constrain potential recovery; (6) identifying effective restoration or mitigation options, accompanied by (7) provision of funds and technical support to implement

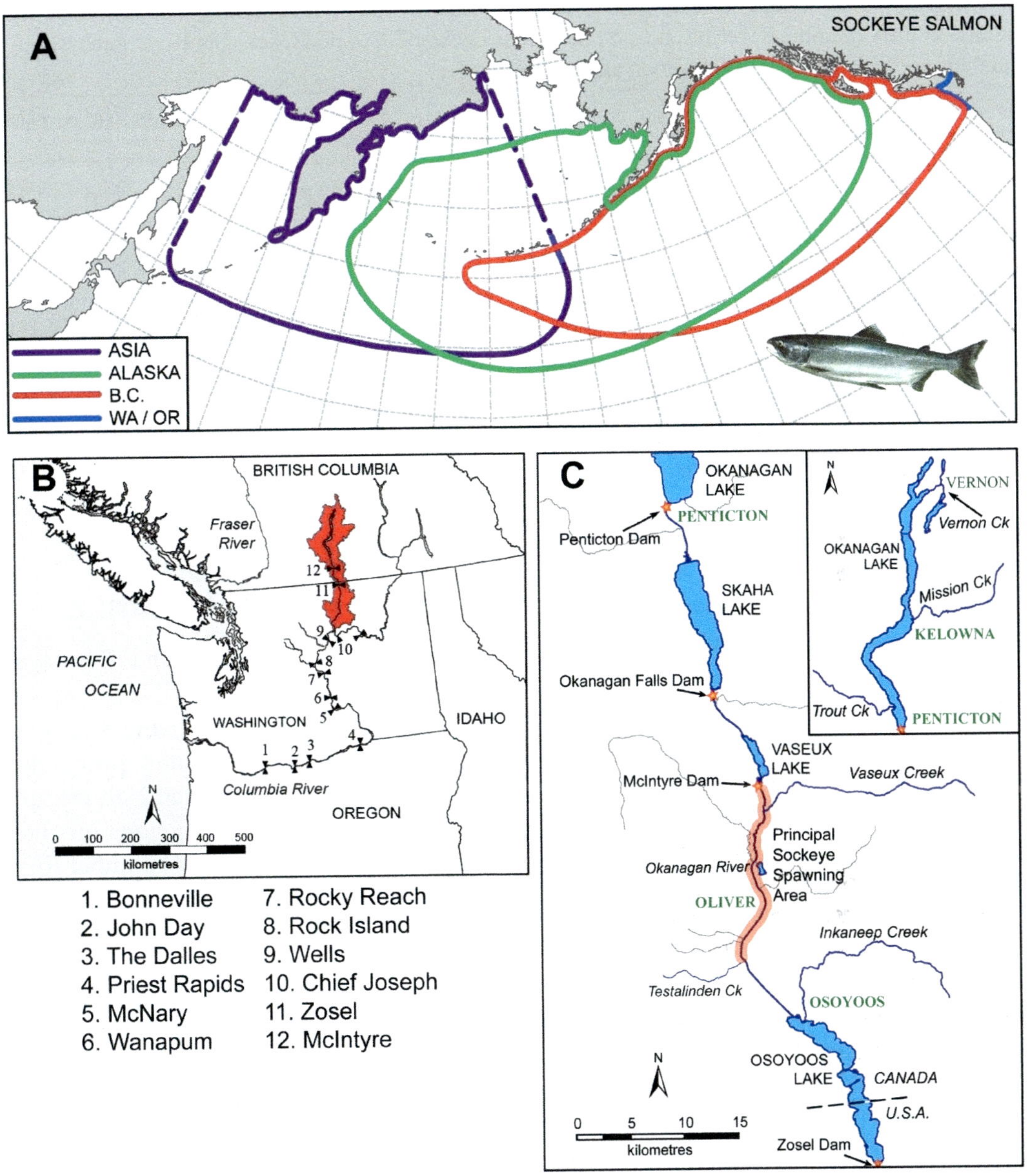

FIGURE 1. Spatial domain of Sockeye Salmon during **(A)** 1–3 years of ocean residence in the northeast Pacific Ocean, **(B)** 8–16-week interval of river migration by salmon smolts or adults, respectively, and **(C)** 1–2-year interval of residence in Osoyoos Lake and/or Okanagan River (Burgner 1991; Hyatt and Rankin 1999; Hyatt et al. 2003, 2015).

priority restoration work; and (8) implementation and assessment of restoration activities for responses by a species and its associated ecosystem. Execution of steps 1–3 to support salmon-and-habitat restoration may require investment of only a few hours to a few days of effort. However, assembling sufficient information, expertise, and financial resources for execution of even a single cycle of steps 4–7 often involves years of sustained effort to

TABLE 1. Estimated historic (late 1800s) and current (late 1990s–2000s) run size(s) of wild salmon in the Columbia River basin (modified from Lackey 2000) and Okanagan River subbasin. All numbers are in millions of wild salmon and are rounded.

Area	Historic run size	Current run size	% of historic run size
Columbia basin	11–15	0.11–0.33	1.7
Okanagan subbasin	0.002–0.150	0.075–0.530	>100.0

overcome knowledge and financial resource limitations on success.

Okanagan Sockeye Salmon return through U.S. waters of the Columbia River, but their freshwater breeding and rearing habitats lie entirely in Canada (Figure 1B, 1C). The Canadian Okanagan Basin Technical Working Group (COBTWG) represents Fisheries and Oceans Canada (DFO), the Okanagan Nation Alliance (ONA), and the British Columbia Ministry of Forests, Lands, and Natural Resource Operations (BC-FLNRO). The COBTWG parties are responsible for joint planning, implementation, and assessment of restoration efforts to facilitate rebuilding of salmon populations in Canadian portions of the Okanagan subbasin.

Canadian Okanagan Basin Technical Working Group experience suggests that close attention to each of the key steps of restoration planning is necessary for success. The current chapter documents our progression through these steps in three sections (Figure 2). In the first section, we briefly discuss the origins of a common salmon restoration vision along with the statutory and collaborative governance context required to support that vision. In the second section, we deal with assembly, review, and analysis of historic information on factors controlling trends exhibited by salmon. We also describe the status and trends shown by salmon in the Okanagan River basin as the foundation of a state of knowledge synthesis to identify (1) limiting factors and current threats, (2) a priority list of potentially effective salmon-and-habitat restoration options, and (3) an integrated, multijurisdictional plan to guide implementation of salmon restoration. Finally, the third section of the chapter documents the extent of restoration work implementation and outcomes achieved along with lessons learned from both successes and failures experienced by the COBTWG and its partners since 1997.

SECTION 1: VISION, POLICY, STATUTES, AND GOVERNANCE

Historically, the Columbia and Okanagan rivers supported substantial U.S. commercial, treaty tribal, nontreaty, and Canadian First Nations fisheries. Salmon production declined drastically during the 20th century, resulting in continuous reductions for most commercial fisheries in the Columbia River basin from the early 1900s to present (Williams et al. 1999; Lichatowich 1999). These declines precipitated the virtual elimination of First Nations fisheries in Canadian portions of the Columbia River basin for all salmon species over the past 60 years. In 1997, the Okanagan Nation Alliance (ONA), a First Nations Council in Canada, convened a workshop of salmon experts to discuss

1. Factors contributing to decades of declining salmon returns to the Okanagan River;
2. The ONA vision and cultural imperative to restore anadromous salmon populations throughout their ancestral range in the Okanagan subbasin;
3. The feasibility of rebuilding existing and/or restoring extirpated populations of anadromous salmon in the Okanagan; and

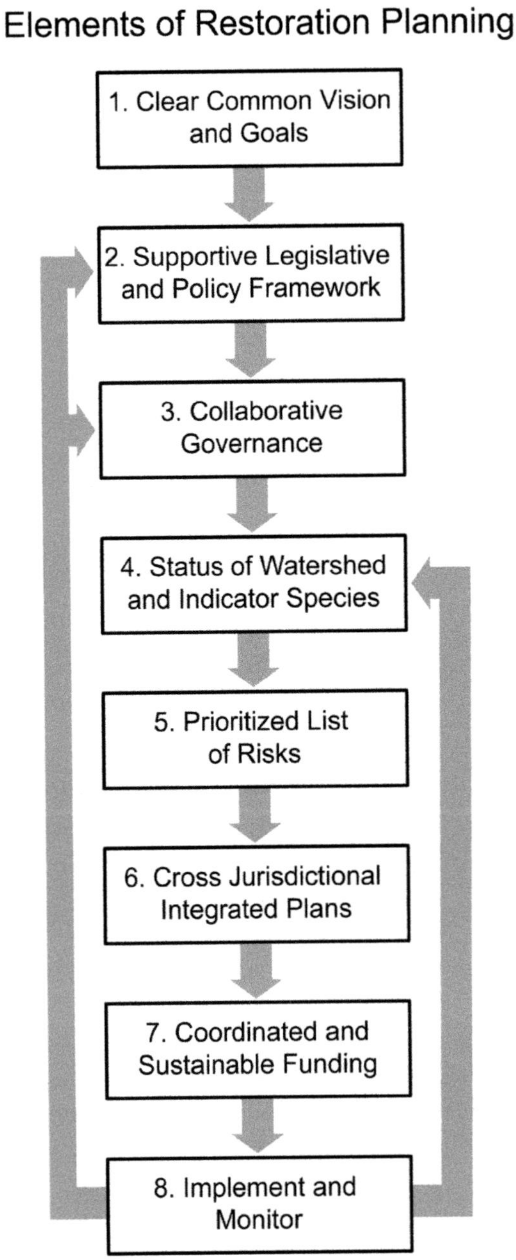

FIGURE 2. Key planning elements that facilitate successful restoration initiatives. The idealized sequence of elements and limited feedback cycles illustrated here represent a highly simplified process that can engage proponents for months to decades of time (adapted from materials at www.polisproject.org).

4. Obligations of Canadian fisheries management agencies to support an aboriginal right, defined in Canada's 1982 Constitution Act, to fish for food, social, and ceremonial purposes (Law and Fréchette 1993).

Okanagan National Alliance elders, at the 1997 workshop, spoke passionately about the significance of ancestral fisheries for Chinook Salmon *Oncorhynchus tshawytscha*, Sockeye Salmon, and kokanee (nonanadromous Sockeye Salmon). Further, the ONA chiefs and council confirmed their intent to commit multiple generations of ONA leaders and people to the objective of achieving success in restoring depleted and extirpated anadromous salmon to the Okanagan subbasin.

The Okanagan River subbasin is the last international watershed (Figure 1B) in the Columbia River system supporting anadromous salmon populations. All other anadromous salmon were extirpated from transboundary watersheds of the upper Columbia River with the construction of Grand Coulee Dam by the United States in 1939 (Ortolano and Cushing 2000). Legislative, policy, and institutional frameworks that facilitate management of Okanagan salmon returning to Canada through U.S. waters are especially complicated due to their multi-jurisdictional character (Figure 3). From a Canadian perspective, the Fisheries Act (Department of Justice Canada 2016), the Species at Risk Act (2002), DFO's Wild Salmon Policy (DFO 2005), and the 1982 Constitution Act provide overarching authority and guidance for DFO to serve as the principal regulatory agency responsible for salmon conservation and management. Fisheries and Oceans Canada is also obligated to protect First Nations' constitutional right to fisheries for food, social, and ceremonial purposes. However, because water management within Canada is an area of provincial, rather than federal, jurisdiction, formal agreements (e.g., the 1976

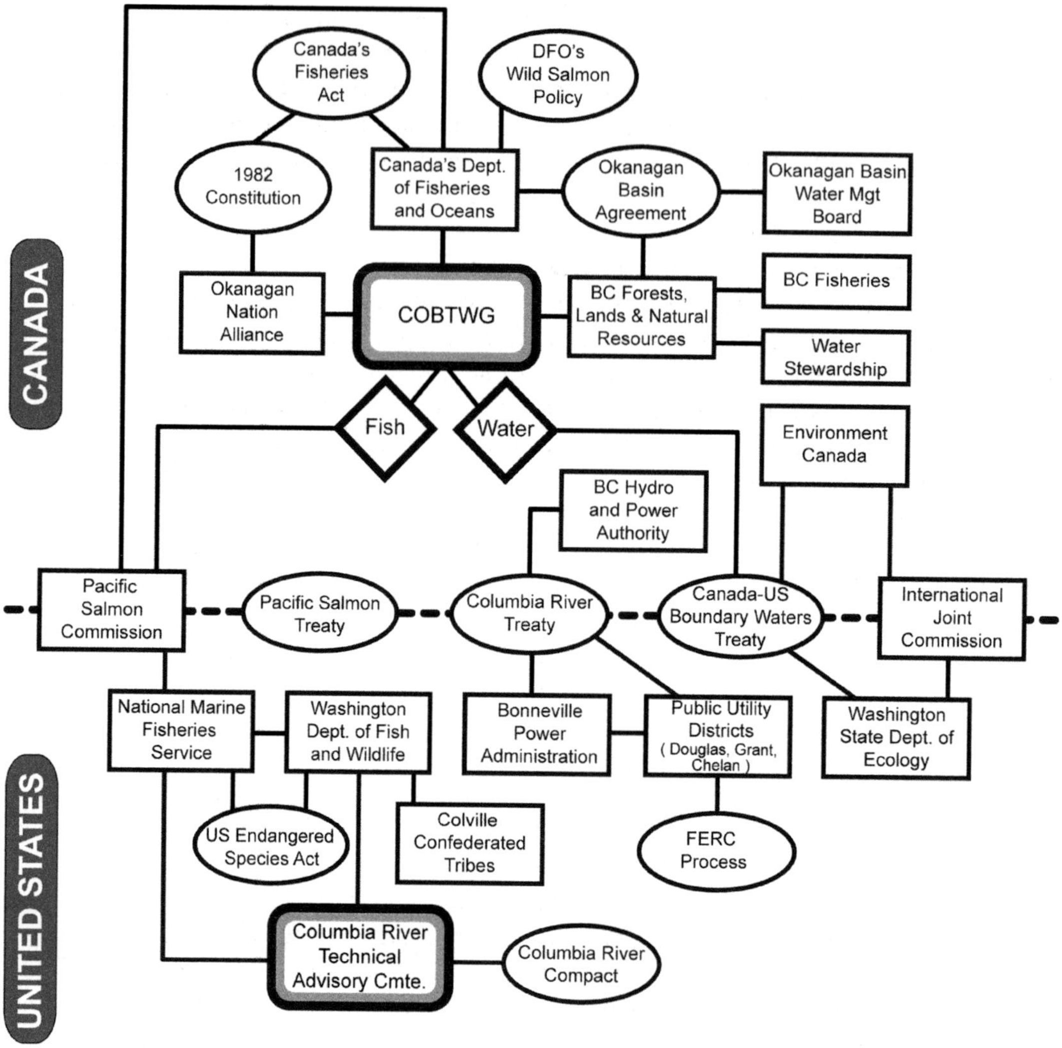

FIGURE 3. Conceptual summary of the complex statutory, policy (ovals), and institutional (rectangles) entities governing conservation and management of salmon populations and associated habitat in Canadian and U.S. portions of the transboundary Okanagan/Okanogan River basin (see text for details).

Canada–British Columbia Okanagan Basin Implementation Agreement) that include provisions to protect fisheries values have also emerged as part of the domestic regulatory framework (Hyatt et al. 2015a). Consequently, fisheries and water stewardship sectors within the BC-FLNRO are important participants in any initiatives involving management of water to sustain anadromous or resident fish in freshwater ecosystems. Finally, management issues affecting salmon originating from transboundary watersheds are governed in part by no fewer than three major Canada–United States treaties (1909 Boundary Waters Treaty, 1964 Columbia River Treaty, 1985 Pacific Salmon Treaty) implemented by two international commissions (International Joint Commission and Pacific Salmon Commission). A parallel system of U.S. statutes, policies, and institutions are involved in conservation and management of these same salmon populations that briefly occupy U.S. waters on their return to the Okanagan River basin (Figure 3).

Despite the wealth of agency expertise, legislation, international treaties, and commissions, the sheer complexity of this transboundary salmon and water governance system poses a major challenge to overcome fragmented or conflicted approaches to effective management of human activities (e.g., habitat disturbance, hydrological regulation, and fish harvest) contributing, on both sides of the border, to production losses of transboundary salmon stocks. Consequently, 20–30 years after establishment of the Canada–United States Pacific Salmon and Columbia River treaties respectively, no examples existed of successful restoration of Canadian-origin salmon returning through the Columbia River. By contrast, specific examples of failure to achieve objectives by groups in Canada and the United States due to jurisdictional fragmentation were apparent. For example, Canada and the United States have repeatedly failed to negotiate formal harvest sharing or co-management arrangements within the 1985 and subsequent versions of the Pacific Salmon Treaty (Shepard and Argue 2005) for Okanagan River salmon of Canadian origin. Similarly, fragmented jurisdictional control over fish, water supplies, and salmon habitat protection within Canada greatly reduced the effectiveness of isolated attempts by DFO, BC-FLNRO, and the ONA prior to the late 1990s to identify limiting factors and to implement successful salmon restoration initiatives. Perhaps not coincidentally, the pattern of multi-decadal declines of resident (Rae 2005) and anadromous salmon (Hyatt and Rankin 1999) within Canadian portions of the Okanagan River basin continued from 1990 to 2000, with kokanee populations in Okanagan Lake and anadromous Sockeye Salmon returns to the Okanagan River reaching historically unprecedented lows.

A shared recognition of the problem of jurisdictional fragmentation emerged among various groups in the years immediately after the 1997 salmon workshop hosted by the ONA. Consequently, in 1998, personnel from the ONA, DFO, and BC-FLNRO agreed to pool resources and expertise to form the three-party COBTWG dedicated to sustainable management and restoration of salmon populations and freshwater ecosystems in the Okanagan subbasin. Signoff on formal terms of reference for COBTWG by senior management occurred in 2003 after 5 years of elevated engagement among the parties. Recognition of issues created by fragmentation of statutory, policy, and institutional approaches to salmon restoration between Canadian and U.S. entities also led to the formation of a Bilateral Okanagan Basin Technical Working Group (BOBTWG), with participation at the discretion of interested parties. Bilateral Okanagan Basin Technical Working Group functions are limited to information exchange between interested parties in Canada and the United States through an annual meeting. Establishment of the COBTWG, and to a lesser extent BOBTWG, created the minimum critical mass of expertise, information sharing, and common purpose to initiate effective restoration and cooperative management of Okanagan salmon (discussed below).

SECTION 2: ANALYSIS OF FACTORS CONTROLLING STATUS AND TRENDS OF OKANAGAN BASIN SALMON

The COBTWG is primarily a technical working group dedicated to identification of (1) factors accounting for declines in abundance of anadromous and resident salmon or aquatic species, and (2) evidence-based options to halt, and if possible reverse, such declines.

Identifying geopolitical constraints to sustainable management or restoration of anadromous salmon is complex enough (Figure 3), but the ecological context of anadromous salmon presents additional challenges. Okanagan salmon life history events follow a trajectory completed within a nested hierarchy of ecosystem (and implicitly geopolitical) units involving local to global spatial scales (Figure 1A–C) and hours to years of elapsed time. Production variations of salmon are controlled by incompletely understood cumulative impacts of natural and human-induced events or regimes (Augerot 2005) operating within the space-time range bounding the life history of a particular salmon population (Figure 4).

Although management operates with highly imperfect knowledge of the specific interactions controlling the status and trends shown by a given population of salmon, systematic assembly of information on known life history and production events plus associated environmental observations serve as an evidence-based foundation for salmon restoration. Given sufficient attention, this approach improves prospects for identifying limiting factors that may be either resolved by or refractive to human intervention(s). For example, atmospheric processes drive climate variation and change that influence biophysical processes in freshwater and oceanic ecosystems (steps P1 to P6 in Figure 4). These processes ultimately affect production outcomes for salmon rearing in the North Pacific (Mantua et al. 1997), including those originating in the Columbia River basin (Anderson 2000). Although this knowledge is useful for forecasting salmon returns and for explanations of the causes of salmon production varia-

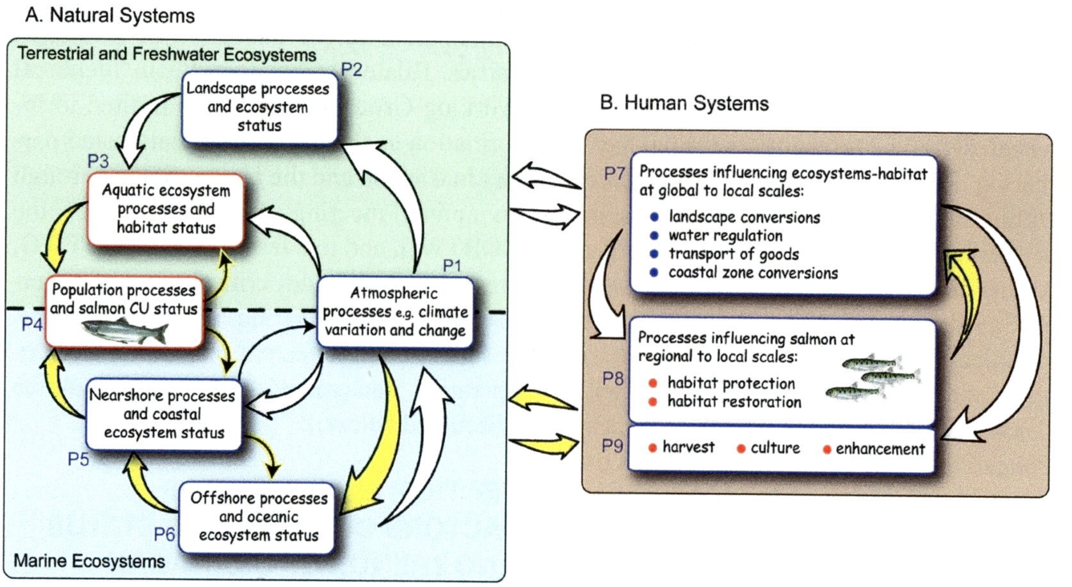

FIGURE 4. Summary of process drivers in **(A)** natural or **(B)** human systems that control salmon conservation unit (CU) status and trends. Yellow arrows highlight the subset of elements/activities for which Fisheries and Oceans Canada (DFO) is principally responsible (P3, P4, P5, P6, P8, and P9). Remaining arrows identify elements/activities dealt with by others (e.g., P1—Environment Canada; P2, P3, P7—British Columbia Ministry of Forests, Lands, and Natural Resource Operations). Red bullets identify the limited local to regional-scale activities over which DFO may exert direct control. Differences in arrow sizes suggest differences in the magnitude of process driver influences on salmon.

tions, regional management agencies have no control over such processes and have limited options other than to adjust salmon harvest to changes in ocean production regimes when known. Considered together, the complexities of governance frameworks (Figure 3) interacting with ecological process constraints within natural (Figure 4A) and human systems (Figure 4B) will inevitably act, sometimes severely, to limit the range of effective salmon management and restoration interventions.

Beechie et al. (2010) have pointed out, as a matter of general principle, that process drivers in natural and human systems are frequently more potent at inducing changes in target species and associated ecosystems from coarse-to-fine temporal and spatial scales while feedback processes from fine-to-coarse scales are much weaker (Figure 4). Unfortunately, the majority of activities implemented to restore depleted salmon populations by fishery management agencies and aligned interest groups operate under an assumption that interventions at fine scales will effectively counter the influence(s) of drivers operating at coarse scales. We do not suggest that most salmon restoration projects will inevitably fail, but point out that careful scoping of drivers and processes operating at several scales will increase prospects for success. The COBTWG used this approach to identify if viable salmon restoration options existed within the Okanagan subbasin (Figure 1C; Bull 1999) despite Canada's limited statutory mandate (Figure 3) and influence over factors affecting Okanagan salmon within the Columbia River basin (Figure 1B) and North Pacific Ocean (Figure 1A; Williams et al. 2014; Hyatt et al. 2015b).

Geographic and Ecological Context of the Okanagan River Basin

The 8,000-km^2 Okanagan basin is defined by a long north–south trench (about 200 km long and 20 km wide) between the Columbia and Cascade Mountain ranges in the Interior Plateau of British Columbia (Merritt et al. 2006; Figure 1B). It extends north from the Columbia Plateau, in Washington State, to the height of land separating the drainage basins of the Columbia and Fraser rivers. The valley bottom contains Okanagan Lake in the north and center, with the Okanagan River and Osoyoos Lake straddling the Canada–United States border in the south (Figure 1B, 1C). Prior to the mid-1800s, a rich complex of lakes, streams, wetlands, and riparian floodplain habitats supported large populations of several species of migratory salmon, resident fish, and high levels of general biodiversity in Canadian portions of the Okanagan subbasin. This area still contains many unique species and more species of resident plants and animals than in most areas of British Columbia and Canada (SOSCP 2000). However, human population growth (details below) and associated landscape conversions have produced road densities of 2.1–3.0 km/km^2 that are the highest in British Columbia (Austin et al. 2008) in addition to cumulative losses of native grassland, riparian, and wetland plant communities of between 41% and 92% (Lea 2008). One-third of all plant and animal species listed as being at imminent risk of extinction in British Columbia are located in the Okanagan basin (Bezener et al. 2004) and the area represents one of the four most endangered ecosystems in Canada (www.biodiversitybc.org).

Human-Induced Disturbance of Habitat and Salmon Status and Trends

A relatively rich scientific record exists of historic to contemporary threats to aquatic ecosystems and their salmon populations in the Okanagan. Although to review these in detail is beyond the scope of this chapter, a brief overview of human-induced disturbance events and processes that have shaped the status and trends exhibited by Okanagan salmon is necessary to clarify the difficult challenges that restoration of these populations posed.

Human Population Growth

Over the past two centuries, the principal driver of landscape and aquatic ecosystem change in the Okanagan has been the growth of human populations. Humans have inhabited the Okanagan valley for at least 6,000 years. The original Okanagan peoples, or the Syilx, were hunter-gatherers moving from camp to camp throughout their territory, depending on food availability. At the time of first contact with Europeans, one of the most important salmon fishing areas for the Syilx was located at Okanagan Falls (Figure 1C), where fish aggregated before moving upstream into Skaha Lake (Kennedy and Bouchard 1998; Cannings and Durance 1998). Between 1840 and 1859, roughly 400 Europeans settled in the Okanagan valley. By 1971, the population of permanent residents had increased to 114,000 and by 2016 nearly quadrupled to 441,000 permanent residents plus another 500,000 annual tourists. Projections suggest that this area will contain 550,000 permanent residents by 2031 (Northcote 1996).

Dams, Channelization, and Hydrology

Given the difficult access afforded by mountainous terrain, European settlement in the Okanagan valley (e.g., at Penticton, Kelowna, Oliver, and Osoyoos) developed first on readily accessible floodplain areas of the valley bottom. Development required river channelization, low-head dams, and water distribution systems of ever-increasing complexity (see Symonds 2000) to respond to navigation needs, agricultural demands, and growing urban population demands for water where the Okanagan has the lowest per-capita water supply of any basin in Canada (Summit Environmental Consultants 2010). Repeated cycles of design, construction, and subsequent modification of low-head dams at the outlets of four of the largest lakes in the valley bottom occurred between 1914 and 2016 (Table 2). The arid nature of the area and water storage demands of a growing human population also resulted in a proliferation of dams on virtually all Okanagan tributaries (Chapman et al. 1956) from 11 in 1913 to 147 by 1998 (Hall et al. 2001).

Highly variable winter storage and spring melt of snowpack at high elevations dominate Okanagan River hydrology that includes naturally occurring flood and drought events. Between 1921 and 2002, Okanagan Lake received net annual inflows differing by more than an order of magnitude, from 78×10^6 to 1.4×10^9 m^3 of water, inducing large floods (1928, 1942, 1948, and 2017) and acute droughts (1930–1932 and 2001–2003) (Symonds 2000; Alexander et al. 2008). Spring freshets typically eliminate winter snowpack during April to June and account for as much as 90% of annual inflows to the Okanagan lakes system in this arid region (Dobson 2004). The 1942 and 1948 floods were especially damaging to agricultural and urban infrastructure. Consequently, from 1950 to 1958, the Okanagan River was altered into a straight, flood-control channel that reduced the river's unconfined, natural length by 50% (Hourston et al. 1954; Hyatt et al. 2015a). This channelization also reduced the area of original floodplain, riparian, and wetland ecosystems (Figure 5A) by 96% (Rae 2005) due to channel simplification, diking, and wetland infilling, although outlines of the original river channel and off-channel wetland habitat are still perceptible nearly 40 years after channelization (e.g., compare Figure 5B and 5C). In addition, seasonal regulation of impounded waters focused principally on meeting agricultural and urban rather than aquatic ecosystem or fisheries needs. Consequently, dams, dikes, culverts, and seasonal

TABLE 2. Summary of the construction dates, purpose, impacts, and current status of the four main dams on main-stem segments of the Okanagan Lake and River system.

Valley bottom dams	Date of first completion	Purpose and impacts	Current status
Penticton Dam at the outlet of Okanagan Lake	1914–1915	Reduce flooding, control Okanagan Lake level, and dredge Okanagan River channel for navigation by small draft vessels. Altered seasonal hydrograph including reduced peak discharge and increased base flows in summer. Fish passage from the Okanagan River into Okanagan Lake blocked by this and successive versions of dam to present. Audit indicated low compliance with fisheries flow provisions of Canada–British Columbia Okanagan Basin Implementation Agreement (Bull 1999).	Development of fish-and-water management tools environmental decision support system greatly improved compliance levels with fisheries flow provisions of the Canada–British Columbia Okanagan Basin Implementation Agreement (Hyatt et al. 2015a).
McIntyre Dam at the outlet of Vaseux Lake	1919–1921	Control Vaseux Lake elevation to support water withdrawals for irrigation. Largely impassable to upstream movement by all fish except within a limited range of high discharge levels. Unscreened intakes diverted fish moving downstream into irrigation works resulting in elevated fish mortalities.	Floodgates re-engineered in 2009 to ease upstream passage by anadromous salmon. irrigation intakes screened in 2010.
Zosel Dam at the outlet of Osoyoos Lake	1927	Flood control and regulation of Osoyoos Lake levels. Initial and subsequent Zosel dams engineered with fishway(s) for upstream passage of anadromous fish into Canadian portions of the Okanagan valley.	Seasonal water levels in Osoyoos Lake prescribed by the Osoyoos Lake Board of Control under International Joint Commission authority.

TABLE 2. Continued.

Valley bottom dams	Date of first completion	Purpose and impacts	Current status
Okanagan Falls Dam at the outlet of Skaha Lake	1950–1958	Dam constructed as part of Okanagan system flood-control works. Engineering design included a fishway that was never activated due to concerns regarding need to deny upstream access to exotics (e.g. Smallmouth Bass *Micropterus salmoides*).	Fishway modified and opened for anadromous fish passage after 2011. Passage remains difficult at discharge <10 m^3/s, and at >15 m^3/s, fish pass preferentially through the dam floodgates and avoid the fishway.

water withdrawals frequently blocked free passage of anadromous salmon and resident fish to and from spawning and rearing habitat in the valley (Rae 2005).

Pollutants, Invasive Species, and Climate Change

From the time of the earliest water quality studies in the Okanagan (Rawson 1939) to present, the human population of the Okanagan valley has roughly doubled every 30 years. Consequently, by 1970, the inherently oligotrophic to mesotrophic lakes of the valley bottom (Rawson 1939; Cumming et al. 2015), were showing clear signs of eutrophication (Stockner and Northcote 1974; Jensen and Epp 2002) from agricultural, urban, and industrial effluents. Although implementation of tertiary treatment in the 1970s greatly reduced cultural eutrophication of valley bottom lakes (Jensen and Epp 2002; Cumming et al. 2015), a range of contaminants still enter Okanagan aquatic ecosystems from urban sewer systems, agricultural lands (Hall et al. 2001), and mass wastage events associated with earthen dam failures (Tannant and Skermer 2013). Thus, water-quality degradation remains a long-term concern for maintenance and restoration of salmonid fishes given their requirements for high-quality water (Bjornn and Reiser 1991; Solbe 1997).

Aquatic ecosystems of the Okanagan valley also have been subject to successive introductions and invasions of nonnative species in association with approximately 150 years of European settlement and development activities. Twenty-five indigenous fish species were present in the Okanagan River prior to 1900 and largely reflected the coldwater species complex originating from the Columbia glacial refuge. European settlement of the Okanagan valley facilitated the invasion of many new species. These species include aquatic plants (e.g., water milfoil *Myriophyllum spicatum*), the planktivorous macroinvertebrate *Mysis diluviana* (freshwater shrimp formerly known as *M. relicta*), and 13 species of nonnative fish occupying a variety of trophic levels (e.g., piscivorous Smallmouth Bass, benthic feeding Brown Bullhead *Ictalurus nebulosus*, zooplankton feeding Black Crappie *Pomoxis nigromaculatus*, Yellow Perch *Perca flavescens*, and omnivorous Common Carp *Cyprinus carpio* (Wright et al. 2002).

The absence of good quantitative data concerning aquatic ecosystem state changes, accompanying introductions and invasions

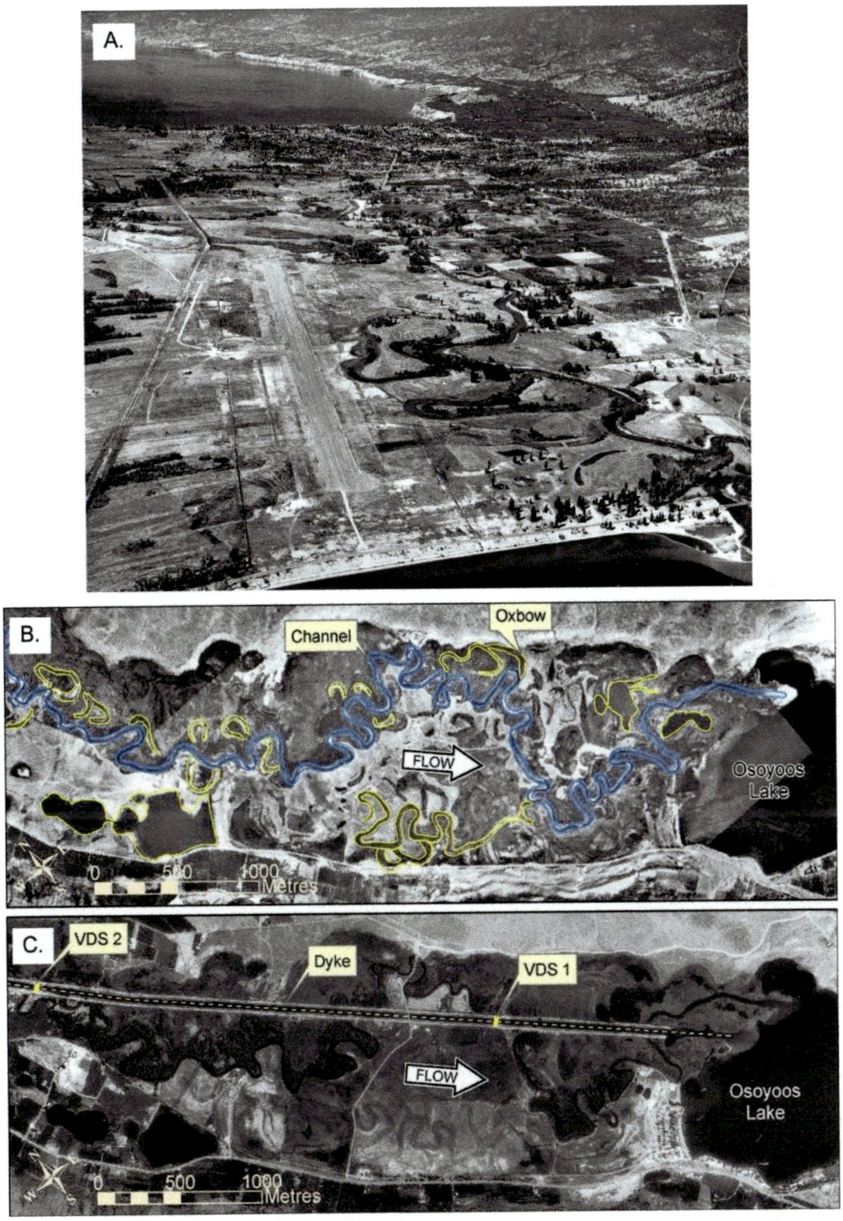

FIGURE 5. Aerial photos (BC Land Management Bureau; see Figure 1C for map orientation) of **(A)** Okanagan floodplain between Okanagan Lake (top of photo) and Skaha Lake (bottom right of photo) in 1950. The Okanagan River with intact meanders and oxbows is readily apparent east of the newly constructed Penticton airstrip (for comparison, Figure 13A indicates the extent of urban and industrial development along the Okanagan River in the same general area as of 2017). **(B)** Edited image of a segment of the original Okanagan River channel (blue highlight), associated side channels, oxbows (yellow highlights), and wetlands immediately north of Osoyoos Lake as they appeared in 1938. **(C)** Same area north of Osoyoos Lake as it appeared in 1996 after channelization and diking to convert the river course into a flood-control channel. VDS 1 and VDS 2 are concrete structures engineered to reduce river velocity for adult salmon access to spawning areas (Figure 1C) 12–20 km upstream of Osoyoos Lake. Figure 5A photo source, Penticton MuseumArchives, AP #9.

of so many species, severely limits our understanding of cumulative impacts of exotics on the indigenous species complex. However, detailed studies on some of the invasive species noted above suggest significant impacts on native species. For example, water milfoil first appeared in the Okanagan in the 1970s and subsequently established extensive littoral-zone beds in all major valley bottom lakes (Dearden 1983). Research in the Okanagan and elsewhere indicates that milfoil can suppress development of native aquatic plants and invertebrates (Nichols and Mori 1971; Keast 1984; Madsen 1999) in addition to providing cover for other invasive species such as Yellow Perch, which now dominates the littoral-zone fish communities of Osoyoos and Skaha lakes (Naito 2000). These changes have likely reduced the littoral-zone capacity of Okanagan lakes to support the juvenile life stages of indigenous species such as Coho Salmon *Oncorhynchus kisutch*, Chinook Salmon, and resident Rainbow Trout *O. mykiss*.

The freshwater shrimp *M. diluviana* is a second invasive species that has become well established in Wood, Kalamalka, Skaha and Osoyoos lakes (DFO 2017; Hyatt et al. 2018b) following its intentional introduction into Okanagan Lake in 1966 (Northcote 1991). Various studies suggest that mysids have the potential to induce disruptive effects on food webs (Spencer et al. 1991; Hyatt et al. 2005, 2018a; Schindler et al. 2012) and the associated collapse of pelagic fish populations in lakes (see references in Nesler and Bergersen 1991). Both outcomes have been observed in Okanagan lakes, where mysids have potentially caused reductions of up to 43% in the daily ration of juvenile Sockeye Salmon rearing in Osoyoos Lake (Hyatt et al. 2018a) and the collapse of pelagic kokanee in Okanagan Lake (Andrusak et al. 2008).

Finally, during the past century, British Columbia has warmed significantly in all seasons, and climate-change projections suggest further seasonal warming, wetter conditions in winter–spring, and drier conditions during summer in the south and on the coast (Walker and Sydneysmith 2008). Climate change is an overarching concern that complicates the maintenance and restoration of Okanagan salmon (e.g., see Hyatt et al. 2003, 2015b) due to conflicts between managing water to satisfy human-system versus fisheries and aquatic systems needs in times of drought (Payne et al. 2004; Sellars and Smith 2005; Cohen et al. 2006).

Given the information above, the COBTWG understood the magnitude and challenging nature of historic impacts and contemporary threats to be addressed (Figure 6) if Okanagan salmon restoration was to be successful. Individually, the COBTWG parties had a history of implementing (DFO, BC-FLNRO) or facilitating (ONA) interventions to resolve environmental impacts on the maintenance or restoration of fish populations within their separate jurisdictions (e.g., eutrophication control, Jensen and Epp 2002; environmental flow allocations, Okanagan Basin Implementation Board 1982). However, the 2003 terms of reference for COBTWG formally committed the parties to work together on restoration action(s) to accelerate recovery of beleaguered Okanagan salmon populations. To begin, the parties to COBTWG reviewed traditional environmental knowledge and early written records to identify historic versus contemporary abundance levels of various salmon species and the status of associated habitat within baseline (Figure 7A, 7B) versus contemporary periods (Figure 7C, 7D) in the Okanagan. This information provided an initial means to identify which species and habitat elements to focus on for the greatest restoration success in the short term. These comparisons confirmed that precipitous declines of native salmon in the Okanagan were of similar

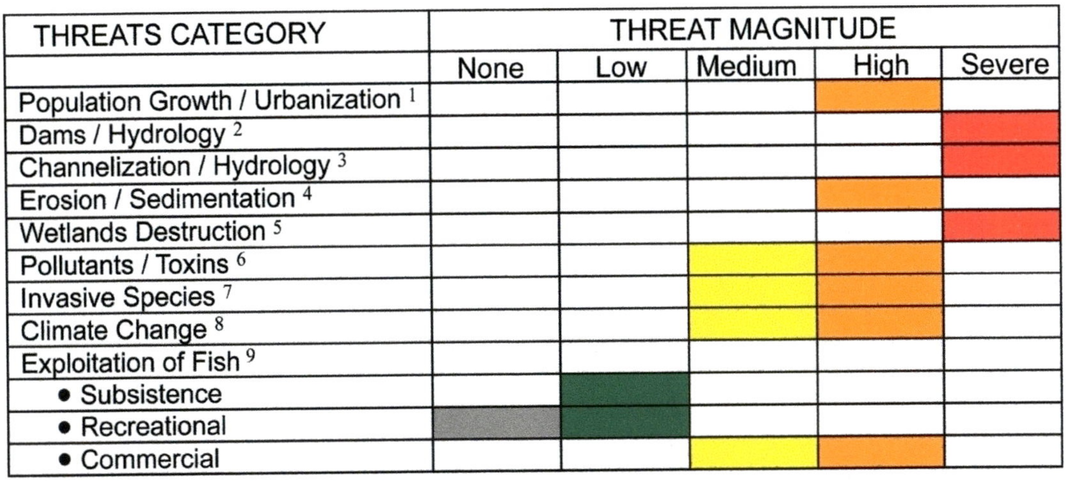

FIGURE 6. Summary results from a five-class risk assessment* by major category of human-origin threats to Okanagan salmon. Results represent a semi-quantitative synthesis of risk characterizations from 2 to 3 years of workshops and discussions among Canadian Okanagan Basin Technical Working Group members. Supporting documents include 1. Northcote 1996; 2. Dobson 2004, Rae 2005; 3. Hourston et al. 1954; Symonds 2000, Hall et al. 2001, Rae 2005; 4. Tannant and Skermer 2013; 5. British Columbia Ministry of Environment 1998; 6. Stockner and Northcote 1974, Jensen and Epp 2002; Cumming et al. 2015; 7. Wright et al. 2002; Rae 2005; 8. Leith and Whitfield 1998, Payne et al. 2004, Cohen et al. 2006, Merritt et al. 2006; and 9. Hyatt and Rankin 1999. *Standard ratings recommended by Canada's Treasury Board to be used for risk assessment work by federal departments.

magnitude to other subbasins of the Columbia River (e.g., Chapman 1986; Lichatowich 1999). Only anadromous Okanagan Sockeye Salmon, Chinook Salmon, steelhead (anadromous Rainbow Trout) plus kokanee, and Rainbow Trout have persisted in quantifiable numbers in the contemporary period (Figure 7D). Chinook Salmon, steelhead, and Coho Salmon all rely on main-channel and off-channel habitat complexes in streams and littoral-zone waters of lakes for spawning and especially juvenile rearing (see species details in Groot and Margolis 1991; Quinn 2005). Efforts to restore these species were considered to have less potential for short-term (<20 years) success due to the extent of critical habitat loss as well as its increased use by a host of invasive predators and competitors. Consequently, restoration efforts initially focused on anadromous Sockeye Salmon (and kokanee) that spawn in streams but rear in the pelagic zone of lakes where historic biophysical alterations appeared less limiting.

Simultaneous with the formation of the COBTWG, several Public Utility Districts (PUDs) in the United States were searching for new opportunities to mitigate salmon losses caused by their dams on the Columbia River as a component of dam-operator relicensing by the U.S. Federal Energy Regulatory Commission (FERC). At the time, Douglas County PUD and the Colville Tribe, in Washington State, were attempting to increase Sockeye Salmon returns to the Okanogan River via introductions of hatchery-origin Sockeye Salmon fry to rear and overwinter in the south basin of Osoyoos Lake (Rick Klinge, Douglas County PUD, personal communication). The attempt failed due to chronically marginal environmental conditions in this basin for Sockeye Salmon fry survival (Hyatt and Rankin 1999; Simmatis et al. 2018). Discussions between

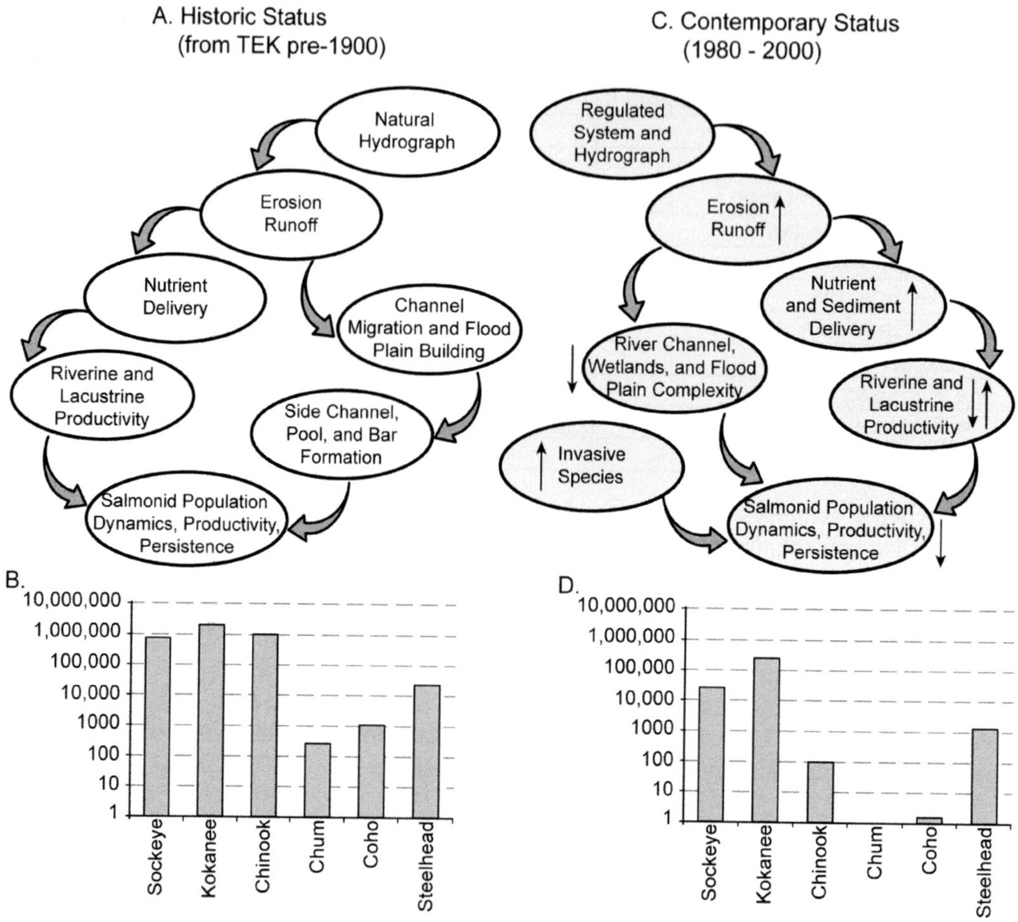

FIGURE 7. Conceptual summary of **(A, B)** historic (<1900) versus **(C, D)** contemporary (1980–2000) biophysical process controls in aquatic ecosystems of the Okanagan River basin. Log-scale declines in salmon abundance are associated with losses of historic aquatic ecosystem integrity. Large arrows (7A, 7C) indicate directional influence of major environmental process drivers. Smaller arrows (7C) identify direction(s) of change in process drivers (e.g., erosion has generally increased; freshwater productivity has shown increases and decreases) relative to the historic, base case (7A). Pre-1900 fish abundance (7A) inferred from traditional environmental knowledge of Okanagan Nation Alliance informants, ethnographic or archeological sources (e.g., Kennedy and Bouchard 1998), and habitat capacity assessments (e.g., Bradford et al. 1997; Hyatt and Rankin 1999). Contemporary abundance estimates (7D; 1980–2000) from direct assessments (e.g., Hyatt and Rankin 1999; COSEWIC 2006; Andrusak et al. 2008).

COBTWG and Douglas County PUD personnel identified mutual interests in Okanagan salmon restoration, where the COBTWG had the authority and technical expertise to implement restoration work in Canada, and the Douglas County PUD had the financial resources and motivation from their FERC licensing process to become a funding partner and recipient of mitigation credits from successful projects. Subsequent investments in additional assessment work by the parties to the COBTWG (Hyatt and Rankin 1999; Bull et al. 2000a; Rae 2005; Andrusak et al. 2008; Hyatt et al. 2018a, 2018b; Rivard-Sirois et al. 2018) and the Douglas County PUD (Bull 1999) on local habitat factors limiting Okanagan salmon and resident fish production led to the identification of a suite of restoration projects (Table 3). The parties to the COBTWG have guided the identification, implementation, and assessment of outcomes from these projects over the past 20 years and have built an enviable record of success along the way.

SECTION 3: EVIDENCE OF SUCCESS OF THE COBTWG RESTORATION STRATEGY

The COBTWG restoration strategy to date has focused on three major initiatives supported by a series of implementation activities (Table 3). The initiatives, briefly reviewed here, include (1) rebuilding the wild-origin population of Sockeye Salmon that rears in Osoyoos Lake; (2) expanding the contemporary range of Sockeye Salmon to Skaha and Okanagan lakes, upstream of Osoyoos Lake (Figure 1C); and (3) improving the quality of off-channel and in-channel habitat in the Okanagan River, its flood-plain, and associated tributaries for use by all native species of anadromous and resident fish.

Rebuilding Wild-Origin Sockeye Salmon Abundance

In an analysis of Okanagan Sockeye Salmon production and the state of their habitat, Hyatt and Rankin (1999) concluded that the quantity and quality of accessible spawning and rearing habitat could support at least a threefold increase in salmon production if spawner abundance were increased by the same amount (i.e., from a 1990–1999 mean of 20,000 to around 60,000 fish). Consequently, they recommended formally increasing Canada's domestic escapement objective from roughly 25,000 to 66,000 adults, as enumerated passing Wells Dam, assuming that no catch occurs above that location. The goal of elevating the escapement objective was to encourage fishery managers to begin testing the upper limits of the Okanagan River system to produce wild-origin Sockeye Salmon. Two impediments stood in the way of achieving this objective. The first was that harvest management rules affecting interception of Okanagan Sockeye Salmon in U.S. waters were not stock-specific, but rather treated the three populations of Sockeye Salmon returning to the Columbia River (Okanagan, Wenatchee, and Redfish Lake–Snake River) as a single aggregate. For example, decision rules to either open or close fisheries on Okanagan and Wenatchee Sockeye Salmon in the mid-Columbia River depended only on achieving a total escapement equal to 65,000 fish past Priest Rapids Dam (Figure 1B) with any surplus to that value considered available for harvest by treaty tribal and commercial fisheries. Accordingly, between 1953 and 2000, exploitation rates on Okanagan Sockeye Salmon averaged 53% for the 17 years in which total Okanagan returns exceeded 50,000 fish (Table 1 of Hyatt and Rankin 1999). Therefore, Okanagan Sockeye Salmon escapement from these fisheries reached levels above 50,000 at Wells Dam further upriver in only 6 out of 46 years between 1953 and 1999 (Hyatt and Rankin 1999). However, the listing of Redfish Lake Sockeye Salmon under the U.S. Endangered Species Act in 1991 (Bevan et al. 1993; Kline et al. 2019, this volume) significantly constrained harvest of Sockeye Salmon in traditional fishing areas of the lower Columbia River.[1] Consequently, in 6 of 7 years of large returns between 2008 and 2014, Okanagan Sockeye Salmon escapement to terminal spawning grounds increased to levels in excess of 60,000 adult fish.

In theory, Okanagan Sockeye Salmon escapement increases provided an opportunity

[1] "Combined non-Indian impacts on ESA-listed Snake River Sockeye Salmon will be minimized, and shall not exceed 1% of the run entering the Columbia River" (Joint Columbia River Management Staff 2016).

TABLE 3. Major initiatives and activities (1998–2017) comprising the Canadian Okanagan Basin Technical Working Group strategy for restoration of anadromous salmon to Canadian portions of the Okanagan/Okanogan River basin (note the two versions of spelling for this basin represent the forms used in Canada and the United States, respectively, with reference to the basin).

Initiative	Specific activities	Status of activities	Documentation
(1) Rebuild the population of wild-origin Osoyoos Lake Sockeye Salmon (focus on Sockeye Salmon with secondary benefits for Okanagan Lake kokanee)	a. Increase wild Sockeye Salmon escapement to make use of available freshwater spawning and rearing habitat.	Fisheries and Oceans Canada formally revised its domestic escapement objective upwards in 1999.	Hyatt and Rankin 1999; Hyatt et al. 2018a
	b. Develop tools to improve water management compliance with flow provisions of the Canada–British Columbia Okanagan Basin Implementation Agreement to control density-independent losses of fry production due to flood-and-scour or drought-and-desiccation events.	A model-based decision support system (fish-and-water management tools) was designed, built, tested, and deployed between 2001 (design) and fall 2003 (deploy) for annual operational use.	Hyatt et al. 2001, 2015a; Alexander et al. 2008
(2) Expand current range of Sockeye Salmon (focus on Sockeye Salmon with collateral benefits for all species)	a. Reintroduce Sockeye Salmon into lakes with impassable barriers (i.e., Skaha and Okanagan lakes, Figure 2C).	Hatchery-origin Sockeye Salmon fry introduced into (1) Skaha Lake between spring 2004 and present, and (2) Okanagan Lake in spring 2016. A naturalized population has now been established in Skaha Lake. Assessment of outcomes and impacts on resident fish currently underway.	Parnell et al. 2003; Wright and Smith 2004; Alexander and Hyatt 2015; Veale and Russello 2016; DFO 2017; Hyatt et al. 2018b
	b. Re-engineer passage facilities at McIntyre and Okanagan Falls dams (Figure 2C) to permit upstream passage of anadromous and resident salmonids.	Upstream fish passage at McIntyre Dam restored via installation of re-engineered floodgates and activation/modification of a fishway at Okanagan Falls Dam.	Bull 1999

TABLE 3. Continued.

Initiative	Specific activities	Status of activities	Documentation
(2) Continued	c. Remove or modify barriers to upstream passage by fish on smaller tributaries.	Upstream passage restored at former barriers on Shingle, Ellis, Vaseux, and Inkaneep creeks.	K. Alex, Okanagan Nation Alliance, personal communication
(3) Improve quantity and quality of off-channel and in-channel habitat	a. Re-engineer Okanagan River flood control channel with Newbury riffles, river meanders, spawning platforms, reconnections to side channels, and off-channel wetlands.	Okanagan River Restoration Initiative projects completed between 2008 and 2017 include installation of rock riffles, spawning platforms, Okanagan River re-meander (Figure 11), and wetland connectors.	Bull et al. 2000a, 2000b; Rivard-Sirois et al. 2018

to test for associated increases in production previously proposed by Hyatt and Rankin (1999), as long as the influence of exceptional flood-and-scour or drought-and-desiccation impacts on incubating Sockeye Salmon eggs and alevins could be controlled. However, historic patterns of seasonal flow regulation in the Okanagan River to achieve flood protection and irrigation benefits frequently placed incubating salmon eggs and alevins at risk of loss. Fisheries provisions of the 1982 Canada–British Columbia Okanagan Basin Implementation Agreement (OBIA) specified seasonal lake level and discharge ranges to accommodate various species or life history stages of fish (Sockeye Salmon and kokanee). However, an audit of 1982–1999 water management performance (Bull 1999) indicated frequent noncompliance of regulated flows relative to OBIA-recommended flow ranges for migrating Sockeye Salmon adults, spawning adults, incubating eggs, and alevins (Figure 8). Flood-and-scour events (discharge >28 m^3/s) or drought-and-desiccation events (discharge <5 m^3/s) had the potential to induce density-independent losses of eggs and alevins (Figure 8C; Alexander et al. 2008). Production losses due to the absence of fish-friendly flows had the potential to disrupt benefits expected from increased escapement in more than half of all historic return years. Fortunately, results from the 1999 OBIA flow audit led to coincidental actions taken by the COBTWG to facilitate increased compliance of water managers with fish-friendly flows.

Development and deployment of an environmental decision support system (EDSS) to provide real-time fish and water management tools (FWMT) to decision makers offered a potential means to improve water management decisions affecting human and natural systems in the Okanagan (Hyatt et al. 2001). Accordingly, the COBTWG and the Douglas County PUD entered into a partnership to build and deploy such a system. The resultant FWMT-EDSS described in detail by Hyatt et al. (2015a) included a coupled set of four biophysical models of critical relationships among climate, fish, and water that interact with a fifth water management rules

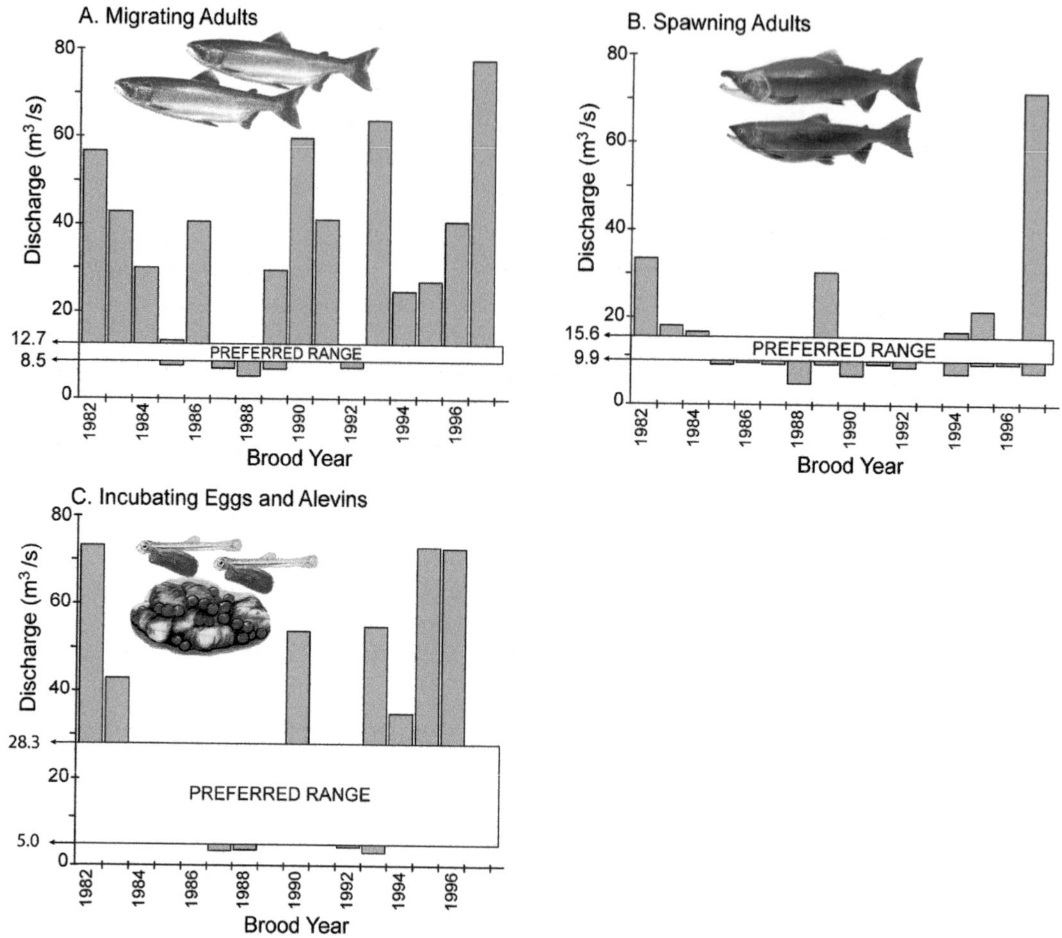

FIGURE 8. Discharge ranges recommended as fish-friendly flows (horizontal rectangles) relative to regulated flows observed annually (gray bars) during 1982–1997 for **(A)** migrating adults, **(B)** spawning adults, and **(C)** incubating eggs/alevins of Sockeye Salmon (adapted from Bull 1999).

model used to predict potential consequences of decisions for fish and other water users. Representatives from government, industry, and local communities use FWMT software to conduct risk assessments of water-management decisions to satisfy human system versus natural system needs (e.g., irrigation demands versus fisheries conservation flows) on a daily to weekly basis. Fish and water management tools software allows users to (1) assess complex biophysical process interactions, (2) examine hydrological impacts on property and fish at multiple locations (e.g., river flow impacts on riparian infrastructure, Sockeye Salmon eggs in the Okanagan River, kokanee eggs in Okanagan Lake), (3) anticipate socioeconomic outcomes of water management decisions (e.g., risk of flood damage to property or shortages of water for irrigation), and (4) facilitate increased cooperation among water users to improve fish-and-water management (e.g., see Hyatt and Stockwell 2013). Comparisons of observations from pre-FWMT "control" versus FWMT deployment years (pre-2003 and post-2003, respectively) indicated that significant improvements from 20% to 90% in all years of Okanagan River regulated flow compliance with OBIA guidelines to protect salmon during critical egg-to-fry emergence stages (see Hyatt et al. 2015a for details).

The COBTWG assessed outcomes of simultaneous increases in the frequency of large escapements and fish-friendly flows on annual production of Sockeye Salmon juveniles before and after the 2003 deployment of the FWMT-EDSS. In six of those years, Sockeye Salmon escapements exceeding 60,000 spawners have occurred, and in five of those years, juvenile Sockeye Salmon production increased significantly (Figure 9), as anticipated by Hyatt and Rankin (1999). It is noteworthy that in two of the six cases, fry recruitment and, subsequently, spring smolt production were much lower than expected. However, in both cases, the anomalously low production of juvenile Sockeye Salmon had clear linkages to other events driving losses of eggs and alevins—impact of a 2010 Testalinden Creek (Figure 1C) mass-wastage event on fry production in Osoyoos Lake and reduced fry production in 2011 from an escapement of >200,000 Sockeye Salmon that exceeded spawning ground capacity (K. D. Hyatt, unpublished observations). The three- to fivefold increase in average annual production of Okanagan Sockeye Salmon smolts leaving Osoyoos Lake after 2007 has been matched by a parallel increase in Okanagan adult Sockeye Salmon returns to the Columbia River after 2009 (Figure 10). Although hatchery-origin fry introduced into Skaha Lake have also contributed to stock rebuilding during the same interval (see text below), they comprise, on average, less than 10% of total adult returns. Thus, the wild-origin run

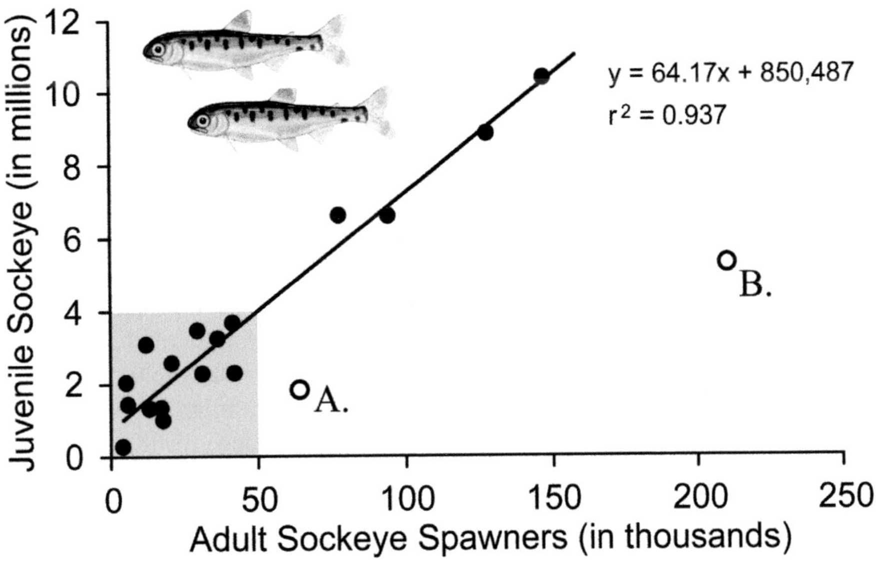

FIGURE 9. Relationship between annual variations in abundance expressed as area-under-the-curve (AUC) estimates of adult Sockeye Salmon spawning in the Okanagan River during 1998–2015 and subsequent numbers of juveniles the following year. Open data points represent 2 years of anomalously low juvenile production associated with (A) an earthen dam failure and effluent input to Osoyoos Lake in spring 2010 (i.e., 2009 brood year adults) and (B) a record number of spawners in 2010 that resulted in redd superimposition and reduced fry recruitment in summer 2011. Direct observations of Sockeye Salmon fry production prior to 2008 were restricted to escapements of less than 50,000 adults (shaded area). Regression relationship excludes 2010 and 2011 outliers. Survey methods and source observations from Hyatt et al. 2015c.

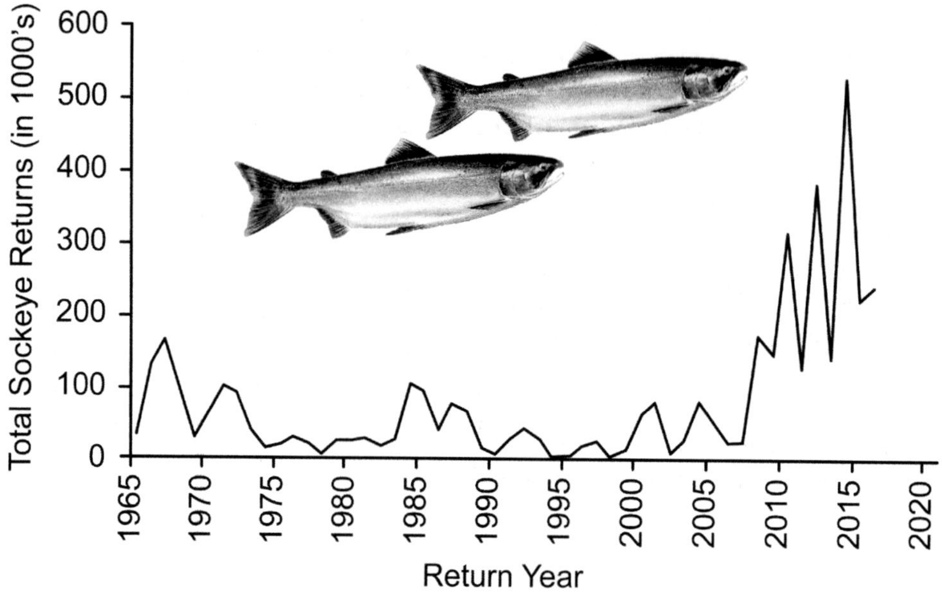

FIGURE 10. Total returns of adult Okanagan-basin Sockeye Salmon taken each year as catch in the Columbia River plus counts of escapement through Wells Dam during1965–2016.

of Sockeye Salmon returning to the Okanagan River basin has not only been restored, but the all-year average return for the most recent decade has exceeded the maximum single-year return (Figure 10) during the period of record from 1965 to 2008 by a substantial margin!

EXPANDING THE CONTEMPORARY RANGE OF SOCKEYE SALMON

From prior to the formation of the COBTWG, the ONA First Nations government, representing the collective interests of its eight member communities, had been investigating the feasibility of expanding the range of salmon in the Okanagan basin (Wright and Smith 2004) beyond McIntyre Dam (Figure 1C). As early as 1920–1921, this dam had become the virtual upstream limit of access for anadromous salmon (Table 2). Concerned with the depressed state of the Sockeye Salmon population in the Okanagan basin in the 1990s, the ONA adopted a policy in 1997 to restore anadromous salmon to their historic range throughout the basin. In the same year, the ONA and Colville Confederated Tribes (CCT; the Okanagan Nation's tribal relatives in Washington State) hosted a workshop to explore the possibility of restoring Sockeye Salmon to historic numbers in the Okanagan valley. The workshop included participants from (1) DFO, (2) the British Columbia Ministry of Environment, (3) the ONA and CCT, and (4) select scientific experts to discuss the potential risks and benefits of potential reintroductions of Sockeye Salmon above the barrier to upstream migration at McIntyre Dam. Due to the concern for the depressed state of kokanee populations in Okanagan Lake (Andrusak et al. 2008), they agreed that if a reintroduction of Sockeye Salmon was to occur, it should initially be restricted to Skaha Lake (Figure 1C), immediately downstream from Okanagan Lake. After 3 years (2000–2003) of risk assessment and project design, including a pilot fry release in 2003, the COBTWG provided approval in principle for an experimental program to reintroduce Sockeye Salmon into Skaha Lake (Peters et al. 1998).

Successful engagements, discussed above, of the COBTWG and DCPUD on joint salmon assessment and restoration work initiated in 1998, increased the confidence of other interested parties (Grant and Chelan PUDs) in the United States regarding the utility of transboundary investments in salmon restoration work in the Okanagan. Their interests, like those of the Douglas County PUD, pertained to FERC relicensing requirements to identify Sockeye Salmon restoration measures to mitigate for losses of this species associated with operation of Priest Rapids, Wanapum, Rocky Reach, and Rock Island dams on the Columbia River. Following assessment of the Skaha Sockeye Salmon Project design (Wright and Smith 2004), the Chelan and Grant PUDs agreed to fund the project, and in 2004, the ONA, with oversight from DFO and BC-FLNRO, began a 12-year experiment (2004–2015; brood years 2003–2014) to reintroduce Sockeye Salmon into Skaha Lake. The objectives of the experiment were to assess Sockeye Salmon production and returns for Skaha Lake and to increase understanding of the interactions among juvenile Sockeye Salmon, kokanee, the introduced shrimp species *M. diluviana*, and their zooplankton forage base. The management goals for the Skaha Lake experimental Sockeye Salmon reintroduction program were to

1. Restore Sockeye Salmon to a portion of their original range within the Okanagan basin and enhance Sockeye Salmon abundance to support tribal ceremonial and subsistence fisheries;
2. Determine whether Skaha Lake Sockeye Salmon could be produced in sufficient numbers to assist the Grant and Chelan PUDs in meeting their Sockeye Salmon smolt production requirements (up to 1.73 million per annum);
3. Determine the effects that hatchery-reared Sockeye Salmon may have on abundance, productivity, and angling quality of Skaha Lake kokanee; and
4. Assess the impact of invasive mysids in Skaha Lake on pelagic fish production.

In October 2003, the ONA initiated wild adult broodstock collection from the Okanagan River (Osoyoos nursery stock), raised the eggs in a hatchery, and, in May 2004, released 352,500 Sockeye Salmon fry into Skaha Lake. During the next 8 years (2005–2012), the hatchery program expanded and a detailed monitoring program was implemented (McQueen et al. 2014). Between 2005 and 2012, numbers of hatchery fry introduced into Skaha Lake have varied by more than a factor of 3 from 0.44 million in 2007 to 1.5 million in 2010. This variation has provided opportunities to assess introduced Sockeye Salmon fry impacts on kokanee, the resident form that completes its entire life cycle in freshwater. Key performance measures, monitored annually, include rates of growth and survival for Skaha Lake wild and hatchery age-0 Sockeye Salmon, resident age-0 and older kokanee, numbers of wild and hatchery Sockeye Salmon smolts, and the abundance of spawning kokanee (Alexander and Hyatt 2015).

Of greatest importance, the project demonstrated that hatchery-origin fry stocked into Skaha Lake survived at, or close to, projected levels to produce an annual smolt run without signs of any significant retention of residual Sockeye Salmon maturing to adulthood in Skaha Lake. Smolt-to-adult survival rates of hatchery-origin Sockeye Salmon from Skaha Lake do not differ significantly from wild-origin Sockeye Salmon from Osoyoos Lake (McQueen et al. 2014). The Skaha project currently contributes an all-year average of around 8% of total smolt and adult production of Okanagan Sockeye Salmon. As of 2009, Sockeye Salmon bound for Skaha Lake have relatively free passage due to engineered modifications to McIntyre

Dam (Tables 2 and 3). Further, adult Sockeye Salmon have demonstrated an ability to bypass Okanagan Falls Dam through floodgates to access Skaha Lake, given above-average summer flows (e.g., 2011 and 2012). These fish spawned in portions of Penticton Channel above Skaha Lake and produced fry that recruited naturally to Skaha Lake. This preliminary result involves too few years of observation to claim total success in establishing a self-sustaining population of Sockeye Salmon in Skaha Lake, but given progress to date, and plans to improve fishway passage at Okanagan Falls Dam, that goal would appear to be within reach.

IMPACTS OF OKANAGAN SOCKEYE SALMON RESTORATION ON FISHERIES

Multiple variables have contributed to rapid rebuilding of Okanagan Sockeye Salmon returning through the Columbia River. The annual average return since 2008 is more than 200,000 adult Sockeye Salmon (Figure 10). Consequently, the Okanagan River population has accounted for greater than 80% of all Sockeye Salmon returning to the Columbia River during the current decade. The recent resurgence of Okanagan River Sockeye Salmon has allowed not only the restoration of tribal ceremonial and subsistence fisheries, but also the creation of valuable recreational and commercial fisheries in Canada and the United States (Figures 11 and 12). These fisheries contrast starkly with circumstances prior to the early 2000s when declining abundance of Okanagan Sockeye Salmon was associated with losses of tribal and commercial Sockeye Salmon fisheries in the United States and the virtual elimination of Okanagan Sockeye Salmon harvest by the ONA in Canada for more than 35 years. Average annual catches of approximately 50,000 Sockeye Salmon of Okanagan River origin occurred in U.S. and Canadian fisheries between 2008 and 2014 (Figure 11), in addition to annual escapements of around 100,000 spawners. The relatively modest exploitation rates of roughly 30% are due to the continued constraint of Endangered Species Act (ESA) protection of Redfish Lake-origin Sockeye Salmon in the United States that return through traditional fishing areas of the lower Columbia River at the same time as the Okanagan and Wenatchee River stocks. Consequently, entirely new fisheries have had to develop in the mid-Columbia River above McNary Dam (Figure 1B) and, in Canada, in Osoyoos Lake and the Okanagan River. Regardless, new recreational, tribal, First Nations, and commercial fisheries in Canada and the United States (Figure 12) have experienced sustainable benefits over the past 10 years from this Okanagan Sockeye Salmon restoration success.

FUTURE IMPROVEMENT OF OFF-CHANNEL AND IN-CHANNEL HABITAT

Okanagan Basin Technical Working Group terms of reference and operational guidelines direct the group to take an ecosystem-based management approach to restoration in which projects focused on one valued species do not result in major, collateral damage to others. In addition, the vision of ONA elders and leaders is one in which the diversity of anadromous salmon (e.g., Chinook Salmon, steelhead, and Coho Salmon) that once existed in Canadian portions of the Okanagan basin (Figure 7) is eventually restored. Given the obstructed state of most small tributaries and the almost total loss of side-channel, off-channel, and main-channel habitat complexes (Figure 13A) that these species depend on (see references in Naiman and Bilby 1998), such restoration will be a monumental task for which short-term success, such as that achieved for Sockeye Salmon restoration, is unlikely. Despite this prognosis, the parties to the COBTWG have committed to a long-

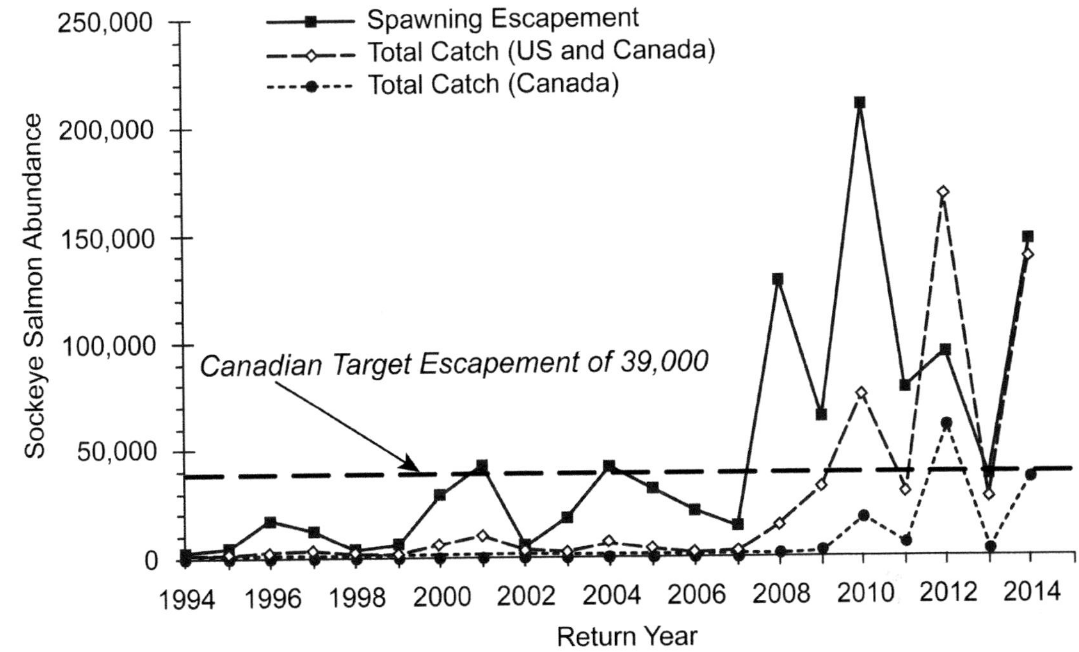

FIGURE 11. Annual abundance of Okanagan Sockeye Salmon taken as catch or escaping to spawn (1994–2014). Note the target escapement of 39,000 fish (area-under-the-curve [AUC] metric) in terminal spawning areas of the Okanagan River is equivalent to 66,612 passing Wells Dam (direct count metric), assuming no catch upstream of that point. The AUC conversion to direct counts at Wells is AUC = 0.576 Wells + 1,096 ($r^2 = 0.89$, $p < 0.001$, $n = 27$; authors' unpublished data).

term, incremental restoration approach that involves identification and implementation of opportunities to restore habitat access for salmon and resident fish (Bull et al. 2000a, 2000b). These actions include (1) removing or modifying water storage and diversion structures on small, obstructed tributaries, (2) re-engineering the current flood-control channel by installing fish-passable riffles (Newbury 2013), and, where feasible, (3) relocating dike segments to provide fish and flowing-water access to relict side channels (Rivard-Sirois et al. 2018) and associated river meanders (Figure 13B, 13C). This work involves large investments of expertise (hydrologists, biologists, engineers, and heavy equipment operators), finances (roughly half a million to a million Canadian dollars per kilometer), and time (years to decades). However, it represents the only approach with any potential to alter the current dismal state of production by anadromous salmon species, other than Sockeye Salmon, in Canadian portions of the Okanagan River basin.

LESSONS LEARNED

Restoration of wild salmon clearly does not always constitute futile pursuit of a mere illusion. Instead, success was possible and achieved! This chapter documented the remarkable level of success experienced over the past 20 years by the COBTWG in the joint execution of several projects (e.g., Testing Habitat Capacity Limits, Fish-and-Water Management Tools, the Skaha Sockeye Reintroduction Project, and the Okanagan River Restoration Initiative), which have effectively contributed to restoration of Sockeye Salmon to the Okanagan River basin. The magnitude

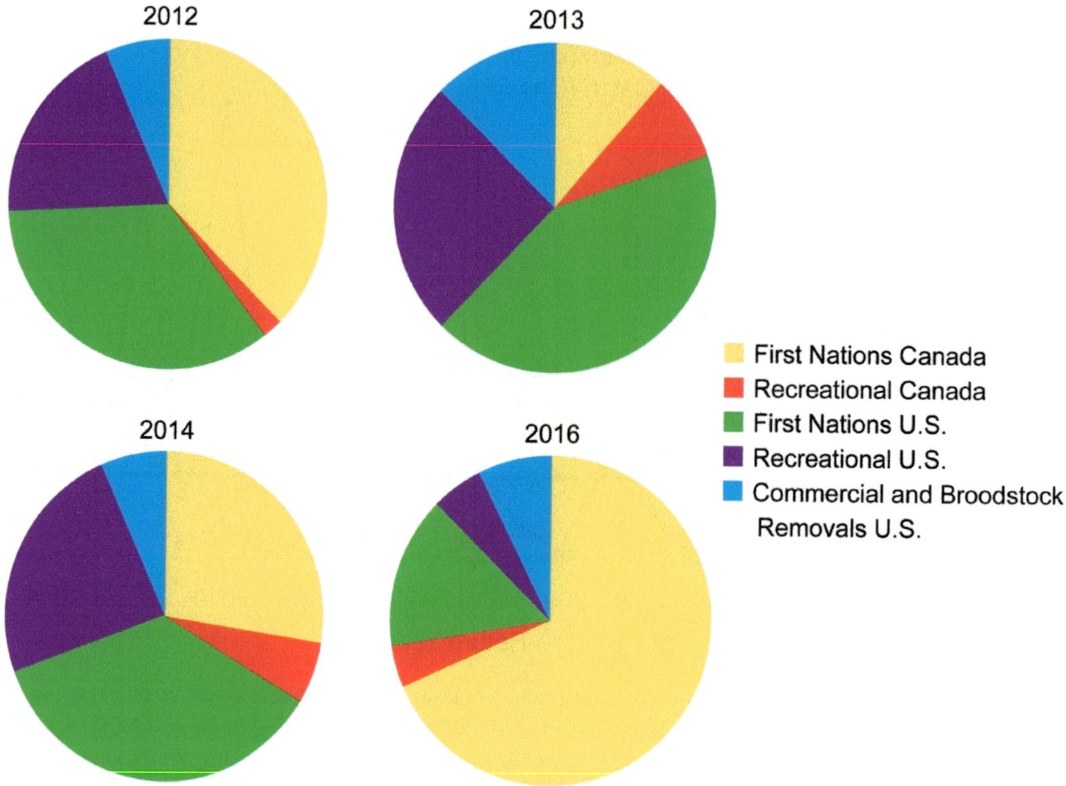

FIGURE 12. Distribution of Okanagan Sockeye Salmon catch among fisheries executed in the United States and Canada in 2012–2016 (authors' unpublished data).

and speed of this success are exceptional relative to the initial projections by COBTWG members (Hyatt et al. 2001). Consequently, one may question whether the outcomes achieved were a consequence of good planning or just good luck! We acknowledge that both have been involved but suggest that our success has been principally due to adherence to a recognized process (Figure 2) that increases the likelihood of restoration success.

The emergence of leadership and a vision served as the starting point. In the current example, the ONA, for whom salmon restoration represents a cultural imperative, provided both. Absent their efforts to bring federal and provincial agencies to the table, the state of salmon in the Okanagan River basin would differ little today from that observed during the 1990s (Figure 10) when Sockeye Salmon extirpation seemed imminent. Cultural (ONA) and institutional (DFO) mandates dictated that avoiding extirpation was a minimum but powerful common objective to which all could readily agree. Canada's Species at Risk Act (2002) and DFO's adoption of a region-wide Wild Salmon Policy (DFO 2005) reinforced the agency's commitment to this objective. The vision, objectives, and call for joint action for implementation of the 2005 Wild Salmon Policy not only matched, but also added to the sense of commitment by DFO to support ONA aspirations for salmon restoration. Effective leadership has also been a key ingredient for implementation of the separate components making up the COBTWG restoration strategy (Table 3), and in each case, specific individuals have served, for periods of a decade or more, as

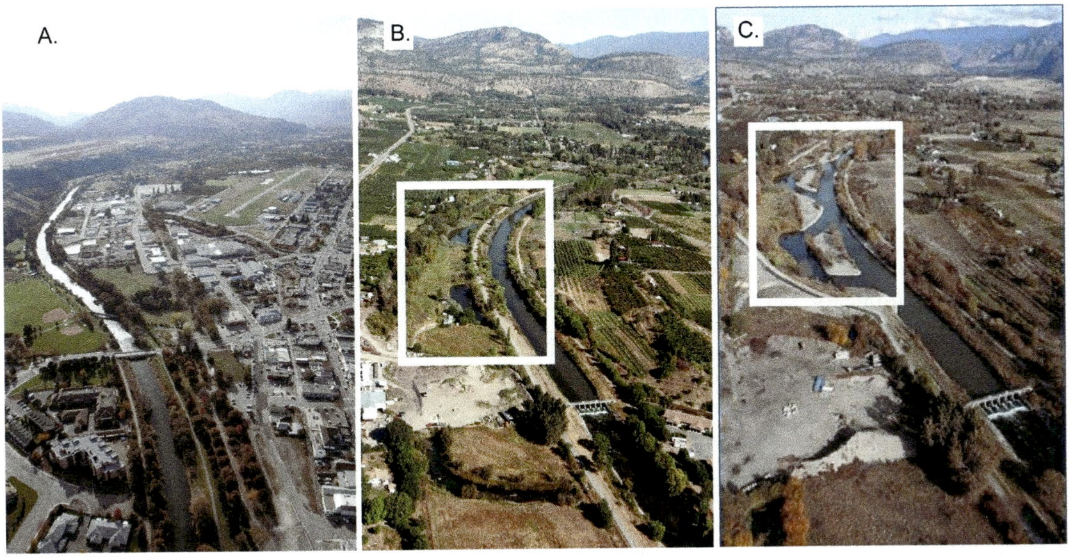

FIGURE 13. Aerial photographs (Okanagan Nation Alliance) of **(A)** the extent of urban development along the channelized Okanagan River, **(B)** a potential meander and side channel restoration site in the Okanagan River near Oliver, British Columbia, and **(C)** the meander and side-channel complex reconnected to the Okanagan River main channel in 2008–2009 after dike breaching and setback (Bull et al. 2000a, 2000b; Rivard-Sirois et al. 2018).

champions of particular projects. Of equal importance, as individual projects achieved success, each success added to the sense of common purpose, positive chemistry, and trust among COBTWG members.

As a general principle, restoration actions need to reflect not only the potential for a biophysical intervention to stimulate an effective response from a fish population, but also the potential for local actors to effectively initiate and sustain such actions. From this perspective, the formation of the three-party COBTWG represented the minimum set of actors with sufficient technical expertise, institutional authority, and fiscal resources to collectively reach a critical mass capable of supporting sustained and effective salmon restoration work in Canadian portions of the Okanagan basin. The limited, three-party membership of the restoration-planning group created advantages and disadvantages. Because the group's terms of reference included making decisions by consensus, the principal advantage of the simple, three-party structure of the group was to minimize delays between planning discussions, decisions, and implementation of actions. The principal disadvantage was that the limited Canadian membership of the COBTWG largely eliminated restoration opportunities to resolve production issues controlled by out-of-basin events or regimes in the Columbia River or on the high seas.

Anadromous Pacific salmon follow life history trajectories lived out over months to years of time spent in spatial "patches" (Figure 1) ranging from a few thousand square meters (e.g., freshwater spawning areas) to millions of square kilometers (e.g., ocean rearing areas). Numerous natural and human-induced events or regimes interact cumulatively throughout this space–time trajectory to shape life history and production outcomes for salmon populations (Figure 4). The difficulty this complexity creates for identifying effective actions to reverse declines in wild salmon populations is more likely to be overlooked than overstated in restoration planning, where solutions to problems emerging at multiple scales are too frequently sought at just the local scale with which restoration proponents are most familiar (Beechie et al. 2010). The COBTWG acknowledged this problem, sought expert advice, and either commissioned or solicited analysis and reporting (e.g., references in Table 3) to realistically identify local opportunities for biophysical restoration actions to provide an effective response by the subject salmon populations. Together, these assessments indicated that the highly disturbed state of Okanagan River basin aquatic ecosystems (Figure 7C versus 7A) would preclude rapid recovery of depleted populations of Chinook Salmon and steelhead, extirpated populations of Coho Salmon (all greatly affected by channelization), or Chum Salmon *Oncorhynchus keta* (Figure 7D). Thus, developing a deeper understanding of the salmon–ecosystem–habitat complex, shaped by interactions among environmental drivers emerging from natural and human systems (Figure 4), is essential to improving planning and the odds of success for either restoration or maintenance of wild salmon populations in the Okanagan River and elsewhere.

A common observation is that "success has many fathers while failure is an orphan." However, it would be a mistake to ignore the role of good luck in this story. At least two lucky events contributed to the success discussed here. The first involved ESA listing of Redfish Lake Sockeye Salmon in Idaho (Kline et al. 2019), which resulted in (1) dramatic curtailment of tribal and commercial fisheries for Sockeye Salmon in the lower Columbia River and (2) large increases of escapement of adult Sockeye Salmon in terminal spawning areas of the Okanagan. Absent these events, historic underutilization of spawning ground and lake-rearing potential by Sockeye Salmon (Hyatt and Rankin 1999) might still be awaiting resolution. The second fortuitous event involved the highly variable influence of ocean conditions on survival trends exhibited by the Columbia River and west coast Vancouver Island Sockeye Salmon populations. Hyatt et al. (2015b) have noted that southern Sockeye Salmon stocks with sea entry into the northern California Current upwelling domain (Okanagan and Barkley Sockeye Salmon) have shown rapid rebuilding (2009–2013) to record (Okanagan) or near-record (Barkley) returns in 2014, due in part to consecutive years of favorable marine survival conditions. The impact of the latter more than doubled average smolt-to-adult return rates from 4% to 11% for the 2006–2010 sea-entry years relative to the 1992–2005 sea-entry years for Okanagan Sockeye Salmon (Hyatt, unpublished data). However, if continued harvest had capped escapement at its 45-year average of 20,000 Sockeye Salmon adults (pre-2007) and fry production (Figure 9) had remained under the frequent influence of flood and drought losses, the maximum average return, even at an 11% smolt-to-adult "marine" survival, would have been well below 100,000 adults for the 2008–2012 return years. This contrasts sharply with observed average returns of more than 230,000 Sockeye Salmon adults during this interval. Thus, improved marine survival (Williams et al. 2014; Hyatt et al. 2015b) may have contributed a maximum of 43% to the observed increase in adult Sockeye Salmon after 2007, but the increase was insufficient to account for the unprecedented magnitude of Okanagan Sockeye Salmon increase documented here (Figure 10).

Although we have already commented on the direct benefits of Okanagan Sockeye Salmon restoration to fisheries in the United States and Canada, several additional benefits warrant comment. Of special significance, the process of collaborative salmon stock restoration facilitated significant capacity building by the ONA First Nation's fisheries group. In 1997, the latter was comprised of four personnel consisting of a policy adviser, a biologist, and two seasonal technicians. However, as of 2017, the ONA fisheries group has grown to more than 40 personnel, including at least 20 full-time staff (e.g., biologists, hydrologists, and fish culturists) and an equivalent number of seasonal employees. To be fair, not all of these positions work exclusively on Okanagan River basin salmon issues, but the latter continues to represent the principal focus of ONA's fisheries work. As participating agencies, BC-FLNRO and DFO have also received tangible benefits. The water-management, decision-support system (FWMT-EDSS) has been embraced by BC-FLNRO managers as an indispensable set of modeling tools to identify seasonally changing solutions to satisfy competing objectives among fisheries (production), agricultural (irrigation), urban (flood protection), and recreational (navigation) water users (Hyatt et al. 2015a). Finally, given DFO's mandate for conservation and sustainable management of wild salmon in Canada's Pacific region, participation in the restoration of Okanagan River Sockeye Salmon represents an uncommon success story. In addition, the highly cooperative path taken and outcomes achieved with the ONA, BC-FLNRO and several industry funding sponsors (in particular, the Douglas County, Grant, and Chelan PUDs) are entirely in line with the model envisaged by DFO's forward-looking Wild Salmon Policy (DFO 2005).

SUMMARY OF LESSONS LEARNED

1. Acknowledging and satisfying each of the several key elements of restoration planning (Figure 2) will improve the odds of implementation success and its outcome(s).
2. Leadership, vision, and persistent support for a common cause are essential ingredients to develop trust, as well as the ever-elusive, positive interpersonal chemistry.
3. Restoration objectives and implementation actions need to be realistically assessed within the context of developing a deep understanding of salmon–ecosystem–habitat associations shaped by complex interactions with drivers in natural and human systems (Figures 4 and 7).
4. Careful scoping to identify root causes of fish population declines and current limiting factors is an important step towards designing successful recovery initiatives. Failing to do so virtually guarantees more failure than success in salmon restoration and management.
5. The scale (space and time) of restoration actions pursued as solution(s) for limiting factor(s) must match the scale at which the factor(s) exert their influence(s).
6. Restoration actions need to reflect not only the local potential for a biophysical action to provide an effective response by the fish population, but also the potential for local actors to effectively initiate and sustain such actions.
7. Clearly identifiable and readily measurable outcomes within a finite interval are important to establish as part of any effective restoration initiative(s).
8. Sustained funding is essential, and although not essential, a bit of good luck can be helpful.
9. Despite the incredible challenge that restoring depleted populations of wild salmon poses, given their complex life histories and wide ranging habits, restoration success is possible!

ACKNOWLEDGMENTS

We gratefully acknowledge ONA fisheries program managers, including Byron Louis, Dawn Machin, Deana Machin, and Howie

Wright, who have provided vision, leadership, and persistent efforts to bridge the many cultural and technical gaps among First Nations, government agencies, and industry partners involved in Okanagan salmon restoration work. All past and present members of the COBTWG deserve our thanks for their dedication and exceptional commitment to overcome the difficult challenge of restoring severely depressed salmon population in the Okanagan River basin. Many organizations have provided funding for COBTWG restoration projects, but the Douglas County, Grant County, and Chelan County PUDs of Washington State, Fisheries and Oceans Canada, and the BC Habitat Conservation Trust deserve special mention.

REFERENCES

Alexander, C. A. D., K. Hyatt, and B. Symonds. 2008. The Okanagan fish/water management tool (v.2.4000): guidelines for apprentice water managers. Prepared for the Canadian Okanagan Basin Technical Working Group, Kamloops, British Columbia by ESSA Technologies Ltd., Vancouver.

Alexander, C. A. D., and K. D. Hyatt, editors. 2015. Proceedings of the peer review workshop for the Okanagan Sockeye reintroduction experiment. Prepared for the Okanagan Nation Alliance and Canadian Okanagan Basin Technical Working Group. Available from the Okanagan Nation Alliance, Westbank, British Columbia.

Anderson, J. J. 2000. Decadal climate cycles and declining Columbia River salmon. Pages 467–484 *in* E. E. Knudsen, C. R. Steward, D. D. MacDonald, J. E. Willaims, and D. W. Reiser, editors. Sustainable fisheries management: Pacific salmon. Lewis Publishers, Boca Raton, Florida.

Andrusak, H., G. Andrusak, S. Matthews, A. Wilson, T. White, P. Askey, D. Sebastain, G. Scholten, P. Woodruff, J. Webster, L. Vidmanic, and J. Stockner. 2008. Okanagan Lake Action Plan years 11 (2006) and 12 (2007) report. British Columbia Ministry of Environment, Ecosystems Branch, Fisheries Project Report 123, Victoria.

Augerot, X. 2005. Atlas of Pacific salmon. The first map-based assessment of salmon in the North Pacific. University of California Press, Berkeley.

Austin, M. A., D. A. Buffet, D. J. Nicolson, G. G. E. Scudder, and V. Stevens, editors. 2008. Taking natures pulse: the status of biodiversity in British Columbia. Biodiversity BC, Victoria, British Columbia.

Beechie, T. J., D. A. Sear, J. D. Olden, G. R. Pess, J. M. Buffington, H. Moir, P. Roni, and M. M. Pollock. 2010. Process-based principles for restoring river ecosystems. BioScience 60:209–222.

Bevan, D. E., P. K. Bergman, T. C. Bjornn, J. A. Crutchfiled, J. P. Harville, P. C. Klingeman, and J. W. Litchfield (Snake River Salmon Recovery Team). 1993. Draft Snake River salmon recovery plan recommendations. National Marine Fisheries Service, Portland, Oregon.

Bezener, A., M. Dunn, H. Richardson, O. Dyer, R. Hawes, and T. Hayes. 2004. South-Okanagan Similkameen conservation program: a multi-partnered, multi-species, multi-scaled approach to conservation of species at risk. In T. D. Hopper, editor. Species at risk 2004 conference proceedings [online publication]. Available: www.arlis.org/docs/vol1/69415913/bezener_edited_final_feb_8.pdf. (January 2019).

Bjornn, T. C. and D. W. Reiser. 1991. Habitat requirements of salmonids in streams. Pages 83–138 *in* W. R. Meehan, editor. Influence of forest and rangeland management on salmonid fishes and their habitats. American Fisheries Society, Special Publication 19, Bethesda, Maryland.

Bradford, M. J., G. C. Taylor, and J. A. Allan. 1997. Empirical review of Coho Salmon smolt abundance and the prediction of smolt production at the regional level. Transactions of the American Fisheries Society 126:49–64.

British Columbia Ministry of Environment. 1998. Habitat atlas for wildlife at risk. British Columbia Ministry of Environment, Victoria

Bull, C., M. Gaboury, and R. Newbury. 2000a. Okanagan River habitat restoration feasibility. Prepared for the Ministry of Environment, Lands and Parks, Penticton, British Columbia.

Bull, C., M. Gaboury, and R. Newbury. 2000b. Setback dyke and meander restoration designs for the Okanagan River. Report submitted to Ministry of Environment, Lands and Parks, Kamloops, British Columbia. Available: http://a100.gov.bc.ca/appsdata/acat/documents/r21384/Setback_1296683789105_3c2947f5334658c3a6a1dc311dd893ab71af9e49ebea08f1f91581081231ec34.pdf. (April 2018).

Bull, C. J. 1999. Fisheries habitat in the Okanagan River phase 2 – investigation of selected options. Prepared for Public Utility District No. 1 of Douglas County, Washington.

Burgner, R. L. 1991. Life history of Sockeye Salmon (*Oncorhynchus nerka*). Pages 1–117 *in* C. Groot and L. Margolis, editors. Pacific salmon life histories. UBC Press, Vancouver.

Cannings, R. J. and E. Durance. 1998. Human use of natural resources in the South Okanagan and Lower Similkameen valleys. *In* I. M. Smith and G. G. E. Scudder, editors. Assessment of species diversity in the Montane Cordillera Ecozone. Ecological Monitoring and Assessment Network, Burlington, Ontario.

Chapman, D. W. 1986. Salmon and steelhead abundance in the Columbia River in the nineteenth century. Transactions of the American Fisheries Society 115:662–670.

Chapman, J. D., D. B. Turner, A. L. Farley, and R. I. Ruggles. 1956. British Columbia: atlas of resources. Natural Resources Conference 1956, Vancouver, British Columbia.

Cohen, S. J., D. Neilsen, S. Smith, T. Neale, T. Taylor, B. Barton, W. Merritt, W. S. Younes, A. Shepherd, P. McNeill, R. Tansey, J. Carmichael, and S. Langsdale. 2006. Learning with local help: expanding the dialogue on climate change and water management in the Okanagan region, British Columbia, Canada. Climatic Change 75:331–358.

COSEWIC (Committee on the Status of Endangered Wildlife in Canada) 2006. COSEWIC assessment and status report on the Chinook Salmon *Oncorhynchus tshawytscha* (Okanagan population) in Canada. COSEWIC, Ottawa.

Cumming, B. F., K. R. Laird, I. Gregory-Eaves, K. G. Simpson, M. A. Sokal, R. N. Nordin, and I. R. Walker. 2015. Tracking past changes in lake-water phosphorous with a 251-lake calibration data set in British Columbia: tool development in multi-proxy assessment of eutrophication and recovery of Osoyoos Lake, a transboundary lake in western North America. Frontiers in Ecology and Evolution [online serial] 3:84.

Dearden, P. 1983. Anatomy of a biological hazard: *Myriophyllum spicatum* L. in the Okanagan basin, British Columbia. Journal of Environmental Management 17:47–61.

DFO (Fisheries and Oceans Canada). 2005. Canada's policy for conservation of wild Pacific salmon. DFO, Vancouver. Available: http://waves-vagues.dfo-mpo.gc.ca/Library/315577.pdf. (December 2018).

DFO (Fisheries and Oceans Canada). 2017. Review of potential impacts associated with recent and proposed Okanagan Sockeye Salmon fry introductions to Skaha and Okanagan lakes. Fisheries and Oceans Canada, Canadian Science Advisory Secretariat Science Response 2017/031, Vancouver. Available: http://waves-vagues.dfo-mpo.gc.ca/Library/40625783.pdf. (December 2018).

Department of Justice Canada. 2016. Fisheries Act. R.S.C., 1985, c. F-14. Minister of Justice. Available: http://laws-lois.justice.gc.ca/PDF/F-14.pdf. (August 2018).

Dobson, D. 2004. Hydrology and watershed management. Pages 71–83 *in* M. A. Roed and J. D. Greenough, editors. Okanagan geology: British Columbia, 2nd edition. Kelowna Geology Committee, Kelowna, British Columbia.

Groot, C., and L. Margolis, editors. 1991. Pacific salmon life histories. University of British Columbia Press, Vancouver.

Hall, K., J. Stockner, H. Schreier, and R. Bestbier. 2001. Nutrient sources and ecological impacts on Okanagan Lake. Institute for Resource and Environment, University of British Columbia, Vancouver. Available: http://a100.gov.bc.ca/pub/acat/public/viewReport.do?reportId=16579. (December 2018).

Hourston, R., C .H. Clay, E. W. Burridge, K. C. Lucas, D. R. Johnson, H. T. Heg, W. R. McKinley, J. T. Barnaby, L. A. Fulton, and A.A. Gentry. 1954. The salmon problems associated with the proposed flood control project on the Okanagan River in British Columbia, Canada. Prepared by the technical staffs of the U.S. Fish and Wildlife Service, Washington, D.C.; the Washington State Department of Fisheries, Olympia; and the Department of Fisheries, Canada, Vancouver.

Hyatt, K. D., C. A. D. Alexander, and M. M. Stockwell. 2015a. Testing the utility of a decision-support system for improving compliance with 'fish friendly' flows in the regulated Okanagan Lake and River system of British Columbia. Canadian Water Resources Journal 40:87–110.

Hyatt, K. D., E. Fast, M. Flynn, D. Machin, S. Matthews, and B. Symonds. 2001. Water management tools to increase production of Okanagan Sockeye Salmon. Fish-water management proposal (June 21, 2001) submitted to Douglas County Public Utility Division No. 1, Wenatchee, Washington.

Hyatt, K. D., D. J. McQueen, and A. D. Ogden. 2018a. Have invasive mysids (*Mysis diluviana*) altered the capacity of Osoyoos Lake, British Columbia to produce Sockeye Salmon (*Oncorhynchus nerka*)? Open Fish Science Journal [online serial] 11:3–28.

Hyatt, K. D., D. J. McQueen, D. P. Rankin, M. M. Stockwell, H. Wright, S. Lawrence, and A. Stevens. 2015c. Methods and summary data for limnology and food-web structure in Osoyoos Lake, B.C. (2005–2013). Canadian Data Report of Fisheries and Aquatic Sciences 1258.

Hyatt, K. D., C. Ramcharan, D. J. McQueen, and C. L. Cooper. 2005. Trophic triangles and competition among vertebrate (*Oncorhynchus nerka*, *Gasterosteus aculeatus*) and macroinvertebrate (*Neomysis mercedis*) planktivores in Muriel Lake, British Columbia, Canada. Ecoscience 12:11–26.

Hyatt K. D. and D. P. Rankin. 1999. A habitat based evaluation of Okanagan Sockeye Salmon escapement objectives. Fisheries and Oceans Canada, Canadian Stock Assessment Secretariat Research Document 99/191, Ottawa. Available: http://waves-vagues.dfo-mpo.gc.ca/Library/244185.pdf. (December 2018).

Hyatt, K. D., H. Stiff, M. M. Stockwell, and R. Ferguson. 2015b. Sockeye salmon indicator stocks: regional overview of trends, 2014 returns, and 2015–2016 outlook. Pages 118–123 *in* P. Chandler, S. King, and I. Perry, editors. State of the physical, biological, and selected fishery resources of Pacific Canadian marine ecosystems in 2014. Canadian Technical Report of Fisheries and Aquatic Sciences 3131.

Hyatt, K. D. and M. M. Stockwell. 2013. Fish and Water Management Tools project assessments: record of management strategy and decisions for the 2007–2008 water year. Canadian Manuscript Report of Fisheries and Aquatic Sciences 3022.

Hyatt, K. D., M. M. Stockwell, and D. P. Rankin. 2003. Impact and adaptation responses of Okanagan River Sockeye Salmon (*Oncorhynchus nerka*) to climate variation and change effects during freshwater migration: stock restoration and fisheries management implications. Canadian Water Resources Journal 28:689–711.

Hyatt, K., R. Withler, and K. Garver. 2018b. Review of recent and proposed Okanagan Sockeye Salmon (*Oncorhynchus nerka*) fry introductions to Skaha and Okanagan lakes: history, uncertainties, and implications. Fisheries and Oceans Canada, Canadian Stock Assessment Secretariat Research Document 2018/018, Nanaimo, British Columbia.

Jensen, V., and P. F. Epp. 2002. Water quality trends in Okanagan, Skaha, and Osoyoos lakes in response to nutrient reductions and hydrologic variation. Ministry of Water, Land, and Air Protection, Penticton, British Columbia. Available: http://a100.gov.bc.ca/pub/acat/public/viewReport.do?reportId=11842. (December 2018).

Joint Columbia River Management Staff. 2016. 2016 Joint staff report: stock status and fisheries for spring Chinook, summer Chinook, Sockeye, steelhead, and other species, and miscellaneous regulations. Oregon Department of Fish and Wildlife, Salem and Washington Department of Fish and Wildlife, Olympia.

Keast, A. 1984. The introduced aquatic macrophyte, *Myriophyllum spicatum*, as habitat for fish and their invertebrate prey. Canadian Journal of Zoology 62:1289–1303.

Kennedy, D., and R. T. Bouchard. 1998. Northern Okanagan, Lakes and Colville. Pages 620–640 *in* D. E Walker, editor. Handbook of North American Indians. Smithsonian Institute, Washington, D.C.

Kline, P. A., T. A. Flagg, C. C. Kozfkay, D. J. Baker, D. J. Maynard, and W. C. McAuley. 2019. Snake River Sockeye Salmon—a perspective on twenty-five years of effort to recover a population in central Idaho, USA. Pages 101–123 *in* C. C. Krueger, W. W. Taylor, and S. Youn, editors. From catastrophe to recovery: stories of fishery management success. American Fisheries Society, Bethesda, Maryland.

Lackey, R. T. 2000. Restoring wild salmon to the Pacific Northwest: chasing an illusion? Pages 91–143 *in* P. Koss and M. Katz, editors. What we don't know about Pacific Northwest fish runs: an inquiry into decision-making. Portland State University, Portland, Oregon.

Law, J. A. and Fréchette, J-D. 1993. The aboriginal fisheries and the Sparrow decision. Library of Parliament, Background Paper BP-341E, Ottawa, Ontario.

Lea, T. 2008. Historical (pre-settlement) ecosystems of the Okanagan valley and Lower Similkameen valley of British Columbia: pre-European contact to present. Davidsonia 19:3–36.

Leith, R., and P. Whitfield. 1998. Evidence of climate change effects on the hydrology of streams in south-central BC. Canadian Water Resources Journal 23:219–230.

Lichatowich, J. 1999. Salmon without rivers. Island Press, Washington, D.C.

Madsen, J. D. 1999. Predicting the invasion of Eurasian watermilfoil into northern lakes. U.S. Army Engineers, Waterways Experiment Station, Technical Report A-99-2, Vicksburg, Mississippi.

Mantua, N. J., S. R. Hare, Y. Zhang, J. M. Wallace, and R. Francis. 1997. A Pacific interdecadal climate oscillation with impacts on salmon production. Bulletin of the American Meteorological Society 78:1069–1079.

McQueen, D., K. Hyatt, H. Wright, R. Bussanich, A. Stevens, M. Stockwell, and P. Rankin. 2014. Evaluation of the experimental introduction of Sockeye Salmon (*Oncorhynchus nerka*) into Skaha Lake and Sockeye rearing in Osoyoos Lake: synthesis of the first 8 years. Okanagan Nation Alliance Report for the Canadian Okanagan Basin Technical Working Group, Westbank, British Columbia.

Merritt, W. S., Y. Alila, M. Barton, B. Taylor, S. Cohen, and D. Neilsen. 2006. Hydrologic response to scenarios of climate change in sub watersheds of the Okanagan basin, British Columbia. Journal of Hydrology 326:79–108.

Naiman, R. J. and R. E. Bilby, editors. 1998. River ecology and management. Springer-Verlag, New York.

Naito, G. 2000. Presence and distribution of exotic fish species in the Okanagan watershed. Report prepared for the Okanagan

Nation Fisheries Commission, Westbank, British Columbia. Available: http://a100.gov.bc.ca/pub/acat/public/viewReport.do?reportId=35431. (December 2018).

National Research Council. 1996. Upstream: salmon and society in the Pacific Northwest. The National Academies Press, Washington, D.C.

Nehlsen, W., J. E. Williams, and J. A. Lichatowich. 1991. Pacific salmon at the crossroads: stocks at risk from California, Oregon, Idaho, and Washington. Fisheries 16:4–21.

Nesler, T. P., and E. P. Bergersen. 1991. Mysids in fisheries: hard lessons in headlong introductions. American Fisheries Society, Symposium 9, Bethesda, Maryland.

Newbury, R. W. 2013. Designing fish-passable riffles as gradient controls in Canadian streams. Canadian Water Resources Journal 38:232–250.

Nichols, S. A., and S. Mori. 1971. The littoral macrophyte vegetation of Lake Wingra; an example of a *Myriophyllum spicatum* invasion in a southern Wisconsin lake. Transactions of the Wisconsin Academy of Sciences Arts and Letters 59:107–119.

Northcote, T. G. 1991. Successes, problems, and control of introduced mysid populations in lakes and reservoirs. Pages 5–16 *in* T. P. Nesler and E. P. Bergersen, editors. Mysids in fisheries: hard lessons from headlong introductions. American Fisheries Society, Symposium 9, Bethesda, Maryland.

Northcote, T. G. 1996. Effects of human population growth on the Fraser and Okanagan River systems, Canada: a comparative inquiry. GeoJournal 40:127–133.

Okanagan Basin Implementation Board. 1982. Summary report on the Okanagan Basin Implementation Agreement (OBIA). Okanagan Basin Implementation Board of Canada. Available: http://a100.gov.bc.ca/pub/acat/public/viewReport.do?reportId=13372. (June 2012).

Ortolano, L., and K. K. Cushing. 2000. Grand Coulee Dam and the Columbia Basin Project USA. Prepared for the World Commission on Dams (WCD), Cape Town, South Africa. Available: http://dina.centre-cired.fr/IMG/pdf/F9_GranCouleeDam.pdf. (December 2018).

Parnell, I. J., C. N. Peters, and D. R. Marmorek. 2003. Evaluate alternative experimental strategies for reintroducing Sockeye Salmon to Skaha Lake. Prepared for Okanagan Nation Alliance Fisheries, Westbank, British Columbia by ESSA Technologies Ltd., Vancouver.

Payne, J. T., A. W. Wood, A. F. Hamlet, R. N. Palmer, and D. T. Lettenmaier. 2004. Mitigating the effects of climate change on the water resources of the Columbia River basin. Climatic Change 62:233–256.

Peters, C. N., D. P. Bernard, and D. R. Marmorek. 1998. Should Sockeye be reintroduced to Okanagan Lake? An exploration of potential benefits, impacts, and uncertainties: Okanagan Nation Fisheries Commission. Available from the Okanagan Nation Alliance, Westbank, British Columbia.

Quinn, T. P. 2005. The behavior and ecology of Pacific salmon and trout. American Fisheries Society, Bethesda, Maryland and University of Washington Press, Seattle.

Rae, R. 2005. The state of fish and fish habitat in the Okanagan and Similkameen basins. Prepared for the Canadian Okanagan Basin Technical Working Group, Westbank, British Columbia.

Rawson, D. S. 1939. Physical and chemical studies, plankton, and bottom fauna of Okanagan Lake B.C. in 1935 with appended data from adjacent smaller lakes. Pages 3–26 *in* W. A. Clemens, D. S. Rawson, and J. L. McHugh, editors. A biological survey of Okanagan Lake, British Columbia. Fisheries Research Board of Canada Bulletin 56.

Rivard-Sirois, C., K. Alex, and C. Louie. 2018. Aquatic monitoring of the Okanagan River Restoration Initiative (ORRI) – post-construction 2017. Prepared by Okanagan Nation Alliance Fisheries Department, Westbank, British Columbia.

Schindler, D. E., J. L. Carter, T. B. Francis, P. J. Lisi, P. J. Askey, and D. C. Sebastian. 2012. *Mysis* in the Okanagan Lake food web: a time-series analysis of interaction strengths in an invaded plankton community. Aquatic Ecology 46:215–227.

Sellars, C. D., and R. Smith. 2005. Application of the water use plan approach to resolve water management issues on Trout Creek in Summerland. Pages 163–176 *in* Water – our limiting resource: towards sustainable water management in the Okanagan. Canadian Water Resource Association, Ottawa, Ontario.

Shepard, M. P., and A. W. Argue. 2005. The 1985 Pacific Salmon Treaty: sharing conservation burdens and benefits. UBC Press, Vancouver.

Simmatis, B., A. Jeziorski, A. Zemanek, D. T. Selbie, K. Hyatt, J. K. Fryer, B. F. Cumming, and J. P. Smol. 2018. Long-term reconstruction of deep-water oxygen conditions in Osoyoos Lake (British Columbia, Canada): implications for Okanagan River Sockeye Salmon. Lake and Reservoir Management 34:392–400.

Slaney, T. L., K. D. Hyatt, T. G. Northcote, and R. J. Fielden. 1996. Status of anadromous salmon and trout in British Columbia and Yukon. Fisheries 21:20–35.

Solbe, J. 1997. Water quality for salmon and trout. The Atlantic Salmon Trust, Pitlochry, Scotland.

SOSCP (South Okanagan-Simikameen Conservation Program) 2000. South Okanagan-Simikameen Conservation Program: a prospectus. Conserving Canada's Desert Country. Environment Canada. Available: https://soconservationfund.ca/wp-content/uploads/2017/07/SOSCP-Prospectus-Final2.pdf. (December 2018).

Spencer, C. N., B. R. McClelland, and J. A. Stanford. 1991. Shrimp stocking, salmon collapse, and eagle displacement. BioScience 41:14–21.

Species at Risk Act. 2002. Bill C-5, an act respecting the protection of wildlife species at risk in Canada. Available: http://laws-lois.justice.gc.ca/eng/acts/s-15.3/. (December 2018).

Stockner, J. G., and T. G. Northcote. 1974. Recent limnological studies of Okanagan basin lakes and their contribution to comprehensive water resource planning. Journal of the Fisheries Research Board of Canada 31:955–976.

Summit Environmental Consultants. 2010. Okanagan water supply and demand project: phase 2 summary report. Prepared by Summit Environmental Consultants for the Okanagan Basin Water Board, Kelowna, British Columbia.

Symonds, B. J. 2000. Background and history of water management of Okanagan Lake and River. British Columbia Ministry of Environment Lands and Parks, Water Management, Penticton. Available: http://a100.gov.bc.ca/pub/acat/public/viewReport.do?reportId=18253. (September 2014).

Tannant, D. W., and N. Skermer. 2013. Mud and debris flows and associated earth dam failures in the Okanagan region of British Columbia. Canadian Geotechnical Journal 50:820–833.

Veale, A. J., and M. A. Russello. 2016. Sockeye salmon repatriation leads to population re-establishment and rapid introgression with native kokanee. Evolutionary Applications 9:1301–1311.

Walker, I. J., and R. Sydneysmith. 2008. British Columbia. Pages 329–386 *in* D. S. Lemmen, F. J. Warren, J. Lacroix, and E. Bush, editors. From impacts to adaptation: Canada in a changing climate 2007. Government of Canada, Ottawa. Available: www.nrcan.gc.ca/sites/www.nrcan.gc.ca/files/earthsciences/pdf/assess/2007/pdf/full-complet_e.pdf. (December 2018).

Williams, J. G., S. G. Smith, J. K. Fryer, M. D. Scheuerell, W. D. Muir, T. A. Flagg, R. W. Zabel, J. W. Ferguson, and E. Casillas. 2014. Influence of ocean and freshwater conditions on Columbia River Sockeye Salmon *Oncorhynchus nerka* adult return rates. Fisheries Oceanography 23:210–224.

Williams, R. N., P. A. Bisson, D. L. Bottom, L. D. Calvin, C. C. Coutant, M. W. Erho, Jr., C. A. Frissell, J. A. Lichatowich, W. J. Liss, W. E. McConnaha, P. R. Mundy, J. A. Stanford, and R. R. Whitney. 1999. Return to the river: scientific issues in the restoration of salmonid fishes in the Columbia River. Fisheries 24:10–19.

Wright, H., K. Hyatt, and C. Bull. 2002. Possible impacts of exotic species on the indigenous aquatic communities of the Okanagan valley mainstem lakes. Okanagan Nation Alliance, Westbank, British Columbia.

Wright, H., and H. Smith. 2004. Management plan for experimental reintroduction of Sockeye Salmon into Skaha Lake; proposed implementation, monitoring and evaluation. Bonneville Power Administration, Report DOE/BP-00005136-6, Portland, Oregon. Available: www.osti.gov/servlets/purl/889735. (December 2018).

Snake River Sockeye Salmon—A Perspective on Twenty-Five Years of Effort to Recover a Population in Central Idaho, USA

Paul A. Kline*

Idaho Department of Fish and Game
Post Office Box 25, Boise, Idaho 83707, USA

Thomas A. Flagg[1]

National Oceanic and Atmospheric Administration
National Marine Fisheries Services, Northwest Fisheries Science Center
Environmental and Fisheries Sciences Division
7305 Beach Drive East, Port Orchard, Washington 98366, USA

Christine C. Kozfkay and Danny J. Baker

Idaho Department of Fish and Game
1800 Trout Road, Eagle, Idaho 83616, USA

Desmond J. Maynard and William C. McAuley

National Oceanic and Atmospheric Administration
National Marine Fisheries Services, Northwest Fisheries Science Center
Environmental and Fisheries Sciences Division
7305 Beach Drive East, Port Orchard, Washington 98366, USA

Abstract.—In November 1991, the U.S. National Marine Fisheries Service listed Snake River Sockeye Salmon *Oncorhynchus nerka* as endangered under the U.S. Endangered Species Act. The last remnants of the Snake River stock return to Redfish Lake in the Sawtooth Valley in central Idaho, a 1,448-km freshwater migration through the Columbia, Snake, and Salmon rivers. In May 1991, about 6 months prior to formal listing, a decision was made by the Idaho Department of Fish and Game, National Marine Fisheries Service, and Shoshone-Bannock Tribes to collect Redfish Lake out-migrating Sockeye Salmon smolts and to retain any anadromous adults that returned to begin the Snake River Sockeye Salmon Recovery Program (hereafter, "Program"). In the ensuing 25 years, many actions have been taken to conserve the Redfish Lake population and to rebuild the wild, natural spawning fish. These include captive broodstock gene-rescue technologies for Sockeye Salmon, carrying-capacity evaluations, and the development of reintroduction strategies. Overall, the Program has resulted in more than 40 published papers that helped advance the science for recovery of this species and provided general guidance applicable to recovery attempts for other species. The Program reduced extinction risk for Redfish Lake Sockeye Salmon by amplifying adult returns from 16 fish during the decade of the 1990s to more than 7,000 fish in 2017

* Corresponding author: paul.kline@idfg.idaho.gov

[1] Retired.

and by retaining the majority of genetic diversity (>95%) of the founder population. In this chapter, we describe the actions taken for success and the myriad of legal and political processes we encountered and navigated during the 25-year history of the project. Nonscientific issues created challenges equally as formidable as biological challenges faced. Lessons learned include the importance of clearly stated goals and objectives being frequently communicated to decision makers, the necessity of science panels for review and advice, the establishment of effective fish culture and biosecurity protocols, the establishment of a partnership of scientists and stakeholders, and conducting monitoring and evaluation to interpret the success of Program actions and to adaptively manage Program efforts. Last, we learned the importance of becoming engaged and vigilant over the legal, policy, and funding frameworks required for the Program to remain successful.

INTRODUCTION

A two-part story exists behind recovery efforts for Snake River Sockeye Salmon *Oncorhynchus nerka*. This stock of fish is recognized as the most depleted anadromous stock on the West Coast of the United States and was the first salmon stock listed as endangered under the U.S. Endangered Species Act (ESA; November 1991) by the National Marine Fisheries Service (NMFS; NMFS 1991). The last known remnant of the Snake River Sockeye Salmon stock returns annually to Redfish Lake in the Sawtooth Valley in central Idaho (Figure 1). The first part of the story is of the oft-told and well-documented science behind the development of captive broodstock gene-rescue efforts for the species. To date, the Snake River Sockeye Salmon Recovery Program (hereafter, "Program") has been successful in amplifying adult returns from a total of 16 fish during the decade of the 1990s to more than 7,000 fish to date, and in retaining a majority of population diversity (>95%). These science-based efforts have been central to reducing the extinction risk for the stock. In this chapter, we provide an overview of the gene-rescue fish-culture efforts to give readers a basic understanding of the development of the recovery science for Redfish Lake Sockeye Salmon. More detailed information can be found in the more than 40 peer-reviewed publications listed in Box 1, including carrying-capacity dynamics, fish culture, genetics, and reintroduction strategies. Much of the overview of the gene rescue science presented is an update of information previously reported by Kline and Flagg (2014).

The second part of the story describes the highs and lows we encountered in making the Program acceptable in an administrative, political, and stakeholder sense. This portion includes the heretofore untold history of the bureaucratic and regulatory processes we navigated to bring the recovery program to its current state, where progression towards recovery has begun. Information is provided on these processes, including authorization to implement recovery actions, establishment of funding channels, and initiation of a technical and coordinating body to advise project implementation. Additionally, permitting processes and required reviews are described, including multiphase science reviews and the many steps that resulted in finalization of a federal recovery plan in 2015. Amazingly, it took 24 years after ESA listing to develop a plan agreeable to all concerned parties (e.g., state, federal, and tribal scientists; federal regulatory agencies; and citizen stakeholders; NMFS 2015).

The chapter includes a description of the logic for the present-day restoration strategies and describes efforts underway to expand hatchery production to more effectively

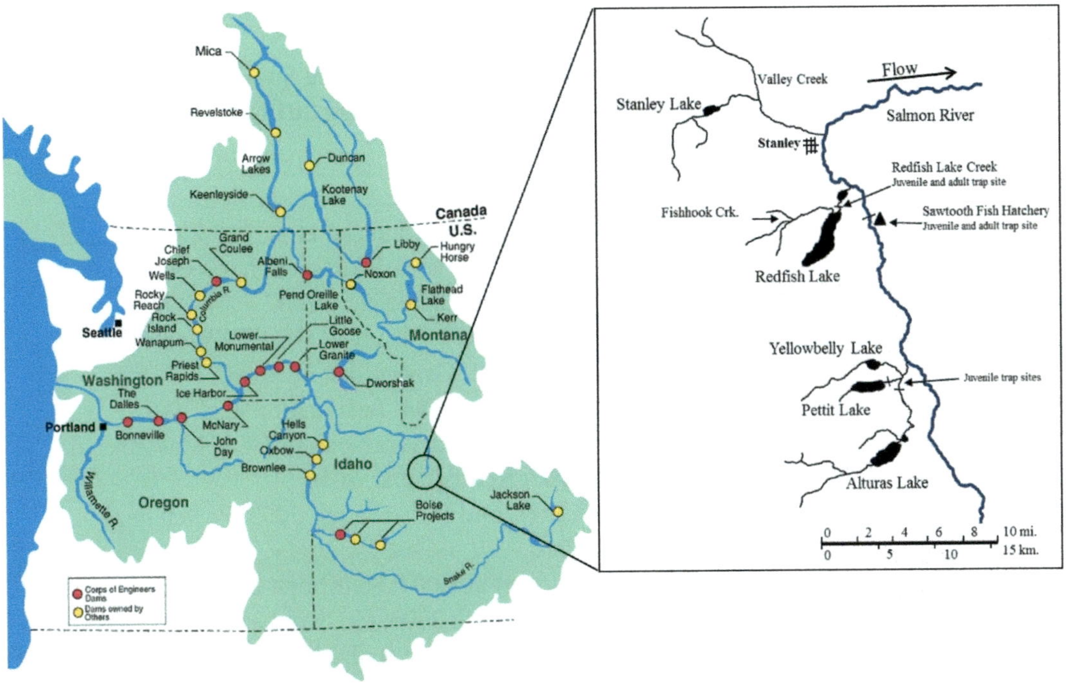

FIGURE 1. Map of the Columbia River basin in the United States and Canada, locations of the Columbia River and Snake River hydropower dams, and Snake River Sockeye Salmon recovery habitat in the upper Salmon River basin, Idaho.

address recolonization and local adaptation goals. What follows is the story of the 25-year process to conserve the Redfish Lake population and to begin rebuilding numbers of natural spawning fish in the wild, a story of substantial progress on the road to full recovery.

CAUSES OF DECLINE

Sockeye Salmon were once relatively widespread in the Snake River basin of Idaho and Oregon (Figure 1). Prior to development of the American West, Sockeye Salmon runs to the Snake River basin were estimated to have been upwards of about 150,000 fish (NPCC 1986). Initial causes of population decline mirror the development of the West, beginning with the inception and intensification of commercial fisheries in the lower Columbia River and Pacific Ocean in the mid-to-late 1800s. As human development increased, many common habitat issues affected this river system, including changes in the landscape from farming and grazing practices and the establishment of towns and cities, all of which had an impact on the water quality and quantity of the river system. Additionally, many of the watersheds were mined using hydraulic or conventional dredge methods, destroying fish habitat and the productive capacity of the river due to gravel displacement, channel alteration, and habitat destruction (NMFS 2015).

Construction of small power and irrigation dams throughout the Snake River watershed had, by the 1950s, blocked many migratory corridors and markedly reduced the geographic distribution of Snake River Sockeye Salmon to Redfish Lake, a single lake in the Sawtooth Valley at the headwaters of the Salmon River in Idaho (Figure 1). In the late 1950s and early 1960s, barriers to upstream migration were installed on outlet streams of

BOX 1. Published articles by researchers involved with the Snake River Sockeye Salmon Recovery Program (updated from Kline and Flagg 2014)

Subject	Reference
Habitat/limnology evaluations	Budy et al. 1995; Luecke et al. 1996; Gross et al. 1998; Pilati and Wurtsbaugh 2003; Sawatzky et al. 2006; Selbie et al. 2007
Lake-fertilization evaluations	Gross et al. 1997; Budy et al. 1998; Wurtsbaugh et al. 2001; Griswold et al. 2003
Predator–prey and life-history evaluations	Rieman et al. 1994; Beauchamp et al. 1997; Steinhart and Wurtsbaugh 1999; Massee et al. 2007; Kendall et al. 2010
Juvenile fish growth and survival evaluations	Steinhart and Wurtsbaugh 2003; Powell et al. 2010
Juvenile and adult migration studies	Hebdon et al. 2004; Steinhorst et al. 2004; Keefer et al. 2008; Griswold et al. 2011; Axel at al. 2013; Tucker et al. 2015; Kozfkay et al. 2019
Genetics	Winans et al. 1996; Cummings et al. 1997; Kozfkay et al. 2008; Waples et al. 2011; Kalinowski et al. 2012; O'Reilly and Kozfkay 2014; Nichols et al. 2016
Gene-rescue hatchery methodologies	Flagg et al. 1995, 2004; Schiewe et al. 1997; Flagg and Mahnken 2000; Berejikian et al. 2004; Pollard and Flagg 2004; Heindel et al. 2005; Swanson et al. 2008; Maynard et al. 2012; Kline and Flagg 2014
Fish health	Purcell et al. 2009; Kurath et al. 2011

three Sockeye Salmon nursery lakes in the Sawtooth Valley (Yellowbelly, Pettit, and Stanley lakes) and lakes were chemically treated. This action was taken by the Idaho Department of Fish and Game (IDFG) to promote recreational sport fisheries for resident salmonid species as anadromous Sockeye Salmon were no longer returning to these three lakes (NMFS 2015). Access to an additional Sawtooth Valley lake, Alturas Lake, was hampered by the presence of a privately installed barrier in place to divert water for irrigation use. This left Redfish Lake (Figure 1) as the only Sawtooth Valley lake available to migrating anadromous salmonids. Further complicating the picture, eight major hydroelectric dams were constructed on the Columbia River and lower Snake River system between 1934 and 1975, which interfered with migration to and from the ocean.

Historically, perhaps 25,000–30,000 Sockeye Salmon may have returned to the five nursery lakes in the Sawtooth Valley (Selbie et al. 2007). By the time of modern record-keeping in the 1950s, Sockeye Salmon returns to Redfish Lake were reduced to less than 5,000 adults annually (Bjornn et al. 1968). By the 1960s, returns had fallen into the hundreds, and by the late 1980s, adult return numbers were reduced to the teens. In the decade of the 1990s, only 16 wild adults returned to Idaho (Figure 2). Clearly, the population was in crisis!

IMPLEMENTING THE GENE-RESCUE PHASE OF THE PROGRAM

In simple terms, the sole goal of the Program at start-up was to use fish culture to help prevent extinction and to maintain as much population genetic diversity as possible, much in the same way zoos function to conserve at-risk populations of birds and mammals. Initial project actions focused on the development of redundant captive broodstock programs at IDFG and NMFS hatcheries to prevent extinction and preserve remaining population genetic diversity. This approach was highly controversial because at that time, application of captive broodstock technology to Pacific salmon was considered experimental. Previous attempts with other salmon species had been marginally successful (e.g., low survival of fish in culture through maturation and low subsequent egg viability; Flagg et al. 1995, 2004).

The Sockeye Salmon returning to Redfish Lake were considered ecologically unique in that they travel a greater distance from the Pacific Ocean (1,448 river kilometers [RKM]) to a higher elevation (2,138 m) and to the southernmost habitat than any other Sockeye Salmon population in the world. Additionally, this population had no indication of genetic dilution caused by the stocking of other populations (Waples et al. 2011). No other nearby Sockeye Salmon population with similar attributes was avail-

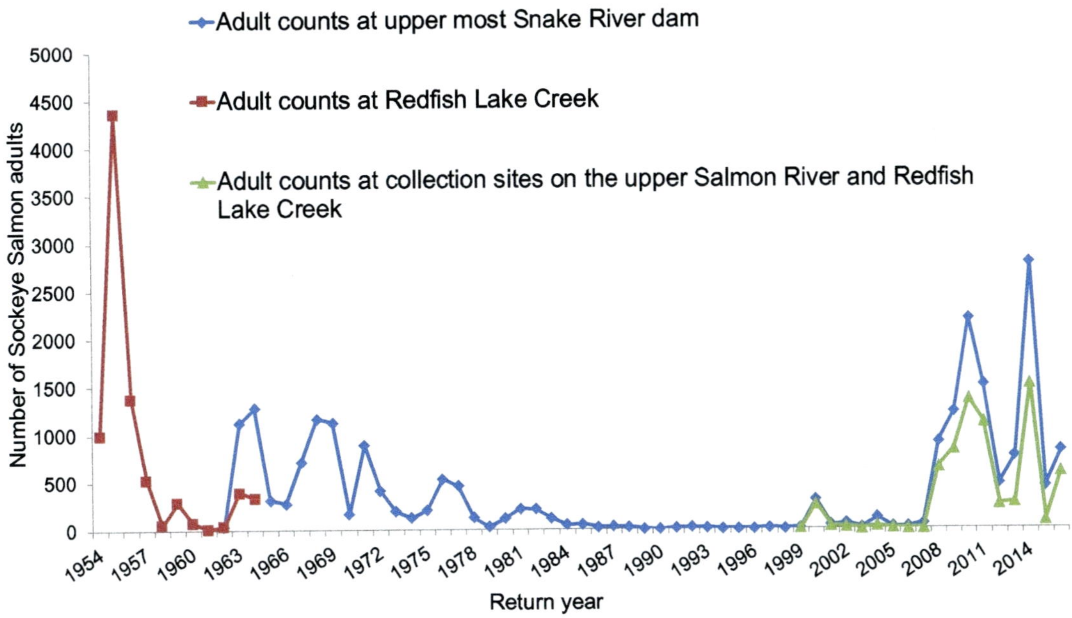

FIGURE 2. Sockeye Salmon adult counts at main-stem dams on the Snake River and at trapping sites in the Sawtooth Valley, Idaho near Redfish Lake for years 1954 through 2016. Note: 1954–1965 counts in red are for wild Sockeye Salmon at a trap on Redfish Lake Creek, the outlet to Redfish Lake; 1962–2016 counts in blue are for wild-, natural- and hatchery-origin Sockeye Salmon at dams on the Snake River; 1999–2016 counts in green are for wild, natural- and hatchery-origin Sockeye Salmon at trap sites on the Salmon River and Redfish Lake Creek.

able as a source of broodstock from which to initiate the Program. Next nearest geographic neighbor populations occurred approximately 1,200 RKM away in tributaries of the upper Columbia River but were genetically and ecologically distinct from Redfish Lake Sockeye Salmon (Winans et al. 1996).

In May 1991, about 6 months prior to formal declaration of ESA listing, a decision was made by IDFG, NMFS, and the Shoshone-Bannock Tribes to collect out-migrating Redfish Lake Sockeye Salmon smolts and to retain any anadromous adults that returned to begin the Snake River Sockeye Salmon Recovery Program. As discussed above, this action was in many ways a last ditch effort to save the remnant population. The only other alternative at the time appeared to be extinction!

Captive Broodstock Fish-Culture Practices and Facilities

As noted above, to address demographic risk and the possibility of hatchery-rearing system failure, redundant broodstock fish-culture facilities were established by IDFG near Boise, Idaho and by NMFS near Seattle, Washington. Because of the small number of fish involved at the start of the Program in 1991, existing rearing containers at IDFG and NMFS facilities (vats and raceways) were initially used for fish rearing. Over time, fish propagation systems were modernized to include flexible rearing facility layouts that better accommodated fish-culture activities ranging from spawning and incubation to adult rearing. Different-sized circular and semicircular rearing tanks were used at IDFG and NMFS facilities to rear Sockeye Salmon through maturation (see Maynard et al. 2012 for review of facility designs). Stringent biosecurity plans and disease containment procedures were adopted early on. Fish-culture methods used in the Program followed accepted standard practices (for an overview of standard methods, see Leitritz and Lewis 1976; Maynard et al. 2012).

Currently, NMFS operates both a fresh well-water facility for juvenile rearing (egg to smolt) and adult spawning and a pumped and ultraviolet-sterilized seawater system for smolt-to-adult rearing on Washington's Kitsap Peninsula. The Idaho Department of Fish and Game operates a captive broodstock facility in Eagle, Idaho supplied with pumped well water. Combined, these captive broodstock facilities can readily sustain more than 3,000 adults and provide up to 1.2 million eggs annually for restoration.

At the outset of the Program, we did not envision the need for rearing space beyond that needed to maintain the captive population. As the Program grew and after we had successfully met our near-term goal of preventing population extinction, we began to consider the need to address population-rebuilding objectives in the habitat. Accordingly, we secured egg incubation and juvenile rearing space at the IDFG Sawtooth Fish Hatchery near Stanley, Idaho to rear limited numbers of presmolts and smolts. Additional smolt rearing space was later secured at Bonneville and Oxbow hatcheries—two Oregon Department of Fish and Wildlife hatcheries located on the lower Columbia River. Combined, these facilities provided space to rear approximately 300,000 smolts annually. In 2013, IDFG completed construction of the Springfield Hatchery near American Falls, Idaho for the purpose of rearing up to 1 million Sockeye Salmon smolts annually.

Genetics and the Spawning Program

As indicated, an extremely small founder population existed that included only 16 anadromous Sockeye Salmon that returned to Redfish Lake over the decade of the 1990s (11 males and 5 females), several hundred Redfish Lake out-migrating smolts trapped between 1991 and 1993, and 26 residual

Sockeye Salmon adults trapped in Redfish Lake in the early 1990s (Burgner 1991; Table 1). Maintenance of the genetic integrity of the brood-lines was a primary consideration in spawning fish and rearing offspring. Individual and family representation and relatedness among captive broodstock adults and returning anadromous adults was consistently taken into account to guide spawning designs. Through 2005, pedigree information (e.g., known origin/lineage of individual members of the population) was used to guide the development of annual spawning designs. As genetic techniques evolved from 2006 forward, a suite of 7–16 microsatellite loci have been used to accomplish pedigree management, as well as to provide information on population-level genetics. Maximum avoidance of inbreeding matrices was developed and implemented from 2005 to 2012 to identify kinship among all possible mate combinations (O'Reilly and Kozfkay 2014). From these analyses, modified factorial mating schemes based on genetic distance were developed to guide mate selection on spawning days. Starting in 2012, a modified mean kinship approach using genetic pedigree data was implemented to minimize relatedness in the Program and maximize retention of genetic diversity (Ballou and Lacy 1995). This genetic maintenance strategy was shown to be highly successful. The genetic focus of the Program and adherence to various central tenets of conservation aquaculture has enabled Program managers to retain approximately 95% of the original founding genetic variability of the population (Kalinowski et al. 2012).

Release Strategies

Although the initial focus of the Program was the gene-rescue effort described above, extra eggs and fish were produced each year as a result of maintaining captive broodstock populations. Eggs, juveniles, and adults not needed to maintain captive broodstocks were released using various release strategies with outcomes evaluated. Through 2016, approximately 1.1 million eyed eggs, 1.6 million presmolts, 2.9 million smolts, and 13,564 prespawn adults were released back to rearing lakes and outlet streams in the Sawtooth Valley. The success of these releases were evaluated for brood years 2004 through 2011 (adult return years 2007 through 2016). More than 500 adult Sock-

TABLE 1. Wild adult and juvenile Sockeye Salmon collected to develop the Snake River Sockeye Salmon Recovery Program.

Collection year	Anadromous adults	Residual adults[a]	Out-migrating smolts
1991	4 (3 male, 1 female)		759
1992	1 male	5 (4 male, 1 female)	79
1993	8 (6 male, 2 female)	18 (16 male, 2 female)	48
1994	1 female		
1995		3 male	
1996	1 female		
1997			
1998	1 male		
Total	16 (11 male, 5 female)	26 (23 male, 3 female)	886

[a] Residual Sockeye Salmon are genetically similar to anadromous Sockeye Salmon but complete their lifecycle in freshwater (see Burgner 1991 for a review).

eye Salmon returned to collection sites in the Sawtooth Valley in five return years, and in return years 2010, 2011, and 2014, more than 1,000 adult Sockeye Salmon returned (Figure 2). Smolt-to-adult return rates from the ocean for Sockeye Salmon produced from juveniles released as smolts were more than 3.5 orders of magnitude greater in some years than rates for adults produced from presmolt releases to Redfish Lake, while egg box releases produced almost no adult returns (Figure 3). Across all reported brood years, more than 75% of returning adults originated from full-term smolt releases while presmolt releases accounted for a relatively small proportion (approximately 5%) of all returning adults (Figure 4). Additionally, smolts produced from in-lake spawning events produced approximately 13% of adult returns for this period of time (Figure 4).

Gene Rescue Program Results

As discussed above, taking all remaining Redfish Lake Sockeye Salmon adult fish into captivity was a last ditch effort to save the remnant population. The decade of the 1990s bore this out as an appropriate action as only 16 adult Sockeye Salmon returned to Redfish Lake over this period of time, way too few to have provided sufficient natural reproductive success to sustain the population. As described by Kline and Flagg (2014), if the five female adult fish incorporated in the broodstock in the decade of the 1990s (Table 1) had not been taken into captivity and had instead been left to spawn naturally, they may have generated at most three adult returns from the ocean, spread over a 4- to 10-year return window, or less than a fraction of an adult per year. Based on these figures, without the captive-broodstock intervention, extinction would have been all but certain.

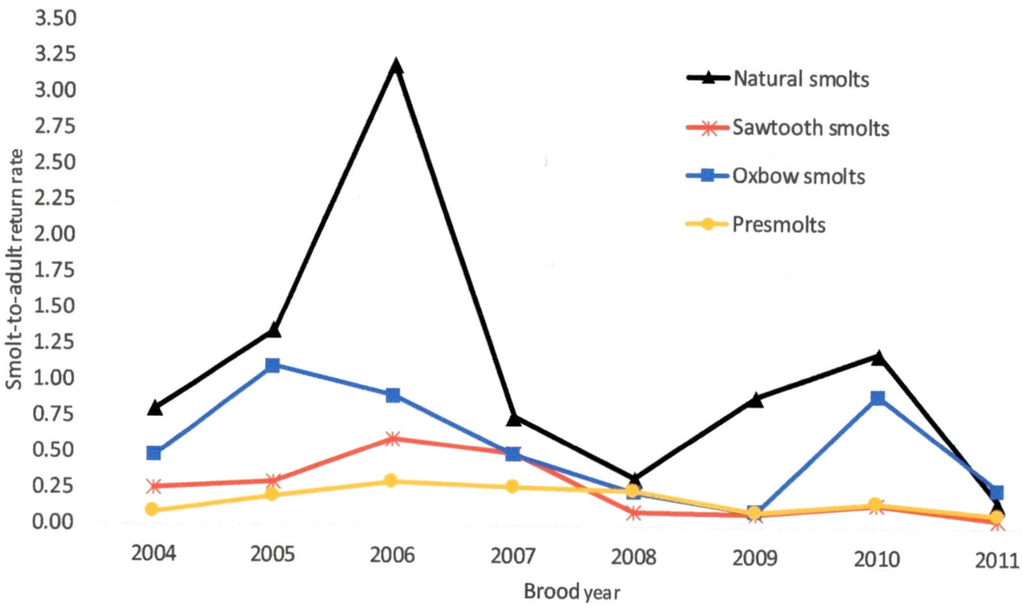

FIGURE 3. Smolt-to-adult return rates (Sawtooth Valley to Sawtooth Valley) by brood year for Snake River Sockeye Salmon produced from natural spawning (black line), smolt releases (red and blue lines) and presmolt releases (yellow line). Naturally produced smolts were the result of successful spawning events in Redfish Lake. Smolts were produced at the Oregon Department of Fish and Wildlife's Oxbow Fish Hatchery and the Idaho Department of Fish and Game's Sawtooth Fish Hatchery. Presmolts were produced at the Idaho Department of Fish and Game's Sawtooth Fish Hatchery.

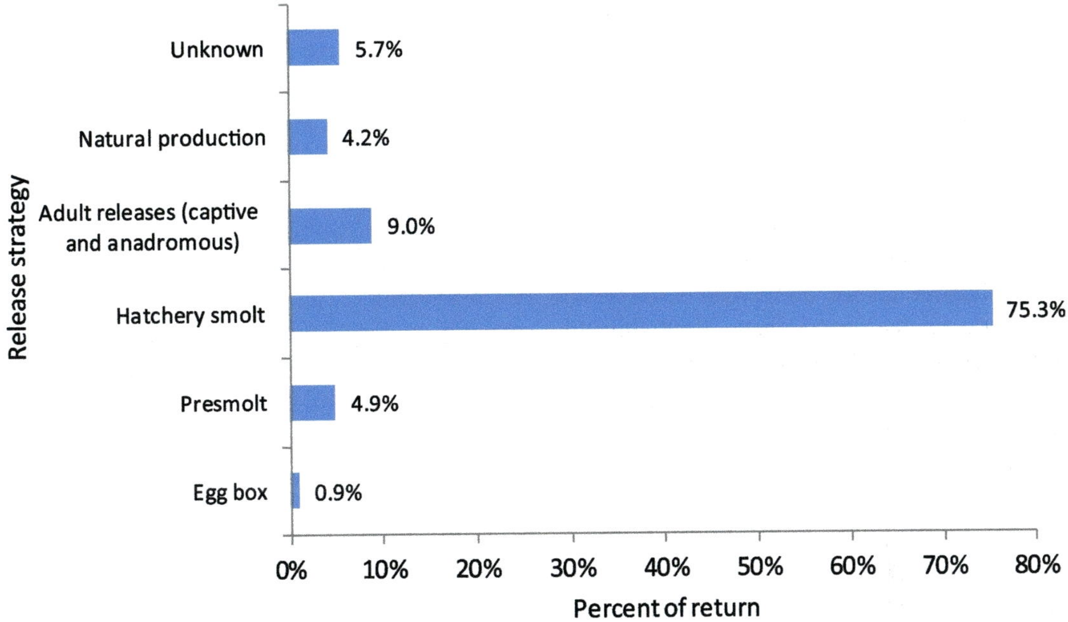

FIGURE 4. Proportion of adult Sockeye Salmon returns to Sawtooth Valley trap sites from natural production in Redfish Lake, and from hatchery smolt, presmolt, and egg box releases for return years 2008–2016.

Although easily overlooked, a major Program accomplishment was simply the development of fish culture protocols for rearing captive Sockeye Salmon full term to maturation (Kline and Flagg 2014). Earlier attempts to rear various species of Pacific salmon and steelhead (anadromous Rainbow Trout *Oncorhynchus mykiss*) had resulted in low egg survival to the eyed-stage of development (commonly only 30–60% compared to more than 80% for wild cohorts), egg-to-adult survival rates generally below 30%, and reduced size at maturity (30–50% smaller than wild fish; Schiewe et at. 1997). Overall, the Redfish Lake Sockeye Salmon Recovery Program has been more productive than earlier programs. Egg survival to the eyed-stage of development has improved greatly to levels that exceed 80%. Fry-to-maturation survival for captive broodstocks is now routinely at least 60% (see Kline and Flagg 2014 for review).

For the Snake River Sockeye Salmon Recovery Program, for every two eggs retained in captivity, approximately one adult fish survives to contribute to the next generation. Because each adult female has a fecundity of more than 2,000 eggs, a large number of adult offspring are generated in a short period of time (in 3–4 years) for use in recovery. The overall, survival advantage afforded by the captive broodstock program has been in excess of 2,000% compared to not intervening (i.e., not initiating the Program). The amplification of the Redfish Lake Sockeye Salmon population from the initial 16 adults captured in the 1990s to the tens of thousands of adults already produced, and the ability to produce millions more in the future, is a remarkable conservation achievement.

Likewise, the genetic maintenance strategy employed by the Program has been highly successful. In fact, the Redfish Lake captive broodstock program has retained a greater proportion of founder genetic diversity than any other known conservation program (Kalinowski et al. 2012).

The release of excess fish not needed to maintain breeding lines for the captive broodstock has also produced useful results. Adult Sockeye Salmon produced in the Program and released to Redfish Lake to volitionally spawn were first observed building redds in 1999, 8 years after the initiation of the Program. Importantly, naturally produced smolts accounted for more than 13% of the adult returns for brood years 2004 through 2011. Brood year 2006 smolt-to-adult return rates for fish produced from natural spawning events exceeded 2.0%, the level estimated to be the required minimum for population self-sustainability (Figures 3 and 4).

Over the course of the Program, critics suggested that the Redfish Lake Sockeye Salmon population was too compromised (demographically and genetically) to have the fitness required to be self-sustaining (ISRP 2006). However, our evaluations described above indicate that properly scaled hatchery-reared smolt releases have the potential to produce sufficient adult returns to fully seed Redfish Lake spawning habitat. Additionally, juveniles produced from in-lake spawning events appear to possess improved fitness attributes, including increased smolt-to-adult return rates, sufficient to match or exceed the level of self-sustainability required for recovery to advance.

The captive broodstock gene-rescue phase of the Program has been a clear success. In the sections below, we describe the processes now underway associated with Program transition from pursuing simple gene rescue to full recolonization of the species in its habitat.

RECOVERY PLANNING

The NMFS recovery plan for Snake River Sockeye Salmon (NMFS 2015) identified biological goals of 1,000 naturally spawning adult fish returning to Redfish Lake and 1,500 fish combined returning to two other Sawtooth Valley lakes (2,500 total; about 10% of historic maximum returns). The approaches developed to meet recovery goals were outlined in the Program's master plan (IDFG 2010, 2012) as well as in the NMFS recovery plan (NMFS 2015). The plans took into account results from the release strategy evaluations described above, which indicated that full-term smolt releases were the most successful approach to recolonization because they produced the largest number of adult returns from the ocean in the shortest time frame possible. We used a modification of the Congressionally authorized Columbia River basin Hatchery Scientific Review Group's (HSRG) recommendations (Paquet et al. 2011) to determine the level of smolt releases needed to consistently generate adequate adult returns sufficient to meet the recovery plan's numerical goals described above. Smolt-to-adult return rates from our evaluation of brood year 2004–2006 returns were then used to determine how returns from natural spawning events would fluctuate over time. Our calculations indicated that releasing 1 million hatchery-reared smolts could, under ideal circumstances, produce approximately 5,000 returning adult fish annually.

Spawning plans for these projections were structured to properly adjust the proportion of natural-origin spawners taken into the broodstock (pNOB)[1] and the proportion of hatchery-origin (pHOS)[1] to natural-origin (pNOS)[1] adult Sockeye Salmon released to the habitat for natural spawning. Emphasis was placed on increasing the use of fish returning from the ocean, as opposed to captive-reared Sockeye Salmon in hatchery-spawning designs. Emphasis was also placed on releasing increased numbers of natural-origin adults to the habitat for natural spawning to increase the adaptive

[1] See Paquet et al. (2011) for review of population metrics such as pNOB, pHOS, pNOS, proportionate natural influence (PNI) factors.

potential and fitness of the population relative to the full-term captive-reared adults that have been used. We used sliding-scale logistics (Box 2) to develop rulemaking for managing the distribution of natural- and hatchery-origin adults. In years of low numbers of fish returning from the ocean, pHOS guidelines could be relaxed to address potential demographic concerns. In years of high abundance of fish returning from the ocean, pHOS would be reduced and focus would be on allowing natural (wild) fish to make up most of the spawning component.

Planning beyond the first gene-rescue phase of the program emphasized two "next phases" of the project: (1) establish parameters for expanding the project and producing enough fish to recolonize the historic habitat, and (2) provide for development of local adaption and the rebuilding of natural population structure. We also developed targets for phasing out both the captive broodstock and, ultimately, all hatchery intervention components of the Program.

Recolonization Phase—Expanding the Program

The current recolonization phase (initiated in 2013 with the completion of Springfield Hatchery) focuses on producing and releasing increased numbers of smolts, returning more adults from the ocean to Idaho collection sites, incorporating more ocean-return adults in hatchery-spawning designs, and releasing more ocean-return adults to spawn naturally in the wild. In the recolonization phase of the Program, an average of 4,984 anadromous Sockeye Salmon (637 and 4,347 natural- and hatchery-origin adults, respectively; IDFG 2010, 2012) are projected to return annually from the ocean to the Sawtooth Valley. In this phase, emphasis will be placed on increasing the use of ocean-returning Sockeye Salmon in hatchery-spawning designs, as well as in the habitat for natural spawning. In-hatchery spawning plans will be structured to include a minimum number of wild (e.g., naturally reared) adults (approximately 10%). During this phase of the Program, the proportion of hatchery-origin captive and ocean-return adults spawning in the wild will not be restricted.

The recolonization phase includes a provision to release hatchery-origin adults returning from the ocean in Pettit Lake (Figure 1), a historical Sockeye Salmon lake in the Sawtooth Valley. The barrier on the outlet of Pettit Lake was removed in the mid-1990s. This release could occur when the combined hatchery- and natural-origin return-from-ocean escapement to Redfish Lake exceeds 5,000 fish. Allowing adult escapement to reach this level before allocating adults returning from the ocean to other lakes will provide managers the opportunity to observe Redfish Lake fish-habitat use and productivity over a range of adult-spawning densities. Maturing captive broodstock adults will continue to be released to Redfish and Pettit lakes during the recolonization phase of the Program. Ocean-return adult Sockeye Salmon captured at the Sawtooth Fish Hatchery weir (Figure 1) and identified as Pettit Lake origin will be transferred to Pettit Lake for natural spawning. Alturas Lake (Figure 1), a third potential recovery lake in the Sawtooth Valley, will not receive Redfish Lake-stock prespawn adults. Program cooperators and NMFS are discussing reintroduction strategies that include the possible development of an independent, localized Alturas Lake broodstock based on resident nonanadromous Sockeye Salmon that reside in the lake. However, anadromous adults trapped at the Sawtooth Hatchery weir and identified as Alturas-origin will be transferred and released in Alturas Lake.

Local Adaptation Phase—Rebuilding Natural Populations

The final phase for the Program emphasizes the importance of local adaptation and the

BOX 2. Scenarios for sliding scale logistics for managing hatchery versus wild escapements in high versus low wild fish abundance years for Redfish Lake Sockeye Salmon (modified from HSRG 2009).

Variable sliding scales are useful for managing salmon abundance so that in low abundance years, more hatchery-origin fish of the appropriate population component are allowed to reach the spawning grounds to reduce demographic risk to the respective populations. As an example,

1. Each year, depending on NOR[a] run size, pNOB[b] and pHOS[c] are allowed to "float" or slide. Managers should establish an acceptable level of removal of NORs for use in the hatchery brood. This will be a fixed percentage of the total NOR return (say 40%) and will not change, regardless of NOR return.
2. In years of high NOR abundance, this 40% could make up 100% of the needed hatchery brood (pNOB = 100%). In that case, no HORs would be used in the hatchery brood. Hatchery fish can be allowed to reach the spawning ground (pHOS) if needed to achieve an appropriate number of fish spawning naturally (for demographic benefit and maximum use of available habitat). This approach, however, would not be required during years of very high NOR returns as both objectives (pNOB and natural spawning) may be meet with NORs.
3. In years of low NOR abundance, the same 40% of the NOR return would be removed for use in the hatchery brood (pNOB). However, in these years, that 40% may make up only a small part of the needed brood (i.e., pNOB 10%). In these years, enough HORs should be used to achieve needed hatchery brood, and additional HORs should be allowed to spawn naturally (pHOS) to achieve the minimum acceptable level of natural spawning.
4. The goal of this sliding scale is to achieve an average PNI[d] over time of the desired level (0.67 or 0.5) depending on the population designation, even though it may not be achieved in any one year. PNIs of >0.5 are believed necessary for the natural environment to influence evolutionary processes.

A good way to determine the level of NORs that should be removed each year (see above) is to review the return of NORs over a long time frame and iterate what level (30%, 40%, or 50%) is needed, on average, to achieve the desired PNI.

---

[a] NOR is the numerical natural origin return for wild fish in the watershed.
[b] pNOB is the proportion of natural-origin spawners taken into the broodstock.
[c] pHOS is the proportion of hatchery-origin spawners in the natural habitat.
[d] PNI = pNOB/(pNOB+pHOS) and is the proportional mean fitness of the integrated population relative to the natural population.

potential of natural-origin fish to increase population viability to a point where population abundance, productivity, spatial structure, and diversity are sufficient to no longer need safeguards of the ESA (NMFS 2015). In this phase of the Program, an average of 6,719 anadromous Sockeye Salmon (1,647 and 5,072 natural- and hatchery-origin adults, respectively) are projected to return annually from the ocean to the Sawtooth Valley (IDFG 2010, 2012). The local adaptation phase of the Program will be initiated after performance objectives for the recolonization phase have been satisfied. As the number of anadromous (hatchery- and natural-origin) adults returning from the ocean and spawning in Redfish Lake increases, it follows that increased numbers of natural-origin adults will be produced over time and return to collection sites in the Sawtooth Valley. As sufficient numbers of natural-origin adults return, sliding-scale management will be employed to guide how natural- and hatchery-origin adults are used in hatchery-spawning designs, as well as in the wild, following guidelines developed by the HSRG. The population will be managed to ensure the environment drives the fitness of the composite population. The HSRG identified that this generally occurs when the number of natural-origin adults incorporated in hatchery spawning designs is greater than the number of hatchery-origin adults released to the habitat to spawn volitionally (HSRG 2009; Paquet et al. 2011).

Once the local adaptation phase is fully initiated, captive broodstock programs will have been scaled back considerably or terminated. The end result of the local adaption phase will ideally be a self-sustaining, natural Sockeye Salmon population in Redfish Lake that can reproduce and survive on its own without hatchery intervention. This state of existence defines the ultimate success of the program. However, we recognize that the landscape we live in today has changed considerably from the pre-Western development state and that periodic hatchery assistance may be necessary to offset downturns in population productivity and production that result from variable natural environmental patterns. Developing a strategy for maintenance of recovery that includes hatcheries is a novel concept not normally used in ESA recovery programs—one that will certainly require more discussion at the regional level between regulatory and management entities. In the near term, as population fitness and natural-origin returns to Redfish Lake increase, efforts to establish a self-sustaining population in Pettit Lake and possibly Alturas Lake will expand.

NAVIGATING THE RECOVERY PROCESS

Developing the science for the Snake River Sockeye Salmon Recovery Program was critical for proper Program implementation; just as important was navigating all of the rules, regulations, reviews, and other such processes required for this endeavor. Fisheries scientists are trained to apply and follow principals of the scientific method, including the incorporation of an adaptive management feedback loop. For us, planning, implementing, and managing the technical aspects of the Program have been largely intuitive. However, as we learned, satisfying regulatory processes and conditions associated with funding and stakeholder expectations proved to be time-consuming and, at times, frustrating. We believe the section that follows will be a useful primer to inform prospective conservation program biologists and administrators about the processes that must be navigated to build a successful program.

Any action involving ESA-listed fish requires federal approval. These actions include research, enhancement (propagation and stocking), or incidental harassment activities that may result in the take (e.g., handling of an organism for biological sampling or incidental/deliberate mortality) of a threatened or endangered species, whether or not deliberate. The basic permits and consultations associated with the Snake River Sockeye Salmon Recovery Program are described in Box 3.

Funding and Program Review

Each of the permit-related steps described in Box 3 required considerable staff time and involvement beyond that required to implement a conventional hatchery operation. Adding rigorous chain-of-custody (chronological documentation or paper trail of control) permit compliance requirements where fish health and fish survival outcomes need to be documented and reported adds additional operational costs. How to fund such endeavors is a continuing challenge that included deep and active engagement in Pacific Northwest regional fisheries forums. The two overarching groups involved in fisheries funding policy and action decisions in the Columbia River basin are the Northwest Power and Conservation Council (hereafter, "Council") and the Bonneville Power Administration (BPA).

In 1980, the Pacific Northwest Electric Power Planning and Conservation Act

BOX 3. State and federal authority, regulatory process and permitting required to implement Snake River Sockeye Salmon recovery efforts.

Permitting authority	Description
State of Idaho	In-state and Interstate egg and fish transfer permits required.
Federal regulatory agencies U.S. Forest Service, Bonneville Power Administration	Compliance with the U.S. Endangered Species Act required. Operational plans and hatchery and genetic management plans developed to satisfy Section 10(a)(1)(A) of the ESA. Incidental as well as direct take of listed species authorized.
Federal regulatory agencies U.S. Forest Service, Bonneville Power Administration	Compliance with the U.S. National Environmental Policy Act required. Biological assessments developed to support federal biological opinions to permit Snake River Sockeye Salmon Recovery Program implementation.
Federal regulatory agencies U.S. Forest Service, Bonneville Power Administration	Compliance with the National Historic Preservation Act required. Determination of no impact to cultural and historic resources obtained.
Federal regulatory agencies U.S. Forest Service	Compliance with U.S. Forest Service land use policy. Special use permit to operate field installations on U.S. Forest Service property required.

(hereafter, "Act") authorized Idaho, Montana, Oregon, and Washington to develop a regional power plan and fish and wildlife program to balance the Pacific Northwest's environment and energy needs (NPCC 1980). Governors of these states appoint two members each as Council members. One major activity of the Council is to act as a recognized, credible, and objective hub for regional fish and wildlife planning information and analysis and to improve the efficiency and cost-effectiveness of the region's restoration actions through fish and wildlife program amendments and project reviews. In 1996, the Act was amended and the Council was directed to create an 11-member Independent Scientific Review Panel (ISRP) to conduct reviews of projects proposed for funding through the Council's Columbia River basin Fish and Wildlife Program. Annually, the Council provides science review on more than US$250 million of Fish and Wildlife Program activities in the Columbia basin (NPCC 2014). The Bonneville Power Administration uses its funds to address Council priorities identified in its Fish and Wildlife Program. Part of the U.S. Department of Energy, BPA is a nonprofit federal power marketer based in the Pacific Northwest. Along with coordinating power distribution, BPA funds the majority of fish and wildlife programs throughout the Pacific Northwest. The Bonneville Power Administration has a large staff of biologists and program managers to help guide these activities. The Bonneville Power Administration has provided the majority of the funding support for the Snake River Sockeye Salmon Recovery Program.

One of the first and most important activities of the Snake River Sockeye Salmon Recovery Program was to immediately convene a working group, the Stanley Basin Sockeye Technical Oversight Committee (TOC), made up of state, tribal, federal, and nongovernmental organization partners. From the onset, the TOC was chaired by a BPA fish and wildlife manager. The TOC meets at least quarterly; consistent membership has included biologists and fish culturists from IDFG, NMFS, and the Shoshone-Bannock Tribes. Additionally, over time, membership has included academicians; federal regulatory agency personnel; leading Pacific Northwest experts in areas such as fish culture, ecology, and genetics; and local interested landowners and conservation groups. The group has remained together and functioning for more than 25 years and has been central in maintaining credibility and broad-based support for the Program. The cohesive structure and availability of expertise has been foundational in guiding science and technical issues for the Program.

Continuation of funding for the project has been partially dependent on multiple reviews of Program science conducted by the Council's ISRP. Reviews have been time-consuming and have required considerable staff attention to address successfully. Early reviews conducted by the ISRP (1999, 2001, 2004) indicated that the Snake River Sockeye Salmon Recovery Program met science standards and should receive continued funding. In 2004, the Council asked for a programmatic review of all conservation aquaculture programs funded through its Fish and Wildlife Program by BPA (referred to as Programmatic Issue 12). This review was primarily conducted to provide the Council with some basis for making funding recommendations to BPA for the growing use of hatcheries and hatchery fish to supplement wild populations of salmon and steelhead. The response we received for our Issue 12 submittal was inconsistent with past science reviews, suggesting that deleterious genetic affects (e.g., inbreeding effects and domestication selection) were likely outcomes of continued Program implementation and that adult returns to date indicated that the Snake River

Sockeye Salmon population may no longer be viable. At the other end of the opinion spectrum, the 2004 Federal Columbia River Power System (FCRPS) biological opinion released by NMFS endorsed the Snake River Sockeye Salmon Recovery Program and recommended that juvenile production space be increased to more effectively address population rebuilding goals.

In its 2006 review, the Council's science panel concluded that the Program was not scientifically justified and recommended that funding be terminated (ISRP 2006). This opinion was our wake-up call that not all reviews were strictly science-based or well informed. The review concluded that the Program had failed to produce adequate adult returns to sustain the population and that Snake River Sockeye Salmon no longer had the fitness necessary to be sustainable, a recommendation made despite the fact that genetic results supporting their assertion did not exist. Additionally, derived measurements of productivity for Snake River Sockeye Salmon, such as smolt-to-adult return rates, were comparable to those observed for spring Chinook Salmon O. tshawytscha released from the IDFG Sawtooth Fish Hatchery, located on the Salmon River in the Sawtooth Valley near Redfish Lake.

The rationale the ISRP used to recommend terminating the Program stemmed from the fact that we had always expressed two goals for the Program. Our near-term goal addressed preventing extinction and maintaining population genetic diversity. Our stated long-term goal was to recover the population and to provide sport and tribal harvest opportunity. While the ISRP acknowledged that we had successfully addressed the near-term goal, they made the determination that the Program had no ability (due to low population fitness) to address the long-term goal of recovery. The ISRP's proposal to defund was taken seriously by the Council, and the Snake River Sockeye Salmon Recovery Program was almost defunded. The Program and fish were saved from termination/extinction through efforts of Idaho Council members and importantly, by then Idaho Governor James Risch (now a U.S. senator from Idaho). Risch addressed the full Council in June 2006 and exclaimed that this Program was not going to be terminated on his watch and that he would not tolerate the extinction of Sockeye Salmon in Idaho! This display of support had great impact on the Council and BPA, who, after review, made the decision to continue funding the Program and to discount the recommendation of the science panel, which appeared to be more politically than scientifically driven.

Even though Program funding was maintained, we realized that we needed to prioritize the development of smolt-rearing space and to focus on meeting a growing political and stakeholder interest to fast-track population expansion and rebuilding. The NMFS 2004 FCRPS biological opinion included language to support (fund) the development of smolt rearing space as well as associated expansion of adult holding and spawning space at IDFG and NMFS facilities. Between 2004 and 2008, improvements were made to facilities to address federal biological opinion recommendations. The outcome was the development of hatchery space to rear approximately 150,000 smolts at an Oregon Department of Fish and Wildlife facility, as well as a continuing commitment to fund the rearing of approximately 150,000 smolts at the IDFG Sawtooth Fish Hatchery annually. Building on this success, the 2008 NMFS FCRPS biological opinion included a recommendation to construct a new smolt-rearing hatchery in Idaho capable of producing 500,000–1 million annually. Funding to construct the Springfield Hatchery came through the 2008 Columbia Basin Fish Accords process, a series of funding agreements developed between BPA and Columbia basin state and tribal jurisdictions to strengthen BPA's commitment to address mitigation responsibilities identified in the 2008 FCRPS biological opinion for the construction and operation of Columbia and Snake River hydropower dams (State of Idaho 2008).

Additional science reviews were conducted by the Council in 2010 and 2012. In their 2010 review, the ISRP concluded that the Program now met scientific review criteria and acknowledged that they "better appreciate that the project was conceived and funded as a gene maintenance program to prevent extinction and that the rather minimal adult returns over project years 1994 to 2006 were very much anticipated" (ISRP 2010). The 2012 review was also favorable (ISRP 2012). Following direction provided in the 2008 FCRPS biological opinion, we initiated Council process to gain approval to fund and construct the Springfield Hatchery. We produced a master plan for Program expansion and new hatchery construction and received approval to construct the Springfield Hatchery in 2010. The master plan articulated the three-tiered recovery strategy described above (IDFG 2010, 2012).

CHALLENGES MOVING FORWARD

As we pursue population rebuilding (recolonization phase) and meeting expectations to consistently return thousands of anadromous Sockeye Salmon from the ocean to spawning locations in the Sawtooth Valley each year, we are discovering new challenges. Some lie out of our control but can negatively affect survival within the salmon life cycle. Ocean conditions, specifically the Pacific Decadal Oscillation and the North Pacific Gyre Oscillation have been shown to strongly influence salmon survival during their ocean phase of residency (Kilduff et al. 2015). When ocean temperatures rise and cause food web shifts, the availability of desirable prey species for juvenile and adult salmon can be greatly reduced, resulting in poor fish survival. Unfortunately, ocean conditions that support high salmon survival have been in decline since 2013 and are forecasted to continue to be negative through 2018 and possibly beyond. Whether poor ocean conditions have impacted the number of Sockeye Salmon returning to Idaho is difficult to determine, but we did observe a sharp drop-off in adult return numbers in 2015 and 2016 compared to 2014 (Figure 2). In 2015, we observed the lowest returns of adults to the Sawtooth Valley since 2008. Low adult returns in 2015 were primarily attributed to a different environmental anomaly—high river temperatures during adult upstream migration caused by seasonally low snow packs and lower than average river discharge. In 2015, we estimated that 4,069 adult Snake River Sockeye Salmon passed Bonneville Dam, the first hydropower project on the Columbia River (Figure 1). This passage was the highest estimate of Snake River Sockeye Salmon movement past this location since the inception of the Program. Unfortunately, only 50 adults successfully made the long migration back to Redfish Lake, resulting in an unprecedented low conversion rate of only 1.25%. Average survival of adult Sockeye Salmon between Bonneville Dam and the Sawtooth Valley is typically about 39% (Kline and Flagg 2014). If 2015 upstream migration conditions had been more favorable, returns to the Sawtooth Valley could have exceeded 1,500 adults.

Such setbacks are unfortunate but perhaps to be expected over the course of long-running restoration efforts. We raise this point to provide perspective for other researchers, managers, and fish culturists that short-term disruptions are, while very frustrating at the time, intermittent and not long-lasting. Success for species recovery programs such as the one discussed in this chapter will require the application of good science, adherence to best management practices, and a healthy dose of patience by the human "restorers" working to address such a long-term recovery goal. As someone else said, Rome was not built in a day. However, constant vigilance is required so as not to be caught, as Nero did, fiddling while Rome burned!

LESSONS LEARNED

The chief aim of this chapter has been to describe the development and implementation of hatchery-based gene-rescue activities, review present-day release strategies and associated adult returns, and describe a new effort underway to expand salmon production to more effectively address recolonization and local adaptation goals. A focus on conservation aquaculture and genetic principles has enabled Program managers to retain approximately 95% of the original founding genetic variability of the population while, at the same time, producing large numbers of offspring for use in restoration. Attention to administrative detail, policy interactions, and coping with environmental uncertainty inherent in managing an endangered stock that must travel more than 1,400 river kilometers to reproduce have been as much the key to success as developing the science. We believe that we have put as much effort into developing information to address Program reviews and regulatory process needs as we have into fish culture and monitoring and evaluation activities. Lessons learned include the need to

- Define clearly stated goals and objectives for the program and ensure that policymakers and technical reviewers are frequently reminded of them;
- Rapidly identify and implement methods to address the immediate extinction threat;
- Implement a suite of actions that include careful stepwise efforts that address multiple priorities and limiting factors, including habitat improvements and carrying-capacity issues; and
- Implement science-driven efforts aimed at understanding population demographics and dynamics (e.g., robust monitoring and evaluation support).

As indicated, we might have lost the Program and the potential for saving ESA-listed endangered Snake River Sockeye Salmon in 2006 if not for the intervention of the governor of Idaho. Winning such small, unexpected battles has turned out to be of equal or greater importance than all of the science accomplishments the Program has achieved. Continuous outreach and collaboration with all local, state, tribal, federal, and nongovernmental organization partners involved was essential to Program success. Thus, several other human-dimension lessons learned include the need to

- Incorporate communication experts and people that can deal with bureaucracy into the program;
- Become engaged and vigilant over the legal, policy, and funding frameworks required of your program;
- Become participatory in local and regional forms that involve all agencies and stakeholders; and
- Consistently communicate about your program and its progress to all those interested in achieving successful outcomes.

Additionally, at the onset of the Program, we convened the Stanley Basin Sockeye Technical Oversight Committee to provide technical guidance for the Program. This cohesive structure has been foundational and is requisite for projects of this kind. Thus, another key lesson learned is to

- Establish a scientific and technical committee to provide advice and oversight of your program.

The careful, stepwise efforts carried out by the Snake River Sockeye Salmon Recovery Program can be seen as a model for future endeavors. We believe that we are one of the first to report on the use of standards developed by the HSRG (Paquet et al. 2011) to help balance both hatchery and developing natural components of a critically at-risk population. Our hope is that by sharing the Program's science direction and successes, our description of the policy challenges we encountered, and the above lessons learned, future workers will be guided when facing last ditch population interventions of this type.

ACKNOWLEDGMENTS

The Idaho Department of Fish and Game, National Marine Fisheries Service, and Oregon Department of Fish and Wildlife provided fish culture, fish transportation, and monitoring and evaluation support for the Program. Primary funding for the Program was provided by the Bonneville Power Administration through the Northwest Power and Conservation Council's Columbia Basin Fish and Wildlife Program.

REFERENCES

Axel, G. A., M. Peterson, B. P. Sanford, E. E. Hockershith, B. J. Burke, K. E. Frick, J. J. Lamb, M. G. Nesbit, and N. D. Dumdei. 2013. Characterizing migration and survival between the upper Salmon River basin and lower Granite Dam for juvenile Snake River Sockeye Salmon. Report of research by the National Marine Fisheries Service to Bonneville Power Administration, Division of Fish and Wildlife, Portland, Oregon.

Ballou, J. D., and R. C. Lacy. 1995. Identifying genetically important individuals for management of genetic diversity in pedigreed populations. Pages 76–111 *in* J. D. Ballou, M. Gilpin, and T. J. Foose, editors. Population management for survival and recovery. Columbia Press, New York.

Beauchamp, D., C. Luecke, W. A. Wurtsbaugh, H. G. Gross, P. Budy, S. Spaulding, R. Dilenger, and C. P. Gubala. 1997. Hydroacoustic assessment of abundance and diel distribution of Sockeye Salmon and kokanee in the Sawtooth Valley lakes, Idaho. North American Journal of Fisheries Management 17:253–267.

Berejikian, B. A., P. Kline, and T. A. Flagg. 2004. Release of captively reared adult anadromous salmonids for population maintenance and recovery: biological trade-offs and management considerations. Pages 233–245 *in* M. J. Nickum, P. M. Mazik, J. G. Nickum, and D. D. MacKinlay, editors. Propagated fish in resource management. American Fisheries Society, Symposium 44, Bethesda, Maryland.

Bjornn, T. C., D. R. Craddock, and D. R. Corley. 1968. Migration and survival of Redfish Lake, Idaho, Sockeye Salmon, *Oncorhynchus nerka*. Transactions of the American Fisheries Society 97:360–375.

Budy, P., C. Luecke, and W. A. Wurtsbaugh. 1998. Adding nutrients to enhance the growth of endangered Sockeye Salmon: trophic transfer in an oligotrophic lake. Transactions of the American Fisheries Society 127:19–34.

Budy, P., C. Luecke, W. A. Wurtsbaugh, and H. P. Gross. 1995. Limnology of Sawtooth Valley lakes with respect salmon-derived nutrients in two lakes from Kodiak Island, Alaska. Journal of Paleolimnology 30:35–53.

Burgner, R. L. 1991. Life history of Sockeye Salmon *Oncorhynchus nerka*. Pages 3–117 *in* C. Groot and L. Margolis, editors. Pacific salmon life histories. University of British Columbia Press, Vancouver.

Cummings, S. A., E. L. Brannon, K. J. Adams, and G. H. Thorgaard. 1997. Genetic analyses to establish captive breeding priorities for endangered Snake River Sockeye Salmon. Conservation Biology 11:662–669.

Flagg, T. A., and C. V. Mahnken. 2000. Captive broodstocking to aid recovery of endangered species. Pages 290–292 *in* R. R. Stickney, editor. The encyclopedia of aquaculture. Wiley, New York.

Flagg, T. A., C. V. W. Mahnken, and K. A. Johnson. 1995. Captive broodstocks for recovery of Snake River Sockeye Salmon. Pages 81–90 *in* H. L. Schramm, Jr. and R. G. Piper, editors. Uses and effects of cultured fishes in aquatic ecosystems. American Fisheries Society, Symposium 15, Bethesda, Maryland.

Flagg, T. A., C. W. McAuley, P. Kline, M. S. Powell, D. Taki, and J. C. Gislason. 2004. Application of captive broodstocks to preservation of ESA-listed stocks of Pacific salmon: Redfish Lake Sockeye Salmon case example. Pages 387–400 *in* M. J. Nickum, P. M. Mazik, J. G. Nickum, and D. D. MacKinlay, editors. Propagated fish in resource management. American Fisheries Society, Symposium 44, Bethesda, Maryland.

Griswold, R., A. Koler, and D. Taki. 2011. Survival of endangered Snake River Sockeye Salmon smolts from three Idaho lakes: relationships with parr size at release, parr growth rates, smolt size, discharge, and travel time. North American Journal of Fisheries Management 31:813–825.

Griswold, R., D. Taki, and J. S. Stockner. 2003. Redfish Lake Sockeye Salmon: nutrient supplementation as a means for restoration. Pages 197–211 *in* J. Stockner, editor. Nutrients in salmonid ecosystems: sustaining production and biodiversity. American Fisheries Society, Symposium 34, Bethesda, Maryland.

Gross, H. P., W. A. Wurtsbaugh, and C. Luecke. 1998. The role of anadromous Sockeye Salmon *Oncorhynchus nerka* in the nutrient loading and productivity of Redfish Lake, Idaho. Transactions of the American Fisheries Society 127:1–18.

Gross, H. P., W. A. Wurtsbaugh, C. Luecke, and P. Budy. 1997. Fertilization of an oligotrophic lake with a deep chlorophyll maximum: predicting the effect on primary productivity. Canadian Journal of Fisheries and Aquatic Sciences 54:1177–1189.

Hebdon, J., P. Kline, D. Taki, and T. Flagg. 2004. Evaluating reintroduction strategies for Redfish Lake Sockeye Salmon captive broodstock progeny. Pages 401–413 *in* M. J. Nickum, P. M. Mazik, J. G. Nickum, and D. D. MacKinlay, editors. Propagated fish in resource management. American Fisheries Society, Symposium 44, Bethesda, Maryland.

Heindel, J. A., D. J. Baker, K. A. Johnson, P. A. Kline, and J. J. Redding. 2005. A simple isolation incubator for specialized rearing of salmonid eggs and first-feeding fry. North American Journal of Aquaculture 67:13–17.

HSRG (Hatchery Scientific Review Group). 2009. Report to Congress on Columbia River basin hatchery reform. Available: http://hatcheryreform.us/wp-content/uploads/2016/05/HSRG-2009-Report-to-Congress.pdf. (August 2018).

IDFG (Idaho Department of Fish and Game). 2010. Springfield Sockeye hatchery master plan for the Snake River Sockeye program, volume 1: master plan. IDFG, Boise.

IDFG (Idaho Department of Fish and Game). 2012. Springfield Hatchery Sockeye program combined step 2 and step 3 submittal, volume 2: appendices. IDFG, Boise.

ISRP (Independent Scientific Review Panel). 1999. Volume 1: review of the Columbia River basin Fish and Wildlife Program for fiscal year 2000 as directed by the 1996 amendment to the Northwest Power Act. Report ISRP 99-2 for the Northwest Power and Conservation Council, Portland, Oregon. Available: https://nwcouncil.org/sites/default/files/isrp99-2.pdf. (August 2018).

ISRP (Independent Scientific Review Panel). 2001. Final review of fiscal year 2002 proposals for the Mountain Snake and Blue Mountain provinces. Report ISRP 2001-12a to the Northwest Power and Conservation Council, Portland, Oregon. Available: https://nwcouncil.org/fw/reviews/2002province/isrp2001-12a/. (August 2018).

ISRP (Independent Scientific Review Panel). 2004. Review of captive propagation program elements: Programmatic Issue 12 for the Mountain Snake and Blue Mountain provinces. Report ISRP 2004-14 for the Northwest Power and Conservation Council, Portland, Oregon. Available: www.nwcouncil.org/sites/default/files/isrp2004_14.pdf. (August 2018).

ISRP (Independent Scientific Review Panel). 2006. ISRP final review of proposals submitted for fiscal years 2007–2009 funding through the Columbia River Basin Fish and Wildlife Program. Report ISRP 2006-6 for the Northwest Power and Conservation Council, Portland, Oregon. Available: https://nwcouncil.org/fw/isrp/isrp2006-6/. (August 2018).

ISRP (Independent Scientific Review Panel). 2010. Final RME and artificial production categorical review report. Report ISRP 2010-44 for the Northwest Power and Conservation Council, Portland, Oregon. Available: https://nwcouncil.org/fw/isrp/isrp2010-44/. (August 2018).

ISRP (Independent Scientific Review Panel). 2012. Snake River Sockeye Springfield Hatchery step 2/step 3 review (project #200740200). Report ISRP 2012-7 for the Northwest Power and Conservation Council, Portland, Oregon. Available: https://nwcouncil.org/fw/isrp/isrp2012-7/. (August 2012)

Kalinowski, S. T., D. M. Van Doornik, and C. C. Kozfkay. 2012. Genetic diversity in the Snake River Sockeye Salmon captive broodstock program as estimated from broodstock records. Conservation Genetics 13:1183–1193.

Keefer, M. L., C. A. Perry, and M. J. Heinrich. 2008. Temperature-mediated en route migration mortality and travel rates of endangered Snake River Sockeye Salmon. Ecology of Freshwater Fishes 17:136–145.

Kendall, N. W., A. H. Dittman, and T. Quinn. 2010. Comparative maturation schedules of two Columbia River Sockeye Salmon *Oncorhynchus nerka* populations. Endangered Species Research 13:51–61.

Kilduff, D. P., E. DiLorenzo, L.W. Botsford, and S. L. H Teo. 2015. Changing central Pacific El Niños reduce stability of American salmon survival rates. Proceedings of the National Academy of Sciences of the United States of America 112:10962–10966.

Kline, P. A., and T. A. Flagg. 2014. Putting the red back in Redfish Lake: 20 years of progress toward saving the Pacific Northwest's most endangered salmon population. Fisheries 39:488–500.

Kozfkay, C. C., M. R. Campbell, J. A. Heindel, D. J. Baker, P. Kline, M. S. Powell, and T. Flagg. 2008. A genetic evaluation of relatedness for broodstock management of captive, endangered Snake River Sockeye Salmon, *Oncorhynchus nerka*. Conservation Biology 9:1421–1430.

Kozfkey, C. C., M. Peterson, B. P. Sandord, E. L. Johnson, and P. Kline. 2019. The productivity and viability of Snake River Sockeye Salmon hatchery adults released into Redfish Lake, Idaho. Transactions of the American Fisheries Society 148:308–323.

Kurath, G., K. A. Garver, S. E. LaPatra and M. K. Purcell. 2011. Resistence and protective immunity in Redfish Lake Sockeye Salmon exposed to M type infectious hematopoietic necrosis virous (IHNV). Journal of Aquatic Animal Health 22:129–139.

Leitritz, E., and R. C. Lewis. 1976. Trout and salmon culture (hatchery methods). California Department of Fish and Game Fish Bulletin 164.

Luecke, C., W. Wurtsbaugh, P. Budy, H. Gross, and G. Steinhart. 1996. Simulated growth and production of endangered Snake River Sockeye Salmon: assessing management strategies for the nursery lakes. Fisheries 21:18–25.

Massee, K. C., J. Kim, B. A. Berejikian, and R. W. Hardy. 2007. Prey selection and efficiency of naive and experienced juvenile Sockeye Salmon. Journal of Fish Biology 70:1213–1223.

Maynard, D. J., T. A. Flagg, W. C. McAuley, D. A. Frost, B. Kluver, M. Wastel, J. Colt, and W.

W. Dickhoff. 2012. Fish culture technology and practices for captive broodstock rearing of ESA-listed salmon stocks. NMFS Technical Memorandum NMFS-NWFSC-117.

Nichols, K. M., C. C. Kozfkay, and S. R. Narum. 2016. Genomic signatures among *Oncorhynchus nerka* ecotypes to inform conservation management of endangered Sockeye Salmon. Evolutionary Applications 9:1285–1300.

NMFS (National Marine Fisheries Service). 1991. Endangered and threatened species; endangered status for Snake River Sockeye Salmon. Federal Register 56:224(20 November 1991):58619–58624.

NMFS (National Marine Fisheries Service). 2015. ESA recovery plan for Snake River Sockeye Salmon (*Oncorhynchus nerka*). National Oceanic and Atmospheric Administration, NMFS, West Coast Region, Portland, Oregon. Available: https://repository.library.noaa.gov/view/noaa/16001. (August 2018).

NPCC (Northwest Power and Conservation Council). 1980. Pacific Northwest Electric Power Planning and Conservation Act. NPCC, Portland, Oregon. Available: www.nwcouncil.org/reports/poweract/. (August 2018).

NPCC (National Power and Conservation Council). 1986. Compilation of information on salmon and steelhead losses in the Columbia River basin. NPCC, Portland, Oregon. Available: https://permanent.access.gpo.gov/lps88417/index.htm. (August 2018).

NPCC (National Power and Conservation Council). 2014. 2014 Columbia River Basin Fish and Wildlife Program. NPCC, Portland, Oregon. Available: www.nwcouncil.org/reports/2014-columbia-river-basin-fish-and-wildlife-program. (August 2018).

O'Reilly, P. T., and C. C. Kozfkay. 2014. Use of microsatellite data and pedigree information in the genetic management of two long-term salmon conservation programs. Reviews in Fish Biology and Fisheries 24:819–848.

Paquet, P., T. Flagg, A. Appleby, J. Barr, L. Blankenship, D. Campton, M. Delarm, T. Evelyn, D. Fast, J. Gislason, P. Kline, D. Maynard, L. Mobrand, G. Nandor, P. Seidel, and S. Smith. 2011. Hatcheries, conservation, and sustainable fisheries—achieving multiple goals: results of the hatchery scientific review group's Columbia River basin review. Fisheries 36:542–561.

Pilati, A., and W. A. Wurtsbaugh. 2003. Importance of zooplankton for the persistence of a deep chlorophyll layer: a limnocorral experiment. Limnology and Oceanography 48:249–260.

Pollard, H. A., and T. A. Flagg. 2004. Guidelines for use of captive broodstocks in recovery efforts for Pacific salmon. Pages 333–345 *in* M. J. Nickum, P. M. Mazik, J. G. Nickum, and D. D. MacKinlay, editors. Propagated fish in resource management. American Fisheries Society, Symposium 44, Bethesda, Maryland.

Powell, M. S., R. W. Hardy, T. A. Flagg, and P. A. Kline. 2010. Proximate composition and fatty acid differences in hatchery-reared and wild Snake River Sockeye Salmon overwintering in nursery lakes. North American Journal of Fisheries Management 30:530–537.

Purcell, M. K., K. A. Garver, C. Conway, D. G. Elliott, and G. Kurath. 2009. Infectious haematopoietic necrosis virus genogroup-specific virulence mechanisms in Sockeye Salmon, *Oncorhynchus nerka* (Walmaum), from Redfish Lake, Idaho. Journal of Fish Diseases 32:619–631.

Rieman, B. E., D. L. Meyers, and R. L. Nielsen. 1994. Use of otolith microchemistry to discriminate *Oncorhynchus nerka* of resident and anadromous origin. Canadian Journal of Fisheries and Aquatic Sciences 51:68–77.

Sawatzky, C. L., W. A. Wurtsbaugh, and C. Luecke. 2006. The spatial and temporal dynamics of deep chlorophyll layers in high-mountain lakes: effects of nutrients, grazing and herbivore nutrient recycling as growth determinants. Journal of Plankton Research 28:65–86.

Schiewe M. H., T. A. Flagg, and B. A. Berejikian. 1997. The use of captive broodstocks for gene conservation of salmon in the western United States. Bulletin of the Natural Research Institute of Aquaculture, Supplement 3:29–34.

Selbie, D. T. B. A. Lewis, J. P. Smol, and B. P. Finney. 2007. Long-term population dynamics of the endangered Snake River Sockeye Salmon: evidence of past influences on stock decline and impediments to recovery, Transactions of the American Fisheries Society 136:800–821.

State of Idaho. 2008. 2008 Columbia Basin Fish Accords memorandum of agreement between the state of Idaho and FCRPS action agencies. Available: Available: www.salmonrecovery.gov/Files/BiologicalOpinions/ID_MOA_Final.pdf. (August 2018).

Steinhart, G. B., and W. A. Wurtsbaugh. 1999. Under-ice diel vertical migrations of *Onchorynchus nerka* and their zooplankton prey. Canadian Journal of Fisheries and Aquatic Sciences 56(Supplement 1):152–161.

Steinhart, G. B., and W. A. Wurtsbaugh. 2003. Winter ecology of kokanee: implications for salmon management. Transactions of the American Fisheries Society 132:1076–1088.

Steinhorst, K., Y. Wu, B. Dennis, and P. Kline. 2004. Confidence intervals for fish outmigration estimates using stratified trap efficiency methods. Journal of Agricultural, Biological and Environmental Statistics 9:284.

Swanson, P., B. Campbell, K. D. Shearer, J. Dickey, B. R. Beckman, D. A. Larsen, L. K. Park, and B. Berejikian. 2008. Application of reproductive technologies to captive breeding programs for conservation of imperiled stocks of Pacific salmon. Cybium 32(Supplement 2):279–282.

Tucker, S., M. E. Thiess, J. F. T. Morris, D. Mackas, W. T. Peterson, J. R. Candy, T. D. Beacham, E. M. Iwamoto, D. J. Teel, M. Peterson, and M. Trudel. 2015. Coastal distribution and consequent factors influencing production of endangered Snake River Sockeye Salmon. Transactions of the American Fisheries Society 144:107–123.

Waples, R. S., P. B. Aebersold, and G. A. Winans. 2011. Population genetic structure and life history variability in *Oncorhynchus nerka* from the Snake River basin. Transactions of the American Fisheries Society 140:716–733.

Winans, G. A., P. B. Aebersold, and R. S. Waples. 1996. Allozyme variability of *Oncorhynchus nerka* in the Pacific Northwest, with special consideration to populations of Redfish Lake, Idaho. Transactions of the American Fisheries Society 125:645–663.

Wurtsbaugh, W. A., H. P. Gross, P. Budy, and C. Luecke. 2001. Effects of epilimnetic versus metalimnetic fertilization on the phytoplankton and periphyton of a mountain lake with a deep chlorophyll maxima. Canadian Journal of Fisheries and Aquatic Sciences 58:2156–2166.

Restoring the Vindel River, Sweden for Atlantic Salmon and Brown Trout

Gustav Hellström*
Department of Wildlife, Fish and Environmental Studies
Swedish University of Agricultural Sciences
Skogsmarksgränd, Umeå SE-907 36, Sweden

Christer Nilsson
Department of Ecology and Environmental Science, Umeå University
Tvistevägen 48, Umeå SE-901 87, Sweden

Stig Westbergh
The Ume/Vindel River Fishery Advisory Board
Vindelvägen 3, Sorsele SE-924 94, Sweden

Daniel Palm and Kjell Leonardsson
Department of Wildlife, Fish and Environmental Studies
Swedish University of Agricultural Sciences
Skogsmarksgränd, Umeå SE-907 36, Sweden

Daniel Holmqvist and Stefan Ågren
The Ume/Vindel River Fishery Advisory Board
Vindelvägen 3, Sorsele SE-924 94, Sweden

Johanna Gardeström
Vindelälven-Juhtatdahka Biosphere Candidate
Kommunalhusvägen 11, Vindeln SE-922 81, Sweden

Anders Alanärä and Hans Lundqvist
Department of Wildlife, Fish and Environmental Studies
Swedish University of Agricultural Sciences
Skogsmarksgränd, Umeå SE-907 36, Sweden

Abstract.—The Vindel River (Vindelälven), Sweden, is 450 km long with a mean annual discharge of 190 m^3/s and runs through sparsely populated areas in northern Sweden, joining the Ume River near the Baltic Sea. A severe decline in Atlantic Salmon *Salmo salar* and Brown Trout *S. trutta* during the past century was caused by (1) intense timber-floating activities starting in the mid-1800s, which degraded stream habitat; (2) hydropower development in the early 1900s, which limited or blocked upstream migration; and (3) an expanded coastal/ocean fishery after 1950, which overharvested anadromous populations. Restoration efforts during the 1970s and 1980s failed due to low efficacy of stocking programs and insufficient habitat restoration. A legislative reform to unify the fishery right owners (FROs) along the river helped initiate restoration efforts in the

* Corresponding author: gustav.hellstrom@slu.se

late 1990s, focusing on improving fish migration past a hydropower station and restoring degraded habitat. Sweden's membership in the European Union made large funding for restoration projects possible. The number of returning Atlantic Salmon increased significantly after migration conditions improved around the hydropower station and with stricter regulations on the offshore fishery. Successful habitat restoration was based upon gaining trust from landowners and FROs via extensive communication to gain access to their land. Restoration work was adaptive and experiences gained were incorporated into restoration guidelines. Involvement of universities, as a provider of expertise and as an unbiased interpreter of data, provided support to legal processes and when evaluating restoration measures. During the course of the restoration work, managers learned that stocking often did not produce satisfactory results. By studying historical documents from the timber-floating era, managers learned that the scope of modifications of the tributaries had been much greater than previously thought and that habitat restoration needed to be extensive. In many tributaries, the number of juvenile Brown Trout increased significantly after habitat restoration, sometimes dramatically exceeding expectations, which made the managers question the validity of established production estimates for northern boreal streams. The experience and knowledge gained from the Vindel River restoration served as the catalyst for many other major restoration projects in rivers emptying into the Baltic Sea.

INTRODUCTION

Streams and rivers are among the fish habitats most extensively harmed by human activities worldwide. The nature and severity of the habitat degradation and the threat to the native fish communities vary widely among rivers, and consequently, so do the approaches and practices needed for successful restoration. Atlantic Salmon *Salmo salar* is considered threatened throughout its natural distribution range (Hindar et al. 2010). The problems facing many Atlantic Salmon populations are, however, far from uniform. Even within geographically contained populations of Atlantic Salmon residing in the Baltic Sea (i.e., the Baltic Salmon), threats and restoration needs differ strongly among salmon rivers depending on geographical location as well as the economic and cultural history of the catchment.

This chapter tells the story behind the restoration of Atlantic Salmon and Brown Trout *S. trutta* in the Vindel River, a large free-flowing river in the northern part of the Baltic Sea catchment, Sweden. In their effort to restore the entire catchment and its salmon and trout populations, managers needed to tackle several parallel challenges. Some of the challenges, such as degraded connectivity due to hydropower or the depletion of stocks due to overfishing, are general and recognizable across the salmon distribution range. Others, such as habitat degradation due to extensive timber floating or navigating the specific environmental and constitutional legislations opposing restorations, are more regional to Scandinavian rivers of the northern Baltic Sea catchment.

The Vindel River is one of several large (>150 m^3/s) rivers draining the northeastern part of the Scandinavian peninsula into the Baltic Sea. The river originates in the alpine regions of the Scandes mountain range, 60 km east of the Atlantic Ocean, and flows 450 km in a southeasterly direction across northern Sweden, entering into the heavily regulated Ume River 30 km from the Gulf of Bothnia (Figure 1). The 12,650-km^2 catch-

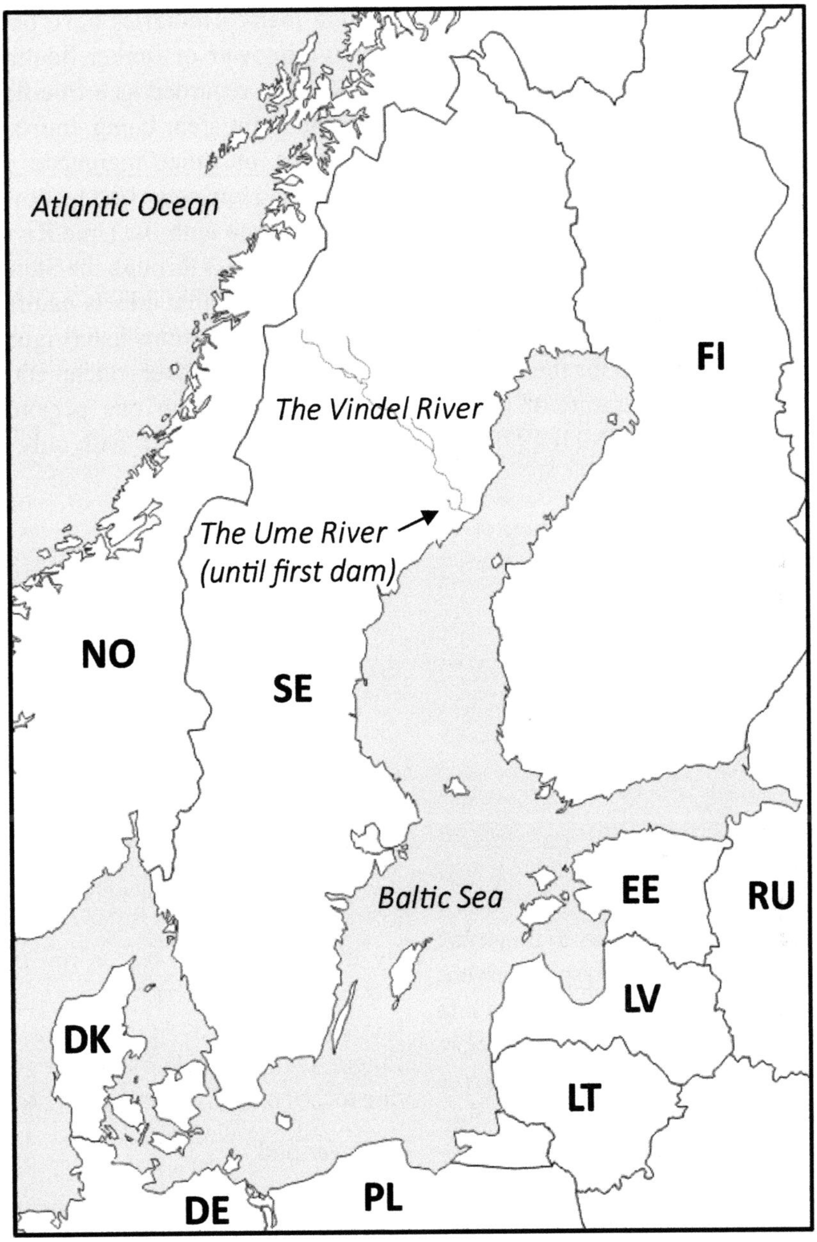

FIGURE 1. The Baltic Sea region, and Vindel River. The Vindel River Atlantic Salmon are generally believed to use the southern part of the main basin of the Baltic Sea as a feeding area. Abbreviations: NO = Norway, SE = Sweden, FI = Finland, RU = Russia, EE = Estonia, LV = Latvia, LT = Lithuania, PL = Poland, DE = Germany, and DK = Denmark.

ment consists of a dense network of tributaries and lakes and spans across a geomorphic gradient formed by ice sheets and ancient seas. Glacial and legacy sediments characterize the upper part of the river, whereas silt, sand, and gravel sediment dominates close to the sea (Nilsson et al. 2017). Falling almost 1,000 m between the headwaters and the river mouth, the Vindel River has a steep vertical profile with more than 40 large rapids in the main stem alone. The largest tributary, the Lais River, connects high up in the system, resulting in the Vindel River becoming a seventh-order river already in its upper parts. The mean annual discharge is 190 m^3/s close to the confluence with the Ume River, but seasonal variation is considerable and flow can be as low as 15 m^3/s during cold winters months and greater than 2,000 m^3/s during extreme flooding events (SMHI 1979). The mean annual air temperature for the region is 3°C, with an average temperature of 13°C in July and –12°C in January (SMHI 1979). Although many tributaries have been dammed for hydropower or timber floating, the Vindel River is regarded as a free-flowing river, with the main stem being entirely free from any dams or other manmade impounding structures. However, 10 km downstream of the confluence with the Ume River, the combined rivers pass through the Stornorrfors hydropower plant that affects natural migration routes for anadromous fish (Figure 2).

The Vindel River catchment is sparsely populated (less than one person per square kilometer; SCB 2012), with only a few small

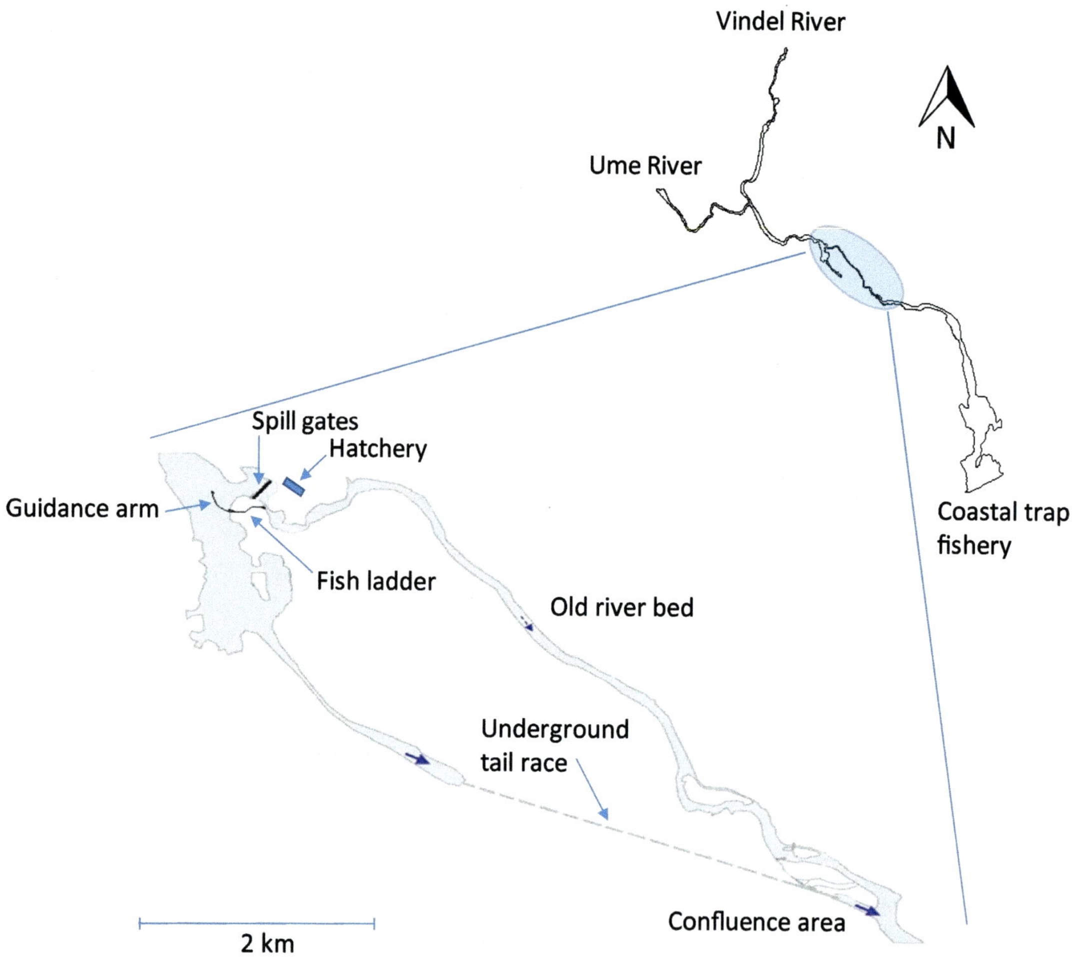

FIGURE 2. Overview of the lower part of the Ume River and the Stornorrfors hydropower complex. The Ume River is dammed with no fish passage.

population centers along the river. The catchment is within the jurisdiction of two provinces and six municipalities. Each administrative region has extensive rights to define its own policies and action plans regarding exploitation and management of the river and its fishery. Private citizens and large forest companies dominate as landowners, each owning approximately 50% of the land. In Sweden, land ownership often also includes the right to fish in affected waters. More than 4,000 FROs occur along the river; the vast majority are private citizens, with jurisdictional authority to regulate fishing within their land. The native Sami people have no restrictions regarding harvest limits or the types of fishing gears used in areas where they conduct reindeer husbandry, which is along the majority of the river catchment, including the coastal parts.

The fish fauna in the Vindel River is typical for northern rivers in Sweden, with Atlantic Salmon, Brown Trout, Grayling *Thymallus thymallus*, Eurasian Perch *Perca fluviatilis*, Northern Pike *Esox lucius*, whitefish *Coregonus* spp., and European Bullhead *Cottus gobio* found throughout most of the river and Arctic Char *Salvelinus alpinus* restricted to upper parts of the system. Two species of nonnative char, Brook Trout *S. fontinalis* and Lake Trout *S. namaycush*, have established reproducing populations in the catchment after introductions in the 1960s in an effort by the government to improve the recreational fishery. Brook Trout are found in the headwaters of small creeks in the lower parts of the river. Lake Trout occur in deep lakes close to the mountains. The dispersal of these species seems slow, and their potential impact on native fish in the catchment has been regarded as limited. Hence, the incentive among managers to try to remove them has been weak. However, it is important to acknowledge that little actual data exist on either the dispersal dynamics or impact of these nonnative char species in the catchment.

The Vindel River salmon stay as juveniles 2–4 years in the river, after which time they migrate to the sea and spend 1–4 years in the southern Baltic Sea (average 2 years) before returning to spawn (Alanärä et al. 2017). All Baltic Salmon are genetically distinct from populations in the eastern Atlantic Ocean and little genetic introgression occurs over the Danish straits (Nilsson et al. 2001). The Vindel River sea trout (anadromous Brown Trout) seem more flexible in anadromy and life history pattern compared to the salmon, but in general, sea trout smolt at age 3 and spend an average of 2–3 years along the coast of the Gulf of Bothnia where they may overwinter in nearby rivers before returning to the Vindel River to spawn (Andersson 1988; Degerman et al. 2012). Upstream of the Stornorrfors power station, salmon and sea trout are free to migrate almost the entire length of the Vindel River and the Lais River, excluding the uppermost alpine reaches.

This chapter describes the history of successful management of the Vindel River salmon and trout populations. The chapter first reviews the background behind population declines, starting several hundred years ago, and then in a chronological order describes the 40-year-long restoration process that has led to the current, improved conditions. The chapter ends with reflections and conclusions about the restorations and tries to identify which management actions worked and which did not.

HISTORICAL ABUNDANCE OF FISH

The Vindel River has, like most large rivers in the Baltic Sea catchment, had abundant populations of sea-running salmonids before the age of extensive human exploitation (Gisler 1751; Christensen and Larsson 1979). Five-thousand-year-old rock carvings indicated prehistorical settlements close to the site of the Stornorrfors hydropower plant, and although these petroglyphs only portray moose and not salmon, the location implies a reliance on fish, likely salmon aggregating below the adjacent waterfall (Ramqvist et al. 1985). Estimating the past abundance of salmon in the Vindel River is difficult. Historically, salmon were used as a tax revenue by the state, and early tax records from mid-16th century indicated that salmon were an important and plentiful resource for early settlers along the river (Öhman 2007). Historically, the salmon fishery used fixed trap-net installations in the river, as well as beach seines and gill nets, and was likely extensive along the lowermost parts of the Vindel and Ume rivers. Early signs of overexploitation of salmon were already evident at the end of 16th century in the form of reduced catches and complaints about bad fishing (Öhman 2007). A riverwide trap closing off the entire migration route was established in the lower parts of Vindel River in the late 1600s, which severely restricted salmon from reaching their upstream spawning grounds. This barrier drastically reduced the salmon population, and by early 1700s, all commercial and tax-based fisheries had ceased due to low returns, leaving only subsistence fishing remaining (Öhman 2007). Although migration statistics from this period remain very sparse, some evidence suggests that large runs occurred sporadically during the 19th century (McKinnell and Karlström 1999; Öhman 2000), especially during high floods (Zetterstedt 1833), which may have provided salmon with migration pathways past the fishery. More reliable migration statistics became available in the first half of the 20th century with the construction of the hydropower dam at Norrfors (Montén 1988).

MORE RECENT DECLINES OF FISH

Timber Floating

Although the historical in-river fishery likely had negative effects on the salmon population in the Vindel River as early as 400 years ago, the first major structural destruction of salmon and trout habitat started in the mid-19th century with the establishment of an extensive network of floatways to transport timber (tree logs) from the remote roadless inland forests to the sawmills and pulp mills on the Baltic Sea coast (Törnlund and Östlund 2006). Driven by an increasing demand for wood and paper from industrialized western continental Europe, the forest industry quickly became the main employer in the region. As logging expanded all over the catchment, massive efforts were undertaken to channelize streams to facilitate timber transport (Törnlund 2002). Such modifications were modest in the beginning and limited by the inefficiencies of engineering technology of the time. However, as the forest industry grew and timber output increased, river modifications gradually intensified. By the mid-1950s, log-floating companies effectively used bulldozers and excavators to channelize streams into floatways all over the catchment (Törnlund and Östlund 2006; Figure 3).

Timber floating and floatway construction profoundly altered the environmental conditions in the streams throughout the catchment. The removal of boulders decreased structural heterogeneity of the riverbed (Polvi et al. 2014), making the habitat less capable of supporting diverse and abundant communities of fish, invertebrates, and plants. Channelization and narrowing of the streams by lining the banks with stone piers and bulldozed landfills homogenized flow and increased the water velocity, flushing out lighter materials such as salmonid spawning substrate and dead wood used for juvenile habitat (Palm et al. 2007; Figure 3). Side channels and eddies were cut off by wing dams and piers, completely draining or dewatering large and often productive areas of the natural river channel (Nilsson et al. 2005). The retention capacity of channelized streams decreased significantly, as did land–

FIGURE 3. Historical photographs depicting the timber-floating era in northern Sweden. **(A)** A rare photo of a pristine tributary before channelization. Such photos were used as reference for restoration targets. **(B)** The same tributary after channelization. Boulders have been removed or bulldozed to the edges to facilitate transport of timber, **(C)** the use of bulldozer to channelize streams, and **(D)** a forester standing on the riverbank of a newly channelized stream.

water interactions and the riparian ecotone, all contributing to reduced inflow and accessibility of allochthonous detritus and leaf litter vital for production of macroinvertebrates in these nutrient poor northern boreal waters (Nilsson et al. 2005). A multitude of splash dams were constructed in the tributaries to control the water flow during the timber-floating season, and many of them came to act as distinct barriers for migrating fish.

For fish such as salmon and trout, these habitat alterations affected all life stages. Loss of habitat complexity made suitable feeding territories and resting pools sparser for juvenile and adult fish (Palm 2007). Lack of access to side channels and natural shallow areas severely limited the amount of nursery areas for juveniles (Palm et al. 2010). Vanishing or substantially degraded spawning beds had negative influences on trout populations in the tributaries, as did the splash dams, which hindered spawners' ability to reach many spawning grounds. Increased erosion and transport of fine sediments clogged the spawning beds with sand and suffocated eggs, and the lack of deep water and other refuges increased the risk of the riverbed freezing solid during cold winters, killing eggs and juveniles (Palm 2007).

By the end of the 1960s, log-floating companies had converted close to 100% of all high-gradient regions in the catchment into floatways (Nilsson et al. 2005). A few years later, with motorized transports (trucks) becoming more readily available, the industry abandoned timber floating due to the high costs and inconvenience compared with trucking timber (Törnlund and Östlund 2002). Although less invasive to instream habitat, the shift to trucking dramatically expanded the network of forest roads in the catchment. The road expansion further reduced the connectivity in the river system, as many of the road-stream crossings had culverts that came to act as migration barriers for fish (Lundberg et al. 2005).

Exactly how devastating floatway constructions were to the trout and salmon populations of the Vindel River is unknown. However, even during the early timber-floating era, concerns were raised about the negative influence channelization had on the salmonid populations (e.g., Byström 1868). Although no reliable catch or monitoring data exist for premodified sites, juvenile densities of trout and salmon estimated from electrofishing surveys were often appallingly low at most sites throughout the catchment before restoration started, seldom more than a few yearlings per 100 m^2. Such densities are 100-fold lower than pristine boreal streams in northern Karelia, Russia (Andersson et al. 1998), indicating that densities likely do not reflect the potential capacity of the premodified streams in the Vindel River catchment. It is notable that as almost all boreal rivers and streams in northern Sweden had been converted into floatways during the timber-floating era, no good reference sites existed to provide reliable baseline data on juvenile densities in natural streams. During the 1990s, the recommendations from the Swedish National Board of Fisheries stated that if a stream contained more than three Brown Trout juveniles per 100 m^2, no funding for restoration would be defensible as this density was considered typical based on an average of densities from multiple streams and rivers in northern Sweden.

Hydropower Dams

The Vindel River is one of only a few large Swedish rivers completely free from hydropower dams on the main stem. Plans to develop the river for hydropower during the 1960s were met with strong local and national resistance, making the government abandon further planning in 1970. At the time, this decision was unprecedented in Swedish politics and represented one of the first true cases where environmental and conservational values triumphed over the demands of the rapidly expanding modern Swedish society. The decision gave the Vindel River an iconic environmental national status among the general public (Johansson 2008). In the story about the recovery of the salmon and trout in the Vindel River, the hard-fought battle to protect the river from hydropower development during the 1960s was the prerequisite for all further efforts taken to revive its anadromous salmonid populations.

Given that the Vindel River today serves as the national symbol for free-flowing rivers, it is rather ironic that merely 10 km downstream of its confluence with the Ume River, one of the largest hydropower plants in Europe blocks the entire river. Complete damming of the whole river for hydropower began in 1926. The construction of the dam prevented the Vindel River salmon from reaching their spawning grounds for more than 10 years, and while a few spawners were manually transported past the dam (as few as 36 salmon in 1931, of which 22 were females), the naturally reproducing population was likely close to extinction during these years (Montén 1988). To compensate for the damages, up to 1.5 million salmon fry were stocked annually in the river and tributaries above the dam (Montén 1988). The result of these stocking events seemed to be poor at first, but after installation of a fish ladder in 1935 at the dam, numbers of ascending salmon began to increase steadily (Andersson 1988; Figure 4).

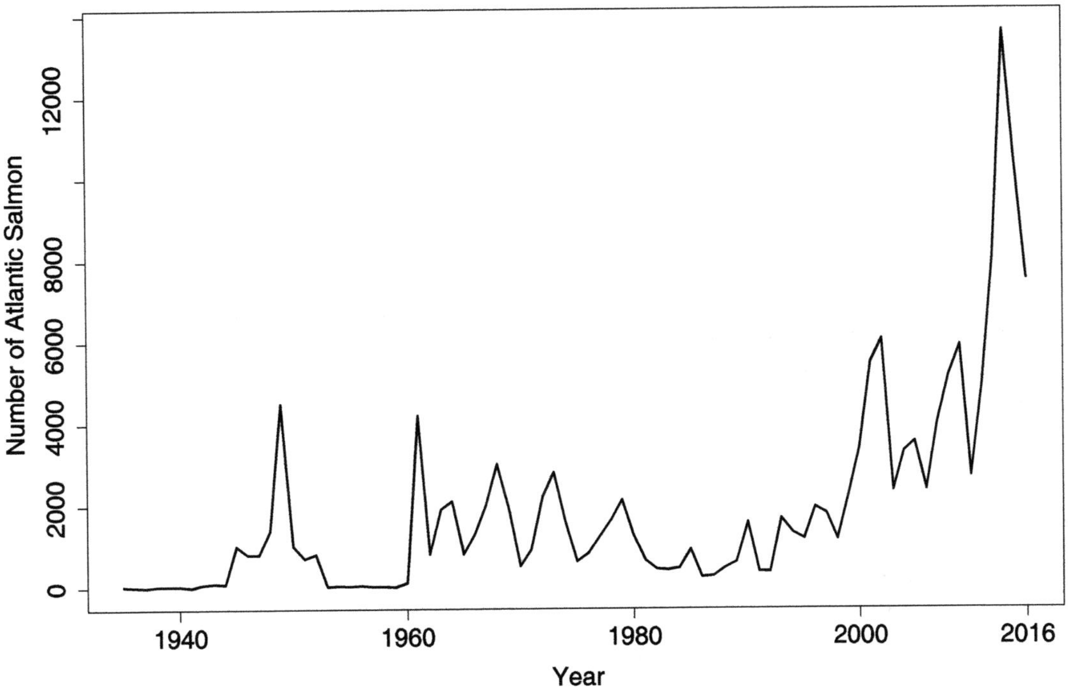

FIGURE 4. Number of Atlantic Salmon ascending the fish ladder at the Stornorrfors Dam, Ume River. The low number during the 1950s may be due to a malfunctioning fish ladder and reconstruction of the dam.

In 1958, a major reconstruction transformed the Norrfors hydroelectric facility into the power plant that is operating today. The new plant, named Stornorrfors, diverts the whole river through a headrace canal into four 75-m-deep vertical tunnels with four large Francis turbines at the bottom of each tunnel (Figure 2). An underground tailrace then reconnects the diverted water with the old river channel 4 km downstream of the turbines. The diversion of water leaves the old river channel dry for 8 km if no water comes through the spillway gates. Thus, Stornorrfors creates a series of migration problems for salmon and trout heading both up- and downstream (Rivinoja 2005; Lundqvist et al. 2008).

As upstream passage through the tailrace canal is impossible, fish have to find their way into the old riverbed, which has a small fraction of the discharge compared to the tailrace, making the channel less attractive as a migration route than the outflow from the turbines (Ferguson et al. 2002; Rivinoja 2005). For the fish finding their way up the old river channel, challenges exist in the form of shallow and narrow waterfalls, which may be difficult to pass depending on the amount of water released through spill gates. Once at the upstream end of the old river channel, fish have to find the entrance to the fish ladder to pass over the remaining 22-m-high dam separating the reservoir upstream and the old river channel. If they successfully navigate the ladder, further migration to the spawning grounds upstream in the Vindel River is unhindered (Figure 2).

The majority of the downstream migrating smolts and kelts (fish that have completed spawning) end up passing through the powerhouse on their way to the sea, experiencing rapid compression and risk of physical trauma as they fall 75 m vertically through the running turbines (Montén 1985; Lundqvist et al. 2014, 2015). Researchers estimated downstream passage mortality to be around 15–20% for smolts and 40–75% for kelts depending on size (Montén 1985; Ferguson et al. 2008; Östergren and Rivinoja 2008).

Fishing and Stocking

In-river subsistence and commercial fisheries were gradually replaced (and eventually completely abandoned) during the 20th century by a coastal and offshore fishery. The offshore fishery primarily occurred in the main basin of the Baltic Sea and targeted mixed-stock salmon on their feeding areas (Karlsson and Karlström 1994; Figure 1). The fishery was predominately based on drift gill nets and longlines, and as vessels and fuel became more affordable after the Second World War, the offshore fleet rapidly expanded (McKinnell 1997). By the 1990s, offshore fishing accounted for 80% of the total catch of Baltic Salmon, with more than 5,000 metric tons caught annually. With no quotas in place to regulate harvest, the Baltic Salmon may well have been the most exploited salmonid in the world at the time (Karlsson and Karlström 1994; McKinnell 1997). How much of the offshore catch consisted of Vindel River salmon is unknown during this period, but likely the Baltic Sea fishery contributed substantially to the decline in number of ascending salmon during the 1980s (Romakkaniemi et al. 2003; Figure 4).

Beginning in the 1950s, large-scale releases of hatchery smolts began occurring in all major regulated rivers draining to the Baltic Sea as a way to compensate for production losses of wild salmon caused by hydropower development (Lindroth 1963). By the 1990s, more than 6 million smolts were released annually in regulated rivers flowing into the Baltic Sea, which may have accounted for as much as 95% of all the salmon in the whole Baltic Sea system, wild and reared combined (Karlsson and Karlström 1994; McKinnell 1997). The stocking became a major threat to the wild salmon, not only because of potential genetic introgression or competition, but also because it sustained extraordinarily high exploitation rates caused by the unregulated offshore fishery, which harvested wild and reared salmon indiscriminately. In the Vindel River, hatchery fry and smolts have been stocked in the river sporadically since the 1920s, first as a way to compensate for the impact of the dam at Norrfors and later as part of various management programs and research projects. However, since 1960, approximately 80,000 hatchery-reared salmon and 20,000 sea trout have been released annually just below the dam in Stornorrfors because of compensation claims from the fishery (Andersson 1988). These fish can be distinguished from the wild fish because their adipose fins have been clipped. The releases made up approximately 50% of all returning salmon ascending the ladder during the 1970s and 1980s (Eriksson et al. 2008). Although most of these fish originated from wild Vindel River parental fish, the hatchery also used breeders from other neighboring regulated rivers during a few years (Andersson 1988).

From the early 1990s, decreased survival of postsmolts was beginning to be observed all over the Baltic Sea. Survival of wild postsmolts during their first year in the sea decreased from about 30% to less than 10% all over the Baltic Sea between 1990 and today (Eriksson et al. 2008; ICES 2016). For hatchery-reared smolts, the corresponding figures have shown a survival decline from about 20% to 5% during the same period (Eriksson et al. 2008). No clear understanding of this trend has yet been laid forth, although altered environmental conditions in the sea with regards to predation and food may be involved (Mäntyniemi et al. 2012). Similarly, return rates have decreased both in wild and hatchery salmon in rivers emptying into the Baltic Sea (Kallio-Nyberg et al. 2006; Mäntyniemi et al. 2012; ICES 2016). In the Vindel River, the return rates of the hatchery-reared salmon have decreased from as high as 30% in the 1960s to below 5% today (Montén 1988; Leonardsson et al. 2013; Alanärä and Persson 2017). Although this trend may be a result of multiple variables, increased post-smolt mortality plays a key role.

EARLY ATTEMPTS TO REVITALIZE FISH POPULATIONS

To make up for the loss of regional construction jobs after the 1971 decision not to exploit the Vindel River for hydropower, a federal compensation package was presented in 1981 with the intention to further develop tourism and recreational fishing along the river (Johansson et al. 1987). This effort was the first directed attempt to revitalize the salmon and trout populations in the river on a major scale and was headed by a broad group of stakeholders, including the County Administrative Board in the county of Västerbotten, FROs, the Swedish National Board of Fisheries, and nongovernmental organizations. The group had two main objectives: (1) implement fisheries management actions to improve fishing along the river valley, and (2) organize and consolidate FROs along the river valley. The project concluded that the major habitat damages due to timber floating were too extensive to solve, and as a result, the direct management actions proposed as a first objective involved stocking of 2- and 3-year-old Brown Trout in the main stem, as well as the creation of alternative fishing spots within the river valley. These new fishing locations were established in small inland lakes using rotenone treatment followed by stocking to create put-and-take trout fisheries. Some less-extensive habitat restorations targeted the heavily channelized tributaries by creating pools and weirs in repetitive patterns with the intention to create easily fishable habitats that could hold large fish. The results from these actions were inconclusive at best (habitat restorations) and clear failures at worst (stocking) (Johansson et al. 1987).

The second objective, consolidation of FROs, proved to be much more successful than the stocking and the creation of new fishing locations. Being one of the first efforts to implement national fisheries legislation that provided a legal framework for individual FROs to unite in local organizations to govern common fishing areas (Swedish Law 1981:533), the restoration project helped create several functional management units along the river and, hence, established the first glimpse of an organizational structure among thousands of individual FROs in the catchment. The consolidation approach went further by establishing an umbrella organization to represent and advise the newly created management units. The organization took the name of the Vindel River Fishery Advisory Board (FAB; later extended to Ume/Vindel River Fishery Advisory Board) and came to be the key player in all future work to revive the salmon and trout populations in the river. Large, wood-crafted information signs designed according to a common template were also installed at all the major rapids along the river, clearly stating the name of the rapid and current fishing regulations. This seemingly simple action was likely important in creating awareness among local residents and tourists that the fishery along the river now was more organized than in the past.

A new regional effort was started in 1988, again headed by the County Administrative Board of Västerbotten. The overall approach was now less about creating attractive and easy fishing opportunities and more toward establishing self-sustaining and viable salmon and trout populations that could withstand some coastal and river fishery harvest (CAB 1995). Although stocking still was the main management action, the majority of the fish stocked were yearlings instead of 3-year-olds, with the rationale of creating more naturalized fish populations. An electrofishing monitoring program for juveniles was also established with the intention to create a long and comparable time-series.

Despite the two fairly large projects running during the 1980s and early 1990s to help revive the salmon and trout in the Vindel River, the population status continued to dwindle (Figure 4). By the late 1980s and early 1990s, just a few hundred salmon passed the fish ladder at Stornorrfors; in 1986, as few as 84 fish made the passage, of which 50% were of hatchery origin (Eriksson et al. 2008). To make the situation even worse, an intensified outbreak of the thiamine deficiency syndrome (M74) hit the Baltic Salmon in the early 1990s, killing as many as 96% of all fry at the Stornorrfors hatchery during 1993 and 1994 and with clear indication that wild fish also were heavily affected (Karlström 1999; ICES 2016). The recruitment during these years sank to record lows, with less than 0.17 yearlings/100 m^2 reported from the electrofishing surveys (Larsson 2015). The situation for the sea trout seemed even more grave, with the number of ascending fish seldom exceeding 30 fish after 1980 (as few as five fish in 1994; Larsson 2015). In their concluding reports, the restorations projects warned that a risk of extinction existed for wild salmon and trout in the Vindel River. This risk was a blow to morale to agencies and local FROs (CAB 1995). Nationally, Sweden concurrently entered into an economic crisis (Englund 1999), and with the resulting large downsizing of the welfare state, the government became less willing to fund conservation work.

TURNING OF THE TIDE

With the restoration projects coming to a close during the late 1990s and funding running out, the road map ahead was unclear. Although overexploitation of salmon in the Baltic Sea fishery had been identified as a contributing factor to declining returns (Karlsson and Karlström 1994), local and regional stakeholders of the Vindel River did not feel empowered to lobby for reduced quotas, especially since the offshore fishery involved a multitude of national fishing fleets and with evidence of high-level corruption and illegal fishing. In 1995, Sweden entered the European Union (EU), transferring much of the control of the Baltic Sea fishery to the EU in Brussels, Belgium even further away from this remote little river valley in northern Sweden. Hence, the focus of the FROs continued to be on problems at the regional and local scales, primarily passage problems at the Stornorrfors power plant and the habitat destruction caused by timber floating. Most urgent was the need to solve the Stornorrfors issue, as the power company in 1996 appealed to the environmental court to have the 40-year-old compensation claim statute-barred.

The Legal Battle

Like most legal cases concerning hydropower, the scope and nature of the biological damage caused by the Stornorrfors power plant have been controversial, as is the level of compensation required for the damage. During the court proceedings connected to the reconstruction of the new dam in 1958, the estimated impact of the dam on the salmon population was not resolved, and hence the appropriate level of compensation could not be set (Montén 1988). The court decided to postpone the ruling until more was known about possible impacts. Since then, despite various attempts to come to an agreement, the case was never settled. In 1996, after a renewed claim for compensation by the FROs, the power company appealed to the environmental court to have the case statute-barred, claiming that no clear impact to the Vindel River salmon populations could be seen and that the case already was too old to be valid for adjudication. The appeal spurred the fighting spirit among the FROs, and in connection with the upcoming court decision, a demonstration was organized outside the gates of the power company, bringing the case to the attention of the media and public. Here, the unifying role of the FAB became valuable and clearly evident as support along the entire catchment quickly could be organized.

Unlike the situation in most other European countries, as well as in the United States and Canada, permits to operate hydropower facilities in Sweden are not reviewed at regular intervals, but instead are considered as permanently valid once decided upon in court (Rudberg et al. 2014). Thus, century-old rulings concerning operational conditions and damage assessments for hydropower are treated as a nonnegotiable baseline by the court when considering any subsequent changes in construction and operation of the plant. In the case of the reconstruction of Stornorrfors in 1958, an old ruling connected to the first damming of the river in 1926 had determined that the damming had caused 20% damage to the salmon fishery (a remarkable conclusion given that the dam made the population go extinct). The legal case came to revolve around whether or not the reconstruction in Stornorrfors had made any further damage to the salmon, apart from the 20% already determined. Two ways were possible for interpreting this question. The power company claimed that as the number of ascending salmon after the reconstruction were approximately equal to, or better than, the number before the reconstruction (Figure 4), the new power plant had not made the situation any worse for the salmon, and hence no compensation measures were needed. The FROs, on the other hand, viewed the ruling to mean that 80% of the salmon run should be unaffected by the dam, regardless of the actual numbers ascending the river. Using such a benchmark, the FROs felt certain that Stornorrfors Dam indeed caused significant disruption to the Atlantic Salmon population, but proving that would require new data and thorough investigations. Support from the government was nowhere to be found. In fact, a few years earlier, the National Board of Fisheries concluded that problems for the ascending salmon around the power plant were negligible (Andersson 1988), and the Swedish Legal, Financial and Administrative Services Agency (Kammarkollegiet) gave their full support to the power company in having the case barred. However, pioneering tracking studies of the salmon migration conducted by researchers at the Swedish University of Agricultural Science (SLU) indicated that problems actually existed around the dam (see below). Increasing involvement of universities now became vital in the case as they provided quality assured, unbiased data collected by neutral actors with high credibility.

In 1999, partly based on data provided by the SLU, which clearly indicated migration problems for the salmon, the court rejected the petition to bar the case and new negotiations started concerning compensation measures. Early on in negotiations, the FAB adopted a policy that any form of compensation from the power company should be directed towards actual measures to improve the salmon population rather than being paid out as financial reparation to individual FROs. This approach was unusual at the time and made claims more complex as investigations had to be made to specify what would be suitable measures and how much such measures would cost. Although not particularly favored by the power company, the altruistic element in this policy, benefitting the environment while forgoing any claims of private financial compensation, may have made the environmental court and, in the end, even the power company, more sympathetic to the cause of the FROs.

In the beginning of 2000, the power company had become more open for discussions. Reasons for this change in approach are difficult to pinpoint. The effort from the FAB to increase awareness of the case among the public may have increased pressure to find a solution that would be acceptable to the public image of the company. During the 1990s, an increased focus on environmental issues and sustainable development in society at large had made the conservation reputation of companies more important (Svedin and Hagerhall Aniansson 2012). At the same time, the EU began formulating common, more environmentally friendly objectives and directives for the management of regulated rivers, such as the EU Water Framework Directive (Directive 2000/60/EC), which potentially could override laxer national legislation. Accumulating evidence of migration problems around the power plant provided by the university (Rivinoja et al. 2001) also made unsubstantiated claims and denials more difficult to sustain in court. Changes among the top leadership and negotiating staff in the power company helped the negotiations, suggesting that personal grievances and poor chemistry among negotiators may have been restricting progress. A clear sign that the power company had changed stance was a trip to the United States in 2001. Organized by the SLU and paid for by the power company, representatives from all stakeholder groups, including top management of the power company, were introduced to solutions and technologies used to solve migration problems of the Pacific salmon in the heavily dammed Columbia River, Washington (Sandström et al. 2002). As a result of that trip, the company officials had seen what state-of-the-art engineering and fishery science could achieve, and this revelation may very well have inspired them when discussing solutions for the Stornorrfors Dam.

During the next couple of years, many things happened in parallel. The SLU continued to lead increasingly detailed studies on the salmon and trout migration around Stornorrfors, getting a better understanding of how, when, and where problems occurred (see below). Based on these studies, actual measures to improve migration now started to be suggested, both from international expertise and from the National Board of Fisheries, which had been consulted to provide qualified advice (Ferguson et al. 2002). To end the increasingly expensive and long-lasting legal process, the power company now pushed for a final judgment by the court on what kind of compensation measures should be made. The FAB, on the other hand, strived for a settlement rather than specific compensation measures. The reason for this stance was that they did not trust the court to provide qualified decisions on what measures should be implemented, especially as the knowledge of the migration complications in the area was still poor. Further, knowing that technical fish passage solutions could be difficult to get functional, they wanted an adaptive process where measures could be tested and adjusted over time. When the National Board of Fisheries recommended an increase in the discharge in the spillways by 400%, which would have cost the power company several million euros annually (Larsson and Sparrevik 2009), the company finally signaled they wanted a settlement, possibly to avoid a court decision that would favor such a recommendation. In August 2006, a settlement was reached. Staying committed to keep a collaborative environment between stakeholders, the FAB suggested establishing a committee to head further work, consisting of representatives from the FROs, the power company, regional and local governmental agencies, and universities. The committee was given €2.5 million, which they could spend as they chose. An additional €2.5 million were also donated to a foundation predominately consisting of local FROs with the aim to support small local-based fishery projects along the river valley.

Getting the Facts Right

Ever since the first dam construction at Norrfors in 1926, the impact of the dam on the salmon and trout populations in the Vindel River has been much debated. A reoccurring difficulty has been verification of the various claims and statements that have been made from different stakeholders. Variations in catch data from the salmon fishery have been used by both sides to prove their points, but such data may have been unrepresentative due to the shifting inclination of the commercial fishery to report their harvest, as well as the by the influence of the stocking programs constantly artificially boosting the fishery with hatchery-reared fish (Montén 1988; Öhman 2000). Although the installation of the fish ladder has allowed collection of more precise data on the spawning migration since the 1930s, no reliable predam data were available to compare with more recent data. Hence, any quantitative estimation of the impact of the dam has been difficult to achieve, both with regards to the number of ascending and descending fish in the system and the migration behavior of these fish. Various claims, such as the dam having a minor role in affecting the salmon population dynamics compared to other variables (e.g., Baltic Sea fishery harvest, ocean conditions, and timber floating) or that the potential migration problems observed below the dam are natural due to the steep and complex bathymetry of the area, have been difficult to disprove.

Since the building of the Stornorrfors power plant in 1958, more ambitious attempts to understand and quantify potential impacts of the dam on the survival and migration of the Vindel River salmon have been done. The lack of effective tracking technology made many of these early studies more anecdotal than quantitative in nature. Although the methods were often quite ingenious, such as tagging salmon with a string attached to a floating light bulb to be able to study the behavior in the confluence area or packaging smolt in small wooden boxes to be able to collect them after being released through the turbines to estimate mortality rates (Lindroth 1970; Montén 1985), most of these studies did not contribute to the understanding of how salmon migration had been affected. However, a few studies did produce reliable data. A series of skillfully planned and executed experiments in the 1960s aimed at estimating smolt mortality through the turbines using externally tagged smolts that were released in large numbers above the dam and later reported back as catches by the fishery concluded that an average of 16–26% were lost due to turbine mortality (Montén 1985), a result that still today provides one of the best estimates of smolt mortality through the power plant.

With the advancement of telemetry technology, researchers at SLU started to track the salmon passing through the Stornorrfors complex with much higher resolution than before. A series of initial studies in the 1990s provided the first hard evidence that salmon indeed were delayed at the confluence area, and also that they had apparent difficulties passing through rapids in the old river channel (Rivinoja et al. 2001). Due to the sensitive issue, the studies had to be rigorous in their design and execution to hold up as evidence in court discussions, and aspects such as tagging effects and nonrepresentative behavior by potential strayers were controlled for by various double-tagging schemes and by genetic considerations (Rivinoja et al. 2006; Lundqvist et al. 2008). The results showed that salmon took 1.5–2 months to pass the first 30 km of the river, from the estuary through the Stornorrfors complex to the fish ladder, while they managed the remaining 250 km to their spawning grounds in the Vindel River in less than a quarter of that time (reviewed in Rivinoja 2005; Lundqvist et al. 2008). Some studies also showed that only as few as 14% of the salmon entering the river from the estuary made it past the fish ladder, the remaining either turning back to sea or staying in the old river channel after considerable migration delays (Rivinoja 2005). Apart from providing evidence of migration difficulties around the dam complex, the ever-more-extensive tracking studies also managed to identify where and why migration delays were happening, including how different levels of discharge through the spillway affected the migration behavior (Leonardsson et al. 2005). An innovative program was initiated in the late 1990s when the public were invited to bet on the migration performance of the tagged salmon by sponsoring a transmitter (H. Lundqvist, Swedish University of Agricultural Sciences, personal communication). Tagged salmon were then tracked as they migrated up the river, and their positions were updated on a Web page. If a sponsored salmon migrated the farthest or fastest, the sponsor won a trophy. The program, named "the salmon race," became popular and even the Swedish royal family contributed by sponsoring a salmon. Apart from generating valuable management data, this largely crowd-funded program also effectively reinforced the Vindel River as a salmon river among the public by highlighting the spectacular migrations occurring in the river every year.

Undoing the Timber-Floating Era

At the end of the timber-floating era, some efforts were made to restore the heavily modified floatways. These early restoration projects, conducted by the forest companies according to a "polluter pays" agreement, only returned a fraction of the removed substrate back into the streams (FAB 2015). More ambitious restoration followed during the 1980s, now under the guidance of the County Administrative Board, but these habitat restoration projects tackled only a few kilometers of the more than 15,000 km of floatways in the catchment. Restoration efforts at this time were largely synonymous with creating sites tailored for recreational fishing (e.g., by constructing weirs and pools in repetitive patterns). Besides looking artificial and unnatural, such measures did not address the more critical damages caused by channelization, such as loss of juvenile nursery areas and adult spawning grounds (Palm 2007). With no evaluation program in place and with limited funding, the restoration projects were patchy and haphazard. In the mid-1990s, a recognition occurred that much more comprehensive restoration efforts were needed to fully rehabilitate the extensive damage done by 150 years of timber floating in the Vindel River catchment. The resources for conducting such major habitat restoration, however, were nowhere to be found in the post-banking crisis in Sweden. This changed with Sweden's entrance into the EU.

In the EU, each member state pays an annual fee and the resulting pool of money is then redistributed to different countries and sectors as needed. One such sector is fisheries, and support for fisheries-related projects may be applied for through the European Fisheries Fund (EFF), a multi-billion-euro instrument renewed every 5 years. Although the application and the accounting/reporting are bureaucratically challenging, the EFF (and other EU funds and programs) allows funding for very large projects that would be unlikely to be approved based solely on national funding (Hagen et al. 2013). The Vindel River Fishery Advisory Board seized upon this opportunity and prepared an application at the end of the 1990s. Key here was organization and the ability to build a large coalition of stakeholders acting together. Via the unique role of the FAB, the collective will of all FROs along the river was credibly represented in one single organization. This aspect, together with regional and local governmental agencies and the universities, provided a powerful alliance that fulfilled many of the collaborative requirements sought by the EU.

Umeå University and SLU took the lead in designing an extensive evaluation program to assess the effectiveness of the restoration actions. The involvement of the universities also directed restoration projects toward more ecological goals rather than creating sites primarily for anglers. The aim was to restore habitats to a pristine state and hence to recreate habitat characteristics comparable to the pre-timber-floating era. This ecological focus was a new approach at the time, and as such, no manuals, guidelines, or experienced expertise were available to guide the work (Nilsson et al. 2007; Gardeström et al. 2013). Further, as almost all of the northern boreal forest streams in Sweden had been exploited for timber floating, few unaffected sites were available as comparative reference sites. A collaboration with historians of timber floating proved valuable to get access to old and rare records and descriptions made by the timber companies to describe what modifications had been made at specific sites (Nilsson et al. 2005). Sometimes photographs were found for stream sites before modifications had been made. These photographs often showed a highly complex, seemingly random structure of large boulders mixed with large wood, a radical departure from the straight weirs commonly created in earlier restoration projects (Figure 3). After approval of the EU-application in 2001, and with a budget of €3 million, an extensive habitat inventory was made during 2002 to document all timber-floating modifications done in the entire catchment. With some sites being modified almost 150 years ago, nature had begun to mask many of the alterations, with stands of large trees now covering old piers, closed-off channels, and excavated walls (Figure 5A, 5C). Even to the trained eye, some modifications were difficult to detect, and here again the insights from the historical records provided by the historians of timber floating proved helpful. Key sources of knowledge were financial records from timber-floating companies, which in detail documented reimbursements for the work that had been done in floatways. Following inventory, a list of prioritized sites to be restored was compiled based mainly on accessibility and ability to get permission from landowners (FAB 2015).

Before a restoration project could start, all needed permissions had to be obtained. By referring to environmental legislation, many restoration projects could technically be performed without the landowners' explicit permission. However, to foster acceptance and cooperation among affected local residents, a policy was adopted that no work or contact with licensing agencies would be made before having full approval of landowners and FROs. This meant thousands of hours spent in meetings and consultations with FROs and private landowners along the river, the majority during evenings when people were home from work (Vindel River LIFE 2016). Gaining approvals was at times tough. Skepticism existed among local residents about the benefits of radically altering their streams and riparian forest, tearing down old piers and dams, and significantly changing flow regimes. Many residents had personal memories attached to certain sites or had been involved in building the often well-crafted timber-floating constructions now considered by many as cultural landmarks. Here, listening to the landowners was the key. To achieve approval from landowners, providing facts about the anticipated ecological benefits that would be achieved after restoration came second to just listening

FIGURE 5. Pictures of floatways before and after restoration. **(A)** A channelized tributary. Note the relatively homogenous flow as well as the lack of any major impounding structures. Also, note how the riparian forests mask the bulldozed walls on each side of the stream, **(B)** the same site after being restored, pictured from exactly the same angle as in the previous photo. The bulldozed walls have been removed, and the substrate returned into the stream. A pier further upstream has been removed, allowing water to enter an overgrown side channel. The difference in heterogeneity and width of the streambed is striking compared to the site before restoration, **(C)** a large stone pier intended to regulate and canalize the river to improve timber transportation, and **(D)** the same site after being restored.

and respectfully responding to the views and wishes of the landowner. Ultimately, it often came down to winning landowner trust and respect at a personal level. The value of having staff that were fully engaged in the topic, with extensive knowledge and interest in the local history and fishery, cannot be overstated in this process. Being able to let go of certain objectives when encountering too much opposition and instead turning the energy to other sites and work often proved to be the correct strategy. Sometimes, initially negative landowners would change their attitude over time as positive results from restoration projects in neighboring areas became known.

After having permission from landowners, all restoration projects had to be approved by the responsible County Administrative Board. Even though not always necessary, most restoration cases were also prepared for the environmental court to be able to state a declaratory judgment, which released managers from any legal liability if unforeseen negative consequences of restoration projects occurred (Vindel River LIFE 2016). All this preparatory work initially took a lot of effort and time, and sometimes years were required to obtain all permissions needed to start a restoration. However, as managers, as well as the County Administrative Board and the court, became more experienced in handling such tasks, the process became quicker and smoother than in the beginning. A recurrent frustration was opposition from the cultural history departments of the County Administrative Boards, which, in their ambition to protect the historical legacy of the timber-floating era, frequently opposed restoration projects, especially if they meant tearing down old piers and stonemasonry (Lundberg et al. 2005). Here, restoration managers often took their case to the court to get the final verdict, although in the end trying to seek constructive dialogue and finding compromises was often the most effective way forward to achieve permissions.

Instream Restoration

Stream habitat restoration came to focus on four major actions; (1) returning removed substrate back into the streams, (2) opening up closed-off side channels by removing piers, (3) building new spawning beds, and (4) removing migration barriers in the form of old splash dams. To work efficiently in the main river and the tributaries, large excavators were used, directed by a restoration supervisor on the ground (Nilsson 2007). An experienced driver and supervisor could restore up to 100 m of stream habitat in a day. Although efforts were made to keep year-to-year continuity in personnel, this proved to be particularly difficult with regards to drivers as the work was hard on the machines and their drivers. Work was done during late summer and early autumn when water levels were low. The overall aim was to make the restored habitat look as natural as possible, meaning that straight lines and repeated patterns were avoided when placing the boulders. Natural weirs were reinforced, as were natural curves and meanders. Blasted pieces of rock that were left over from timber-floating constructions were placed deep under water to hide any traces of human activity.

An early insight was the extraordinary amount of boulders that had been removed from the streams and placed along the edges of the channels. When the boulders were returned back into the water, the appearance of the stream was often radically altered in a way that strongly departed from the traditional preconception of what a natural boreal stream in northern Sweden should look like. For example, in some cases the amount of added boulders was so large that it came to cover the entire stream channel, making the riverbed almost seem dry as all water seeped beneath the boulders. Here, the restoration managers had to readjust their original image of what they thought the final restoration would look like and trust that the new appearance represented the natural state of the stream. Boulder placement was complemented by the addition of large wood, often from trees that were overturned into the stream. The combination of multiple boulders breaking the water's surface and large amount of large wood lying in the water often made fishing more difficult at restored sites. Managers had to be responsive to such complaints from FROs and landowners and sometimes needed to adapt the restoration projects to avoid clashes with recreational fishing interests (FAB 2015). For example, the amount of large wood added to a restored site seldom came close to what has been found at pristine reference sites in neighboring Russia (Nilsson et al. 2007).

As experience and successes grew, restoration practices became bolder and more advanced. With an additional €3 million from the EU granted in 2009, some sites received what was referred to as "enhanced" restoration (Gardeström et al. 2013; Nilsson et al. 2017). Habitat characteristics that previously had been regarded as forever lost to modified streams were now being recreated. Very large boulders, for example, were almost entirely lacking from the channelized streams as they often were the first obstacles blasted out of the streambed when constructing the floatway. Hence, huge boulders were now transported from the nearby terrain and placed in the streams, slowing down flows and raising water levels. The streams were allowed to overflow large areas of productive forestland, creating wetlands as well as initiating new erosion dynamics. With an increasingly pervasive policy to return all bulldozed substrate back into the river, a restored site often looked like a battlefield once finished, with excavated land and broken trees being scattered all over. However, after only one or two large floods, such ugly scars from restoration activities were erased, something that likely reduced complaints considerably from the landowners (Figure 4B, 4D).

The natural spawning substrate for trout and salmon (gravel and cobble-sized stone) was lacking in most of the channelized floatways, either due to having been flushed downstream or having been buried under boulders (Palm 2007). Restoration of spawning habitats hence required either that substrate had to be added to the site or that the buried substrate had to be uncovered. Most restoration of spawning habitats occurred in the tributaries and demanded a level of finesse that only manual labor could bring. Although physically laborious, the work was rewarding as large fish often found and used the new spawning beds the same autumn they were constructed (FAB 2015). Establishing new spawning grounds hence became a restoration activity that engaged local FROs, and several educational courses were provided to help local landowners know how to construct and maintain spawning beds. Perhaps better than any other restoration action, the efforts to establish and restore spawning beds clearly reflected the new approach by managers to create self-sustaining, natural populations. This activity was partly influenced by the restoration goals promoted by the collaborating universities, but it was also a consequence of the fact that years of stocking of yearlings and smolts had not been able to compensate or reverse the decline of returning spawners or low juvenile densities (CAB 1995; Eriksson et al. 2008)

One of the most effective restoration measures in terms of gaining new productive habitat was to open up closed-off side channels by removing piers and walls that blocked the flow of water into these channels. These side channels often had not been channelized and hence had been left in relatively pristine condition in terms of habitat heterogeneity and geomorphology. As many of the piers were tediously crafted by stonemasons, they became particularly sensitive structures to remove from a cultural and historical perspective. Compromises were sometimes negotiated by leaving the piers in place while the water was diverted into the cut-off side channel through pipes or culverts that could be built into the piers.

Much emphasis was put on dissemination of information about habitat restoration to the landowners. Information leaflets were sent out to 6,000 stakeholders along the river, informing them about ongoing activities, and multiple meetings, seminars, conferences and courses were arranged to keep all stakeholders involved and up to date (Vindel River LIFE 2016). Films were also produced and spread through the Internet, showcasing not only restoration projects, but also the potential for fishing and recreation in the catchment. A special effort was made to involve schoolchildren in the region by arranging theme-based education about the river and its restoration (Vindel River LIFE 2016). Progress and news were constantly updated on Web sites, and experiences and results were regularly summarized in technical and scientific reports and manuals, something that helped get the Vindel River restoration widely recognized outside of Sweden and also facilitated acquisition of new funding. All restoration work followed adaptive principles in which goals and best practices were adjusted as new knowledge was gained. All sites were also regularly visited after they had been restored to check on how boulder placements and structures held up and to make adjustments if needed (Vindel River LIFE 2016).

A NEW DAWN

With the legal process concerning the Stornorrfors hydropower plant concluded, work to improve migration past the dam now proceeded according to the new organizational and financial conditions established by the legal settlement. After reviewing all the reports and recommendations coming out from a decade of migration studies, the committee swiftly acted to reduce the migration problems in the old river channel by adding weirs and pools to lessen the challenges provided by particular rapids in the river stretch and by altering the spill discharge regime to increase attraction using pulses of high and low flows in cycles (reviewed in Leonardsson et al. 2013). All measures were evaluated by multi-year studies under the direction of SLU, by now highly experienced in fish telemetry as well as in hydraulic modeling and fish passage issues.

In 2010, the power company undertook reinforcement of the dam to prepare for predicted increased peak discharge from the Vindel River due to climate change, and this work meant that the fish ladder had to be reconstructed. Here the power company, under the advice of the committee, proposed a completely new 300-m-long fish ladder adapted for both upstream and downstream passage and equipped with advanced monitoring and fish collection technology. The €20 million facility became operational in 2010 and represents one of the most extensive and advanced migration and monitoring stations in Europe (Lundqvist et al. 2014; Lindberg et al. 2016). To increase attraction to the ladder entrance, most of the spill water is directed to the lower end of the ladder to act as auxiliary water, but the water will first pass through a small turbine to generate power, reducing the financial loss for the company. In 2012, a 100-m-long semi-floating guidance arm was installed above the intake to the ladder to direct downstream migrating smolts and kelts through the ladder and into the old river channel on their way to the sea, instead of having them pass through the turbines (Lundqvist et al. 2014). Evaluation programs running since 2010 have continuously monitored and adjusted the functionality of various sections of the ladder complex, gradually improving the upstream and downstream effectiveness (Leonardsson et al. 2013; Lundqvist et al. 2014).

Coinciding with the efforts to improve passage in Stornorrfors, stricter regulation of the commercial fisheries in the Baltic Sea started to be enforced by the EU. Implementation of a quota system during the 1990s helped curb the unchecked and excessive offshore fisheries, and since then management has strived to achieve a maximum-sustained-yield-based harvest goal under the guidance of scientific expertise from the International Council for the Exploration of the Sea (ICES; Romakkaniemi et al. 2003; ICES 2016). These actions may have reduced the total exploitation rate by more than half over the past 20 years, from as high as 95% of the total Baltic Sea salmon stock being caught by the commercial fishing during the 1980s to approximately 40% today (ICES 2016). In 2007, a key regulation to ban drift-net fishing was implemented by the EU. Although being justified as a measure to save the threatened Baltic sea harbour porpoise Phocoena phocoena (which risked being entangled in the drift nets) rather than protecting the Baltic Salmon, this regulation, in combination with regulations by Finland and Sweden to simultaneously also ban the longline fishery, effectively transformed the commercial salmon fishery from a mixed-stock offshore fishery on the feeding grounds to a coastal trap-net fishery on salmon entering the rivers. This change allowed for more regional-based harvest quotas, which could be more fine-tuned to the population status of specific salmon rivers. The opening of the commercial season was also delayed to protect the earliest part of the salmon run, which often contained a higher proportion of wild fish, to try to shift the bulk of the catch to stocked salmon, which ascend later than the wild fish in many rivers (Siira et al. 2009).

Consequences of the stricter regulations started to be apparent in many rivers entering the Baltic Sea by the late 1990s (Romakkaniemi et al. 2003), including the Vindel River where the number of ascending salmon through the ladder gradually began to increase (Figure 4). With the migration problems in the old river channel improved and the new fish ladder becoming operational, tracking studies started to report a higher proportion than in the past of ascending salmon passing up the fish ladder, as well as shorter run times through the old river channel (Leonardsson et al. 2013). In 2013, 52% of all fish tagged at the coast ($n = 350$) successfully passed through the ladder, traversing the old river channel 30% faster than previously seen (Leonardsson et al. 2013). That same year, almost 15,000 salmon ascended the ladder, which represented a 150-fold increase compared to the early 1990s (Larsson 2015). Only a few percent of these fish were of hatchery origin compared to almost 40% during the 1990s, despite stocking levels being the same (Eriksson et al. 2008). As the Atlantic Salmon entered the Vindel River, they now met thousands of hectares of restored habitat to serve as proper spawning and nursing grounds. After 14 years of restoration projects at a cost of more than €6 million, 90% of almost 1,600 km of channelized floatways were now considered as being set in a trajectory towards a preindustrial state (Lundberg et al. 2005; Vindel River LIFE 2016). Dozens of dams had been removed and multiple side-channels opened up, making available vast new areas for reproduction, juvenile production, and migration. More than 1,000 new spawning beds were created in tributaries all over the catchment.

In 2014, more than 40 yearlings/100 m^2 were reported from the annual monitoring programs, a 40,000% increase since the early 1990s (Larsson 2015). Monitoring programs also started to find salmon at sites previously seen as predominately Brown Trout habitat, such as in small tributaries and at reaches far up the river close to the natural migration barriers, where Atlantic Salmon had not been seen for decades. Going from being a marginalized salmon river in the 1980s and 1990s, the Vindel River was now the third-most important river for wild Atlantic Salmon production in the Baltic Sea (ICES 2016), despite being considerably smaller than other important rivers. The ability of the new fish ladder to accurately monitor both up- and downstream migrating fish made the Vindel River suitable as an index river for Baltic Sea salmon, and statistics from the river now contribute heavily to the annual management recommendations of the Baltic Salmon provided by ICES (ICES 2016).

Habitat Restoration Evaluation and the Aftermath

Throughout the habitat restoration, the universities were deeply involved in assessing, evaluating, and guiding the work. Sites were monitored to map effects on everything from habitat heterogeneity, geomorphology, and flow dynamics to the abundance and diversity of the biotic communities, such as riparian plants, insects, and fish. Fish were mostly monitored using electrofishing, focusing primarily on juvenile Atlantic Salmon and Brown

Trout. Common for many habitat restoration projects, proper evaluations required several years of monitoring to allow time for the biotic community to respond. A few years of "before and after" data at a restored site often proved too sparse to serve as the basis for reliable conclusions (Nilsson et al. 2016). Although monitoring and evaluation are an important and well-defined issue, obtaining funding from the EU for monitoring programs that lasted more than only a couple of years has been difficult (Nilsson et al. 2015). A more fundamental problem for the evaluation programs was the lack of proper control data for comparison purposes. Interannual variation in the population dynamics of salmonids can be considerable in boreal streams, potentially masking any trend after restoration at a site. To account for this problem, nonrestored sites, preferably similar in morphology and habitat to the restored site, had to be monitored parallel to the restored site. This meant leaving out sites or streams from the restoration program to serve as control sites, a sensitive issue as such decisions could be seen as primarily favoring scientific reporting rather than the rehabilitation of the catchment. Also, as the goal was to achieve pristine conditions, sites that had not been used for timber floating would ideally serve as true reference sites, but such sites were almost completely lacking in the entire catchment (Nilsson et al. 2007). Finally, even though neighboring tributaries may appear similar, all watersheds are truly unique in their capability to produce and sustain salmonid populations, making any comparisons in restoration responses between streams complex and potentially misleading. General conclusions about the effectiveness of habitat restoration have hence been difficult to make.

Although profound changes to the physical habitat can easily be verified after restoration, decisive positive responses in the biotic community have been more difficult to determine (Nilsson et al. 2015, 2017). With regards to juvenile salmonids, some sites have responded spectacularly well, with almost 20-fold increases in yearling densities per 100 m^2 compared to before restorations. Other sites have not responded equally well, and in some sites, densities even seem to decrease a few years after restoration. The reasons for these results are not entirely clear, but likely more time is needed to allow weak populations to gradually build up in numbers. Indications for this interpretation are newly compiled monitoring data showing tributaries restored in the beginning of the millennium just now starting to respond strongly in terms of increased densities of juvenile trout (D. Holmqvist, personal communication). To allow populations to build up after restoration, they should be spared from fishing harvest. However, stricter regulations may not be popular among FROs, which sometimes expected fishing to become much better immediately after restoration. Implementation of stricter regulations (e.g., size and bag limits, banning bait fishing, or complete fishing closure) has been one of the most challenging messages to communicate to FROs. The experience suggests that regulations have to be firmly accepted by the FROs to be effective; otherwise, they will not be obeyed and hence will be ineffective. Getting the FROs to understand why the regulations are needed has been key, and the best scenario has been to have the FROs suggest the regulations themselves, preferably before the restoration has even begun. A general view by the managers have been to consider many tributaries primarily as nursery habitat for juveniles and spawning habitats for migrating adult trout, and hence, because of their critical sensitive nature to the population, essentially wanting to ban fishing in them altogether. In a few cases, small, restored tributaries have been able to sustain an exclusive trophy-based fishery, but so far, this circumstance has been very limited throughout the catchment. Policies such as catch and release have gradually gained momentum among the FROs and are increasingly being implemented. The most successful example is a fishery on large adfluvial Brown Trout in the alpine region of the Vindel River, where a strict no-kill fishery likely has helped sustain a rapidly expanding tourism focused on sportfishing, which generates substantial income and employment in the region.

REFLECTIONS AND LESSONS LEARNED

For the young generation growing up along the Vindel River during the 1980s and 1990s, the fishing was likely considered to be fairly good. They knew that their river was unique and one of the few large free-flowing rivers left in northern Europe, and they had heard all about the famous fight to save the river from damming during the 1960s. Since then, local and regional governments constantly communicated about the potential to expand sportfishing tourism as a way to generate new jobs for the region. The catchment was in good shape according to the National Board of Fisheries. Many of the tributaries had Brown Trout densities around 3–5 yearlings/100 m^2, which was considered normal for northern boreal streams. Few salmon were caught in the river. A few old men caught some dark-colored salmon every autumn by letting their gill nets drift slowly over deep pools. Instead, youths fished for Grayling and Brown Trout, often by standing with their fishing rods on the large stone piers and dipping their baited hook down the steep edges, just as their fathers and grandfathers had done before them. Fishing was cheap, and on a good day they could catch a dozen fish between 15 and 20 cm, just enough to fill the fishing bag. Sometimes they could also catch large trout, but to do so they needed to fish just a few days after a stocking event, before all the fish mysteriously disappeared. This was the normal fishery at the time, what it has always been like in recent memory. Twenty years later, the situation is very different. The stone piers are gone, the gill nets and the baited hooks banned. Fishing in the tributaries is either prohibited or strictly regulated; in some cases, it is not allowed to keep a single fish. A license, which can be bought over the Internet, may cost up to €70 for 1 d of fishing, and text messages remind the angler to report the catch after every fishing trip. A good day of fishing may involve landing a few 3–4-lb (1.4–1.8 kg) resident wild Brown Trout or Grayling or battling a much larger Atlantic Salmon or adfluvial Brown Trout. This new situation is a result of 40 years of restoration work, with the most positive changes occurring during the past 20 years. The key factors instrumental in making this change possible are summarized below.

Share Your Vision, but Be Ready to Compromise

An overall effort in the restoration work has been trying to convince the local residents to embrace and accept that their river and the fishery should change. The arguments from the managers have become more and more convincing as positive results from the ongoing restoration efforts in the river or tributaries can be presented as examples. Listening to landowners, and not challenging their decisions by referring to environmental legislation, has helped reduced conflicts and complaints, even though it sometimes may have resulted in delayed or suboptimal restoration at some locations. The strategy has been to temporarily skip sites when met by strong opposition from FROs and to instead focus on sites where approval was easily obtained. This approach has been effective as it has reduced conflicts and complaints. Another challenge has been to engage local residents to support the vision that the Vindel River could become a productive Atlantic Salmon river. The number of salmon that entered the river during the 1970s and 1980s had been so few that the notion of the Vindel River as a major salmon river was slowly fading among the public and politicians. The "salmon-race" in the late 1990s brought that back by creatively making the public aware of, and engaged in, the spectacular spawning migration that happened every year. A shared vision among managers and FROs has been important for the FAB to credibly represent FROs in court, and when lobbying toward politicians and agencies. The process to get FROs and local residents on board to support the transformation of the Vindel River would not have been possible without massive dissemination of information and thousands of hours of meetings with local residents on the ground at their land.

Striving to establish a collaborative environment with the power company meant accepting the fact that the power station would not be torn down and that hydropower plays an increasingly important role as the world gradually converts to fossil-free energy production. The goal has never been to drastically reduce the power production in favor of salmon, but rather to find best possible solutions for how largescale hydropower can coexist with viable salmon runs. Adopting such policy has definitely helped building a better relationship with the power company over the years and helped restoration work move forward.

Aim for Complete Restoration

The habitat restoration conducted during the 1980s proved insufficient and misguided. The ecological and more holistic approach to restoration came first after the universities became involved, which influenced the work by sharing the latest theories and philosophies from restoration sciences. These ecological restoration projects became increasingly aggressive as the managers learned more about the scale and effects of the timber-floating modifications and received increased support from the landowners. Often, the old restoration projects from the 1980s had to be redone as they had not placed enough excavated substrate back into the river. Later restoration projects hence aimed to be complete, not requiring supplementary future restoration efforts.

Organize to Become Stronger

Although early restoration efforts (1980s) largely can be considered failures in terms of improving the habitat for fish, they did initiate the organizational groundwork for unifying residents into FROs along the river. This achievement came to be key for much of the later work and was essential to the success seen today. The creation of the FAB, which came to represent thousands of united FROs along the river, has been particularly important.

The Dubious Value of Stocking

The early restoration projects put large effort and resources into stocking programs to boost weak populations, and millions of yearlings and smolts of trout and salmon were stocked in the catchment over the years. Most of stocking seem to have had little, if any, lasting effects on the fish populations in terms of abundances or genetic introgression (D. Palm, personal data). Today, most stocking in the catchment is related to research-based programs and only involve fry or fertilized eggs. Very recent success with stocking fertilized eggs deposited in gravel-filled plastic crates, which are deployed in the river during early spring to promote a natural timing of hatching, has so far been promising in terms of first-year survival, especially in tributaries where trout populations have been slow to respond to habitat restoration. Although this new approach clearly suggests that stocking could have a value when using early life stages, it is still too early to evaluate the method with regards to promoting viable, self-reproducing populations.

Perseverance Will Be Rewarded

Despite the bad prospects confronting the managers in the beginning of the 1990s, with two major restoration projects unable to stop the dwindling decline of the Vindel River salmon, a few committed individuals and organizations kept up the energy and pushed toward renewed efforts to restore salmon and trout populations. This steady push brought the Vindel River to public and governmental attention and kept the organizational network intact during this difficult time. Swedish membership in the EU, the result of a very close national election, completely transformed conditions overnight, not only with regards to habitat restoration funding, but also in the way the offshore fishery and hydropower could be regulated. Thanks to keeping the local organizational structure intact and active, managers could quickly act on the new opportunities EU membership provided and became pioneers in whole river restoration work.

Choose Your Management Battles

Taking on a whole river restoration project can seem daunting as it often involves a multitude of aspects, especially when the target is anadromous salmon where the ocean phase may be in geographical regions way beyond control of the local fishery. The decision to keep focusing on issues regional to the Vindel River, such as passage problems around the dam and the habitat restoration of floatways, may seem narrow-minded, especially as the offshore fishery prevented most Atlantic Salmon from ever returning back to the river. Still, apart from gradually improving the situation in the river, the work kept the public and policymakers aware of the river and the problem with its fish. When the offshore fishery finally started to be regulated, the Vindel River was ready to host the returning fish, quickly boosting it to one of the most productive salmon rivers in the Baltic Sea. The apparent injustice in spending millions of euros to improve the habitat in the river while the majority of Atlantic Salmon were never able to return to the river due to offshore fishery could likely have been used more effectively as an argument to win sympathy to the cause. Today, the coastal trap-net fishery is the only commercial fishery still targeting salmon in Sweden. Instead of actively confronting these fishermen to have them reduce their harvest, managers have started to negotiate buying their quota to allow all returning salmon to reach the river.

Adapt Your Restoration Goals and Practices to New Knowledge

Examples of sites increasing 20-fold in juvenile Brown Trout densities after habitat restoration instantly made the old guidelines of 3–5 fish/100 m^2 from the National Board of Fisheries obsolete and triggered a debate about what the actual production maximum could be and, hence, what goals should be adopted. In the past couple of years, reports of very high densities from certain sites along the river (>200 yearlings/100 m^2) have been surprising and suggest that we may have been too cautious when speculating about maximum production capacities in the catchment. Throughout the restoration work in the Vindel River, the lack of solid reference information regarding the prehistoric state of fish and habitat has forced managers to constantly adjust their views and preconceptions of what to expect from the river after restoration. Questioning established facts has been the first step in such a process, critically reviewing statements and so-called "best practices" being promoted by agencies and by scientists. Being at the forefront of ecological restoration of boreal rivers, no one really had the answers to the problems and questions that would arise during the work. Adopting an adaptive approach, where restoration goals and practices were updated as new knowledge was gained, has been a successful strategy. Regular meetings where managers, researchers, and County Administrative Board officials discussed current issues about ongoing restoration has also been vital as input to the adaptive process.

The Value of Research

The involvement of the universities has been valuable throughout the restoration process of the Vindel River. Researchers can often provide up-to-date knowledge of modern restoration practices and have extensive understanding of how to design and conduct evaluation programs. Hence, engaging universities a few years before a restoration project begins may be advisable so solid evaluations can be planned and conducted. Funding may also sometimes be easier to acquire if universities are involved as partners. For the universities, participating in the various Vindel River restoration projects has been rewarding, not only due to the challenging and relevant research questions posed, but also because it provided opportunities to acquire funding for students and other researchers. For the Vindel River projects, the close collaboration with regional universities also gave credibility when combating unsubstantiated statements by the power company.

The Ripple Effects from Pioneering Projects

The Vindel River restoration projects represent pioneering efforts affecting much of the river restoration work elsewhere in Sweden. The experience coming from these pioneering projects is now instrumental in helping guide current efforts to restore other rivers in northern Sweden. Also, the skills gained to apply for and manage EU-based projects have today been transferred to a new generation of managers and mastered to such a degree that the new river restoration projects in the region have budgets unheard of in northern Europe.

CONCLUDING REMARKS

When summarizing the restoration work done in the Vindel River over the past 40 years and the positive development of the salmon and trout populations that has followed, the work can rightly be called a success story. Comparing the status of fish and habitat 40 years ago and today, the transformation of the catchment is remarkable by any standard. Nothing has been more important for the success than the dedication and spirit of a handful of committed individuals and organizations working relentlessly to improve fish and habitat conditions in the river. In the annals of the Vindel River, their work should be regarded equally important for the preservation of the beauty and soul of the river as the heroes involved in saving the river from hydropower development in the 1960s. Still, the work is not yet complete, if ever. As most habitat has been restored and the upstream migration problems around the dam are reaching a satisfactory solution, attention is now directed to other aspects. Establishing safe downstream passage of kelts and smolts through the dam is the next challenge, and solutions will likely not be cheap. Recent work is also focusing on softer targets, such as increasing the diversity in the FRO-organizations along the river by engaging more youths, women, and immigrants in the fishery and restoration work.

New threats are also rising on the horizon. Worrying signs of increased salmon die-offs due to disease, possibly related to a renewed outbreak of M74 syndrome, has negatively affected the runs over the past 2 years. In the absence of a commercial offshore fishery, an unregulated recreational trolling fishery has quickly expanded on the feeding areas in the Baltic Sea, potentially catching as many salmon as the entire commercial quota annually (ICES 2016). Post-smolt mortality is still very high, and the underlying cause is unknown. In addition, the

Vindel River sea trout have not responded as positively as salmon to the restoration work, and at the present we lack sufficient knowledge of what the problem is. Further, we have yet to see any true economic boom resulting from the restoration work in the form of increased tourism and employment in the majority of the catchment. Cost–benefit studies have hence concluded that even slight reductions of the power production at Stornorrfors to further enhance the Atlantic Salmon run may not result in a net benefit for society at large (Håkansson 2009). The response to these issues should be to continue working tirelessly to improve the situation for the salmon and trout amid an uncertain future. Hopefully, when the 30-year revision of this chapter is published, we will report that the Vindel River is approaching its maximum production capacity with tens of thousands of Atlantic Salmon ascending the river every year and that the management is now so well understood that one no longer manages the Vindel River salmon as a single population, but instead as multiple different subpopulations spatially and temporally distributed along the river. Hopefully, in the future we will be able to tell how the downstream migrating smolts and kelts now safely pass the hydropower dam in Stornorrfors and how the current work on re-establishing sea trout by large-scale egg stocking proved to be the key to get the production going in slowly responding tributaries. Perhaps we will also report about legislative battles that followed when private landowners tried to disband the FRO organization to gain more direct profit from the economically valuable recreational salmon fishery that developed on their property, a scenario that may very well occur in the wake of a growing salmon population. Until then, we can stay assured that managers and researchers will do their very best to let the Vindel River reach its full potential as a beautiful and productive river.

ACKNOWLEDGMENTS

We thank Umeå municipality, Vattenfall AB, and the County Administrative Board of Västerbotten for providing data and giving input on the chapter. We also thank the editors and reviewers for improving the chapter with valuable edits and comments.

REFERENCES

Alanärä, A., and L. Persson. 2017. Slutrapport projekt "RETURLAX". [Final report for project "Returning Salmon" project.] Energiforsk, Stockholm.

Andersson, T. 1988. Havsöring. [Seatrout.] Swedish Board of Fisheries, PM 1988 part 1, Härnösand. (In Swedish.)

Andersson, M., P. Molin, B. Nilsson, and K. Söderkvist. 1998. Expedition Varzuga – Naturlaxens referensälv? [Varzuga exedition – should the Varzuga River in Russia be used as a reference river when restoring rivers for wild salmon in Sweden?] County Administrative Board, Västerbotten, Umeå, Sweden.

Byström, C. 1868. Om fisket i Umeå elf och skärgård. [On fishing in the Ume River and the coastal archipelago.] Lantbruks-Akademiens Tidskrift (The Journal of the Swedish Agricultural Academy), Stockholm.

CAB (County Administrative Board, Västerbotten). 1995. Projekt Västerbottenslax. [The "Salmon in Vasterbotten, Sweden" project] County Administrative Board, Final Report 1988-1994, Message 8:1995, Västerbotten, Umeå, Sweden.

Christensen, O., and P. O. Larsson. 1979. Review of Baltic salmon research: a synopsis compiled by the Baltic Salmon Working Group. International Council for the Exploration of the Sea, Cooperative Research Report 89, Charlottenlund, Denmark.

Degerman, E., K. Leonardsson, and H. Lundqvist. 2012. Coastal migrations, temporary use of neighbouring rivers, and growth of sea trout (*Salmo trutta*) from nine northern Baltic Sea rivers. ICES Journal of Marine Science 69:971–980.

Eriksson, L., P. Rivinoja, J. Östergren, I. Serrano, and S. Larsson. 2008. Smolt quality and survival of compensatory stocked Atlantic Salmon and Brown Trout in the Baltic Sea. Swedish University of Agricultural Sciences, Department of Wildlife, Fish and Environmental Studies, Report 62, Umeå, Sweden. (In Swedish.)

Englund, P. 1999. The Swedish banking crisis: roots and consequences. Oxford Review of Economic Policy 15:80–97.

FAB (Ume/Vindel River Fishery Advisory Board). 2015. Vindel River LIFE. Vunna erfarenheter av vattendragsrestaurering. [The "Vindel River Life" project: what we have learned from habitat restorations in rivers.] Report to the European Commission. Umeå Universitet, Umeå, Sweden.

Ferguson, J. W., J. G. Williams, and E. Meyer. 2002. Recommendations for improving fish passage at the Stornorrfors Power Station on the Umeälven, Umeå, Sweden. National Oceanic and Atmospheric Administration, National Marine Fisheries Service, Northwest Fisheries Science Center, Seattle, Washington.

Ferguson, J. W., G. R. Ploskey, K. Leonardsson, R. W. Zabel, and H. Lundqvist. 2008. Combining turbine blade-strike and life cycle models to assess mitigation strategies for fish passing dams. Canadian Journal of Fisheries and Aquatic Sciences 65:1568–1585.

Gardeström, J., D. Holmqvist, L. E. Polvi, and C. Nilsson. 2013. Demonstration restoration measures in tributaries of the Vindel River catchment. Ecology and Society [online serial] 18(3):8.

Gisler, N. 1751. Rön om laxens natur och fiskande i de norrländska älfvarna. [Observations of salmon life, and salmon fishing, in the northern rivers of Sweden.] Kongl. Vetenskaps Academiens Handlingar (Documents from the Royal Academy of Science), Stockholm.

Hagen, D., K. Svavarsdottir, C. Nilsson, A. K. Tolvanen, K. Raulund-Rasmussen, Á. L. Aradóttir, A. Fosaa, and G. Halldorsson. 2013. Ecological and social dimensions of ecosystem restoration in the Nordic countries. Ecology and Society [online serial] 18(4):34.

Håkansson, C. 2009. Costs and benefits of improving wild salmon passage in a regulated river. Journal of Environmental Planning and Management 52:345–363.

Hindar, K., J. A. Hutchings, O. H. Diserud, and P. Fiske. 2010. Stock, recruitment and exploitation. Pages 299–325 *in* O. Aas, S. Einum, A. Klemetsen, and J. Skurdal, editors. Atlantic Salmon ecology. Wiley-Blackwell, Hoboken, New Jersey.

ICES (International Council for the Exploration of the Sea). 2016. Report of the Baltic Salmon and Trout Assessment Working Group (WGBAST), 30 March–6 April 2016, Klaipeda, Lithuania. ICES, ICES CM 2016/ACOM:09, Copenhagen, Denmark.

Johansson, E. S. 2008. Debatten om Vindelälven. [The debate concerning the Vindel River.] Västerbotten 2008(2):53–58.

Johansson, S., H. Hedin, L. Nilsson, B. Bengtsson, Ö. Karlström, G. Laestander, B. Grönlund, and L. Rosenqvist. 1987. Fiskprojekt Vindelälven Laisälven. Verksamhetsrapport för 1981–1986 och förslag till handlingsprogram för 1987–1991. ["The Vindel River and the Lais River" fishery project. Activity report for 1981–1986 and proposal for a plan of action for 1987–1991.] Fishery Board, Report 1987, Västerbotten County, Umeå, Sweden.

Kallio-Nyberg, I., E. Jutila, E. Jokikokko, and I. Saloniemi. 2006. Survival of reared Atlantic Salmon and sea trout in relation to marine conditions of smolt year in the Baltic Sea. Fisheries Research 80:295–304.

Karlsson, L., and Ö. Karlström., 1994. The Baltic Salmon (*Salmo salar* L.): its history, present situation and future. Dana 10:61–85.

Karlström, Ö. 1999. Development of the M74 syndrome in wild populations of Baltic Salmon (*Salmo salar*) in Swedish rivers. Ambio 28:82–86.

Larsson, S. 2015. Västerbottens vildlaxälvar. [Wild salmon rivers in Västerbotten.] County Administrative Board, Message 2:2015, Västerbotten, Umeå, Sweden.

Larsson, M., and E. Sparrevik. 2009. Återskapande av vandringsmöjligheter för havsvandrande fisk: ekologiska effekter och verksamhetspåverkan. [Restoration of migration routes for anadromous fish: ecological effects and impact on hydropower operation. Vattenfall AB, Report 2792700, Solna, Sweden.

Leonardsson, K., Y. Belyaev, P. Rivinoja, and H. Lundqvist, 2005. Modelling upstream migration of Atlantic Salmon as a function of environmental variables. Chapter III *in* P. Rivinoja, editor. Migration problems of Atlantic Salmon (*Salmo salar* L.) in flow regulated rivers. Doctoral dissertation. Swedish University of Agricultural Science, Umeå.

Leonardsson, K., R. Karlsson, J. Nilsson, and H. Lundquist. 2013. Uppströmsvandring i den restaurerade gamla älvfåran samt funktionskontroll av ny fisktrappa i Norrfors (Umeälven) under laxens vandringssäsong 2012–2013. [Upstream migration in the restored old river bed, and evaluation of the new fish ladder in Norrfors (Ume River) during the salmon migration 2012–2013.] Swedish University of Agricultural Sciences, Report 7, Umeå.

Lindberg, D.-E., K. Leonardsson, and H. Lundqvist. 2016. Path selection of Atlantic Salmon (*Salmo salar*) migrating through a fishway. River Research and Applications 32:795–803.

Lindroth, A. 1963. Salmon conservation in Sweden. Transactions of the American Fisheries Society 92:286–291.

Lindroth, A. 1970. Inverkan av Stornorrfors kraftstation på lax- och havsöringsbestånden i Ume älv t.o.m. 1969. [The effect of the Stornorrfors hydropower station on the populations of salmon an dsea trout in the Ume River up until 1969.] Vattenfall AB, PM 16/2 1970, Solna, Sweden.

Lundberg, P., D. Holmqvist, S. Westbergh, J. Östergren, and E. Törnlund. 2005. Miljöåterställning av nationalälvarna Vindel- och Piteälven. [Restoration of the national rivers Vindel and Pite.] Vindel River Fishery Advisory Board, Final Report 2002-2005, Vindeln, Sweden.

Lundqvist, H., K. Leonardsson, D. Lindberg, S. Westbergh, Å. Forssén, and G. Hellström. 2014. Laxens nedströmsvandring mot fiskavledare till Stornorrfors fisktrappa i Umeälvens nedre del. [The downstream migration of salmon towards guidance structure at the Stornorrfors fish ladder in the lower part of the Ume River.] Swedish University of Agricultural Sciences, Department of Wildlife, Fish and Environmental Studies, Report 1:2014, Umeå.

Lundqvist, H., K. Leonardsson, J. Williams, J. Östergren, G. Hellström, and Å. Forssén. 2015. Problematiken kring flergångslekare av lax, öring och steelhead i vildlaxförande stora flödesreglerade vattendrag. [The problem concerning conservation of multispawners of wild salmon, trout and steelhead in large regulated rivers.] Swedish University of Agricultural Sciences, Department of Wildlife, Fish and Environmental Studies, Report 8:2015, Umeå.

Lundqvist, H., P. Rivinoja, K. Leonardsson, and S. McKinnell. 2008. Upstream passage problems for wild Atlantic Salmon (*Salmo salar* L.) in a regulated river and its effect on the population. Hydrobiologia 602:111–127.

Mäntyniemi, S., A. Romakkaniemi, J. Dannewitz, S. Palm, T. Pakarinen, H. Pulkkinen, A. Gårdmark, and O. Karlsson. 2012. Both predation and feeding opportunities may explain changes in survival of Baltic Salmon post-smolts. ICES Journal of Marine Science 69:1574–1579.

McKinnell, S. M. 1997. A retrospective on Baltic Salmon (*Salmo salar* L) biology and fisheries. Nordic Journal of Freshwater Research 73:73–88.

McKinnell, S. M., and Ö. Karlström. 1999. Spatial and temporal covariation in the recruitment and abundance of Atlantic Salmon populations in the Baltic Sea. ICES Journal of Marine Science 56:433–443.

Montén, E. 1985. Fish and turbines; fish injuries during passage through power station turbines. Vattenfall AB, Stockholm, Sweden.

Montén, E. 1988. Fiskodling och vattenkraft: en bok om kraftutbyggnadernas inverkan på fisket och hur man sökt kompensera skadorna genom främst fiskodling. [Aquaculture and hydropower: the effect of hydropower on fisheries, and how to mitigate negative impacts of fish stocking programs.] Vattenfall, Vällingby, Sweden.

Nilsson, C. 2007. Återställning av älvar som använts för flottning: en vägledning för restaurering. [Guidelines for restoration of rivers that have been used as floatways.] Swedish Environmental Protection Agency, Report 5649, Stockholm.

Nilsson, C., A. L. Aradottir, D. Hagen, G. Halldorsson, K. Høegh, R. J. Mitchell, K. Raulund-Rasmussen, K. Svavarsdóttir, A. Tolvanen, and S. D. Wilson. 2016. Evaluating the process of ecological restoration. Ecology and Society [online serial] 21(1):41.

Nilsson, C., R. Jansson, B. Malmqvist, and R. J. Naiman. 2007. Restoring riverine landscapes: the challenge of identifying priorities, reference states, and techniques. Ecology and Society [online serial] 12(1):16.

Nilsson, C., F. Lepori, B. Malmqvist, E. Törnlund, N., Hjerdt, J. M. Helfield, D. Palm, J. Östergren, R. Jansson, E. Brännäs, and H. Lundqvist. 2005. Forecasting environmental responses to restoration of rivers used as log floatways: an interdisciplinary challenge. Ecosystems 8:779–800.

Nilsson, C., L. E. Polvi, J. Gardeström, E. Hasselquist, L. Lind, and J. M. Sarneel. 2015. Riparian and in-stream restoration of boreal streams and rivers: success or failure? Ecohydrology 8:753–764.

Nilsson, C., J. M. Sarneel, D. Palm, J. Gardeström, F. Pilotto, L. E. Polvi, L. Lind, D. Holmqvist, and H. Lundqvist. 2017. How do biota respond to additional physical restoration of restored streams? Ecosystems 20:144–162.

Nilsson, J., R. Gross, T. Asplund, O. Dove, H. Jansson, J. Kelloniemi, K. Kohlmann, A. Löytynoja, E. E. Nielsen, T. Paaver, and C. R. Primmer. 2001. Matrilinear phylogeography of Atlantic Salmon (*Salmo salar* L.) in Europe and postglacial colonization of the Baltic Sea area. Molecular Ecology 10:89–102.

Öhman, B. 2000. Lax- och nättingfisket i Umeälven. [The salmon and lamprey fishery in the Ume River.] Västerbotten 2000(1):18–25.

Öhman, B. 2007. Avradslax och husbehovsfisk : en studie över fiskets betydelse för de äldsta fasta bosättningarna vid Vindelälven. [The importance of salmon and subsistence fishing for the earliest settlements along the Vindel River.] Unpublished report available from http://libris.kb.se/bib/10556507. (In Swedish.)

Östergren, J., and P. Rivinoja. 2008. Overwintering and downstream migration of sea trout (*Salmo trutta* L.) kelts under regulated flows—northern Sweden. River Research and Applications 24:551–563.

Palm, D. 2007. Restoration of streams used for timber floating. Doctoral dissertation. Swedish University of Agricultural Sciences, Umeå.

Palm, D., E. Brännäs, F. Lepori, K. Nilsson, and S. Stridsman. 2007. The influence of spawning habitat restoration on juvenile Brown Trout (*Salmo trutta*) density. Canadian Journal of Fisheries and Aquatic Sciences 64:509–515.

Palm, D., F. Lepori, and E. Brännäs. 2010. Influence of habitat restoration on post-emergence displacement of Brown Trout (*Salmo trutta* L.): a case study in a northern Swedish stream. River Research and Applications 26:742–750.

Polvi, L. E., C. Nilsson, and E. M. Hasselquist. 2014. Potential and actual geomorphic complexity of restored headwater streams in northern Sweden. Geomorphology 210:98–118.

Ramqvist, P., M. Backe, and L. Forsberg. 1985. Hällristningar vid Stornorrfors. [The rock carvings at Stornorrfors.] Västerbotten 1985(2):29–49.

Rivinoja, P. 2005. Migration problems of Atlantic Salmon (*Salmo salar* L.) in flow regulated rivers. Doctoral dissertation. Swedish University of Agricultural Sciences, Umeå.

Rivinoja, P., K. Leonardsson, and H. Lundqvist. 2006. Migration success and migration time of gastrically radio-tagged v. PIT-tagged adult Atlantic Salmon. Journal of Fish Biology 69:304–311.

Rivinoja, P., S. McKinnell, and H. Lundqvist. 2001. Hindrances to upstream migration of Atlantic Salmon (*Salmo salar*) in a northern Swedish river caused by a hydroelectric power-station. River Research and Applications 17:101–115.

Romakkaniemi, A., I. Perä, L. Karlsson., E. Jutila, U. Carlsson, and T. Pakarinen. 2003. Development of wild Atlantic Salmon stocks in the rivers of the northern Baltic Sea in response to management measures. ICES Journal of Marine Science 60:329–342.

Rudberg, P. M., M. Escobar, J. Gantenbein, and N. Niiro. 2014. Mitigating the adverse effects of hydropower projects: a comparative review of river restoration and hydropower regulation in Sweden and the United States. Georgetown International Environmental Law Review 27:251–273.

Sandström, H., K. A. Wallin, B. Svensson, S. Westbergh, and H. Lundqvist. 2002. Förbättring av lax-och havsöringbestånden i Vindelälven – ett åtgärdsförslag. [Proposal to restore the salmon and sea trout populations in the Vindel River.] Vindel River Fishery Advisory Board, Vindeln, Sweden.

SCB (Statistics Sweden). 2012. Official statistics of Sweden. Available: www.scb.se. (September 2018).

Siira, A., J. Erkinaro, P. Jounela, and P. Suuronen. 2009. Run timing and migration routes of returning Atlantic Salmon in the northern Baltic Sea: implications for fisheries management. Fisheries Management and Ecology 16:177–190.

SMHI (Swedish Meteorological and Hydrological Institute). 1979. Streamflow records of Sweden. Liber, Stockholm.

Svedin, U., and B. Hagerhall Aniansson. 2012. Society and the environment: a Swedish research perspective. Springer, Dordrecht, Netherlands.

Swedish Law 1981:533. Lag om fiskevårdsområden. [Law concerning fishery right owners.] Government Offices of Sweden, Svensk författningssamling 1981:533, Stockholm.

Törnlund, E. 2002. Flottningen dör aldrig: bäckflottningens avveckling efter Ume- och Vindelälven 1945–70. [Timber floating will never die: the abolition of timber floating in the Ume and Vindel River tributaries 1945–70.] Doctoral dissertation. Umeå University, Umeå, Sweden.

Törnlund, E., and L. Östlund. 2002. Floating timber in northern Sweden: the construction of floatways and transformation of rivers. Environment and History 8:85–106.

Törnlund, E., and L. Östlund. 2006. Mobility without wheels: the economy and ecology of timber floating in Sweden 1850–1980. The Journal of Transport History 27:48–70.

Vindel River LIFE. 2016. LIFE08NAT/S/266. Vindel River LIFE, final report. Umeå University, Department of Ecology and Environmental Science, Umeå, Sweden.

Zetterstedt, J. H. 1833. Resa genom Umeå lappmarker i Vesterbottens län, förrättad år 1832. [Journey through Swedish Lapland in Umea, Vasterbotten County 1832.] N. M. Lindh, Örebro, Sweden.

Recovery of Willamette River (Oregon, USA) Fish Assemblages: Successes and Remaining Threats

Robert M. Hughes*
Amnis Opes Institute and Oregon State University
2895 SE Glenn Street, Corvallis, Oregon 97333, USA

Brian L. Bangs
Oregon Department of Fish and Wildlife
28655 Highway 34, Corvallis, Oregon 97333, USA

Stanley V. Gregory
Department of Fisheries and Wildlife, Oregon State University
104 Nash Hall, Corvallis, Oregon 97331, USA

Paul D. Scheerer
Oregon Department of Fish and Wildlife (retired)
24886 Grange Hall Road, Philomath, Oregon 97370, USA

Randall C. Wildman
Department of Fisheries and Wildlife, Oregon State University
104 Nash Hall, Corvallis, Oregon 97331, USA

Jeffrey S. Ziller
Oregon Department of Fish and Wildlife
3150 East Main Street, Springfield, Oregon 97478, USA

Abstract.—Fish assemblages in the Willamette River basin (Oregon) were once substantially degraded by water pollution, channelization, dams, nonnative fish, and conversion of natural forest and savanna to agriculture and urbanization. Restoration actions have included basinwide waste treatment, physical habitat rehabilitation, recovery of the Oregon Chub *Oregonichthys crameri* to stable status, and stocking reductions of nonnative fish to protect native fish. State and federal sewage treatment regulations and funding, federal endangered species regulations and funding, and reduced funding and support for stocking nonnative trout led to those rehabilitated fish assemblages. Periodic fish and habitat monitoring has documented the following improvements in fish assemblages: (1) decreased occurrences of pollution-tolerant species and increased occurrences of pollution-sensitive species and native main-stem species, (2) increased number of abundant Oregon Chub populations, and (3) persistence of resident native Rainbow Trout *Oncorhynchus mykiss*. Notably, no known extinctions of native fish species have occurred in the Willamette River, water quality index scores in the lower river have improved from poor to fair, and water quality in the upper

* Corresponding author: hughes.bob@amnisopes.com

river remains good to excellent. In conclusion, enactment of laws and regulations for environmental protection and the collective actions of state and federal agencies, tribes, municipal governments, universities, land trusts and conservation groups, watershed councils, and private landowners have led to a substantially rehabilitated river. However, population and economic growth, climate change, nonnative fish, winter steelhead (anadromous Rainbow Trout) and spring Chinook Salmon *O. tshawytscha* listings, a superfund site, channel alterations, toxic substances, poor fish passage at dams, and altered flow regimes remain challenges. Four lessons learned are that (1) pollution control and improved water quality and flows are essential to the recovery and persistence of native fish populations, (2) recovery of endangered species is achievable but requires knowledge of their life history needs, (3) the greater ecological fitness of native stocks facilitates their persistence, and (4) research and monitoring, combined with public communication and collaboration, are essential for habitat and native fish assemblage rehabilitation.

INTRODUCTION

The Willamette River is Oregon's largest river with a 29,800-km^2 basin area and a 680 m^3/s mean annual discharge (Figure 1). Its runoff per unit drainage area is the greatest of all large U.S. rivers (USGS 1949); typical summer and winter discharges are 250 and 1,800 m^3/s, respectively. The basin holds 70% of Oregon's human population, and its three largest cities are located along the river (Hulse et al. 2002).

Before European settlement in the 1840s, the main-stem Willamette River drained a 70 km by 200 km mix of wet and dry prairie and oak savanna that was maintained by Calapooia Indian fires (Johannessen et al. 1971). The river was bordered by floodplain forests ranging from 2 to 9 km wide, and the channel was anastomosing and meandering, incorporating multiple shaded channels, alcoves, oxbow lakes, islands, regularly flooded floodplains, and many snags and log jams (Sedell and Froggatt 1984; Benner and Sedell 1997; Hulse et al. 2002). During summers, cool water flowed from mountain tributaries, as well as hyporheic (subsurface) inflows along the floodplain that were driven by winter flooding and floodplain aquifer recharging (Hughes et al. 2005; Wohl et al. 2017). Water was clear and aquatic primary production was low because of relatively low nutrient concentrations, extensive wetlands, and extensive channel shading along upper-river side channels (Sedell and Froggatt 1984; Hughes et al. 2005; Wohl et al. 2017). In winter, the floodplain, backwaters, side channels, and snags provided fish with refuge from frequent large floods, as well as rearing and feeding sites in other seasons (Benke et al. 1985; Lehtinen et al. 1997; Wohl et al. 2017). The floodplain also was ideal habitat for the North American beaver *Castor canadensis* (Scott 1924). Such conditions were once common in and along many floodplain rivers throughout the conterminous United States prior to European settlement and river channelization (Gammon 2005; Wohl et al. 2017).

Beavers were largely removed by the 1840s by fur trappers, and intensive agriculture in the fertile Willamette Valley and Willamette River channel alterations followed shortly thereafter. To increase navigation and commerce, the U.S. Army Corps of Engineers removed more than 68,000 snags and had channelized the river before 1910 (Benner and Sedell 1997; Hulse et al. 2002). More than 26% of the total length of all channel types was lost (Sedell and Froggatt 1984; Gregory 2008), area of the active channel decreased by 22% (Gregory et al. 2002a; Greg-

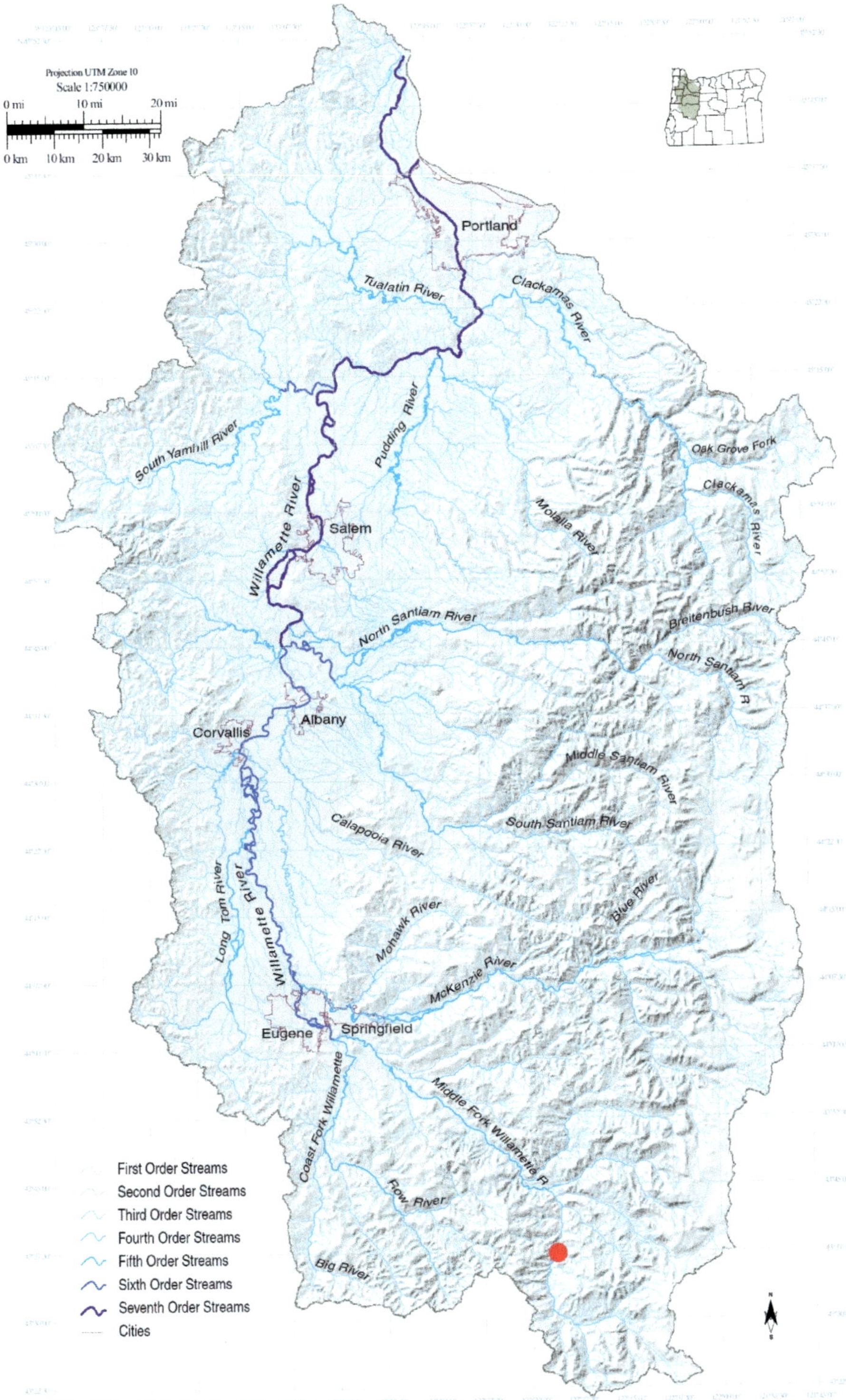

FIGURE 1. Willamette River basin depicting the main stem and its major tributaries and cities (Hulse et al. 2002). A red dot indicates the location of the Hills Creek Dam.

ory 2008), 26% of the channel was rip-rapped (Gregory et al. 2002b; Hughes et al. 2005), and much of the lower main-stem shoreline was lined with bulkhead (Hughes and Gammon 1987). Using digitized historical U.S. Geological Survey topographic maps, Dykaar and Wigington (2000) estimated an 80% reduction in island and bar area in the upper river from 1910 to 1988. The lower reaches of the Willamette River's major tributaries experienced similar transformations. Agriculture became the dominant land use in the valley and was followed by canneries, meat packing plants, sawmills, pulp and paper mills, and urbanization, all of which disposed untreated wastes into the river and its tributaries. In addition, agriculture and urbanization have altered or removed many headwater streams and wetlands in the valley by ditching, tiling, and burial, as is common elsewhere (Colvin et al. 2009; Hughes et al. 2014).

By 1950, the river experienced critically low dissolved oxygen levels, extensive filamentous bacteria *Sphaerotilus natans* beds, and floating and benthic sludge deposits in the summer months. From the 1940s to the 1960s, large dams were built on the Willamette's major tributaries for hydropower and for flood control to replace the lost floodplain storage, reducing flood discharges by 30–40% (Gregory et al. 2007). Those dams trapped sediment and reduced sediment delivery from major tributaries to the main-stem Willamette River by 60–90% (O'Connor et al. 2014) and altered channel and floodplain formation (Wallick et al. 2013). By 1938, ambient toxicity tests resulted in the death of juvenile Chinook Salmon *Oncorhynchus tshawytscha* in less than 2 min when placed in Willamette River water (Smith 1940). Loss of appropriate backwater spawning and rearing habitats, altered flow regimes, and the introduction of nonnative centrarchid sport fishes led to the Oregon Chub *Oregonichthys crameri*, a native endemic species, being federally listed as an endangered species in 1993. Additionally, dams on the major tributaries, major federal and state salmon stocking programs, and massive stocking of hatchery Rainbow Trout *Oncorhynchus mykiss* threatened the persistence of native resident trout in the upper Middle Fork Willamette River in the 1960s and 1970s as well as salmon and steelhead (anadromous Rainbow Trout) in the upper Willamette River and its major tributaries.

Together with remaining threats, this chapter provides a review of the problems encountered and actions taken by the collective work of state and federal agencies, municipal governments, tribes, universities, land trusts, conservation groups, watershed councils, and private landowners to successfully rehabilitate the Willamette River.

GOALS AND OBJECTIVES FOR RESTORATION

In response to the watershed changes described above, Oregon fishery biologists collectively pursued several fishery rehabilitation goals: (1) restore main-stem and major tributary fish assemblages to support all native fishes known to have ranged historically in those rivers, as mandated by Oregon's Native Fish Conservation Policy (Oregon Administrative Rule 635-007-0502); (2) restore Oregon Chub, winter steelhead, fall Chinook Salmon, and Coho Salmon *Oncorynchus kisutch* populations to be sufficient to delist them from federal endangered or threatened status; and (3) recover productive resident native Rainbow Trout and Bull Trout *Salvelinus confluentus* fisheries in the upper Middle Fork Willamette River.

PROBLEMS ENCOUNTERED

Main-Stem Willamette Fish Assemblages

The Willamette River cleanup was not easily attained. The river was an open sewer like many others across the United States in the pre-Clean Water Act (1972) era (Robbins 2004). Sewage plants, pulp and lumber mills, food processing plants, and woolen mills dumped untreated wastes directly into the Willamette River and major tributaries. Dissolved oxygen decreased and bacterial counts increased abruptly in the lower 120 km below Salem (Rogers et al. 1930). The absence of dissolved oxygen hindered or temporally stopped migration of salmon and steelhead through the 1950s (Fish and Wagner 1950). The cleanup pitted Oregon's mythical promises of limitless freedom and economic opportunity, followed by industrial capitalism and jobs, against environmental quality, environmental justice, environmental restraints, and public health (Robbins 2004). Installation of municipal and industrial waste treatment systems required millions of dollars of infrastructure and dedicated waste treatment plant managers.

Additionally, nonnative Midwestern U.S. sport fishes were introduced into the Willamette River basin to provide fisheries for anglers (Lampman 1946). Many of these Midwestern fishes thrived in the off-channel habitats and flood-control reservoirs and reduced or extirpated native fishes through competition and predation in some of those habitats (Hughes and Herlihy 2012). Once established, these sport fishes became part of the fish assemblage and fisheries that continue today. Concerted efforts are ongoing to modify streamflows and reconnect currently isolated floodplain habitats to favor native fishes over nonnative fishes. Currently, popular fisheries exist for nonnative Smallmouth Bass *Micropterus dolomieu* and Largemouth Bass *M. salmoides* and other nonnative warmwater species on the river and its tributaries. These fisheries have created conflict because anglers focused on nonnative fishes oppose actions that would reduce those fisheries in favor of native fish assemblages, including salmonids. Management agencies are tasked with the difficult tradeoffs of balancing the mandated protection of wild native fishes with public demand for popular sport fisheries for hatchery salmonids and nonnative game fishes (Maas-Hebner et al. 2016). One way was to select a set of basins from which hatchery fish will be eliminated and to which no further introductions will occur (Maas-Hebner et al. 2016).

Oregon Chub

Oregon Chub (Figure 2) was historically widespread in the Willamette basin (Long 1982; Long and Bond 1983; Pearsons 1989). Loss of suitable off-channel habitat resulted from the channel alterations, dams, and wetland drainage for agriculture. Additionally, many off-channel habitats contained nonnative fish predators and competitors. Those factors led to the listing of Oregon Chub as endangered under the Endangered Species Act (ESA) in 1993 (U.S. Office of the Federal Register 1993). Lack of targeted surveys prior to the late 1970s resulted in incomplete knowledge of their modern and historical distribution, habitat, and life history requirements.

Resident Trout in the Upper Middle Fork Willamette River

Prior to the construction of Hills Creek Dam (Figure 1) in 1957 to 1962, native resident Rainbow Trout were common in the upper Middle Fork Willamette River and provided a popular sport fishery (Oregon Department of Fish and Wildlife [ODFW], unpublished data). When the dam and reservoir were proposed, the U.S. Fish and Wildlife Service and the Oregon State Game Commission both recognized that nongame fish species would likely proliferate in the reservoir. Therefore, a broad-spectrum piscicide (rotenone) treatment project to remove unwanted Largescale Sucker *Catostomus macrocheilus* and North-

FIGURE 2. Oregon Chub (photo by Dave Herasimtschuk, Freshwaters Illustrated 2011; Corvallis, Oregon)

ern Pikeminnow *Ptychocheilus oregonensis* was conducted prior to closing the gates of the dam. This project was funded by the Oregon State Game Commission and the U.S. Army Corps of Engineers and was implemented in the fall of 1960.

Although the chemical treatment project did not completely remove all fish species, it was very successful in removing most individuals of the large main-stem native species such as Largescale Sucker, Northern Pikeminnow, Rainbow Trout, and Bull Trout. After the removal project, the reservoir and tributaries were heavily stocked with hatchery-reared Rainbow Trout because of its popularity as a sport fish. The Middle Fork Willamette River and Hills Creek received an average of more than 60,000 large (catchable size) trout per year during the 1960s and 1970s. Releases of hatchery-reared trout escalated angling pressure and probably resulted in excessive harvest of remaining native trout. Angling regulations during this era did not distinguish between wild and hatchery trout, allowing both to be included as part of the harvest limit. Releases of hatchery trout from other basins not only encouraged harvest of wild trout, they also facilitated potential genetic introgression (gene transfer) between native and introduced fish stocks. In 2001, ODFW began introducing sterile triploid Rainbow Trout in the upper Middle Fork Willamette River. The cumulative negative effects of the fish removal project, heavy fish stocking with subsequent heavy angler pressure, and potential for introgression with hatchery fish left fish managers with little hope for the survival of the native Rainbow Trout of the Middle Fork Willamette River above Hills Creek Dam.

MANAGEMENT SOLUTIONS AND ACTIONS

Main-Stem Willamette Fish Assemblages

The Willamette River cleanup was driven by citizen outrage and individual leadership. The Isaac Walton League, the Oregon Wildlife Federation, and fisheries and water resource professors at Oregon State College (now Oregon State University) gave presentations about Willamette River water quality to Willamette Valley community groups and the Oregon legislature in the 1930s. Many citizens wrote letters to the editors of Willamette Valley newspapers. Portland Mayor Joseph Carson marched with youths demanding clean rivers. The Oregon Legislature passed a river cleanup bill, but Governor Charles Martin vetoed it under industry pressure. As a result, a citizen-initiated referendum was passed by 78% approval in 1938 to establish the Oregon Sanitary Authority and create water quality standards, pollution controls, and waste treatment plants to improve water quality and fisheries. The Isaac Walton League and Oregon Wildlife Federation also produced a film on pollution in the Willamette River and presented it to public audiences along the Willamette River and other parts of the state (Smith 1940). The visual effects of seeing juvenile salmon die quickly when placed in Willamette River water angered many citizens, but political intransigence and World War II delayed full implementation of the 1938 referendum until 1950. In 1950 and 1960, the Oregon Sanitary Authority mandated wastewater treatment by mills and secondary sewage treatment by cities along the lower river. A journalist, Tom McCall, developed and presented a persuasive television documentary in 1962 about the river that propelled him to the Oregon governorship in 1966. Under McCall's progressive Republican leadership, Oregon established the Oregon Department of Environmental Quality, strengthened pollution controls and basinwide secondary waste treatment, and obtained state and federal infrastructure funding, and Willamette River water quality improved significantly by the late 1960s (Starbird and Georgia 1972; Robbins 2004).

Oregon's success with the Willamette River was a model for the 1972 federal Clean Water Act. The Willamette River Greenway Act was passed in 1973, leading to increased protection of riparian zones. In 1991, Northwest Environmental Advocates sued the city of Portland because its stormwater overflows contaminated the lower Willamette River with raw sewage. That successful lawsuit led to collection, storage, and treatment of urban stormwater runoff by most major Willamette Valley cities, aided by local, state, and federal funding (Robbins 2004).

More recently, recovery of main-stem fish assemblages has focused on increased numbers of citizens, agencies, and conservation groups collaborating to rehabilitate river channels and floodplain forests, improve land uses, and implement regulatory programs to upgrade waste treatment and improve physical habitat structure. The Willamette Basin Task Force formed in 1997 leading to the Willamette Restoration Initiative (1998) and the Willamette Special Investment Partnerships (2006).

In the late 1990s, a number of distinct spring Chinook Salmon and winter Steelhead populations were listed as threatened under the federal ESA, including two in the Willamette River. The National Marine Fisheries Service subsequently issued a biological opinion for the U.S. Army Corps of Engineers, Bonneville Power Administration, and U.S. Bureau of Reclamation ("the Action Agencies"; NMFS 2008) prescribing a number of actions to prevent further jeopardizing those populations.

Currently more than 100 agencies, watershed groups, land trusts, and nongovernmental organizations contribute to aquatic conservation and rehabilitation of the main-stem Willamette River. Without a guiding vision, such disparate attempts to rehabilitate rivers and streams were not likely to succeed because they were economically and ecologically uncoordinated (Palmer et al. 2005). Even after the major changes in river management resulting from the 1938 referendum and the McCall administration in the 1960s, native fish assemblages of the lower river were still highly altered and water quality was poor.

In 2005, only one floodplain rehabilitation project existed along the entire length of the main-stem Willamette River; however, in 2008, a funding partnership was formed between the Oregon Watershed Enhancement Board and the Meyer Memorial Trust, a philanthropic organization based in Oregon. In 2009, the model Watersheds Program, with additional funding from the Bonneville Power Administration, joined the partnership to rehabilitate seven major watersheds in the basin. Between 2008 and 2016, those entities invested more than US$40 million in rehabilitating degraded habitats, protecting ecologically healthy habitats, and monitoring status and trends of physical, chemical, and biological resources in the main-stem Willamette River. The Meyer Memorial Trust convenes a Within Our Reach conference biennially to share information and coordinate activities among the partners and interested citizens. Presently, 20 major rehabilitation projects occur along the main-stem Willamette River, some of which include more than 2,000 acres (809.3 ha), and 56 smaller rehabilitation projects are occurring in the tributaries. Because of overlapping linkages, more than 1,300 projects are funded by the partnership. In 2010, only 270 individual landowners were participating in rehabilitation efforts; in 2018, the partnership now includes more than 1,150 landowners.

Proposals for these projects are ranked on their contribution to recovering native fish assemblages, ESA-listed fish populations, and ecosystem processes such as nutrient and sediment retention, riparian vegetation growth, and large wood deposition. The National Marine Fisheries Service biological opinion (NMFS 2008) required the Action Agencies to mitigate the effects of their dams (migration barriers, habitat degradation, and reduced genetic diversity from hatcheries). Those mitigation measures included upstream and downstream fish passage at dams, coldwater releases from dams, naturalized downstream flows, projects to increase habitat complexity, screening of irrigation diversions, and increased genetic diversity of hatchery fish. Recently, it was reported that brief annual draining of a reservoir on the Middle Fork Willamette River facilitated juvenile Chinook Salmon passage and eliminated nonnative predatory fish from the reservoir (Murphy et al. 2019). In 2012, the International River Foundation awarded the Theiss International Riverprize to the Willamette River Partnership in recognition of the innovative approaches to collaborative river rehabilitation described above (International River Foundation 2012).

Periodic fish assemblage sampling has been conducted on the main stem since 1944 (Dimick and Merryfield 1945). Additional sampling sites after that study were located near those used by Dimick and Merryfield (1945), notably above and below major pollution sources and tributaries, and were to achieve sampling intervals of 20 km or less. Such a sampling design facilitated detection of anomalous conditions and spatial ecological patterns in an economically efficient manner (Hughes and Gammon 1987; Hughes et al. 2012). Fish were sampled in the summer after adult migrating spring Chinook Salmon had passed upstream so as to limit harm to these fish (Dimick and Merryfield 1945; Hughes and Gammon 1987; Farr and Ward 1993; Tetra Tech 1994; Friesen 2005; Hughes et al. 2005; LaVigne et al. 2008; Hughes and Herlihy 2012; Gregory et al. 2014; Williams 2014). The 1944 study collected fish by "seining, angling, set lines, and even in a few cases by dynamiting" (Dimick and Merryfield 1945); subsequent sampling was by boat and backpack electrofishing or by seining.

The differences in sampling sites, sampling gear, and sampling effort preclude the comparison of species abundance data in this chapter. Instead, native species richness, nonnative species richness, sensitive species richness, percent nonnative species, and percent tolerant species were compared (discussed below). Nonetheless, these metrics can also vary as a result of sampling differences and sampling effort (Cao et al. 2001; Dauwalter and Pert 2003; Hughes and Herlihy 2007; Hughes and Peck 2008; LaVigne et al. 2008). Origin and tolerance classifications were based on Zaroban et al. (1999) and Whittier et al. (2007).

Oregon Chub

No targeted surveys for Oregon Chub were conducted prior to the late-1970s; however, surveys in the late-1970s indicated that Oregon Chub distribution had apparently been severely reduced (Long 1982; Long and Bond 1983; Pearsons 1989). Consequently, this species' historical distribution is based on incidental catches from the surveys of Snyder (1908) and Dimick and Merryfield (1945) and from specimens in the Oregon State University Fish Collection. In response to a petition for ESA listing, Oregon State University (Markle et al. 1991) and ODFW (Scheerer 2002) initiated surveys of historically occupied habitats and other similar, suitable habitats in the Willamette River basin.

Recovery efforts were guided by the pre-ESA listing Oregon Chub Conservation Agreement (USFWS 1998) and the Oregon Chub Recovery Plan (USFWS 1998). The conservation agreement, which was signed by ODFW, the U.S. Fish and Wildlife Service (USFWS), U.S. Army Corps of Engineers, Bureau of Land Management, U.S. Forest Service, and Oregon Parks and Recreation Department, detailed tasks and criteria for species recovery. Those documents detailed habitat enhancement and protection activities, species and habitat monitoring, landowner legal protections, and adaptive management approaches. In addition, because of their interest in preserving all native fishes, tribal governments purchased Oregon Chub habitat from private landowners and set it aside for recovery efforts.

Since 1992, fish surveys have been conducted at 1,700 unique habitats to monitor Oregon Chub abundance and distribution and to locate undocumented populations. As a result, annual abundance estimates were recorded for more than two decades in habitats occupied by Oregon Chub, and life history information was collected to guide recovery and restoration efforts (Scheerer and McDonald 2003). Additionally, Oregon Chub was reintroduced into suitable habitats within its historical range.

Prior to conducting reintroductions on private properties, USFWS issued a conservation agreement in 1994 and two safe harbor agreements in 2002 and 2008 to private landowners agreeing to reintroduce Oregon Chub on their properties. Those three agreements protected landowners from legal actions under the ESA. In 2009, USFWS developed a programmatic safe harbor agreement permitting ODFW to enroll multiple landowners as cooperators under a single agreement (U.S. Office of the Federal Register 2009). The programmatic agreement simplified and expedited the process of establishing new populations of Oregon Chub and enabled ODFW to establish five additional populations.

Two additional research initiatives were implemented to facilitate Oregon Chub recovery. In 2009, ODFW initiated a floodplain study to evaluate the relationship between the regulated tributary discharges and temperatures and the water levels, water temperatures, connectivity, and fish assemblages in off-channel Oregon Chub habitats (USFWS 2008). The U.S. Fish and Wildlife Service studied the levels of genetic variation and population structuring among Oregon Chub populations (DeHaan et al. 2012) and assessed genetic diversity and similarity between introduced Oregon Chub populations and their donor populations (DeHaan et al. 2016).

Resident Trout in the Upper Middle Fork Willamette River

As noted above, native Rainbow Trout above Hills Creek Dam were in danger of being replaced by hatchery Rainbow Trout stocks through gene transfer and overharvest, especially during the first few decades after dam construction. Therefore, in 2008, ODFW developed a reintroduction program for Rainbow Trout above Hills Creek Dam. The program objectives were to (1) determine if Rainbow Trout that recolonized the basin above the dam since the 1960s were of hatchery or wild origin, (2) determine locations in the upper Willamette River basin where native Rainbow Trout continue to persist, and (3) develop a genetic management plan for Rainbow Trout above Hills Creek Dam.

If the results of this study suggested that naturally produced Rainbow Trout in the upper Middle Fork Willamette basin were the result of the colonization of hatchery fish, then the feasibility of re-establishing native Rainbow Trout genetic material from an adjacent basin was to be assessed. Genetic samples of naturally produced Rainbow Trout in the upper Willamette River basin and from hatchery stocks known to have been released into the upper Willamette River basin were collected and provided to a genetics laboratory for analysis. For each fish, genomic DNA was examined for 22 microsatellite loci (Williamson et al. 2010).

Similarly, the Bull Trout population of the Middle Fork Willamette River was exposed to the chemical fish removal project and potential for interactions with hatchery Rainbow Trout releases. Although the remnant Bull Trout population was subjected to heavy angling pressure, the population persisted in the Middle Fork Willamette River for several decades before disappearing around 1990.

ASSESSMENT AND EVALUATION

Main-Stem Willamette River Fish Assemblages

The occurrence and numbers of native fish species and sensitive fish species have increased throughout the main stem since 1945 (Hughes and Gammon 1987; Hughes et al. 2005; Gregory et al. 2014; Williams 2014). The Willamette River main stem now supports 29 native species and 24 nonnative species (Appendix). Total native fish species richness in 1945 was 0–1 and 1–12 at sites in the lower and upper river, respectively (Dimick and Merryfield 1945), in contrast to 6–9 and 12–19 now (Figure 3).

Pollution-sensitive fish species now occur downriver of river kilometer 160, whereas none were found there in 1945. The percent of tolerant species has decreased systemwide since 1945; however, nonnative species richness occurrence at sites and the percent of nonnative fish species have increased since 1945. Juvenile Chinook Salmon rear along the full length of the main stem, and the subyearling life history now makes a substantial contribution to returning adults in some years (Schroeder et al. 2016). Bayley and Baker

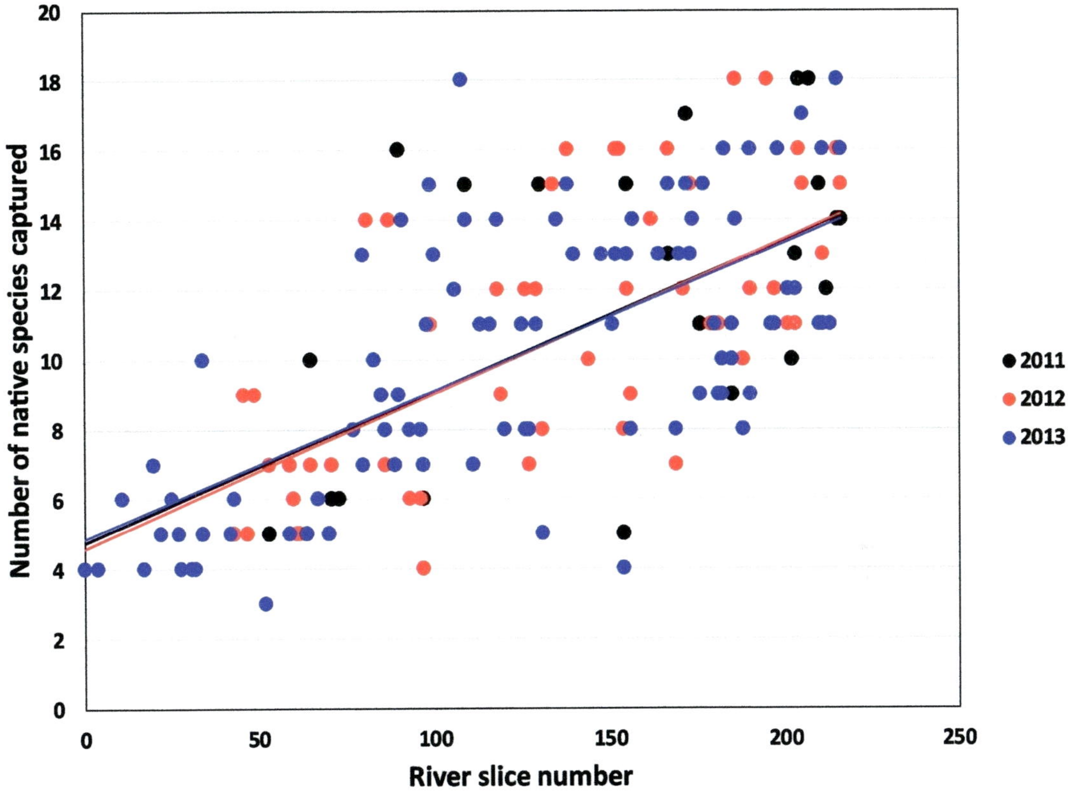

FIGURE 3. Main-stem Willamette River native fish species richness as a function of river kilometer (river slice) and recent year. Lower river sites are at the bottom left. Note, one might expect greater fish species richness in the lower river based on its greater discharge (McGarvey and Terra 2015). However, the lower Willamette flows slowly through Coast Range basalts in an incised, mostly deep channel with bed substrates dominated by clay and sand—versus the more complex meandering floodplain channel dominated by gravel and cobble substrates of the more rapidly flowing middle and upper river (Hughes and Gammon 1987; Wallick et al. 2013). Those differences in natural and current physical habitat structure likely explain some of the reduced species richness and the different species of the lower Willamette River (Pringle et al. 1988; Fausch et al. 2002).

(2002) reported that gravel mines on regularly inundated floodplains provided highly productive juvenile Chinook Salmon rearing habitat when graded and off-channel ponds are connected to the river. A resident native Cutthroat Trout *Oncorhynchus clarkii* and Rainbow Trout fishery now also exists in the upper main stem. In addition, various forms of water contact recreation such as swimming and boating are commonly pursued in this reach.

Unfortunately, listed salmonid species (Chinook Salmon, Coho Salmon, and steelhead) have declined, whereas encounter frequencies of nonnative Common Carp *Cyprinus carpio*, Smallmouth Bass, and Largemouth Bass have increased. The increased occurrence frequency of Smallmouth Bass in the past 20 years creates a potential predation hazard to native fish (Friesen 2005; Hughes and Herlihy 2012). Smallmouth Bass were not encountered in the middle or upper main-stem Willamette River or its tributaries in 1983 but is now a dominant species in the lower Calapooia River, common to other

Willamette River tributaries, and occasionally encountered in the upper and middle main stem. Though the number of nonnative taxa has increased, native species comprise most of the total fish caught in the main-stem Willamette River. In sampling from 2011 to 2013, native species comprised 93% of the total fish captured (Gregory et al. 2014; Williams 2014).

We are aware of no other changes during the period of study, so the likely explanations for the observed fish assemblage changes are improved water quality, more accurate species identifications, and increased sampling effectiveness. However, the greater efficiency of electrofishing is highly unlikely to have doubled the species that were encountered, and concurrent electrofishing and beach seining in 1999 produced nearly the same species (authors' unpublished data). Recovery of native species has occurred!

Regulatory requirements, municipal improvements, and better land-use practices have improved water quality in the Willamette River in recent decades. The Oregon water quality index (OWQI), a composite of several traditional water quality parameters, has been monitored at nine stations along the length of the main stem since 1980. The OWQI integrates eight subindices based on water temperature, dissolved oxygen, biochemical oxygen demand, pH, ammonia+nitrogen, total phosphorus, total solids, and *Escherichia coli* into a single number (Cude 2001, 2005). Each subindex converts a measurement into a rating from 10 (worst) to 100 (ideal) based on a nonlinear regression curve (Cude 2001). Water quality in the lower river has improved from poor to fair, whereas water quality in the upper river remains good to excellent (Figure 4; Habitat Technical Team of the Willamette Action Team for Ecosystem Restoration 2015; Mulvey 2015). Again, good news over the recent past for Willamette River fishes.

The trends for improved water quality in the lower and middle river are likely to increase with continued improvements in stormwater runoff controls, sewage treatment, and industrial discharge treatment. In addition, diffuse pollution control actions in urban areas (Molina 2014; Yeakley and Dunham 2014) together with Oregon's statewide land use planning regulations (ODLCD 2010), its forest practices regulations (ODF 2012), and the Oregon plan for salmon and watersheds (Oregon Government 2010) continue to improve watershed conditions, especially in smaller tributaries. Implementation of rehabilitation actions and pollution abatement as part of the Lower Willamette River Superfund Program by the U.S. Environmental Protection Agency also will increase the likelihood that future water quality throughout the Willamette River will better support native fish assemblages.

Oregon Chub

Recently, previously unknown populations of Oregon Chub have been discovered and new ones have been established. Between 1992 and 2013, ODFW discovered 59 previously undocumented Oregon Chub populations and successfully introduced Oregon Chub into 19 locales within its historical range (Bangs et al. 2013). Half of the introduced populations are on privately owned property where voluntary agreements guide management. At the time of ESA delisting of the Oregon Chub in 2015, ODFW documented 42 abundant Oregon Chub populations (≥500 adult fish as defined in the recovery plan), up from just a handful at the time of listing in 1992 (Figure 5). The floodplain study provided data relating regulated tributary discharges and temperatures with off-channel habitat conditions (temperatures, water levels, and connectivity) and fish assemblages, identifying locales where Oregon Chub can coexist with nonnative fishes (Bangs et al. 2014).

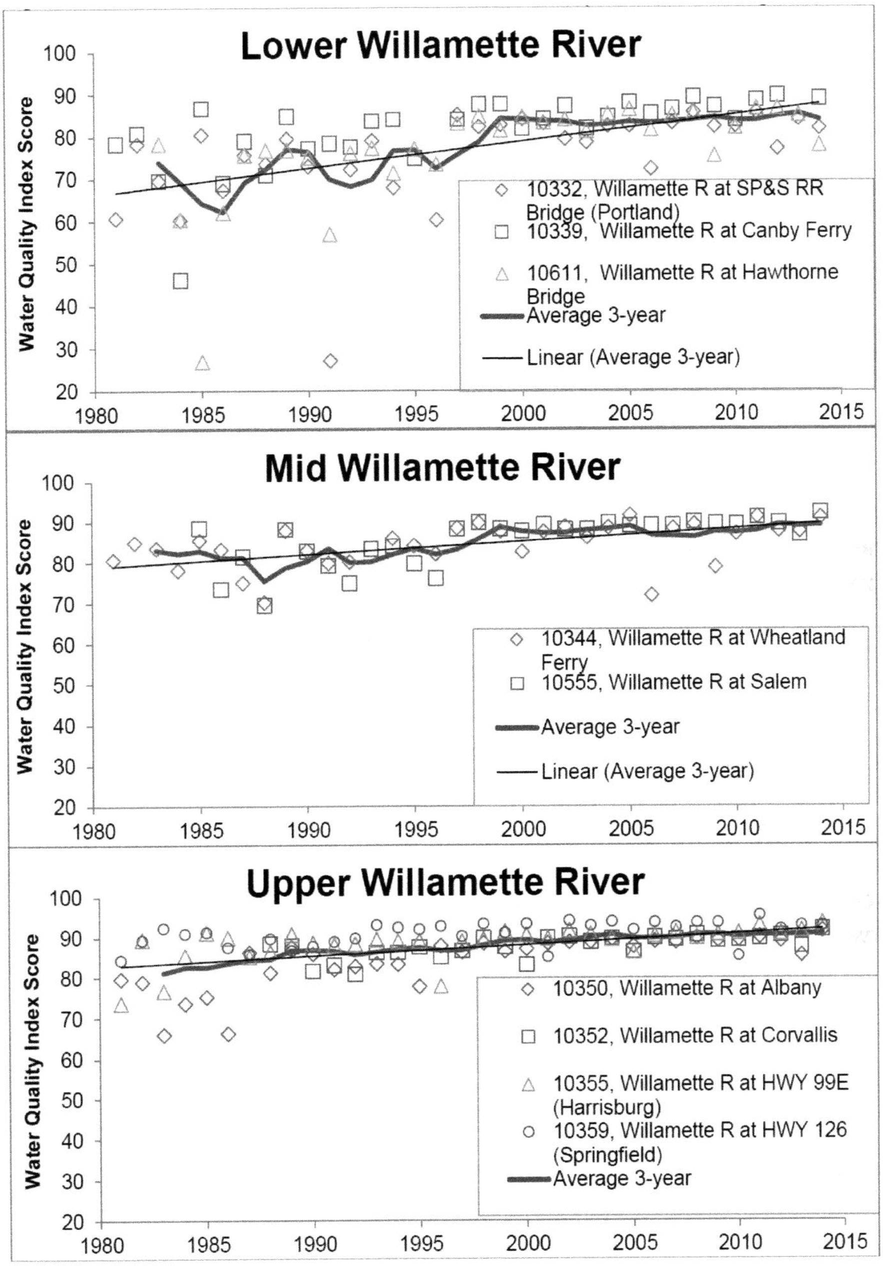

FIGURE 4. Water quality index scores for the lower, middle, and upper main-stem Willamette River (Mulvey 2015).

The genetics of naturally occurring Oregon Chub indicated that populations were organized at the tributary level and that most populations did not have reduced levels of genetic variability, despite their isolation for 50–70 years since the flood control dams were constructed (DeHaan et al. 2012). The genetic variability of introduced populations was similar to that of the abundant, naturally occurring donor populations in the same subbasin. Also, when multiple donor populations were used, genetic variability

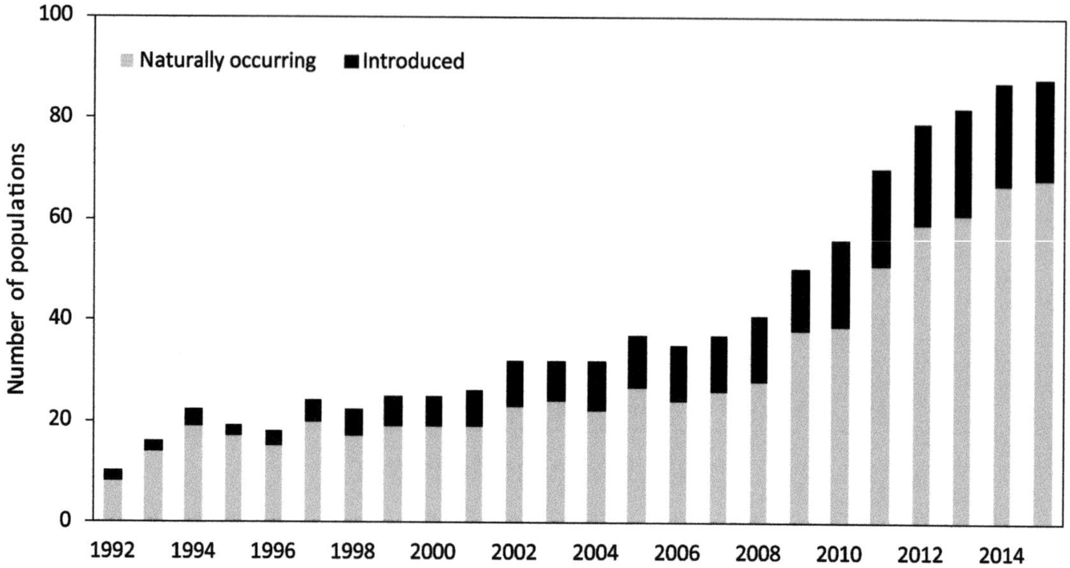

FIGURE 5. Number of known naturally occurring (gray) and introduced (black) Oregon Chub populations in the Willamette River basin, 1992–2015.

was higher in the introduced populations than in naturally occurring populations (DeHaan et al. 2016).

Resident Trout in the Upper Middle Fork Willamette River

Naturally reproduced and hatchery Rainbow Trout were genetically distinct based on comparisons of collections made by ODFW biologists and volunteers (Williamson et al. 2010; Figure 6). This analysis provided evidence that hatchery trout released into the Middle Fork Willamette River and other upper Willamette River tributaries had not introgressed with the native populations. A similar finding was reported for the Deschutes River basin by Matala et al. (2009).

The lack of introgression likely resulted from the high reproductive success of the locally adapted native stocks and the lower reproductive and survival traits of hatchery stocks poorly adapted for natural environments (Lynch and O'Hely 2001; Ford 2002; Marchetti and Nevitt 2003). Nonnative Rainbow Trout encroachment also exhibited "considerable ebb and flow in steep-gradient tributaries" to which native trout were evolutionarily adapted (Larson et al. 1995; Moore and Kulp 2019, this volume). In addition, native stocks and hatchery stocks of Rainbow Trout are separated in spawn timing (Kinunen and Moring 1978). For example, in the upper Middle Fork Willamette River, native Rainbow Trout spawn April–May, whereas hatchery stocks spawn October–December.

To further protect these native populations, ODFW has implemented a management plan that allows only sterile hatchery Rainbow Trout releases into the upper Middle Fork Willamette River basin. This strategy should ensure that future genetic introgression will not occur. Reintroduction of Bull Trout to the waters above Hills Creek Dam was initiated in 1996 and continues to this day. The project has been successful in bringing back two populations of reproducing wild Bull Trout to the watershed, totaling 20–40 adults per year (ODFW 2018). In addition, the current Northwest Forest Plan (FEMAT 1993) provides protections for streams draining federal forestlands by shifting manage-

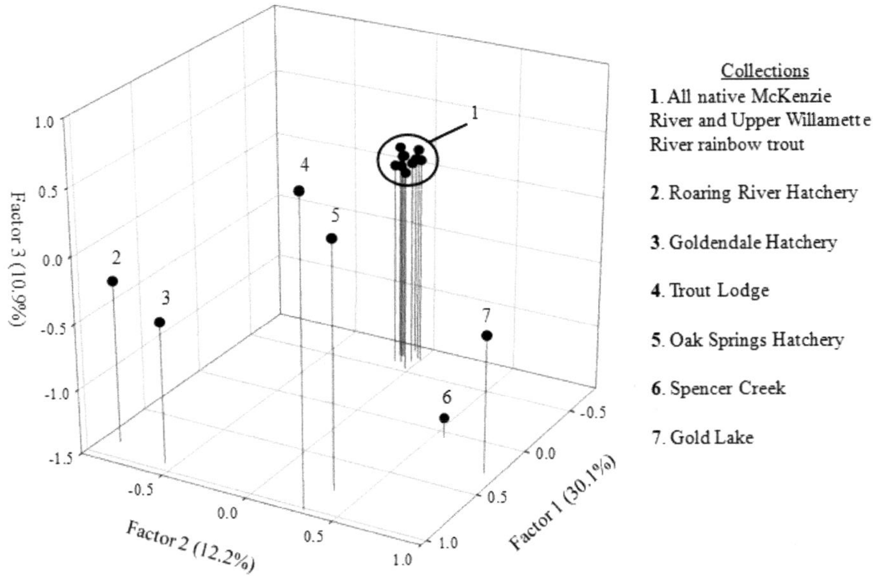

FIGURE 6. Factorial correspondence analysis plot of genetic composition variance (distance) in hatchery stocks known to have been planted versus natural origin collections of Rainbow Trout sampled from the McKenzie and Middle Fork Willamette rivers (from Williamson et al. 2010). The non-native stock basin origins were Roaring River (Santiam), Goldendale (Yakima), Trout Lodge (Hood), Oak Springs (Deschutes), Spencer Creek (Klamath), and Gold Lake (Rogue).

ment focus from primarily timber extraction to more of a science- and landscape-based ecosystem approach that includes an aquatic conservation strategy. Riparian reserves protect all floodplains, more than 100 m on both sides of perennial streams, and more than 50 m along intermittent streams. In addition, large blocks of forest are managed as late-successional reserves. These reserves protect the majority of Bull Trout habitat in the Willamette basin.

REMAINING THREATS

Unfortunately, some serious problems remain to challenge native fish assemblages and species of interest in the Willamette basin. Channel-forming high flows have been reduced from five per decade to one per decade (Hughes et al. 2005), leading to simplified physical habitat structure and sub-optimal conditions for native species. The Portland harbor continues to have low water quality with diffuse municipal, industrial, and agricultural pollutants present; the dissolved oxygen criterion was not met in the lower river in four of the past nine summers; and the main stem fails to meet temperature criteria for salmonid spawning and rearing. The key ESA biological opinion mitigation prescriptions for salmonids have not been fully implemented by the Action Agencies to mitigate the effects of their dams and hatcheries, thereby limiting wild Chinook Salmon populations (Schroeder et al. 2016) as well as winter steelhead and Pacific Lamprey *Entosphenus tridentatus* populations (Poirier et al. 2017). Many of the current permits regulating major industry and municipal polluters to the Willamette River require modifications to lessen threats from elevated temperatures and unsafe levels of toxics, microbes, pesticides, pharmaceuticals, and personal care products (Maas-Hebner et al.

2014; Yeakley et al. 2014; Segerstrom et al. 2017). A fish consumption advisory exists because of high levels of mercury in piscivorous Northern Pikeminnow, Smallmouth Bass, and Largemouth Bass (Peterson et al. 2002). Thornbrugh et al. (2018) classified the mostly agricultural watersheds and urban areas of the Willamette Valley as having moderate to low integrity; they are likely to remain sources of diffuse pollutants because agricultural controls are largely voluntary (ODA 2017; ODEQ 2017).

Last but far from least, climate change and population and economic growth are ticking time bombs for the native ichthyofauna during this century (Limburg et al. 2011; Mote et al. 2014; Ripple et al. 2017). In the northwestern United States, mean annual air temperatures are projected to increase by up to 5°C, summer precipitation is projected to decrease by 30%, and the mountain snowpack is projected to be reduced by 60% between the present and the 2090s (Mote et al. 2014; Jaeger et al. 2017). Such changes will lead to markedly reduced summer water availability and higher water temperatures, with severe consequences for salmonid health and ranges (Mantua et al. 2010; Isaac et al. 2018). Equally important, human societies in general are poorly suited for adapting to such changes because of conflicts and coordination gaps among differing governments, limited understanding and stalled social learning by most citizens, and multiple financial constraints and high transaction costs for allocating reduced amounts of water (Jaeger et al. 2017; Oberlack and Eisenack 2017). Finally, little recognition exists that economic and population growth drive climate change and other pressures leading to biodiversity losses (Limburg et al. 2011; Crist et al. 2017; Ripple et al. 2017). Clearly, the amazing capacity of human cultures to adapt will be tested over the next few decades.

LESSONS LEARNED

Main-Stem Willamette River Fish Assemblages

Pollution control and improved water quality and flow are essential to the recovery of native fish populations. The most critical step in recovering Willamette River fish assemblages was the installation and maintenance of municipal sewage and industrial discharge treatment plants, along with subsequent installation of stormwater collection, holding, and treatment. Together with dams providing low flow supplementation, the pollution abatement infrastructure greatly improved main-stem water quality. Knowledge of the water quality needs of sensitive fish species was essential for providing the flows and waste treatment needed to meet those requirements. However, the pollution history of recent decades and repeated calls for restoring the Willamette River in the 1930s, 1960s, and 2000s clearly indicate that continued vigilance and improvements regarding both diffuse and point source waste discharges and enforcement are needed. Future improvements must be generated by citizens and nongovernmental organizations within collaborative partnerships with agencies because currently insufficient legislative and agency support and funding exists as a result of competing demands for limited government funds (Hughes and Noss 1992; Hughes 1997; USFWS 1998; Palmer et al. 2005; Wood 2013; Meyer Memorial Trust 2015; Jaeger et al. 2017; Oberlack and Eisenack 2017; Thomas et al. 2019, this volume).

Research and monitoring, combined with public communication, are essential for native fish assemblage rehabilitation. The 1938 toxicity tests in the Willamette River were filmed and widely distributed to the public, leading to support for basinwide sewage treatment. Subsequent fish assemblage sampling along the main stem by Oregon State University, combined with ODFW and Oregon Department of Environmental Quality communication with the public (e.g., Mulvey et al. 2009; Mulvey 2015), helped provide support for installation of sewage and municipal stormwater runoff collection and treatment. Those studies also indicated the need for greater controls than were currently in place on agricultural diffuse pollutants. Differences in fish assemblage monitoring protocols reinforced the need to establish cost-effective and consistent methods among institutions to facilitate temporal and spatial comparisons of the fish assemblages over time (LaVigne et al. 2008; Curry et al. 2009; Mulvey et al. 2009). Such protocols and regular river monitoring have been implemented for decades by the states of Ohio (Yoder et al. 2005; Yoder et al. 2019, this volume) and Maryland (Klauda et al. 1998) and nationally by the U.S. Environmental Protection Agency (Hughes and Peck 2008; USEPA 2016).

A review by Palmer et al. (2005) identified five critical requirements for successful rehabilitation: (1) a specified guiding image of a more dynamic, healthy river; (2) measurable improvement of the river's ecological condition; (3) a self-sustaining river with minimal need for follow-up maintenance; (4) no lasting harm during project implementation; and (5) publicly available information from pre- and post-assessments and monitoring. Consistent with these requirements, rehabilitation partners in the Willamette River have developed several guiding visions of past change, current conditions, and future trajectories of a healthier river (Hulse et al. 2002; Wallick et al. 2013; Habitat Technical Team of the Willamette Action Team for Ecosystem Restoration 2015).

Although the requirements described above are critical for resource managers, scientists, and informed practitioners, they can be confusing or technically complicated for the general public. The Meyer Memorial Trust developed a Willamette River report card to describe the health of the main-stem Willamette River more clearly and interestingly for the citizens, communities, and schools along the river and in the region (Meyer Memorial Trust 2015). The report card developed a score for the upper, middle, and lower river based on 15 metrics of ecological health that were available from monitoring programs and considered relevant to the general public—water quality index, toxics, temperature, native fish richness, proportion of nonnative fish, juvenile Chinook Salmon abundance, bald eagle abundance, channel complexity, floodplain forest extent and condition, flow targets, peak flows, fish consumption advisories, tribal fishery condition, *Escherichia coli* occurrence, and harmful algal blooms. The upper and middle reaches received a grade of B and the lower reach received a grade of C+. The overall grade for the main-stem Willamette River was a B–. Though technical challenges and concerns exist with converting quantitative monitoring data into a letter grade, the communication value for the broader public is substantially greater when the letter grade is linked with explanations of the associated technical data. The report card is designed to be repeatable at decadal intervals if partners and agencies continue to collect the ecological information on which it is based, and this requirement provides a stimulus for future support of river monitoring efforts. A similar approach is employed by the U.S. Environmental Protection Agency in its National Aquatic Resource Surveys (e.g., Paulsen et al. 2008; USEPA 2016).

Oregon Chub

Recovery of endangered species is achievable, but for species having little background information, recovery requires research to understand their life history needs. Detailed studies of the natural variability in abundance of Oregon Chub, their habitat requirements and life history characteristics, the population responses of introduced Oregon Chub populations, relationships between regulated river flows and habitat conditions and fish assemblages in off-channel habitats, and genetic variability within and among Oregon Chub populations all contributed to its recovery. The factors that caused the declines of endangered species occurred incrementally over 150 years, and recovery from the cumulative impacts of habitat loss, impaired water quality, and nonnative species will be a gradual process.

Partnerships and collaboration among numerous institutions and private landowners were essential to recovery success. The development of agreements between state and federal agencies and private landowners strengthened working relationships on private lands (USFWS 1998). This collective effort by a very strong public–private partnership in the Willamette Valley clearly demonstrates that listed species can be recovered and delisted in a highly populated, extensively altered landscape—a testament to the resilience of the species and the innovative and cooperative thinking of biologists and other stakeholders.

Resident Trout in the Upper Middle Fork Willamette River

The greater ecological fitness of native stocks facilitates their persistence. Given all of the transgressions land and fisheries managers have imposed on the native Rainbow Trout of the upper Willamette River basin, we were fortunate. The native stock appears to have persisted in spite of chemical removal, dam building, instream habitat degradation, and introductions of the less-fit, nonnative hatchery fish. At the present time, most of the upper Middle Fork Willamette River basin is being managed for wild fish production and is also protected as national forest and wilderness. The outlook for these Rainbow Trout is optimistic as long as managers remain protective of this unique fish stock in the future by maintaining and improving fish habitat and avoiding stocking hatchery fish. Similar approaches are needed wherever wild fish conservation is the goal (Lynch and O'Hely 2001; Mclean et al. 2003; Moore and Kulp 2019).

CONCLUSIONS

Despite several remaining threats, the fish assemblages of the Willamette River main stem, a once-endangered species, and a valued wild trout fishery have recovered. Key characteristics of all successful rehabilitation and management programs are detailed adaptive management and effective monitoring plans. Such plans must (1) clearly identify current knowledge and major uncertainty gaps, (2) clearly express management expectations and uncertainties, (3) plan and apply monitoring programs for obtaining knowledge to resolve the identified uncertainty gaps, (4) improve predictive models with the improved monitoring information, (5) communicate widely in effective ways, and (6) modify forthcoming management actions based on the latest knowledge regarding the managed resources, the pressures and stressors affecting them, and citizen inputs (Runge 2011; Peters et al. 2014; Maas-Hebner et al. 2016). Those adaptive management plans must clearly state what shall and shall not happen when previously listed trigger points for decisions are attained. Monitoring designed to reduce critical knowledge gaps will not only help reduce those gaps, but also facilitate more efficient use of always-limited monitoring staff, staff time, and funds because only those assessments that improve long-term management will be a priority. Improved management can best occur if the new knowledge is applied to re-examine management alternatives, improve stakeholder engagement and collaboration, re-evaluate ambiguities, and alter future management choices as appropriate. Those future management choices are subsequently monitored, and the cycle resumes (Maas-Hebner et al. 2016).

ACKNOWLEDGMENTS

This chapter is dedicated to the memory of Kathy Maas-Hebner who facilitated our research and writing in innumerable ways. We also thank the countless public citizens, scientists, and managers that cared enough about the Willamette River ecosystem to contribute freely to its rehabilitation in myriad aspects. Critical comments of three anonymous reviewers and the copy editing of Debby Lehman improved the chapter contents and appearance.

REFERENCES

Bangs, B. L., P. D. Scheerer, and S. Clements. 2013. 2013 Oregon Chub investigations. Oregon Department of Fish and Wildlife, Fish Research Project EF-13 Annual Progress Report, Corvallis.

Bangs, B. L., P. D. Scheerer, and S. Clements. 2014. Effects of U.S. Army Corps of Engineers Willamette Project operations on Oregon Chub and other floodplain fishes (2009–2012). Oregon Department of Fish and Wildlife, Annual Progress Report, Corvallis.

Bayley, P. B., and C. Baker. 2002. Fish population observations in Endicott and Truax floodplain areas under restoration following aggregate mining in the Willamette River basin, Oregon (1998-01). Report to Willamette River Gravel Removal Restoration Fund Program by Oregon State University, Department of Fisheries and Wildlife, Corvallis.

Benke, A. C., R. L. Henry, III, D. M. Gillespie, and R. J. Hunter. 1985. Importance of snag habitat for animal production in southeastern streams. Fisheries 10(5):8–13.

Benner, P. A., and J. R. Sedell. 1997. Upper Willamette River landscape: a historical perspective. Pages 23–47 *in* A. Laenen and D. A. Dunnette, editors. River quality: dynamics and restoration. CRC Press, Boca Raton, Florida.

Cao, Y., D. P. Larsen, and R. M. Hughes. 2001. Evaluating sampling sufficiency in fish assemblage surveys: a similarity-based approach. Canadian Journal of Fisheries and Aquatic Sciences 58:1782–1793.

Colvin, R. G. R. Giannico, J. Li, K. L. Boyer, and W. J. Gerth. 2009. Fish use of intermittent watercourses draining agricultural lands in the upper Willamette River valley, Oregon. Transactions of the American Fisheries Society 138:1302–1313.

Crist, E., C. Mora, and R. Engelman. 2017. The interaction of human population, food production, and biodiversity protection. Science 356:260–264.

Cude, C. G. 2001. Oregon water quality index: a tool for evaluating water quality management effectiveness. Journal of the American Water Resources Association 37:125–137.

Cude, C. G. 2005. Accommodating change of bacterial indicators in long term water quality datasets. Journal of the American Water Resources Association 41:47–54.

Curry, R.A., R.M. Hughes, M. McMaster, and D. Zafft. 2009. Coldwater fish in rivers. Pages 139–158 *in* S. A. Bonar, W. A. Hubert, and D. W. Willis, editors. Standard methods for sampling North American freshwater fishes. American Fisheries Society. Bethesda, Maryland.

Dauwalter, D. C., and E. J. Pert. 2003. Effect of electrofishing effort on an index of biotic integrity. North American Journal of Fisheries Management 23:1247–1252.

DeHaan, P., B. Adams, P. Scheerer, and B. Bangs. 2016. Influence of introduction history on genetic variation in introduced populations: a case study of Oregon Chub. North American Journal of Fisheries Management 36:1278–1291.

DeHaan, P., P. D. Scheerer, R. Rhew, and W. R. Ardren. 2012. Analyses of genetic variation in populations of Oregon Chub, a threatened floodplain minnow in a highly altered environment. Transactions of the American Fisheries Society 141:533–549.

Dimick, R. E., and F. Merryfield. 1945. The fishes of the Willamette River system in relation to pollution. Oregon State College, Engineering Experiment Station Bulletin Series 20, Corvallis.

Dykaar, B. B., and P. J. Wigington, Jr. 2000. Floodplain formation and cottonwood colonization patterns on the Willamette River, Oregon, USA. Environmental Management 25:87–104.

Farr, R. A., and D. L. Ward. 1993. Fishes of the lower Willamette River, near Portland, Oregon. Northwest Science 67:16–22.

Fausch, K. D., C. E. Torgersen, C.V. Baxter, and H. W. Li. 2002. Landscapes to riverscapes: bridging the gap between research and conservation of stream fishes. BioScience 52:483–498.

FEMAT (Forest Ecosystem Management Assessment Team). 1993. Forest ecosystem management: an ecological, economic, and social assessment. U.S. Government Printing Office, Washington, D.C. Available: www.blm.gov/or/plans/nwfpnepa/FEMAT-1993/1993_%20FEMAT_Report.pdf. (January 2018).

Fish, F. F., and R. Wagner. 1950. Oxygen block in the main-stem Willamette River. U.S. Fish and Wildlife Service Special Scientific Report Fisheries 41.

Ford, M. J. 2002. Selection in captivity during supportive breeding may reduce fitness in the wild. Conservation Biology 16:815–825.

Friesen, T. A. editor. 2005. Biology, behavior, and resources of resident and anadromous fish in the lower Willamette River. Oregon Department of Fish and Wildlife, Salem.

Gammon, J. R. 2005. Wabash River fishes from 1800 to 2000. Pages 365–382 *in* J. N. Rinne, R. M. Hughes, and B. Calamusso, editors. Historical changes in large river fish assemblages of the Americas. American Fisheries Society, Symposium 45, Bethesda, Maryland.

Gregory, S. 2008. Historical channel modification and floodplain forest decline: implications for conservation and restoration of a large floodplain river—Willamette River, Oregon. Pages 763–777 *in* H. Habersack, H. Piegay, and M. Rinaldi, editors. Gravel-bed rivers 6: from process understanding to river restoration. Elsevier, Amsterdam, Netherlands.

Gregory, S., L. Ashkenas, and C. Nygaard. 2007. Summary report to assist development of ecosystem flow recommendations for the middle fork and coast fork of the Willamette River, Oregon. Institute for Water and Watersheds, Oregon State University, Corvallis.

Gregory, S., L. Ashkenas, D. Oetter, P. Minear, and K. Wildman. 2002a. Historical Willamette River channel change. Pages 18–24 *in* D. Hulse, S. Gregory, and J. Baker, editors. Willamette River basin atlas. Oregon State University Press, Corvallis. Available: www.fsl.orst.edu/pnwerc/wrb/Atlas_web_compressed/3.Water_Resources/3c.historic_chl_web.pdf. (June 2017).

Gregory, S., L. Ashkenas, D. Oetter, R. Wildman, P. Minear, S. Jett, and K. Wildman. 2002b. Revetments. Pages 32–33 *in* D. Hulse, S. Gregory, and J. Baker, editors. Willamette River basin atlas. Oregon State University Press, Corvallis. Available: www.fsl.orst.edu/pnwerc/wrb/Atlas_web_compressed/3.Water_Resources/3g.revetments_web.pdf. (June 2017).

Gregory, S. V., R. C. Wildman, and J. E. Williams. 2014. Monitoring of native and non-native fish communities in the Willamette River. Oregon State University, Department of Fisheries and Wildlife, technical report, Corvallis.

Habitat Technical Team of the Willamette Action Team for Ecosystem Restoration. 2015. Tracking progress in restoring the Willamette River floodplain. Available: https://digital.osl.state.or.us/islandora/object/osl:19087. (August 2018).

Hughes, R. M. 1997. Do we need institutional

change? Pages 559–568 *in* D. J. Stouder, P. A. Bisson, and R. J. Naiman, editors. Pacific salmon and their ecosystems. Chapman and Hall, New York.

Hughes, R. M., S. Dunham, K. G. Maas-Hebner, J. A. Yeakley, C. B. Schreck, M. Harte, N. Molina, C. C. Shock, V. W. Kaczynski, and J. Schaeffer. 2014. A review of urban water body rehabilitation and mitigation challenges and approaches: (1) rehabilitation and remediation. Fisheries 39:18–29.

Hughes, R. M., and J. R. Gammon. 1987. Longitudinal changes in fish assemblages and water quality in the Willamette River, Oregon. Transactions of the American Fisheries Society 116:196–209.

Hughes, R. M., and A. T. Herlihy. 2007. Electrofishing distance needed to estimate consistent index of biotic integrity (IBI) scores in raftable Oregon rivers. Transactions of the American Fisheries Society 136:135–141.

Hughes, R. M., and A. T. Herlihy. 2012. Patterns in catch per unit effort of native prey fish and alien piscivorous fish in 7 Pacific Northwest USA rivers. Fisheries 37:201–211.

Hughes, R. M., A. T. Herlihy, W. J. Gerth, and Y. Pan. 2012. Estimating vertebrate, benthic macroinvertebrate and diatom taxa richness in raftable Pacific Northwest rivers for bioassessment purposes. Environmental Monitoring and Assessment 184:3185–3198.

Hughes, R. M., and R. F. Noss. 1992. Biological diversity and biological integrity: current concerns for lakes and streams. Fisheries 17(3):11–19.

Hughes, R. M., and D. V. Peck. 2008. Acquiring data for large aquatic resource surveys: the art of compromise among science, logistics, and reality. Journal of the North American Benthological Society 27:837–859.

Hughes, R. M., R. C. Wildman, and S. V. Gregory. 2005. Changes in fish assemblage structure in the mainstem Willamette River, Oregon. Pages 61–74 *in* J. N. Rinne, R. M. Hughes, and B. Calamusso, editors. Historical changes in large river fish assemblages of the Americas. American Fisheries Society, Symposium 45, Bethesda, Maryland.

Hulse, D., S. Gregory, and J. Baker. 2002. Willamette River basin planning atlas: trajectories of environmental and ecological change. Oregon State University Press, Corvallis.

International River Foundation. 2012. Thiess International Riverprize winner: Meyer Memorial Trust, Willamette River, USA. Available: http://riverfoundation.org.au/our-programs/riverprize/international-riverprize/. (February 2018).

Isaac, D. J., C. H. Luce, D. L. Horan, G. L. Chandler, S. P. Wollrab, and D. E. Nagel. 2018. Global warming of salmon and trout rivers in the northwestern U.S.: road to ruin or path through purgatory? Transactions of the American Fisheries Society 147:566–587.

Jaeger, W. K., A. Amos, D.P. Bigelow, H. Chang, D.R. Conklin, R. Haggerty, C. Langpap, K. Moore, P.W. Mote, A.W. Nolin, A.J. Plantinga, C.L. Schwartz, D. Tullos, and D. P. Turner. 2017. Finding water scarcity amid abundance using human-natural system models. Proceedings of the National Academy of Sciences of the United States of America 114:11884–11889.

Johannessen, C. L., W. A. Davenport, A. Millet, and S. McWilliams. 1971. The vegetation of the Willamette Valley. Annals of the Association of American Geographers 61:286–302.

Kinunen, W., and J. R. Moring. 1978. Origin and use of Oregon Rainbow Trout brood stocks. The Progressive Fish-Culturist 40(3):87–89.

Klauda, R., P. Kazyak, S. Stranko, M. Southerland, N. Roth, and J. Chaillou. 1998. Maryland Biological Stream Survey: a state agency program to assess the impact of anthropogenic stresses on stream habitat quality and biota. Environmental Monitoring and Assessment 51:299–316.

Lampman, B. H. 1946. The coming of the pond fishes: an account of the introduction of certain spiny-rayed fishes, and other exotic species, into the waters of the lower Columbia River region and the Pacific coast states. Binfords and Mort, Hillsboro, Oregon.

Larson, G. L., S. E. Moore, and B. Carter. 1995. Ebb and flow of encroachment by nonnative Rainbow Trout in a small stream in the southern Appalachian Mountains. Transactions of the American Fisheries Society 124:613–622.

LaVigne, H. R., R. M. Hughes, R.C. Wildman, S.V. Gregory, and A. T. Herlihy. 2008. Summer distribution and diversity of nonnative fishes in the main-stem Willamette River, Oregon, 1944–2006. Northwest Science 82:83–93.

Lehtinen, R. M., N. D. Mundahl, and J. C. Madejczk. 1997. Autumn use of woody snags by fishes in backwater and channel border habitats of a large river. Environmental Biology of Fishes 49:7–19.

Limburg, K. E., R. M. Hughes, D.C. Jackson, and B. Czech. 2011. Population increase, economic growth, and fish conservation: collision course or savvy stewardship? Fisheries 36:27–35.

Long, J. J. 1982. Status of the Oregon Chub on the Lowell Ranger District, Willamette National Forest. Oregon State University, Final Report 43-04R-1-00049, Corvallis.

Long, J. J., and C. E. Bond. 1983. Distribution of Oregon Chub *Hybopsis crameri*. Oregon State University, Department of Fish and Wildlife, Corvallis.

Lynch, M., and M. O'Hely. 2001. Captive breeding and the genetic fitness of natural populations. Conservation Genetics 2:363–378.

Maas-Hebner, K. G., R. M. Hughes, and C. Schreck. 2014. Wild salmonids in the urban environment: lethal and sublethal effects. Pages 169–182 *in* J. A. Yeakley, K. G. Maas-Hebner, and R. M. Hughes, editors. Wild salmonids in the urbanizing Pacific Northwest. Springer, New York.

Maas-Hebner, K. G., C. B. Schreck, R. M. Hughes, J. A. Yeakley, and N. Molina. 2016. Scientifically defensible fish conservation and recovery plans: addressing diffuse threats and developing rigorous adaptive management plans. Fisheries 41:276–285.

Mantua, N., I. Tohver, and A. Hamlet. 2010. Climate change impacts on streamflow extremes and summertime temperature and their possible consequences for freshwater salmon habitat in Washington State. Climate Change 102:187–223.

Marchetti, M. P., and G. A. Nevitt. 2003. Effects of hatchery rearing on brain structures of Rainbow Trout, *Oncorhynchus mykiss*. Environmental Biology of Fishes 66:9–14.

Markle, D. F., T. N. Pearsons, and D. T. Bills. 1991. Natural history of *Oregonichthys* (Pisces: Cyprinidae), with a description of a new species from the Umpqua River of Oregon. Copeia 1991:277–293.

Matala, A. P., S. Marx, and T. G. Wise. 2009. A genetically distinct wild redband trout (*Oncorhynchus mykiss gairdneri*) population in Crane Prairie Reservoir, Oregon, persists despite extensive stocking of hatchery Rainbow Trout (*O. m. irideus*). Conservation Genetics 9:1643–1652.

McGarvey, D. J., and B. F. Terra. 2015. Using river discharge to model and deconstruct the latitudinal diversity gradient for fishes of the Western Hemisphere. Journal of Biogeography 43:1436–1449.

Mclean, J. E., P. Bentzen, and T. P. Quinn. 2003. Differential reproductive success of sympatric, naturally spawning hatchery and wild steelhead trout (*Oncorhynchus mykiss*) through the adult stage. Canadian Journal of Fisheries and Aquatic Sciences 66:443–440.

Meyer Memorial Trust. 2015. Willamette River report card 2015: connecting water, wildlife, and people. Meyer Memorial Trust, Portland, Oregon. Available: https://ecoreportcard.org/report-cards/willamette-river/. (June 2017).

Molina, N. M. 2014. Regulatory and planning approaches to protecting salmonids in an urbanizing environment. Pages 31–45 *in* J. A. Yeakley, K. G. Maas-Hebner, and R .M.

Hughes, editors. Wild salmonids in the urbanizing Pacific Northwest. Springer, New York.

Moore, S. E., and M. A. Kulp. 2019. Overcoming challenges to native fish restoration in a national park. Pages 45–64 *in* C. C. Krueger, W. W. Taylor, and S. Youn, editors. From catastrophe to recovery: stories of fishery management success. American Fisheries Society, Bethesda, Maryland.

Mote, P., A. K. Snover, S. Capalbo, S. D. Eigenbrode, P. Glick, J. Littell, R. Raymond, and S. Reeder. 2014. Northwest. Pages 487–513 *in* J. M. Melillo, T. C. Richmond, and G. W. Yohe, editors. Climate change impacts in the United States: the third national climate assessment. U.S. Global Change Research Program, Washington, D.C. Available: http://nca2014.globalchange.gov/report/regions/northwest. (January 2018).

Mulvey, M. 2015. More information about the Willamette River report card water quality indicator. Oregon Department of Environmental Quality, Hillsboro. Available: http://www.oregon.gov/deq/FilterDocs/WillwqRpt.pdf. (June 2017).

Mulvey, M., R. Leferink, and A. Borisenko. 2009. Willamette basin rivers and streams assessment. Oregon Department of Environmental Quality, DEQ 09-LAB-016, Portland. Available: www.oregon.gov/deq/FilterDocs/WillametteBasinAssessment2009.pdf. (June 2017).

Murphy, C. A., G. Taylor, T. Pierce, I. Arismendi, and S. L. Johnson. 2019. Short-term reservoir draining to streambed for juvenile salmon passage and non-native fish removal. Ecohydrology, doi: 10.1002/eco.2096.

NMFS (National Marine Fisheries Service). 2008. 2008–2023 Willamette River basin project biological opinion. National Oceanic and Atmospheric Administration, NMFS, Northwest Region, F/NWR/2000/02117. Seattle, Washington.

O'Connor, J. E., J. F. Mangano, S. W. Anderson, J. R. Wallick, K. L. Jones, and M .K. Keith. 2014. Geologic and physiographic controls on bed-material yield, transport, and channel morphology for alluvial and bedrock rivers, western Oregon. Geological Society of America Bulletin 126:377–397.

Oberlack, C., and K. Eisenack. 2017. Archetypical barriers to adapting water governance in river basins to climate change. Journal of Institutional Economics 14(Special Issue 3):527–555.

ODA (Oregon Department of Agriculture). 2017. Agricultural water quality resources. Available: www.oregon.gov/ODA/programs/NaturalResources/AgWQ/Pages/AgWQResources.aspx. (January 2018).

ODEQ (Oregon Department of Environmental Quality). 2017. Water quality status and trends analysis. Available: https://www.oregon.gov/deq/wq/programs/Pages/wqstatustrends.aspx. (August 2018).

ODF (Oregon Department of Forestry). 2012. Oregon Forest Practices Act: adaptable and informed by sound science. Available: www.oregon.gov/ODF/Documents/WorkingForests/FPAFactSheet.pdf. (January 2018).

ODFW (Oregon Department of Fish and Wildlife). 2018. Middle Fork Willamette Bull Trout Rehabilitation Project. Available: http://odfw.forestry.oregonstate.edu/willamettesalmonidrme/mid-willamette-bull-trout-rehabilitation-project. (January 2018).

ODLCD (Oregon Department of Land Conservation and Development). 2010. Oregon's statewide planning goals & guidelines. ODLCD, Salem. Available: www.oregon.gov/lcd/docs/goals/compilation_of_statewide_planning_goals.pdf. (January 2018).

Oregon Administrative Rule 635-007-0502. Native fish conservation policy. Revised September 12, 2003. Oregon Fish and Wildlife Commission. Salem, Oregon.

Oregon Government. 2010. Oregon plan for salmon and watersheds. Available: www.oregon.gov/OPSW/Pages/archived.aspx. (January 2018).

Palmer, M. A., E. S. Bernhardt, J. D. Allen, P. S. Lake, G. Alexander, S. Brooks, J. Carr, S. Clayton, C. N. Dahm, J. Follstad Shah, D. L. Galat, S. Gloss, P. Goodwin, D. D. Hart, B. Hassett, R. Jenkinson, K. M. Kondolf, R. Lave, J. L. Meyer, T. K. O'Donnell, L. Pagano, and E. Sudduth. 2005. Standards for ecologically successful river restoration. Journal of Applied Ecology 42:208–217.

Paulsen, S. G., A. Mayio, D. V. Peck, J. L. Stoddard, E. Tarquinio, S. M. Holdsworth, J. Van Sickle, L. L. Yuan, C. P. Hawkins, A. T. Herlihy, P. R. Kaufmann, M. T. Barbour, D. P. Larsen, and A. R. Olsen. 2008. Condition of stream ecosystems in the US: an overview of the first national assessment. Journal of the North American Benthological Society 27:812–821.

Pearsons, T. N. 1989. Ecology and decline of a rare western minnow: the Oregon Chub (*Oregonichthys crameri*). Master's thesis. Oregon State University, Corvallis.

Peterson, S. A., A. T. Herlihy, R. M. Hughes, K. L. Motter, and J. M. Robbins. 2002. Level and extent of mercury contamination in Oregon, USA, lotic fish. Environmental Toxicology and Chemistry 21:2157–2164.

Peters, R. J., J. J. Duda, G. R. Pess, M. Zimmerman, P. Crain, Z. Hughes, A. Wilson, M. C. Liermann, S. A. Morley, J. R. McMillan, K. Denton, D. Morrill, and K. Warheit. 2014. Guidelines for monitoring and adaptively managing restoration of Chinook Salmon (*Oncorhynchus tshawytscha*) and steelhead (*O. mykiss*) on the Elwha River. U.S. Fish and Wildlife Service, Washington, D.C. Available: https://walrus.wr.usgs.gov/elwha/PetersElwhaMonAdapt.pdf. (April 2017).

Poirier, J., A. Gray, and B. Clemens. 2017. Pacific Lamprey 2017 regional implementation plan for the Lower Columbia/Willamette regional management unit: Willamette sub-unit. U.S. Fish and Wildlife Service, Washington, D.C. Available: www.fws.gov/pacificlamprey/Documents/RIPs/2017.06.13%20WillametteRIP.pdf. (January 2018).

Pringle, C. M., R. J. Naiman, G. Bretschko, J. R. Karr, M. W. Oswood, J. R. Webster, R. L. Welcomme, and M. J. Winterbourne. 1988. Patch dynamics in lotic systems: the stream as a mosaic. Journal of the North American Benthological Society 7:503–524.

Ripple, W. J., C. Wolf, T. M. Newsome, M. Galetti, M. Alamgir, E. Crist, M. I. Mahmoud, and W. F. Laurance. 2017. World scientists' warning to humanity: a second notice. BioScience 67:1026–1028.

Robbins, W. G. 2004. Landscapes of conflict: the Oregon story, 1940–2000. University of Washington Press, Seattle.

Rogers, H. S., C. A. Mockmore, and C. D. Adams. 1930. A sanitary survey of the Willamette Valley. Oregon State Agricultural College, Engineering Experiment Station, Corvallis. Available at Oregon State University Library, Corvallis.

Runge, M. C. 2011. An introduction to adaptive management for threatened and endangered species. Journal of Fish and Wildlife Management 2:220–233.

Scheerer, P. D. 2002. Implications of floodplain isolation and connectivity on the conservation of an endangered minnow, Oregon Chub, in the Willamette River, Oregon. Transactions of the American Fisheries Society 131:1070–1080.

Scheerer, P. D., and P. J. McDonald. 2003. Age, growth, and timing of spawning of an endangered minnow, the Oregon Chub (*Oregonichthys crameri*), in the Willamette basin, Oregon. Northwestern Naturalist 84:68–79.

Schroeder, R. K., L. D. Whitman, B. Cannon, and P. Olmsted. 2016. Juvenile life-history diversity and population stability of spring Chinook Salmon in the Willamette River basin, Oregon. Canadian Journal of Fisheries and Aquatic Sciences 73:1–14.

Scott, L. M., editor. 1924. John Work's journey from Fort Vancouver to Umpqua River and return in 1834. Oregon Historical Quarterly 24:236–268.

Sedell, J. R., and J. L. Froggatt. 1984. Importance of streamside forests to large rivers: the isolation of the Willamette River, Oregon, USA, from its floodplain by snagging and streamside forest removal. Internationale Vereinigung für Theoretische und Angewandte Limnologie Verhandlungen 22:1828–1834.

Segerstrom, C., K. Smith, and E. Carey. 2017. DEQ has Oregon in dirty hot water. Eugene Weekly, Eugene, Oregon. Available: https://www.eugeneweekly.com/2017/02/23/deq-has-oregon-in-dirty-hot-water/. (April 2017).

Smith, W. J. 1940. Pollution in the Willamette (film). Oregon State University Archives, Corvallis. Available: http://archiveswest.orbiscascade.org/ark:/80444/xv77310/pdf. (June 2014).

Snyder, J. O. 1908. The fishes of the coastal streams of Oregon and northern California. U.S. Bureau of Fisheries Bulletin 27:153–189.

Starbird, E. A., and L. J. Georgia. 1972. A river restored: Oregon's Willamette. National Geographic 141:816–835l.

Tetra Tech. 1994. Willamette River basin water quality study. Phase II: biological sampling data report. Prepared for the Oregon Department of Environmental Quality by Tetra Tech Incorporated, Redmond, Washington.

Thomas, J. A., J. G. Schulte, P. A. Tennant, and D. R. Argo. 2019. Recovery of a great river fishery: the story of the Ohio River. Pages 211–231 *in* C. C. Krueger, W. W. Taylor, and S. Youn, editors. From catastrophe to recovery: stories of fishery management success. American Fisheries Society, Bethesda, Maryland.

Thornbrugh, D. J., S. G. Leibowitz, R. A. Hill, M. H. Weber, Z. C. Johnson, A. R. Olsen, J. E. Flotemersch, J. L. Stoddard, and D. V. Peck. 2018. Mapping watershed integrity for the conterminous United States. Ecological Indicators 85:1133–1148.

U.S. Office of the Federal Register. 1993. Endangered and threatened wildlife and plants; determination of endangered status for the Oregon Chub. Federal Register 58:199(19 February 2015):53800–53804.

U.S. Office of the Federal Register. 2009. Proposed programmatic safe harbor agreement for Oregon Chub, Willamette Valley, OR; Notice of availability; receipt of application. Federal Register 74:95(19 May 2009):23431–23432.

USEPA (U.S. Environmental Protection Agency). 2016. National rivers and streams assessment 2008–2009: a collaborative survey. USEPA, EPA/841/R-16/007, Washington, D.C.

USFWS (U.S. Fish and Wildlife Service). 1998. Oregon Chub (*Oregonichthys crameri*) recovery plan. Portland, Oregon. Available: https://ecos.fws.gov/docs/recovery_plans/1998/980903b.pdf. (June 2017).

USFWS (U.S. Fish and Wildlife Service). 2008. Biological opinion on the continued operations and maintenance of the Willamette River Basin project and effects to Oregon Chub, Bull Trout, and Bull Trout critical habitat designated under the Endangered Species Act. As proposed by: U.S. Army Corps of Engineers, Bonneville Power Administration, and Bureau of Reclamation. Available: www.nwp.usace.army.mil/Portals/24/docs/environment/biop/USFWS_WillametteBO-Final_071108.pdf. (June 2017).

USGS (U.S. Geological Survey). 1949. Large rivers of the United States. U.S. Geological Survey Circular 44. Available: https://pubs.usgs.gov/circ/1949/0044/report.pdf. (June 2017).

Wallick, J. R., K. L. Jones, J. E. O'Connor, M. K. Keith, D. Hulse, and S. V. Gregory. 2013. Geomorphic and vegetation processes of the Willamette River floodplain, Oregon: current understanding and unanswered questions. U.S. Geological Survey, Open-File Report 2013-1246, Reston, Virginia.

Whittier, T. R., R. M. Hughes, G. A. Lomnicky, and D. V. Peck. 2007. Fish and amphibian tolerance values and an assemblage tolerance index for streams and rivers in the western USA. Transactions of the American Fisheries Society 136:254–271.

Williams, J. E. 2014. Habitat relationships of native and non-native fishes of the Willamette River, Oregon. Master's thesis. Oregon State University, Corvallis.

Williamson K., P. DeHaan, and D. Hawkins. 2010. Genetic origin of *Oncorhynchus* mykiss collected from the upper Willamette River basin, OR. U.S. Fish and Wildlife Service, Abernathy Fish Technology Center Report, Longview, Washington.

Wohl, E. A., K. B. Lininger, and J. Baron. 2017. Land before water: the relative temporal sequence of human alteration of freshwater ecosystems in the conterminous United States. Anthropocene 18:27–46.

Wood, M. C. 2013. Nature's trust: environmental law for a new ecological age. Cambridge University Press, New York.

Yeakley, J. A., and S. Dunham. 2014. Watershed and landscape scale actions for mitigating impacts on urban salmonids. Pages 227–241 *in* J. A. Yeakley, K. G. Maas-Hebner, and R. M. Hughes, editors. Wild salmonids in the urbanizing Pacific Northwest. Springer, New York.

Yeakley, J. A., K. G. Maas-Hebner, and R. M. Hughes. 2014. Wild salmonids in the urbanizing Pacific Northwest. Springer, New York.

Yoder, C. O., E. T. Rankin, V. L. Gordon, L. E. Hersha, and C. E. Boucher. 2019. Degradation and recovery of the Scioto River (Ohio, USA) fish assemblage from presettlement to present-day conditions. Pages 233–265 *in* C. C. Krueger, W. W. Taylor, and S. Youn, editors. From catastrophe to recovery: stories of fishery management success. American Fisheries Society, Bethesda, Maryland.

Yoder, C. O., E. T. Rankin, M. A. Smith, B. C. Alsdorf, D. J. Altfater, C. E. Boucher, R. J. Miltner, D. E. Mishne, R. E. Sanders, and R. E. Thoma. 2005. Changes in fish assemblage status in Ohio's nonwadeable rivers and streams over two decades. Pages 399–429 *in* J. N. Rinne, R. M. Hughes, and B. Calamusso editors. Historical changes in large river fish assemblages of the Americas. American Fisheries Society, Symposium 45, Bethesda, Maryland.

Zaroban, D. W., M. P. Mulvey, T. R. Maret, R. M. Hughes, and G. D. Merritt. 1999. Classification of species attributes for Pacific Northwest fishes. Northwest Science 73:81–93.

APPENDIX: WILLAMETTE FISHES

TABLE A.1. Fishes recorded in or near the main-stem Willamette River (listed alphabetically by family and by scientific name within family).

Species name	Occurrence	Location	Tolerance
White Sturgeon *Acipenser transmontanus*	rare	lower-mid	sensitive
Cascadian Sucker *Catostomus bondi*	occasional	mid-upper	
Largescale sucker *C. macrocheilus*	common	all	
Green Sunfish *Lepomis cyanellus*[a]	common (sloughs)	all	tolerant
Pumpkinseed *L. gibbosus*[a]	common (sloughs)	all	tolerant
Warmouth *L. gulosus*[a]	occasional (sloughs)	all	tolerant
Bluegill *L. macrochirus*[a]	common (sloughs)	all	tolerant
Redear Sunfish *L. microlophus*[a]	rare	all	tolerant
Smallmouth Bass *Micropterus dolomieu*[a]	occasional	all	
Largemouth Bass *M. salmoides*[a]	occasional	all	tolerant
White Crappie *Pomoxis annularis*[a]	rare	all	tolerant
Black Crappie *P. nigromaculatus*[a]	rare	all	tolerant
American Shad *Alosa sapidissima*[a]	occasional	lower	
Oriental Weatherfish *Misgurnus anguillicaudatus*[a]	occasional (clumped)	lower	tolerant tolerant
Prickly Sculpin *Cottus asper*	common	all	sensitive
Paiute Sculpin *C. beldingii*	occasional	mid-upper	sensitive
Reticulate Sculpin *C. perplexus*	common	all	
Torrent Sculpin *C. rhotheus*	common	all	sensitive
Chiselmouth *Acrocheilus alutaceus*	common	all	
Goldfish *Carassius auratus*[a]	rare	lower	tolerant
Grass Carp *Ctenopharyngodon idella*[a]	rare	lower	tolerant
Common Carp *Cyprinus carpio*[a]	occasional	all	tolerant
Peamouth *Mylocheilus caurinus*	common	all	
Golden Shiner *Notemigonus crysoleucas*[a]	rare	lower	tolerant
Oregon Chub *Oregonichthys crameri*	rare	floodplain	
Northern Pikeminnow *Ptychocheilus oregonensis*	common	all	
Longnose Dace *Rhinichthys cataractae*	common	mid-upper	
Leopard Dace *R. falcatus*	occasional	mid-upper	
Speckled Dace *R. osculus*	common	all	
Redside Shiner *Richardsonius balteatus*	common	all	
Banded Killifish *Fundulus diaphanus*[a]	occasional	lower-mid	tolerant
Threespine Stickleback *Gasterosteus aculeatus*	common (edges)	all	
Amur Goby *Rhinogobius brunneus*[a]	rare	lower	tolerant
White Catfish *Ameiurus catus*[a]	rare	lower	tolerant

TABLE A.1. Continued.

Species name	Occurrence	Location	Tolerance
Yellow Bullhead *A. natalis*[a]	rare	lower	tolerant
Brown Bullhead *A. nebulosus*[a]	rare	all	tolerant
Channel Catfish *Ictalurus punctatus*[a]	rare	all	tolerant
Eulachon *Thaleichthys pacificus*	rare	lower	
Yellow Perch *Perca flavescens*[a]	occasional (sloughs)	all	
Walleye *Sander vitreus*[a]	rare	lower-mid	
Sand Roller Percopsis *transmontana*	occasional	all	tolerant
Western River Lamprey *Lampetra ayresii*	rare	all	sensitive
Pacific Brook Lamprey *L. pacifica*	occasional	all	sensitive
Pacific Lamprey *L. tridentata*	common	all	sensitive
Starry Flounder *Platichthys stellatus*	rare	lower	sensitive
Western Mosquitofish *Gambusia affinis*[a]	common (edges)	all	tolerant
Cutthroat Trout *Oncorhynchus clarkii*	occasional	upper	sensitive
Chum Salmon *O. keta*[b]	rare	lower	sensitive
Coho Salmon *O. kisutch*[b]	occasional	lower	sensitive
Rainbow Trout O. mykiss[b]	occasional	all	sensitive
Sockeye Salmon *O. nerka*[b]	rare	lower	sensitive
Chinook Salmon *O. tshawytscha*[b]	occasional	all	sensitive
Mountain Whitefish *Prosopium williamsoni*	common	all	sensitive
Red-bellied Pacu *Piaractus brachypomus*[a]	rare	lower-mid	tolerant

[a] Nonnative species.

[b] Federally threatened or endangered population or species.

Restoration of Anadromous Sea Lamprey to the Connecticut River and Adjacent Watersheds

Stephen R. Gephard*

Connecticut Department of Energy and Environmental Protection
Fisheries Division, Marine Headquarters
Post Office Box 719, Old Lyme, Connecticut 06371, USA

Abstract.—Anadromous Sea Lamprey *Petromyzon marinus* spawning runs in the Connecticut River and other streams in the state of Connecticut, USA were decimated by the construction of dams, which were built from 1720 to 1920 for a variety of reasons, notably hydropower. Many of these dams blocked migratory routes to spawning grounds. Government fish management programs begun in the 1960s and 1970s to restore other anadromous species to the Connecticut River did not initially target Sea Lamprey for restoration. The installation of fishways intended for Atlantic Salmon *Salmo salar* and American Shad *Alosa sapidissima* at barrier dams resulted in coincidental upstream passage of Sea Lamprey. The public knew about the destructive invasion by Sea Lamprey into the Great Lakes and initially questioned the wisdom of letting Sea Lamprey move upstream into areas where they were not present. The Connecticut Department of Energy and Environmental Protection (CTDEEP) engaged in a public education campaign to inform about the benefits of Sea Lamprey restoration, which include increased biodiversity and forage base for other fish species, importation of marine-derived nutrients into the freshwater ecosystems, and beneficial physical habitat modification through nest building. By the 1990s, the species was officially targeted for restoration, reflecting evolving public perceptions and state agencies' embrace of increased biodiversity as a program goal. The CTDEEP pursued Sea Lamprey restoration in other watersheds in addition to the Connecticut River watershed. The main strategies for restoration have been provision of fish passage, transplantation, and public education. Restoration efforts have succeeded in increasing the number of adult Sea Lampreys in specific watersheds and expanding their geographical range within these watersheds. Monitoring and research have advanced our knowledge of anadromous Sea Lamprey, including details of the juvenile and adult migrations. Results in Connecticut suggest that individual Sea Lampreys will move upstream beyond any pheromone signal when the run has been previously established in that river below the dam, but they tend not to enter a river from the ocean without a pheromone signal. Many reasons exist why Sea Lamprey runs expanded into vacant habitat more quickly than other anadromous species, but the fact that Sea Lampreys do not home to natal rivers has limited the subsequent increase of the Connecticut River annual run. Lessons learned include specific tools for restoring Sea Lamprey, how to combat negative public perceptions of a parasitic animal, and how to promote support for restoring runs of ecologically valuable but uncharismatic species.

* Corresponding author: steve.gephard@ct.gov

INTRODUCTION

Sea Lamprey *Petromyzon marinus* is an ancient, eel-like, jawless fish that parasitizes other fish species during its juvenile phase (Scott and Crossman 1973). Sea Lamprey is anadromous, which means that it begins its life as an embryo in freshwater, spends its larval phase in freshwater, and migrates as a juvenile to the ocean where it feeds on large marine fish off the Atlantic Continental Shelf, grows, and matures, and returns to freshwater as an adult to spawn (McDowell 1988). Anadromous Sea Lampreys generally parasitize fish while in the ocean (Scott and Crossman 1973), but feeding in freshwater rivers by migrating juveniles has been reported (Hardisty and Potter 1971), particularly in estuaries (Silva et al. 2013). Native anadromous Sea Lampreys in North America range from Florida, USA in the south to Newfoundland, Canada in the north, and in Europe from Turkey in the Mediterranean Sea, around the Iberian Peninsula and north to Finland, including the British Isles (Erdogan et al. 2015; Hansen et al. 2016). Waldman et al. (2004, 2006) suggested that the species was native to Lakes Ontario and Champlain, but that view is disputed by others (Eshenroder 2009). The species has experienced declines throughout much of its native range (Limburg and Waldman 2009; Hansen et al. 2016), but restoration efforts are underway in some watersheds, including the Connecticut River (Gephard and McMenemy 2004). The Connecticut River is one of the largest rivers on the East Coast of the United States, with a mean monthly discharge of 333 m^3/s, a length of 660 km, and a drop of 800 m (Merriman and Thorpe 2004). It enters Long Island Sound, an estuary of the North Atlantic Ocean, in the state of Connecticut (Figure 1). Dozens of other smaller streams enter Long Island Sound to the west and east of the Connecticut River, and some of them have also been targeted for Sea Lamprey restoration.

Sea Lamprey eggs deposited among rocks in streams hatch in spring (in Connecticut and the Connecticut River watershed), and larvae known as ammocoetes emerge and drift downstream until they find suitable habitat in which to settle. Preferred habitat is sand or silt, often with organic debris such as sticks and leaves (Potter 1980). The larvae burrow into the stream substrate vertically with their heads near the surface of the streambed where they feed on alga (primarily diatoms), organic detritus, and bacteria (Manion 1967). Larvae remain buried in feeding burrows for their entire larval phase, which averages 5 years in Connecticut (Aarrestad 1992). During this time, they have no eyes and lack the oral disk used for parasitism. The mouths of larvae are small, without teeth, and folded inward (Scott and Crossman 1973). By the time Sea Lampreys are ready to migrate to sea, they may range in length from 109 to 130 mm (Potter et al. 1978; Quintella et al. 2003). The transformation from a freshwater fish to a saltwater fish generally begins in Connecticut in late July or August of the fish's fifth year (Aarrestad 1992). Eyes form and the mouth enlarges to form an oral disk with hard teeth arranged in a circular pattern. At this time, the juvenile fish is referred to as a transformant or transformer. Seaward migration is generally passive with the fish being carried by streamflow. The migration is believed to range from September to mid-November in New England (Beamish 1980; Aarrestad 1992). The initiation of the migration appears to coincide with decreasing water temperature and increasing streamflow after the trees have dropped their leaves and autumn rains raised water levels. Insufficient data are available to determine the timing of the end of the annual juvenile migration in the Connecticut River, but transformers have been reported moving downstream in December.

Little is known of the Sea Lamprey's biology or distribution while in the ocean. Captures of solitary juvenile Sea Lamprey or those feeding on a host fish in the ocean

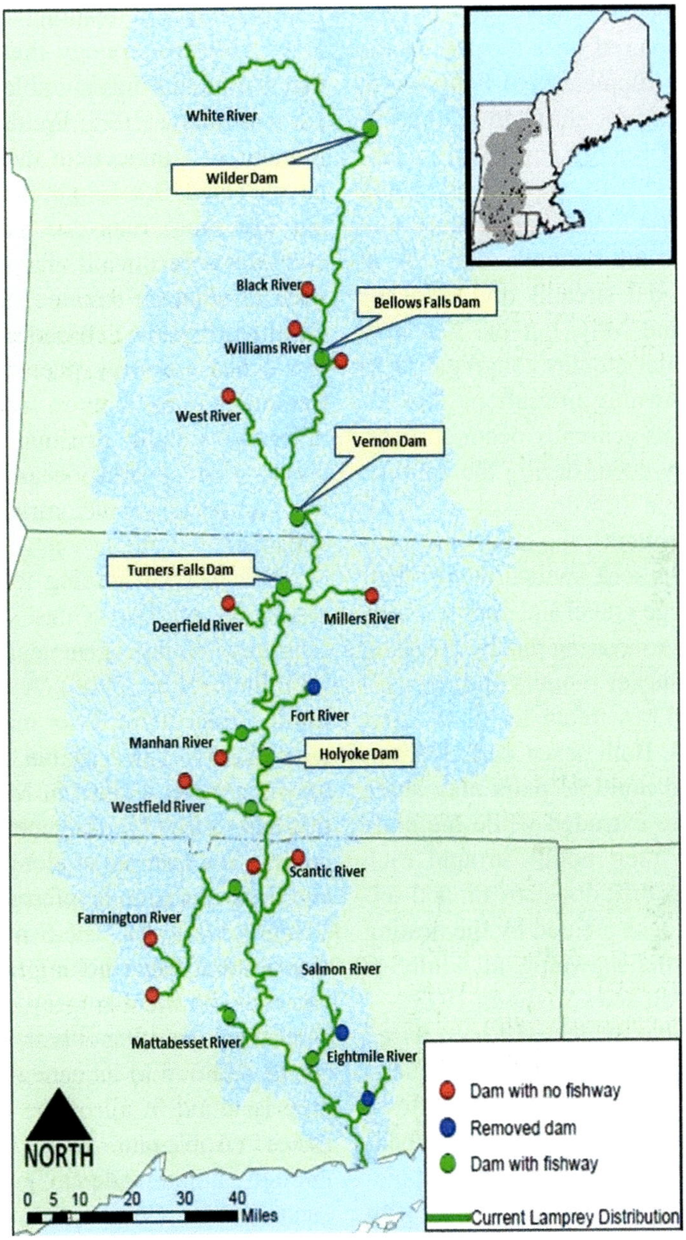

FIGURE 1. Map of the Connecticut River watershed showing the current distribution of Sea Lamprey and the location of major dams, referenced in this paper. (Map courtesy of the Connecticut River Atlantic Salmon Commission.)

are rare (Halliday 1991). Most Sea Lampreys spend 12–16 months at sea feeding prior to returning to freshwater to spawn as adults that can be as long as 860 mm (Scott and Crossman 1973). Little is known of the exact time of river entry, but adult Sea Lamprey are known to enter rivers in Portugal (at a similar latitude as Connecticut) in December and January (Andrade et al. 2007). No directed efforts to sample for Sea Lamprey have occurred in the Connecticut River estuary, but one adult was captured near the top of the salt wedge (river kilometer 10) February 5, 1996 while trawling for sturgeon (Connecticut Department of Energy and Environmental Protection [CTDEEP], unpublished data). Sea Lampreys cease feeding upon entry into freshwater. Adults are typically not seen in Connecticut nontidal streams and fishways until late April and May, but the fish may spend the late winter months congregating in the estuaries. Spawning migrations into inland stream habitats generally occur at night, but spawning may occur during the daytime because by the time they are ready to spawn (in June in Connecticut), the adults are generally blind. Adults seek swift-flowing, shallow water over large gravel and cobble where they create nests, excavating pits by dragging rocks with their sucker mouths and depositing them at the downstream lip of the nest (Applegate 1950). Both sexes participate in nest building, and multiple pairs may share nests. Gametes are extruded while the male and female wrap their bodies around each other, and the eggs drift downstream and settle in the pile of rocks created by the nesting activity. Shortly after spawning, all adult Sea Lampreys die (Scott and Crossman 1973).

Preferred habitat, which will be referred to throughout this paper, consists of clean cobble and gravel with swift water flows (spawning) and soft sand or silt with embedded organic material (larval rearing; Applegate 1950; Potter 1980). Stream reaches with these attributes constitutes high-quality Sea Lamprey habitat.

Most other anadromous fishes imprint to natal river water odor as juveniles and home to the same streams as adults, using their olfactory senses (McDowell 1988). Sea Lampreys do not imprint and home, but rather seek out freshwater streams entering the ocean more or less randomly. Adults are attracted by pheromones released by Sea Lamprey larvae residing in rivers, which is an adaptive strategy that assures the fish that the stream has suitable habitat (Bergstedt and Seelye 1995; Waldman et al. 2008). Streams in Connecticut that were known to have hosted Sea Lamprey runs historically but lost those runs due to industrialization continue to be devoid of adult Sea Lampreys even as adjacent streams' run size increases (Whitworth 1996). Without the scent of the larval Sea Lamprey pheromones, adult Sea Lampreys seem slow to locate and colonize otherwise suitable streams. While straying to unoccupied streams occasionally occurs in Sea Lamprey, it is uncommon in Connecticut streams.

The lack of homing to specific streams results in all East Coast anadromous Sea Lamprey being genetically homogenous (Waldman et al. 2008). A fish that hatched in the Connecticut River may spawn 7 years later in a river in Virginia, and its offspring may spawn in a river in Maine. Over time, this has resulted in no genetic differentiation among Sea Lampreys along the East Coast, and the population is referred to as panmictic (King et al. 2006). The Sea Lampreys in the Connecticut River belong to the same population as the Sea Lampreys in the Delaware, Merrimack, and Penobscot rivers.

In addition to its native range, Sea Lamprey is found in all of the Laurentian Great Lakes (Applegate 1950), having entered through human-made navigational canals that connect East Coast anadromous runs with tributaries of Lake Erie (Scott and Crossman 1973; Eshenroder 2014). The Great Lakes population, which is separate from the East Coast population, spawns in streams that flow into the Great Lakes, and the fish spend their juvenile parasitic phase in the lakes instead of the Atlantic Ocean. Sea Lamprey parasitized Lake Trout *Salvelinus namaycush* and many other native species in the Great Lakes, which resulted in severe declines of many of the host species (Hansen 1999). A binational multi-million dollar annual program to control Sea Lamprey in the Great Lakes began in the 1950s and continues to the present (Smith and Tibbles 1980; Great Lakes Fishery Commission 2017). The decimation of valuable fish stocks and the subsequent ambitious control program became much better known to the average person in North America than the relatively small and covert spawning runs along the East Coast, and because of that, Sea Lamprey generally has been considered negatively by the public. The Great Lakes Sea Lamprey Control Program has spent considerable effort educating the public about the threat of Sea Lamprey in the Great Lakes and justifying large government expenditures to control the population in the Great Lakes. The literature and Internet is full of references to Great Lakes Sea Lamprey, but scant information is available about the East Coast anadromous Sea Lamprey runs. For example, a fact sheet posted by the National Oceanic and Atmospheric Administration (NOAA 2017) never mentions East Coast Sea Lamprey, even though the agency is a strong supporter of East Coast efforts such as the Connecticut River restoration program.

The CTDEEP has sought to restore anadromous Sea Lamprey runs in the Connecticut River in association with the four-state and federal cooperative Connecticut River Atlantic Salmon Commission (CRASC), as well as in selected coastal watersheds in Connecticut that also flow into Long Island Sound (LIS), as part of a comprehensive effort to restore the native suite of diadromous fish species to support increased biodiversity and natural trophic relationships.

This chapter reviews these restoration efforts for Sea Lamprey, which have been a fish management success story, summarizes the results of restoration efforts, reviews lamprey behavior and response to management actions, and shares some lessons learned that may be applicable to other species restoration programs.

WATERSHEDS

The Connecticut River Watershed

The Connecticut River is the longest river in New England at 660 km long, beginning near the Quebec border in New Hampshire and entering LIS in Connecticut at 41°16’51” N–72°20’22” W (Figure 1). The 28,176-km^2 watershed is diverse, draining spruce-fir northwoods, the White and Green Mountains, and the pastoral hills with deciduous forests of southern New England. The lower 80 km are tidal, and there are 45 major tributaries (Figure 1). Much of the lower 450 km has a sand-mud substrate that will not support Sea Lamprey spawning due to lack of gravel, but most of the tributaries have abundant spawning substrate. Sea Lamprey was reported to have been able to ascend past Bellows Falls (river kilometer 450; Figure 1), which blocked most anadromous species (Scarola 1973).

Dam building in the watershed began early during the colonial period with many small tributaries dammed by the early 1700s for local needs such as sawmills, gristmills, and ice ponds. Later, the dams supported industrial mills for manufacturing and, beginning in the late 1800s, for hydroelectric generation. During this time, much of the spawning habitat in the tributaries became inaccessible to adult Sea Lampreys. The first dam across the main-stem Connecticut River was built near Montague, Massachusetts (river kilometer 198) in 1798 and blocked most of the upper half of the watershed to Sea Lamprey (Gephard and McMenemy 2004). By 1900, Sea Lampreys could access only 139 km of the 660-km main-stem river and only short reaches of tributary habitat below the first dams. Additionally, water pollution had become so severe during the 1900s that it likely caused the death of many larval Sea Lampreys that lived below the lowermost dams. Due to the early date at which the population was decimated, no data exist describing the size of the historic run of Sea Lamprey in the watershed.

An interstate–federal cooperative effort to restore anadromous fish to the Connecticut River began in 1967 with the creation of the Policy Committee for the Fisheries Management of the Connecticut River (Foster 1991), which later turned into CRASC (Gephard and McMenemy 2004). Initially, the efforts of CRASC were focused on Atlantic Salmon *Salmo salar* and American Shad *Alosa sapidissima*, but later the CRASC promoted the restoration of all diadromous fish species, including Sea Lamprey. The expansion of mission was a result of changing times, perceptions, and the realization that all species have value even if they are not valued as sport or food.

Other Watersheds that Are Tributaries to Long Island Sound

The same history of dam building, habitat degradation, and reduction of Sea Lamprey described for the Connecticut River was repeated in other streams in Connecticut that flow into LIS. Sea Lamprey likely entered most of the streams that entered LIS prior to dam building (Whitworth 1996). Notable runs occurred in the Housatonic, Naugatuck, and Shetucket rivers (Figure 2), referenced later in this paper. Evidence of the species' presence in the Shetucket River watershed occurs in historical accounts that describe a war that was waged between the Nipmuck and Narragansett Indian tribes along the banks of the Quinebaug River, a tributary of the Shetucket River, because Narragansett guests were served whole (internal organs intact) Sea Lampreys by their Nipmuck hosts at a location 64 km upstream from LIS. The Narragansetts traditionally ate Sea Lampreys with the organs removed and were so offended by the act that they killed their hosts (De Forest 1851).

By the 1970s, too much habitat had been lost and too many dams were present to initially consider restoring salmon and shad in all Connecticut streams. Attention of the fisheries managers in Connecticut during this time was focused on the larger, more-intact,

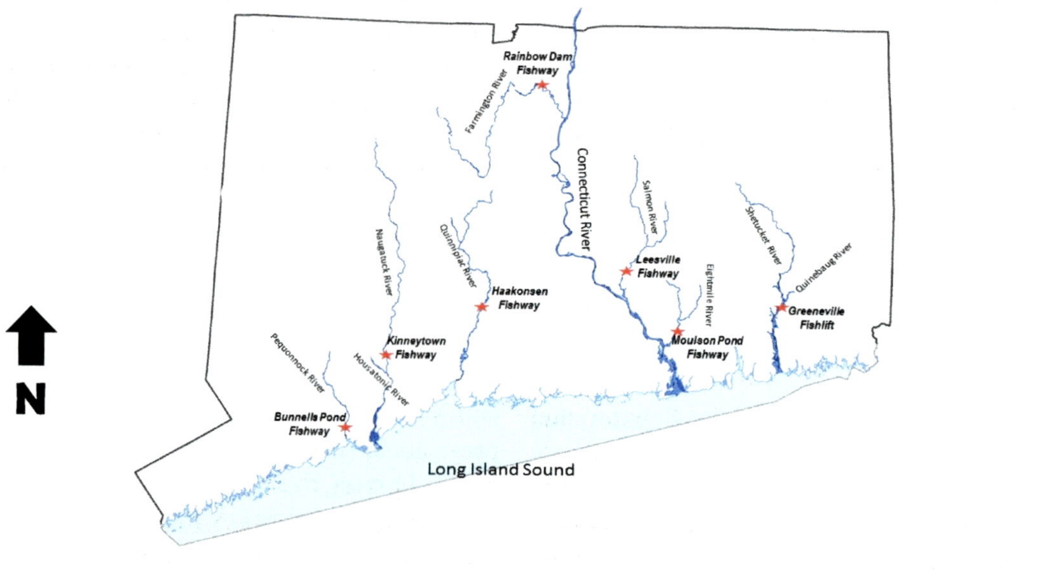

FIGURE 2. Map of the state of Connecticut showing watersheds and fishways referred to in the text.

and better-known Connecticut River. The Connecticut River is shared by four states and presented the opportunity for interstate–federal cooperation. By the 1990s, however, attention turned to other watersheds in Connecticut, and the CTDEEP began to work to build fishways on these streams. This expansion was due in part by the desire of local residents who wished to see the benefits of increased fish runs seen in the Connecticut River to occur also in their streams and also by the opportunity for fishway construction through relicensing of hydroelectric dams by the Federal Energy Regulatory Commission. By this time, the public had learned through public outreach efforts of the CTDEEP and partners that Sea Lampreys were not destructive in Connecticut as they were in the Great Lakes, and as a result, the species was chosen for restoration, along with American Shad, Alewife *Alosa pseudoharengus*, and Blueback Herring *A. aestivalis*. Experiences with Sea Lamprey restoration in these other watersheds in addition to the Connecticut River are included because they offer additional insight and lessons.

METHODS

The Connecticut River Basin

The key strategy for restoration of Sea Lamprey runs was the provision of fish passage at the barrier dams. Sea Lamprey was never extinct from the lower Connecticut River due to the presence of small amount of available spawning and nursery habitat below the first dam above the heads-of-tide, and remnant runs remained in both the main-stem river and some tributaries low in the watershed. When fishways for Atlantic Salmon and American Shad were built at barrier dams, Sea Lampreys used the fishways to migrate upstream into previously blocked habitat. Daily counts were made of all migrating anadromous fishes, including Sea Lamprey, which moved past viewing windows at the first fishways. Initially, counts were made by workers physically present at the fishways in underground counting houses with a viewing window. Unfortunately, this technique missed most Sea Lampreys migrating up fishways at night. The slender and persistent Sea Lampreys were able to swim through fishway gates with narrow bar gaps designed to block undetected passage of larger fish such as Atlantic Salmon or American Shad during overnight hours when counting houses were not staffed. Later, video cameras were installed in counting houses and counts were conducted 24 h each day, allowing the documentation of passage overnight and other times when staff were not present. Count data allowed workers to document the total number of Sea Lamprey that ascended fishways by day and year and also movement patterns of the species through the fishways.

Some fishways on tributaries had no counting facilities and Sea Lamprey were able to pass undetected and uncounted. To estimate the run size on two such Connecticut River tributaries, the CTDEEP conducted annual Sea Lamprey nests surveys in habitat upstream of the fishways. Staff and trained volunteers walked the portions of the Salmon River (upstream of the Leesville Fishway) and the Eightmile River (upstream of the Moulson Pond Fishway) watersheds (Figures 1 and 2) that were accessible to Sea Lamprey to enumerate nests and generate a run size estimate based upon a formula that equated two adults for each single nest and five adults to each community nest (Gephard et al. 1992). Annual nest surveys were also conducted on the Fort Hill River, a Connecticut River tributary in Amherst, Massachusetts (B. Kynard and M. Horgan, paper presented at the American Fisheries Society annual meeting, 2011). Tracking the numbers of Sea Lamprey adults in selected tributaries allowed comparison with such counts at fishways on the main-stem river and provided a more comprehensive picture of each year's adult run. Additional insight into the behavior of Sea Lampreys was gained through snorkeling surveys of selected spawning habitat, mostly conducted below the Leesville Dam and Fishway on the Salmon River (Figures 1 and 2).

Downstream fish passage facilities were constructed at dams to provide a means for downward migrating diadromous fish to avoid passing through hydroelectric turbines where mortality can be high. Sea Lamprey adults die after spawning so downstream passage of adults was not a concern. Small Sea Lamprey transformants were likely able to pass through the turbines without harm, although no evaluations were conducted. Transformants are known to have used some of the downstream fish passage facilities designed for Atlantic Salmon and American Shad and other species, but the extent of such use is unknown. Most such facilities (surface spill gates near turbine intakes that pass water down a pipe or sluiceway to below the dam) were operated without any monitoring, so few observations and little data exist for Sea Lamprey. At the Rainbow Dam on the Farmington River, a tributary of the Connecticut River in Windsor, Connecticut (Figure 2), the CTDEEP operated a sampling station within the downstream passage facility, where water was diverted into a tank and salmon and other species could be removed, examined, and inspected. Sampling of salmon smolts occurred weekly from mid-April until early June each year from the late 1990s until 2012. Observations of other fish species, including Sea Lamprey transformants, were recorded.

Besides fishways, no other restoration strategies were employed for Sea Lamprey rehabilitation in the Connecticut River basin. Substantial reduction of water pollution was achieved by the four states from 1970 to 2000, due to both the Connecticut Clean Water Act of 1967 (State of Connecticut 1967; Focht 1969) and 1972 amendments to the Federal Water Pollution Control Act (Federal Water Pollution Control Act Amendments of 1972; Copeland 2016), and while the abatement of water pollution was pursued for its own merits, the water-quality improvement benefitted the Sea Lamprey population by increasing survival of the larvae through increased oxygen and reduced contaminant loads in the substrate where they burrowed.

After many years of informally promoting the restoration of the Sea Lamprey, CRASC adopted a Sea Lamprey management plan for the Connecticut River (CRASC 2017), perhaps the first of its kind. The mission of CRASC is to restore many native diadromous fish species, and watershed restoration plans have been developed for Atlantic Salmon and American Shad. The Sea Lamprey Restoration Plan is the latest is a series of single-species management plans from CRASC, but relies heavily upon restoration initiatives (like fishways) used for other species. The plan states why the species should be restored, identifies targeted habitat within the watershed, and lists restoration strategies, but it does not state a numerical goal for run size because insufficient data exist on run increases in response to newly opened habitat on which to base a targeted run size. This establishment of a numerical goal is complicated by the fact that adult Sea Lampreys do not home to natal streams (as discussed in more detail later), thus disrupting any cause-and-effect relationship between increased juvenile production and subsequent adult runs. The plan does not establish a goal or mechanism for commercial harvest, for which there has been no demand.

Other Coastal Watersheds of Connecticut

Inclusion of Sea Lamprey restoration work conducted on selected Connecticut tributaries of LIS is instructive and expands our understanding of Sea Lamprey behavior and biology in ways that may guide future restoration work. The restoration of the species to the Pequonnock, Naugatuck, Quinnipiac, and Shetucket rivers (Figure 2) also focused on fish passage and benefitted from experience gained from the Connecticut River program, which preceded efforts in these streams. One major benefit from the Connecticut River program was the public acceptance of Sea Lamprey restoration. Also, it was understood from the Connecticut River program that if fishways were designed for other targeted species, Sea Lampreys would use them without any additional special consideration.

Unfortunately, less than 10 of the 36 coastal streams in Connecticut possessed runs of Sea Lamprey by 1980, and those that did hosted only small runs, perhaps numbering less than 20 fish annually. As a result, transplantation of adult Sea Lamprey was used as a reintroduction tool for some of these rivers. The CTDEEP tested the transplantation method with an experiment at two coastal rivers. The Naugatuck and Shetucket rivers (Figure 2) each had a dam near the head of tide, and no Sea Lamprey were found in these rivers. The Greeneville Fish Lift was built on the Shetucket River in 1996, and the Kinneytown Dam Fishway was built on the Naugatuck River in 1998. Both dams were the lowermost on the rivers. For the first years of operation, no Sea Lamprey were documented (by videography at a viewing window) ascending either fishway (Starr 2008; CTDEEP, unpublished data). In 2004, 110 adult prespawn Sea Lamprey adults of both sexes were transplanted into tributaries of the Naugatuck River upstream of the Kinneytown Dam (Starr 2008), but no such transplantation occurred in the Shetucket River. This transplantation was used as a paired test to monitor the rate of restoration: one river with transplantation and one river without. The donor source for the transplantation was the Farmington River, where more than 1,000 Sea Lampreys migrate up the Rainbow Dam Fishway (Figures 1 and 2) in a typical spring. Sea Lampreys captured in the Rainbow Dam Fishway were released into selected tributaries with suitable habitat, upstream of temporary blocking nets deployed to keep the lamprey from dropping out of the stream before they became acclimated. After several days, the nets were removed, and subsequent nest and electrofishing surveys confirmed successful reproduction (Starr 2008). These actions were repeated a second year (Starr 2008), and the presence of adult Sea Lampreys was monitored at the Greeneville and Kinneytown fishways during every subsequent year, to date.

The Pequonnock River never had a reported Sea Lamprey run, even after the Bunnells Pond fishway was built in 2002 (Figure 2). Sea Lampreys were transplanted from the Farmington River to the Pequonnock River upstream of the Bunnells Pond Dam Fishway for the first time in 2016 using techniques similar to those used for the Naugatuck River. In the Quinnipiac River (Figure 2), Sea Lampreys had ascended to the first dam without interruption but never ascended any farther due to a lack of a fishway. In 2011, the Haakonsen Fishway was built at that dam and allowed Sea Lampreys to continue upstream, as documented by a viewing window and videography beginning the first year of operation.

RESULTS

The Connecticut River Basin

Adult Sea Lampreys moved up Connecticut River fishways during the very first migratory season of the operation of all fishways in rivers where the remnant runs had been present prior to fishway construction. No prior upstream releases, pheromone application, or other strategies were needed to invoke their use of the fishways, which included a variety of designs: steeppass (Moulson Pond), Denil

(Leesville), vertical slot (Rainbow), modified Ice Harbor pool and weir (Turners Falls), and fish lift (Holyoke; Figure 1). Counts of adult Sea Lampreys at these fishways increased over time (Figure 3). Sea Lampreys were counted at four main-stem Connecticut River fishways, and the numbers counted decreased with each additional upstream fishway encountered (Figure 3). The number of Sea Lamprey nests counted in the Salmon and Eightmile rivers, where no fishway counts were possible, also documented immediate use of upstream habitat (Figure 4).

The Norton Dam on the Jeremy River (Salmon River drainage) blocked Sea Lamprey migration from about 1830 to 2016 (Figure 2). The dam was removed in 2016 and the Sea Lamprey Nest Survey in 2017 documented 56 nests upstream of the former dam site (CTDEEP, unpublished data).

The use of videography to accomplish around-the-clock counting has been essential for accurate enumeration of runs. Fish typically ascend fishways on tributaries during the dark hours during the beginning of the run, but as the run progresses, an increased percentage of migrants pass up fishways during the daylight hours (CTDEEP, unpublished data). Haro and Kynard (1997) reported that adult migrants are active after dark with the peak of migratory activity 1–2 h after midnight. However, Hardisty and Potter (1971) found that the eyes of the Sea Lamprey deteriorate as the migration progresses, and later in the migration, the fish are no longer photo-sensitive. Binder and McDonald (2008) found that dermal photoreceptors in the Sea Lamprey tail may play an important role in light avoidance, and this mechanism also weakens with increasing water temperature, which would occur as the

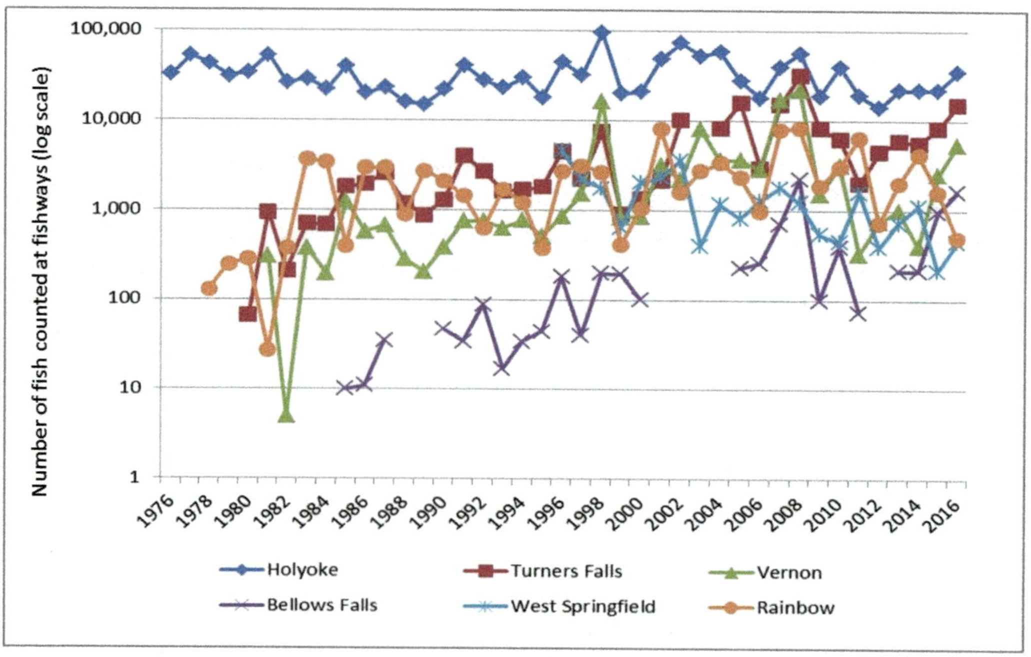

FIGURE 3. Annual counts of upstream migrating Sea Lampreys at major fishways in the Connecticut River watershed. Rainbow and West Springfield are the first dams on separate tributaries. Other fishways are on the main stem and are listed in downstream to upstream order. The Holyoke Fish Lift began operation in the 1950s but Sea Lampreys were counted previous to these dates only in 1967 when 46 were counted. (Graph courtesy of the Connecticut River Atlantic Salmon Commission.)

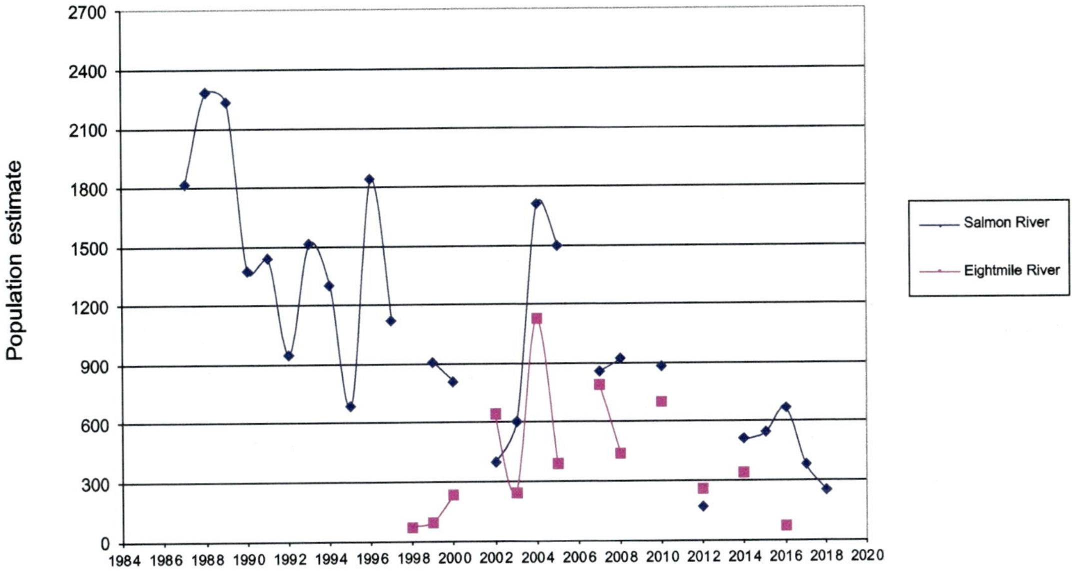

FIGURE 4. Sea Lamprey population estimates for the Salmon and Eightmile River watersheds, 1986–2017, based upon nest counts made during watershed surveys. Counts were not made on some years due to high water events.

adult migration progressed. These progressive trends explain the shift from a strictly nocturnal use of fishways at the beginning of the run to the use of fishways at all hours later in the migration season.

At the Rainbow Dam Fishway (Figure 2), for the years 1978–2016, mean dates were as follows: beginning of run = April 28, peak of run = May 18, end of run = June 23, and length of run = 57 d (CTDEEP, unpublished data).

Other Coastal Watersheds of Connecticut

The transplantation experiment to the Naugatuck (Sea Lampreys transplanted) and Shetucket (no Sea Lampreys transplanted) rivers resulted in an immediate establishment of a sustained Sea Lamprey run to the Naugatuck River but not to the Shetucket River (Starr 2008; Figure 5). A small number of Sea Lampreys were counted at the Greeneville Fish Lift on the Shetucket River immediately before and after the beginning of the experiment, but no Sea Lampreys have been counted there after 2007 (Figure 5). The reasons for this brief appearance of the species are not understood. The run to the Naugatuck River (Kinneytown Fishway counts) has continued to the present day (Figure 5).

In the Pequonnock River, adult Sea Lampreys were released upstream of the Bunnell Pond Fishway (Figure 2) for the first time in 2016, and an adult Sea Lamprey was counted migrating up the fishway for the first time days later.

In the Quinnipiac River, adult Sea Lampreys began using the Haakonsen Fishway during its first year of operation (Figures 2 and 5).

The Role of Public Perception of Sea Lamprey Restoration

When the Connecticut River restoration program for Atlantic Salmon and American Shad began, fisheries agencies gave little thought to the fact that Sea Lampreys might ascend the fishways. When the species ascended the

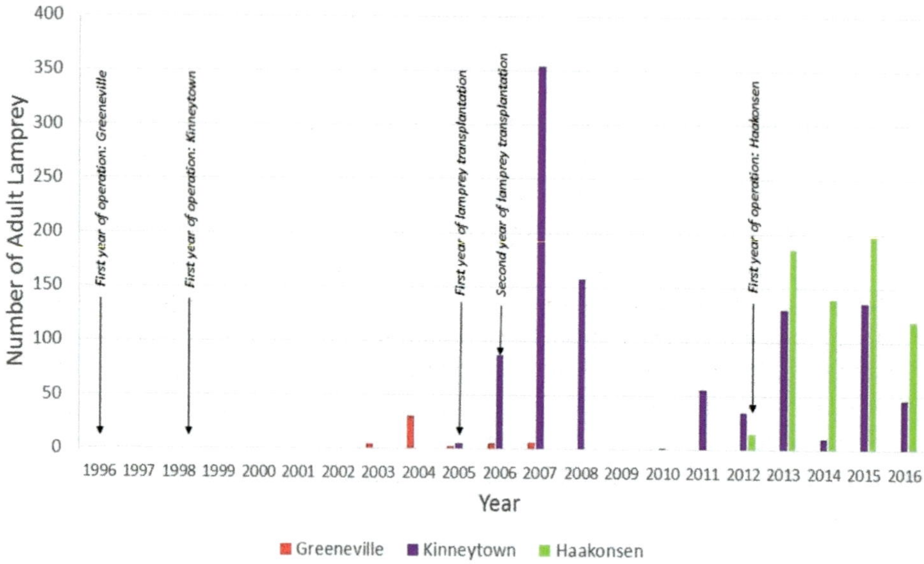

FIGURE 5. Numbers of Sea Lamprey counted at three selected fishways in coastal Connecticut. Greeneville is located on the Shetucket River downstream of the confluence with the Quinebaug River. Kinneytown is located on the Naugatuck River. Haakonsen is located on the Quinnipiac River (see Figure 2).

fishways, agency biologists who knew that it did not parasitize other fish in these streams did not express concern over its presence or relay their lack of concern to the public. As numbers of Sea Lamprey began to increase and migrate into parts of the watershed where it had not been observed during anyone's lifetime, members of the public began to notice them, particularly during their conspicuous nesting behavior in shallow, clear waters. This prompted complaints, mostly from anglers who often killed them (Desbonnet 2007). The public knew of Sea Lamprey only from the Great Lakes experience and never knew the species existed within their home community. Resistance to Sea Lamprey restoration occurred even in parts of the professional fish management community. At this time, biologists in Maine and other states were killing Sea Lampreys that were found in fishways (Ed Baum, formerly of the Maine Sea-Run Salmon Commission, personal communication; Williams 2017). In response, the CTDEEP and later CRASC engaged in active and deliberate public outreach to educate the public and promote Sea Lamprey restoration through talks, articles (e.g., Gephard 1984), Web pages (USFWS 2017), brochures, interviews, and personal interactions along streams and at fishways. With information provided by state and federal biologists, opposition to Sea Lamprey restoration faded. It is now uncommon to read an unsupportive letter or newspaper article or have a member of an audience question the wisdom of letting Sea Lampreys go upstream. After an annual open house at the Rainbow Dam Fishway was advertised in a newspaper with a photograph of a Sea Lamprey attached to the viewing window, public attendance set a record for that day, with many visitors expressing their desire to see the Sea Lamprey. Public acceptance of Sea Lampreys spread to other states including Vermont, Massachusetts, and Maine. In June of 2017, a conservation group hosted a streamside event in Massachusetts to watch Sea Lampreys spawn in a shallow

Connecticut River tributary. A large number of people attended and some were moved to tears of joy (Andrew Fisk, Connecticut River Conservancy, personal communication). Most successful fisheries management programs have focused on species that have strong public support because of the perceived value of the species for recreation, food, or employment benefits. This may be one of the few fish restoration programs that targets a species that offers none of those benefits, but instead, supports increased biodiversity. Public support for the program would have not occurred without strong public outreach by fisheries professionals and allied conservationists.

DISCUSSION

The effort to restore anadromous Sea Lampreys to the Connecticut River has been an ongoing success as defined by increasing numbers of fish to the watershed and increasing the geographic range of the species in the watershed. A potential exists for further increases in Sea Lamprey as new projects provide fish passage around the remaining barrier dams and as existing fishways are modified to improve fish passage efficiency. Efforts to restore anadromous Sea Lampreys to other Connecticut watersheds have also been successful but, to varying degrees, mostly dependent upon the efficiency of the existing fishways and the time elapsed since restoration efforts have been initiated. Future transplantation efforts may accelerate the restoration process.

The traditional expectation for restoration programs for anadromous fishes is that as fish passage facilities reconnect runs with upstream habitat, the numbers of juveniles produced in freshwater will increase, and therefore, the number of adults that subsequently return to the same river will increase. Future expectations for Sea Lamprey restoration remain unclear because the species does not home to natal streams. Even though the production of seaward-migrating juvenile Sea Lampreys increased in the Connecticut River watershed, it was not clear if that would result in more adults subsequently entering the river to spawn because those adults could chose to spawn in other rivers. Data show that the annual numbers of adult Sea Lampreys at most dams increased after the provision of fish passage. The likely explanation is that with more larval Sea Lamprey in the Connecticut River, more pheromones are produced, therefore creating a stronger attraction to maturing adult Sea Lampreys in the coastal marine waters.

The number of adult Sea Lampreys fluctuates annually, and the trend is not consistently upward at all streams. One explanation may be that sea survival of Sea Lamprey varies like other anadromous fish species and the number of adults counted at Connecticut River fishways reflects the varying rate of marine survival. Another possibility is that as more habitat is opened in the upper portion of the Connecticut River watershed, and more larvae are produced, the pheromones coming down the river from these upstream areas overwhelm the pheromones produced from lower watershed tributaries and adults overshoot the weaker pheromone plumes. A comparison of the Salmon River, lower tributary to the Connecticut River, estimates (Figure 4) with the main-stem fishway counts (Figure 3) suggest that the Salmon River may be losing spawners to the upper Connecticut River watershed. The fishway at Bellows Falls (the furthest upstream on the main stem; Figure 2) is the newest. The numbers of counted Sea Lampreys at Bellows Falls on the Connecticut River is increasing (Figure 3) while the number to the Salmon River is decreasing (Figure 4), which may represent a redistribution of fish within the watershed. Localized impacts, such as floods that scour the streambed, may temporarily reduce the larval Sea Lamprey population in specific subdrainages.

Adult Sea Lampreys appear to be able to ascend most types of fishways built along the Connecticut River and elsewhere in Connecticut. Differences in the rates of passage of Sea Lampreys occur with different styles of fishways. Castro-Santos et al. (2017) found that many Sea Lampreys below main-stem dams on the Connecticut River were not able to pass through two fishways and those that passed experienced migratory delay due to repeated attempts. Underwater observations of Sea Lampreys in small tributary fishways of different designs suggest that they are able to pass through these structures much easier, although controlled experiments with tagged fish have not been conducted. Large number of Sea Lamprey nests upstream of these small fishways confirmed significant upstream passage. Sea Lampreys that do not pass through some fishways may be able to spawn downstream of the fishways. Therefore, inefficiencies of some fishways may restrict upstream dispersal but perhaps not overall freshwater production within the watershed.

Nest surveys on the Salmon River and other Connecticut River tributaries confirm that once adult Sea Lamprey pass through fishways, they are able to distribute throughout the watershed, find suitable spawning habitat, build nests, and successfully spawn. Subsequent electrofishing surveys have documented Sea Lamprey larvae in both widely scattered and concentrated areas of suitable habitat downstream of nesting areas. Nest surveys reveal that some areas of the watershed can be expected to support spawning each year, but the relative numbers of nests in specific areas vary considerably annually. During some years, most of the nests are concentrated in the lower part of the Salmon River watershed, whereas in other years, most of the nests are concentrated in the headwaters. The factors that influence such distribution patterns are unknown but could be related to the timing of river entry relative to the timing of spawning and the environmental conditions (e.g., stream discharge, water temperature).

Observations of downstream migrating juvenile Sea Lampreys at fishways have expanded our knowledge of the species' emigration to sea. The Rainbow Dam downstream migrant bypass on the Connecticut River was not operated in the fall when most juvenile Sea Lampreys are believed to migrate to sea. Springtime operation to sample migrating smolts of Atlantic Salmon also recorded the occasional presence of juvenile Sea Lampreys. In some years, Sea Lamprey transformants were found impinged in the wedge wire screen at the sampling station, and numerous salmon smolts had transformants attached to them. These observations were noteworthy because Sea Lamprey transformants usually migrate to the ocean in the fall and only rarely in the spring (Beamish 1980), and they are not known to attack freshwater fish such as salmon smolts. Examination showed these transformants to be exceptionally thin yet not feeding on the salmon smolts. In fact, the skin was never broken by their rasping teeth. These observations invariably occurred only during springs that followed very dry autumns, an observation also reported by Swink and Johnson (2014). Sea Lamprey transformants are relatively weak swimmers and rely on the downstream flow of the river to aid their seaward migration. It is speculated that if elevated stream flows do not occur in the fall before water temperatures become too low, transformants will spend the winter in the stream substrate and initiate their seaward migration in the spring (Applegate 1950). At this point, the nonfeeding transformant has a reduced condition factor and is likely less capable of swimming to the sea under its own locomotion. The ability of Sea Lampreys to locate and attach to migrating Atlantic Salmon smolts appears to be an evolutionary adaptation for these fish to get to the ocean as quickly as possible. The lack of feeding behavior on these smolts by these hitchhiking Sea Lampreys may have evolved so as not to weaken their mode of transportation. Only one report from Connecticut has occurred during the past 35 years of a transformant attaching to any other species other than Atlantic Salmon smolts and one sea-run Brown Trout *Salmo trutta* smolt. None of these juvenile Sea Lampreys attached to Atlantic Salmon smolts were found to be feeding. However, others have reported some feeding by juveniles in freshwater (Hardisty and Potter 1971; Silva et al. 2013). However, feeding juvenile Sea Lamprey in northwestern Spanish rivers were documented by electrofishing upstream habitat and many host fish were adult anadromous fish. The authors state that there was no way to determine if the Sea Lampreys attacked the hosts in the upriver habitat or did so in the saltwater estuary and were carried back upstream by the migrating host. In Connecticut, feeding juvenile Sea Lampreys were occasionally observed on upstream migrating anadromous fishes (e.g., Atlantic Salmon and American Shad) in fishways. This also occurs mainly in springs after a dry autumn. It is suggested that this is inadvertent freshwater feeding by juvenile Sea Lamprey because they attacked their host in estuarine or coastal waters but their host was an adult anadromous fish returning to its natal stream to spawn and carried the juvenile Sea Lamprey back to freshwater. In contrast, attachment of juvenile Sea Lamprey to Atlantic Salmon smolts in upriver, freshwater areas is deliberate but without the intent of immediate feeding. Possibly, the Sea Lampreys might initiate feeding once the salmon smolt reaches saltwater. No other examples of juvenile Sea Lamprey feeding in freshwater exists in Connecticut, even though the streams in which transforming Sea Lampreys are found are full of three species of trout that presumably would be as suitable hosts as Atlantic Salmon smolts.

The results of work on the Pequonnock, Naugatuck, and Shetucket rivers that flow directly into LIS clearly demonstrated the species' reluctance to enter a river without a larval population emanating pheromones (e.g., Shetucket River—Greeneville Fishway). Yet when the species is already present in a river, Sea Lamprey will ascend into a new stretch of unpopulated river (with no pheromones) when made available by a fishway or dam removal (e.g., Quinnipiac River—Haakonsen Fishway; Jeremy River—Norton Dam removal). Hogg et al. (2013) documented the same occurrence in a tributary of the Penobscot River and speculated that adventurous males, perhaps pushed out of downstream habitat by other territorial males, penetrated beyond the extent of the pheromone source. The pheromones released by these males will then attract females upstream to find receptive males (Siefkes et al. 2005). Based upon results in these watersheds, I hypothesize that if Sea Lampreys are already in a river for spawning purposes, they will continue upstream in that same river into newly opened habitat without a pheromone signal, but they will not leave the ocean and enter river without the pheromone signal. Fine et al. (2004) found that all species of lampreys use a common migratory pheromone that makes use of bile acids. However, only a few small populations exist of another (nonanadromous) lamprey the American Brook Lamprey *Lampetra appendix* in the Connecticut River watershed, and neither the American Brook Lamprey nor any other species of lamprey exist in the other referenced watersheds (Jacobs and O'Donnell 2009), so the presence of such a common pheromone is discounted. Clearly, some straying to vacant rivers will occur as demonstrated by the fact that a handful of Sea Lampreys were counted at the Greeneville Fishway on the Shetucket River, 2003–2007 (Figure 5). However, the fact that most streams that flow into LIS that lost their runs of Sea Lamprey more than 100 years ago remain vacant of the species suggests that the incidence of adventurous entry into a vacant river is rare. The transplantation of adults on the Naugatuck River demonstrated that when pheromones are added (via restored reproduction), adults will enter the previously vacant stream. The experience on the Pequonnock River offers a slightly different lesson. One adult Sea Lamprey ascended the Bunnells Pond Fishway shortly after prespawned adults were stocked many miles upstream of the fishway. Brant et al. (2016) reported that female Sea Lampreys will enter rivers occupied by sexually mature male Sea Lampreys that release a bile salt that functions as a sexual pheromone. In the case of the Pequonnock River, the first adult Sea Lamprey to use the fishway did not wait until larval Sea Lamprey populations became established, but was likely attracted upstream by the presence of the transplanted (male) adults.

Relatively little research has been directed at anadromous Sea Lampreys as part of the restoration program, particularly when compared to the efforts expended toward Atlantic Salmon and American Shad. The effort to restore Sea Lamprey to both the Connecticut River and other tributaries of LIS was conducted by management agencies and not organized as scientific research. As more Sea Lampreys entered the river, researchers became more interested in the species and began directed research to further guide managers involved in the restoration effort. Aarrestad (1992), Yergeau (1983), and Steir and Kynard (1986a) described life history traits in Connecticut River tributaries, and Steir and Kynard (1986b) described the migration as the species recolonized historic tributaries. Similar work has more recently been done in Maine (Hogg et al. 2013; Weaver et al. 2016). For a summary of research in the Connecticut River basin, see also Kynard and Horgan (2011) and Nislow and Kynard (2009). Recent studies of Sea Lampreys using the fishways at large, main-stem dams revealed that the dams and fishways created migratory delays of 13.6–14.6 d (median values), and percent passage at the dams ranged between 29% and 55% (Castro-Santos et al. 2017). However, the body of work conducted on Sea Lampreys continues to be sparse because most funds are spent on more charismatic and endangered species like Atlantic Salmon or Atlantic Sturgeon *Acipenser oxyrinchus* or on developing means for controlling Sea Lampreys in the Great Lakes. More research on East Coast anadromous Sea Lampreys would expand our knowledge of the species and help support restoration.

Value of Sea Lamprey

Several studies have been conducted on the ecological benefits of Sea Lamprey populations. Most have focused on the transfer of marine-derived nutrients and providing forage for other species (Johnston 1994; Nislow and Kynard 2009; Weaver et al. 2015, 2016, 2018). My own unpublished observations while snorkeling provide additional insight. During spawning of Sea Lamprey, many species, including Spottail Shiner *Notropis hudsonius*, Common Shiner *Luxilus cornutus*, Fallfish *Semotilus corporalis*, and American Eel *Anguilla rostrata*, were observed feeding on Sea Lamprey eggs that failed to settle into protected crevices of the gravel nests. Some of these cyprinids were observed spawning over the tailspill of the Sea Lamprey nests, which is consistent with other reports (Johnston 1994). Sea Lamprey carcasses (postspawned) that collect on the streambed and in woody debris were observed covered by grazing larvae of caddisflies (Trichoptera). Anglers and scientists fishing or sampling the Connecticut River estuary in the fall have reported that the stomachs of Striped Bass *Morone saxatilis* were full of Sea Lamprey transformants. Walleyes *Sander vitreus* in the upper Connecticut River have been reported with stomachs full of larval or juvenile Sea Lampreys (Gabe Gries, U.S. Fish and Wildlife Service, personal communication). Sea Lampreys have also been described as "ecological engineers" for the physical impact their nesting has on the stream substrate (Hogg et al. 2014) and the resulting increase in macroinvertebrate abundance when compared to undisturbed substrate (Weaver et al. 2018). Habitat disturbed and cleaned (or conditioned) by nesting Sea Lampreys is often used by salmonids and other stream-dwelling fishes as spawning habitat (Saunders et al. 2006; F. W. Kircheis, retired biologist, unpublished report). These observations and studies document the ecological benefits of Sea Lamprey restoration that are less tangible than benefits realized by a community when economically valuable species like salmon are restored to a river.

Why Has Sea Lamprey Restoration Fared Better than Restoration Programs for Other East Coast Anadromous Species?

1. Sea Lampreys in North America have no directed harvest, commercial or recreational, in either fresh or saltwater. Most other East Coast diadromous species are harvested, even if the harvest is bycatch. This means there is one less source of mortality that can impede population growth.
2. Sea Lampreys were able to persist in some streams by spawning below many lowermost dams for decades. Some other species, such as Atlantic Salmon, were not. This allowed for immediate recolonization of reconnected streams without the need to re-establish a local run.
3. The panmictic nature of the Sea Lamprey population and the persistence of fish across its North American range likely prevented the loss of coadapted gene complexes that has hampered the restoration of other anadromous fishes, notably Atlantic Salmon. The panmictic nature of the population means that fish from one stream are adapted to succeed in other vacant streams. This breed ing structure is unlike many anadromous species that home to natal streams, and streams host uniquely adapted strains of fish that may be maladapted to other streams. The panmictic nature of Sea Lamprey expedites the restoration of vacant streams.
4. Implied in both points (2) and (3) is the fact that no hatchery program was required for Sea Lamprey restoration as has been for some restoration programs for other anadromous species such as Atlantic Salmon and American Shad. While hatcheries are valuable tools for some restoration and recovery programs, they introduce risks (Young 2013).
5. Sea Lampreys have been able to readily use any kind of fishway that was provided at barrier dams, even if not all migrants are able to fully ascend every fishway.
6. Pheromone communication allows even small numbers of upstream migrants to find each other to spawn (Teeter 1980). Small numbers of radio-tagged Atlantic Salmon that were released above dams migrated to many different upstream tributaries and never found a mate.
7. The quantity and quality of Sea Lamprey habitat upstream of barrier dams that were made passable were suitable for supporting increased freshwater production of Sea Lamprey.
8. The life history of Sea Lamprey larvae includes a longer freshwater phase than other East Coast anadromous species. However, the fact that the larvae remain burrowed and therefore not vulnerable to most predation throughout the dura-

tion of the freshwater phase results in higher freshwater survival than some other species, notably Atlantic Salmon juveniles, which are subject to high rates of predation by birds and resident fishes in streams during their 2- or 3-year residency (Mills 1989).

9. Larval Sea Lampreys feed on alga and bacteria (Manion 1967), which are likely abundant and subject to less foraging from competitors than the aquatic invertebrates fed upon by other juvenile diadromous fish species.
10. Sea Lampreys are highly fecund, with individual females reported to produce 172,000 eggs (Maitland 2003), which is considerably more than most nest builders in streams, such as Atlantic Salmon, which in Maine has been reported to produce between 5,000 and 12,000 eggs per female (Baum 1997).

Why Have Lampreys Not Increased More in Response to Restoration Efforts?

1. Passage inefficiency of adult Sea Lampreys at large main-stem Connecticut River dam fishways may be limiting the number of fish that spawn upstream in vacant habitat and therefore the growth of the Sea Lamprey spawning run.
2. Members of anadromous fish species share the common experiences of the marine ecosystem with their conspecifics from other rivers. However, river-specific impacts affect only the fish that enter each river. If pollution, dam construction, or overharvest occurs within the freshwater zone of one watershed, fish that enter a different nonimpacted watershed are not affected by those variables. Similarly, if restoration activities such as dam removal and the closure of fisheries occur within one watershed, only fish that enter that river experiences those benefits (typically, population increases). However, the increases to the run of Sea Lampreys to a river such as the Connecticut River, where marked increases in available habitat has occurred, have not been as great as might have been expected, possibly due to the fact that Sea Lampreys do not imprint and home to natal streams. They enter streams based on their attraction to larval Sea Lamprey pheromones (Teeter 1980). A full understanding on the exact mechanisms of such migration from the Atlantic Ocean is lacking, but this method of attraction is different from the way anadromous fish that imprint and home to natal streams. It is likely that many adult Sea Lampreys that hatch in the Connecticut River will not return to the Connecticut River, but instead will migrate into other East Coast rivers. This results in a collective benefit to the entire coastwide population, not just to the run in the river where the habitat improvements occurred. This limits the population increase to the run of the restoration river in the next generation, even after more larvae are produced in the river.

Consider a hypothetical situation in which several barrier dams in "River A" were made passable to all anadromous fishes by either removal or the installation of effective fishways. Suppose these actions opened up 1,000,000 m² of upriver riverine habitat. Under some conditions, this action might be expected to eventually generate an additional 200 adult Atlantic Salmon to River A due to increased freshwater production of imprinted juveniles. Although the relationship between habitat and juvenile production of Sea Lampreys is not understood, nor is the marine mortality rate, for this exercise, let us assume that habitat in River A will also produce an additional 200 adult Sea Lampreys that will enter some river from the ocean due to an increase in the number of transformants emigrating from River A. Those 200 adult Sea Lampreys did not imprint to River A and instead can be expected to distribute themselves among other rivers. Suppose that there are four other rivers located near to River A that have similar numbers of juveniles producing pheromones. If those "new" 200 adult Sea Lampreys that originated in River A are distributed equally among the five rivers, River A would only receive 20% of the increased production, or 40 fish.

This oversimplified exercise underscores the breeding structure of Sea Lamprey along the East Coast of North America, may explain why run sizes have not rebounded as quickly as other species, and argues for a regional approach to Sea Lamprey management. To realize the full benefit of restoration, all parties managing Sea Lampreys in various watersheds along the East Coast should pursue habitat connectivity and contribute to the increase of adult Sea Lampreys in the ocean. If some managing parties failed to do so, their watersheds would be realizing the collective benefit from restoration efforts elsewhere without contributing to the effort.

LESSONS LEARNED

The lessons learned from decades of restoring Sea Lamprey to the Connecticut River and other Connecticut tributaries of LIS are somewhat unique because most of the success stories of fish management typically involve rebuilding the population sizes of species of highly sought after (and overexploited) fish. Sea Lampreys in the United States are not highly valued and not impacted by any significant exploitation.

Lesson 1: All Native Species Have Value, even if There Does Not Seem to Be a Direct Benefit to Humans

Species of low interest to stakeholders are rarely studied and therefore their ecological services are often unknown. Early values are based on whether the species can be caught for fun, eaten, or sold. The ecological value of Sea Lamprey became apparent only after the species was studied by scientists. "The last word in ignorance is the man who says of an animal or plant, 'What good is it?' If the land mechanism as a whole is good, then every part is good, whether we understand it or not." (Leopold 1953). When the restoration effort began, managers adopted Aldo Leopold's philosophy on blind faith. As the effort continued and researchers began investigations of the species, the eyes of biologists were opened along with those of the general public.

Lesson 2: Public Misconceptions and Outdated Values Must Be Actively and Publicly Challenged by Scientists and Conservationists if Changes Are to Occur

Many state agencies refused to support Sea Lamprey restoration because of public opposition. Connecticut was the first state to publicly contradict the commonly held belief about Sea Lamprey and took every opportunity to educate the public and promote restoration. No erroneous statements or misrepresentations were left unchallenged. The opposition to Sea Lamprey disappeared quickly in Connecticut, followed by the other Connecticut River states and finally most of New England. The fish managers in Vermont have been particularly outspoken as they helped residents understand why it was good to expend much time and money to control Sea Lamprey in Lake Champlain in western Vermont (nonnative invasive species) but promote Sea Lamprey restoration in the Connecticut River basin in eastern Vermont (native species) (Vermont Fish and Wildlife 2017). Public engagement on conservation efforts has been critical restoration success elsewhere, such as in Sweden (Hellström et al. 2019, this volume) and British

Columbia (Hyatt and Stockwell 2019, this volume)

Lesson 3: A Remnant Run of the Target Species with an Intact Genome Has High Value to Restoration Programs

Experience has shown that restoring anadromous fish to a river from which the native run has been extirpated is very difficult due to the loss of stream-specific adapted traits. Even runs that still persist below the first barrier dam but are reduced to small numbers may suffer from genetic bottlenecks (reduction of allelic diversity) that can hamper a full restoration. In the case of Sea Lamprey in the Connecticut River and other Connecticut streams, some streams had remnant runs below the first dams while other streams were devoid of the species. Because of the lack of homing to natal streams, all East Coast anadromous Sea Lampreys belong to one population, and thus, those returning to U.S. East Coast streams share the same genome. Even when the annual run of Sea Lamprey below a dam shrinks from hundreds to four, those four fish belonging to one population has not suffered the extreme genetic bottlenecking as would a river-specific Atlantic Salmon population whose numbers had shrunk to four. This lesson may have little value to future Sea Lamprey restoration efforts but may be very instructive for efforts seeking to restore other anadromous species.

Lesson 4: A Combination of Fish Passage at Migratory Barriers and Good Quality and Quantity of Upstream Habitat Aids Restoration

In the case of Sea Lamprey in the Connecticut River, studies have shown that some fish cannot get through large main-stem fishways; however, sufficient upstream natural reproduction occurred to prompt population increases. Improved passage into upstream habitat will continue to be incremental. Managers will likely build fishways at barrier dams that cannot be removed, even if such fishways are not highly effective. This will begin the recolonization process with the understanding that as time progresses, studies and adaptive management will result in improvement of passage of adult Sea Lampreys.

Lesson 5: Transplantation of Adults into Spawning Areas Can Be Beneficial in Accelerating Recovery of Populations

In the case of Sea Lamprey in the Naugatuck and Pequonnock rivers, the release of prespawned Sea Lampreys into a stream previously devoid of Sea Lamprey can support natural reproduction and help establish larvae upstream that will attract more adults and lead to the restoration of a run. This technique results in establishing a larval population in previously vacant streams without the use of a hatchery, which can have inadvertent deleterious impacts to a species' population. Transplantation has been effectively used elsewhere, such as with Brook Trout *Salvelinus fontinalis* (Moore and Kulp 2019, this volume).

Based on the experiences to date and the trend of improved stream connectivity by dam removal and fishway construction, the number of Sea Lamprey in the Connecticut River and other LIS tributaries will likely continue to increase. The extent of these increases may be in part dependent upon actions taken in neighboring watersheds as well as factors in the marine ecosystem. Other diadromous fish species have shown signs of recovery in response to improvements in freshwater ecosystems only to experience subsequent declines due to unknown variables in the ocean. More information about Sea Lampreys in the ocean is needed to monitor trends of ocean mortality, assess the future impacts of climate change, and counteract excessive mortality rates if they occur.

ACKNOWLEDGMENTS

Many people assisted in the collection and analysis of the data presented in this paper. Special thanks go to Tim Wildman (including assistance with the figures); Bruce Williams, Dave Ellis, and Peter Aarrestad of the CTDEEP; Kelly Starr (Connecticut River Coastal Conservation District); Ken Sprankle and Melissa Grader of the U.S. Fish and Wildlife Service; the many workers counting Sea Lamprey at fishways over the years; and the many volunteers who counted Sea Lamprey nests each summer. This paper benefitted greatly from the suggestions of two anonymous reviewers.

REFERENCES

Aarrestad, P. J. 1992. Distribution and life history characteristics of anadromous Sea Lamprey (*Petromyzon marinus* L.) in the Salmon River drainage basin, Connecticut. Master's thesis. University of Connecticut, Storrs.

Andrade, N. O., B. R. Quintella, J. Ferreira, S., I. Póvoa, S. Pedro, and P. R. Almeida. 2007. Sea Lamprey (*Petromyzon marinus* L.) spawning migration in the Vouga River basin (Portugal): poaching impact, preferential resting sites and spawning grounds. Developments in Hydrobiology 195:121–132.

Applegate, V. 1950. Natural history of the Sea Lamprey (*Petromyzon marinus*) in Michigan. U.S. Fish and Wildlife Service Special Scientific Report Fisheries 55.

Baum, E. 1997. Maine Atlantic Salmon: a national treasure. Atlantic Salmon Unlimited, Hermon, Maine.

Beamish, F. W. H. 1980. Biology of the North American anadromous Sea Lamprey (*Petromyzon marinus*) in New Brunswick. Journal of Zoology 177:57–72.

Bergstedt, R. A., and J. G. Seelye. 1995. Evidence for lack of homing by Sea Lampreys. Transactions of the American Fisheries Society 124:235–239.

Binder, T. R., and D. G. McDonald. 2008. The role of dermal photoreceptors during the Sea Lamprey (*Petromyzon marinus*) spawning migration. Journal of Comparative Physiology A 194:921–928.

Brant, C. O., N. S. Johnson, K. Li, T. J. Buchinger, and W. Li. 2016. Female Sea Lamprey shift orientation toward a conspecific chemical cue to escape a sensory trap. Behavioral Ecology 27:810–819.

Castro-Santos, T., X. Shi, and A. Haro. 2017. Migratory behavior of adult Sea Lamprey and cumulative passage performance through four fishways. Canadian Journal of Fisheries and Aquatic Sciences 74:790–800.

Copeland, C. 2016. Clean Water Act: a summary of the law. Congressional Research Service, CRS Report 7-5700, Washington, D.C. Available: https://fas.org/sgp/crs/misc/RL30030.pdf. (July 2018)

CRASC (Connecticut River Atlantic Salmon Commission). 2017. Connecticut River Anadromous Sea Lamprey management plan. CRASC, Sunderland, Massachusetts. Available: www.fws.gov/r5crc/pdf/CRASC-sea-lamprey-plan-final-2018-11-26-18.pdf. (December 2018).

De Forest, J. W. 1851. The history of the Indians of Connecticut: from the earliest known period to 1850. Reprint 1964, Shoestring Press, Hamden, Connecticut.

Desbonnet, A. 2007. Aquatic vampires. On the Water (June), East Falmouth, Massachusetts.

Erdogan, C., S. S. Birecikligil, and R. Fricke. 2015. Freshwater fishes of Turkey: a revised and updated annotated checklist. Biharean Biologist 9:141–147.

Eshenroder, R. L. 2009. Comment: mitochondrial DNA analysis indicates that Sea Lamprey are indigenous to Lake Ontario. Transactions of the American Fisheries Society 138:1178–1189.

Eshenroder, R. L. 2014. The role of the Champlain Canal and Erie Canal as putative corridors for colonization of Lake Champlain and Lake Ontario by Sea Lampreys. Transactions of the American Fisheries Society 143:634–649.

Federal Water Pollution Control Act Amendments of 1972. Public Law 92-500, 92nd Congress, 2nd session (18 October 1972).

Fine, J. M., L. A. Vrieze, and P. W. Sorensen. 2004. Evidence that petromyzontid lampreys employ a common migratory pheromone that is partially comprised of bile acids. Journal of Chemical Ecology 30:2091–2110.

Foster, C. H. W. 1991. Yankee salmon. CIS, Cambridge, Massachusetts.

Focht, T. H. 1969. Connecticut's administrative control of water pollution: the fluid administrative process. Connecticut Institute of Water Resource, University of Connecticut, Special Report 8, Storrs.

Gephard, S. 1984. Sea Lampreys—how bad are they? The Citizen's Bulletin (June):15–16. Connecticut Department of Energy and Environmental Protection, Hartford.

Gephard, S., G. Hamley, and J. Ravita. 1992. Anadromous fish enhancement and restoration, 1992. Connecticut Department of Energy and Environmental Protection, Fisheries Division, Annual Performance Report, Old Lyme.

Gephard, S. and J. R. McMenemy. 2004. An overview of the program to restore Atlantic Salmon and other diadromous fishes to the Connecticut River with notes on the current status of these species in the river. Pages 287–317 *in* P. M. Jacobson, D. A. Dixon, W. C. Leggett, B. C. Marcy, and R. R. Massengill, editors. The Connecticut River ecological study (1965–1973) revisited: ecology of the lower Connecticut River 1973–2003. American Fisheries Society, Monograph 9, Bethesda, Maryland.

Great Lakes Fishery Commission. 2017. Sea Lamprey control in the Great Lakes: a remarkable success! Available: www.glfc.org/control.php. (July 2018).

Halliday, R. G. 1991. Marine distribution of the Sea Lamprey in the northwest Atlantic. Canadian Journal of Fisheries and Aquatic Sciences 48:832–842.

Hansen, M. J. 1999. Lake Trout in the Great Lakes: basinwide stock collapse and binational restoration. Pages 417–453 *in* W. S. Taylor and C. P. Ferreri, editors. Great Lake fisheries policy and management: a binational perspective. Michigan State University Press, East Lansing.

Hansen, M. J., C. P. Madenjian, J. W. Slade, T. B. Steeves, P. R. Almedia, and B. R. Quintella. 2016. Population ecology of the Sea Lamprey (*Petromyzon marinus*) as an invasive species in the Laurentian Great Lakes and an imperiled species in Europe. Reviews in Fish Biology and Fisheries 26:509–535.

Hardisty, M. W., and I. C. Potter. 1971. The general biology of adult lamprey. Pages 127–206 *in* M. W. Hardisty and I. C. Potter, editors. The biology of lamprey, volume 1. Academic Press. London.

Haro, A., and B. Kynard. 1997. Video evaluation of passage efficiency of American Shad and Sea Lamprey in a modified Ice Harbor fishway. North American Journal of Fisheries Management 17:981–987.

Hellström, G., C. Nilsson, S. Westberg, D. Palm, K. Leonardsson, D. Holmqvist, S. Ågren, J. Gardeström, A. Alanärä, and H. Lundqvist. 2019. Restoring the Vindel River, Sweden for Atlantic Salmon and Brown Trout. Pages 125–156 *in* C. C. Krueger, W. W. Taylor, and S. Youn, editors. From catastrophe to recovery: stories of fishery management success. American Fisheries Society, Bethesda, Maryland.

Hogg, R., S. M. Coghlan, Jr. and J. Zydlewski. 2013. Anadromous Sea Lampreys recolonize a Maine coastal river tributary after dam removal. Transactions of the American Fisheries Society 142:1381–1394.

Hogg, R. S., S. M. Coghlan, Jr., J. Zydlewski, and K. S. Simon. 2014. Anadromous Sea Lampreys (*Petromyzon marinus*) are ecosystem engineers in a spawning tributary. Freshwater Biology 59:1294–1307.

Hyatt, K. D., and M. M. Stockwell. 2019. Chasing an illusion? Successful restoration of

Okanagan River Sockeye Salmon in a sea of uncertainty. Pages 65–100 *in* C. C. Krueger, W. W. Taylor, and S. Youn, editors. From catastrophe to recovery: stories of fishery management success. American Fisheries Society, Bethesda, Maryland.

Jacobs, R. P., and E. B. O'Donnell. 2009. A pictorial guide to the freshwater fishes of Connecticut. Connecticut Department of Environmental Protection, Bulletin 42, Hartford.

Johnston, C. E. 1994. The benefit to some minnows of spawning in the nests of other species. Environmental Biology of Fishes 40:213–218.

King, R. C., W. D. Stansfield, and P. K. Mulligan. 2006. A dictionary of genetics, 7th edition. Oxford Press, New York.

Leopold, A. 1953. Round river: from the journals of Aldo Leopold. L. B. Leopold, editor. Oxford University Press, New York.

Limburg, K. E., and J. R. Waldman. 2009. Dramatic declines in North Atlantic diadromous fishes. BioScience 59:955–965.

Maitland, P. S. 2003. Ecology of the River, Brook and Sea lamprey: *Lampetra fluviatilis*, *Lampetra planeri* and *Petromyzon marinus*. Conserving Natura 2000 Rivers Ecology Series No. 5. English Nature, Peterborough, UK.

Manion, P. J. 1967. Diatoms as food of larval Sea Lampreys in a small tributary of northern Lake Michigan. Transactions of the American Fisheries Society 96:296–298.

McDowell, R. M. 1988. Diadromy in fishes. Timber Press, Portland, Oregon.

Merriman, D., and L. M. Thorpe. 2004. Introduction. Pages 3–25 *in* P. M. Jacobson, D. A. Dixon, W. C. Leggett, B. C. Marcy, and R. R. Massengill, editors. The Connecticut River ecological study (1965–1973) revisited: ecology of the lower Connecticut River 1973–2003. American Fisheries Society, Monograph 9, Bethesda, Maryland.

Mills, D. 1989. Ecology and management of Atlantic Salmon. Chapman and Hall, London.

Moore, S. E., and M. A. Kulp. 2019. Overcoming challenges to native fish restoration in a national park. Pages 45–64 *in* C. C. Krueger, W. W. Taylor, and S. Youn, editors. From catastrophe to recovery: stories of fishery management success. American Fisheries Society, Bethesda, Maryland.

Nislow, K. H., and B. E. Kynard. 2009. The role of anadromous Sea Lamprey in nutrient and material transport between marine and freshwater environments. Pages 485–494 *in* A. J. Haro, K. L. Smith, R. A. Rulifson, C. M. Moffitt, R. J. Klauda, M. J. Dadswell, R. A. Cunjak, J. E. Cooper, K. L. Beal, and T. S. Avery, editors. American Fisheries Society, Symposium 69, Bethesda, Maryland.

NOAA (National Oceanic and Atmospheric Administration). 2017. What is a Sea Lamprey? Available: https://oceanservice.noaa.gov/facts/sea-lamprey.html. (July 2018).

Potter, I. C., G. W. Wright, and J. H. Youson. 1978. Metamorphosis in the anadromous Sea Lamprey, *Petromyzon marinus* L. Canadian Journal of Zoology 56:561–570.

Potter, I. C. 1980. Ecology of larval and metamorphosing lampreys. Canadian Journal of Fisheries and Aquatic Sciences 37:1595–1615.

Quintella, B. R., N. O. Andrade, R. Espanhol, and P. R. Almeida. 2003. Distribution, larval stage duration and growth of the Sea Lamprey ammocoetes, *Petromyzon marinus* L., in a highly modified river basin. Ecology of Freshwater Fish 12:286–293.

Saunders, R., M. A. Hachey, and C. W. Fay. 2006. Maine's diadromous fish community: past, present, and future implications for Atlantic Salmon recovery. Fisheries 31:537–555.

Scarola, J. F. 1973. Freshwater fishes of New Hampshire. New Hampshire Fish and Game Department, Nashua.

Scott, W. B., and E. J. Crossman. 1973. Freshwater fishes of Canada. Fisheries Research Board of Canada Bulletin 184.

Siefkes, M. J., S. R. Winterstein, and W. Li. 2005. Evidence that 3-keto petromyzonal sulphate specifically attracts ovulating female Sea Lamprey, *Petromyzon marinus*. Animal Behaviour 70:1037–1045.

Silva, S., M. J. Servia, R. Vieira-Lanero, and F. Cobo. 2013. Downstream migration and hematophagous feeding of newly metamorphosed Sea Lampreys (*Petromyzon marinus* Linnaeus, 1758). Hydrobiologia 700:277–286.

Smith, B. R., and J. J. Tibbles. 1980. Sea Lamprey (*Petromyzon marinus*) in Lakes Huron, Michigan, and Superior: history of invasion and control, 1936–78. Canadian Journal of Fisheries and Aquatic Sciences 37:1780–1808.

Starr, K. 2008. Life history of the anadromous Sea Lamprey (*Petromyzon marinus*) colonizing vacant habitat in three tributaries of the Naugatuck River. Master's thesis. Southern Connecticut State University, New Haven.

State of Connecticut. 1967. Connecticut's Clean Water Act of 1967, an analysis of Public Act 57. Clean Water Task Force, Hartford, Connecticut.

Steir, K., and B. Kynard. 1986a. Abundance, size, and sex ratio of adult sea-run Sea Lamprey, *Petromyzon marinus*, in the Connecticut River. U.S. National Marine Fisheries Service Fishery Bulletin 84:476–480.

Steir, K., and B. Kynard. 1986b. Movements of sea-run Sea Lamprey, *Petromyzon marinus*, during the spawning migration in the Connecticut River. U.S. National Marine Fisheries Service Fishery Bulletin 84:749–753.

Swink, W. D., and N. S. Johnson. 2014. Growth and survival of Sea Lampreys from metamorphosis to spawning in Lake Huron. Transactions of the American Fisheries Society 143:380–386.

Teeter, J. 1980. Pheromone communication in Sea Lamprey (*Petromyzon marinus*): implications for population management. Canadian Journal of Fisheries and Aquatic Sciences 37:2123–2132.

USFWS (U.S. Fish and Wildlife Service). 2017. Species conservation: migratory fish restoration benefits everyone! Available: www.fws.gov/r5crc/species.html. (September 2018).

Vermont Fish and Wildlife. 2017. Sea Lamprey in Connecticut River drainage. Available: https://vtfishandwildlife.hosted.civiclive.com/cms/One.aspx?portalId=73163&pageId=8132090. (July 2018).

Waldman, J. R., C. Grunwald, N. K. Roy, and I. I. Wirgin. 2004. Mitochondrial DNA analysis indicates Sea Lamprey (*Petromyzon marinus*) are indigenous to Lake Ontario. Transactions of the American Fisheries Society 133:950–960.

Waldman, J. R., C. Grunwald, and I. Wirgin. 2006. Evaluation of the native status of Sea Lamprey *Petromyzon marinus* in Lake Champlain based on mitochondrial DNA sequencing analysis. Transactions of the American Fisheries Society 135:1076–1085.

Waldman, J., C. Grunwald, and I. Wirgin. 2008. Sea Lamprey *Petromyzon marinus*: an exception to the rule of homing in anadromous fishes. Biology Letters 4:659–662.

Weaver, D. M., S. M. Coghlan, Jr., J. Zydlewiski, R.S. Hogg, and M. Canton. 2015. Decomposition of Sea Lamprey *Petromyzon marinus* carcasses: temperature effects, nutrient dynamics, and implications for stream food webs. Hydrobiologia 760:57–67.

Weaver, D. M., S. M. Coghlan, Jr., and J. Zydlewiski. 2016. Sea Lamprey carcasses exert local and variable food web effects in a nutrient-limited Atlantic coastal stream. Canadian Journal of Fisheries and Aquatic Sciences 73:1616–1625.

Weaver, D. M., S. M. Coghlan, Jr., and J. Zydlewski. 2018. Effects of Sea Lamprey substrate modification and carcass nutrients on macroinvertebrate assemblages in a small Atlantic coastal stream. Journal of Freshwater Ecology 33:19–30.

Whitworth, W. E. 1996. Freshwater fishes of Connecticut. Connecticut Department of Energy and Environmental Protection, State Geological and Natural History Survey, Bulletin 114, Hartford.

Williams, T. 2017. Cool green science: recovery: why Sea Lampreys need to be recovered and killed. Cool Green Science, The Nature Conservancy. Available: https://blog.nature.org/science/2017/12/11/recovery-why-sea-lampreys-need-to-be-restored-and-killed/. (July 2018).

Yergeau, K. M. 1983. Population demography, riverine movement and spawning habitat of the Sea Lamprey, *Petromyzon marinus*, in the Connecticut River. Master's thesis. University of Massachusetts, Amherst.

Young, K. A. 2013. The balancing act of captive breeding programmes: salmon stocking and angler catch statistics. Fisheries Management and Ecology 20:434–444.

Recovery of a Great River Fishery: The Story of the Ohio River

Jeff A. Thomas*[1], Jerry G. Schulte[1], and Peter A. Tennant[1]
Ohio River Valley Water Sanitation Commission (retired)
5735 Kellogg Avenue, Cincinnati, Ohio 45230, USA

D. Ryan Argo
Ohio River Valley Water Sanitation Commission
5735 Kellogg Avenue, Cincinnati, Ohio 45230, USA

Abstract.—Great river systems (>5,180 km^2 drainage area or >3,000 m^2 average annual discharge), due to their sheer size, wide variety of uses, and cross-jurisdictional watersheds, represent unique challenges for fishery managers. Multi-agency collaborations involving direct and consistent communication of all stakeholders, including government and nongovernment organizations, are essential for the successful natural resources management of these systems. Recovery of the severely degraded fishery of the Ohio River is an excellent illustration of the power of direct engagement of large numbers of stakeholders. The river and its tributaries were in such poor condition immediately after the industrial revolution that by the late 19th and early 20th centuries, the main stem had effectively become an industrial and municipal sewer rather than the beautiful stream it had once been. Important steps in the recovery of the system involved the creation of the Ohio River Valley Water Sanitation Commission in 1948 and the U.S. Environmental Protection Agency in 1970. Collaborations between these two entities and many other partners, including state natural resource and water quality agencies, other federal agencies, industry, academia, and nongovernmental organizations, coupled with the enactment of the Clean Water Act in 1972, have been responsible for improving the Ohio River from its lowest point in the 1930s to the thriving resource that it is today, sustaining upwards of 160 fish species. Threats still exist to the aquatic communities of the basin, and recent data sets show reasons that more work remains to be done to maintain this valuable resource.

INTRODUCTION

The great rivers of the United States, whether defined as rivers draining more than 5,180 km^2 (Simon and Emery 1995) or the more selective definition, those with a total average discharge greater than 3,000 m^3/s (Angradi et al. 2009), have played a vital role in the settlement and development of the country. They have served as routes for exploration, thoroughfares for commodities ranging from beaver pelts to refined chemicals, and sources of food sustaining people and native cultures for thousands of years. These same rivers still support a multitude of industrial and recreational uses. They are unique environmental and ecological resources that present resource managers with equally unique management challenges due to their jurisdictional and geopolitical boundaries as well as their sheer size and hydrologic regimes (Simon and Emery 1995).

* Corresponding author: jthomas@epri.com
[1] Retired.

Of these resource management challenges, severe hydrologic and morphological changes due to the construction of water control structures (e.g., dams) that result in significant alterations to the ecological and biological processes of many great river systems are often among the most substantial. Another issue is that the water bodies are so wide that water-quality conditions, physical habitats, and biological communities can differ, not only upstream and downstream, but also laterally (from shore to shore) at the same location. Additionally, water velocity, stage, and discharge are highly dynamic and can fluctuate within hours, absent local precipitation, due to the size and development of subwatersheds. Great rivers are a continuum providing incredibly complex ecological services from their headwaters to their terminus (the confluence of another great river or, ultimately, the ocean), resulting in a highly complicated system for management.

Great rivers also serve multiple anthropogenic interests and uses, with most having undergone significant morphological changes to support these various uses. Industrial and municipal needs place withdrawal demands on the resource to provide cooling water, process water, and potable water while simultaneously serving as the receiving stream for industrial and municipal wastewater discharges. The abundance of these industrial facilities on the larger rivers within a great river system does not preclude the need for these same corridors to retain their often-majestic landscapes, providing esthetic and economic value as well as maintaining ecosystem services. Concurrently, these systems also generally experience recreational uses such as boating, angling, and swimming that must be balanced with the need to support commercial navigation. Additionally, these systems must be managed to support harvesting of fish for both commercial and recreational uses.

The Ohio River may exemplify this complexity and delicate balance better than any stream in the United States, supporting all of the above activities and more without interruption (Figure 1). Based on any definition, the Ohio River truly is a great river. However, among other great river systems in the nation, due to the incredible abundance of natural resources such as water, timber, and coal available to support industrialization combined with a naturally rich biodiversity, the story of the Ohio River and its fishery is unique.

Formed in Pittsburgh by the confluence of the Allegheny and Monongahela rivers, the Ohio River flows in a generally southwest direction for 1,579 km, during which it forms the border of, or is entirely contained within, six states (Figure 1). The Ohio River drains a watershed of 524,200 km^2, which includes parts of 14 states (and likely at times a 15th, South Carolina, assuming some minor precipitation runoff) and has a discharge (almost 8,000 m^3/s) that is second in the United States only to that of the Mississippi River into which it flows (www.orsanco.org). The volume of water flowing down the Ohio River is staggering. At their confluence, the Ohio frequently triples the flow of the Mississippi River. This volume of water, coupled with the seemingly inexhaustible supply of natural resources (e.g., timber, coal), has supported and sustained industrial development along the Ohio River for well over a century. Heavily influenced by natural topographical variation, the industrial development of the Ohio River has differed greatly between the upper and lower reaches of the river. At 981 mi (1,579 km) in length, the midpoint of the system is found near Cincinnati, Ohio. Roughly at this point, a significant transition in morphology of the river occurs due to historical glacial activity during the last ice age (Wisconsin Glaciation) in North America 10,000 years ago.

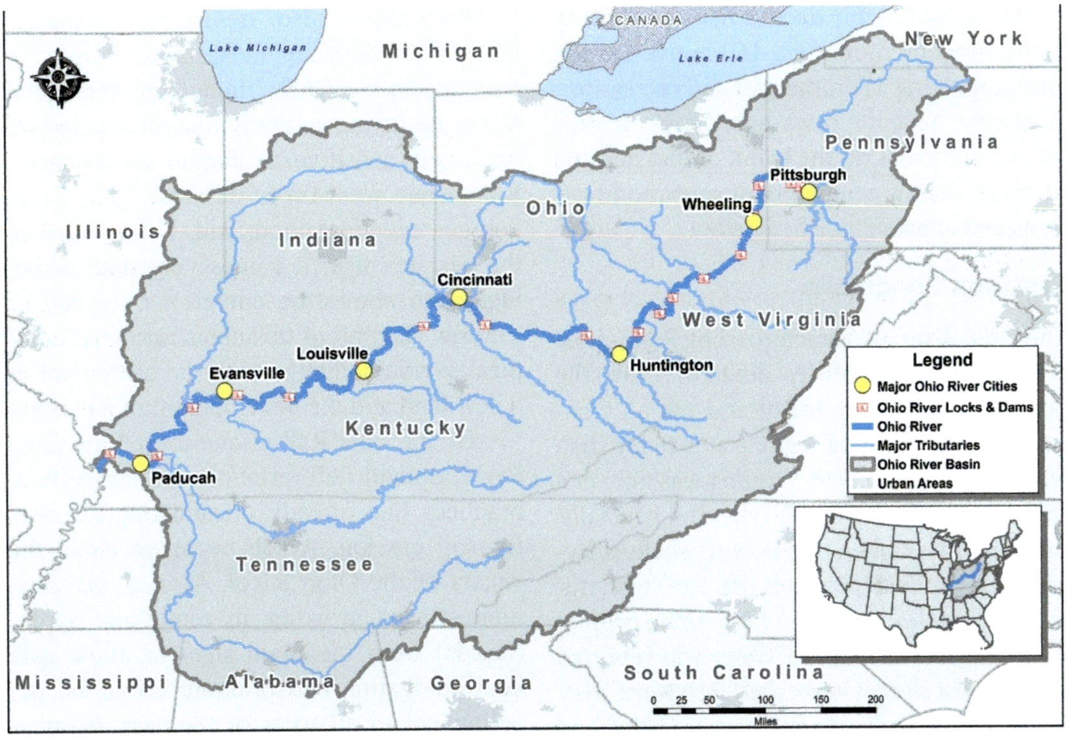

FIGURE 1. The Ohio River drainage in the United States.

The lower half of the Ohio River was affected by this glaciation, leaving the "valley" broad, wide, and flat. By comparison, the upper Ohio River valley is highly dissected with a fairly steep-walled river valley. The abundance of coal, water, and this steep-walled topography historically promoted tremendous industrial development in the upper reaches, the likes of which did not occur in the lower half of the river, which has been used more heavily for agriculture.

Water resource management on a great river like the Ohio River is challenging, to say the least. Classical water quality or fishery management techniques are typically not directly applicable without substantial modification, often due to the sheer size of the water body involved and diversity of uses (Simon and Emery 1995; Angradi et al. 2009). Integrated, cooperative approaches on these vast, dynamic systems generally show the greatest success. Past initiatives that have proven most effective at improving conditions for aquatic communities in the Ohio River over the past century have consisted of multi-agency collaborations, which have relied heavily on consistent and direct communication to achieve their objectives. The remainder of this chapter will tell the tale of the degradation of the Ohio River and highlight successes of several multi-agency initiatives responsible, at least in part, for the dramatic recovery of the system. Many valuable lessons that have been learned during this process will also be conveyed.

DEGRADATION

The recovery of the Ohio River fishery is indeed a contemporary success story of cooperation and collaboration among several nongovernmental organizations and multiple state and federal resource managers and management paradigms. Today, the Ohio River is a thriving and highly diverse fish community represented by more than 160 native species and supporting a commercial and recreational fishery. Seventy years ago, however, this same fishery was on the brink of disaster due to more than a century of severe modifications and abuses described below.

Pre-Dam Conditions

The Ohio River as characterized by Rafinesque (1820) was substantially different from the body of water we see today, and not only because the basin drains more states today than what he reported (West Virginia was not yet a state, and he neglected Maryland). Originally, the Ohio River was a free-flowing shallow river (mean depth at low water was 1 m) with numerous "ripples" (rapids) where native peoples walked across its breadth. Water was reported to be almost clear at low stages. Fish were "very abundant" according to Rafinesque (1820) and were "taken sometimes by thousands with the seines." He reported that the grunting of Freshwater Drum *Aplodinotus grunniens* was so common that "every navigator of the Ohio is well acquainted with it." Over the course of his 2-year tour, he described 100 species of fish from the river, recognizing that this did not "include the whole number existing in the Ohio" but should make up a "great proportion of them." While as many as a dozen of these "new species" were based on completely fabricated illustrations by John James Audubon as part of a practical joke (Markle 1997), Rafinesque was still responsible for giving us our best comprehensive survey of the magnificent diversity of the Ohio River prior to the pending severe water-quality impacts. This diversity that Rafinesque encountered, including species with such colorful names such as toters (Northern Hog Sucker *Hypentelium nigricans*), bubblers (Freshwater Drum), and ribbonfish (most likely gar *Lepisosteus*), allows modern researchers at least a rough benchmark with which to compare current conditions.

Rafinesque also described "muscles" or "Unio" (meaning mussels) as being "so common in the Ohio," indicating very good water quality since many mussel species are extremely sensitive to human disturbances. Rafinesque went on to describe Ohio River fish as "excellent food, and, if not equal to the best sea fish, it comes very near it, being much above the common river fish of Europe." But all of this abundance and complexity would change with the settlement of a new land and the development of a nation.

By the time Rafinesque toured the Ohio River, upland deforestation and agricultural practices had already created opportunities for soil erosion, which began to cloud the waters of the Ohio River. Also at this time, interest was growing in removing ripples (rapids) from the main stem to allow safe boat navigation and promote economic development to all parts of the river (Cramer 1824). Thus began the modern evolution of the Ohio. Pearson and Krumholz (1984) did a thorough review of activities undertaken to improve the river for navigation. Snag removal began in 1824 to begin to address the multitude of navigational hazards described by Rafinesque. The height of the steamboat era on the Ohio River occurred between 1850 and 1920 and was preceded by ample political pressure to ensure safe navigation year-round. The first major modification to accommodate the growing shipping industry occurred in 1830 with the completion of a navigation canal around the Falls of the Ohio River in Louisville, Kentucky. This first canal was followed in the late 19th and early 20th century with the construction of a series of 46 low head locks and dams (original plan of 54 structures was later reduced) to facilitate commercial navigation throughout the river (Pearson and Krumholz 1984). By 1929, the canalization of the Ohio River was complete, forever changing the river's morphology and hydrology.

Post-Dam Conditions

Over the past century, as the lock and dam system converted from 46 low head dams to 19 high lift dams (the 19th is currently under construction), the mighty Ohio River evolved into a series of pools carefully managed for navigation, each with a constant minimum depth of 3 m throughout and an average depth of 8 m. Significant modifications have also occurred throughout the watershed, with countless thousands of dams constructed on tributary streams to support navigation, provide flood control, or to raise water levels for mills and other similar activities.

The canalization of the system accelerated shore-based industrial, agricultural, and urban development throughout the watershed. On the main stem of the larger rivers of the basin, and especially the Ohio River itself, steel mills, coking plants, chemical plants, and electric generating facilities populated the shorelines. Little concern at this time existed about the effects on water quality or the aquatic life of the developing industry within the Ohio River valley. Many believed that the quantity of water in the Ohio River and its tributaries was so vast that wastes discharged to the rivers were readily diluted. Environmental regulations were all but nonexistent, allowing all forms of pollutants and toxins to be discharged into the waterways. As recently as 1950, less than 1% of the wastes discharged to the Ohio River and its tributaries received any treatment.

By the late 1950s, the upper 100 mi (160.9 km) of the river were "characterized by heavy industrial pollution" (Williams et al. 1960), with the river overall being described as "not a "clean stream" (ORSANCO 1962a). Although conditions at this time were reported to improve going downstream, many tributaries throughout the river were described by Williams et al. (1960) as having an "unpleasant appearance," having "an odor of sewage-laden streams," or producing "a strong sickening odor." Colors of tributaries and effluents alike were described as murky gray, rust brown, chalky white, or deep yellow. Some effluents were visible for more than 100 m, had measured pH values as low as 2.4 and low dissolved oxygen (DO) content, and often also had high copper and iron levels. The water quality of the tributaries was as bad or worse than the main-stem Ohio River, with one tributary near the city of Cincinnati having a DO of 0.6 mg/L, which produced a measurable effect on the main stem for miles below. The water of another tributary below Cincinnati was described as becoming progressively "more noxious" going upstream and had a uniform DO throughout of 1.1–1.3 mg/L. The Kanawha River in West Virginia was described as having "chemical odors so intense that it was difficult to take samples without becoming nauseated" (Williams et al. 1960; see description of the Scioto River in Yoder et al. 2019, this volume).

The fishery and ecology of the Ohio River and its basin were being stressed as never before due to increasing discharges of untreated wastes. Massive fish kills were observed for days as dead fish passed along the riverfront of Huntington, West Virginia. Large sections of the Kanawha River were reported by locals to not have been fished for years due solely to pollution (Williams et al. 1960). In a sampling event in an Allegheny River tributary that had a pH of 3.5, no fish at all were collected. The other half of the Ohio River origin was in even worse condition as two sampling events at two lock chambers on the main stem of the Monongahela River in 1957 and 1968 using a fish toxin (rotenone) produced a grand total of two and one fish, respectively (all Bluegill *Lepomis macrochirus*; Ohio River Valley Water Sanitation Commission Fish Population Data, www.orsanco.org/data/fish-population).

The fish diversity that Rafinesque (1820) originally described was severely diminished, with the likelihood that an unknown number of species present at that time would never return. Lachner (1956) reported that the fish fauna in many areas of the Ohio River basin had been "completely eliminated." In the main stem, fish species tolerant to low DO, such as the Black Bullhead *Ameiurus melas*, became abundant, particularly in the upper reaches of the river where even Channel Catfish *Ictalurus punctatus* could not thrive (Williams et al. 1960). Gone too was the delectable nature of the fish. Taste surveys of Flathead Catfish *Pylodictis olivaris*, Channel Catfish, and Freshwater Drum from the late 1950s nearly always resulted in reports of the fish having "strong, disagreeable tastes and odors," "a slight petroleum flavor," or "an oily taste that was not desirable" (Williams et al. 1960). It is no wonder then that fish offered to lock personnel during sampling events were very rarely accepted. One particular meal prepared for a family was scarcely touched, although "the remainder was fed to the dogs and they found them to be quite tasty" (Williams et al. 1960).

As bad as things appeared to be as observed and reported by the Aquatic Life Resources Project commissioned by the Ohio River Valley Water Sanitation Commission (ORSANCO) in the late 1950s (Williams et al. 1960), conditions may have already been improving from their lowest points of the mid–late 1930s (Pearson and Krumholz 1984). In fact, a series of outbreaks of waterborne diseases at that time finally raised widespread concern about water-quality conditions of the Ohio River and the need to take action.

RECOVERY—WATER QUALITY

Due to conditions described above, governors of eight states in the Ohio River valley appointed representatives to negotiate a compact that would pledge their cooperative efforts in reducing and preventing pollution of interstate waters of the Ohio River basin. The Ohio River Valley Water Sanitation Compact was signed by the governors of the states of Illinois, Indiana, Kentucky, New York, Ohio, Pennsylvania, Virginia, and West Virginia on June 30, 1948. The compact committed the signatory states to faithful cooperation in the control of water pollution, set objectives for beneficial uses of the waters of the valley and set treatment requirements for sewage and industrial wastes. The compact also created ORSANCO as the organization empowered to implement its provisions and to ensure that the waters of the Ohio River basin were "capable of maintaining fish and other aquatic life."

The compact established a minimum level of treatment for sewage (essentially primary treatment, or the removal of settleable suspended solids through sedimentation) and a narrative treatment requirement for industrial wastes. This reflected the state of knowledge at the time when the compact was written. While processes for secondary treatment of sewage (biological process to remove dissolved and suspended organic compounds) were known, large-scale applications were not in existence. Treatment methods for many industries had not been developed. For those methods that had been used, operational data were too scant to allow establishment of numerical standards. In recognition of this lack of knowledge, the compact gave authority to ORSANCO to establish additional "rules, regulations and standards" for the treatment of sewage and industrial wastes and to issue orders on individual dischargers to comply.

For its first 10 years of operation, ORSANCO's priorities included a campaign with the member states to build support for expenditures for municipal sewage treatment and efforts to determine treatment methods for wastewaters from the various industries in the valley. To accomplish the latter, ORSANCO invited representatives from the various industries (e.g., chemical, manufacturing, and power-producing) to form industry action committees to study the generation of wastes and identify solutions. The work products from those committees not only led to improved water quality in the Ohio River basin, but also made important contributions to efforts to abate pollution from industrial wastes in the United States and worldwide as other programs referenced the ORSANCO model. The Ohio River Valley Water Sanitation Commission did not have sufficient resources to conduct water quality monitoring in its early years. Instead, ORSANCO relied primarily on data collected by municipal and industrial water users to provide some characterization of conditions (ORSANCO 1957). The drawback to this approach was that the data were inconsistent due to numerous variations in collection and reporting methods. Each user measured parameters needed for the particular use, and monitoring locations were not selected in any systematic manner. As a result of these inconsistencies, characterization of the water-quality conditions of the Ohio River is difficult to establish at the onset of ORSANCO's pollution control activities. In 1960, ORSANCO developed an electronic monitor that could provide continuous measures of temperature, pH, conductivity, and DO (ORSANCO 1960, 1962b). Over the next 2 years, monitors were placed at key locations along the Ohio River and certain major tributaries. This system of electronic monitors operated until 1986.

In 1972, Congress enacted the Clean Water Act, which was the capstone of a growing national resolve to restore and protect water resources. The act provided increased regulatory authorities to require wastewater treatment and grants to support construction of municipal treatment facilities. Research on treatment methods increased, which allowed development of standards for each industrial category (chemical, power-producing, manufacturing, and municipal). The act also provided operating grants for state and interstate water pollution control agencies. The Ohio River Valley Water Sanitation Commission began to receive an annual federal grant that allowed expansion of monitoring activities to better understand water quality conditions and to prescribe actions (generally implemented by state water quality agencies) needed to improve them.

By the late 1980s, secondary treatment had been attained for all municipal discharges to the Ohio River, and all industrial discharges were receiving some level of treatment. While combined sewer systems in old urban areas continued to present problems during and after wet weather, water quality criteria for pH and DO were met the vast majority of the time over the entire Ohio River. In addition, the U.S. Environmental Protection Agency released its Combined Sewer Overflow (CSO) Control Policy in 1994 (USEPA 1994) to establish a national approach for controlling discharges from CSOs.

Water quality trends in the upper Ohio River from 1977 to 1987 indicated decreasing levels of cyanide, phenolics, lead, and zinc. The results of all of these efforts to control water pollution in the basin were documented by Cavanaugh and Mitsch (1989). The Ohio River Valley Water Sanitation Commission (ORSANCO 1990) also showed decreasing trends riverwide from 1975 to 1986 for many parameters, including total phosphorus, ammonia, Kjeldahl nitrogen, total nitrogen, copper, lead, and zinc.

RECOVERY—FISH COMMUNITIES

Ohio River fish community response to these improved water quality conditions was remarkable, with pollution-tolerant species declining and native species diversity and abundance increasing (Thomas et al. 2005). Thomas et al. (2005) analyzed ORSANCO's long-term fish monitoring data, collected via rotenone sampling at main-stem lock chambers from 1957 to 2001 (Ohio River Valley Water Sanitation Commission Fish Population Data, www.orsanco.org/data/fish-population), to determine temporal trends in fish populations, species guilds, and various community metrics. The study documented evidence of positive trends correlated with water quality improvements at all three levels (i.e., species, guild, and assemblage). The authors showed decreasing populations of some pollution-tolerant species such as bullhead catfishes (Yellow Bullhead *Ameiurus natalis*, Black Bullhead, and Brown Bullhead *A. nebulosus*) along with a decrease in the overall percentage of all individuals classified as pollution-tolerant. In the other direction, the number of native species encountered increased significantly over time. Likewise, the number of species classified as intolerant to pollution increased over time, driven by the populations of several species, including Mooneye *Hiodon tergisus*, Smallmouth Redhorse *Moxostoma breviceps*, Smallmouth Bass *Micropterus dolomieu*, and Logperch *Percina caprodes*.

At the assemblage level, corresponding increases were observed using the modified index of well-being, which incorporates abundance and biomass of native fish species (Thomas et al. 2005). While the majority of the trends observed in the study corroborated improvements in the water quality of the river, Thomas et al. (2005) also pointed out that populations of two pollution-tolerant species were significantly increasing (Common Carp *Cyprinus carpio* and Green Sunfish *Lepomis cyanellus*), and a pollution-intolerant species (Goldeye *Hiodon alosoides*) was shown to be becoming less abundant over time. In addition, the percentage of individuals of two different guilds (simple lithophils and invertivores) both seemed to decrease over time. These species and guild trends all indicated that though vast improvements have been made in the system since the 1950s, threats such as nonpoint source pollution and invasive species still exist. Last, Thomas et al. (2005) also pointed out that a key aspect of maintaining a healthy sport and commercial fishery in a system like the Ohio River is managing for water quality improvements to benefit the entire aquatic biological assemblages.

When considering the recovery of the Ohio River fish assemblage, questions commonly arise related to fish diversity and abundance today relative to before the start of the Industrial Revolution, which began as early as the 1790s in the United States. It is extremely difficult to address these topics with any confidence, primarily due to the lack of quantitative and standardized sampling methods employed by Rafinesque and others two centuries ago. Observations by the early naturalists were purely anecdotal, with few estimates of abundance and observations occurring intermittently over long time periods. The pattern of observations in the past stands in stark contrast to the intense, quantitative sampling now conducted by a multitude of entities on the Ohio River, including long-term monitoring efforts of the Electric Power Research Institute (Lohner and Dixon 2013) and Thomas More College (Lorentz et al. 2006). Also, a drastic difference from the past is the ease of communication among modern investigators, allowing instant sharing of data and new discoveries. Last, the taxonomy of fishes has been refined greatly over the past 200 years, allowing better identification of species today.

For these reasons, a current tally of native species encountered in the Ohio River in recent times would likely be greater than a list created from historical records, even though surely some species have been extirpated from the river. Pearson and Krumholz (1984) reported eight species not recorded in the river since 1920 and an additional five not recorded since 1970. However, of the former, two species (Mud Darter *Etheostoma asprigene* and Gilt Darter *Percina evides*) have since been collected in the main stem. Of the latter since 1970, only Southern Redbelly Dace *Chrosomus erythrogaster* has not been documented in the past 20 years. Additionally, Pearson and Pearson (1989) reported three native species as being gone from the river (Lake Sturgeon *Acipenser fulvescens*, Alabama Shad *Alosa alabamae*, and Crystal Darter *Crystallaria asprella* [also known as *Ammocrypta asprella*]). Of those three species, Lake Sturgeon has been documented since that study; however, no records of the latter two species in the river have occurred (Compton et al. 2004).

While encouraging when an "extirpated" species from the main stem is discovered again, determining their relative abundance is challenging, and it is especially difficult to compare estimates to presettlement days. For instance, although Lake Sturgeon has been documented several times since the 1980s, no quantitative density estimates exist for Ohio River fishes before 1957; however, it is likely safe to assume that this species and many others were much more common historically than today. If water quality continues to improve throughout the system, some other species will probably return to the Ohio River and increase in density, although the changed habitat of the river due to its impoundments will be a major deterrent for some native species dependent on riffles and faster-flowing water for their life history. For example, Paddlefish *Polyodon spathula* were slow to recover in the Ohio River even after water quality improvements. Their delay in returning was likely caused at least in part by their migratory behavior and the series of navigational dams along the main stem.

ROLE OF STATE FISHERY AGENCIES

Parallel to the significant efforts to improve the water quality conditions of the basin, an important part of the story of the recovery of the Ohio River fishery is the role of the state fishery agencies. Six fishery management agencies have jurisdiction along the main stem of the Ohio River and a role in the management of the Ohio River fishery. Over the past quarter of a century, these main-stem states have coordinated efforts to monitor and actively manage the complex Ohio River sport and commercial fishery.

Classical fish management techniques, though not as simple to implement or measure as in smaller systems or reservoirs, can sometimes be effective in cross-jurisdictional, great river systems, provided ample coordination occurs among vested agencies. For some taxa, coordinated stocking programs likely had a positive effect on species reestablishment or represented successful new introductions to the system (i.e., hybrids). Blue Catfish *Ictalurus furcatus*, historically thought to inhabit the entire Ohio River main stem, showed a decline in abundance in the upper half of the river from 1970 to 1980, to the point that the species became rare in the upper third of the river (Pearson and Krumholz 1984). According to Frank Jernejcic (personal communication), a 45-year veteran (retired) with the West Virginia Department of Natural Resources, recent Blue Catfish stocking efforts in the West Virginia section of the river have resulted in the species becoming a highly desirable sport fish, and since the stocking of these fish, the West Virginia angling size record for the species has increased almost annually due to large individuals caught from the Ohio River. Similarly, nonnative hybrid Striped Bass (Striped Bass *Morone saxatilis* × White Bass *Morone chrysops*) were introduced into the main stem Ohio River and have also become a popular species for anglers. Populations of Walleye *Sander vitreus* (found almost exclusively in the upper river) and Paddlefish were documented to have increased significantly from 1957 to 2001 (Thomas et al. 2005), likely due to a combination of water quality improvements and stocking programs initiated in the 1990s.

While helpful for a few species, or in certain sections of the river, stocking programs in the Ohio River for other species have generally proven to be too costly to ensure that adequate numbers survive in such a large system, even with coordination among multiple state agencies. Although Herrala (2014) documented the value of Largemouth Bass *Micropterus salmoides* population enhancement in two pools in the middle section of the river, stocking programs for two black bass species could not be shown to have a measurable impact on sport fishing success in the upper Ohio River (Jernejcic, personal communication). Catch per unit effort (CPUE) rates have been calculated annually from 1975 to 2013 for Ohio River bass fishing tournament events in West Virginia (West Virginia Department of Natural Resources, unpublished data). While significant increases in the number of tournaments and tournament hours occurred over that time period and significant fluctuations in CPUE were observed over the years, no positive correlations, either spatially or temporally, occurred between stocking efforts and increases in CPUE. The increases in tournaments and tournament hours more likely represented increasing angling pressure, possibly due in part to improving perception of the river by the public, more so than enhanced black bass populations. While seemingly not successful for black bass species, the management activities implemented in the upper half of the river are believed to be responsible for observed population increases of all four of the other taxa mentioned above due to a combination of fish stocking initiatives along with water quality management programs.

In addition to direct fishery management activities, another important role served by the fishery management agencies has been that of advocacy. Sport fish are often the primary, if not the only, visible component of an aquatic ecosystem to which the general public, particularly anglers, relate (Jernejcic, personal communication). Fishery resource agencies have been particularly integral in parlaying this relationship between the general public and the resource into federal funding directed towards fish restoration and management plans and projects. This effort has been most directly evidenced via the fish management agencies' continued encouragement of regular reauthorizations of the Federal Aid in Sport Fish Restoration Act since it was enacted in 1950. This type of funding has greatly helped accelerate the recovery and use of the river. As the sport fishery improved, infrastructure improvements were also made with the construction of boat and angler access sites, which further helped to rebuild and promote the connection of Ohio River valley residents with the rivers and streams flowing through their neighborhoods. The combination of advocacy, the improved fishery, and public education has helped encourage anglers and others to appreciate the resource of the Ohio River and has likely inspired the political will for watershed protection and other advocacy groups throughout the basin, among many other intangible benefits.

In 1990, another important development occurred that increased the legislative advocacy power of the six state agencies responsible for the Ohio River fishery: the formation of the Ohio River Fish Management Team (ORFMT). Initiated by the state fishery chiefs, the ORFMT, for the first time, established direct lines of communication between its members and partners riverwide surrounding fish management issues and actions, including representation on ORSANCO's Biological Water Quality Subcommittee, ensuring a direct link between water quality objectives and fishery resources. This communication created the forum for coordination of Ohio River fishery management practices, the development of joint management projects, and the establishment of standardized, multi-state, sport fish regulations. The ORFMT, working collaboratively, was able to submit more unified, holistic, and coordinated permit reviews related to sections 10 and 404 of the Clean Water Act to the U.S. Army Corps of Engineers, enhancing the reduction and minimization of environmental impacts. Working together cooperatively saved valuable resources, improved regulation and management of the fishery, and enabled the six states to create a unified voice when soliciting funding opportunities from state and federal sources. In addition, the enhanced communication and coordination due to the formation of the ORFMT surely had a positive influence on the effectiveness of fishing regulations such as size and creel limits. Though enacted for various species in the Ohio River as early as the late 1800s, more unified regulations likely played a more important role in preserving the fishery than disjunct ones, particularly for certain species.

RECENT FISH POPULATION TRENDS

In the early 1990s, ORSANCO began employing electrofishing as a technique for assessing fish species presence and their abundance. A nonlethal alternative to rotenone lock chamber surveys, this technique allowed crews to sample nearly any area along the mainstem Ohio River shoreline. Since its first use in 1990, ORSANCO crews have conducted nearly 3,000 electrofishing surveys, including targeted point sources, randomly selected sites, and a network of long-term research locations using this single, repeatable, assessment method. From 1990 to 2016, methods have remained relatively constant, although some minor variation has occurred. Standardization includes time of year (within an index period of July 1 to October 31), specific electrofishing unit used, number of netters (one or two), and other details related to the electrofishing setup (e.g., number of booms, sphere anodes versus wire anodes).

The fish species sampled during these events have distinct life histories and varying responses to habitat and water quality changes, making their presence or absence an excellent indication of the Ohio River's ecological status. By examining recent trends in the abundance of individual species, we can determine the current status of the Ohio River and attempt to identify emerging concerns. When observing fish populations across time, trends can be clouded by a few extreme values or an overabundance of similar (often null) values. To reduce the influence of these values we performed 75th quantile regressions when examining species trends across all electrofishing data collected by ORSANCO from 1991 to 2016 (2,734 sites, each 0.5 km along the Ohio River shoreline). The summary of these analyses can be seen in the appendix.

Species groups that showed positive trends since 1991 include suckers, sunfishes, and shiners. A prior study of Ohio River fish trends showed a return of pollution-intolerant round-bodied suckers as water quality improved from 1957 to 2001 (Thomas et al. 2005). While our analyses showed that intolerant Smallmouth Redhorse and Shorthead Redhorse *Moxostoma macrolepidotum* populations remained steady, the moderately intolerant species from this group (i.e., Silver Redhorse *M. anisurum* and Golden Redhorse *M. erythrurum*) continued their historical positive trend in abundance. The increase in sunfish populations likely has less to do with water quality and more to do with changes to instream habitat. Since the mid-2000s, the Ohio River has experienced a drastic increase in submerged aquatic vegetation (SAV). Small invertivorous sunfish species use SAV for both cover and forage. Not surprisingly, a marked increase in Bluegill, Longear Sunfish *Lepomis megalotis*, and Orangespotted Sunfish *L. humilis* has simultaneously occurred over the past decade. The increased SAV may also partially explain the increased abundance of small shiners, minnows, and some other small species (e.g., Emerald Shiner *Notropis atherinoides*, Channel Shiner *N. wickliffi*, Bluntnose Minnow *Pimephales notatus*, Trout-perch *Percopsis omiscomaycus*) as they either favor SAV or the fine substrates that settle around the vegetation beds for forage and cover.

One alarming decrease observed over the past quarter century was shown in the darter group. Darters are typically good indicators of river and stream condition as they require higher-quality water and substrates for spawning and foraging success. Decreasing River Darter *Percina shumardi* and Channel Darter *P. copelandi*, both designated as great river species, were observed, from double-digit records to infrequent observations per year, and are worrisome trends. These decreases may be tied to SAV proliferation and increased sedimentation limiting their preferred spawning and feeding habitats or limiting field crews' access to their preferred habitat immediately along the shorelines. Sauger *Sander canadensis*, a member of the same family as darters (Percidae), showed a similar decrease over the past decade. However, variable stocking efforts since 1990 by state agencies on the Ohio River and its major tributaries make conclusions difficult as to the cause and validity of this decline.

The trends shown in the aforementioned groups are directly reflected in the species and abundance metrics of ORSANCO's unpublished modified Ohio River fish index

TABLE 1. Spearman rank correlations ($\alpha = 0.05$) results for the 13 metrics of the modified Ohio River fish index (*m*ORFIn; $N = 2,362$) and two instream variables versus year from 1991 to 2016. For Secchi depth and fine/sandy substrate correlations, $N = 1,899$ and 2,166, respectively. Abbreviations: r_s = Spearman rank correlation statistic; Trend, + = increasing and – = decreasing; NS = not significant; DELT = deformities, erosions, lesions, and tumors.

Metric	r_s	Trend
Fish		
Number of species	0.29	+
Number of sucker species	0.25	+
Number of sunfish species	0.19	+
Number of intolerant species	0.17	+
Number of great river species	0.15	–
Number of DELT anomalies	0.17	–
% individuals as tolerant species	0.18	NS
% individuals as nonnative species	0.11	–
% individuals as simple lithophils	0.17	–
% individuals as detritivores	0.10	+
% individuals as invertivores	0.28	+
% individuals as piscivores	0.16	–
Catch per unit effort	0.17	+
Instream		
Secchi depth	0.06	–
% fine/sandy substrates	0.10	+

(*m*ORFIn; Table 1), which updated the existing Ohio River fish index (ORFIn) developed by Emery et al. (2003). Temporal trends in the 13 metrics of the *m*ORFIn were examined using Spearman rank correlations. The strongest trends were observed in the invertivores metric and native species metrics. Both of these metrics were highly influenced by the marked increase in the minnows, shiners, and sunfish species observed per site. All of these species, excluding Bluntnose Minnow, are invertivores (Table 1).

The decline of Silver Chub *Macrhybopsis storeriana*, River Darter, and Channel Darter (Appendix), along with infrequent recent encounters of Goldeye and Mooneye, contributed to the decreasing trend observed in the great river species metrics. The decrease in these species, along with the lower Sauger catches, also explains the decreasing trend observed in the simple lithophils metric, a breeding guild that requires clean substrates for spawning. Superficially, this decrease in many simple lithophil species, of which the majority are classified as intolerants, makes the increasing trend in the intolerants metric counterintuitive. However, the intolerants metric is based on the abundance of individuals, not simply species diversity. Thus, the large numbers of Channel Shiner collected per site may often offset the decreased abundance of other intolerant species.

Trend analysis of the Ohio River fish species would be incomplete without the examination of exotic species. Exotic species are of great concern when tracking the health of a fishery as they can disrupt native communities in a multitude of ways by competing for food and habitat or even facilitating the invasion of additional exotics. The Ohio River has a long history of exotic species. Common Carp, brought to North America in the 1800s, are now found throughout the basin and have been followed by several relatives (e.g., Goldfish *Carassius auratus*, Grass Carp *Ctenopharyngodon idella*, Silver Carp *Hypophthalmichthys molitrix*, and Bighead Carp *H. nobilis*). Striped Bass (a marine species) and its hybrid as well as Rainbow Trout *Oncorhynchus mykiss* are commonly stocked throughout the basin for recreation and subsequently find their way into the main stem. White Perch *Morone americana* was apparently becoming established in the mid-2000s but then has seemingly disappeared while Banded Killifish *Fundulus diaphanus* seems to just recently be gaining a foothold. Other introduced invasive fish species are knocking at the door of the Ohio River, with recent collections occurring near the main stem (Black Carp *Mylopharyngodon piceus* and Round Goby *Neogobius melanostomus*). Though difficult to prove establishment, by the authors' estimate likely six nonnative fish species occur that could be considered established in the Ohio River out of a total of 17 nonnative species encountered.

While not a fish species, the invasion of nonnative zebra mussel *Dreissena polymorpha* provides a prime example of typical population growth by an exotic species (Figure 2). Upon first arrival in the mid-1990s, zebra mussel abundance exploded as they initially invaded the Ohio River. Their numbers subsequently crashed potentially due to predation by native fauna (e.g., Freshwater Drum) or as a result of inter- and/or intraspecific competition. The muted fluctuations observed since the early 2000s are typical of a species that has achieved their carrying capacity within a system. Conversely, the trend observed in White Perch, a species native to the Atlantic Continental Slope, is a prime example of a failed invasion (Figure 2). First observed in 2003, potentially caused by a bait bucket or contaminated stocking introduction, the number of total individuals per sample riverwide never exceeded 10 before its last recording in 2011.

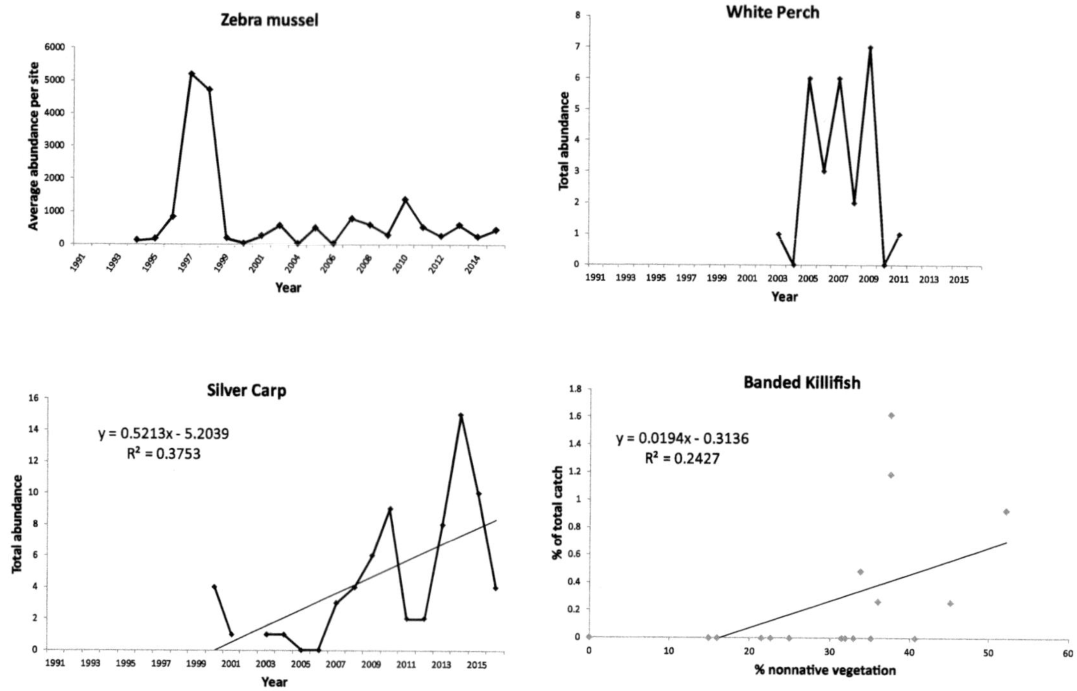

FIGURE 2. Nonnative and invasive species trends and relationships for four species from 1991 to 2016.

Silver Carp first arrived in the Ohio River in the late 1990s (Figure 2). These large fish filter-feed on phytoplankton and can spawn multiple times in a season. They also show an extreme flight response to any disturbance, including electrofishing equipment, and have earned the nickname of "flying fish." Our data therefore severely underestimate the number of individuals present during assessment electrofishing. Nonetheless, the data collected for Silver Carp indicate that they are continuing to increase in abundance. Likewise, Banded Killifish, another Atlantic Continental Slope species, has continued to increase in abundance over the past decade (Figure 2). Historically an infrequent component of the Ohio River community, Banded Killifish numbers have increased since 2010. The Ohio River Valley Water Sanitation Commission is currently investigating the hypothesis that their recent success may be linked to the increased prevalence of nonnative vegetation, hydrilla *Hydrilla verticillata* and Eurasian watermilfoil *Myriophyllum spicatum*.

CURRENT AND FUTURE THREATS

Although the Ohio River fish community has truly recovered from the brink of disaster, as evidenced by Thomas et al. (2005) and the trends in the fish data described in the section above, more work remains to be done to continue recovery and ensure a stable fishery for generations to come. Continued development of the basin associated with population growth, agricultural practices, and natural resources extraction will likely threaten aquatic organisms throughout the watershed for the foreseeable future through nonpoint source pollution and altered hydrologic regimes. Among other effects, this development will continue to contribute sediment loads to the river, resulting in increased siltation of the substrate and higher turbidity levels. Though not measured in this study, these effects could be behind observed declines in some of the smaller insectivores such as darters and minnows. It is also possible that the recent boom of natural resource extraction in the upper basin in recent years may have led to altered water chemistry in some of the upper river tributaries, thereby affecting source populations necessary to sustain main-stem populations.

Importantly, the Ohio River fish communities recovered from 1957 to 2003 despite the rapid human population growth (from just over 20 million in 1970 to more than 25 million in 2010) and associated development of the basin over that time period. Water quality conditions apparently were so drastic that the abatement of in-stream pollution counteracted the negative effects of adding 5 million humans to the basin. This water quality effect may have been best documented by Krumholz and Minckley (1964), who reported the occurrence of "clean-water fishes" within 11 d of temporary steel mill closures caused by a worker strike. This type of response caused by temporary closures illustrates what likely happened when wastewaters associated with steel production in the upper Ohio River corridor diminished indefinitely in the early 1980s, and that undoubtedly contributed greatly to fish community improvements.

Given these much-improved water quality conditions of today, however, Ohio River fish communities are likely much more sensitive to watershed disturbances than 50 years ago. While stream and watershed restoration can be effective in protecting and restoring aquatic communities in small systems, these types of projects generally are less effective when undertaken directly on rivers the size of the Ohio River. However, the improvements in water quality in the Ohio River have had positive effects on the landscape of the main stem, after property values and green spaces along the river began to increase when local residents once again began to appreciate the river as a valuable resource. Today's Ohio River aquatic communities are, however, certainly at the mercy of future landscape activities throughout its cumulative watershed.

Climate change is another factor that not only may have significant impact on Ohio River fish populations in the near future, but also may already be affecting the community now. The temperature of the Ohio River main stem appears to be increasing since the early 1990s based on unpublished data collected by ORSANCO every other month primarily at Ohio River locks and dams. Spearman rank correlations for temperature data from 1990 to 2010 for July and August from all Ohio River sample locations show significantly increasing trends for both months ($\alpha = 0.05$, $r_s = 0.126$ and 0.522, respectively). Several factors may be responsible, such as cumulative effects as tributaries warm due to loss of shading as riparian zones are developed or possibly loss of connectivity of streams in the basin to cool groundwater aquifers, among others. But, these effects may already be symptoms of a steadily increasing global climate (Hansen et al. 2006). Besides increasing average stream temperatures, climate change is also expected to severely affect aquatic communities through an ever increasing pattern of more frequent and more severe extremes (i.e., droughts and floods) throughout the basin (Drum et al. 2017).

Changing stream conditions due to climate change may also accelerate the spread of another threat to Ohio River fish communities. Invasive species are likely responsible for some of the trends observed in the recent electrofishing data set from the river. These species will likely be able to gain larger footholds as current stream conditions are altered from the conditions preferred by native species. Not only will nonnative species already in the basin continue to adversely affect native species, but also the introduction of new invasive species through human intervention must also be viewed as a reality.

Another current and future threat to the Ohio River ecosystem is that of spills of various chemicals that are either transported on (barges) or near (trains and trucks) the Ohio River and its tributaries or are stored near waterways throughout the basin in above- and below-ground storage tanks. Ironically, this threat may be thought of as an emerging threat simply because a spill in the 1950s may not have resulted in a fish kill because of the lack of pollution-intolerant fish species in the vicinity. Today, that same spill would likely be devastating to the more-sensitive fish community inhabiting the river. Despite the best efforts of all involved, hazardous spills to the Ohio River and water bodies throughout the world will continue to be a reality so long as materials are used, stored, and transported in or around those water bodies.

Last, though we have come a long way since the Industrial Revolution, we must not forget that contaminants capable of being bioaccumulated are still being introduced into the river, both through point and nonpoint sources as well as atmospheric deposition. Some of these contaminants, like mercury, can cause the consumption of some fish to be unsafe for humans, or other animals for that matter. While we would expect declines in the levels in fish of some contaminants like polychlorinated biphenyls and pesticides such as chlordane and dichlorodiphenyltrichloroethane (DDT), other poorly understood or as yet unidentified contaminants may be on the rise. Some of these compounds, often known as endocrine-disrupting chemicals, can disrupt the natural reproductive cycles of fish and include a broad suite of chemicals from sources like common pharmaceutical and personal care products. While we have made great strides in protecting the valuable natural resources of the Ohio River over the past quarter of a century, these emerging contaminants remind us to be ever vigilant.

LESSONS LEARNED

Over the course of the recovery of the river, many valuable lessons have been learned that may be of use in other aquatic ecosystems across the world. First and foremost, collaboration among agencies was essential. No single agency or organization could undo the effects of more than a century of abuse to a system as large as the Ohio River. Success came through a collaboration of efforts beginning in the 1930s between multiple state and federal agencies and, shortly thereafter, a newly created interstate agency. Through the interstate agency, beneficial resource objectives were established and actions taken to achieve those objectives. These objectives were accomplished with open and direct communication and coordination among all stakeholders (government agencies, industry, academia, and nongovernmental organizations) as the keys to success.

A second lesson to be learned from this story is to not be overwhelmed. Get the ball rolling to start a program of restoration, even in small steps initially. Even a task as monumental as improving the condition of the Ohio River when observed at its worst began with something as simple as starting the conversation among affected jurisdictions, organizations, and industry that something must be done. A wise man once said, "Unless someone like you cares a whole awful lot, nothing is going to get better. It's not" (Seuss 1971). Someone has to get things started.

Third, for restoration to be successful, nonnative invasive species must be controlled and actions must be taken to avoid future introductions. Starting in the 1800s with the importation of Common Carp, the Ohio River main stem and corridor are forever changed due to invasions of exotic plants and animals. Kudzu *Pueraria montana*, Japanese knotweed *Fallopia japonica*, Chinese honeysuckle *Combretum indicum*, and a myriad of other alien plants line the riparian areas of the Ohio River and its tributaries. Hydrilla and Eurasian watermilfoil are beginning to dominate the upper sections of the main stem. Four nonnative carp species have strong footholds throughout the basin and main stem with another threatening to invade the lower river. Zebra mussels, Asiatic clams, and rusty crayfish *Orconectes rusticus* are among the host of invertebrates challenging native species for the right to exist in the river. While this picture may seem dismal, native species hopefully can weather the storm and once again thrive when the new threats have reached equilibrium. Further, new techniques may become available to control species like Silver Carp, although efforts to control nonnative species use up financial and personnel resources that could have been used elsewhere. Government agencies and local citizens alike should take all necessary steps to avoid future introductions.

In summary, we cannot forget that the Ohio River has been, is, and likely always will be a working, populated, industrialized river. Once an industrial sewer, the Ohio River today supports numerous recreational activities ranging from major national bass tournaments, to regional and local catfish, crappie, and Walleye tournaments, to boat regattas, as well as paddling and swimming competitions. The story of the success of the Ohio River fishery is a story of resource management in its most holistic form, with underlying water quality improvements being primarily responsible for the observed recovery of the fishery. This case study demonstrates that recovery of a fish community to the point of supporting a viable fishery is possible. However, maintenance of this success is in many ways more challenging than the recovery in that the threats are less obvious and require different types of control actions than the traditional collect and treat approach used for sewage and industrial wastes. This story of the Ohio River fishery spans more than 100 years. Resource utilization, management, and improvements over the next 100 years will hopefully be as successful.

ACKNOWLEDGMENTS

The authors thank Frank Jernejcic and John Arway for their discussions involving the role of state fishery agencies in the recovery of the Ohio River. Many thanks also to the various members of the ORSANCO community over the years, including staff, committee members, and commissioners. Last, the editors and two anonymous reviewers provided valuable comments that greatly helped improve the manuscript.

REFERENCES

Angradi, T. R., D. W. Bolgrien, T. M. Jicha, M. S. Pearson, B. H. Hill, D. L. Taylor, E. W. Schweiger, L. Shepard, A. R. Batterman, M. F. Moffett, C. M. Elonen, and L. E. Anderson. 2009. A bioassessment approach for mid-continent great rivers: the upper Mississippi, Missouri, and Ohio (USA). Environmental Monitoring and Assessment 152:425–442.

Cavanaugh, T. M., and W. J. Mitsch. 1989. Water quality trends of the upper Ohio River from 1977 to 1987. Ohio Journal of Science 89(5):153–163.

Compton, M. C., D. J. Eisenhour, R. R. Cicerello, L. E. Kornman, A. Surmont, Jr., and E. L. Laudermilk 2004. Distributional records of selected Kentucky fishes. Journal of the Kentucky Academy of Sciences 65(2):76–84.

Cramer, Z. 1824. The navigator, 12th edition. Cramer and Spear, Pittsburgh, Pennsylvania.

Drum, R. G., J. Noel, J. Kovatch, L. Yeghiazarian, H. Stone, J. Stark, P. Kirshen, E. Best,

E. Emery, J. Trimboli, J. Arnold and D. Raf. 2017. Ohio River basin: formulating climate change mitigation/adaptation strategies through regional collaboration with the ORB Alliance. U.S. Army Corps of Engineers, Institute for Water Resources, Civil Works Technical Report CWTS 2017-01, Alexandria, Virginia.

Emery, E. B., T. P. Simon, F. H. McCormick, P. L. Angermeier, J. E. Deshon, C. O. Yoder, R. E. Sanders, W. D. Pearson, G. D. Hickman, R. J. Reash, and J. A. Thomas. 2003. Development of a multimetric index for assessing the biological condition of the Ohio River. Transactions of the American Fisheries Society 132:791–808.

Hansen, J., M. Sato, R. Ruedy, D. W. Lo, and M. Medina-Elizade. 2006. Global temperature change. Proceedings of the National Academy of Sciences of the United States of America 103:14288–14293.

Herrala, J. 2014. Evaluation of a supplemental stocking program to improve the Largemouth Bass fishery in the Markland and Meldahl pools of the Ohio River. Kentucky Department of Fish and Wildlife Resources, Fisheries Bulletin 115, Frankfort.

Krumholz, L. A., and W. L. Minckley. 1964. Changes in the fish population in the upper Ohio River following temporary pollution abatement. Transactions of the American Fisheries Society 93:1–5.

Lachner, E. A. 1956. The changing fish fauna of the upper Ohio basin. Pages 64–78 *in* C. A. Tryon, Jr. and M. A. Shapiro, editors. Man and the waters of the upper Ohio basin. Pymatuning Laboratory of Field Biology, Special Publication 1, Pittsburgh, Pennsylvania.

Lohner, T. W., and D. A. Dixon. 2013. The value of long-term environmental monitoring programs: an Ohio River case study. Environmental Monitoring and Assessment 185:9385–9396.

Lorentz, C. N., D. T. Saalfeld, and S. T. Saalfeld. 2006. Status and changes of Ohio River fish assemblages around William H. Zimmer Power Plant, Moscow, Ohio. Journal of the Kentucky Academy of Science 67:39–46.

Markle, D. F. 1997. Audubon's hoax: Ohio River fishes described by Rafinesque. Archives of Natural History 24:439–447.

ORSANCO (Ohio River Valley Water Sanitation Commission). 1957. Ohio River Valley Water Sanitation Commission: a report on the interstate crusade for clean streams. 9th annual summary. ORSANCO, Cincinnati. Available: www.orsanco.org/wp-content/uploads/2016/07/annualreport1957.pdf. (August 2018).

ORSANCO (Ohio River Valley Water Sanitation Commission). 1960. Ohio River Valley Water Sanitation Commission: a report on the interstate crusade for clean streams. Chronicle of the 12th year. ORSANCO, Cincinnati. Available: www.orsanco.org/wp-content/uploads/2016/07/annualreport1960.pdf. (August 2018).

ORSANCO (Ohio River Valley Water Sanitation Commission). 1962a. Aquatic life resources of the Ohio River. ORSANCO, Cincinnati.

ORSANCO (Ohio River Valley Water Sanitation Commission). 1962b. Yesterday…today and tomorrow: 14th annual report on the interstate crusade for clean streams. ORSANCO, Cincinnati. Available: www.orsanco.org/wp-content/uploads/2016/07/annualreport1962.pdf. (August 2018).

ORSANCO (Ohio River Valley Water Sanitation Commission). 1990. Water quality trends: Ohio River and its tributaries. ORSANCO, Cincinnati.

Pearson, W. D., and L. A. Krumholz. 1984. Distribution and status of Ohio River fishes. Oak Ridge National Laboratory, ORNL/SUB/79-7831/1, Oak Ridge, Tennessee.

Pearson, W. D., and B. J. Pearson, 1989. Fishes of the Ohio River. The Ohio Journal of Science 89:181–187.

Rafinesque, C. S. 1820. *Ichthyologia Ohiensis*. W. G. Hunt, Lexington, Kentucky.

Seuss, Dr. 1971. The lorax. Random House, New York.

Simon, T. P., and E. B. Emery. 1995. Modification and assessment of an index of biotic integrity to quantify water resource quality in great rivers. Regulated Rivers: Research and Management 11:283–298.

Thomas, J. A., E. B. Emery, and F. H. McCormick. 2005. Detection of temporal trends in Ohio River fish assemblages based on lockchamber surveys (1957–2001). Pages 431–450 *in* J. N. Rinne, R. M. Hughes, and B. Calamusso, editors. Historical changes in large river fish assemblages of the Americas. American Fisheries Society, Symposium 45, Bethesda, Maryland.

USEPA (U.S. Environmental Protection Agency). 1994. Combined sewer overflow (CSO) control policy. USEPA, Office of Water, EPA 830-B094-001, Washington, D.C.

Williams, J. C., J. E. Hannegan, and W. M. Clay. 1960. ORSANCO–University of Louisville surveillance study of the Ohio River, 1960–61 final report. Ohio River Valley Water Sanitation Commission, Cincinnati.

Yoder, C. O., E. T. Rankin, V. L. Gordon, L. E. Hersha, and C. E. Boucher. 2019. Degradation and recovery of the Scioto River (Ohio, USA) fish assemblages from presettlement to present-day conditions. Pages 233–265 *in* C. C. Krueger, W. W. Taylor, and S. Youn, editors. From catastrophe to recovery: stories of fishery management success. American Fisheries Society, Bethesda, Maryland.

APPENDIX: TEMPORAL TRENDS OF OHIO RIVER FISH SPECIES

TABLE A.1. Results of temporal trends analyses for 100 Ohio River fish species collected from 1991 to 2016 (2,734 sites) with guild assignments. Slopes (m) and p values, when significant ($\alpha = 0.05$), are provided for 75th quantile regressions of species abundance per 500-m site versus year. Great river species (GRS) and simple lithophils (SL) are denoted with an "X." For tolerance (Tol), T = tolerant, M = moderately intolerant, and I = intolerant.

Species	Slope	*p*	Trophic	GRS	SL	Tol
Ohio Lamprey *Ichthyomyzon bdellium*			piscivore			I
Silver Lamprey *I. unicuspis*			piscivore			
Paddlefish *Polyodon spathula*			planktivore	X	X	I
Spotted Gar *Lepisosteus oculatus*			piscivore			
Longnose Gar *L. osseus*	0.11	<0.001	piscivore			
Shortnose Gar *L. platostomus*			piscivore			
Bowfin *Amia calva*			piscivore			
Goldeye *Hiodon alosoides*			invertivore	X	X	I
Mooneye *H. tergisus*			invertivore	X	X	I
Skipjack Herring *Alosa chrysochloris*			piscivore	X		
Gizzard Shad *Dorosoma cepedianum*			herbivore			
Threadfin Shad *D. petenense*	−1.08	0.038	plankivore			
Central Stoneroller *Campostoma anomalum*			herbivore			
Goldfish *Carassius auratus*			detritivore			T
Grass Carp *Ctenopharyngodon idella*			herbivore			T
Spotfin Shiner *Cyprinella spiloptera*	0.33	<0.001	omnivore			
Steelcolor Shiner *C. whipplei*			invertivore			
Common Carp *Cyprinus carpio*			detritivore			T
Common Carp × Goldfish			detritivore			
Streamline Chub *Erimystax dissimilis*			invertivore			

TABLE A.1. Continued.

Species	Slope	*p*	Trophic	GRS	SL	Tol
Mississippi Silvery Minnow *Hybognathus nuchalis*			detritivore	X		
Silver Carp *Hypophthalmichthys molitrix*			herbivore			T
Bighead Carp *H. nobilis*			omnivore			T
Striped Shiner *Luxilus cerasinus*			invertivore			
Silver Chub *Macrhybopsis storeriana*	−0.36	0.00002	invertivore	X	X	
River Chub *Nocomis micropogon*			invertivore			
Golden Shiner *Notemigonus crysoleucas*			omnivore			T
Emerald Shiner *Notropis atherinoides*	1.0	0.01152	planktivore			
River Shiner *N. blennius*	0.36	<0.001	invertivore	X		
Spottail Shiner *N. hudsonius*	−0.21	0.00905	omnivore			
Silver Shiner *N. phtogenis*	−0.77	0.03288	invertivore			
Rosyface Shiner *N. rubellus*			invertivore			I
Silverband Shiner *N. shumardi*			invertivore			
Sand Shiner *N. stramineus*			invertivore			
Channel Shiner *N. wickliffi*	2.85	<0.001	invertivore	X		I
Suckermouth Minnow *Phenacobius mirabilis*			invertivore			
Bluntnose Minnow *Pimephales notatus*	0.55	<0.001	detritivore			T
Fathead Minnow *P. promelas*			detritivore			T
Bullhead Minnow *P. vigilax*			omnivore			
Creek Chub *Semotilus atromaculatus*			invertivore			
River Carpsucker *Carpiodes carpio*	0.333	<0.001	detritivore			
Quillback *C. cyprinus*			detritivore			
Highfin Carpsucker *C. velifer*			detritivore			
White Sucker *Catostomus commersonii*			detritivore		X	T
Blue Sucker *Cycleptus elongatus*	−0.4	0.021	invertivore	X	X	I
Northern Hog Sucker *Hypentelium nigricans*			invertivore	X	I	
Smallmouth Buffalo *Ictiobus bubalus*	0.083	0.042	detritivore			
Bigmouth Buffalo *I. cyprinellus*			detritivore			
Black Buffalo *I. niger*	0.05	0.024	detritivore			
Spotted Sucker *Minytrema melanops*			invertivore	X		
Silver Redhorse *Moxostoma anisurum*	0.25	<0.001	invertivore	X	M	
River Redhorse *M. carinatum*			invertivore	X	I	
Black Redhorse *M. duquesnei*			invertivore	X	I	
Golden Redhorse *M. erythrurum*	0.231	0.0009	invertivore	X	M	
Shorthead Redhorse *M. macrolepidotum* or Smallmouth Redhorse *M. breviceps*			invertivore	X	I	
Blue Catfish *Ictalurus furcatus*			omnivore	X		
Channel Catfish *I. punctatus*			omnivore			

TABLE A.1. Continued.

Species	Slope	*p*	Trophic	GRS	SL	Tol
Flathead Catfish *Pylodictis olivaris*			piscivore			
Northern Pike *Esox lucius*			piscivore			
Muskellunge *E. masquinongy*			piscivore			
Trout-perch *Percopsis omiscomaycus*	0.421	0.018	invertivore			
Brook Silverside *Labidesthes sicculus*			invertivore			I
Mississippi Silverside *Menidia audens*			invertivore			
Banded Killifish *Fundulus diphanus*			invertivore			
Morone sp.			piscivore			
White Perch *Morone americana*			invertivore			
White Bass *M. chrysops*			piscivore			
Yellow Bass *M. mississippiensis*			invertivore			
Striped Bass *M. saxatilis*	−0.231	0.008	piscivore			
Hybrid striper			piscivore			
Rock Bass *Ambloplites rupestris*	0.062	0.012	piscivore			
Lepomis hybrid			invertivore			
Green Sunfish *Lepomis cyanellus*			invertivore			T
Pumpkinseed *L. gibbosus*			invertivore			
Warmouth *L. gulosus*			invertivore			
Orangespotted Sunfish *L. humilis*	0.588	0.015	invertivore			
Bluegill *L. macrochirus*	0.364	0.014	invertivore			
Longear Sunfish *L. megalotis*	0.167	0.006	invertivore			
Redear Sunfish *L. microlophus*	0.077	0.033	invertivore			
Smallmouth Bass *Micropterus dolomieu*			piscivore			I
Spotted Bass *M. punctulatus*			piscivore			
Largemouth Bass *M. salmoides*			piscivore			
White Crappie *Pomoxis annularis*			piscivore			
Black Crappie *P. nigromaculatus*			invertivore			
Mud Darter *Etheostoma asprigene*			invertivore			
Greenside Darter *E. blennioides*			invertivore			I
Rainbow Darter *E. caeruleum*	−0.091	0.0005	invertivore		X	M
Fantail Darter *E. flabellare*			invertivore			
Johnny Darter *E. nigrum*			invertivore			
Banded Darter *E. zonale*			invertivore			I
Yellow Perch *Perca flavescens*			invertivore			
Logperch *Percina caprodes*	−0.1	0.02	invertivore		X	I
Channel Darter *P. copelandi*	−0.435	<0.001	invertivore	X	X	I
Slenderhead Darter *P. phoxocephala*			invertivore		X	I
Dusky Darter *P. sciera*			invertivore		X	
River Darter *P. shumardi*	−0.1	0.022	invertivore	X	X	
Sauger *Sander canadensis*	−0.222	0.017	piscivore		X	
Walleye *S. vitreus*	0.071	0.054	piscivore		X	
Saugeye (Sauger × Walleye)			piscivore		X	
Freshwater Drum *Aplodinotus grunniens*	−0.6	0.0002	invertivore			

Degradation and Recovery of Scioto River (Ohio, USA) Fish Assemblages from Presettlement to Present-Day Conditions

Chris O. Yoder,* Edward T. Rankin, Vickie L. Gordon, and Lon E. Hersha

Midwest Biodiversity Institute, Center for Applied Bioassessment and Biocriteria Post Office Box 21561, Columbus, Ohio 43221, USA

Charles E. Boucher

Ohio Environmental Protection Agency, Division of Surface Water Ecological Assessment Section 4675 Homer Ohio Lane, Groveport, Ohio 43125, USA

Abstract.—A 37-year series of standardized fish assessments in the Scioto River (Ohio, USA) since 1979 coupled with historical information documents a near complete recovery from heavily polluted conditions in the late 19th and early to mid-20th centuries. Nearly 100 fish species were extirpated downstream from the city of Columbus (Ohio, USA) by sewage and industrial pollution. The 1972 amendments to the Federal Water Pollution Control Act (Clean Water Act) mandated the control of sewage and industrial pollution. Reductions in loadings of untreated or poorly treated sewage were incremental. Full recovery to near-prepollution composition and abundance took more than two decades after advanced wastewater treatment was achieved. Unpolluted tributaries served as recolonization sources for populations of extirpated species. These positive changes extended across all fish assemblage members as evidenced by increased values of the Ohio index of biotic integrity; modified index of well-being; native species richness, density, and biomass; and the reduced incidence of external anomalies on fish. These restoration successes and their documentation were facilitated by the Clean Water Act that set forth the goals for water quality standards and treatment technology for reducing water pollution and conducting baseline and follow-up monitoring. An important lesson learned was that serious doubts that existed in the 1970s about the feasibility of advanced wastewater treatment technology and the attainability of water quality standards in an effluent dominated river were completely erased by the demonstrated improvements in the fish and macroinvertebrate assemblages in the Scioto River. The extent of improvements in recreational opportunities have tracked that of the biota by an increased use for fishing, canoeing, kayaking, and related forms of recreation. However, maintaining these improvements will require continuation of high levels of wastewater treatment and water quality standards. A growing human population that is forecast to increase by one-half million persons by 2050 makes maintaining the currently high levels of biological integrity a continuing challenge. Given the lessons learned with the mosaic of stressors in the Scioto River over the past 150 years, we believe this challenge can be met successfully.

* Corresponding author: cyoder@mwbinst.com

INTRODUCTION

This chapter focuses on the central Scioto River main stem, a major tributary to the Ohio River, located in Franklin and Pickaway counties between Columbus and Circleville, Ohio, USA (Figure 1). The study area includes a 65 km (40 mi) reach of the main stem from within the city limits of Columbus to just downstream from Circleville. Our focus is on the changes in the fish assemblages that occurred between the pre-European settlement period of the late 1700s to the present. Information about fish assemblages through that time period initially included the observations of pioneering naturalists who inventoried the aquatic fauna to the first treatises of fishes in Franklin County (Williamson and Osburn 1898) and Ohio (Osburn 1901) at the turn of the 20th century. The documentation of changes to the fish fauna and the factors and conditions that contributed to those changes through 1977 were described in detail by Trautman (1957, 1977, 1981) in *The Fishes of Ohio* and *The Ohio Country from 1950–1977*. The data on the distribution of fishes for that time period are held within The Ohio State University Museum of Biodiversity (OSUMB) fish collections database. However, the OSUMB records of that time period are comprised of presence–absence of species without standardized relative abundance data.

The increase in water quality monitoring spurred by the rapid proliferation of government agencies tasked with environmental protection and management in the early 1970s eventually included biological monitoring of aquatic assemblages such as macroinvertebrates and fish. Whereas water quality monitoring had focused on selected chemical parameters even after passage of the 1972 Federal Water Pollution Control Act (Clean Water Act [CWA]), the biological integrity and propagation of fish, shellfish, and wildlife provisions[1] spurred a renewed interest in biological monitoring and assessment (Davis and Simon 1995). Whereas the U.S. Environmental Protection Agency (U.S. EPA) initially required only pilot programs, the Ohio Environmental Protection Agency (Ohio EPA) initiated a statewide program of intensive main-stem river and watershed surveys of fish and macroinvertebrate assemblages in combination with chemical and physical monitoring in 1979 (Yoder and Rankin 1995a, 1998; Ohio EPA 2011). In Ohio, monitoring information is used to set the water quality standards (WQS) that determine the requirements for pollution discharges to individual rivers and streams such that the latter are based on actual data as opposed to an assumed basis (i.e., a stream or river is assumed to be protected at the minimum required by the CWA and without any supporting data). As a result, the Ohio CWA programs have been guided and affirmed by the results of comprehensive monitoring and assessment data for nearly 40 years.

The main stem of the Scioto River between Columbus and Circleville (the central Scioto River) was monitored annually beginning in 1979 with the goal of providing detailed information about year-to-year variations in the biological assemblages, measured by species counts and relative abundance, and about the need for and responses to what were then seen as increasingly stringent controls for wastewater discharges. Serious doubts existed at that time about the feasibility of wastewater treatment yielding the stringent quality needed to meet WQS. In addition, wastewater flows from municipal sewage treatment plants throughout Ohio and the Midwest comprised the majority of the flow in rivers during the summer

[1] Section 101(a) for biological integrity and 101(a)(2) for the protection and propagation provisions.

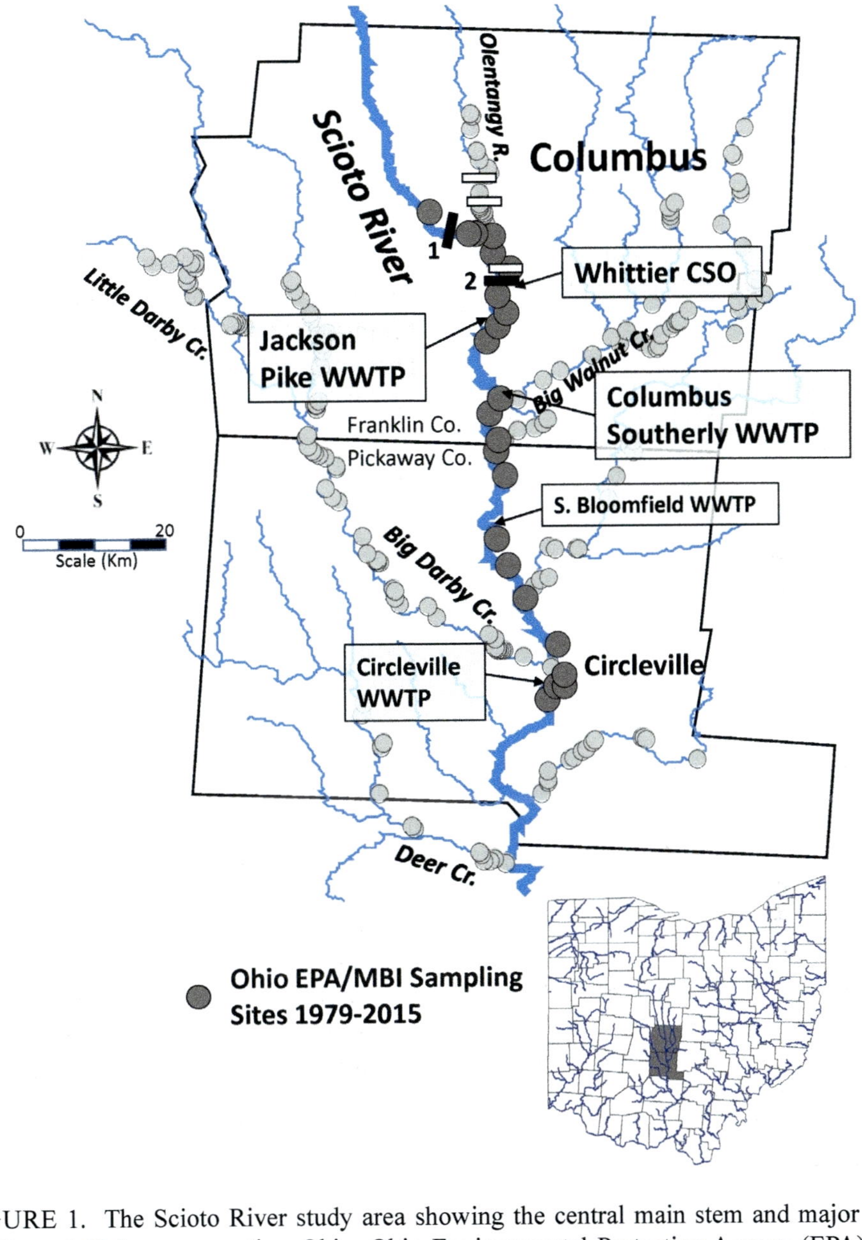

FIGURE 1. The Scioto River study area showing the central main stem and major tributaries in Franklin and Pickaway counties, Ohio; Ohio Environmental Protection Agency (EPA) and Midwest Biodiversity Institute (MBI) fish sampling sites (1979–2015); major wastewater treatment plants (WWTPs); removed dams (open bars); and remaining dams (solid bars; 1—Dublin Road Water Treatment Plant Dam; 2—Greenlawn Dam). CSO = combined sewer overflow.

and fall months. Termed "effluent dominated flows," the attainability of CWA-mandated WQS under the extant philosophy of "dilution is the solution to pollution" was widely questioned. Serious proposals to reclassify such river segments as limited for aquatic life were advanced to preclude what were then seen as technologically and economically unrealistic levels of treatment needed to meet CWA goals for water quality. This issue and the close proximity to the Ohio EPA facilities in the city of Columbus, Ohio made the central Scioto River main stem an ideal site for long-term monitoring of the response of the fish and other aquatic assemblages to pollution abatement practices. The Ohio EPA program was designed at the outset to provide meaningful biological endpoints, as opposed to single chemical pollutants, as the primary measures of pollution control success; thus, the monitoring and assessment program was integrated with the WQS and permitting programs early on (Yoder and Rankin 1998). This approach is not always the case, especially in states where the production of a statewide list of impaired waters is the singular focus of monitoring and assessment (Yoder and Barbour 2009). The Ohio approach not only satisfied the statewide objective, but also offered a way to scale assessment to individual river reaches and sources of pollution.

This chapter focuses on the decline and recovery of fish assemblages using the central Scioto River main stem and concludes with lessons learned about the successes and shortcomings of water pollution control over the past century.

ENVIRONMENTAL SETTING

The Scioto River basin drains 16,879 km^2 (6,517 mi^2) and is the second-largest watershed contained entirely within Ohio (16% of all land area in Ohio; Figure 1). The main stem is 372 km (231 mi) in length (Ohio DNR 1960), originating in Auglaize County in north-central Ohio and entering the Ohio River at Portsmouth, Ohio in Scioto County. The river has the longest length of unimpounded free-flowing riverine habitat in Ohio, being open from the Greenlawn Dam in Columbus, Ohio to the confluence with the Ohio River, a distance of 208 km (129.5 mi). The geology of the basin transitions from the glaciated Eastern Corn Belt Plains to the unglaciated Western Allegheny Plateau ecoregion (Omernik 1987; Woods et al. 1995) just south of Chillicothe, Ohio. The main stem and several of its tributaries were formed within outwash valleys after the end of the Wisconsinan age of glaciation. As a result, the Scioto River and its tributaries had continuous base flows comprised of groundwater inputs and contained coarse substrates of gravel and cobbles with slab boulders at points along the main stem where outcroppings occur of Devonian limestone. The climate is temperate with four distinct seasons (spring, summer, autumn, and winter) and 95.5 cm (37.6 in) of average annual precipitation, which includes an average of 80 cm (32 in) of annual snowfall. The elevation at Columbus is 238 m (781 ft) above sea level.

The Scioto River valley was originally home to several aboriginal cultures, and the name Scioto is derived from the Wyandot word skɛnǫ·tǫ' (deer). Permanent European settlement was initiated after the American Revolution in the late 1700s and early 1800s, displacing the aboriginal populations soon thereafter. Of the 11 major cities that now border the Scioto River, the city of Columbus has by far the largest population. When Columbus was chartered in 1834, its population was 3,500 people. The population grew rapidly after 1900, and especially through the latter half of the 20th century, reaching just over 2 million in 2015 across the greater metropolitan region. Unlike many of the larger cities in Ohio and neighboring states, Columbus lacks heavy industry, and as the state capitol, its economic base is comprised mostly of government and nonindustrial businesses. The Scioto River serves as the principal water supply for the Columbus metropolitan area, with two public water supply reservoirs, Griggs and O'Shaughnessy, located on the main stem upstream and two others (Alum Creek and Hoover reservoirs) on major Scioto River tributaries in the Big Walnut Creek watershed to the northeast of Columbus.

HYDROLOGICAL AND POLLUTION HISTORY

Documentation of water pollution in the Scioto River dates to 1886 by the Ohio State Board of Health (Sharp 1887) with descriptions of pollution serious enough to elicit numerous complaints by the public in nearly every city along the main stem. Wastewater from municipalities and industries were discharged without any treatment during that time. Leighton (1903) commented, "The river is little more than a dumping ground for refuse and its misuse affords a good example of the wanton destruction of a valuable resource." The description of polluted conditions in Ohio rivers and streams by Trautman (1933, 1977, 1981) adds the degradation caused by extensive changes in the landscape beginning with the near deforestation of Ohio and the conversion of the naturally occurring forests and prairies to agricultural and urban land uses into the early and mid-20th century. The accumulation of these changes with a rapidly increasing population and human activity altered the fish fauna throughout Ohio by greatly reducing or even eliminating many species of fish from entire regions of the state (Trautman 1981). Trautman (1977) observed that general littering and exploitation of natural resources had reached its zenith in Ohio in the 1960s and that such polluted conditions could no longer be ignored.

Sewage pollution in the study area has existed since the initial settlement of Columbus in the first half of the 19th century. The first public sewer was constructed in 1841, and in 1872, the first waterworks came into service for delivering potable water to homes and businesses and also for building and maintaining sanitary sewers (Lenz 1993). By 1880, nearly 0.29 million m^3/d (2.2 million gal/d [MGD]) of water was supplied, with the majority collected by sewers and discharged as raw, untreated sewage to receiving streams, all of which were tributaries to the Scioto River. Williamson and Osburn (1898) provided the first evidence of adverse effects to the aquatic biota, stating that the Scioto River downstream from Columbus was so polluted in the summer of 1897 that only a "few species of fish" could be obtained by seining. This account of a degraded fish assemblage was a result of the discharges of raw sewage and manufacturing wastes from Columbus and communities upstream. The first treatment of sewage was proposed by the city of Columbus in 1898 but was rejected by the Ohio Department of Health as being inadequate to resolve the problems posed by raw sewage discharges. After a redesign, the first treatment works using septic tanks and spray filtration was initiated in 1908, primarily in response to the threat of waterborne diseases.

Between that time and the 1970s, population growth and sewage flows continually exceeded the capacity of the treatment system such that an almost continual process of latent treatment upgrades occurred. During this time, Trautman (1977) specifically noted visual evidence of the resulting pollution below Columbus in the form of "globs of suds" that were more than 2 m (6 ft) high, which completely enveloped the boat he was using to navigate the main stem.

Known first as the Improved Sewage Works, what became the Jackson Pike Wastewater Treatment Plant (WWTP) was constructed in 1908 as the first dedicated sewage treatment facility, located 8 km (5 mi) downstream from downtown Columbus (Figure 1). The plant transitioned from primary to secondary treatment in the late 1950s with flows of nearly 11 million m^3/d (82 MGD). In 1967, the Southerly WWTP was constructed to handle the rapid growth in sewage flows, discharging to the main stem 19 km (12 mi) downstream from Columbus (Figure 1).

The eventual recognition that secondary treatment would be inadequate to meet water quality requirements mandated by the 1972 CWA led to the installation of advanced treatment in 1988. Termed "water quality-based limitations," effluent concentrations for 5-d biochemical oxygen demand (BOD_5), total suspended solids, and ammonia nitrogen (NH_3-N) were required to meet instream WQS based on the warmwater habitat aquatic-life-use designation.[2] Use designations and chemical water quality criteria are the key parts of the Ohio WQS that underwent several revisions in the 1970s and 1980s in response to the recommendations of the first of a series of biological and water quality surveys conducted by the Ohio EPA during and after that same time (Yoder et al. 1981; Ohio EPA 1986, 1999b, 2012).

The first Ohio EPA biological assessments followed nearly three decades of a singular focus on chemical measures of the water quality of receiving rivers and for setting treatment requirements for wastewater discharges. Furthermore, chemical assessments prior to the late 1970s were focused entirely on oxygen demanding wastes as the awareness and technology to measure and assess effects of toxicants lagged in their development. The earliest biological assessments of the Scioto River were focused on macroinvertebrates and represented their pioneering use as indexes of water quality in the 1960s by Olive and Smith (1975) and the Federal Water Pollution Control Administration (FWPCA) in 1965–1968 (FWQA 1970). Both studies concluded that

> the primary cause of water quality degradation was the excessive amount of oxygen demanding material introduced by municipalities and industries...which was particularly true of that portion of the Scioto River downstream from Columbus where the benthos reflected significant degradation of the Scioto River for a distance of 60–75 mi (97–120 km).

The FWQA (1970) included a compilation of fish kills during the same time frame, which were common occurrences involving thousands to tens of thousands of dead fish. One kill of more than 300,000 fish occurred in 1967. During that time, bypasses of untreated or partially treated sewage were commonplace because the treatment plants did not yet have the capacity to treat all sewage flows (Yoder et al. 1981). The raw sewage bypass at the Southerly WWTP was identified as a major cause of fish assemblage impairment as it comprised more than 78% of the total BOD_5 loading discharged in 1979 (Yoder et al. 1981).

Whereas the awareness of water pollution raised by these early assessments helped spur improved water pollution controls, alone they were inadequate to understand and address the needs for fully restoring aquatic assemblages to levels expected by the CWA. The Ohio WQS eventually filled this void by adding an initial list of toxic chemical parameters in 1978 and expanding it in coverage and scientific adequacy by 1990. The advent of biological criteria consisting of direct measures of the fish and macroinvertebrate assemblages (Ohio EPA 1987a, 1987b, 1989a, 1989b, 2015) and a process for determining the attainment and attainability of tiered aquatic life uses (Yoder 1995; Yoder and Rankin 1995a, 1998) completed gaps left by chemical assessments and criteria alone. Taken together, the integration of WQS and monitoring and assessment provided a firmer basis for requiring advanced wastewater treatment via National Pollutant Discharge Elimination System (NPDES) permitting, resulting in the eventual attainment of CWA goals in the Scioto River main stem.

Today, the most significant impacts from the Columbus sewer system include the Whit-

[2] Use designations and water quality criteria are defined in the Ohio Administrative Code (OAC 3745-1-07).

tier Street Combined Sewer Overflow (CSO) at river kilometer (RK) 208 (river mile [RM] 129.5), the Jackson Pike WWTP at RK 204.5 (RM 127.1), and the Southerly WWTP at RK 190.4 (RM 118.4; Figure 1). Together, the Jackson Pike and Southerly WWTPs can discharge approximately 21.4 million m^3/d (160 MGD) of treated wastewater, comprising 90% or more of the flow in the main stem during dry weather periods. Bypasses that were the most frequent in the 1970s and 1980s have been virtually eliminated by composting sewage sludge, expanding treatment capacity, and enlarging the interceptor sewer lines ahead of the WWTPs. Columbus has 31 CSO locations on the Scioto and lower Olentangy rivers between the Olentangy River confluence at RK 213 (RM 132.3) and the Whittier Street CSO. Another CSO is located on Alum Creek, which is in the tributary network of Big Walnut Creek and well removed from the Scioto River main stem. Approximately 90% of all CSO volume flows from the Whittier Street CSO and another 7% from the Alum Creek CSO, and these are currently being addressed by a variety of measures, including reducing the infiltration of precipitation into the sewer system. Two smaller sewage plants are located on the Scioto main stem at South Bloomfield and Circleville, each comprising less than 1% of the Columbus WWTPs by volume.

The Scioto River, through and just downstream from downtown Columbus, was widened and dammed in response to a record flood that decimated parts of the city in 1913. The remnants of the original channelization are still evident in an overwidened channel in an 8.4 km (5.2 mi) reach between the Olentangy River confluence and the Jackson Pike WWTP. The removal of the 93-year-old Main Street Dam, also in downtown Columbus, was completed in 2014. This dam impounded 3.7 km (2.3 mi) of the Scioto River, artificially enlarging the channel width to an average of 150 m (492 ft). Riffles and pools were restored in an effort to provide better habitat for native plant, fish, and mussel species (Stantec 2012). Two other dams were recently removed in the lower Olentangy River just upstream from the confluence with the Scioto River in downtown Columbus. The two remaining impoundments and barriers to upstream fish movement in the study area are the Dublin Road Water Treatment Plant Dam and the Greenlawn Dam (Figure 1). The Greenlawn Dam is the downstream-most barrier to upstream fish movement, leaving 208 km (129.5 mi) of the main stem open downstream to the Ohio River at Portsmouth, Ohio.

METHODS

Given that insufficient relative abundance data were collected prior to 1979, the Ohio EPA results from that same year are regarded herein as the pre-CWA pollution control baseline. Conditions during this time period were first assessed by Yoder et al. (1981) with a standardized pulsed DC electrofishing methodology that provided consistent relative abundance data, including numbers and biomass across all fish species.

For the Ohio EPA and Midwest Biodiversity Institute (MBI) data sets, all methods for capturing, identifying, and processing electrofishing samples followed those developed by the Ohio EPA (1989b, 2006, 2015) and the rationale described by the Ohio EPA (1987a, 1987b) and Yoder and Smith (1999). Fish were collected via standardized boat-mounted electrofishing (Ohio EPA 1989b, 2015). Boats 4.8 m in length were rigged with a hinged aluminum boom mounted on a bow platform. Four 6.25-mm-diameter by 0.9-m-long woven-steel cables served as anodes and were spaced evenly on a cross-member extending 2.75 m in front of the bow. Four 25-mm-diameter by 1.8-m flexible-steel cathodes were hung directly from the bow. Pulsed DC was produced and transformed by Smith-

Root type VI-A and 5.0 GPP electrofishing units. Power output was varied depending on relative conductivity that ranged from 500 to 900 μS/cm to produce 12–20 A of output.

Electrofishing was conducted during daylight within a June 15–October 15 seasonal index period along the shoreline for a lineal distance of 0.5. km at each site (Yoder and Smith 1999; Ohio EPA 2015). Time was recorded and sampling continued for a minimum of 2,400 s to ensure sufficient intensity of sampling effort. All habitats (pools, runs, riffles, shoals, and backwaters) that occur along shore were thoroughly electrofished. The boat was maneuvered by outboard motors (9–15 hp) and/or manually by pushing in shallow water where motoring was not effective. Stunned and immobilized fish were netted by one person standing on the bow platform and placed in an aerated live well. This method of sampling was effective for a wide spectrum of river fish, including smaller benthic and riffle dwelling species, large bottom dwellers, pool dwellers, and deep, fast water inhabitants. This sampling increased the known ranges for several Ohio fish species beyond those documented in Trautman (1981). Each sample typically produced 20–35 species (maximum = 50) and 250–500 fish (maximum >1,000), provided that chemical quality and physical habitat were not limiting. Captured fish were enumerated by species, weighed, examined for external anomalies, and released. Voucher specimens were preserved in 10% formalin and later transferred to 70% ethyl alcohol. Lengths were taken on selected species; otherwise, individuals were classified as adults, juveniles (1+), or young of year (0+) and recorded by those classes. Fish less than 15–20 mm total length (excepting adults of small species) were not included in the samples, following the recommendation of Angermeier and Karr (1986).

A qualitative habitat assessment was conducted during each sampling pass using the qualitative habitat evaluation index (QHEI; Ohio EPA 1989b; Rankin 1989, 1995) and as applied to nonwadeable rivers. The QHEI is a visual assessment of the quality, composition, amount, and extent of substrate, cover, channel complexity, riparian vegetation, current types, pool/run/riffle, and gradient variables. The QHEI corresponds to key attributes of fish assemblage quality (Rankin 1989, 1995) and is an important tool in determining the appropriate and attainable use classification for Ohio rivers and streams (Rankin 1995; Yoder and Rankin 1995a). These data were entered and stored in the Ohio ECOS data management system.

Macroinvertebrates were collected less often than fish but frequently enough to fulfill the dual assemblage approach employed by the Ohio EPA. Macroinvertebrates were sampled using modified Hester-Dendy artificial substrate samplers (quantitative sample) and a qualitative dip net/hand pick method in accordance with Ohio EPA macroinvertebrate assessment procedures (Ohio EPA 1989b, 2015). The artificial substrates were exposed for a colonization period of 6 weeks during a July 1–September 30 seasonal index period and placed to ensure adequate flow velocity over the substrates (minimum of 9.1 cm/s). A qualitative sample collected with a triangular frame dip net and hand picking was obtained at the time of substrate retrieval. All samples were initially preserved in a 10% solution of formaldehyde. Substrates were then transferred to the laboratory, disassembled, sieved (standard no. 30 and 40 sieves), and transferred to 70% ethyl alcohol. Laboratory sample processing of quantitative and qualitative samples included an initial scan and prepick for large and rare taxa followed by subsampling procedures in accordance with the Ohio EPA (1989b, 2015). Identifications were performed to the lowest taxonomic resolution possible for the commonly encountered orders and families, which is genus/species for most organisms. From these results, the density of macroinvertebrates per square foot is determined as well as taxonomic richness and an invertebrate community index (ICI; Ohio EPA 1987b; DeShon 1995) score for the quantitative samples and a narrative assessment for the standalone qualitative samples.

Sampling sites in the study area were selected using a spatially intensive design to comprehensively assess chemical and physical impacts within a main-stem river reach (Yoder et al. 2005). This design employed multiple sampling sites (Figure 1) so that results can be analyzed and displayed in a longitudinal context. The Ohio EPA river mile index system was used to designate individual sampling sites and calculate the distance of river between sampling sites and determine the length of river reaches. Data were stored in and retrieved from the Ohio EPA and MBI data management systems, and calculations of relative abundance and indices were performed using the available routines in each.

The sampling design, results, and interpretation relative to pollution sources was based on the concepts advanced by Bartsch (1948) and Doudoroff and Warren (1951) to facilitate detection and quantification of degradation and recovery from pollution influences along a river reach (i.e., pollution impact reaches). Sampling sites were located upstream from major sources of disturbance, in areas of immediate impact and potentially acute effects, through reaches of increasing and lessening degradation and reaches of recovery. The goal was to determine the role of specific sources and the cumulative effects from multiple sources. Large rivers were treated as a single study unit to understand how changes take place along a longitudinal continuum with respect to both natural and anthropogenic influences. This approach yielded a detailed assessment of impairments, the extent and severity of indicator responses in a particular river reach, and temporal changes (Yoder and DeShon 2003; Yoder et al. 2005). An index of biotic integrity (IBI; Ohio EPA 1987b; Yoder and Smith 1999) modified in accordance with Karr et al. (1986) and calibrated for boat electrofishing samples from large rivers, the modified index of well-being (MIwb; Gammon 1976; Gammon et al. 1981; Ohio EPA 1987b; Yoder and Smith 1999), and the ICI (Ohio EPA 1987b; DeShon 1995) were used to assess the status of the fish and macroinvertebrate assemblages in accordance with the biological criteria in the Ohio WQS.[3] Fish sampling efforts in the same reach of river between the Griggs Dam in Columbus downstream to Circleville have been conducted during 28 of the 37 years since 1979, mostly at the same level of effort in terms of individual sampling locations within a mid-June to mid-October seasonal index period. Macroinvertebrates were sampled over fewer years but at a rate of approximately once every 5 years between 1981 and 2015. The Ohio EPA has produced three major assessment reports since 1979 (Yoder et al. 1981; Ohio EPA 1986, 1999b, 2012), the most recent based on a 2010 survey. These data were also used to develop and update the Ohio integrated report (Ohio EPA 2016), which is a requirement of CWA Section 303[d] to report on the extent of impaired surface waters by the states on a biennial basis.

The status of the fish assemblages in the central Scioto River main stem over the 37-year period of 1979–2015 is presented herein as the compilation of results for selected Ohio fish assemblage measures (IBI, MIwb, and species richness, density, and biomass),

[3] Ohio Administrative Code Chapter 3745-1-07(B) Table 7-15.

the incidence of DELT[4] anomalies, and the area of degradation/area of attainment (ADV/AAV; Yoder and Rankin 1995b; Yoder et al. 2005) values from the main-stem reach between the Greenlawn Dam and an abandoned railroad trestle at RK 208 (RM 99.5) near Circleville comprises the data for assemblage level analyses.

Two species-level analyses were also included to supplement the assemblage-based IBI and MIwb assessments. Criticisms of such indices have been partially rooted in single index values obscuring important changes in species populations. Although aggregations of species by taxa similarity, function, and tolerance to pollution are factored into a properly developed and regionally calibrated IBI (Simon and Evans 2017), an index value alone is insufficient to directly highlight changes at the species level. We included this aspect by examining the size of selected large-bodied species in terms of individual fish species biomass and changes in the occurrence of species in the main stem over time to better reveal species level changes before and after water pollution controls.

Historical fish data prior to 1979 were obtained from the OSUMB Fish Division (https://fish-division.osu.edu) database and as displayed on the species distribution maps in Trautman (1981). The Ohio EPA and MBI data developed after 1978 are part of the current OSUMB database and comprise the vast majority of species occurrence records for the study area for the 1979–2015 time period. The OSUMB fish data prior to 1979 were comprised of presence–absence data with no relative abundance data; thus, it was restricted to species mapping and for making general comparisons about pre-1979 species occurrences in the study area. These data included adjacent portions of major tributaries to help explain the origin of species ingress into the main stem during and following water-quality-based permitting and pollution controls through 2015.

The frequency of occurrence within eight discrete time periods across the main-stem sites with fish sample records was used as the indicator of change through time. The time intervals included 1979–1983, 1984–1988, 1989–1993, 1994–1998, 1999–2003, 2004–2008, 2009–2013, and 2014–2016. We then followed the approach used by Buckwalter et al. (2017) by using R statistical software (R Core Team 2014) to evaluate trends in species occurrences. The lm function was used to fit the null and linear models, and the nlsLM function in the "minpack.lm" package (Elzhov et al. 2013) was used to fit the exponential model. Beta 1 confidence limits were computed with the confint function. The AICc function in the "AICcmodavg" package (Mazerolle 2015) was used to compute AICc values. We selected the strongest output between the linear and exponential functions and classified each of the 88 species and 3 hybrids with sufficient occurrences as exhibiting a strong increase, moderate increase, weak increase, decline, or no change across the eight sequential time periods. Differences in species occurrence between the pre-CWA pollution control period of 1979–1988 and the pollution recovery period of 2009–2015 were also analyzed using a chi-square test and reported for species that did not have significant relationships from the curve fitting analysis. We did this to account for variable results between the eight time intervals that may have masked an otherwise important trend.

RESULTS

Fish Species Occurrence

The pre-1979 historical record in Trautman (1981) and the Ohio EPA/MBI and OSUMB

[4] DELT = deformities, eroded fins, lesions, and tumors on fish, after the methodology of the Ohio EPA (1989a) and Sanders et al. 1999.

databases of 1979–2016 revealed that the central Scioto River now has one of the most species-rich fish assemblages and is among the strongest recoveries from previously polluted conditions of any river in Ohio. Eight previously unrecorded species were found in the central Scioto River study area by the Ohio EPA and MBI during 1979–2015 and, when added to the historical (Trautman 1981) list of 100 fish species, yields a current total of 108 species (Table 1). Two species, the Harelip Sucker *Moxostoma lacerum* and the Scioto Madtom *Noturus trautmani* were considered extinct and not counted in these totals. Three species are nonnative, with two species, Common Carp *Cyprinus carpio* and Goldfish *Carassius auratus*, introduced shortly after European settlement and the third, Striped Bass *Morone saxatilis*, as a managed species for fishery purposes in the 1980s. Hybrids have also been purposely stocked to enhance recreational fishing mainly in tributary reservoirs. These include Sauger *Sander canadensis* × Walleye *S. vitreus* and Striped Bass *Morone saxatilis* × White Bass *M. chrysops*. The native Muskellunge *Esox masquinongy* was also stocked for the same purpose and some individuals have entered the Scioto River main stem and been collected by electrofishing and caught by anglers on a regular basis. The effect of the Greenlawn Dam located at RK 208 (RM 129.5) as the downstream-most barrier to upstream fish movement was noteworthy in that 26 of the 99 native fish species that occurred downstream were not recorded upstream during 1979–2015 (Table 1). Those 26 species include some that have shown strong increases in occurrence downstream from the dam but are precluded from recolonizing their former ranges upstream because of their inability to bypass dams.

Ammonia-Nitrogen Loadings

Loadings of ammonia nitrogen (NH_3-N) from the two Columbus WWTPs and concentrations measured periodically at a fixed station located less than 5 km downstream from the Southerly WWTP markedly declined during 1988–2016 due to the implementation of advanced wastewater treatment in 1988 (Figure 2). Mass loadings of

TABLE 1. Number of fish species recorded in the Scioto River study area based on Ohio Environmental Protection Agency and Midwest Biodiversity Institute sampling from 1979 to 2015.

Reach description	River kilometer (river mile)	Native species	Nonnative species	Hybrids
Griggs Dam to Circleville	225 to 160 (140 to 99.5)	100	5	19
Griggs to Greenlawn Dam	225 to 208 (140 to 129.5)	73	3	15
Greenlawn Dam to Circleville	208 to 160 (129.5 to 99.5)	99	5	18
Absent above Greenlawn Dam	225 to 208 (140 to 129.5)	26	2	3
Absent below Greenlawn Dam	208 to 160 (129.5 to 99.5)	1	0	0
Species with one individual above Greenlawn Dam	225 to 208 (140 to 129.5)	10	0	0

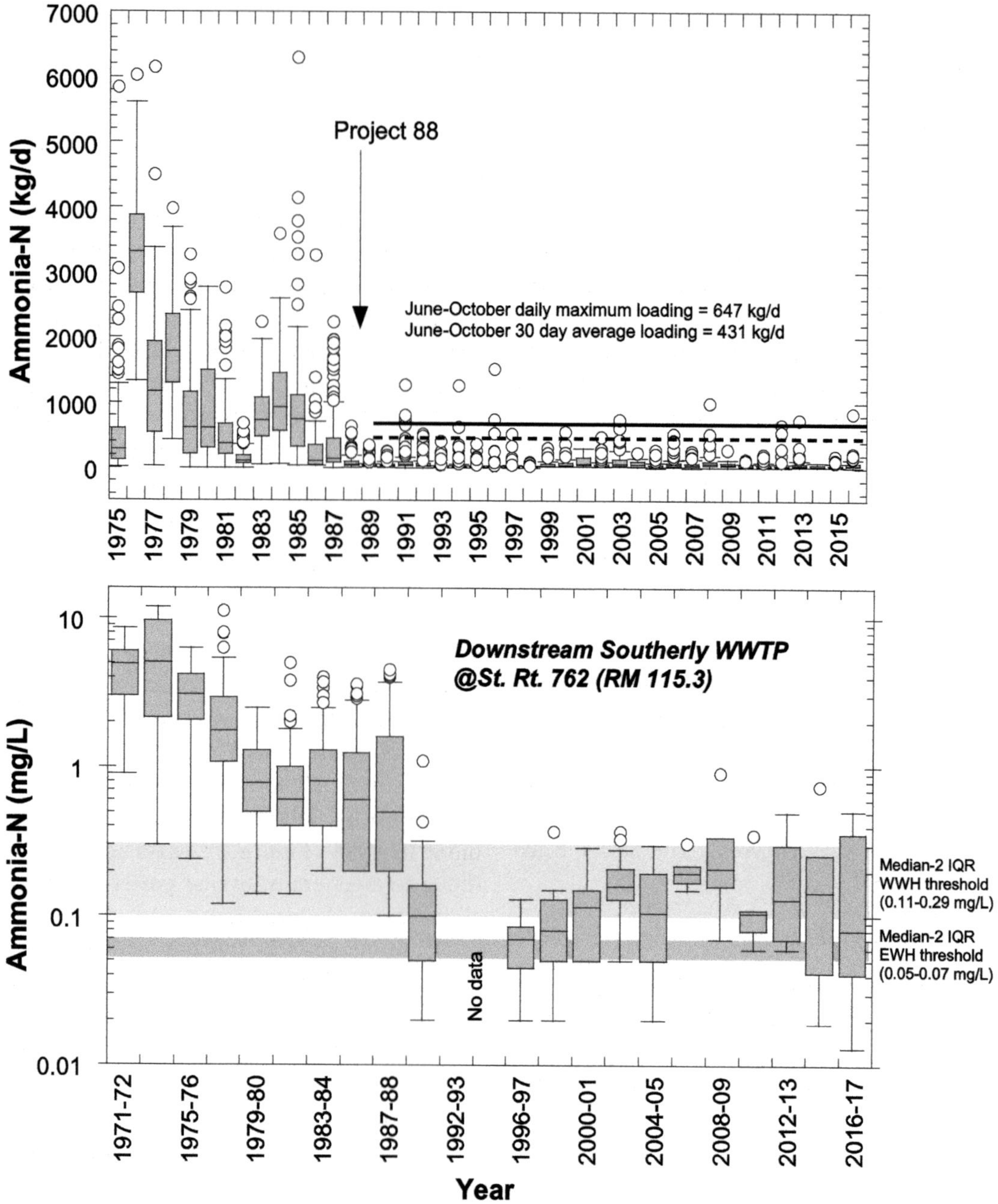

FIGURE 2. Box-and-whisker plots of June–October 1975–2016 ammonia-nitrogen loadings discharged by the Columbus Southerly Wastewater Treatment Plant (WWTP; upper graph) and 1971–2016 ammonia-nitrogen concentrations measured at a fixed monitoring location three miles downstream (lower graph). The solid and dashed lines in the upper graph are the permitted loading limitations. The shaded areas in the lower graph are the biological effect thresholds of ammonia-nitrogen consistent with the warmwater habitat (WWH) and exceptional warmwater habitat (EWH) use designations as reported by the Ohio Environmental Protection Agency (Ohio EPA 1999). IQR = interquartile range; RM = river mile.

NH_3-N declined from a median range of 500–2,000 kg/d prior to 1988 to less than 25–50 kg/d after 1988, well below water-quality-based NPDES permit limitations. The reduced loadings were matched by declining instream concentrations of total NH_3-N consistent with attainment of biological criteria for warmwater habitat use (Ohio EPA 1999a). Loadings of BOD_5, total suspended solids, and other pollutants followed a similar pattern (Ohio EPA 2012). The reduced levels of NH_3-N effluent loadings and improved instream water quality have generally been maintained through 2016. Loadings at the Jackson Pike WWTP and instream water quality downstream followed a similar pattern of improved quality (Ohio EPA 2012). Also important was the elimination of treatment system bypasses at the Southerly WWTP shortly after the installation of advanced treatment, a milestone that coupled with improved wastewater treatment fulfilled recommendations made by the Ohio EPA in 1981 (Yoder et al. 1981).

Assemblage Level Indicators—IBI, MIwb, ICI, DELT Anomalies, and ADV/AAV

Longitudinal and temporal trends in the IBI, MIwb, and macroinvertebrate ICI (DeShon 1995) show the biological effects of reduced pollutant loadings (Figure 3). The 1979 results reflected the severity of impairments that existed between downtown Columbus and Circleville, which resulted from excessive loadings from WWTPs, WWTP bypasses, and CSOs. By 1988, improvement was evident and attributed to reductions in the extensive bypassing of untreated sewage at the Southerly WWTP by the composting of sewage sludge as early as 1981 (Figure 4). After the installation of advanced treatment in 1988, the fish assemblage recovery continued into the mid-1990s and was sustained through 2015. The macroinvertebrate assemblage lagged behind the fish assemblage in the segment between the Whittier CSO and the Jackson Pike WWTP. In that segment, the assimilative capacity is diminished by a combination of diverted flows at the Dublin Road Water Treatment Plant and the aforementioned channel modifications made after the 1913 flood. The fish MIwb preceded the IBI in attaining the warmwater habitat biocriteria, a pattern that has been widely observed in rivers across the state (Yoder and Smith 1999; Yoder et al. 2005). The incidence of DELTs was not substantially reduced from extremely elevated levels (>3.0%) until 1996; it reemerged above that level in 2004 and 2005 and subsided thereafter. This pattern indicated that detectable incidences of DELT anomalies can be expected to occur in an effluent-dominated river under current WWTP permitting and treatment technology (Figure 4). Improvements in other fish assemblage measures such as catch per unit effort (density, numbers/km minus tolerants), biomass (kg/km minus tolerants), and native species richness followed a generally similar pattern of overall increases (Figure 5).

The increases in the fish and macroinvertebrate indices resulted in full attainment of the warmwater habitat biological criteria below the Jackson Pike WWTP to Circleville after implementation of advanced wastewater treatment in 1988 and the elimination of bypasses at the WWTPs by 1991. Consistent attainment of the IBI and MIwb biological criteria occurred only after the bypasses were eliminated, and this was reflected in near-zero or zero ADVs and increased AAVs (Figure 4).

Change in Maximum Weight of Selected Fish Species

Based on observations made by field personnel over the 37-year span of the central Scioto

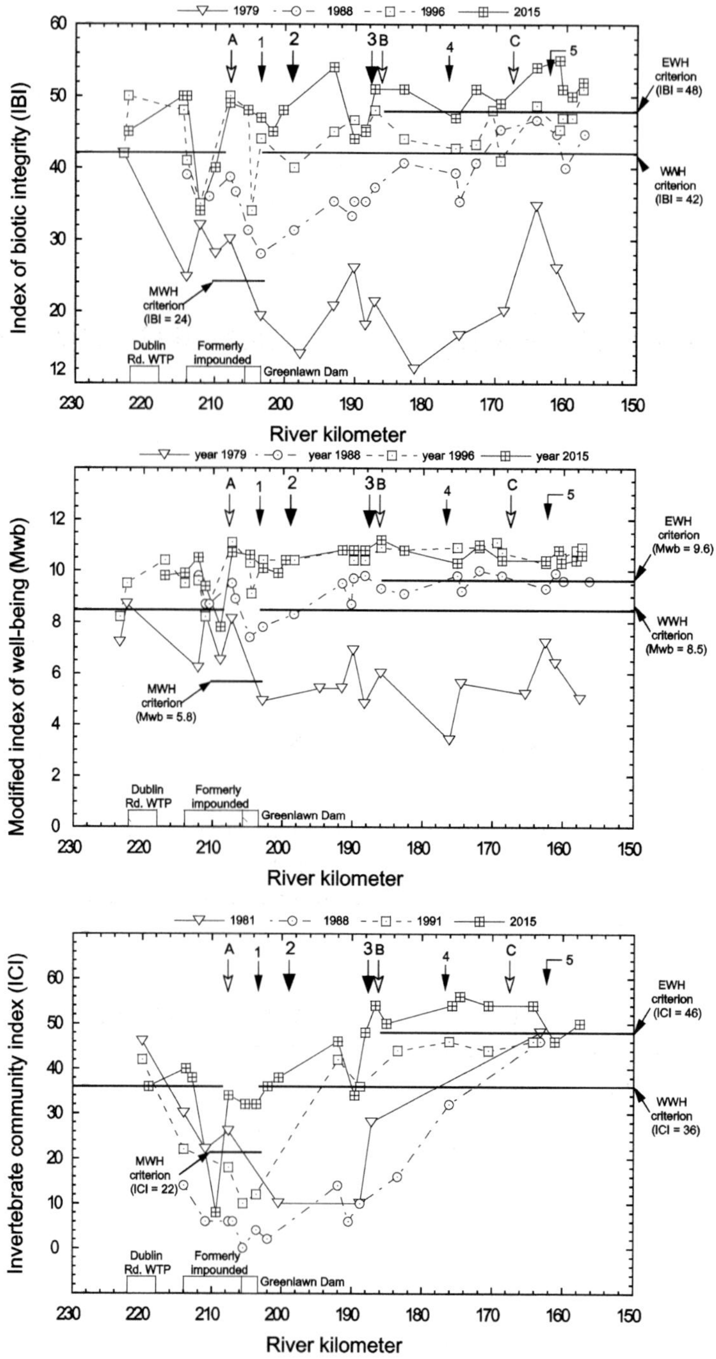

FIGURE 3. Fish and macroinvertebrate assemblage index results at four time intervals at multiple locations. Use designations: warmwater habitat = WWH, modified warmwater habitat = MWH, and exceptional warmwater habitat = EWH. Major pollution sources: 1 = Whittier Street Combined Sewer Overflow, 2 = Jackson Pike Wastewater Treatment Plant (WWTP), 3 = Southerly WWTP, 4 = South Bloomfield WWTP, and 5 = Circleville WWTP. Different font sizes indicate proportional pollutant loading. Major tributaries: A = Olentangy River, B = Big Walnut Creek, and C = Big Darby Creek. Dams and their impoundments: shaded boxes along the *x*-axis. WTP = water treatment plant.

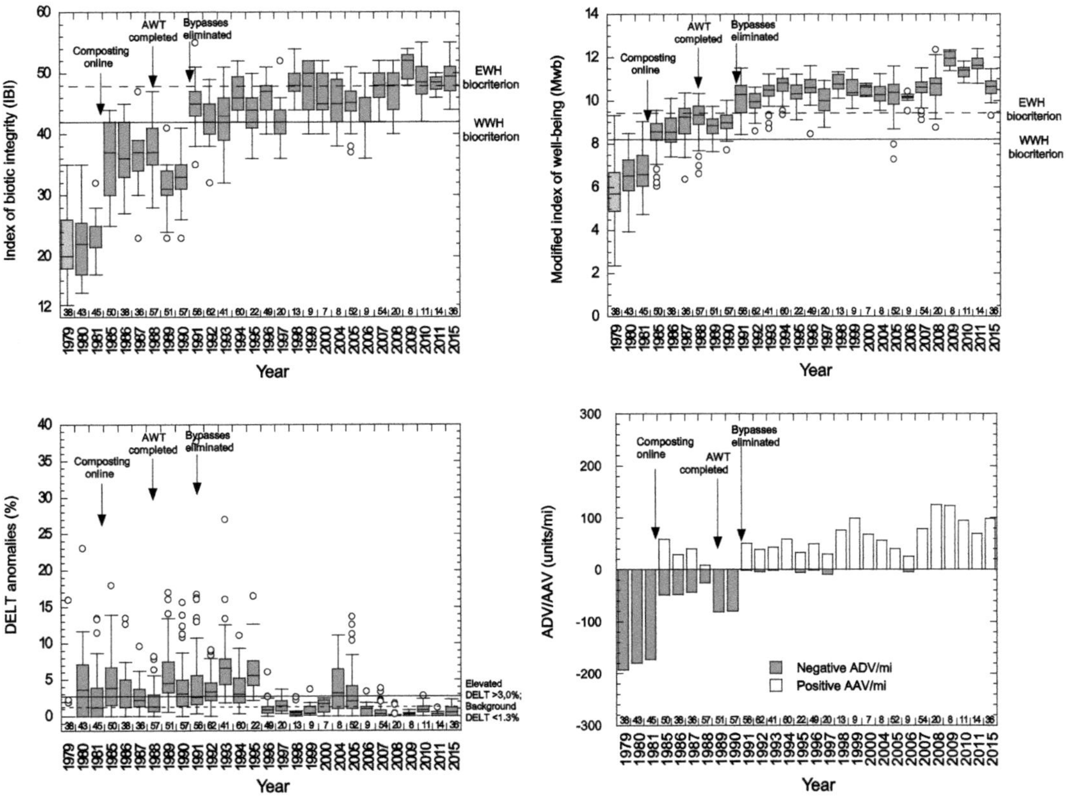

FIGURE 4. Box-and-whisker plots of the Ohio index of biotic integrity (IBI; upper left), the modified index of well-being (MIwb; upper right), the incidence of DELT anomalies (deformities eroded fins, legions, and tumors; lower left), and the area of degradation/area attainment (ADV/AAV) values (lower right) between 1979 and 2015.

River main-stem fish assemblage surveys, larger individual fish than in the past were being collected as water quality improved. To validate this casual observation, we examined trends in individual fish size across eight time intervals using the biomass of the largest individuals in samples for six large-bodied fish species (Figure 6). The results show a general increase through time in maximum weight for Smallmouth Bass *Micropterus dolomieu*, Sauger, Black Buffalo *Ictiobus bubalus*, River Redhorse *Moxostoma carinatum*, Longnose Gar *Lepisosteus osseus*, and Flathead Catfish *Pylodictis olivaris*, consistent with a recovered fish assemblage. All six of these species are relatively long-lived, reaching maturity over multiple years.

Evidence of Attainment of Biological Criteria

The assemblage level results for fish and macroinvertebrates showed that consistent recovery occurred within a few years after the installation of advanced treatment in 1988 (Figure 5). The most sustained attainment of the biological criteria occurred in 1991 and for all of the years through 2015. The most consistent measure of attainment of the IBI biological criterion for the warmwater habitat use was the ADV/AAV, which takes into account the extent to which IBI scores at each site meet or surpass the warmwater habitat biocriterion over the entire reach of the central Scioto River. The virtually complete at-

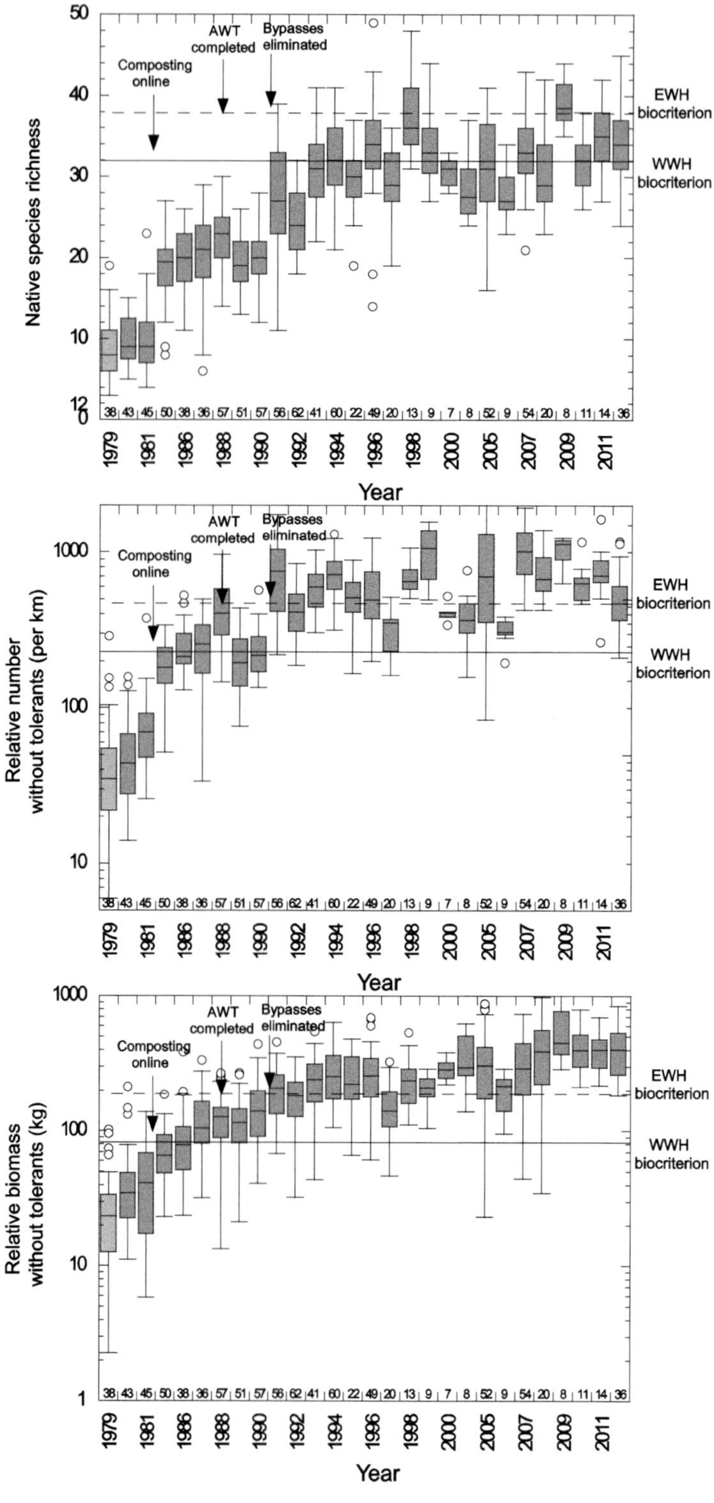

FIGURE 5. Box-and-whisker plots of fish native species richness, fish catch per unit effort (numbers/km) minus highly tolerant species (upper) and fish biomass (kg/km) minus highly tolerant species between 1979 and 2015. EWH = exceptional warmwater habitat; WWH = warmwater habitat.

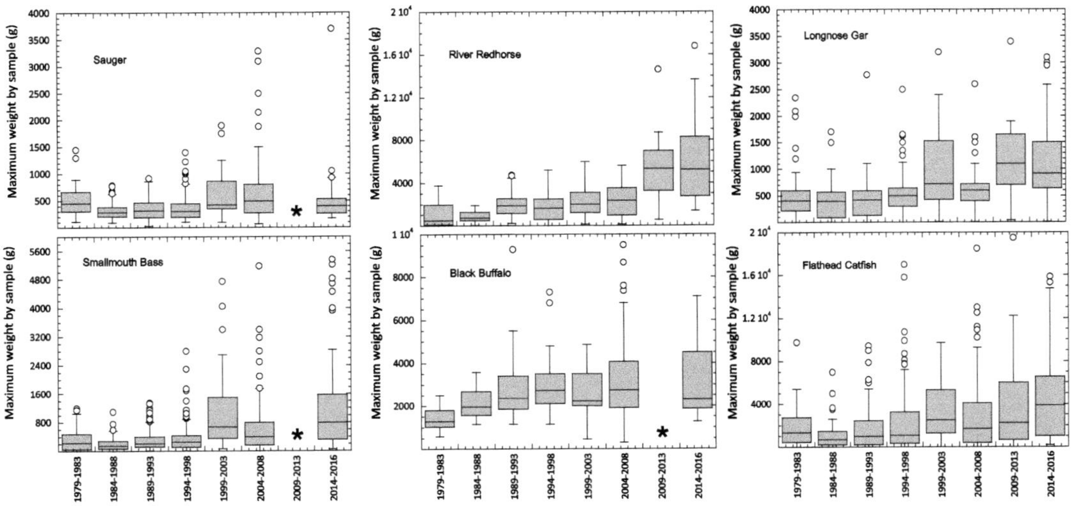

FIGURE 6. Box-and-whisker plots of the biomass (g) of the largest fish for six selected species in each Scioto River main-stem sample in which that species occurred over eight time intervals during 1979–2015 (* individual weight data not recorded).

tainment of the IBI biological criterion was first evident in 1991 and every year thereafter, with zero or near zero ADVs and the AAV generally increasing, which means IBI scores exceeded the minimum warmwater habitat biological criterion over an increased distance of the main stem (Figure 4). Prior to the installation of advanced treatment in 1988 and the elimination of raw sewage bypasses in 1991, the ADVs were indicative of a severe and embedded impairment across the entire main-stem reach from Columbus to Circleville, as was the AAV of zero, the diametric opposite of the results that occurred after 1991. Beginning in 1994, an increasing number of samples met the IBI and MIwb biological criteria for exceptional warmwater habitat (EWH) use downstream from Big Walnut Creek (RK 188.7 [RM 117.3]) less than 2 km downstream from the Southerly WWTP. The attainment demonstrated by the fish assemblage was accompanied by concurrent improvements in the macroinvertebrate ICI, partially attaining the EWH biological criterion in 1996 and completely attaining it by 2015. This confirms the attainability of the EWH designation because the Ohio EPA requires a showing of attainment by both assemblages to designate a stream or river as such in the Ohio WQS.

Change in Occurrence and Distribution of Fish Species

Increases in the frequency of occurrence in the main stem were significant for 35 fish species and one hybrid in the curve fitting analysis and an additional 34 species and one hybrid in the chi-square analysis (Tables 2 and 3). Of the species increasing in occurrence by curve fitting, 19 species plus the hybrid showed a strong increase, 10 showed a moderate increase, and 7 showed a weak increase (Table 2). Among the increasing species, 20 are classified by the Ohio EPA (1987b) as highly or moderately intolerant and 12 species as intermediate. The top nine species with strong increases in terms of how they were ranked in Table 2 (by linear slope coefficient) were all highly or moderately intolerant. Two species classified as moderately tolerant and one as tolerant also showed increases. Of the two species and one hybrid that showed sig-

TABLE 2. Linear and exponential curve fit results for 37 species and 2 hybrids occurring at more than five sites in the study area at various times based on % occurrence at all sampled sites. The direction and significance of the linear slope coefficients and the significance of the exponential (Exp.) coefficients were used to determine if a species had exhibited a strong increase (SI), moderate increase (MI), weak increase (WI), or decline (D) or was unchanged (null) over the eight time intervals. Species are ranked by the linear slope coefficient; the origin of ingress is provided and for species with significant increased or decreased % occurrence. Tolerance: I = highly intolerant; M = moderately intolerant; N = intermediate; P = moderately tolerant; T = highly tolerant; na = not assigned.

	% occurrence—year interval (number of sites sampled)											
Species	1979–1983 (71)	1984–1988 (105)	1989–1993 (67)	1994–1998 (100)	1999–2003 (61)	2004–2008 (91)	2009–2013 (36)	2014–2015 (61)	Model	Trend	Tolerance	Origin of ingress
Banded Darter *Ethostoma zonale*	10.0	29.8	19.3	44.4	54.2	45.8	80.9	75.4	Linear	SI	I	Large tributaries
Logperch *Percina caprodes*	8.3	16.7	25.4	48.4	64.4	69.1	72.1	86.0	Linear	SI	M	Large tributaries
Sand Shiner *Notropis stramineus*	16.1	47.6	32.6	67.2	78.0	61.4	92.7	69.3	Linear	SI	M	Large tributaries
Central Stoneroller *Campostoma anomalum*	19.4	44.6	46.7	76.9	82.2	61.0	82.4	82.5	Linear	SI	N	All tributaries
Greenside Darter *Etheostoma blennioides*	20.3	47.4	39.8	67.2	77.1	53.4	72.1	87.7	Linear	SI	M	Large tributaries
Streamline Chub *Erimystax dissimilis*	6.4	6.4	8.8	9.1	42.4	38.1	36.8	57.9	Exp.	SI	I	Big Darby Creek
Sauger *Sander canadensis* × Walleye *S. vitreus*	0.0	9.2	37.6	50.0	37.3	42.4	52.9	53.5	Linear	SI	na	Reservoir stocking
Rainbow Darter *Etheostoma caeruleum*	10.6	27.3	19.1	44.1	58.5	29.2	55.9	70.2	Linear	SI	M	All tributaries
Emerald Shiner *Notropis atherinoides*	3.9	17.0	27.1	41.9	39.0	57.2	61.8	35.1	Exp.	SI	N	Lower main stem
Mimic Shiner *Notropis volucellus*	1.4	4.3	13.0	13.8	32.2	37.7	30.9	43.9	Linear	SI	I	Main stem, tributaries
Smallmouth Redhorse *Moxostoma breviceps*	20.6	32.8	59.4	68.1	48.3	70.8	73.5	62.3	Linear	SI	I	Mainstem, tributaries

TABLE 2. Continued.

Species	% occurrence—year interval (number of sites sampled)								Model	Trend	Tolerance	Origin of ingress
	1979–1983 (71)	1984–1988 (105)	1989–1993 (67)	1994–1998 (100)	1999–2003 (61)	2004–2008 (91)	2009–2013 (36)	2014–2015 (61)				
Freshwater Drum *Aplodinotus grunniens*	18.1	29.8	64.4	62.8	50.9	64.4	79.4	53.5	Linear	SI	P	Lower main stem
Bluebreast Darter *Etheostoma camurum*	2.8	8.2	3.9	10.0	24.6	11.0	38.2	48.3	Exp.	SI	I	Big Darby, Deer Creek
Northern Hog Sucker *Hypentelium nigricans*	45.8	70.6	72.9	91.3	93.2	83.1	97.1	91.2	Linear	SI	I	Large tributaries
Tippecanoe Darter *Etheostoma tippecanoe*	2.2	8.8	3.9	10.6	21.2	14.8	30.9	47.4	Exp.	SI	I	Big Darby, Deer Creek
Rosyface Shiner *Notropis rubellus*	13.9	20.6	15.8	23.8	42.4	23.3	35.3	61.4	Linear	SI	I	Large tributaries
Quillback *Carpiodes cyprinus*	16.7	13.1	35.4	42.5	30.5	33.5	66.2	42.1	Linear	SI	N	Main stem
Gravel Chub *Erimystax x-punctatus*	0.0	5.2	28.2	48.4	17.8	38.6	41.2	32.5	Linear	SI	I	Big Darby, Deer Creek
Smallmouth Buffalo *Ictiobus bubalus*	21.7	21.5	52.8	48.8	25.4	57.6	61.8	51.8	Linear	SI	N	Main stem
Bigeye Chub *Hybopsis amblops*	0.0	0.0	0.0	0.6	9.3	4.2	11.8	34.2	Exp.	MI	I	Scippo Creek
Flathead Catfish *Pylodictis olivaris*	10.8	10.5	34.5	32.2	22.9	45.3	36.8	42.1	Linear	MI	N	Lower main stem
Johnny Darter *Etheostoma nigrum*	9.4	16.5	10.8	23.1	33.1	21.2	26.5	51.8	Linear	MI	N	Small tributaries
Bluntnose Minnow *Pimephales notatus*	55.6	72.1	67.7	84.7	94.9	76.3	97.1	85.1	Linear	MI	T	All
Bluegill *Lepomis macrochirus*	60.8	49.8	80.9	80.9	82.2	80.1	92.7	82.5	Linear	MI	P	Main stem
Smallmouth Bass *Micropterus dolomieu*	58.6	70.0	88.1	89.1	93.2	90.7	92.7	91.2	Linear	MI	M	Main stem

TABLE 2. Continued.

Species	% occurrence—year interval (number of sites sampled) 1979–1983 (71)	1984–1988 (105)	1989–1993 (67)	1994–1998 (100)	1999–2003 (61)	2004–2008 (91)	2009–2013 (36)	2014–2015 (61)	Model	Trend	Tolerance	Origin of ingress
River Redhorse *Moxostoma carinatum*	8.6	2.8	11.1	17.2	20.3	20.8	26.5	35.1	Exponential	MI	I	Main stem, large tributaries
Slenderhead Darter *Percina phoxocephala*	0.0	0.4	2.2	10.3	11.0	19.9	27.9	17.5	Linear	MI	I	Lower main stem
Black Buffalo *Ictiobus niger*	6.1	3.2	18.2	26.6	15.3	36.4	36.8	16.7	Linear	MI	N	Lower main stem
Largemouth Bass *Micropterus salmoides*	26.7	38.0	52.5	50.0	42.4	46.2	61.8	54.4	Linear	MI	N	Impoundments
Sauger *Sander canadensis*	13.6	28.1	47.2	37.5	46.6	48.7	47.1	34.2	Linear	MI	N	Main stem
Orangespotted Sunfish *Lepomis humilis*	24.4	18.7	20.7	25.3	24.6	33.1	26.5	29.0	Exp.	MI	N	Impoundment
Dusky Darter *Percina sciera*	0.8	0.9	0.3	6.9	5.9	11.0	11.8	14.0	Linear	WI	M	Large tributaries
Golden Redhorse *Moxostoma erythrurum*	76.9	70.6	90.3	92.5	89.0	87.7	94.1	86.8	Linear	WI	M	Large tributaries
Mottled Sculpin *Cottus bairdii*	0.0	0.6	1.1	14.7	5.9	6.8	4.4	14.0	Linear	WI	I	Small tributaries
Redear Sunfish *Lepomis microlophus*	0.0	1.5	0.0	0.0	2.5	4.2	1.5	7.0	Linear	WI	N	Main stem
Goldeye *Hiodon alosoides*	0.0	0.2	0.0	0.3	0.0	0.0	1.5	0.9	Linear	WI	I	Lower main stem
Common Carp *Cyprinus carpio* × Goldfish *Carassius auratus*	5.8	1.9	8.6	6.3	0.9	0.4	0.0	0.0	Linear	D	T	
Pumpkinseed *Lepomis gibbosus*	8.3	14.2	14.6	14.4	12.7	3.8	1.5	4.4	Linear	D	N	
Goldfish *Carassius auratus*	11.9	8.2	6.6	3.8	3.4	0.0	1.5	0.9	Exp.	DT		

TABLE 3. Overall change in % occurrence for 51 species and 1 hybrid occurring at more than five sites in the study area that did not exhibit significant linear slope coefficients ranked by chi-squared value (increase or decrease) based on the % occurrence between 1979–1988 and 2004–2015. The origin of ingress is provided for species with significant increases based the chi-squared value ($P < 0.0005$ of 12.12). Tolerance: I = highly intolerant; M = moderately intolerant; N = intermediate; P = moderately tolerant.

Species	% occurrence—year interval (number of sites sampled)					
	1979–1988 (176)	2009–2015 (188)	Chi square	% change	Tolerance	Origin of ingress
Longear Sunfish *Lepomis megalotis*	81.72	85.71	416.08	4.00	N	Main stem, tributaries
Spotfin Shiner *Cyprinella spiloptera*	72.28	89.56	391.82	17.28	N	Main stem, tributaries
Green Sunfish *Lepomis cyanellus*	65.62	73.63	340.85	8.01	T	Main stem, tributaries
Gizzard Shad *Dorosoma cepedianum*	64.77	67.03	328.73	2.26	P	Main stem
Channel Catfish *Ictalurus punctatus*	35.59	67.58	282.28	31.99	N	Main stem
River Carpsucker *Carpiodes carpio*	47.09	60.44	260.23	13.35	N	Main stem
Silver Shiner *Notropis photogenis*	30.75	57.69	240.29	26.94	I	Main stem, tributaries
Striped Shiner *Luxilus chrysocephalus*	29.30	52.75	217.91	23.45	N	Tributaries
Spotted Bass *Micropterus punctulatus*	37.05	48.90	208.20	11.86	N	Lower main stem, tributaries
Silver Redhorse *Moxostoma anisurum*	36.44	45.60	198.58	9.16	M	Main stem, large tributaries
Black Redhorse *Moxostoma duquesnei*	26.51	43.41	177.79	16.89	I	Big Darby Creek, Deer Creek
Steelcolor Shiner *Cyprinella whipplei*	34.38	37.36	176.76	2.98	M	Main stem, large tributaries
Bullhead Minnow *Pimephales vigilax*	8.23	27.47	153.32	19.24	N	Lower main stem
Fantail Darter *Etheostoma flabellare*	8.48	27.47	150.43	19.00	N	Tributaries
Variegate Darter *Etheostoma variatum*	23.73	35.71	147.05	11.99	I	Big Darby Creek, Deer Creek
Suckermouth Minnow *Phenacobius mirabilis*	19.86	34.62	142.36	14.76	N	Main stem
Stonecat *Noturus flavus*	18.64	30.22	123.78	11.58	I	Larger tributaries
White Sucker *Catostomus commersonii*	22.52	22.53	113.50	0.01	T	Main stem, tributaries
Creek Chub *Semotilus atromaculatus*	15.13	23.08	94.86	7.94	T	Tributaries
Spotted Darter *Etheostoma maculatum*	2.18	10.99	80.41	8.81	I	Big Darby Creek
Brook Silverside *Labidesthes sicculus*	12.59	18.68	77.08	6.09	M	Main stem, tributaries
Blackside Darter *Percina maculata*	14.29	17.03	76.02	2.75	N	Main stem, large tributaries

nificant declines, two are classified as highly tolerant and the other moderately tolerant. Eight species that showed either a strong increase, a moderate increase, or a decline are illustrated in Figure 7.

The chi-square analysis (Table 3) included species such as the Spotted Darter, which occurred at low frequencies in the pre-CWA pollution control period of 1979–1988 and which increased in the 2009–2015 pollution recovery period but not consistently enough to yield significant curve fit. This result seemed to be true of species that were either absent or in low occurrence before 1988 but that showed increases during some but not all of the post-1988 time intervals. Grouping the intervals into longer time periods helped to amplify not only the overall increase in occurrence of 35 species and 1 hybrid, but also the declines among 10 species, including the highly tolerant Common Carp and Golden Shiner *Notemigonus crysoleucas*, several species that are affiliated with impounded habitats, and small stream species that occurred sporadically in the main stem throughout the 37-year time period.

The role and importance of tributaries in the recovery of several species in the main stem is exemplified by the occurrence of Tippecanoe Darter in the main stem and tributaries over three intervals: pre-1979, 1979–1999, and 2000–2016 (Figure 8). This species likely occupied most of the study area before the Scioto River, and some tributaries became heavily polluted in the late 19th century, becoming extirpated at all except a few tributary refuges, such as Big Darby and Deer creeks located south of Columbus. The 1979–1999 period results are the first evidence of its reappearance in the main stem at sites in close proximity to the confluence with Big Darby and Deer creeks. The 2000–2016 period results reflect a near-complete return to the known historical range in Franklin and Pickaway counties, including tributaries upstream from the discharges of sewage pollution that had reduced the fish fauna to only a few species (Williamson and Osburn 1898). The only exception is that Tippecanoe Darters have not returned to the pre-1900 locations in the lower Olentangy River (Figure 8; Trautman 1957, 1981). Greenlawn Dam is apparently serving as a barrier to their further upstream movement, which explains their absence from these two historic locations.

Discussion

All of the fish-assemblage and species-level indicators examined showed clear trends of improvement through time, which corresponds to the decreased loadings of sewage pollution discharged after 1988. The fish assemblage indices showed a steady improvement that has exceeded the minimum expectations of CWA-mandated WQS and effluent limitations. Species-level analyses likewise showed an increase in occurrence of formerly extirpated or reduced species that comprised nearly 65% of the total species recorded in the study area. In turn, five highly tolerant species and one hybrid have seen significant declines. All of these indicators of improvement coincide with the reduction of water pollution from Columbus WWTPs and a commensurate improvement in river water quality.

The return of the Tippecanoe Darter to the Scioto River main stem and other tributaries from which it was likewise extirpated (all of which are downstream from the Greenlawn Dam) most likely emanated from Big Darby and Deer creeks where their populations have remained unchanged. This recovery was similar to recovery patterns of other species that relied on unpolluted tributaries as refuges and repopulation epicenters and the central Scioto River main stem as a route of ingress to the main stem and then to tributaries located within and upstream from the pollution

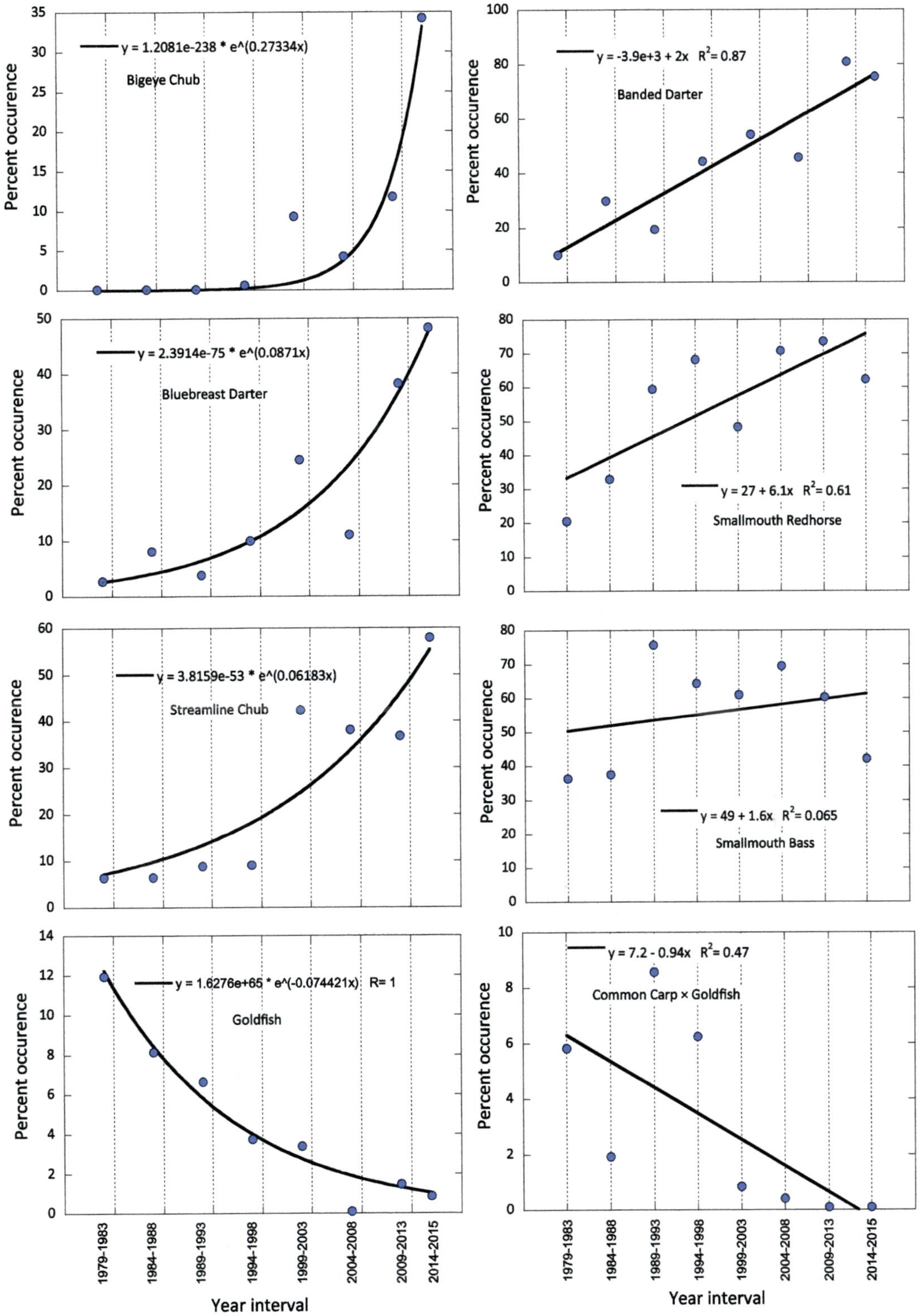

FIGURE 7. Percentage of occurrence of eight fish species at sampled sites for eight time intervals with significant linear and exponential coefficients.

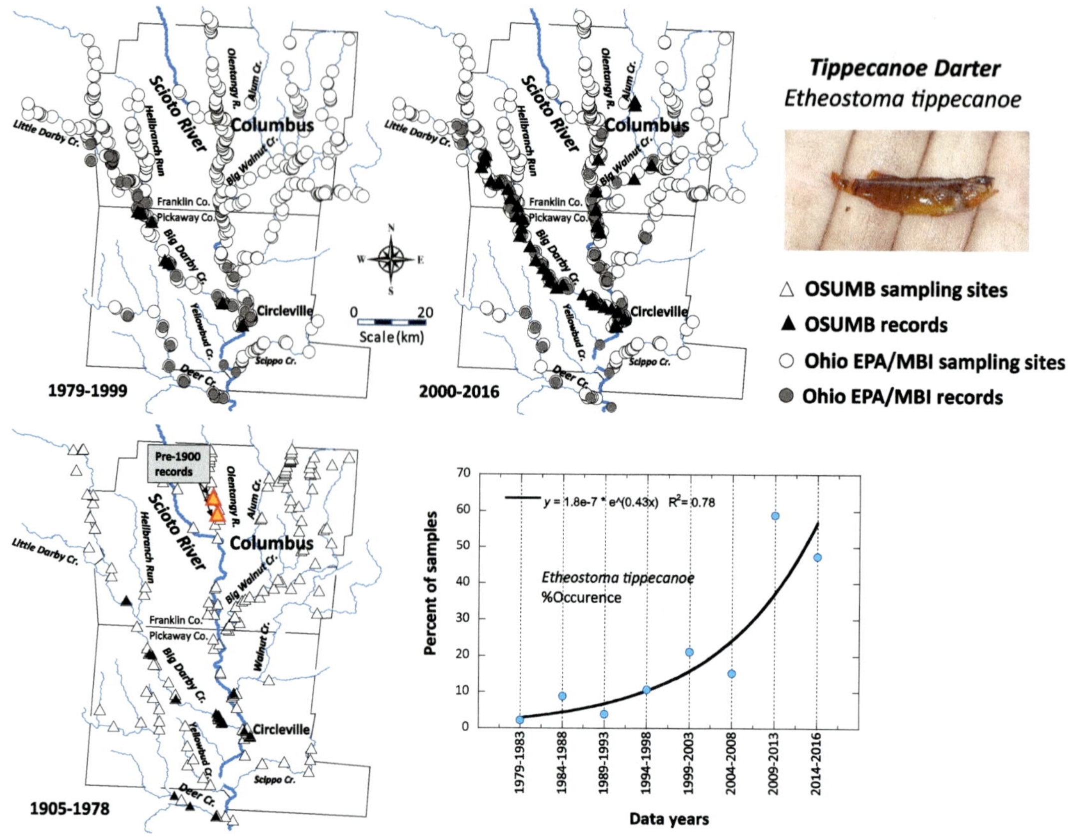

FIGURE 8. Change in occurrence of Tippecanoe Darter between 1905 and 2016. Open symbols are sampling locations and solid symbols denote Tippecanoe Darter occurrence. The exponential function (lower right) shows the percentage occurrence at sampled sites for 1979–1983, 1984–1988, 1989–1993, 1994–1998, 1999–2003, 2004–2008, 2009–2013, and 2014–2016 with the r^2 of 0.78 at $P < 0.01$.

footprint of the sewage discharges. Recovery from naturally occurring refuges has also been reported elsewhere, such as Brook Trout *Salvelinus fontinalis* within the watershed of Honnedaga Lake in the Adirondack region of New York (Kraft 2019, this volume). Honick et al. (2017) reported that before 1981, the Tippecanoe Darter in Ohio occurred at only 27 sites in the Scioto and Muskingum River basins of which 20 sites were in Big Darby Creek which enters the Scioto River in the southern part of our study area. We believe that the WWTP upgrades in 1988 essentially diminished a water quality barrier to their upstream and downstream ingress via the Scioto River main stem. During 2011–2015, Honick et al. (2017) documented 187 records of Tippecanoe Darter, some of which were located in the central Scioto River main stem and tributaries, including Walnut, Big Walnut, and Alum creeks. New records were also documented outside of the central Scioto River basin, including the lower Scioto River and two tributaries, Paint Creek and Salt Creek, and the Ohio River main stem, all located south of the central Scioto River but within areas that were heavily polluted by Columbus and other sources since the late 1800s. Similar recoveries of Bluebreast Darter and Spotted Darter were also documented by Honick et al. (2017) within and outside of the central Scioto River. Most of the new records for these three species, documented by Honick et al. (2017), were from large rivers sampled with benthic trawling and electrofishing, which suggests that they may have been missed by early sampling that employed little more than seines to collect fish. Still, all of these water bodies experienced serious pollution during the same period as the Scioto River; thus, these new records likely better document the original ranges of these benthic fish species.

The correspondence of the species-level results to improved fish IBI scores through this same time period is logical in that several IBI metrics are positively affected by increases in sensitive species and declines in tolerant species. R. Miltner (paper presented at the National Water Quality Monitoring Conference, 2014) documented similar increases in sensitive fish species occurrences in small Ohio streams altered primarily by nonpoint sources, including Bigeye Chub, which showed a moderate increase in the central Scioto River main stem, occurring at no locations before 1999 but increasing to 35% by 2014–2015. This recovery is another example of a highly pollution-intolerant species that was reduced to a few relict populations statewide by 1978 and that used main-stem rivers as routes to reoccupy its former range as water quality in the latter improved in response to CWA-mandated pollution controls. While the alleviation of severe point-source pollution was the principal controlling factor in the return of numerous species to the central Scioto main stem, recent successes in controlling some forms of nonpoint-source pollution (Miltner 2015) has also benefited some of these sensitive species.

Based on the most recent 2015 survey, the central Scioto River main stem is in full attainment of the warmwater habitat designated use based on the biological criteria for fish and macroinvertebrates downstream to the Southerly WWTP and full attainment of the EWH designated use downstream to Circleville. Demonstrating such complete attainment of WQS in the main stem is the strongest evidence that compliance with water-quality-based effluent limits actually achieved an important goal of the CWA. The sequence of administrative actions (WQS and NPDES permitting) forcing the installation of advanced wastewater treatment and as evidenced by reduced loadings of key pollutants resulted in improved water quality that allowed the full recovery of the biological assemblages. The framework for this sequence of measurable milestones was originally described as the hierarchy of water indicators by the U.S. EPA (1995a, 1995b) and was further detailed and used by Yoder and Rankin (1998), Karr and Yoder (2004), and Yoder et al. (2005). The framework has been used extensively by the Ohio EPA and includes a hierarchical sequence of six levels of indicators, listed below.

- Level 1—actions taken by regulatory agencies (permitting, enforcement, grants)
- Level 2—responses by the regulated entity (treatment works, pollution prevention)
- Level 3—changes in discharged quantities (pollutant loadings)
- Level 4—changes in ambient conditions (chemical/physical water quality, habitat)
- Level 5—changes in uptake and/or assimilation (tissue contamination, biomarkers, assimilative capacity)
- Level 6—changes in health, ecology, or other effects (ecological condition, human and wildlife health)

In this framework, the results of administrative actions (levels 1 and 2) should bring changes to water quality (levels 3, 4, and 5) that in turn translate to a response (level 6) that is measured by the indicators of aquatic assemblage health and well-being. The administrative steps taken by the Ohio EPA (with U.S. EPA oversight) to issue water quality-based NPDES permits (level 1) and the steps taken by the permit holder (level 2), the city of Columbus, are easily described and documented. Quantifying the resulting changes in the loadings of key pollutants (level 3) involved accessing monthly operating report data submitted by Columbus to the Ohio EPA for the key stressors that affect water quality and assimilative capacity in the central Scioto River (Figure 2, upper). The reduction of effluent loadings resulted in improved ambient water quality (level 4) as evidenced by reduced instream concentrations of NH_3-N (Figure 2, lower) and a better understanding of how the WWTPs are taxing the assimilative capacity (level 5) of the main stem. The end result of the above sequence is evident in the response in the biological indicators, which is expressed as attainment or nonattainment of the Ohio biological criteria for the fish and macroinvertebrate assemblages (Figures 3 and 4). Symptoms expressed by the biota beyond the index scores can be useful in aiding causal diagnosis as a feedback loop in the hierarchy of indicators process. The incidence of DELT anomalies on fish, which showed variable results through the postadvanced treatment years (Figure 4), provides an example of the type of feedback that might signal new problems requiring more-detailed follow-up investigations and potential interventions. Our conclusion was that the control of WWTP effluent quality and the cessation of bypasses were most responsible for the biological improvements meeting the goals of the Ohio WQS and the CWA. The recovery of sensitive and intermediate fish species to occupy their presettlement distributions in the central Scioto River main stem provided additional evidence of the success of water quality management, including the better control and reduction of both point and nonpoint sources of impact. Thomas et al. (2019, this volume) observed a similar recovery of the fish assemblages in the Ohio River main stem, to which the Scioto River is a major tributary, after abatement of industrial and municipal pollution by state and federal agencies. Similarly, recovery of fish assemblages was reported by Hughes et al. (2019, this volume) for the Willamette River in Oregon based on addressing problems associated with pollution, channelization, dams, and nonnative fish introductions.

LESSONS LEARNED

Water-Quality-Based Permitting

The most important lesson learned is that the CWA-mandated reductions in loadings of sewage pollutants from the Columbus sewer system via water-quality-based permitting resulted in water quality improvement sufficient to allow a nearly full biological recovery. This recovery happened despite past serious doubts about the treatability of sewage and attainability of biological goals of the CWA when they were first stated in 1972. From the first reported evidence of serious water pollution in the Scioto River as far back as the 1880s to the installation and operation of advanced wastewater treatment in 1988, it took more than a century before sufficient actions to reduce pollution actually took place. Part of the delay was due to the costs of wastewater treatment and the almost constant pursuit of the engineering technology that was required to reduce pollutants to the levels necessary to meet CWA goals. These achievements did not come easily or without a significant expenditure of public funds at the federal, state, and local levels. The consequences of the serious doubts about the efficacy of advanced wastewater treatment and the attainability of CWA-mandated WQS in an effluent-dominated river first surfaced when the Scioto River was proposed as a limited aquatic habitat in 1978. That was a lower water quality goal than that prescribed by the 1972 CWA, and it was eventually disapproved by the U.S. EPA, which illustrated the critically important role of federal agency oversight of state or local actions.

Once demonstrated to be effective by the results of sustained biological monitoring, advanced treatment is now the accepted minimum technology for municipal wastewater discharges in Ohio and much of the United States. The successes seen in the central Scioto River were also evident statewide in the attainment of the Ohio biological criteria in 93% of the nearly 2,100 km (1,300 mi) of large inland rivers by 2010 (Yoder et al. 2005; Ohio EPA 2016). However, maintaining these levels of attainment will require equally dedicated vigilance as a slight decline in attainment has already occurred (Yoder 2017).

Last, perhaps the most unheralded part of the demonstrated success of water-quality-based permitting are the contributions by the tens of individuals that were directly involved in different aspects of improved wastewater treatment by the city of Columbus. This includes people in the city, state, and federal agencies that were involved in debating and making the policy decisions, setting the WQS, developing the permits, designing the treatment facilities, and financing the capital improvements, the water utility rate payers, and the advocates for improved water quality, all of whom can take a share of the credit for the resulting improvements. The sum total of these accomplishments are the legacy left to today's counterparts to protect and uphold. The extent of improvements in recreational opportunities have tracked that of the fish and macroinvertebrate assemblages, as evidenced by an increased use for fishing, canoeing, kayaking, and related forms of recreation, including multiple new public access points within the study area.

Biological Improvement Occurred in Increments

The improvements documented in the central Scioto River occurred in increments of time and in response to increments of water pollution control. As early as 1981, some evidence of improvements occurred within the fish-assemblage-level indicators, and these occurred up to and through the benchmark year of 1988 and thereafter until the mid-1990s. At the species level, detectable improvements in the form of expanded occurrences within historical ranges took place well into the 2000s, which resulted in further improvements in the assemblage measures such as the fish IBI. What we have learned is that full recovery takes time and meeting a minimally required level of water quality may well not be the zenith of an improvement. This observation was certainly the case in the Scioto River downstream from the Southerly WWTP where the biological criteria for the exceptional warmwater habitat use designation were not fully attained until 2015, nearly three decades after the installation of advanced wastewater treatment. It takes time for a once-extirpated fish species to re-enter a formerly polluted area, re-establish a reproducing population, extend that population over miles of river and variable pollution gradients, and occur in sufficient abundance to change an index value or show as a significant increase. This is particularly an issue with naturally rare or patchy species (e.g., Kanno et al. 2009; Hughes et al. 2019).

Achieving a Complete Return to Presettlement Conditions Is Unlikely

While an issue for continuing debate, most would agree that the biological integrity provision of the CWA implies a return to pre-European settlement conditions. The U.S. EPA developed the biological condition gradient (BCG; Davies and Jackson 2006; U.S. EPA 2016) with a condition equivalent to "as naturally occurs" being the ultimate state of integrity for an aquatic ecosystem. At the same time, the BCG also describes subsequent levels of condition that each reflect an increasing departure from the naturally occurring state. The CWA implicitly recognized that few if any waters would actually achieve full biological integrity by setting an interim goal for the protection and propagation of fish, shellfish, and wildlife, which is widely regarded as the minimum restoration goal required by the CWA. For the Scioto River, this interim goal is exemplified by biological assemblages that can minimally attain the warmwater habitat use designation, which was accomplished in the Scioto River by the sequence of pollution control actions described in this chapter.

Whereas numerous fish species have been restored to their presettlement ranges, we doubt that all aspects of today's aquatic assemblages are equivalent to those historic conditions. The incidence of DELT anomalies on fish are evidence of remaining sublethal stresses that likely were absent during presettlement times. The current levels of DELT anomalies would not meet the "as naturally occurs" level of the BCG that many equate to full biological integrity. Freshwater mussel species, which are dependent on their fish hosts to reoccupy former ranges, have not experienced the same levels of recovery as the fish or other macroinvertebrate assemblages. The near-constant addition of human-made materials and substances to the aquatic biosphere could well be the cause, and these certainly did not exist in presettlement or before preindustrial times. New-generation pesticides that have only recently been used have been detected throughout Midwestern U.S. rivers, some at levels that are harmful to aquatic life (Van Metre et al. 2016). New generation pollutants such as pharmaceuticals and endocrine-disrupting chemicals present in municipal wastewater effluents pose yet another potential barrier to restoring full biological integrity. Physical factors include an altered flow regime and the loss of connectivity in the form of the two remaining dams in the study area. While riverine habitat is mostly intact downstream from Columbus, agricultural practices continue to encroach on the riparian zone, and levees in some places restrict floodwaters from reaching the floodplain. The impending demands for an expanded public water supply affiliated with the anticipated one-half million increase in population by 2050 could further exacerbate the altered flow conditions in the Scioto River upstream from the WWTPs and contribute to an expanded urban footprint on the watershed as a whole, further threatening tributaries that have historically served as native species refugia. Preserving intact habitat where it exists is as critical to the aquatic assemblages as water quality (Hughes et al. 2019).

Suffice it to say, even the exceptional levels of performance recently attained by the fish and macroinvertebrate assemblages fall short of full biological integrity. However, the observed improvements have certainly exceeded the baseline expectations of the interim goals of the CWA, which is in itself a noteworthy accomplishment.

Monitoring and Assessment Before and After Success Is Essential

Last, continued monitoring and assessment of streams and rivers provide essential feedback to the pollution control and water quality management programs as it can serve as an early warning of new problems or potential lapses in pollution controls. It would be a serious mistake to rest on successes documented in this case study, which required so many years to attain, by reducing the vigilance provided by sustained and routine monitoring and assessment. The history of water pollution documented herein and the decades that it took to finally respond should serve as a lesson to not allow such vigilance to suddenly lapse.

ACKNOWLEDGMENTS

Credit for the work to produce the results reported herein are due to the many Ohio EPA staff members and managers and hundreds of college interns who endured long days in the field and provided the resources to sustain a nearly 40-year program of systematic biological assessment of Ohio's rivers and streams. We thank Brian Zimmerman, OSUMB, for providing the historical fish distribution records and Eric Nygaard, Ohio EPA, for providing a quick turnaround with the effluent data for the Columbus WWTPs. Jack Freda of the Midwest Biodiversity Institute provided important input regarding the 2015 macroinvertebrate results. Last, credit is due to the tens of persons at the U.S. EPA, Ohio EPA, and city of Columbus who were directly involved in debating, setting, and implementing the CWA policies that made the improvements in the Scioto River possible. Those who follow in their footsteps would do well to ensure that these gains are maintained and, if possible, enhanced in the future.

REFERENCES

Angermeier, P. L., and J. R. Karr. 1986. Applying an index of biotic integrity based on stream-fish communities: considerations in sampling and interpretation. North American Journal of Fisheries Management 6:418–427.

Bartsch, A. F. 1948. Biological aspects of stream pollution. Sewage Works Journal 20:292–302.

Buckwalter, J. D., E. A. Frimpong, P. L. Angermeier, and J. N. Barney. 2017. Seventy years of stream-fish collections reveal invasions and native range contractions in an Appalachian (USA) watershed. Diversity Distributions 2017:1–14.

Davies, S. P., and S. K. Jackson. 2006. The biological condition gradient: a descriptive model for interpreting change in aquatic ecosystems. Ecological Applications 16:1251–1266.

Davis, W. T., and T. P. Simon. 1995. Biological assessment and criteria: tools for risk-based planning and decision making. Lewis Publishers, Boca Raton, Florida.

DeShon, J. D. 1995. Development and application of the invertebrate community index (ICI), Pages 217–243 *in* W. S. Davis and T. Simon, editors. Biological assessment and criteria: tools for risk-based planning and decision making. Lewis Publishers, Boca Raton, Florida.

Doudoroff, P., and C. E. Warren. 1951. Biological indices of water pollution with special reference to fish populations. Pages 144–153 *in* Biological problems in water pollution. U.S. Public Health Service, Robt. A. Taft Sanitary Engineering Center, Cincinnati, Ohio.

Elzhov, T. V., K. M. Mullen, A. N. Spiess, and B. Bolker. 2013. Minpack lm: R interface to the Levenberg-Marquardt nonlinear least-squares algorithm found in MINPACK, plus support for bounds. R package version 1.1–8. Available: http://CRAN.R-project.org/package=minpack.lm/. (August 2018).

FWQA (Federal Water Quality Administration). 1970. Benthic biology of the Scioto River basin, Ohio. U.S. Department of Interior, Upper Ohio Basin Office, Work Document 42, Wheeling, West Virginia.

Gammon, J. R. 1976. The fish populations of the middle 340 km of the Wabash River. Purdue University, Water Resources Research Center, Technical Report 86, West Lafayette, Indiana.

Gammon, J. R., A. Spacie, A., J. L. Hamelink, and R. L. Kaesler. 1981. Role of electrofishing in assessing environmental quality of the Wabash River. Pages 307–324 *in* J. M. Bates and C. I. Weber, editors. Ecological assessments of effluent impacts on communities of indigenous aquatic organisms. American Society for Testing and Materials, ASTM STP 730, Fort Lauderdale, Florida.

Honick, A. S., B. J. Zimmerman, J. R. Stauffer, Jr., D. G. Argent, and B. A. Porter. 2017. Expanded distributions of three *Etheostoma* darters (subgenus *Nothonotus*) within the upper Ohio River watersheds. Northeastern Naturalist 24:209–234.

Hughes, R. M, B. L. Bangs, S. V. Gregory, P. D. Scheerer, R. C. Wildman, and J. S. Ziller. 2019. Recovery of Willamette River (Oregon, USA) fish assemblages: successes and remaining threats. Pages 157–184 *in* C. C. Krueger, W. W. Taylor, and S. Youn, editors. From catastrophe to recovery: stories of fishery management success. American Fisheries Society, Bethesda, Maryland.

Kanno, Y., J. C. Vokoun, D. C. Daulwater, R. M. Hughes, A. T. Herlihy, T. R. Maret, and T. M. Patton. 2009. Influence of rare species on electrofishing distance when estimating species richness of stream and river reaches. Transactions of the American Fisheries Society 138:1240–1251.

Karr, J. R., K. D. Fausch, P. L. Angermier, P. R. Yant, and I. J. Schlosser. 1986. Assessing biological integrity in running waters: a method and its rationale. Illinois Natural History Survey, Special Publication 5, Champaign.

Karr, J. R., and C. O. Yoder. 2004. Biological assessment and criteria improve TMDL planning and decision-making. Journal of Environmental Engineering 130:594–604.

Kraft, C. 2019. Adirondack Brook Trout and acid rain: environmental legislation fosters successful restoration. Pages 299–322 *in* C. C. Krueger, W. W. Taylor, and S. Youn, editors. From catastrophe to recovery: stories of fishery management success. American Fisheries Society, Bethesda, Maryland.

Leighton, M. O. 1903. Normal and polluted waters in northeastern United States. U.S. Geological Survey, Water Supply Irrigation Paper 79, Series L, Reston, Virginia.

Lenz, E. 1993. Historical milestones for wastewater treatment in Columbus. Available: https://www.columbus.gov/utilities/about/Historical-Milestones-for-Wastewater-Treatment-in-Columbus/. (December 2018).

Mazerolle, M. J. 2015. AICcmodavg: model selection and multimodel inference based on (Q)AIC(c). R package version 2.0–3. Available: http://CRAN.R-project.org/package=AICcmodavg. (August 2018).

Miltner, R. J. 2015. Measuring the contribution of agricultural conservation practices to observed trends and recent condition in water quality indicators in Ohio, USA. Journal of Environmental Quality 44:1821–31.

Ohio DNR (Ohio Department of Natural Resources). 1960. Gazetteer of Ohio streams. Ohio DNR, Division of Water, Ohio Water Plan Inventory Report No. 12, Columbus.

Ohio EPA (Ohio Environmental Protection Agency). 1986. Central Scioto River mainstem comprehensive water quality report. Ohio EPA, Division of Water Pollution Control and Division of Water Quality Monitoring and Assessment, Columbus.

Ohio EPA (Ohio Environmental Protection Agency). 1987a. Biological criteria for the protection of aquatic life, volume I: the role of biological data in water quality assessments. Ohio EPA, Division of Water Quality Monitoring and Assessment, Surface Water Section, Columbus.

Ohio EPA (Ohio Environmental Protection Agency). 1987b. Biological criteria for the protection of aquatic life, volume II: users manual for biological field assessment of Ohio surface waters. Ohio EPA, Division of Water Quality Monitoring and Assessment, Surface Water Section, Columbus.

Ohio EPA (Ohio Environmental Protection Agency). 1989a. Addendum to biological criteria for the protection of aquatic life, volume II: users manual for biological field assessment of Ohio surface waters. Ohio EPA, Division of Water Quality Planning and Assessment, Surface Water Section, Columbus.

Ohio EPA (Ohio Environmental Protection Agency). 1989b. Biological criteria for the protection of aquatic life, volume III: standardized biological field sampling and laboratory methods for assessing fish and macroinvertebrate communities. Ohio

EPA, Division of Water Quality Monitoring and Assessment, Columbus.

Ohio EPA (Ohio Environmental Protection Agency). 1999a. Association between nutrients, habitat, and the aquatic biota in Ohio rivers and streams. Ohio EPA, Technical Bulletin MAS/1999-1-1, Columbus.

Ohio EPA (Ohio Environmental Protection Agency). 1999b. Biological and water quality study of the middle Scioto River and Alum Creek: Franklin, Delaware, Morrow, and Pickaway counties, Ohio. Ohio EPA, Technical Report MAS/1997-12-12, Columbus.

Ohio EPA (Ohio Environmental Protection Agency). 2006. Methods for assessing habitat in flowing waters: using the qualitative habitat evaluation index (QHEI). Ohio EPA, Division of Surface Water, Ecological Assessment Section, Columbus.

Ohio EPA (Ohio Environmental Protection Agency). 2011. Ohio EPA surface and ground waters monitoring strategy, 2011–2015. Ohio EPA, Technical Bulletin DSW/EAS/2011-4-1, Columbus.

Ohio EPA (Ohio Environmental Protection Agency). 2012. Biological and water quality survey of the middle Scioto River and select tributaries 2010. Delaware, Franklin, Pickaway, and Union Counties. Ohio EPA, Technical Report DSW/EAS 2012-12-12, Columbus.

Ohio EPA (Ohio Environmental Protection Agency). 2015. Biological criteria for the protection of aquatic life (revised June 26, 2015). Volume III: Standardized biological field sampling and laboratory methods for assessing fish and macroinvertebrate communities. Ohio EPA, Technical Report EAS/2015-06-01, Columbus.

Ohio EPA (Ohio Environmental Protection Agency). 2016. Ohio 2016 integrated water quality monitoring and assessment report. Ohio EPA, Division of Surface Water, Final Report, Columbus.

Olive, J. H., and K. R. Smith. 1975. Benthic macroinvertebrates as indexes of water quality in the Scioto River basin, Ohio. Bulletin of the Ohio Biological Survey 5(2).

Omernik, J. M. 1987. Ecoregions of the conterminous United States. Annals of the Association of American Geographers 77:118–125.

Osburn, R. C. 1901. The fishes of Ohio. The Ohio Academy of Science, Special Paper 4, Columbus.

R Core Team. 2014. R: a language and environment for statistical computing. R Foundation for Statistical Computing, Vienna, Austria. Available: www.R-project.org/. (August 2018).

Rankin, E. T. 1989. The qualitative habitat evaluation index (QHEI), rationale, methods, and application. Ohio Environmental Protection Agency, Division of Water Quality Planning and Assessment, Ecological Assessment Section, Columbus.

Rankin, E. T. 1995. The use of habitat assessments in water resource management programs. Pages 181–208 *in* W. Davis and T. Simon editors. Biological assessment and criteria: tools for water resource planning and decision making. Lewis Publishers, Boca Raton, Florida.

Sanders, R. S., R. J. Miltner, C. O. Yoder, and E. T. Rankin. 1999. The use of external deformities, erosions, lesions, and tumors (DELT anomalies) in fish assemblages for characterizing aquatic resources: a case study of seven Ohio streams. Pages 225–248 *in* T. P. Simon, editor. Assessing the sustainability and biological integrity of water resources using fish communities. CRC Press, Boca Raton, Florida.

Simon, T. P., and N. T. Evans. 2017. Environmental quality assessment using stream fishes. Pages 319–334 *in* F. R. Hauer and G. Lamberti, editors. Methods in stream ecology, 3rd edition. Elsevier, Amsterdam.

Sharp, H. J. 1887. Report on the pollution of water courses by straw board factories. Pages 1873–1885 *in* Ohio State Board of Health. Annual report for 1886 to the 67th General Assembly of the state of Ohio, part 1. Myers Brothers, State Printers, Columbus, Ohio.

Stantec. 2012. Main street dam removal and Scioto greenways feasibility study. City of Columbus and Columbus Downtown Development Corporation, Columbus, Ohio.

Thomas, J. A., J. G. Schulte, P. A. Tennant, and D. R. Argo. 2019. Recovery of a great river fishery: the story of the Ohio River. Pages 211–231 *in* C. C. Krueger, W. W. Taylor, and S. Youn, editors. From catastrophe to recovery: stories of fishery management success. American Fisheries Society, Bethesda, Maryland.

Trautman, M. B. 1933. The general effects of pollution on Ohio fish life. Transactions of the American Fisheries Society 63:69–72.

Trautman, M. B. 1957. The fishes of Ohio. The Ohio State University Press, Columbus.

Trautman, M. B. 1977. The Ohio country from 1950 to 1977: a naturalists view. Ohio Biological Survey Biological Notes 10.

Trautman, M. B. 1981. The fishes of Ohio, revised edition. The Ohio State University Press, Columbus.

U.S. EPA (U.S. Environmental Protection Agency). 1995a. A conceptual framework to support development and use of environmental information in decision-making. U.S. EPA, EPA 239-R-95-012, Washington, D.C.

U.S. EPA (U.S. Environmental Protection Agency). 1995b. Environmental indicators of water quality in the United States. U.S. EPA, EPA 841-R-96-002, Washington, D.C.

U.S. EPA (U.S. Environmental Protection Agency). 2016. A practitioner's guide to the biological condition gradient: a framework to describe incremental change in aquatic ecosystems. U.S. EPA, EPA-842-R-16-001, Washington, D.C.

Van Metre, P. C., D. A. Alvarez, B. J. Mahler, L. Nowell, M. Sandstrom, and P. Moran. 2016. Complex mixtures of pesticides in Midwest U.S. streams indicated by POCIS time-integrating samplers. Environmental Pollution 2016:1–10.

Williamson, E. B., and R. C. Osburn. 1898. A descriptive catalogue of the fishes of Franklin County, Ohio. The Ohio State University, Columbus.

Woods, A., J. M. Omernik, C. S. Brockman, T. D. Gerber, W. D. Hosteter, and S. H. Azevedo. 1995. Ecoregions of Ohio and Indiana. U.S. EPA, Corvallis, Oregon.

Yoder, C. 2017. A retrospecive on the Clean Water Act: is today as good as it gets? Midwest Biodiversity Institute, Hilliard, Ohio. Available: https://midwestbiodiversityinst.org/publications/articles/a-retrospective-on-the-clean-water-act-in-ohio-is-today-as-good-as-it-gets. (December 2018).

Yoder, C. O. 1995. Policy issues and management applications for biological criteria. Pages 327–344 *in* W. Davis and T. Simon, editors. Biological assessment and criteria: tools for water resource planning and decision making. Lewis Publishers, Boca Raton, Florida.

Yoder, C. O., P. S. Albeit, and M. A. Smith. 1981. The distribution and abundance of fishes in the mainstem Scioto River as affected by pollutant loadings. Ohio Environmental Protection Agency, Technical Report 81/3, Columbus.

Yoder, C. O., and M. T. Barbour. 2009. Critical elements of state bioassessment programs: a process to evaluate program rigor and comparability. Environmental Monitoring and Assessment 150:31–42.

Yoder, C. O., and J. E. DeShon. 2003. Using biological response signatures within a framework of multiple indicators to assess and diagnose causes and sources of impairments to aquatic assemblages in selected Ohio rivers and streams. Pages 23–81 *in* T. P. Simon, editor. Biological response signatures: indicator patterns using aquatic communities. CRC Press, Boca Raton, Florida.

Yoder, C. O., and E. T. Rankin. 1995a. Biological criteria program development and implementation in Ohio. Pages 109–144 *in* W. Davis and T. Simon, editors. Biological assessment and criteria: tools for water resource planning and decision making. Lewis Publishers, Boca Raton, Florida.

Yoder, C. O., and E. T. Rankin. 1995b. Biological response signatures and the area of degradation value: new tools for interpreting multimetric data. Pages 263–286 *in* W. Davis and T. Simon, editors. Biological assessment and criteria: tools for water resource planning and decision making. Lewis Publishers, Boca Raton, Florida.

Yoder, C. O., and E. T. Rankin. 1998. The role of biological indicators in a state water quality management process. Journal of Environmental Monitoring and Assessment 51:61–88.

Yoder, C. O., and M. A Smith. 1999. Using fish assemblages in a state biological assessment and criteria program: essential concepts and considerations. Pages 17–56 *in* T.P. Simon, editor. Assessing the sustainability and biological integrity of water resources using fish communities. CRC Press, Boca Raton, Florida.

Yoder, C. O., M. A. Smith, B. C. Alsdorf, D. J. Altfater, C. E. Boucher, R. J. Miltner, D. E. Mishne, R. E. Sanders, and R. F. Thoma. 2005. Changes in fish assemblage status in Ohio's nonwadeable rivers and streams over two decades. Pages 399–429 *in* J. N. Rinne, R. M. Hughes and B. Calamusso, editors. Historical changes in large river fish assemblages of the Americas. American Fisheries Society, Symposium 45, Bethesda, Maryland.

Evidence of Recoveries from Tropical Floodplain Fisheries: Three Examples of Management Gains for South American Giant Arapaima

Daniel J. Gurdak*
Department of Environmental and Forest Biology
SUNY College of Environmental Science and Forestry
Syracuse, New York 13210, USA
and
Universidade Federal do Oeste do Pará
Santarém, Pará, Brazil

Caroline C. Arantes
Department of Wildlife and Fisheries Sciences, Texas A&M University
College Station, Texas 77840, USA
and
Center for Global Change and Earth Observations, Michigan State University
East Lansing, Michigan 48823, USA

Leandro Castello
Department of Fish and Wildlife Conservation
Virginia Polytechnic Institute and State University
Blacksburg, Virginia 24061, USA

Donald J. Stewart
Department of Environmental and Forest Biology
SUNY College of Environmental Science and Forestry
Syracuse, New York 13210, USA

L. Cynthia Watson
Environment and Climate Change Canada
Watershed Hydrology and Ecology Research Division, Canada Centre for Inland Waters
Burlington, Ontario L7S 1A1, Canada

Abstract.—Although tropical inland fisheries provide important regional income and food security, these fisheries face severe management challenges that are amplified by distinct ecological, economic, and political factors. The arapaima (genus *Arapaima*) are of particular interest because of their large size (up to +3 m and +200 kg) and their substantial economic and cultural value in South America. Arapaima are among the most historically important and over-exploited fisheries in South America. Unfortunately, traditional fishery management approaches have been ineffective, and arapaima populations have suffered drastic stock depletions and even local extinctions across much of their range.

* Corresponding author: daniel.j.gurdak@gmail.com

Fortunately, over the past 15 years, small-scale co-management efforts have promoted pockets of successful recovery for arapaima fisheries. In this analysis, we begin by introducing the history of arapaima fishing, the difficulties encountered by early management efforts, and subsequent improvements to arapaima management approaches using co-management at a small scale with local residents. We then evaluate the implementation of this promising approach across three regions: (1) Mamirauá Sustainable Development Reserve, Amazonas State, Brazil; (2) Santarém region, Pará State, Brazil; and (3) North Rupununi, Guyana. Specifically, each region was evaluated based on the presence and implementation of eight principles needed for the sustainable governance of common-property natural resources in situations with and without active management: (1) defined boundaries, (2) congruence between appropriation and provision rules and local conditions, (3) collective action arrangement, (4) monitoring, (5) graduated sanctioning, (6) conflict resolution mechanism, (7) minimum recognition and right to organize, and (8) nested enterprises. For each region, we examined how management was implemented and identified current and ongoing challenges. Based on our experience with these fisheries, we offer some lessons that can be instrumental for arapaima management and other inland fisheries: manage even when faced with uncertainties; monitor, evaluate, and adapt management efforts; bridge knowledge systems; foster genuine interest and cooperation of various stakeholders to ensure long-term success; and move toward an ecosystem-based approach. These lessons will be instrumental in overcoming management challenges in the future and could be applied to other fisheries.

INTRODUCTION

Although tropical inland fisheries provide important regional income and food security, these fisheries face severe management challenges that are amplified by distinct ecological, economic, and political factors that set them apart from their temperate counterparts (Pauly et al. 1989; Allan et al. 2005). In contrast to traditional, single-species fishery management, which is characteristic of northern, temperate latitudes, tropical fisheries generally have multiple fishery targets set in extremely biodiverse ecosystems with high biomass and productivity. Tropical fisheries are typically in less affluent societies and often are neglected, not only because of their complexity, but also due to limited human and financial resources needed for effective governance and enforcement (Pauly et al. 1989). As a result, inland tropical fisheries are often data-poor, such that standard fishery management assessments cannot be performed (Pauly et al. 1989; Allan et al. 2005). In general, tropical fisheries have some commercial orientation but typically include a substantial emphasis on small-scale, artisanal subsistence fishing. Tropical fishers are often among the world's poorest and most marginalized peoples and depend on fishing as an essential source of income and food (Pauly 1997). Monitoring and enforcement are extremely difficult because these fishers use a variety of gears, fishing is spatially and temporally dispersed, and government resources are limited. As a result of these distinctions, many tropical fisheries are essentially unregulated and virtually open-access, leading to overfishing and tragic fish population declines that threaten the livelihoods of local peoples.

When facing these challenges, an optimistic future for many tropical fisheries is difficult to envision. Can tropical fishery management stabilize declining fish populations or even promote population recovery? Is there any possibility for management success? Although management challenges appear overwhelming, success can be achieved, and some examples in tropical fishery management do exist. Here, we investigate three case studies that promoted sustainability and management of the South American arapaima (genus *Arapaima*), the continent's most valued and overexploited fishery resource (Veríssimo 1895; Isaac et al. 1993; Fernandes 2005; Castello et al. 2011).

Arapaima possess unique bioecological features and are of substantial economic, nutritional, and cultural importance, which make them an ideal target for management attention. The arapaima are among the largest fishes of South America, growing up to 3 m in length and 200 kg in weight (Arantes et al. 2010; Figure 1) and, depending on region, are commonly called arapaima, paiche, or pirarucu. Although five described species exist in the genus *Arapaima* (Stewart 2013a, 2013b), the group faces incredible data shortfalls that make species identification and distributions uncertain (see Linnean and Wallacean shortfalls defined in Bini et al. 2006). Arapaima naturally occur in river floodplains across two major basins (Amazon and Essequibo), five countries (Brazil, Columbia, Peru, Ecuador, and Guyana), and a variety of freshwater wetlands, ecoregions, and habitat types. Arapaima comprise an ecologically unique group of fishes that build nests, guard young, and must breathe air at regular intervals (Lüling 1964; Castello 2008). Arapaima reproduction coincides with rising water levels, and their life histories are synced with the pulse of rising and falling water on the floodplain (Figure 2). They can grow to almost 1 m total length in 1 year and reproduce between 3 and 5 years of age after growing to about 1.5 m (e.g., Fontenele 1948; Imbiriba 1994; Godinho et al. 2005; Arantes et al. 2010; Gurdak et al., in press). Because their

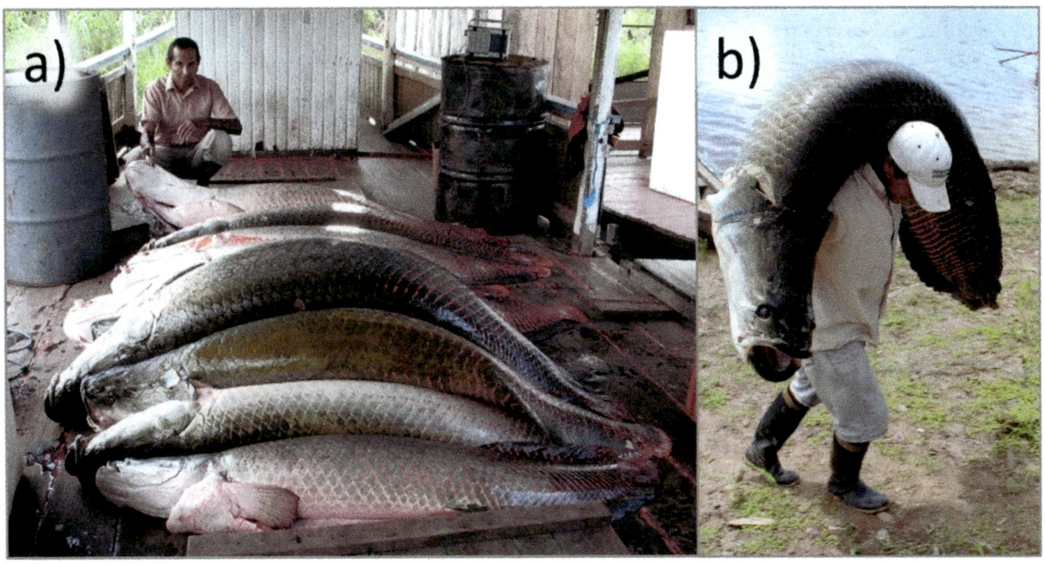

FIGURE 1. Arapaima (genus *Arapaima*) are large, valuable food fish that are an important part of local and regional economies. (a) A day's harvest of legal-sized arapaima at Jarauá Community in the Mamirauá Sustainable Development Reserve, Amazon State, Brazil (photo by CCA). (b) Fisherman carrying ~145-kg arapaima, Pacaya Samiria National Reserve, Peru (photo by DJS).

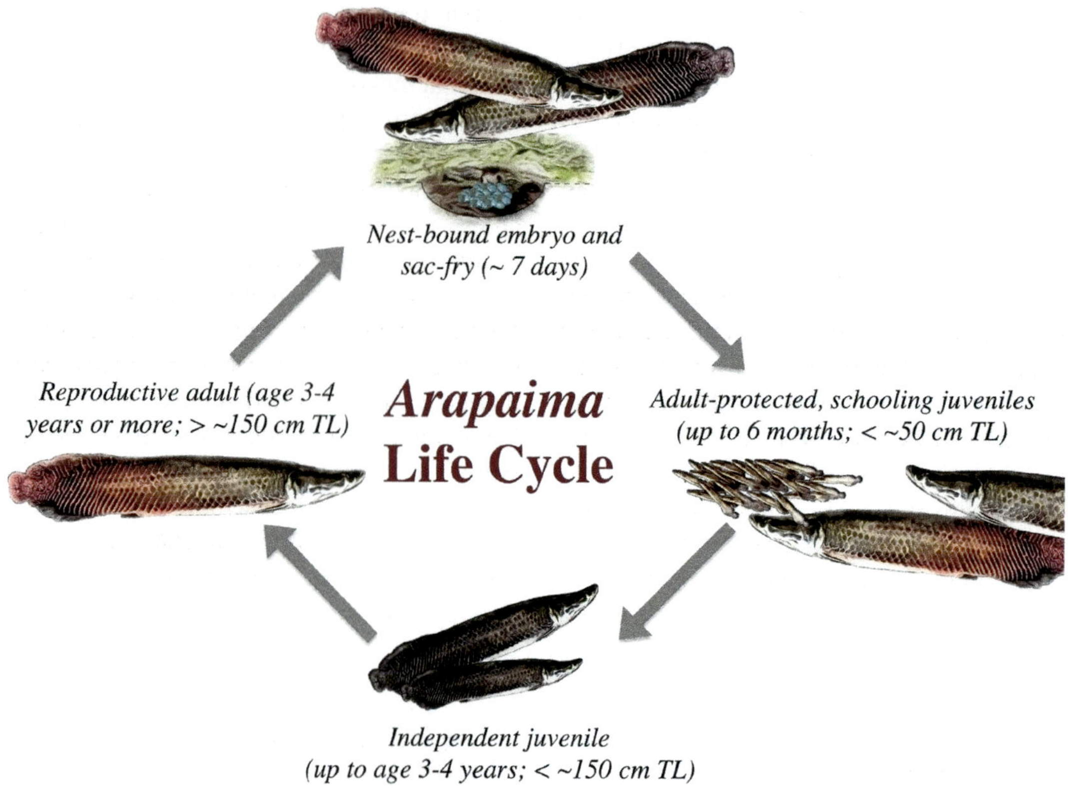

FIGURE 2. The generalized life cycle of the arapaima (genus *Arapaima*) can be divided into four main stages: (1) nest-bound embryo and sac fry, (2) adult-protected, schooling juvenile, (3) independent juvenile, and (4) reproductive adult.

large fillets have good taste and texture without many bones, they are extremely valuable as a food fish and an important part of local and regional economies (Figure 1). Arapaima meat is extremely common in local dishes and restaurants. Arapaima have even captured local, national, and international attention as a cultural symbol and as a sport fish through popular television shows, documentaries, and publications (e.g., Purnell 2017). For South American riverine inhabitants, traditional arapaima fishing is highly specialized and an important cultural and economic activity (Castello et al. 2009; Murrieta 2001). Although gill nets are being used with increasing frequency, traditional fishers use a harpoon or bow and arrow to capture arapaima at the moment it surfaces to breathe air, and they work from a small canoe or while perched in a tree. Fishers also harvest arapaima using a series of baited hooks and lines.

Deterring overexploitation of arapaima has been difficult to achieve using traditional size and season regulations (Castello et al. 2015). As a result of its desirability, arapaima populations have become decimated, overfished, and locally extinct in some areas (Castello et al. 2011, 2015). Fortunately, continuing and improved arapaima management initiatives implemented by governmental and nongovernmental institutions are achieving sustainability of arapaima fisheries across their range (e.g., Castello et al. 2009; Campos-Silva and Peres 2016; Oviedo and Bursztyn 2016; Petersen et al. 2016). Many of these management initiatives have successfully recovered declining arapaima populations while directly engaging local fishers. In some cases, this success has been evaluated using a framework introduced by Ostrom (1990). The framework proposed that people are capable of organizing and self-governing their shared resources if a set of criteria, or design principles, is met (Ostrom 1990; see Table 1). The presence of these eight design principles is crucial in achieving sustainable governance of a collective resource and has helped evaluate management of fisheries across the globe (e.g., Persoon et al. 2003; Castello et al. 2009; Oviedo and Bursztyn 2016).

Implementing arapaima management successfully across its large range has proven difficult. As far as we know, no study has compared or evaluated arapaima management initiatives across multiple regions. A central question is thus, "How can fishery management programs be implemented successfully across a large area that is ecologically, culturally, and politically heterogeneous?" In this chapter, we begin by briefly discussing the history of arapaima fishing and the difficulties of early management efforts and showcase a promising co-management approach involving local residents. We then evaluate the implementation of this developing co-management approach across three regions: (1) Mamirauá Sustainable Development Reserve, Amazonas State, Brazil; (2) Santarém region, Pará State, Brazil; and (3) North Rupununi, Guyana. For each region, we examined how fishery management programs were implemented, evaluated management efforts considering Ostrom's (1990) eight sustainable gover-

TABLE 1. Ostrom's (1990) eight institutional design principles for sustainable governance of common property resources (adapted from Castello et al. 2009).

Principle	Description
1. Defined boundaries	Clear definition exists of both the individuals with rights to exploit the resource and the boundaries of the resource.
2. Congruence between appropriation and provision rules and local conditions.	Appropriation rules are related adequately to ecological and socioeconomic conditions and to provision rules.
3. Collective action arrangement	Individuals affected by the rules can participate in modifying the rules.
4. Monitoring	The monitors auditing the resource and the behavior of appropriators are accountable to the appropriators or are the appropriators.
5. Graduated sanctioning	Appropriators who violate the rules are likely to be assessed graduated sanctions, depending on the seriousness and context of the offenses.
6. Conflict resolution mechanism	Appropriators and officials have rapid access to low-cost arenas to resolve conflict.
7. Minimum recognition and right to organize	The rights of appropriators to devise their own institutions are not challenged by external government authorities.
8. Nested enterprises	Appropriation, provision, monitoring, enforcement, conflict resolution, and governance activities are organized in multiple layers of nested enterprises.

nance design principles of common property natural resources (see Table 1), and identified current and ongoing challenges. Last, we compiled lessons learned based on factors that either promoted success or contributed to ongoing challenges. These lessons will be instrumental in overcoming management challenges in the future and could be applied to other fisheries.

HISTORY OF ARAPAIMA FISHING

The history of arapaima fishing is shaped by unique socioeconomic and environmental variables that span a huge range with various governmental, social, and ecological boundaries (e.g., countries, states, and municipalities), cultural areas, ecological regions, and a variety of flood pulse patterns. The history of arapaima fishing dates to the pre-Columbian era when many South American rivers and river floodplains supported civilizations of high population densities with fertile soils and abundant food fish, other aquatic vertebrates, and various natural resources (Denevan 1966; Meggers 1971; Roosevelt 1989). The rich South American fish diversity, sustained by unique habitats formed by cyclical flooding, provided a reliable supply of nourishment for local populations.

Undoubtedly, arapaima was one among many food resources for early South American peoples. For example, at the mouth of the Amazon River in Brazil, arapaima bones have been found in archeological sites, albeit in lower density than would be expected compared to other fishes (Roosevelt 1991). Large, adult arapaima were likely difficult to capture using pre-Iron Age tools. Possibly because of their capture difficulty, some cultures limited the consumption of arapaima to sporadic and ritualistic use (Murrieta 2001). Even into the 20th century, arapaima fishing was deterred in traditional fishing communities of the Rupununi River basin of Guyana because arapaima were either considered taboo or revered as the "Mother of all Fishes" (Fernandes 2005; Watson et al. 2016).

During the colonial period, fish of South American waterways became an important source of economic activity and sustenance for a growing human population (Veríssimo 1895; Queiroz 1999; Cleary 2001). Early European explorers observed high fish consumption among native peoples and marveled at the amazing diversity of fishes and other aquatic fauna (Papavero 2000). Although predominantly a subsistence activity, fisheries were important to regional economies and provided rural and urban populations with protein (Crampton et al. 2004). Fishing was concentrated in the most accessible areas in freshwaters close to the Atlantic coast and in areas near growing urban centers. Before refrigeration, commercial fisheries preserved fishes by drying and salting. The large size, firm texture, and few bones of arapaima made it ideal for this type of preservation, making it among the most consumed and commercialized fishes. Arapaima was even likened to another important commercial fish, often being referred to as the "freshwater cod." Salted arapaima became a mainstay of the urban poor and a universal item in local and urban markets (McGrath et al. 1993). Salted arapaima was an important source of nutrition that supported various extractive cycles of natural resources such as rubber tapping and Brazil nut collection (Smith 1981; Goulding 1983; McGrath 1986, 1989).

Early South American fisheries were a seasonal activity that supplied dried salted fish to frontier settlements (Goulding 1983; Smith 1985; McGrath et al. 2015). Starting in the middle of the 20th century, the increasing availability of technological advances, such as motor-powered boats, ice, and synthetic fibers to make gill nets, resulted in increased fishing pressures (Castro and McGrath 2003). The broad availability of iced, fresh fish increased demand and exploitation of already moderately exploited fishery resources (e.g., McGrath et al. 1993). As a result, South American fisheries became a year-round activity that supplied urban and export markets with fresh, iced fish (Goulding 1983; Smith 1985; McGrath et al. 2015). Fisheries, including the arapaima, began showing classic signs of overexploitation, such as decreases in landings and sizes of harvested fish (Isaac et al. 1993; Martinelli and Petrere 1999). As arapaima population densities declined due to overexploitation, fishing activities moved to more distant and remote regions. For example, cultural reverence for arapaima in Guyana largely deterred commercial fishing activities until foreign entrepreneurs opened a market for Guyanese arapaima in Brazil in the 1960s (Watson et al. 2016). Today, with increasing interest in opening international markets, arapaima are harvested across their entire range, and there are growing efforts to promote their aquaculture.

MANAGEMENT: CHALLENGES OF EARLY EFFORTS

Due to declines of arapaima fisheries by the middle of the 20th century, these fish became some of the first targets for recovery by management agencies (Isaac et al. 1993). These efforts began with government regulations regarding arapaima harvest. Worried about declining arapaima fisheries in neighboring Brazil, the Guyanese national government banned all arapaima harvest by passing the Fisheries Act in 1952 (Fernandes 2005). The Brazilian Institute of Environment and Renewable Natural Resources (IBAMA) first established a minimum harvest size of 1.5 m total length in 1986 and then, in 1991, a seasonal closure for arapaima harvest from December to May (Isaac et al. 1993; Viana et al. 2004; Castello and Stewart 2010). Some Brazilian states even declared total moratoria on arapaima harvest, like Amazonas (in 1996) and Acre (in 2008) (Castello et al. 2009; Castello and Stewart 2010).

Unfortunately, regulating arapaima fisheries proved nearly impossible due not only to temporal and geographic dispersion of arapaima catch, but also to a lack of urban market controls (Bayley and Petrere 1989; Castello et al. 2011). South American fisheries involve the harvest of many species captured during various seasons, and fishing activity is spread across an enormous area. Limited human and financial resources for enforcement resulted in widespread illegal trade of fish and noncompliance of harvest regulations (Bayley and Petrere 1989; Viana et al. 2004). By the late 1980s, the commercial fishery in Guyana collapsed, despite a moratorium on arapaima fishing, and unregulated arapaima harvest continued with a large focus on subsistence (Fernandes 2005). Various Brazilian states, such as Acre and Amazonas, experimented with complete closures of the arapaima fishery, but illegal fishing continued as a result of limited monitoring and enforcement (Viana et al. 2004; Castello et al. 2009; Oviedo and Bursztyn 2016). In each case, the failure of these regulatory approaches resulted in virtually unregulated, open-access fishing for subsistence or commercial gain. Illegal arapaima that are undersized or out of season could, and still can, be readily found in most urban markets.

The arapaima fishery is increasingly thought to be severely overexploited across most of its range (Isaac et al. 1993; Martinelli and Petrere 1999; Castello et al. 2015). Arapaima population numbers across their range have plummeted to 13% of historic abundance (Castello et al. 2011), and illegal arapaima fishing is extremely common. Recent evaluations of harvest showed that two-thirds of the arapaima harvested from lower Amazon fishing communities was illegal, with many fish captured below minimum size or during seasonal closure (Cavole et al. 2015; Gurdak et al. 2019). In addition, catch surveys from the Araguaia River basin showed limited compliance with fishing gear restrictions, such as prohibiting gill nets (Kirsten et al. 2012). As a result, arapaima populations continue to be decimated by overfishing, leading to localized extinctions (Castello et al. 2015).

MANAGEMENT: A PROMISING APPROACH AT A SMALL SCALE

The failure of early fishery management efforts in South America (i.e., regulations established by state and federal governments) emphasized the need for alternative approaches to manage aquatic resources like the arapaima. Beginning in the 1970s and 1980s, grassroots movements in the Brazilian Amazon led to the development of community management systems and subsequent federal and state fisheries management policies to control commercial fishing pressure in floodplain lake fisheries (see McGrath et al. 2015 for a more detailed history). Initially considered illegal by the Brazilian government, the multi-decade development of co-management institutions and legislation ultimately involved varying inputs from floodplain communities, government management agencies, local nongovernment organizations (NGOs), and national and international donors (McGrath and Castello 2015; McGrath et al. 2015). Due to increasing pressures on floodplain lake-fish populations and the absence of government enforcement, fisheries conflicts erupted throughout the Amazon basin as communities began to organize to prevent nonresident commercial fishers from entering their lakes (Chapman 1989; Hartmann 1989). These isolated efforts grew into regional grassroots organizations that contributed to the development of governmental policies to manage floodplain resources in Brazil (McGrath et al. 2015). Specifically, in the 1980s and early 1990s, growing pressure on federal environmental agencies responsible for fisheries management in Brazil (i.e., IBAMA) led to the negotiation of collective agreements to regulate fisheries in floodplain communities and increasingly sought government recognition and support for enforcement (McGrath et al. 1993). By the early 2000s, Brazil's federal co-management policy was operational through IBAMA and consisted of various legally recognized community fishing agreements across several states (McGrath et al. 2015). Unfortunately, after the Brazilian government restructured its fisheries sector in 2002, further development of federal co-management policies for artisanal fisheries was stalled (McGrath et al. 2015). Instead, individual Brazilian states began to develop fisheries regulations based on federal legislation, but the degree and nature of support varied. These efforts have led to varying implementations of, and outcomes for, Brazilian fisheries management policies from state to state (McGrath et al. 2015; also, see below).

Although not the focus of these co-management efforts, the important and overexploited arapaima fishery was among the fisheries to benefit from this management approach. In addition, management of arapaima is the best example of community management systems in South America (McGrath et al. 2015). The agreements shared basic components: identifying the user group, delineating a management area, and establishing rules for resource use and enforcement (Castro and McGrath 2003). The rules for fishing activities incorporated existing regulations, such as size and season restrictions, and imposed additional fishing gear, season, and area restrictions. For example, Brazilian seasonal closures protect fishes during reproductive activities that coincide with high water levels while, in contrast, many fishing agreements limit fishing activity in the low-water season when lake fishes are isolated and susceptible to harvest (Castro and McGrath 2003). Agreements also included rules for monitoring, usually through surveillance, and enforcement to limit prohibited fishing by local residents and outsiders (Castro and McGrath 2003). The overall management goals of arapaima co-management systems have been to allow local stakeholders to rebuild fishery stocks, improve livelihoods, increase food security, and conserve the ecosystems they depend on.

The techniques for effective co-management of arapaima were enhanced at the Mamirauá Sustainable Development Reserve, where the Mamirauá Institute has pioneered floodplain fisheries management and research activities since the 1990s (Figure 3[b]). Based on the ecological knowledge of expert arapaima fishers, a replicable approach for estimating arapaima populations through counting was introduced (Castello 2004). This method took advantage of the arapaima's unique physiology as an obligate air breather to count the number of arapaima coming up to breathe during a set interval of time. This technique for estimating arapaima populations has proven to be cost-effective and accurate while involving local fishers, thus creating ownership of, and trust in, data collected (Castello 2004). The approach has allowed for replication and the direct assessment of arapaima stocks while providing a method for adaptive management of a floodplain lake fishery (McGrath et al. 2015). Based on annual population assessments, a management plan with specific harvest quotas can be developed (McGrath et al. 2015). Estimates suggest that up to about 30% of adult arapaima can be harvested annually without detriment to the reproductive stock (Castello et al. 2011). In areas where arapaima harvest has been banned (e.g., Guyana, Brazilian states of Acre and Amazonas), harvest exceptions are being granted when management schemes have been implemented. In many cases, effectively implemented arapaima co-management schemes demonstrated the institutional design principles needed for sustainable governance of common property resources like arapaima (Castello et al. 2009; Oviedo and Bursztyn 2016).

MANAGEMENT SUCCESS AND CHALLENGES

Arapaima management has been successfully implemented in several geographically dispersed areas across South America (Figure 3). Here, we selected three cases and examined how management was implemented, evaluated management efforts considering Ostrom's (1990) eight principles (Table 1), and identified ongoing challenges.

Mamirauá Sustainable Development Reserve, Amazonas State, Brazil

The Mamirauá Sustainable Development Reserve was established in the 1990s to promote biodiversity protection through participatory resource management (Figure 3[b]; Castello et al. 2009). The reserve was stratified into two zones: sustainable use, where regulated and sustainable exploitation of resources was allowed; and preservation, where no natural resources use was allowed (Castello et al. 2009). Initial attempts to regulate arapaima harvest were ineffective and illegal fishing was undeterred. In the early 1990s, less than 30% of harvest met the legal size limit, and arapaima were harvested irrespective of season restrictions (Viana et al. 2004). The Mamirauá Institute was founded in the late 1990s to improve sustainable use of natural resources and people's livelihoods and operates under the supervision of the Brazilian federal government (i.e., the Brazilian Ministry of Science, Technology, Innovation and Communication; Castello et al. 2009). In hopes of improving sustainable practices, a management plan was implemented for the reserve in 1996 (Castello et al. 2009). Despite a statewide ban on arapaima fisheries, harvest exceptions were negotiated for those com-

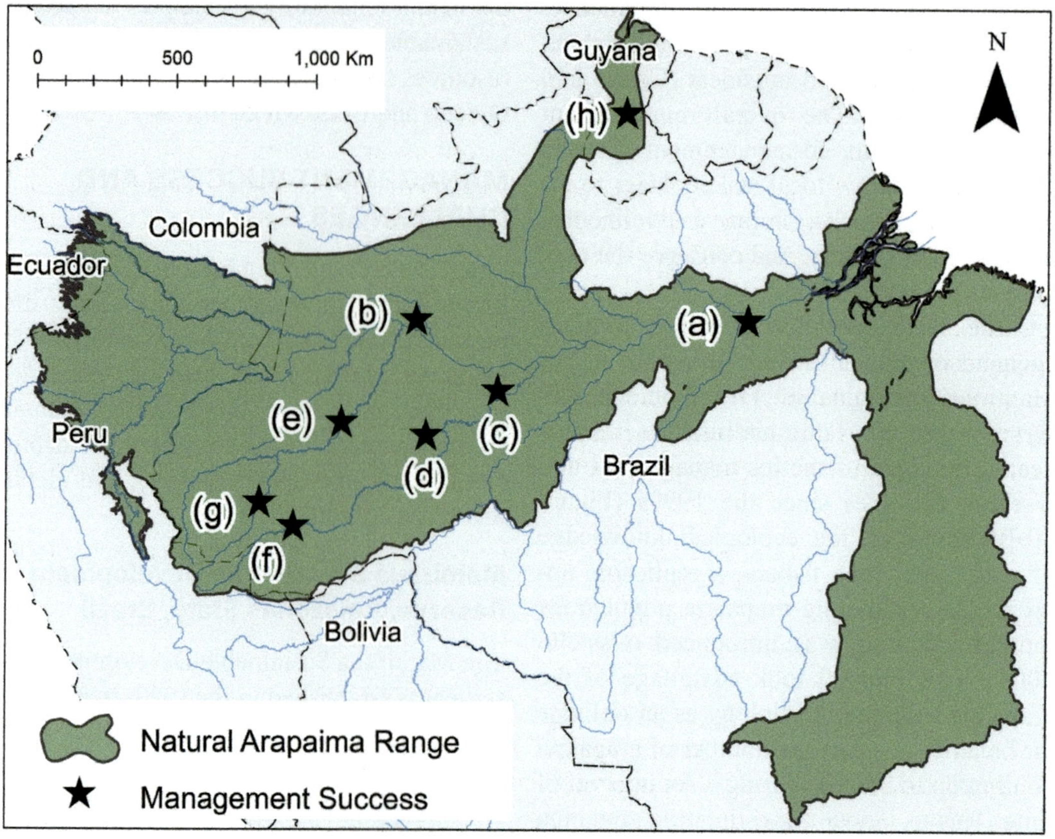

FIGURE 3. Despite drastic population declines, successful arapaima management has been implemented at various sites: (a) lower Amazon, Pará State, Brazil (McGrath et al. 1993); (b) Mamirauá Sustainable Development Reserve, Solimões River, Amazonas State, Brazil (Castello et al. 2009); (c) lower Purus River (including Piagaçu-Purus Sustainable Development Reserve), Amazonas State, Brazil (Petersen et al. 2016); (d) Tapauá River, Amazonas State, Brazil (Petersen et al. 2016); (e) middle Juruá River, Amazonas State, Brazil (Campos-Silva et al. 2016); (f) upper Purus River, Manuel Urbano, Acre State, Brazil (Oviedo and Bursztyn 2016); (g) Envira River, Fiejo, Acre State, Brazil (Oviedo and Bursztyn 2016); and (h) Essequibo River drainage, Guyana.

munities in cooperation with government officials (Castello et al. 2009). After 8 years, success at Mamirauá was observed: arapaima populations increased 9-fold with a 23-fold increase in the adult population, harvest quotas increased 10-fold, the number of fishers participating in the management efforts more than doubled, and per capita income of fishers involved increased 8-fold (Castello et al. 2009). By 2005, arapaima management efforts of the Mamirauá Institute had expanded to 15 other communities and even to fisher's organizations outside of the reserve (Castello et al. 2009).

The Mamirauá Sustainable Development Reserve in Amazonas presents a unique case of sustainable fishery management. In addition to being in a productive and pristine floodplain system with ecological features that favor arapaima production (Arantes et al. 2013), the Mamirauá Institute provided communities on the reserve with opportunities to implement management actions needed for sustainable governance of natural resources like the arapaima (Castello et al. 2009; Table 2). First, management areas were located inside a protected area with well-defined boundaries, resource uses, and users. In addition, fishers largely complied with established management regulations, and arapaima populations showed a remarkable increase in abundance (Castello et al. 2009). With support from the Mamirauá Institute, fishing communities have increasingly been involved and participated, particularly through regular meetings, in collectively modifying rules that affected their fishing practices. These meetings, with participation from the institute, also provided mechanisms for conflict resolution. Communities were allowed to organize and implement programs with support from external institutions and governmental authorities that provided for legal permits to harvest arapaima. In addition, the resource was evaluated by the users themselves; arapaima counts were conducted by fishers, some of whom were instrumental in developing this technique that has become a critical assessment tool of arapaima management (see Castello 2004). The principal mechanism for enforcement and sanctioning of rule offenders on the reserve was accom-

TABLE 2. Evaluation of arapaima management in Mamirauá Sustainable Development Reserve, in the central Amazon, Amazonas State, Brazil, using Ostrom's (1990; see Table 1) eight sustainable governance design principles (adapted from Castello et al. 2009).

Ostrom's design principles	Before management	After management
1. Defined boundaries	Present: zoning system.	Present: as before.
2. Congruence between appropriation and provision rules and local conditions	Absent: size limit and closed season violated.	Present: fishers followed management rules and arapaima populations grew
3. Collective action arrangement	Present: opportunities existed for individuals to participate.	Present: as before with increased participation.
4. Monitoring	Absent: fishers were appropriators but underreported illegal catch, undermining monitoring system.	Present: counts of arapaima were made by the fishers themselves.
5. Graduated sanctioning	Absent: illegal fishing widespread among community fishers.	Present but weak: sanctioning done by reducing fishing quota and/or gear confiscation.
6. Conflict resolution mechanism	Present: regular meetings were attended.	Present: as before.
7. Minimum recognition and right to organize	Absent: fishing of arapaima banned.	Present: sceme provided legal permit to harvest arapaima.
8. Nested enterprises	Present: three-layered scheme with communities, Mamirauá Institute, and the Brazilian Institute of Environment and Renewable Natural Resources (IBAMA).	Present: as before.

plished by volunteer environmental agents called agentes ambientais voluntarios, who are local residents trained and authorized by governmental agencies and receive logistical support from the institute for enforcement activities. In most cases, rule violators can be sanctioned by reducing their fishing quotas and punished by confiscating or destroying gear and boats, but often the application of these tactics has been weak. Ultimately, the success of management and governance in Mamirauá is accomplished due to effective layers of nested enterprises (i.e., communities, Mamirauá Institute, IBAMA), resulting in coordinated management, monitoring, enforcement, conflict resolution, and governance activities.

Arapaima management efforts at the Mamirauá Sustainable Development Reserve were nested in a favorable political climate that fostered the development of effective fisheries co-management efforts throughout the entire state of Amazonas (see McGrath et al. 2015). Although developed later than in other states (e.g., compared to Pará; see below), Amazonas has made a significant commitment to fisheries co-management policies and has implemented legislation that includes procedures for legal recognition, monitoring, and enforcement (McGrath et al. 2015). In Amazonas, grassroots efforts had a major influence on policy development, with the state division of IBAMA, the federal environmental agency, playing a lead role in developing policies and institutional arrangements for fisheries co-management (McGrath et al. 2015). In addition, fisheries and co-management policy was influenced by the state's reserve-based approach to environmental management and sustainable development (McGrath et al. 2015). With further support from state government and nongovernment institutions, comprehensive legislation for statewide fisheries co-management policy was approved, including procedures and criteria for legal recognition, monitoring, and enforcement of collective fishing agreements (McGrath et al. 2015). This favorable climate toward collective fisheries agreements allowed the arapaima management system to develop at Mamirauá and provided a framework for sustainable community-based arapaima management initiatives across the state. Starting in 2004, the state division of IBAMA implemented regulations based on this framework, allowing community groups to submit management proposals based on arapaima population assessments and receive approval for specific harvest quotas (McGrath et al. 2015). To protect chain of custody of managed arapaima, IBAMA provides tags and transport documents to prevent illegal fish from entering the market (McGrath et al. 2015). Harvested arapaima are not filleted, although viscera may be removed, so that total lengths can be confirmed as fish enter markets.

Although management efforts at the Mamirauá Sustainable Management Reserve have been successful, several challenges to long-term sustainability remain. To begin with, existing rules and regulations must be validated through research and adjusted when necessary. For example, although minimum size of capture of arapaima is set to 1.5 m total length and populations have increased based on this regulation, the size at first reproduction (and a more ideal minimum size of capture) has been shown to be at least 1.57 m, 7 cm larger than current regulations (Arantes et al. 2010). As a result, a higher minimum size of capture may be more appropriate to achieve higher sustainable yields from arapaima fisheries on the reserve. In addition, despite the presence of regulatory and enforcement mechanisms, illegal fishing continues, in part because too few volunteer agents have been available to monitor the managed area. In addition, due to the high socioeconomic costs of enforcement activities, not all communities have been able to effectively deter illegal fishing through monitoring and sanctioning of rule offenders (Castello et al. 2009). Illegal fishing inside and outside the reserve supplies local markets with illegal fish, which generates unfair competition with the managed, legally harvested fish. In some communities, arapaima counts can be inaccurate if performed by inexperienced fishers or even intentionally manipulated to inflate harvest quotas (Arantes et al. 2007; Andrade et al. 2011). Mechanisms to assure quality control of counts have been developed but are not always implemented because they require sufficient technical support and monitoring (Arantes et al. 2007; Andrade et al. 2011). Maintaining the financial and human resources of the Mamirauá Institute is essential to support ongoing initiatives and address demands for training and support. Last, despite the state's commitments to co-management efforts and the regulatory systems implemented, government management institutions in Amazonas remain fragile, albeit functional, and the state struggles with limited capacity to meet the regulatory and research needs (McGrath et al. 2015).

Lower Amazon, Pará, Brazil

In the lower Amazon, near the city of Santarém, floodplain inhabitants live in geographically dispersed communities that are outside formally protected areas or reserves (Figure 3[a]). As in the state of Amazonas (see above), arapaima co-management in Pará evolved from grassroots fisheries efforts initiated at the community level (McGrath et al. 1993; Castro and McGrath 2003). The Iara Project, based in Santarém, was initiated in the 1990s by IBAMA through a joint effort between IBAMA and the German government consultancy firm GTZ to lead the development of fisheries co-management policy in the state of Pará (McGrath et al. 2015). This effort involved two interrelated phases to first regulate fishing and then implement a comprehensive land-tenure system (McGrath et al. 2015). First, community fishing agreements, also referred to as fishing accords, were implemented and achieved recognition from the federal government (i.e., IBAMA; McGrath et al. 1993; Castro and McGrath 2003). Based on IBAMA's policies of the 1990s, Pará further developed co-management policy by passing legislation in 2005, followed by slight revisions in 2007 (McGrath et al. 1993). In contrast to the situation in Amazonas, however, local government institutions at the state and municipal levels in Pará did not play a significant role in implementing legislation and provided little recognition or support for existing fishing agreements (McGrath et al. 2015). In an effort to revise and improve the policies and institutional arrangements for co-management, the second phase involved the implementation of a land-tenure system that recognized individual and collective rights of floodplain resources (McGrath et al. 2015). Known as Agro-extractivist Settlement Projects or PAEs, the system proposes that land in the Amazon River floodplains, or várzea, is owned by the federal government and communities are granted settlement concessions in exchange for the commitment to sustainably manage natural resources in their respective areas (McGrath et al. 2008). In addition, the system provided formal recognition of local rules and organizations and provided opportunities for organizing governance activities in multiple layers of nested enterprises (McGrath et al. 2008). Unfortunately, the realization of the PAE system came as the federal government lost interest in fisheries co-management policy (McGrath et al. 2015). In addition, no government institution, federal or state, was appointed or assumed the responsibility for enforcing management plans. Due to these shortfalls, the PAE system, while sound, may never achieve full legalization and functionality. Despite active and ongoing arapaima management efforts in the Santarém region based on the model developed in Mamirauá, these efforts are impeded by a lack of government and institutional support. In contrast to Amazonas, Pará has not implemented substantial efforts to limit illegal exploitation of arapaima or to develop or enforce effective regulations. As a result, managed arapaima from a few sustainable fisheries competes with illegal and unmanaged arapaima.

The majority of communities in the region have no real management, and as a result, arapaima populations are extremely depleted and, in some cases, locally extinct (Castello et al. 2015; Table 3). Although most communities in the region are zoned under the PAE system, conflicts exist regarding specific lakes and rights to use along boundary zones. Compliance with size, season, and gear regulations is nonexistent and much of the fishery is essentially open-access. Most communities have semiregular meetings with varying levels of organization and capacity. Often, rules outlined in fishing agreements have little meaning because of widespread noncompliance. Meetings usually have limited effectiveness in terms of conflict resolution and can even make conflicts worse. Counts of arapaima are not regularly conducted (e.g., Castello et al. 2015) or have never been conducted. Rule violators are not sanctioned (and, as a result, seem to have no fear of alleged sanctions). Community rules and regulations are provided formal recognition via the PAE system; however, even where rules exist, noncompliance with regulations is rampant. Although the PAE system is designed to involve nested enterprises, communities are largely autonomous and receive some input from local NGOs and a fishermen's union but limited government involvement.

Despite these grim generalizations, a few exceptions exist where arapaima management has been successful. About 25% of communities were minimally engaged in arapaima management efforts (Castello et al. 2015). Communities with management and comparable habitat productivity had arapaima densities that were, on average, two orders of magnitude higher when compared to communities without management (Castello et al. 2015). Although the implementation of arapaima management falls along a gradient of efficacy, the iconic example of successful fishery management for the region is the community of Ilha de São Miguel (e.g., McGrath et al. 1993). Here, use of gill nets and seines has been banned for nearly three decades, and only a handful of fish taxa are commercialized, one being the arapaima (McGrath et al. 1993; Castello et al. 2015). Although banning gill nets may negatively affect food and income security, arapaima density and multispecies catch per unit effort at Ilha de São Miguel are actually higher than anywhere else in the region (Castello et al. 2011, 2013, 2015). In addition, this community shows many of Ostrom's (1990) principles (Table 3). First, the community is zoned under the PAE system, and being a floodplain "island," the area has clearly defined boundaries. Although arapaima density in the community is high, low compliance exists with the minimum size of capture regulation, but strong compliance occurs with season and gear regulations. The community has regular meetings to collectively establish fishing agreements and resolve conflict. Although counts of arapaima occur semi-regularly, encouraged by local NGO support, quotas are never established and harvest is primarily limited by gear restrictions. Arapaima are captured using traditional harpoon methods and occasionally by a series of baited hook and lines tied to a stake or tree. Depending on severity of offense, rule violators from within the community can be banned from fishing or even expelled from the community. Unfortunately, the community has a difficult time

TABLE 3. Evaluation of arapaima management in the lower Amazon, Pará State, Brazil, using Ostrom's (1990; see Table 1) eight sustainable governance design principles.

Ostrom's design principles	Without management	With management (e.g., Ilha do São Miguel)
1. Defined boundaries	Present: implementation of zoning system began in 2006. Weakly observed resulting in conflicts regarding lakes and land use.	Present: zoning system. Being a floodplain "island," has well-defined boundaries.
2. Congruence between appropriation and provision rules and local conditions	Absent: low compliance with size and season regulations.	Somewhat present: low compliance with season regulations.
3. Collective action arrangement	Absent or, when present, rules do not mean much.	Present: community collectively establishes fishing agreements.
4. Monitoring	Largely absent: occasional monitoring efforts conducted with support from nongovernmental organizations (NGOs).	Somewhat present: counts of arapaima encouraged by NGOs and conducted semi-regularly.
5. Graduated sanctioning	Largely absent: absent or, when present, very weak.	Present: depending on severity, rule violators can be prohibited from fishing or even expelled. Gear can be confiscated or destroyed.
6. Conflict resolution mechanism	Weak: meetings are attended semi-regularly but are often an ineffective platform for conflict resolution.	Present: regular meetings were atended.
7. Minimum recognition and right to organize	Present but weak: rules and regulations were offered formal recognition.	Present: rules and regulations were offered formal recognition.
8. Nested enterprises	Present but weak: despite scheme present, limited government input in the region leaves communities largely autonomous.	Somewhat present: community is largely autonomous and increasingly insular with some uncoordinated inputs from local NGOs, fishermen's unions, and governments.

punishing rule violators from other communities or nearby urban areas. The community allocates substantial resources for surveillance, regularly patrolling community lakes. Most commonly, the community confiscates, and sometimes destroys, fishing gear of rule violators, even of those from outside the community. Overlooking legal considerations related to punishing violators, community rules and regulations are commonly recognized by local institutions and recognized as satisfactory de facto if not technically de jure management systems. However, nested enterprises within the community remain limited and the community is largely autonomous and increasingly insular, with some uncoordinated inputs from local NGOs, fishermen's unions, and the government.

Despite some success, developing arapaima management in the lower Amazon is a challenge. Currently, nearly 75% of communities in the region do not follow any management rules for arapaima harvest (Castello et al. 2015). The few communities that engage in a fishing agreement, such as Ilha de São Miguel, represent exceptions to the relatively low levels of governance in the region. The majority of communities have not overcome the challenges of implementing arapaima management and, in particular, lack effective systems for monitoring and enforcing rule offenders. Overall estimates suggest that two-thirds of arapaima harvest in the lower Amazon is illegal, being captured either below minimum size limits or during seasonal closures (Cavole et al. 2015; Gurdak et al., in press). Even communities with management lack compliance with certain regulations, principally with minimum size of capture (Cavole et al. 2015). In many cases, communities with successful management initiatives have adopted financially and physically taxing surveillance regimes, especially in lakes that are more exposed or accessible to poachers. In many cases, surveillance is conducted at night by fishers armed with clubs, machetes, and firearms. Communities have no legal authority, however, to punish rule offenders, and government agencies lack the interest and human and financial resources to implement and enforce regulations. To improve management, governance capacity in the region needs to be strengthened and mechanisms for enforcing rules and sanctioning rules offenders must be implemented. In addition, effective policing and market controls that inhibit illegal arapaima from entering the market are needed. Ultimately, assistance and cooperation from the state and local government institutions, while currently limited, is essential to ensure the development and success of arapaima management initiatives in the region.

North Rupununi, Guyana

The North Rupununi region, located in the Essequibo River drainage of southwestern Guyana, is characterized by a mix of savanna and lowland tropical evergreen forests (Figure 3[h]; Lowe-McConnell 1964). The region is inhabited by 13 communities of primarily indigenous and some nonindigenous groups and contains about 200 lakes known to have arapaima (Castello 2002). Despite a federal moratorium on arapaima harvest (established in 1952 through the Fisheries Act), rampant illegal arapaima fishing amounted to open-access conditions. By 2002, regional arapaima populations were severely depleted to less than 500 adult individuals, and commercial fishing collapsed (Castello 2002). In 2001, the ban on arapaima fishing in the region was reinforced by a local community-based organization, the North Rupununi District Development Board. In 2002, through the cooperation of local communities, NGOs, and government institutions, efforts were formalized by developing and approving a management plan (Castello 2002). Based on the management plan, arapaima in the region collectively belonged to all 13 communities. Mechanisms for decision making, management practices, rules, and stakeholder participation were included in the plan. By 2007, the community-based management plan was legally recognized by the Guyanese government and community fishing grounds and fishing regulations officially became enforceable. During 13 years of management in the region, local communities were empowered to protect arapaima fisheries and arapaima populations increased sevenfold (Watson 2011; L. C. Watson, unpublished data).

Implementation of the management plan allowed for improved governance in the region (Table 4), increasing arapaima populations and strengthening local community institutions and participation. Before man-

TABLE 4. Evaluation of arapaima management in North Rupununi, Guyana, using Ostrom's (1990; see Table 1) eight sustainable governance design principles.

Ostrom's design principles	Before management	After management
1. Defined boundaries	Largely absent: only communities with land titles had a zoning system.	Present: zoning system but conflicts exist with regards to which communities have access to what lakes.
2. Congruence between appropriation and provision rules and local conditions	Absent: no compliance with ban on arapaima fishing. Open access resulted in extremely low abundances.	Somewhat present: community members follow rules in some areas and populations are increasing but low compliance in other areas and populations are decreasing.
3. Collective action arrangement	Absent.	Present, but there is often disagreement among communities
4. Monitoring	Mostly absent: some communities with land titles monitored overall presence and absence of arapaima in specific lakes.	Somewhat present: semi-regular monitoring conducted with financial support from various nongovernmental organizations.
5. Graduated sanctioning	Absent: illegal fishing common among communities and outsiders.	Present but very weak.
6. Conflict resolution mechanism	Absent: although mechanisms were present, they were not applied to fisheries.	Somewhat present: meetings attended but not very effective in resolving conflicts.
7. Minimum recognition and right to organize	Absent.	Present: communities have legal permit to harvest arapaima.
8. Nested enterprises	Absent.	Present: three-layered scheme with communities, Iwokrama Centre, and Department of Fisheries.

agement was implemented, all sustainable governance design principles were absent or largely absent. After implementation of management, application was variable across communities in the region; however, the majority of design principles were present, and the others were somewhat present. The management plan provided a zoning system and rules for resource use and graduated sanctioning. Semi-regular meetings are attended but could be more effective in resolving conflicts. Arapaima management is accomplished through a nested, three-layered scheme with input from communities, Iwokrama Centre, and the Guyanese Government Department of Fisheries. The Iwokrama Centre is an NGO created to develop sustainable use practices in the region.

Arapaima harvest has not been allowed by fisheries organizations, and management efforts in the North Rupununi District continue to face several challenges. First, population recoveries have been mixed, principally due to illegal fishing in areas more accessible by roads and boat launches (Watson 2011). In many cases, curbing illegal harvest is difficult for local communities because monitoring remote lakes is costly. In addition, enforcement is complicated when close friends and family are involved in violating rules. Second, the current arapaima population structure (ratio of juveniles to adults) suggests that poaching of young fish is common (Watson 2011). This harvest is a result of bycatch when using gill nets and targeted efforts to catch small, more easily concealed fish. Last, conflicts continue among communities in the region. For example, some communities have land titles that include arapaima lakes while others lack titles for their settlements and surrounding hunting and fishing grounds. Communities with land titles have a greater interest to protect and patrol their zones than communities without land titles. A further complication is that noncommunity members can legally fish for other species of fishes in arapaima lakes, making monitoring and enforcement difficult. Resentment also exists based on geography, with more remote communities claiming limited access to jobs compared to more readily accessible communities. These variables together contribute to cheating within and among communities. Because arapaima in the region are owned as a collective among all 13 communities, a managed harvest has not been agreed upon, except for one trial harvest of 100 fish in 2009. Instead, those communities that have been able to protect arapaima populations and other natural resources are pursuing ecotourism and catch–release sport fishing to generate income in lieu of an arapaima commercial fishery (Purnell 2017; Lennox et al. 2018).

LESSONS LEARNED

Examples of small-scale arapaima management have successfully begun to confront some of the unique sociopolitical and economic challenges associated with managing tropical fisheries (Pauly et al. 1989). In fact, the flexibility of the small-scale, co-management initiatives of arapaima in lake systems is proven by its implementation across distinct contexts. In large part, the model's success is attributed to the arapaima's economic and cultural value along with its unique biology. Arapaima are ecologically well suited to small-scale management because they are not known to migrate long distances and are predominantly lake-dwelling and lake-spawning. In addition, arapaima are obligatory air breathers and the ability to estimate their population in a lake has been drawn from experienced fishers' skill to count the number of arapaima surfacing to breath (Castello 2004). The ability to annually estimate arapaima populations improved confidence and transparency in arapaima management plans while strengthening community capacity for fisheries management (McGrath et al. 2007,

2015). While these details have contributed to the success of arapaima management initiatives, these techniques would not work for all fishes. For example, alternative regulatory and management techniques are needed for fishes that are not as economically valuable or for those that migrate long distances, the latter representing the majority of catch sold in urban markets in the Amazon.

In the cases presented above, arapaima management initiatives improved design principles of sustainable governance (Table 5). On average, 30% of eight design principles were present in unmanaged situations. In contrast, managed situations presented an average of 75% of the eight design principles, with the remainder being at least somewhat present. Although the design principles were useful in evaluating arapaima management initiatives, important distinctions exist between arapaima management systems and other co-management systems that are characterized by the governance design principles. In addition, other factors not accounted by the designs principles can influence the success of the management systems, either directly or indirectly (e.g., leadership presence, specific policies, see Lesson 4 below; Ostrom 2007). Ultimately, our cases also show that success, challenges, and outcomes of arapaima management were not uniform and frequently require unique and context-specific considerations and interventions. Based on our experiences, we identified five specific lessons for achieving sustainable harvests for arapaima, which may be applicable to other fishes.

Lesson 1: Manage Even When Faced with Uncertainties, but Proceed with Caution

Resource management almost always occurs with some degree of uncertainty, even though reaching scientific certainty and consensus before taking action may be desirable. Regardless, management can be effective if uncertainty is recognized and risks to decision making are accepted (Ludwig et al. 1993). In most cases, resource management in the face of uncertainty should be encouraged because suboptimal action is generally preferred to no action at all (Johannes 1998). In the lower Amazon, early grassroots efforts empowered local communities to take control of lake fisheries and recover declining arapaima populations (McGrath et al. 1993). Arapaima management efforts at the Mamirauá Sustainable Development Reserve faced incredible uncertainty, and despite the lack of information, depleted arapaima stocks recovered and set an example for management elsewhere (Castello et al. 2009). These pioneering initiatives provided examples for arapaima management that, despite continued uncertainties, have successfully been implemented in other regions and have led to recovered arapaima populations (Campos-Silva and Peres 2016; Oviedo and Bursztyn 2016; Petersen et al. 2016).

For arapaima, the contrast between areas with and without management is best exemplified by arapaima population densities in the region of the lower Amazon (Castello et al. 2015). The few areas with arapaima populations were the same communities that sought to implement management. In contrast, the vast majority of the region had depleted or locally extinct arapaima populations as a result of limited or nonexistent management efforts. Still, caution should be taken to avoid actions that could result in increased conflict or resentment toward management, thus compromising future conservation efforts. Ultimately, we urge continued management efforts because the alternative of no management could be much worse for the fish and for the local human communities.

Lesson 2: Monitor, Evaluate, and Adapt Management Efforts

Once management actions are implemented, monitoring and evaluation are essential

TABLE 5. Comparison of managed and unmanaged conditions for arapaima fisheries using Ostrom's (1990) eight sustainable governance design principles in (1) Mamiraua Sustainable Development Reserve (MSDR); (2) lower Amazon, Pará State, Brazil; and (3) North Rupununi, Guyana.

	Without management			With management		
Ostrom's design principles	MSDR	Lower Amazon	North Rupununi	MSDR	Lower Amazon	North Rupununi
1. Defined boundaries	Present	Present	Absent	Present	Present	Present
Congruence between appropriation and provision rules and local conditions	Absent	Absent	Absent	Present	Somewhat present	Somewhat present
2. Collective action arrangement	Present	Absent or weak	Absent	Present	Present	Present
3. Monitoring	Absent	Largely absent	Absent	Present present	Somewhat present	Somewhat
4. Graduated sanctioning	Absent	Largely absent	Absent	Present but weak	Present	Present but weak
5. Conflict resolution mechanism	Present	Somewhat present	Absent	Present	Present	Present
6. Minimum recognition and right to organize	Absent	Present	Absent	Present	Present	Present
7. Nested enterprises	Present	Absent	Absent	Present	Somewhat present	Present

to provide feedback about the effectiveness of the actions in achieving management goals (Margoluis et al. 2009; Holland 2010). Short-term successes should not be expected to persist indefinitely, and assessments can guide how management could be modified or adapted to better suit evolving realities. Understanding both management successes and failures provides valuable information to guide future decision making. In most cases, flexibility and adjustments to existing management may be needed due to three reasons described below.

First, uncertainties or incorrect assumptions in the knowledge base may limit the effectiveness of regulations and management. By challenging existing assumptions, research and assessment can reveal new information that may need to be incorporated to adjust existing regulations. For example, the minimum size of capture for arapaima is generally 1.5 m, even though research shows that the minimum length at reproduction (and therefore a better minimum size of capture) should be larger than this length in some parts of the central (~1.57 m) and lower Amazon (~1.7 m) and much larger in Peru (~2 m) (Gurdak et al. 2019). Second, implementing ineffective management protocols or regulations can lead to noncompliance or undesirable outcomes. For example, despite federal and state regulations, most arapaima harvested in Brazil are illegal as a result of inadequate regulatory controls (e.g., Castello and Stewart 2010; Cavole et al. 2015; Gurdak et al. 2019). Arapaima co-management has been a successful alternative that has improved compliance with harvest regulations. Third, management is complicated by the reality that natural systems are dynamic and, at times, unpredictable (Johannes 1998). For arapaima, ecological conditions on floodplains are intrinsically variable, and management regulations established at one time may not be ideal in the long term. For example, annual flood pulses can lead to age-structured fish populations that will respond to management manipulations with time lags related to generation times. As a result, management must consider changes in vulnerability in response to unexpected events and must be capable of being adaptable to changing conditions (Johannes 1998). Regional universities could provide a leading role in gathering needed local information and providing institutional stability for fisheries management in uncertain political climates (McGrath and Castello 2015). Adaptation will be important as climate change affects these already-dynamic freshwater environments and threatens Amazonian fishes even further (Frederico et al. 2016).

Lesson 3. Bridge Scientific and Underused Knowledge Systems

Management depends on applying the best available information to establish harvest regulations (Salo et al. 2013). Although information can be drawn from various sources, evidence-based approaches generally use scientific evidence drawn from research (dos Santos et al. 2015). A growing recognition exists, however, that the complexities of environmental management must draw and incorporate knowledge from many sources (Ludwig 2001). Despite their utility and ability to overcome information deficits, alternative knowledge sources such as individual experience, qualitative observations, and local or indigenous knowledge are generally underused and sometimes ignored. Traditional or local ecological knowledge, for example, is the informal knowledge held by local inhabitants who use natural ecosystems (Johannes et al. 2000; Silvano and Valbo-Jørgensen 2008) and has contributed to managing fisheries around the globe (e.g., Johannes et al. 2000; Valbo-Jørgensen and Poulsen 2000; Castello 2004; Silvano and Valbo-Jørgensen 2008; Brelsford 2009).

Management schemes should consider the use of various knowledge systems and seek to create bridges between systems (Berkes et al. 2006). Encouraging information sharing among key stakeholder groups is not an easy task but can create synergies that improve management (Berkes et al. 2006). Bridging information systems is especially important when data deficits hinder effective management decision making. Bridging knowledge systems has contributed to the success of arapaima management (Castello et al. 2009). The knowledge and skills of local fishers were used to standardize a method to reliably count arapaima and estimate populations (Castello 2004). This method involved local people and allowed them to actively participate in resource assessment and subsequent use while creating ownership and trust in the information being used for decision making (e.g., Castello et al. 2009). The population counts were essential to granting harvest permissions for arapaima in the central Amazon (Castello et al. 2009). In addition, linking government officials and local fishers interested in management efforts allowed for continued government support and improved understanding among stakeholders (Castello et al. 2009). Local knowledge can be used to identify local ecological differences that help optimize regulations and compliance. For example, local knowledge can be used to detect distinctive characteristics and highlight unexpected life histories for a particular fish population (e.g., Gurdak et al. 2019).

Lesson 4. Long-Term Success Depends on the Genuine Interest and Cooperation of Government, Communities, and Other Stakeholders

Resource management typically centralizes authority in governments at regional, national, or international levels. By definition and design, co-management efforts incorporate the active participation of resource users that manage resources in partnership with government and nongovernment institutions. This arrangement strengthens people's participation in governance while distributing and decentralizing power (Pomeroy and Berkes 1997). This integration is essential to co-management and community management efforts where management responsibilities are shared and organized in multiple tiers or nested enterprises (Ostrom 1990; Table 1). At a minimum, these nested enterprises consist of the government and the resource user. Different implementations of arapaima management have, to varying degrees, connected local fishers, various government entities, NGOs, universities, and fisher unions.

The long-term success of this approach can be hindered if the partners' participation does not represent full and genuine engagement. At its foundation, local resource users must be engaged in management discussions and decision making. Arapaima management efforts in each case study above were bolstered by local grassroots initiatives. Identifying fishers' interests was critical to establishing effective management programs (e.g., Castello et al. 2009). The genuine participation of local fishers was essential for arapaima management because harvest and surveillance were in their control, with limited or nonexistent outside enforcement of regulations (Castello et al. 2009). Strong community leadership was critical to ensuring local compliance with management rules.

Government interest and action is vital in creating favorable conditions that allow communities to sustainably co-manage fisheries (McGrath et al. 2015). Specifically, government involvement requires genuine regulation and collaboration with fishers and fisheries industries (Pomeroy and Berkes 1997; McGrath et al. 2015). The role of federal and state governments is exemplified by the outcomes of arapaima management efforts in the Brazilian states of Amazonas and Pará (see above; also, see McGrath et al. 2015). In both states, similar legislation for co-managing floodplain fisheries evolved from grassroots efforts that supported community-based management (McGrath et al. 2015). However, varying support from local and federal governments resulted in two different outcomes. Amazonas was able to implement a comprehensive management system because government agencies had the interest and political support to develop sustainably managed fisheries (McGrath et al. 2015). For arapaima, management meant implementing monitoring and enforcement of regulations and agreements that limited market access for illegal arapaima and enabled arapaima population recoveries in managed areas (Bessa and Lima 2010). In contrast, Pará was not able to implement effective regulatory systems for co-managing fisheries, despite having implemented a legislative framework like that of Amazonas (McGrath et al. 2015). As a result, communities in Pará that manage arapaima are the exception and not the rule, and as is true across most of the Amazon, ineffective fisheries management has led to severely depleted and even locally extinct arapaima populations (Castello et al. 2015; McGrath et al. 2015). Arapaima from managed fisheries compete in a market saturated with illegal arapaima. In addition, communities that manage arapaima face significant external fishing pressure and incur great costs to patrol lakes and deter poaching. Although communities have the legal right to manage their fishers (i.e., minimum recognition and right to organize; Table 1, principle 7), heavy outside poaching pressure renders this recognition insufficient. As a result, integrative nested enterprises with substantial government input and collaboration are needed. Developing effective co-management systems requires the real, long-term commitment of governments, local fisher organizations, and fisheries industries (McGrath et al. 2015).

Lesson 5. Move Toward an Ecosystem-Based Approach

Despite their success, one drawback of arapaima co-management efforts is that they represent a single-taxon focus. Such approaches mainly target fishing effort and, while simplified and easier to implement (Simberloff 1998), could ignore maintenance of conditions that allow for arapaima populations to persist. Currently, the primary threat to arapaima is overfishing (Castello et al. 2013). However, four main drivers of freshwater ecosystem degradation exist that could have negative impacts on arapaima populations: deforestation, construction of dams and navigable waterways, pollution, and overharvesting (Castello et al. 2013). Like many floodplain fishes, the arapaima depends on floodplain habitat and the flood pulse to complete its life cycle. As a result, successful long-term arapaima management depends on preserving habitats, ecosystems, and the natural processes that sustain arapaima populations. Therefore, ecosystem- or catchment-based management strategies can balance human uses and ecosystem integrity (see Castello et al. 2013).

Managing arapaima could serve as a catalyst for expanding fisheries and ecosystem management efforts. For example, managing arapaima populations can raise the political will, funds, and conservation awareness while incorporating management of the ecosystem that arapaima depend on. Arapaima fulfills a surrogacy role (e.g., as a flagship taxon; see Caro 2010) that could, as a result, promote and enhance community management efforts. In addition, arapaima co-management provides an organizational framework to implement ecosystem management (McGrath et al. 2007; McGrath and Castello 2015). The ability to count and monitor arapaima to adapt management plans helps to strengthen community organization, and associated training and capacity building establishes an adaptive learning culture (McGrath et al. 2007; McGrath and Castello 2015). Based on this organizational framework, communities are well suited to build upon arapaima management efforts that include more comprehensive approaches and take into account the floodplain lake ecosystem (McGrath and Castello 2015). As a result, communities may be more inclined to modify economic activities that adversely affect productivity of lake systems, such as limiting cattle herds and restricting their movements to reduce negative impacts on forest and aquatic macrophyte habitat (McGrath et al. 2007).

CONCLUSIONS

Arapaima are the most iconic fishes of South American floodplain rivers. They are economically valued as a food fish because of the highly regarded taste, texture, and size of their fillets. Arapaima are culturally important and symbolic as a flagship species of tropical South American fisheries and floodplain management and conservation. Although arapaima represent a small fraction of total fisheries harvest, their management and conservation could have profound implications for protecting not only their own populations, but also the habitat structure of the critical nurseries for most of the commercial fish species in the Amazon. Here, we showed that managing arapaima fisheries can be successful and directly benefit the environment and local peoples. These experiences could contribute to managing overexploited tropical floodplain fisheries elsewhere. Co-management efforts can have positive impacts on fish stocks and on the participation and income of fishers and fishing communities. Management must go forward even in data-poor situations, but we urge that caution and flexibility be built in, and adaptability to changing environmental circumstances is essential. Last, we stress that the future of arapaima, and other floodplain fishes, is linked to the health and integrity of floodplain ecosystems, and thus, ecosystem-oriented management is critical for sustainable management of arapaima and other floodplain fisheries.

ACKNOWLEDGMENTS

This review was funded in part by the U.S. Institute for International Education Fulbright Scholar Program, the U.S. Environmental Protection Agency under the Science to Achieve Results (STAR) Graduate Fellowship Program, National Geographic Society Young Explorer's Grant and various funding sources from the State University of New York College of Environmental Science and Forestry (Graduate Assistantships in Department of Environmental and Forest Biology, the Robert L. Burgess Graduate Scholarship in Ecology, the Wilford Dence Fellowship, the Tropical Social Forestry Fund, and the Maurice and Annette Alexander Wetlands Research Award). Financial support for CCA was provided by Conselho Nacional de Desenvolvimento Científico e Tecnológico – CNPq (200893/2012-2) and Texas A&M University (i.e., Applied Biodiversity Science Program, Tom Slick Fellowship and Dissertation Fellowship). Research was conducted under authorization granted by the Brazilian National Ministry of Science, Technology and Innovation (Ministério da Ciência, Tecnologia e Inovação, MCTI; CNPq #s 001141/2012-0, 001609/2014-9) and the Brazilian Institute of Environment and Renewable Natural Resources (Instituto Brasileiro do Meio Ambiente e dos Recursos Naturais Renováveis, IBAMA; SISBIO #s 30201 and 45187). This work could not have been accomplished without the work of the following institutions: Mamiraua Sustainable Development Institute, Amazon Environmental Research Institute (Instituto de Pesquisa Ambiental da Amazônia – PAM), and SAPOPEMA (Sociedade para a Pesquisa e Proteção do Meio Ambiente). We appreciate the comments and feedback of those who reviewed drafts of this manuscript and greatly improved it, including anonymous reviewers. Last, but not least, we would like to thank peoples of the fishing communities who work to protect floodplain fisheries. The individuals and organizations mentioned here have not officially endorsed this publication, and the views and opinions expressed here are those of the authors.

REFERENCES

Allan, J. D., R. Abell, Z. E. B. Hogan, C. Revenga, B. W. Taylor, R. L. Welcomme, and K. Winemiller. 2005. Overfishing of inland waters. BioScience 55:1041–1051.

Andrade, L. C. A., E. S. R. Amaral, N. B. Silva, and H. L. Queiroz. 2011. Recount pirarucu: a method for evaluating the quality of the pirarucu countings. UAKARI 7:29–40.

Arantes, C. C., L. Castello, M. Cetra., and A. Schilling. 2013. Environmental influences on the distribution of arapaima in Amazon floodplains. Environmental Biology of Fishes 96:1257–1267.

Arantes, C. C., L. Castello, and D. S. Garcez. 2007. Variações entre contagens de *Arapaima gigas* (Schinz) (Osteoglossomorpha, Osteoglossidae) feitas por pescadores individualmente em Mamirauá, Brasil. [Variations among counts of Arapaima gigas (Schinz) (Osteoglossomorpha, Osteoglossidae) made by individual fishers in Mamirauá, Brazil.] Pan-American Journal of Aquatic Sciences 2:263–269.

Arantes, C. C., L. Castello, D. J. Stewart, M. Cetra, and H. L. Queiroz. 2010. Population density, growth and reproduction of arapaima in an Amazonian river-floodplain. Ecology of Freshwater Fish 19:455–465.

Bayley, P. B., and M. J. Petrere. 1989. Amazon fisheries: assessment methods, current status and management options. Pages 385–398 *in* D. P. Dodge, editor. Proceedings of the international large river symposium (LARS). Canadian Special Publication of Fisheries and Aquatic Sciences 106.

Berkes, F., W. V. Reid, T. J. Wilbanks, and D. Capistrano. 2006. Conclusions: bridging scales and knowledge systems. Pages 315–331 *in* W. V. Reid, F. Berkes, T. J. Wilbanks, and D. Capistrano, editors. Bridging scales and knowledge systems. Island Press, Washington, D.C.

Bessa, J. D., and A. de C. Lima. 2010. Manejo de pesca do pirarucu (*Arapaima gigas*) no Estado do Amazonas: Erros, Acertos e Perspectivas Futuros. [Management of the fishery of Pirarucu (*Arapaima gigas*) in the state of Amazonas: wrongs, rights, and future prospects.] Anais do 1 Seminário Internacional de Ciências do Ambiente e Sustentabilidade na Amazônia, Universidade Federal do Amazonas, Manaus, Amazonas, Brazil.

Bini, L. M., J. A. F. Diniz-Filho, T. F. Rangel, R. P. Bastos, and M. P. Pinto. 2006. Challenging Wallacean and Linnean shortfalls: knowledge gradients and conservation planning in a biodiversity hotspot. Diversity and Distributions 12:475–482.

Brelsford, T. 2009. "We have to learn to work together": current perspectives on incorporating local and traditional/indigenous knowledge into Alaskan fishery management. Pages 381–394 *in* C. C. Krueger and C. E. Zimmerman, editors. Pacific salmon ecology and management of western Alaska's populations. American Fisheries Society, Symposium 70, Bethesda, Maryland.

Campos-Silva, J. V., and C. A. Peres. 2016. Community-based management induces rapid recovery of a high-value tropical freshwater fishery. Scientific Reports 6:34745.

Caro, T. 2010. Conservation by proxy: indicator, umbrella, keystone, flagship, and other surrogate species. Island Press, Washington, D.C.

Castello, L. 2004. A method to count pirarucu Arapaima gigas: fishers, assessment, and management. North American Journal of Fisheries Management 24:379–389.

Castello, L. 2008. Lateral migration of *Arapaima gigas* in floodplains of the Amazon. Ecology of Freshwater Fish 17:38–46.

Castello, L., C. C. Arantes, D. G. McGrath, D. J. Stewart, and F. S. de Sousa. 2015. Understanding fishing-induced extinctions in the Amazon. Aquatic Conservation: Marine and Freshwater Ecosystems 25:587–598.

Castello, L., D. G. McGrath, L. L. Hess, M. T. Coe, P. A. Lefebvre, P. Petry, M. N. Macedo, V. F. Renó, and C. C. Arantes. 2013. The vulnerability of Amazon freshwater ecosystems. Conservation Letters 6:217–229.

Castello, L., and D. J. Stewart. 2010. Assessing CITES non-detriment findings procedures for Arapaima in Brazil. Journal of Applied Ichthyology 26:49–56.

Castello, L., D. J. Stewart, and C. C. Arantes. 2011. Modeling population dynamics and conservation of arapaima in the Amazon. Reviews in Fish Biology and Fisheries 21:623–640.

Castello, L., J. P. Viana, G. Watkins, M. Pinedo-Vasquez, and V. A. Luzadis. 2009. Lessons from integrating fishers of arapaima in small-scale fisheries management at the Mamirauá Reserve, Amazon. Environmental Management 43:197–209.

Castello, L. editors. 2002. Management plan for Arapaima (*Arapaima gigas*) in the North Rupununi, Guyana: a pilot project under the Fisheries Act. Iwokrama International Centre for Rain Forest Conservation and Development, Georgetown, Guyana.

Castro, F., and D. G. McGrath. 2003. Moving toward sustainability in the local management of floodplain lake fisheries in the Brazilian Amazon. Human Organization 622:123–133.

Cavole, L. M., C. C. Arantes, and L. Castello. 2015. How illegal are tropical small-scale fisheries? An estimate for arapaima in the Amazon. Fisheries Research 168:1–5.

Chapman, M. D. 1989. The political ecology of fisheries depletion in Amazonia. Environmental Conservation 16:331–337.

Cleary, D. 2001. Towards an environmental history of Amazonia. Latin American Research Review 36:65–97.

Crampton, W. G. R., L. Castello, and J. P. Viana. 2004. Fisheries in the Amazon várzea: historical trends, current status, and factors affecting sustainability. Pages 76–95 *in* K. Silvius, R. Bodmer, and J. M. V. Fragoso, editors. People in nature: wildlife conservation in South and Central America. Columbia University Press, New York.

Denevan, W. 1966. A cultural-ecological view of the former aboriginal settlement in the Amazon region. Professional Geographer 18:346–351.

dos Santos, J. G., A. C. Malhado, R. J. Ladle, R. A. Correia, and M. H. Costa. 2015. Geographic trends and information deficits in Amazonian conservation research. Biodiversity and Conservation 24:2853–2863.

Fernandes, D. 2005. "'More eyes watching…" Lessons from community-based management of a giant fish, *Arapaima gigas*, in central Guyana. Master's thesis. University of Manitoba, Winnipeg, Manitoba.

Frederico, R. G., J. D. Olden, and J. Zuanon. 2016. Climate change sensitivity of threatened, and largely unprotected, Amazonian fishes. Aquatic Conservation: Marine and Freshwater Ecosystems 26(S1):91–102.

Fontenele, O. 1948. Contribuição para o conhecimento da biologia do pirarucu, *Arapaima gigas* (Cuvier), em cativeiro (Actinopterygii, Osteoglossidae). [Contribution to knowledge of the biology of Pirarucu, *Arapaima gigas* (Cuvier), in captivity (Actinopterygii, Osteoglossidae).] Revista Brasileira de Biologia 8:445–459.

Godinho, H. P., J. E. Santos, P. S. Formagio, and R. J. Guimarães-Cruz. 2005. Gonadal morphology and reproductive traits of the Amazonian fish *Arapaima gigas* (Schinz, 1822). Acta Zoologica 86:289–294.

Goulding, M. 1983. Amazonian fisheries. Pages 89–210 *in* E. Moran, editor. The dilemma of Amazonian development. Westview Press, Boulder, Colorado.

Gurdak, D. J., D. J. Stewart, L. Castello, and C. C. Arantes. 2019. Diversity in reproductive traits of arapaima (*Arapaima* spp., Müller) in Amazonian várzea floodplains: conservation implications. Aquatic Conservation: Marine and Freshwater Ecosystems 29:245–257.

Hartmann, W. 1989. Conflitos de pesca em águas interiores da Amazônia e tentativas para sua solução. [Fishing conflicts in inland waters of the Amazon and attempts for their solution.] Encontro de Ciências Sociais e o Mar 3:1–14.

Holland, D. S. 2010. Management strategy evaluation and management procedures: tools for rebuilding and sustaining fisheries. OOECD Publishing, ECD Food, Agriculture and Fisheries Working Papers, No. 25, Paris.

Imbiriba, E. P. 1994. Reprodução, larva e alevinagem do pirarucu (*Arapaima gigas*). Recomendações básicas, 26. [Reproduction, larvae, and rearing of pirarucu (*Arapaima gigas*). Basic recommendations, 26.] EMBRAPA-CPATU, Belém, Pará, Brazil.

Isaac, V. J., V. L. Rocha, and S. Mota. 1993. Considerações sobre a legislação da "piracema" e outras restrições da pesca da região do Médio Amazonas. [Considerations on the "Piracema" legislation and other fishing restrictions in the middle Amazon region.] Pages 187–211 *in* L. G. Furtado, A. F. Mello, and W. Leitão, editors. Povos das águas: realidade e perspectivas na Amazônia. [Peoples of the waters: reality and perspectives in Amazonia.] Museu Paraense Emilio Goeldi, Belém, Pará, Brazil.

Johannes, R. E. 1998. The case for data-less marine resource management: examples from tropical nearshore finfisheries. Trends in Ecology and Evolution 13:243–246.

Johannes, R. E., Freeman, M. M., and R. J. Hamilton. 2000. Ignore fishers' knowledge and miss the boat. Fish and Fisheries 1:257–271.

Kirsten, I. F., L. R. Puerta, L. A. F. de Mateus, A. C. Catella, and I. S. Lima. 2012. A pesca do pirarucu (*Arapaima* sp.) na bacia do rio Araguaia em Mato Grosso–Brasil. [The Pirarucu (*Arapaima* sp.) fishery in the Araguaia River basin in Mato Grosso–Brazil.] Boletim do Instituto de Pesca, São Paulo 38:131–144.

Lennox, R. J., J. W. Brownscombe, S. J. Cooke, and A. J. Danylchuk. 2018. Post-release behaviour and survival of recreationally-angled Arapaima (*Arapaima* cf. *arapaima*) assessed with accelerometer biologgers. Fisheries Research 207:197–203.

Lowe-McConnell, R. H. 1964. The fishes of the Rupununi Savanna District of British Guiana, South America. Journal of the Linnaean Society Zoology 45:103–144.

Ludwig, D. 2001. The era of management is over. Ecosystems 4:758–764.

Ludwig, D., R. Hilborn, and C. J. Walters. 1993. Uncertainty, resource exploitation, and conservation: lessons from history. Science 260:17–36.

Lüling, K. H. 1964. Zur biologie und ökologie von *Arapaima gigas* (Pisces, Osteoglossidae). [On the biology and ecology of *Arapaima gigas* (Pisces, Osteoglossidae).] Zeitschrift für Morphologie und Ökologie der Tiere 54:436–530.

Margoluis, R., C. Stem, N. Salafsky, and M. Brown. 2009. Design alternatives for evaluating the impact of conservation projects. Pages 85–96 *in* M. Birnbaum and P. Mickwitz, editors. Environmental program and policy evaluation: addressing methodological challenges. New Directions for Evaluation 122:85–96.

Martinelli, N. M. C., and M. J. Petrere. 1999. Morphometric relationships and indirect determination of the length frequency structure of the pirarucu, *Arapaima gigas* (Cuvier), in Brazilian Amazonia. Fisheries Management and Ecology 6:233–240.

McGrath, D. 1986. The animal products trade in the Brazilian Amazon. Final report to the World Wildlife Fund, Washington, D.C.

McGrath, D. 1989. The Paraense traders: small-scale, long distance trade in the Brazilian Amazon. Doctoral dissertation. University of Wisconsin-Madison, Madison.

McGrath, D. G., O. T. Almeida, and F. D. Merry. 2007. The influence of community management agreements on household economic strategies: cattle grazing and fishing agreements on the lower Amazon floodplain. International Journal of the Commons 1:67–88.

McGrath, D. G., Cardoso, A., Almeida, O. T., and J. Pezzuti, J. 2008. Constructing a policy and institutional framework for an ecosystem-based approach to managing the lower Amazon floodplain. Environment, Development and Sustainability 10:677.

McGrath, D. G., F. de Castro, C. Futemma, B. D. de Amaral, and J. Calabria. 1993. Fisheries and the evolution of resource management on the lower Amazon floodplain. Human Ecology 21:167–195.

McGrath, D. G., and L. Castello. 2015. Integrating fishers' ecological knowledge and the ecosystem based management of tropical inland fisheries: an Amazon case study. Pages 127–148 *in* J. Fischer, J. Jorgensen, H. Josupeit, D. Kalikoski, and C. M. Lucas, editors. Fishers' knowledge and the ecosystem approach to fisheries: applications, experiences and lessons in Latin America. FAO (Food and Agriculture Organization of the United Nations) Fisheries and Aquaculture Technical Paper 591.

McGrath, D. G., L. Castello, O. T. Almeida, and G. M. B. Estupiñán. 2015. Market formalization, governance, and the integration of community fisheries in the Brazilian Amazon. Society and Natural Resources 28:513–529.

Meggers, B. J. 1971. Amazonia: man and culture in a counterfeit paradise. Aldine Publishing, Chicago.

Murrieta, R. S. S. 2001. A mística do pirarucu: pesca, ethos e paisagem em comunidades rurais do baixo Amazonas. [The mystique of the pirarucu: fishing, ethos and landscape in rural communities of the lower Amazon.] Horizontes Antropológicos 7:113–130.

Ostrom, E. 1990. Governing the commons: the evolution of institutions for collective action. Cambridge University Press, Cambridge, UK.

Ostrom, E. 2007. A diagnostic approach for going beyond panaceas. Proceedings of the National Academy of Sciences of the United States of America 104:15181–15187.

Oviedo, A. F., and M. Bursztyn. 2016. The fortune of the commons: participatory evaluation of small-scale fisheries in the Brazilian Amazon. Environmental Management 57:1009–1023.

Papavero, N. 2000. O novo Eden: a fauna da Amazônia brasileira nos relatos de viajantes e cronistas desde a descoberta do rio Amazonas por Pinzón (1500) até o tratado de Santo Ildefonso (1777): con transcrição dos principais textos. [The new Eden: the fauna of the Brazilian Amazon in the reports of travelers and chroniclers since the discovery of the Amazon River by Pinzón (1500) until the treaty of Santo Ildefonso (1777): with transcription of the main texts.] Museu Paraense Emílio Goeldi, Belém, Pará, Brazil.

Pauly, D., G. Silvestre, and I. R. Smith. 1989. On development, fisheries and dynamite: a brief review of tropical fisheries management. Natural Resource Modeling 3:307–329.

Pauly, D. 1997. Small-scale fisheries in the tropics: marginality, marginalization, and some implications for fisheries management. Pages 40–49 *in* E. K. Pikitch, D. D. Huppert, and M. P. Sissenwine, editors. Global trends: fisheries management. American Fisheries Society, Symposium 20, Bethesda, Maryland.

Persoon, G., D. M. Van Est, and P. E. Sajise editors. 2003. Co-management of natural resources in Asia: a comparative perspective. NIAS Press, Copenhagen, Denmark.

Petersen, T. A., S. M. Brum, F. Rossoni, G. F. V. Silveira, and L. Castello. 2016. Recovery of *Arapaima* sp. populations by community-based management in floodplains of the Purus River, Amazon. Journal of Fish Biology 89:241–248.

Pomeroy, R. S., and F. Berkes. 1997. Two to tango: the role of government in fisheries co-management. Marine Policy 21:465–480.

Purnell, R. 2017. Guy with a fly rod. How a small village in Guyana became a showcase of fisheries conservation. Fly Fisherman 49:40–47.

Queiroz, H. L. 1999. Artisanal fisheries of pirarucu at the Mamirauá Ecological Station. Pages 83–100 *in* C. Padoch, J. M. Ayres, M. Pinedo-Vasquez, and A. Henderson, editors. Várzea: diversity, development, and conservation of Amazonia's whitewater floodplains. New York Botanical Press, Advances in Economic Botany 13, New York.

Roosevelt, A. C. 1989. Resource management in Amazonia before the conquest: beyond ethnographic projection. Advances in Economic Botany 7:30–62.

Roosevelt, A. C. 1991. Moundbuilders of the Amazon: geophysical archaeology on Marajó Island, Brazil. Academic Press, New York.

Salo, M., A. Sirén, and R. Kalliola. 2013. Knowledge for action and interaction. Pages 319–345 *in* Diagnosing wild species harvest: resource use and conservation. Academic Press, London.

Silvano, R. A., and J. Valbo-Jørgensen. 2008. Beyond fishermen's tales: contributions of fishers' local ecological knowledge to fish ecology and fisheries management. Environment, Development and Sustainability 10:657.

Simberloff, D. 1998. Flagships, umbrellas, and keystones: is single-species management passé in the landscape era? Biological Conservation 83:247–257.

Smith, N. 1981. Man, fishes and the Amazon. Columbia University Press, New York.

Smith, N. J. H. 1985. The impact of cultural and ecological change on Amazonian fisheries. Biological Conservation 32:355–373.

Stewart, D. J. 2013a. A new species of arapaima (Osteoglossomorpha: Osteoglossidae) from the Solimoes River, Amazonas State, Brazil. Copeia 2013:470–476.

Stewart, D. J. 2013b. Re-description of *Arapaima agassizii* (Valenciennes), a rare fish from Brazil (Osteoglossomorpha: Osteoglossidae). Copeia 2013:38–51.

Valbo-Jørgensen, J., and A. F. Poulsen. 2000. Using local knowledge as a research tool in the study of river fish biology: experiences from the Mekong. Environment, Development and Sustainability 2:253–376.

Viana, J. P., J. M. B. Damasceno, L. Castello, and W. G. R. Crampton. 2004. Economic incentives for sustainable community management of fishery resources in the Mamiraua Sustainable Development Reserve, Amazonas, Brazil. Pages 139–154 *in* K. Silvius, R. Bodmer, and J. M. V. Fragoso, editors. People in nature: wildlife conservation in South and Central America. Columbia University Press, New York.

Veríssimo, J. 1895. A Pesca na Amazônia. [The fishery in the Amazon.] Livraria Alves, Rio de Janeiro, Brazil.

Watson, L. C. 2011. Ecology and conservation of arapaima in Guyana: recovery of a giant fish in distress. Doctoral dissertation. SUNY College of Environmental Science and Forestry, Syracuse, New York.

Watson, L. C., D. J. Stewart, and A. M. Kretzer. 2016. Genetic diversity among populations of arapaima in Guyana: implications for conservation of a giant fish. Copeia 104:864–872.

Section II

Lakes

Adirondack Brook Trout and Acid Rain: Environmental Legislation Fosters Successful Restoration

Clifford Kraft*
Department of Natural Resources, Cornell University
Ithaca, New York 14853, USA

Abstract.—Recovery of Brook Trout *Salvelinus fontinalis* in an Adirondack (New York, USA) lake that was subject to anthropogenic acidification provides a remarkable example of fishery improvement in response to environmental regulation. Studies initiated in the 1950s following a steady decline in Brook Trout populations helped document this recovery. Unsuccessful efforts to maintain a fishery in Honnedaga Lake with hatchery-reared fish in the 1950s forced managers to look beyond stocking, the primary approach employed until that time. As a result, fishery scientists collaborated in the 1960s and 1970s with researchers from other disciplines, providing a broad understanding of atmospheric inputs, watershed processes, and chemical interactions influencing lakes and streams. Extensive studies in the 1980s confirmed the connection between Brook Trout mortality and airborne emissions of strong acid nitrogen and sulfur compounds that released toxic inorganic aluminum from increasingly acidic soils. Political debates in that decade focused on federal regulatory efforts to reduce these emissions, which culminated in passage of the Clean Air Act Amendments of 1990. Within the next decade, Brook Trout that took refuge within a few well-buffered, groundwater-fed tributaries began to recolonize Honnedaga Lake as conditions improved in the main lake due to reduced atmospheric deposition of sulfur dioxide and nitrogen oxides. Since then, management of Honnedaga Lake in the 21st century relied upon natural reproduction by wild Brook Trout. Ultimately, social and political decisions made far away from the Adirondack Mountain region developed regulations that fostered recovery of the Honnedaga Lake fishery by restoring necessary water-chemistry conditions. The recovery of Honnedaga Lake highlights three lessons. First, environment and habitat conditions must be suitable before fishery management actions can be effective. This criterion requires a broad understanding of environmental conditions that sustain fisheries, incorporating insights from atmospheric sciences, geology, and limnology. Second, natural reproduction of Brook Trout in Honnedaga Lake successfully increased population abundance without the additional intervention of stocking hatchery-reared fish. Finally, successful management of Honnedaga Lake required political support and regulatory action from beyond the Adirondack region, as well as media attention.

* Corresponding author: cek7@cornell.edu

INTRODUCTION: FISHERIES MANAGEMENT AS AN AGRICULTURAL PRACTICE

The recovery of Brook Trout *Salvelinus fontinalis* populations from acid rain in some Adirondack Mountain region lakes of northern New York highlights an important transition in fishery management from its origin as an agricultural science to its current central—and comfortable—placement in environmental sciences. Fishery management emerged as a profession in the 19th century from a foundation of fish culture and stocking that had deep roots in centuries of food-fish production in pond culture (Neess 1949). Much later, in the mid-20th century, managers of Adirondack fisheries increasingly turned to disciplines in environmental sciences when the traditional agricultural perspective of planting (i.e., stocking) fish could no longer support recreational fisheries that targeted Brook Trout. This transition first required embracing approaches developed by limnologists to understand the physics and chemistry of lakes, then linking these processes to primary productivity and fish production. But even that transition was not sufficient to understand and restore failing Adirondack fisheries until managers developed a more expansive perspective, looking beyond the water bodies in which fish were found. Ultimately, Adirondack Brook Trout management required understanding the atmospheric transport of acidic compounds from distant sources and the underlying geology of watersheds that influenced the chemistry and productivity of lakes (Jenkins et al. 2007). Lakes were no longer viewed in isolation from a much broader landscape. As a result, the widespread impacts of acid rain in the Adirondacks pushed fishery management from a discipline rooted in animal husbandry into an expansive realm of environmental inquiry.

An emphasis on hatchery production of fish to sustain fisheries was signaled in 1870 at the founding of the American Fish Culturists' Association, which in 1885 broadened its scope by changing its name to the American Fisheries Society (AFS; Moffitt et al. 2010). The new charter for the expanded organization described its central mission to "promote the cause of fish culture" (Moffitt et al. 2010). Stocking of cultured fishes and other aquatic organisms was viewed as a way to rehabilitate fisheries degraded and extirpated by overharvest of fish stocks (Whitaker 1892) and destruction of aquatic habitats by timber harvest, mining, and industrial development (Marsh 1965). The fervor for fish stocking is reflected in this 19th century recommendation presented in the *Transactions of the American Fisheries Society* by Herschel Whitaker (Whitaker 1892), a member of the Michigan State Board of Fish Commissioners:

> Weak and erratic efforts made now and then to make good the loss by the planting of a few million fish will not do. With the means at hand and with the information we now have as to fish culture, and with the small outlay of money necessary to carry on the work of artificial propagation, each State should see for itself that every female fish taken during the spawning season in its waters shall have her eggs taken from her, fecundated and after being hatched, properly planted; there is no good reason why this should not be done.

Proceedings from early meetings of the American Fisheries Society often described improvements in fish husbandry similar to other contemporary approaches used to raise large numbers of domesticated animals in confined settings (Stone 1872). This connection to animal husbandry was even apparent in the physical location of the first meeting of AFS that took place in rooms housing the New York Poultry Society (Anonymous 1872).

Early writing by proponents of fish hatcheries focused on how technological innovations could alleviate the disappearance of fish from lakes and streams increasingly influenced by human exploitation and habitat alteration. Stocking continued as the primary tool to manage fisheries into the 1920s and 1930s, at which time ongoing declines in fish abundance and the failure of stocking to sustain many fisheries provided an incentive to seek new approaches. These new perspectives were developed through work conducted by early limnologists pursuing an understanding of why some lakes and streams were more productive than others. Two Wisconsin lake biologists, Edward Birge and Chancey Juday, were at the forefront of efforts examining how lake productivity might influence fish abundance, along with Canadian scientist Donald Rawson (Northcote and Larkin 1966), building upon a foundation described in the late 19th century by Stephen A. Forbes (1887). Most important, these researchers helped promote a recognition that intact fisheries required specific environmental conditions often degraded by human activities. The melding of limnology with traditional fish stocking approaches was first reflected in reports from surveys conducted by state agencies during the 1920s and 1930s that assessed where to stock fish (Moffitt et al. 2010). Although these reports focused on where species of fish were found, they also included early records of the physical and chemical characteristics of lakes (e.g., Greeley 1934; Smith and Moyle 1944). As the middle of the 20th century approached, problems with declining Adirondack fisheries provided a major test of this transition in understanding.

The Adirondack region wake-up call was prompted by extensive, but yet undetected, acidification of surface waters caused by atmospheric deposition of acid pollutants from electric generation utilities and industrial sources in the U.S. Midwest (Driscoll et al. 2001). By the 1950s, fish were disappearing from lakes and streams without any apparent cause. At first, this problem resembled previous fishery declines that resulted from overexploitation of large, mature fish or the destruction of spawning habitat—problems often alleviated by relying upon hatcheries, not nature, to produce young fish. During that era, government agencies were responsible for most fishery management practices. However, anglers began to seek input from fishery scientists aligned with universities as the research questions became more challenging and popular recreational fisheries could no longer be sustained. This new wave of professionals remained true to the founding tradition of fishery science—that is, applying agricultural approaches to support fisheries by stocking hatchery-reared fish—but also began to explore the relevance of limnological insights to fishery management. Eventually, these perspectives converged in the Adirondacks when it became obvious that stocked Brook Trout could no longer survive in strongly acidified lakes and streams.

In the rest of this chapter, the physical and ecological setting for Adirondack fisheries and the cultural importance of Brook Trout is first described, followed by a close look at the decline of the Honnedaga Lake Brook Trout fishery and its recovery. Last, lessons learned from this century-long journey are highlighted.

BROOK TROUT AND THE ADIRONDACKS

Brook Trout have been the heart of the Adirondack angling tradition for more than 150 years. Officially designated the state fish of New York in 1975, Brook Trout belong to the char subgroup of salmonine fishes that is distinct from the true trout and salmon. Originally found only in eastern North America, Brook Trout were described as "ubiquitous" in the first fisheries surveys within the Adirondack region (George 1981). The favored status of Brook Trout at the onset of the fisheries profession was evident in proceedings from the early meetings of the American Fish Culturists' Association, which included many articles describing how to raise this species (Stone 1872).

Not only were Adirondack lakes well suited for Brook Trout, this region's geography also made these fish populations unusually susceptible to acid rain. The underlying geology defining the Adirondack region consists of highly metamorphosed granitic rock thinly covered by infertile, strongly acid glacial till soils (Jenkins et al. 2007). This geological setting provides the Adirondack landscape with little capacity to neutralize the acidity of precipitation, which has a pH of around 5.6 or 5.7 under typical conditions (pH 7 is neutral, and lower values are more acidic). Precipitation is naturally acidic because hydrogen ions are generated by the chemical reaction of naturally occurring levels of atmospheric carbon dioxide and nitric oxide with water. However, the pH of precipitation has become much more acidic in many locations throughout the world from power plants, industrial facilities, and auto emissions that produce strong acid-generating compounds—primarily sulfur dioxide, nitrogen oxides, and ammonia—emitted into the atmosphere (Jenkins et al. 2007).

The geology of the Adirondack Mountain region landscape also presents the first abrupt rise in elevation encountered by prevailing winds from the industrial region of mid-continental North America. This geographical position enhances the precipitation responsible for the abundant lakes and streams in the Adirondacks; both summer rains and winter snowfall are more intense in the Adirondacks than nearby locations at lower elevations (Jenkins et al. 2007). Unfortunately, these same conditions enhance the deposition of acid rain on a landscape that has little capacity to neutralize acidic compounds generated at locations upwind from the Adirondacks.

Geology also influences where people gather, and the mountainous landscape of the Adirondacks provided a refuge since the late 19th century for urban residents pursuing recreational opportunities and a reprieve from their daily lives. The Adirondack region encompasses a 6-million-acre (2.4 million ha) state forest preserve established in 1894 by the New York State Constitution with more than 2,750 individual lakes and ponds larger than 0.4 ha. Initially, the abundant lakes and bold outcrops of rock provided a place where urban dwellers could experience rejuvenation by being closer to nature. Perhaps as important, this higher-elevation region was cooler in summer than nearby cities such as Boston, New York City, and Buffalo. In the 19th century, the Adirondacks attracted and inspired prominent writers and artists such as Ralph Waldo Emerson and Winslow Homer, as well as politicians such as Theodore Roosevelt. Private fishing clubs were also established at that time as Adirondack outdoor recreation typically focused on angling for Brook Trout (Worden 1891).

Almost a century later, a thriving 1950s American economy after World War II provided an opportunity for more recreational anglers to participate in previously established Adirondack angling traditions. The expansion of industry and increasing use of electricity from the booming economy in the 1950s and 1960s also increased atmospheric emissions of nitrous oxides and sulfur dioxide from coal-burning power plants and industrial facilities in the Midwestern United States. Smokestack heights increased during this era to disperse pollutants away from these sources (Schofield 1976; Waxman 1991). The acid rain generated by these facilities followed prevailing wind patterns and geography, reaching the same destination as anglers seeking refuge on Adirondack lakes and streams. As a result, the same economic expansion that provided a broad segment of the American population with leisure time and the financial ability to fish in the Adirondacks also degraded the resource that attracted them to this recreational pursuit.

This contrast between enjoyment and exploitation of natural resources has been a recurring theme in the Adirondacks since European settlers arrived. For example, the July 12, 1888 issue of *Forest and Stream* magazine contained this description of "Adirondack Abominations" (Spears 1888):

> In the North Woods the work goes bravely on. Landlords, guides, tourists, "sportsmen" and railroad managers are doing what their hands find to do in the work of destruction. The State has its elaborate and expensive systems for the conservation of its forests and game and fish, but the destroyers make away in a year with more than these cumbrous and inefficient bureaus can replace in a decade.

This quote illustrates that as the Adirondacks became a destination for wealthy urban residents to escape to a place with abundant natural resources, the region was also subject to extensive exploitation. A broader debate about public and private resource management ensued, culminating in 1894 with the establishment of the "forever wild" provision within the New York State Constitution that specifically prohibited timber harvest on state lands within the Adirondack Forest Preserve (New York State Bar Association 2016). More than a century later, the Adirondack Park encompasses a mix of state and private lands. The constitutional protection of Adirondack lands also established a pattern by which management of the Adirondack landscape was dominated by political decisions made elsewhere (Jacoby 2014). Eventually, this played out favorably for Adirondack Brook Trout fisheries at the end of the 20th century as U.S. and Canadian federal legislation regulated emissions far upwind from the Adirondacks and reduced deposition of sulfur dioxide and nitrogen oxides (Stoddard et al. 1999; Lehmann et al. 2005; Driscoll et al. 2016).

DECLINING FISHERIES

Longstanding angling traditions for Brook Trout have provided records that help document and understand the decline and recovery of Brook Trout in some Adirondack lakes. This long-term perspective also provides an opportunity to track how fishery management became a matter of ecosystem management, which then became the focus of national and international policy debates. These transitions occurred because understanding why Adirondack Brook Trout fisheries disappeared in the mid-20th century required recognizing environmental influences originating far beyond the boundaries of individual lakes and their surrounding watersheds. Prior to the 1950s, it was unimaginable that Brook Trout could be harmed by industrial development hundreds of miles away. This threat was a new type of environmental challenge for which the prevailing science—fisheries or otherwise—was poorly equipped to understand. Few scientists—let alone fishery scientists—recognized that long-distance atmospheric transport of chemical compounds could move acid emissions from distant locations to the Adirondack region. That understanding did not develop until it became obvious in the 1950s that radiation could be detected at locations far downwind from where nuclear weapons were tested (Jessee 2014).

Accounts of angler catches and fish-stocking practices at fishing and hunting clubs within the Adirondack Preserve provide a sketchy record of early fishery management practices in this region. Many accounts describe declining populations of popular inland sport fishes such as Brook Trout, which led fishing clubs to try to boost their fisheries by stocking hatchery-reared fish. But hatchery stocking and the number of anglers dwindled as money became scarce during the Great Depression and through World War II. This period was also a time when large tracts of private land were sold to the state of New York (Comstock and Webster 1990), which moved the responsibility of managing many Adirondack lake-and-stream Brook Trout fisheries to the staff of state agencies. As interest in recreational opportunities returned and expanded after the end of World War II, the legacy of angling at Adirondack fishing clubs provided new incentive for scientists to investigate inland lake fisheries as the benefits of science became evident in other aspects of daily life (Comstock and Webster 1990).

One early example occurred at the Adirondack League Club (ALC), one of the oldest and largest of many Adirondack fishing and hunting clubs established on private lands within the Adirondack Preserve. By the time the ALC was established in 1890, residents of urbanized areas in the northeastern United States were familiar with declining catches in waters near their homes. These declines resulted from unregulated harvest, habitat destruction, and pollution in waterways subject to increasing industrial and urban development. Stories about pack baskets full of fish and fish rising to every cast were probably the exception rather than the rule, but their presence in popular magazines fostered the notion of the Adirondacks as offering a respite from urban ills. Dr. Dwight Webster, a faculty member in the Department of Conservation at Cornell University, started work at the ALC in 1950 to provide management advice for fisheries that primarily targeted Brook Trout (Mark Webster, former president of the ALC, personal communication).

The founders of Adirondack fishing clubs were familiar with the decline of fisheries throughout the industrializing United States that occurred as the American conservation movement was launched (Mitchell and Diamant 2015). Because fish management practices in that era were dominated by stocking, these clubs often constructed fish hatcheries to support fish-stocking programs of native and nonnative fishes. Indeed, hatcheries were viewed as key to conservation throughout the late 1800s and well into the 1900s. For example, in 1874, John Muir (1874) lauded the first federal hatchery at the McCloud River in California for its potential

> to restore wasted waters to their primitive or more than primitive fruitfulness, and … to extend the geographical range of the more important food-fishes, such as shad, salmon and trout, by naturalizing them in new waters.

For the first time in American history, Muir explained, forests and oceans were "limitable." Similar sentiments were clearly in the minds of the vanguard of Adirondack fishery managers who wanted to expand available Adirondack fisheries (George 1981).

THE DECLINE OF BROOK TROUT IN HONNEDAGA LAKE

Although Brook Trout populations were diminished by acid rain in hundreds of Adirondack lakes, focusing on one lake—Honnedaga Lake, from which a unique sequence of data are available—provides an example of the decline and rebirth of a Brook Trout fishery. Honnedaga Lake is a relatively large (310 ha), deep (60 m), and high-elevation (670 m) lake on the property of the ALC in the southwestern Adirondacks, the region of New York State most strongly affected by acid precipitation (Figure 1). Honnedaga Lake and the surrounding land became part of the ALC in 1890 after establishment of a hotel catering to

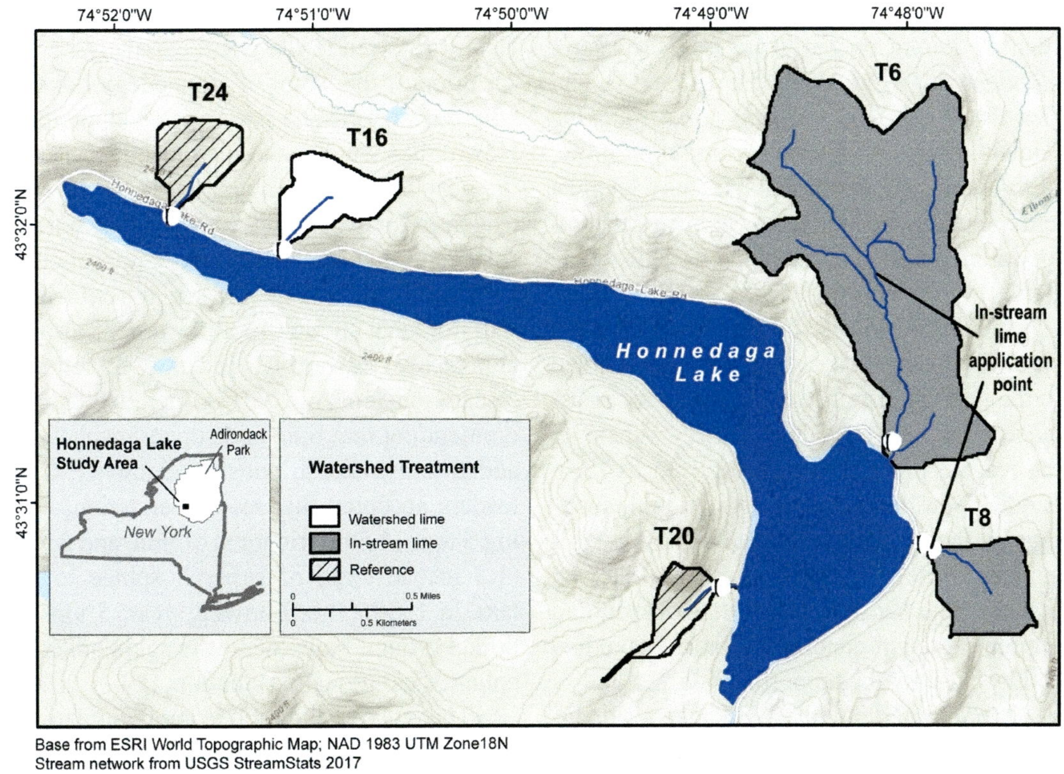

FIGURE 1. Map of Honnedaga Lake, its location in the Adirondack Park (New York, USA), and tributaries where instream and watershed liming has occurred (from George et al. 2018).

anglers at this remote location. A long-standing interest in recreational fishing and stable oversight by the ALC made possible the historic record of the lake's fishery and water chemistry, described below.

Early ALC documents indicated that Brook Trout were the only native fish present in this lake. Several other species were introduced in the late 19th and early 20th centuries, including Lake Trout *Salvelinus namaycush*, White Sucker *Catostomus commersonii*, Round Whitefish *Prosopium cylindraceum*, and Creek Chub *Semotilus atromaculatus*. The latter three species were stocked periodically to provide forage for Lake Trout and were thought to have established self-sustaining populations based on sporadic club-member descriptions. But angler reports largely focused on game fish, and annual reports of the ALC only described Lake Trout caught by anglers from 1896 through 1935, after which they disappeared (Webster 1961). Webster (1961) also noted that interviews with guides and club members reported that White Suckers, Round Whitefish, and Creek Chubs were no longer present in the late 1920s or early 1930s. In retrospect, it is impossible to determine why these introduced fishes disappeared, though perhaps they were only present as a result of stocking and self-sustaining populations were never firmly established.

In addition to soliciting fishery management advice from experts at universities and agencies, angling clubs often solicited advice from individuals who were recognized for their fishing prowess and had a flair for science, technology, and communication. One such individual was Edward R. Hewitt, a renowned Catskill angler and author of books such as *Telling on the Trout* (Hewitt 1926), *Handbook of Fly Fishing* (Hewitt 1933), and *Handbook of Trout Raising and Stocking* (Hewitt 1935). Prior to becoming a promoter of fly-fishing, Hewitt was trained as a chemist and reportedly maintained a chemistry laboratory at his summer home in the Catskills (Van Put 2014). Perhaps that is why Hewitt collected water samples in 1938 for a chemical analysis of Honnedaga Lake water when he traveled to the ALC in response to a request for assistance to improve a declining fishery. Whatever the reason, this analysis (completed by "Mr. Biddick" or "Riddick," given that both spellings are included in the letter documenting this analysis in the ALC historical archive) showed that the lake had low levels of plankton and nutrients. Hewitt concluded that the lake had "a low capacity for supporting fish."

More important, this 1938 analysis documented that the lake's acidity was suitable for Brook Trout, providing a unique snapshot from an era when such measurements were rare. In water samples collected on December 13, 1938, the alkalinity of Honnedaga Lake ranged from 12.5 to 20.0 ppm of calcium carbonate ($CaCO_3$), and the pH ranged from 6.6 to 7.2 (Schofield 1965). Together, these measures indicated that Honnedaga Lake water had enough buffering capacity to keep pH at a level that did not produce conditions toxic to fish and other organisms.

Hewitt's report to the president of the ALC was consistent with fishery management practices during an era that focused on hatchery supplementation. He concluded that the poor Brook Trout fishing was due to the scarcity of food; therefore, Hewitt recommended stocking forage fish and—perhaps less typical—stocking a type of insect. But Hewitt also recommended a less conventional, limnological approach to solve the fishery decline by boosting the lake's productivity. In a 1939 letter describing his thoughts about enhancing Honnedaga Lake fish production, Hewitt referred to the chemical analyst's description of the water of Honnedaga Lake as "the purest water he had ever encountered." Hewitt then went on to note that "the cause of the very low plankton content is the low nitrogen in the water and the very low mineral salts" and that it is "much more important to get food in the lake than to put in trout which will not have enough food for proper growth at present" (Edward R. Hewitt, 1939 letter to Oscar Houston, ALC). He concluded by recommending that lime and superphosphate be added to the lake to boost productivity. Club leaders accepted this recommendation, adding 5 tons (4.5 metric tons) of lime and 2 tons (1.8 metric tons) of superphosphate to the lake in 1939. The following year, 5 tons of boiled ground bone meal was again added to enhance biological production.

The recommendation to apply lime might have reflected some awareness of the growing understanding of limnology. But this approach also built upon practices first employed in China as far back as 500 B.C. to increase the production of fish in ponds (Neess 1949). In a review of European fish-pond fertilization efforts, Neess (1949) pointed out that the traditional agricultural approach to fish culture was at the time of his review in the midst of a transition to manipulating the pond environment instead of attributing poor fish growth to some inherent fish attributes that could be improved by breeding and domestication. Early 17th century efforts included adding mixtures of honey, wine, camphor, and barley gruel to enhance fish growth in ponds. Later, a biochemical perspective showed a connection between plant nutrition and fertilization that subsequently led German scientists to discover that lime additions stimulated carp growth in ponds (Neess 1949). By the early 20th century, lime was commonly applied to European fish ponds in areas where soils had a low calcium content, but the potential influence of humans on soil or water acidity was not considered.

FISHERY MANAGEMENT MEETS LIMNOLOGY

When Dr. Webster arrived at the ALC 5 years after World War II, he heard club members describe a general decline in angling success for the past several decades. He also heard that fisheries on the ALC's 52,000-acre (21,044 ha) preserve had been augmented with hatchery-reared fish since the founding of the club. But what he did not know was even more important. Webster likely did not know that in 1852, Angus Smith identified sulfuric acid in the rain of Manchester, England and attributed this to the burning of coal. As a fisheries scientist with an agricultural mindset, Webster had no reason to wonder about the acidity of rain at a time when few in the scientific community were aware that this might be an important influence on organisms and ecosystems. Webster also had no reason to be aware of work from the vicinity of Manchester, England in the early 1910s showing that the acidity of rain decreased with distance from that city (Crowther and Ruston 1911). Although a previous member of the Cornell University faculty, agronomist Benjamin Wilson, had measured the sulfur content of rainwater in the vicinity of Ithaca, New York (Wilson 1921, 1923), Wilson focused on whether sulfur deposition by precipitation could be beneficial to crops. In the same vein, the first U.S. measurements of rainfall acidity were also made as part of an agricultural investigation and found that precipitation in the state of Tennessee was nearly neutral, which did not raise any concern (MacIntire and Young 1923). In fact, it was not until the 1950s that the acidity of aquatic habitats (in this case, bogs) was linked to the airborne deposition of sulfuric acid by Canadian ecologist Eville Gorham (Gorham 1958). Still, Gorham and others had not yet connected acid rain to lake acidity at a time when Webster was first beginning to pay attention to Adirondack fisheries. Just as important, no one had questioned whether lake acidity could result from human activity. So based on what he knew, Webster applied the tried-and-true approach of stocking hatchery-reared fish to enhance the Honnedaga Lake Brook Trout population.

Within a few years, Webster's annual reports to the Adirondack League Club indicated that things were not working out as expected in this lake with "peculiar water chemistry." Observations of dead fish led him to initiate a series of tests in which hatchery-reared Brook Trout were placed in submerged cages at various locations along the Honnedaga Lake shoreline. These experiments showed extensive mortality of hatchery fish within 12–24 h, as was briefly mentioned in a note published by Webster (1961) describing "An Unusual Lake of the Adirondack Mountains, New York." At the end of that publication, he concluded that "intolerable conditions were acquired (at least currently) within the past few decades" for fish survival in Honnedaga Lake water (Webster 1961). Despite observing that Honnedaga Lake water was noted for "extreme softness and acidity" (D. A. Webster, Cornell University, 1959 unpublished report), Webster treated these "intolerable conditions" as a natural feature of the Adirondack landscape. This attitude prevailed for the next decade until other North American and European scientists began to explore interactions between atmospheric transport of acidic compounds and chemical conditions in lakes and streams.

The foundation for this new understanding of acid rain began when Swedish scientists began documenting increasing acidity of Scandinavian precipitation in 1956 (Odén 1976). That work provided the underpinning for a Swedish Natural Science Research Council publication, *The Acidification of Air and Precipitation and Its Consequences on the Natural Environment* (Odén 1968), which in turn helped prompt Gene Likens, Herbert Bormann, and others to alert the North American scientific community to the destructive influences of acidic precipitation (Likens et al. 1972).

One of Webster's graduate students, Carl Schofield, was key in putting together a North American picture connecting the acidity in precipitation to increasing acidity in lakes and fish mortality. Schofield's Master's thesis, published in 1962 (Schofield 1962), tracked down unpublished results from Hewitt's water chemistry survey conducted on Honnedaga Lake in 1938 that—combined with Schofield's measurements of pH and alkalinity 20 years later—led him to conclude that the chemistry of Honnedaga Lake water had changed. The pH of water samples collected during the fall of 1959 and summer of 1960 ranged from 3.9 to 5.8; $CaCO_3$ alkalinity ranged from 0 to 3.0 ppm (Schofield 1965). This contrasted substantially with the much higher levels of both chemical parameters measured in 1938. However, Schofield's thesis did not yet connect the low pH to the death of Honnedaga Lake Brook Trout, even though he cited an early study of fish in an acid Michigan lake that described "the absence, from Vincent Lake, of a large number of forms [fishes] common in neighboring bodies" (Jewell and Brown 1924). Unfortunately, Schofield and others did not cite other work by Minna Jewell—one of the first prominent female ecologists (Langenheim 1996)—that described acid impacts on fishes (Jewell 1922; Brown and Jewell 1926), and her work was not cited in published results from Schofield's master's thesis (Schofield 1965).

During the 1960s, a time of changing social norms, fishery researchers became more aware of work in other environmental disciplines as ecosystems became a subject of study (Martin 2018). Schofield's examination of Honnedaga Lake water chemistry occurred shortly before Gene Likens moved to Cornell University in 1969 and began to enlist others in describing the influence of acid rain on the chemistry of precipitation samples collected at Hubbard Brook in New Hampshire and near the Cornell University campus in central New York (Likens and Bormann 1974). Fortunately, Schofield's work came to the attention of Likens, whose insights helped reveal that Honnedaga Lake Brook Trout were not surviving due to acidification. Ultimately, a publication in the prominent journal Science (Likens and Bormann 1974) highlighted Schofield's (1965) description of changes in Honnedaga Lake pH from 1938 to 1959 as key evidence that acid rain had altered Adirondack lake water chemistry. This publication by Likens and Bormann (1974) attracted media attention to acid rain and helped place this issue on the public policy agenda in the United States and Canada during a decade when environmental issues received extensive attention by elected officials (Rosenbaum 2016). Their publication, in turn, increased awareness of Schofield's work connecting acidified water to Brook Trout mortality.

Another lake in the southwestern Adirondacks on Adirondack League Club property, Mountain Pond, also contributed substantially to Schofield and Webster's recognition of the link between increasing acidity and Brook Trout mortality. Webster completed surveys of Mountain Pond in the 1950s when he first started working in the Adirondacks. At that time, management of the Brook Trout fishery in Mountain Pond relied upon the familiar historic approach of stocking hatchery-reared fish. But failure of that previously successful approach in the early 1950s prompted Webster to try to modify the lake environment itself, as described in an unpublished Cornell University 1958 report by Webster:

> A recheck on water chemistry indicated an increase in acidity to the critical level

> since 1953 (pH dropped from 5.6 to 4.4). This was verified by mortality of fingerling Brook Trout held in test cages overnight at a depth of 10 ft (3 m; two of the five fish dead, balance not healthy).
>
> On the basis of this information, the 1958 stocking was suspended and 2.5 tons (2.3 metric tons) of limestone was scattered on the ice in March, 1959.

This report provided the first description that acidic conditions were causing death of stocked Brook Trout. This observation led Webster to implement a practice that had been described in some of the first published studies linking fishery management with limnology: adding calcium carbonate to lake water to neutralize acidic conditions. Although this approach reflected the transition of fishery science from its roots as an agricultural endeavor to an environmental science, the agricultural perspective remained. For example, Webster's reports about lime applications in Mountain Pond and other Adirondack lakes for many years continued to describe use of "agricultural" limestone to "fertilize" lakes, not to reduce acidity. The notion that fisheries should be planted, fertilized, and harvested in the manner of an agricultural crop—or that "weedy" species of fish must be removed—still prevailed. However, acid rain would continue to erode that perspective as approaches from other disciplines pushed fishery management into the arms of environmental science.

The intersection of fishery management and limnology had started to develop in work conducted by Arthur Hasler at the University of Wisconsin in the late 1940s. Hasler applied his knowledge of European fish-pond culture practices to the management of fish in northern Wisconsin lakes (Hasler et al. 1951; Elser et al. 1986). Initially, Hasler's group worked in two small bog lakes (<4 ha surface area) in northern Wisconsin where "lime" (calcium and magnesium hydroxide) was added to improve fish productivity (Hasler et al. 1951). Next, Hasler began what has been portrayed as the first long-term ecological experiment (Likens 1985) in a pair of neighboring lakes, Peter and Paul, created by dividing a 5-ha hourglass-shaped lake with an earthen dike. As in Hasler's earlier study, lime was added to Peter Lake to deepen the euphotic-zone of the lake and create more suitable habitat for trout (Johnson and Hasler 1954). The added lime was intended to increase light transparency by reducing concentrations of humic acids that darkened the water to a deep brown, and fish were expected to benefit from the subsequent increase in lake productivity.

These studies provided the foundation for more studies in the following decades as Hasler and his students closed the gap between limnology and fishery management. To a large degree, these early liming efforts were consistent with the prevailing agricultural perspective of fishery management; Hasler stocked lakes with nonnative Rainbow Trout *Oncorhynchus mykiss* and Brown Trout *Salmo trutta* after other fish were exterminated. But the Wisconsin limnologists also paired their fisheries surveys with comprehensive surveys of lake physical and chemical conditions (Johnson and Hasler 1954), adhering to a tradition established at that university by Birge and Juday. Although the Peter and Paul Lake liming experiment was maintained for 25 years (1951–1976), attention to fishery consequences rapidly dwindled when no increase in trout production was observed (Johnson and Hasler 1954; Stross et al. 1961). Still, the Peter and Paul Lake experiment produced useful limnological insights for many years (Elser et al. 1986) that ultimately linked acid rain and fishery management. For example, the influence of the Peter and Paul Lake study was apparent in Gene Likens' approach to experimentally demonstrating the impacts of acid rain, which was not surprising given that Likens completed a Ph.D. with Hasler in the early 1960s.

LINKING ATMOSPHERIC EMISSIONS TO DECLINING BROOK TROUT FISHERIES

The developing link between limnology and fisheries flourished in the 1960s and 1970s as Likens left Wisconsin and moved to Cornell University, then applied the approach pioneered in Hasler's Wisconsin lake studies to the forested Hubbard Brook ecosystem in New Hampshire (Likens 1985). The pillars of this approach were long-term data collection and the manipulation of an entire ecosystem. As experiments at Hubbard Brook and long-term data collection began to foster a broad awareness of acid rain as a serious concern, Carl Schofield was able to link his late 1950s observations of Brook Trout mortality in Honnedaga Lake and other nearby lakes to the broader environmental problem of acid rain, writing in an unpublished 1972 Cornell University report by Webster:

> The long-esteemed quality of environment in the Adirondacks may no longer be inviolable. Air pollution knows no boundaries and is no longer confined to the cities. It is increasingly evident that country air is not as pure as it once was. Two questions urgently need to be answered: (1) Does air pollution constitute a threat to the Adirondacks? (2) What can be done to alleviate the problem? The answer to the latter question is obvious, but the solution will not be obtained readily. Nationwide attempts to curb air pollution are in progress, but it will be some years before significant improvement is obtained. At the moment we have no detailed information on the amounts and composition of air-borne contaminants that fall out over the Adirondack region. There are no programs specifically designed to monitor water quality in headwaters of the Adirondack region. What little information we have obtained from other research programs indicates a definite need for more intensive study of this problem.

In this same report, Schofield provided data documenting the acidification of 14 Adirondack lakes, including Honnedaga Lake, by comparing contemporary pH values with data collected from 1950 to 1953. The ongoing and now-well-documented fishery decline in Honnedaga Lake continued to play an important role in the development of the mid-1970s understanding of acid rain. This role started with Schofield's initial work in Honnedaga Lake showing that "periods of high acidity and heavy metals content, during the summer, were coincident with heavy mortality of Brook Trout" (Schofield 1965). A decade later, Schofield (1976) reported that summer pH values less than 5 were found in 51% of 217 Adirondack lakes at an elevation greater than 610 m and that fish were not present in 90% of these lakes. Overall, this connection between fish mortality and acid rain helped attract public attention to concerns about acid rain. Along with Schofield's work, an understanding of the detrimental impact of acid rain on fish was developing in Norway, Sweden, and Canada (Schofield 1976; Leivestad and Muniz 1976), but the relatively long-term record of pH and fishery conditions in Honnedaga Lake—extending back to 1939—contributed to the developing North American consensus that concerns about acid rain must be addressed.

Although the political consensus about acid rain lagged behind scientific consensus, the type of fishery decline first noted in Honnedaga Lake was at the center of public debate. The opening statement at a February, 1980 U.S. House of Representatives committee hearing implicitly referred to work by Schofield and others documenting the decline of fish in Adirondack lakes (U.S. House of Representatives 1980):

> The brilliant blue of more than 100 Adirondack lakes is misleading. All the life in these remote waters has been killed by acidification; fish can no longer re-

> produce nor survive in them... Fishery biologists are liming lakes in an attempt to neutralize the waters.

Other members of the U.S. Congress also gave testimony at that hearing focusing on Adirondack fisheries, taking advantage of public attention brought to the Adirondacks by the recently completed 1980 Lake Placid Olympic Winter Games. For example, one representative from New York stated,

> Due to their size and geological conditions, Lake Placid and Saranac Lake probably will be able to fend off the worst effects of the acid precipitation but other bodies of water in the Adirondacks, particularly lakes and streams at the higher elevations, will not be so fortunate. Fish will cease to reproduce, aquatic vegetation will perish and the waters will be devoid of life.

Another joint statement by two New York Congressional representatives (McEwen and Mitchell) explicitly referred to Schofield's work by saying,

> A recent survey of 214 lakes at higher elevations shows that more than half are too acidic to support life. Forty years ago, only 4% of these lakes were without fish.

As might be expected, not everyone at the hearing spoke in favor of the scientific consensus about acid rain. The testimony of James McAvoy, director of the Ohio Environmental Protection Agency, cast doubt by saying

> Given the fact that the phenomenon is not new and very little seems to be known for sure, we must raise the question: What has become so critical about the phenomenon as to generate the attention it is now receiving? For what reason have inflammatory and irresponsible headlines about acid rain appeared in the press? What are the reasons, especially when we have been told by Dr. Ralph Perhac, in testimony before the National Commission on Air Quality, the 'purported increase' of acidity in rainfall 'may be more apparent than real.'... There are reports that nature may account for up to two-thirds of the total acid-producing air contaminants... Few people realize that lightning flashes in thunderstorms can create nitric acid. Two such flashes on a 1-square-kilometer mountain lake watershed during a rainfall of 2 centimeters could result in a pH of less than 3.5.

In the end, Congressional representatives were not dissuaded by such presentations of alternative "facts." Instead, they pushed the U.S. government to fund studies supporting a broad array of scientists from many disciplines who confirmed the negative influence of acid rain on Adirondack aquatic ecosystems (Jenkins et al. 2007). These studies provided more insights about the complex mosaic of ecosystem interactions leading to fish mortality from acid rain that puzzled early researchers. For example, scientists confirmed that inorganic aluminum mobilized by acidic soil conditions entered surface waters and was ultimately responsible for the observed toxicity to fish and other aquatic organisms (Driscoll et al. 1980; Baker et al. 1996), which was an idea first suggested by Carl Schofield (Charley Driscoll, Syracuse University, personal communication). As results from those studies produced an overwhelming scientific consensus, political controversy continued about whether government regulations could alleviate this problem (Waxman 1991). In the end, the Clean Air Act Amendments of 1990 included many provisions explicitly designed to reduce acid rain, culminating a decade of political debate in the United States (Waxman 1991). Yet dissenting voices railed against these regulatory measures before and after they were implemented, contending that "the acidification of the lakes in the Adirondack Mountains cannot be stopped by clean air legislation" (Russell

1992). Now, 30 years later, we can evaluate predictions by both the proponents and opponents of government action.

The conclusion is clear. The deniers of acid rain science were wrong in predicting that reductions in acid emissions would not lead to environmental and fishery improvements. The 1990 Amendments to the Clean Air Act provided legal authority for regulations resulting in substantial reductions in sulfur dioxide and nitrogen oxide emissions from coal-fired power plants. Together with similar actions in Canada, these regulations helped reduce acidic deposition in the Adirondacks and across the eastern United States and Canada (Stoddard et al. 1999; Lehmann et al. 2005; Driscoll et al. 2016). These reductions in both sulfur dioxide and nitrogen oxide led to a recovery of pH and acid-neutralizing capacity, and reduction of harmful concentrations of inorganic monomeric aluminum in lakes and streams in eastern North America (Driscoll et al. 2003; Garmo et al. 2014). Although these regulatory measures were not taken solely to manage fisheries in the Adirondacks, documented fisheries declines in Adirondack lakes such as Honnedaga Lake contributed substantially to public and political discussions that led to measures requiring industries and utilities to reduce emissions. Stocking fish was no longer sufficient to cover up environmental degradation. The Honnedaga Lake Brook Trout fishery played an important role in showing that reductions in atmospheric acid emissions were needed to re-establish the environmental conditions in lakes and rivers that attracted people to the Adirondacks.

BROOK TROUT RECOVERY IN HONNEDAGA LAKE

During the 1960s and 1970s, Honnedaga Lake fisheries surveys by Cornell University scientists dwindled as it became apparent that Brook Trout could no longer survive in that lake. Scientists and fishery managers now understood the extent to which acid rain had altered Adirondack ecosystems. After completing annual Honnedaga Lake fish surveys each year from 1960 through 1965, a Cornell University report by Webster of results from the next survey in 1967 described them as "singularly unrewarding." The same document mentioned that members and employees of the Adirondack League Club had recently observed dead Brook Trout along the Honnedaga Lake shoreline, confirming the futility of further stocking. Surveys next conducted in 1972, 1973, and 1976 captured 17, 15 and 3 fish, respectively, and Cornell University reports of these surveys said little more than that conditions in Honnedaga Lake were not suitable for Brook Trout. One final survey of the Honnedaga Lake fishery was completed in 1979, in which six Brook Trout were captured, before a 20-year hiatus ensued. An unpublished 1979 Cornell University report by Webster describing the 1979 survey concluded that "chemical conditions in the lake are extremely unfavorable for fish survival," but "a few small fish also find refuge in the tributary streams." For the next two decades, Honnedaga Lake—a noted angler destination in the late 19th century—was viewed as a clear, blue, but fishless lake.

During the decades of ongoing policy debates and government action to regulate emissions of nitrous oxides and sulfur dioxide, members of the Honnedaga Lake community considered, then rejected, adding lime to the lake because they were concerned that this would reduce the clarity of the lake's water. Infrequent water chemistry surveys in the 1980s and 1990s showed little increase in lake pH (Figure 2, top panel). Throughout that time, club members observed small Brook Trout in some tributaries to the lake; then, at the approach of the new millennium, adult fish were observed rising to the surface of the lake. This prompted a Cornell University fish

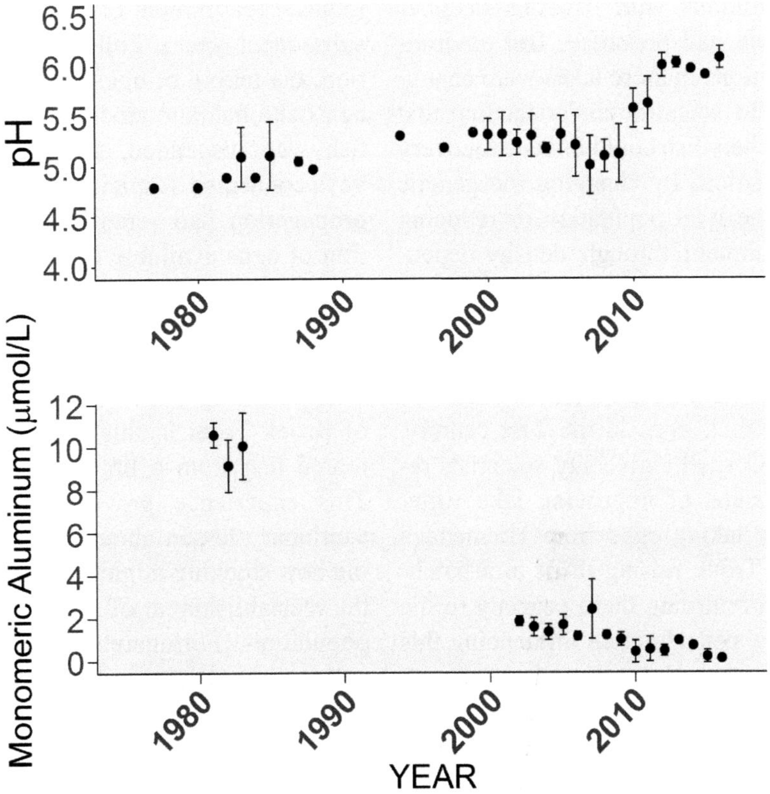

FIGURE 2. Top panel: summer (July and August) surface water pH in Honnedaga Lake. Bottom panel: inorganic monomeric aluminum in surface water (May–October) from Honnedaga Lake.

survey in 2000—using methods similar to those used in the Webster era—that captured 70 Brook Trout ranging from 180 to 460 mm in length. The Honnedaga Lake Brook Trout population had apparently begun to recover from fish that had persisted in tributary refuges with suitable water chemistry conditions, which prompted more comprehensive water chemistry surveys in the lake (Figure 2) and its tributaries (Josephson et al. 2014). Now, Cornell University managers of this fishery were faced with the challenge of responding to this discovery and deciding how to manage or enhance the recovery of the lake's Brook Trout population.

As might have been expected, members of the Adirondack League Club's Fish, Game and Preserve Committee quickly requested a stocking program. The author was responsible for implementing that decision after having recently taken over direction of Cornell University's Adirondack Fishery Research Program. At that time, the author recognized that the stocking request was prompted by decades of management actions that focused on fish stocking as the centerpiece for maintaining the club's fisheries. For more than 30 years, the Adirondack League Club had supported operation of a fish hatchery on the premises of their club. This reinforced the mindset that fisheries were supported by hatcheries, not intact ecosystems, during an era when Adirondack lakes were impaired by acid rain. To a large degree, the notion of hatchery supplementation as a path to restoring self-sustaining fish populations had been lost as frustrations with fisheries declines from acid rain had prevailed. But environmental conditions in these lakes were changing, as was the scientific understanding that stocking hatchery fish could impede recovery of fish populations by changing the genetic makeup of the wild population or reducing natural recruitment through density-dependent processes (Lorenzen et al. 2012; Rand et al. 2012).

The legacy of stocking hatchery-reared fish to sustain a fish population remained so dominant that, even in the 21st century, anglers and Cornell University scientists responded to signs of improving lake water chemistry by taking eggs from Honnedaga Lake Brook Trout, raising them in a hatchery, and then returning these progeny to the lake. One key consideration influencing this decision was that Honnedaga Lake Brook Trout had been designated as one of only a dozen remaining New York "heritage" Brook Trout populations, having never been previously stocked with fish originating beyond that body of water (Keller 1979). So, instead of stocking Honnedaga Lake with fish from other waters, hatchery progeny were raised from eggs collected in Honnedaga Lake, thereby maintaining the genetic integrity of this population.

In fall 2002, 4,000 fingerling Brook Trout raised from eggs collected in fall 2001 surveys were stocked into Honnedaga Lake. This action was repeated the following year. However, a marked decline in the number and proportion of wild fish captured in fall 2004 (Figure 3) led Cornell University staff to question the influence of stocking on the lake's Brook Trout population. Following the 2-year stocking effort, hatchery-reared fish made up a large proportion of yearling and adult Brook Trout, but no increase in the survey catch rate was observed. This suggested that the collection of eggs from spawning adults in 2001 and 2002 had contributed to reduced recruitment of wild Brook Trout in subsequent years. Following that observation, the taking of eggs from wild Honnedaga Lake fish and stocking hatchery-reared fish was suspended, and subsequent surveys confirmed suspicions that the hatchery propagation had removed a large proportion of eggs available from mature females in the wild population without any benefit (Figure 3). Instead, stocking had replaced what would have been naturally reproduced, self-sustaining 2001 and 2002 year-classes of Brook Trout in this lake with hatchery-reared fish from a limited supply of eggs. This experience provides an instructive, cautionary lesson about carefully considering how stocking might negatively influence the reestablishment of a self-sustaining fish population. Fortunately, this management action did not impede the reproduction and recovery of the Brook Trout population (Figure 4), and the impairment of natural reproduction induced by the hatchery effort diminished as stocked fish died and water chemistry conditions continued to improve. Most important, measured levels of toxic inorganic monomeric aluminum in Honnedaga Lake water showed a steady decline after annual water chemistry surveys were initiated in 2003 and remained consistently below the level considered toxic to Brook Trout (2 ɥmol/L; Figure 2, lower panel). Ultimately, the broad ecosystem recovery reflected in the Brook Trout fishery was confirmed by ongoing limnological surveys to document changes in lake water chemistry (Josephson et al. 2014), once again reflecting the influence of ecosystem focused scientists such as Hasler and Likens.

The second decade of the 21st century brought a renaissance of the Honnedaga Lake Brook Trout fishery. Angler interest in the lake skyrocketed after angling was again allowed in 2016, attracted by the opportunity to catch Brook Trout that often achieved lengths

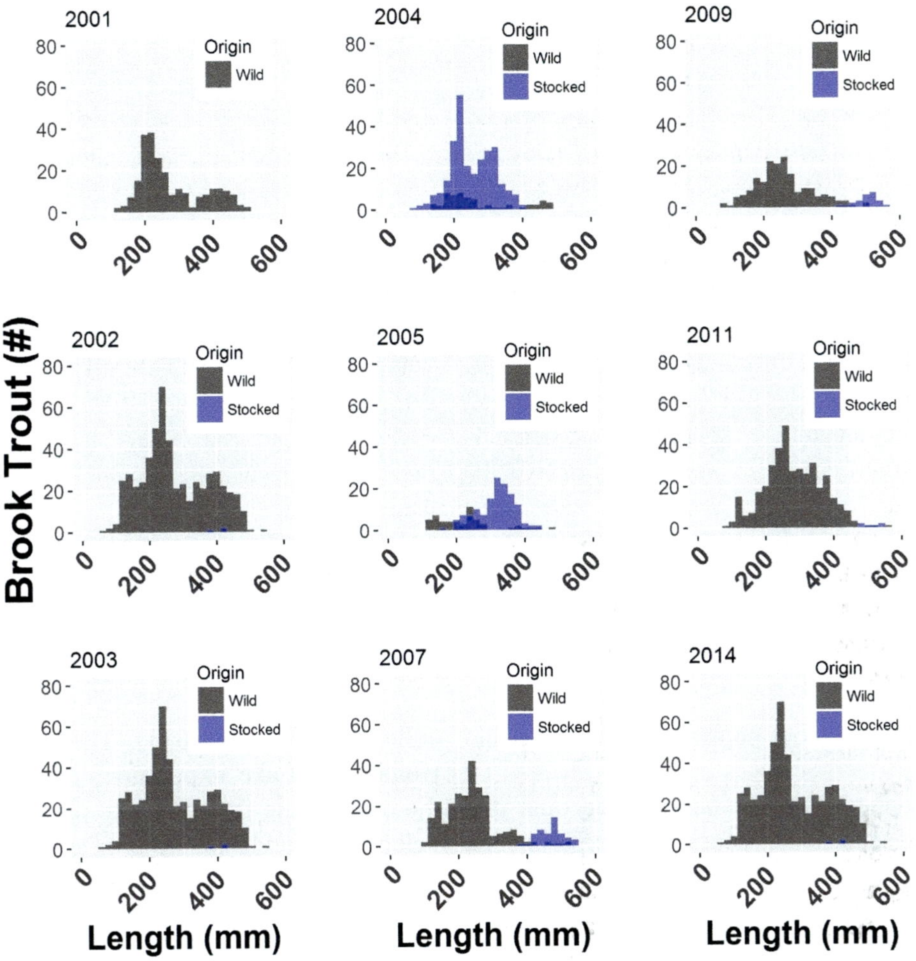

FIGURE 3. Lengths and number of Brook Trout captured in fall trap-net surveys in Honnedaga Lake from 2001 to 2014.

greater than 500 mm. Now, nearly 20 years of data collection has provided information about many different variables contributing to the success of Brook Trout in this lake, such as their reliance on isopods and crayfish as a key prey resource (Jirka and Kraft 2017). Yearly surveys of water chemistry and young-of-the-year fish have identified a small number of lake tributaries that sustain fish reproduction because the tributaries are located in small watersheds with glacial till that has substantial buffering capacity (Driscoll and Newton 1985), though most lake tributaries remain acidic year-round. Additional fall surveys of the number and location of redds (individual fish spawning sites) have shown an increase in reproduction (Cornell Adirondack Fishery Research Program, unpublished data). In addition, 3 years of capture–mark–recapture surveys initiated in 2015 have provided the first firm estimates of Brook Trout population abundance and survival in this lake; this augments information from otolith analyses indicating that Honnedaga Lake Brook Trout are long-lived (maximum age 9 years) and grow slowly (Figure 5). Genetic analyses have also confirmed that Honnedaga Brook Trout are distinct from other Adiron-

FIGURE 4. Brook Trout spawning in Honnedaga Lake. (Photo taken by Eileen Randall).

dack Brook Trout populations (Bruce et al. 2018).

Honnedaga Lake has also become the site of additional ecosystem experiments examining linkages between a lake fishery and its watershed, again reflecting the recent integration of fishery management with a broad array of other environmental sciences (George et al. 2018). This work has included four annual instream point applications of lime to one chronically acidified tributary and one episodically acidified tributary from 2012 to 2015 (George et al. 2018; Figure 1). A watershed-wide liming effort was initiated in October 2013 when 150 tons (136 metric tons) of high-calcium limestone were distributed by helicopter in a pelletized form over a 30-ha watershed encompassing one tributary flowing into Honnedaga Lake; a neighboring watershed served as a reference site (Figure 1). This action at a landscape scale is only the third watershed liming experiment conducted in the United States, one of which was implemented at Hubbard Brook by Likens and colleagues (Likens 2013). The large-scale Honnedaga Lake experiment reflects both the central role that this lake played in developing an understanding of acid rain and the experimental and long-term data collection tradition established by Hasler and Likens. The perspective that they pioneered is reflected in the breadth of the Honnedaga watershed liming study, in which researchers are examining the effects of this experiment on not only Brook Trout, but also soil and soil-water chemistry, tree growth and regeneration, soil fauna, stream chemistry, aquatic invertebrates, and the movement of mercury through food webs.

The recovery of Brook Trout documented in Honnedaga Lake reflects the potential for recovery of Brook Trout populations in waters throughout the Adirondacks as acid rain decreases. Although Baldigo et al. (2016) did not find a change in Brook Trout catch per unit effort in data from 43 Adirondack lakes surveyed three times from 1984 to 2012, Cornell University surveys of other Adirondack lakes and streams have shown

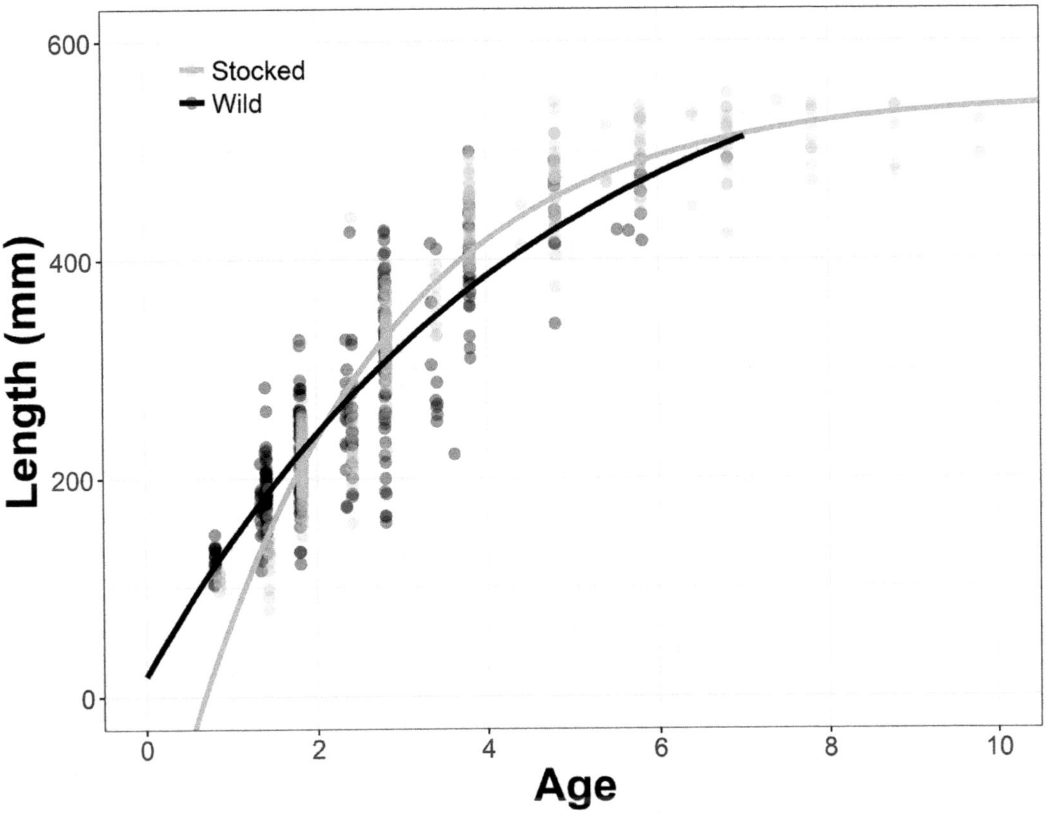

FIGURE 5. Length at age of Honnedaga Lake Brook Trout captured from 2003 to 2015. Ages of wild fish (black) determined by otoliths; ages of stocked fish (gray) determined by fin clip.

increases in Brook Trout abundance similar to that observed in Honnedaga Lake. These other waters have shown an increase in pH and reduction in monomeric aluminum similar to that observed in Honnedaga, allowing them to sustain wild Brook Trout populations that in some instances were obscured by stocking implemented decades ago to support recreational fisheries. Stocking may be required to re-establish self-sustaining Brook Trout populations in lakes where recolonization cannot occur from connected remnant populations, as documented in formerly acidic Brook Trout Lake (Sutherland et al. 2015). But Honnedaga Lake shows that it is just as important to recognize when stocking is no longer or not required to restore a self-sustaining fish population.

CONCLUDING THOUGHTS AND LESSONS LEARNED

As the 20th century progressed, the Adirondacks served as a natural laboratory that showed stocking could not maintain fisheries impaired by human activities occurring far from the lakes and streams where anglers sought refuge. After this was recognized, the Adirondacks provided an example of how a broad environmental understanding of the physical and chemical conditions that sustain fisheries could guide their recovery. But just as important, social and political decisions made far from the Adirondacks were needed to successfully rehabilitate these fisheries.

This chapter has reflected on how what occurred in Honnedaga Lake during the past century helped scientists look forward to understanding current challenges in managing fisheries. Aquatic habitats continue to face physical and chemical degradation from atmospheric transport of chemical constituents—including greenhouse gases released by human activities—that originate far from where the symptoms of declining fisheries are apparent. We hope that reflecting on how science informed management practices—in this case, controls on industrial emissions—that improved Adirondack Brook Trout fisheries will help show how fisheries can be sustained and restored by controlling greenhouse gas emissions and other human-generated impacts.

This chapter also started by looking back at an era in which fisheries were managed like agricultural crops: removing pests, increasing production by fertilizing, and stocking domesticated organisms. Then acid rain prevailed and forced managers of inland fisheries to view them as a resource that depends on environmental conditions, both local, as in the geology of the Adirondacks, and distant, as in industrial emissions hundreds of miles away. Of course, prior generations had recognized obvious human insults caused by habitat degradation and overharvest, but acid rain required more sophisticated scientific analyses and insights that depended on models and forecasts supported by strong empirical data. That helped fishery science become an environmental science, as is now evident in public discussions of declining fisheries throughout the world. Maintaining this comprehensive scientific approach will be important in restoring and sustaining fisheries as we face contemporary environmental challenges that operate at global scales.

Looking forward, science must be respected by government policymakers in the United States and countries throughout the world. This happened in the late 20th century when scientific studies in a corner of the Adirondack region of New York helped citizens and politicians across the nation take successful action based upon a scientific consensus and concern for the environment. Ultimately, the recovery of Adirondack Brook Trout populations provides a notable example of success in which scientists working with public advocates and policymakers together brought about an improved environment and fishery resource.

KEY LESSONS:

- A broad environmental understanding of the physical and chemical conditions that sustain fisheries help guide actions to achieve their recovery.
- Incorporating insights from atmospheric sciences, geology, and limnology into fishery management represents a movement toward ecosystem management of fisheries.
- Social and political decisions made far from the Adirondacks helped rehabilitate Adirondack fisheries.
- Fishery management often requires political support, regulatory action, and media attention—as well as an expansive scientific understanding—to achieve favorable outcomes.
- Once favorable environmental conditions are restored, natural recovery processes can successfully rehabilitate a fishery without the additional intervention of stocking hatchery-reared fish.

ACKNOWLEDGMENTS

I thank Fred Henson, Dan Josephson, Kurt Jirka, Laura Martin, Aaron Sachs, and four anonymous reviewers for helpful comments on a previous draft of the manuscript. Dan also consistently provided key insights during many discussions of this topic that are reflected in the information presented. Eileen Randall and Ben Marcy-Quay provided assistance with figures and photos. For many decades, dozens of additional students and technicians too numerous to list provided necessary help with field work and insights needed to understand ongoing changes in Honnedaga Lake and Adirondack fisheries. Long-standing support provided by the Adirondack League Club and Cornell University was essential to make this work possible.

REFERENCES

Anonymous. 1872. Proceedings of the American Fish Culturists' Association. Transactions of the American Fisheries Society 1:3–21.

Baker, J. P., J. Van Sickle, C. J. Gagen, D. R. DeWalle, W. E. Sharpe, R. F. Carline, B. P. Baldigo, P. S. Murdoch, D. W. Bath, W. a. Krester, H. A. Simonin, and P. J. Wigington. 1996. Episodic acidification of small streams in the northeastern United States: effects on fish populations. Ecological Applications 6:422–437.

Baldigo, B. P., K. M. Roy, and C. T. Driscoll. 2016. Response of fish assemblages to declining acidic deposition in Adirondack Mountain lakes, 1984–2012. Atmospheric Environment 146:223–235.

Brown, H. W., and M. E. Jewell. 1926. Further studies on the fishes of an acid lake. Transactions of the American Microscopical Society 45:20–34.

Bruce, S. A., M. P. Hare, M. W. Mitchell, and J. J. Wright. 2018. Confirmation of a unique and genetically diverse "heritage" strain of Brook Trout *Salvelinus fontinalis* in a remote Adirondack watershed. Conservation Genetics 19:71–83.

Comstock, E. J., and M. C. Webster. 1990. The Adirondack League Club 1890–1990. Adirondack League Club, Old Forge, New York.

Crowther, C., and A. G. Ruston. 1911. The nature, distribution and effects upon vegetation of atmospheric impurities in and near an industrial town. The Journal of Agricultural Science 4:25–55.

Driscoll, C. T., J. P. Baker, J. J. Bisogni, and C. L. Schofield. 1980. Effect of aluminum speciation of fish in dilute acidified waters. Nature (London) 284:161–164.

Driscoll, C. T., K. M. Driscoll, H. Fakhraei, and K. Civerolo. 2016. Long-term temporal trends and spatial patterns in the acid-base chemistry of lakes in the Adirondack region of New York in response to decreases in acidic deposition. Atmospheric Environment 146:5–14.

Driscoll, C. T., K. M. Driscoll, K. M. Roy, and M. J. Mitchell. 2003. Chemical response of lakes in the Adirondack region of New York to declines in acidic deposition. Environmental Science and Technology 37:2036–2042.

Driscoll, C. T., G. B. Lawrence, A. J. Bulger, T. J. Butler, C. S. Cronan, C. Eagar, K. F. Lambert, G. E. Likens, J. L. Stoddard, and K. C. Weathers. 2001. Acidic deposition in the northeastern United States: sources and inputs, ecosystem effects, and management strategies. BioScience 51:180–198.

Driscoll, C. T., and R. M. Newton. 1985. Chemical characteristics of Adirondack lakes. Environmental Science and Technology 19:1018–1024.

Elser, M. M., J. J. Elser, and S. R. Carpenter. 1986. Paul and Peter lakes: a liming experiment revisited. The American Midland Naturalist 116:282–295.

Forbes, S. A. 1887. The lake as a microcosm. Bulletin of the Peoria Science Association 1887:77–87. (Reprinted in Bulletin of the Illinois State Natural History Survey (1925) 15:537–550.)

Garmo, Ø. A., B. L. Skjelkvåle, H. A. de Wit, L. Colombo, C. Curtis, J. Fölster, A. Hoffmann, J. Hruška, T. Høgåsen, D. S. Jeffries, W. B. Keller, P. Krám, V. Majer, D. T. Monteith, A. M. Paterson, M. Rogora, D. Rzychon, S. Steingruber, J. L. Stoddard, J. Vuorenmaa, and A. Worsztynowicz. 2014. Trends in surface water chemistry in acidified areas in Europe and North America from 1990 to 2008. Water, Air, and Soil Pollution 225:1880.

George, C. J. 1981. The fishes of the Adirondack Park. New York State Department of Environmental Conservation, Albany.

George, S. D., B. P. Baldigo, G. B. Lawrence, and R. L. Fuller. 2018. Effects of watershed and in-stream liming on macroinvertebrate communities in acidified tributaries to an Adirondack lake. Ecological Indicators 85:1058–1067.

Gorham, E. 1958. Free acid in British soils. Nature (London) 181:106–106.

Greeley, J. R. 1934. Fishes of the Raquette watershed with an annotated list. Pages 53–108 *in* A biological survey of the Raquette watershed. Supplemental to the twenty-third annual report of the New York State Conservation Department. J. B. Lyon company, Albany, New York.

Hasler, A. D., O. M. Brynildson, and W. T. Helm. 1951. Improving conditions for fish in brown-water bog lakes by alkalization. The Journal of Wildlife Management 15:347–352.

Hewitt, E. R. 1926. Telling on the trout. C. Scribner's Sons, New York.

Hewitt, E. R. 1933. Hewitt's handbook of fly fishing. The Marchbanks Press, New York.

Hewitt, E. R. 1935. Hewitt's handbook of stream improvement. The Marchbanks Press, New York.

Jacoby, K. 2014. Crimes against nature: squatters, poachers, thieves, and the hidden history of American conservation. University of California Press, Berkeley.

Jenkins, J. C., K. Roy, C. Driscott, and C. Buerkett. 2007. Acid rain in the Adirondacks: an environmental history. Cornell University Press, Ithaca, New York.

Jessee, E. J. 2014. A heightened controversy: nuclear weapons testing, radioactive tracers, and the dynamic stratosphere. Pages 152–180 *in* J. R. Fleming and A. Johnson, editors. Toxic airs: body, place, planet in historical perspective. University of Pittsburgh Press, Pittsburgh, Pennsylvania.

Jewell, M. E. 1922. The fauna of an acid stream. Ecology 3:22–28.

Jewell, M. E., and H. Brown. 1924. The fishes of an acid lake. Transactions of the American Microscopical Society 43:77–84.

Jirka, K. J., and C. E. Kraft. 2017. Diet niche width and individual specialization of Brook Trout in Adirondack lakes. Transactions of the American Fisheries Society 146:716–731.

Johnson, W. E., and A. D. Hasler. 1954. Rainbow trout production in dystrophic lakes. The Journal of Wildlife Management 18:113–134.

Josephson, D. C., J. M. Robinson, J. Chiotti, K. J. Jirka, and C. E. Kraft. 2014. Chemical and biological recovery from acid deposition within the Honnedaga Lake watershed, New York, USA. Environmental Monitoring and Assessment 186:4391–4409.

Keller, W. 1979. Management of wild and hybrid Brook Trout in New York lakes, ponds and coastal streams. New York State Department of Environmental Conservation, Albany.

Langenheim, J. H. 1996. Early history and progress of women ecologists: emphasis upon research contributions. Annual Review of Ecology and Systematics 27:1–53.

Lehmann, C. M. B., V. C. Bowersox, and S. M. Larson. 2005. Spatial and temporal trends of precipitation chemistry in the United States, 1985–2002. Environmental Pollution 135:347–361.

Leivestad, H., and I. P. Muniz. 1976. Fish kill at low pH in a Norwegian river. Nature (London) 259:391–392.

Likens, G., and F. Bormann. 1974. Acid rain: a serious regional environmental problem. Science 184:1176–1179.

Likens, G. E. 1985. An experimental approach for the study of ecosystems: the fifth Tansley lecture. Journal of Ecology 73:381–396.

Likens, G. E. 2013. Biogeochemistry of a forested ecosystem. Springer Science & Business Media, Berlin, Germany.

Likens, G. E., F. H. Bormann, and N. M. Johnson. 1972. Acid rain. Environment 14:33–40.

Lorenzen, K., M. C. M. Beveridge, and M. Mangel. 2012. Cultured fish: integrative biology and management of domestication and interactions with wild fish. Biological Reviews 87:639–660.

MacIntire, W. H., and J. B. Young. 1923. Sulfur, calcium, magnesium and potassium content and reaction of rainfall at different points in Tennessee. Soil Science 15:205–227.

Marsh, G. P. 1965. Man and nature. University of Washington Press, Seattle.

Martin, L. J. 2018. Proving grounds: ecological fieldwork in the Pacific and the materialization of ecosystems. Environmental History 23:567–592.

Mitchell, N. J., and R. Diamant. 2015. The necessity of stewardship: George Perkins Marsh and the nature of conservation. The George Wright Forum 32:257–265.

Moffitt, C. M., G. Whelan, and R. Jackson. 2010. Historical perspectives on inland fisheries management in North America. Pages 1–41 *in* W. A. Hubert and M. C. Quist, editors. Inland fisheries management in North America. American Fisheries Society, Bethesda, Maryland.

Muir, J. 1874. Salmon breeding. Daily Evening Bulletin of San Francisco, October 29.

Neess, J. C. 1949. Development and status of pond fertilization in central Europe. Transactions of the American Fisheries Society 76:335–358.

New York State Bar Association. 2016. Reports and recommendations concerning the conservation article in the state constitution (Article XIV) adopted by the Committee on the New York State Association. New York State Bar Association, Albany. Available: www.nysba.org/ArticleXIVreport/. (January 2019).

Northcote, T. G., and R. A. Larkin. 1966. Western Canada. Pages 451–486 *in* D. J. Frey, editor. Limnology in North America. University of Wisconsin Press, Madison.

Odén, S. 1968. The acidification of air and precipitation and its consequences on the natural environment. Swedish National Science Research Council, Ecology Committee Bulletin 1, Stockholm. (In Swedish.)

Odén, S. 1976. The acidity problem—an outline of concepts. Water, Air, and Soil Pollution 6:137–166.

Rand, P. S., B. A. Berejikian, T. N. Pearsons, and D. L. G. Noakes. 2012. Ecological interactions between wild and hatchery salmonids: an introduction to the special issue. Environmental Biology of Fishes 94:1–6.

Rosenbaum, W. A. 2016. Environmental politics and policy, 10th edition. CQ Press, Washington, D.C.

Russell, S. 1992. Potential fall out from the National Acid Precipitation Assessment Program. Brigham Young University Journal of Public Law 6:423–442.

Schofield, C., Jr. 1962. Water quality in a lake of the Adirondack Mountains in relation to the survival of hatchery reared Brook Trout, *Salvelinus fontinalis* (Mitchell). Master's thesis. Cornell University, Ithaca, New York.

Schofield, C. L., Jr. 1965. Water quality in relation to survival of Brook Trout, *Salvelinus fontinalis* (Mitchill). Transactions of the American Fisheries Society 94:227–235.

Schofield, C. L., Jr. 1976. Acid precipitation: effects on fish. Ambio 5:228–230.

Smith, L. L., Jr., and J. B. Moyle. 1944. A biological survey and fishery management plan for the streams of the Lake Superior north shore watershed. Minnesota Department of Conservation, Division of Game and Fish, Technical Bulletin 1, Saint Paul.

Spears, J. R. 1888. Adirondack abominations. Forest and Stream: A Weekly Journal of the Rod and Gun 30:513.

Stoddard, J. L., D. Jeffries, A. Lukewille, T. A. Clair4, P. J. Dillon, C. T. Driscoll, M. Forsius, M. Johannessen, J. S. Kahl, J. H. Kellogg, A. Kemp, J. Mannio, D. T. Monteith, P. S. Murdoch, S. Patrick, A. Rebsdorf, B. L. Skjelkvåle, M. P. Stainton, T. Traaen, H. van Dam, K. E. Webster, J. Wieting, and A. Wilander. 1999. Regional trends in aquatic recovery from acidification in North America and Europe. Nature (London) 401:575–578.

Stone, L. 1872. Trout culture. Transactions of the American Fisheries Society 1:46–56.

Stross, R. G., J. C. Neess, and A. D. Hasler. 1961. Turnover time and production of planktonic crustacea in limed and reference portion of a bog lake. Ecology 42:237–245.

Sutherland, J. W., F. W. Acker, J. A. Bloomfield, C. W. Boylen, D. F. Charles, R. A. Daniels, L. W. Eichler, J. L. Farrell, R. S. Feranec, M. P. Hare, S. L. Kanfoush, R. J. Preall, S. O. Quinn, H. C. Rowell, W. F. Schoch, W. H. Shaw, C. A. Siegfried, T. J. Sullivan, D. A. Winkler, and S. A. Nierzwicki-Bauer. 2015. Brook Trout lake case study: biotic recovery from acid deposition 20 years after the 1990 Clean Air Act Amendments. Environmental Science and Technology 49:2665–2674.

U.S. House of Representatives. 1980. Acid rain: hearings before the Subcommittee on Oversight and Investigations of the Committee on Interstate and Foreign Commerce, House of Representatives, Ninety-Sixth Congress, second session, February 26 and 27, 1980. U.S. Government Printing Office, Washington, D.C.

Van Put, E. 2014. Trout fishing in the Catskills. Skyhorse Publishing, New York.

Waxman, H. A. 1991. An overview of the Clean Air Act Amendments of 1990. Environmental Law 21:1721–1816.

Webster, D. A. 1961. An unusual lake of the Adirondack Mountains, New York Limnology and Oceanography 6:88–90.

Whitaker, H. 1892. Early history of the fisheries on the Great Lakes. Transactions of the American Fisheries Society 21:163–179.

Wilson, B. D. 1921. Sulfur supplied to the soil in rainwater. Agronomy Journal 13:226–229.

Wilson, B. D. 1923. The quantity of sulfur in rain water. Agronomy Journal 15:453–456.

Worden, G. H. 1891. Trouting in the North Woods. Outing: An Illustrated Monthly Magazine of Sport, Travel and Recreation 18:77–79.

Walleye Management in the Red Lakes, Minnesota: Collapse, Recovery, and Cooperative Management

Gary C. Barnard* and Anthony J. Kennedy
Minnesota Department of Natural Resources
2114 Bemidji Avenue, Bemidji, Minnesota 56601, USA

William P. Brown
Red Lake Department of Natural Resources
15761 High School Drive, Red Lake, Minnesota 56671, USA

Donald L. Pereira
Minnesota Department of Natural Resources
500 Lafayette Road, St. Paul, Minnesota, 55155, USA

Abstract.—Fishery management jurisdiction of the Red Lakes is split between the state of Minnesota and Red Lake Band of Chippewa Indians. Historically, the Red Lakes supported productive commercial and recreational Walleye *Sander vitreus* fisheries before crashing in the 1990s from decades of overharvest. In the late 1990s, these management groups began an interagency effort to develop and implement a recovery plan to restore the Walleye population, which included a complete Walleye fishing moratorium with strict enforcement and a short-term fry stocking program. Public involvement occurred early in the planning stages and throughout the implementation process. Public support was essential for effective implementation of restrictive fishing regulations necessary to expedite recovery and to resume harvests. The combination of recovery stockings and a 7-year Walleye fishing moratorium by both jurisdictions was successful, and the Walleye population recovered sufficiently to resume harvest in 2006. Walleye harvest was guided by a jointly prepared "Harvest Plan for Red Lakes Walleye Stocks" (hereafter, "Harvest Plan"; Red Lakes Fisheries Technical Committee, 2006 Harvest Plan for Red Lakes Walleye Stocks, unpublished), which took a conservative approach to harvest management while the Walleye population progressed towards full recovery as defined in the "Recovery Plan for Red Lakes Walleye Stocks" (MOU 1999). The Walleye population reached full recovery thresholds in 2009, and the Harvest Plan was revised in 2015 to allow increased harvest after observing negative density dependent effects on growth and recruitment. Current management focuses on active management of spawning stock biomass for optimal condition as defined by the revised Harvest Plan (2.5–4.5 lb/surface acre [2.8–5.0 kg/ha]; Red Lakes Fisheries Technical Committee, 2015 Harvest Plan for Red Lakes Walleye Stocks, unpublished). This approach strives to manage spawning stock biomass such that wild fry production falls within the optimal range of a density dependent stock–recruitment relationship, and to maximize the likelihood of producing strong year-classes to sup-

* Corresponding author: gary.barnard@state.mn.us

port harvest levels commensurate with the potential productivity of the system. The collapse of the Red Lakes' Walleye fisheries was catastrophic but provided an incredible opportunity to increase our understanding of Walleye population dynamics, the importance of collaborative fishery management, and public involvement in decision-making processes.

INTRODUCTION

The Red Lakes, Minnesota, are comprised of two large, oval-shaped basins known as Upper Red Lake and Lower Red Lake. These lakes combined comprise the largest water body completely within the state's borders (284,264 acres [115,037.6 ha]), with jurisdiction split between the Red Lake Band of Chippewa (hereafter, "Band") and the state of Minnesota. Lower Red Lake lies entirely within the Red Lake Indian Reservation (hereafter, "Reservation"), whereas approximately 60% of Upper Red Lake (72,000 acres [29,137 ha]) is within the jurisdiction of the Band and 40% (48,000 acres [19,425 ha]) under state jurisdiction (Figure 1). The Red Lakes are capable of supporting productive commercial and recreational fisheries with historically documented annual harvests of Walleye *Sander vitreus* approaching 1 million lb (453,592 kg; Smith and Krefting 1954; Brown and Thompson 2004; Figure 2).

In this chapter, we focus on the cooperative recovery efforts between jurisdictions and stakeholders over the past two decades following the Walleye population collapse and culminating in fully restored commercial and sport Walleye fisheries in the Red Lakes. The chapter concludes with lessons learned from our experiences that may be applicable to fishery management elsewhere.

HISTORICAL BACKGROUND

The earliest reports on the Red Lakes suggest that prior to 1917, the only fishing activity was subsistence fishing by Band members (VanOosten and Deason 1957). The state of Minnesota established a commercial Walleye fishery as a food production measure in 1917 as part of a wartime effort to provide affordable meat. Commercial fishing efforts continued after World War I primarily to provide a source of income to Band members who sold fish to the state-run commercial fishery. In 1929, the Minnesota Supreme Court prohibited the state from running for-profit commercial fisheries and the state turned over the commercial fishing operation to the Red Lake Fisheries Association (VanOosten and Deason 1957). Commercial fishing was limited to waters within the Reservation and governed by the U.S. Secretary of the Interior through regulations outlined in the U.S. Code of Federal Regulations (Commercial Fishing on Red Lake Indian Reservation, 25 CFR Part 242). Because Reservation waters are closed to non-Band members, the Red Lake Fisheries Association was comprised entirely of duly-enrolled Red Lake Band members.

From 1929 until 1970, commercial fish production was largely a function of abundance and market demand rather than the imposed 650,000-lb (294,835 kg) quota (Smith and Krefting 1954). In general, annual Walleye production was within 200,000 lb (90,718 kg) of the quota, except the quota was raised to 1 million lb (453,592 kg) from 1941 to 1953 due to increased demand as a result of World War II (Smith and Krefting 1954). During this time, approximately 200 independent tribal fishers earned their living commercial fishing each year. Commercial fishers had to be members of the Red Lake Fisheries Association, which entitled them to the use of eight gill nets per day. In addition, many Band members participated in the subsistence fishery, using one gill net a night

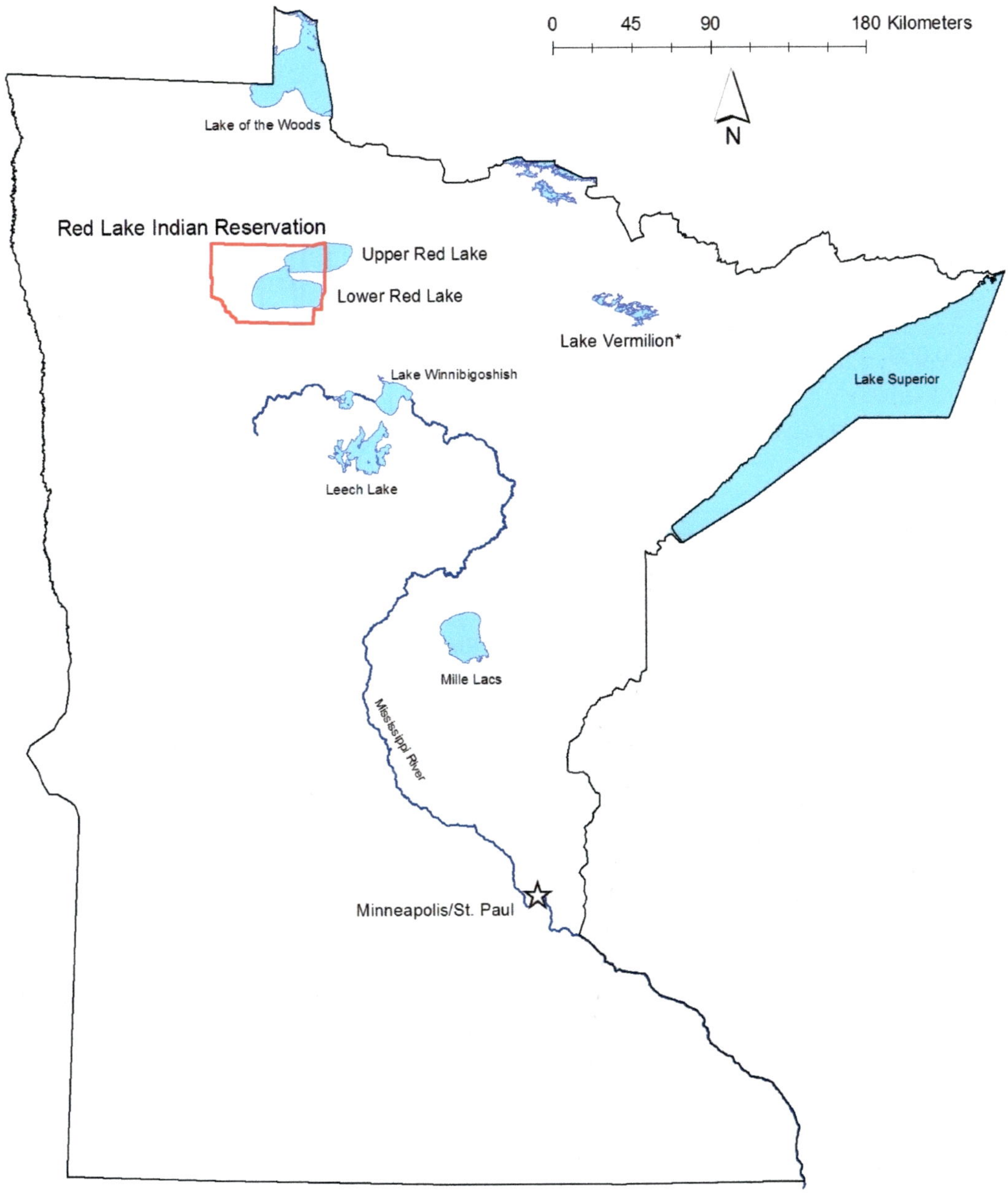

FIGURE 1. Location of the Red Lakes and Red Lake Nation in Minnesota. Lake Vermillion, the donor source of eggs for stocking efforts, is identified by the asterisk.

to provide food for personal use. Subsistence harvest was not recorded at the commercial processing plant and, thus, was not included in the reported harvest.

The Red Lakes have a simple fish community, with Walleye being the most abundant predator followed by a low-density Northern Pike *Esox lucius* population. Primary forage fish are Yellow Perch *Perca flavescens* and cyprinids, although juvenile White Sucker *Catostomus commersonii*, Lake Whitefish *Coregonus clupeaformis*, Freshwater Drum

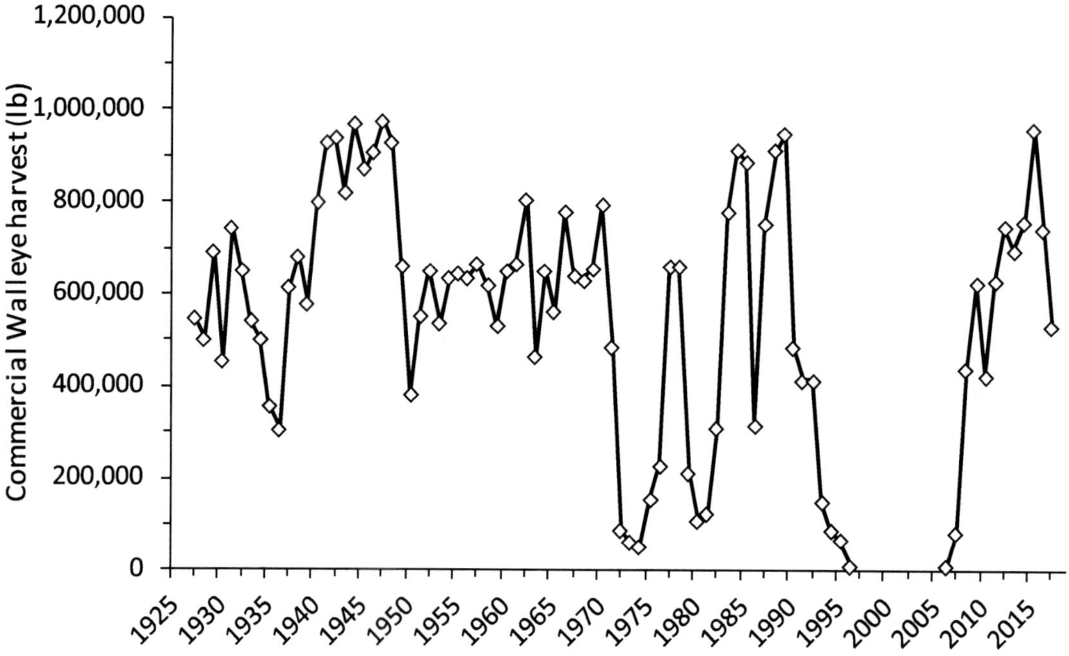

FIGURE 2. Commercial Walleye harvest in the Red Lakes, 1926–2017. Note the Walleye commercial fishing closure from 1997 to 2005.

Aplodinotus grunniens, and Goldeye *Hiodon tergisus* are present in varying abundance and may be important at times. The Red Lakes have a long history of successful Walleye natural reproduction because spawning substrate for lithophilic spawners is abundant along the shorelines. In addition, several tributary rivers experience spawning migrations. Male and female Walleyes in the Red Lakes generally begin to recruit to the spawning population at ages 3 and 4, respectively. Total length when entering the spawning population is approximately 12 in (305 mm) for males and 15 in (381 mm) for females.

Due to the robust Walleye population, the Minnesota Department of Natural Resources (MNDNR) operated a spawning station and hatchery on Upper Red Lake to produce fry for statewide distribution from the late-1920s through the 1970s. A popular recreational fishery developed over this time as well. The state returned a portion of fry produced back into the Upper Red Lake as directed by a policy supporting egg take operations (MNDNR 1982). The Band also operated a spawning station and hatchery with most of those fry returned to Lower Red Lake. As a result, an average of 40 million fry were stocked into the Red Lakes each year.

In the 1970s, annual variation in Walleye abundance increased, with boom-and-bust cycles observed in both the number of eggs taken in spawning operations and commercial harvest production. Egg take operations were suspended in the late 1970s due to the high annual variability in egg production and difficulty in reaching target egg takes during low points in the population cycle. Commercial fishing yields continued to cycle until the population collapsed in the mid-1990s (Figures 2 and 3). The booms were typically the result of recruitment of a single strong year-class to the fishery (Smith 1977). Once these fish were removed, a lag period would occur until the next strong year-class recruited to the fishery. Fishery managers believed pro-

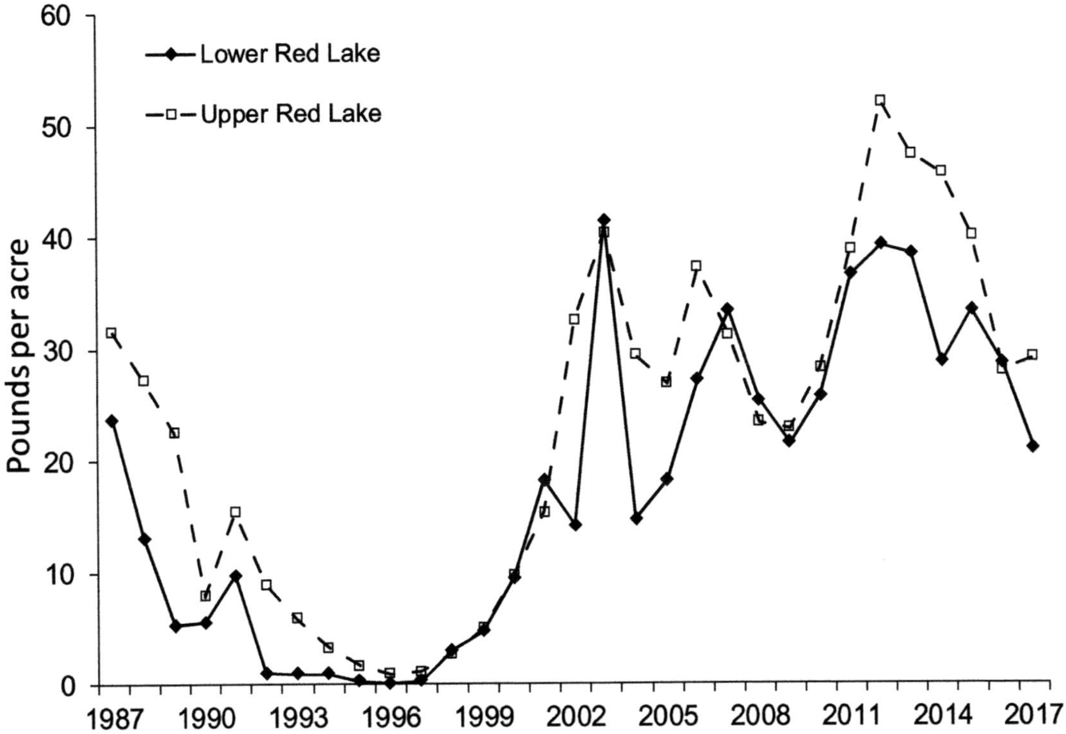

FIGURE 3. Estimated density of Walleyes greater than 8 in in the Red Lakes, 1987–2017.

duction of exceptional year-classes was the result of favorable climatic conditions and independent of spawning stock or routine fry stockings (Smith 1958). However, the cessation in fry stocking did coincide with the onset of boom-and-bust cycles in the 1970s.

Boom-and-bust population cycles continued for more than two decades, during which time the Walleye population always managed to rebound from periods of low abundance. As a result, periods of low Walleye abundance were thought to be normal and it became routine to simply wait for the next population boom. No formal population assessment programs were in place at this time so managers relied on commercial fishing records on Band waters and periodic creel surveys and gill-net assessments on state waters to track population trends. Each jurisdiction began annual independent Walleye population assessments in the 1980s, but data were not shared between the state and Band because each jurisdiction managed their fisheries independently. Thus, as the population began to collapse in the late 1980s and early 1990s, the true extent of the decline was not immediately realized. For example, Walleye catch per unit effort (CPUE) for the fall gill-net assessment in state waters never declined below 3.4 fish/net, whereas CPUE for Band waters reached a record-low 0.1 fish/net in 1996.

Both the Band and the state were overharvesting the Walleye population through commercial and subsistence gill-net fisheries on the Reservation waters and by recreational anglers on state waters. However, commercial fishing gear is more efficient than angling, and commercial fishing effort tends to increase in response to declining fish abundance while recreational fishing effort declines in response to diminishing angling success. The result was that state waters of Upper Red Lake were acting as a refuge of sorts, but the limited refuge effect did not protect the Walleye population from overexploitation by the combination of commercial, subsistence, and sport fisheries. The last strong year-class measured prior to the population collapse was produced in 1985. After the 1985 year-class moved through the fishery and abundance declined to levels that made commercial Walleye fishing no longer economically viable, commercial fishers shifted their efforts to Yellow Perch by reducing the gill-net mesh size. This species was flourishing due to predation relief related to low Walleye abundance, and ready markets were available. However, targeting Yellow Perch with a smaller mesh size than used before for Walleyes resulted in harvesting subsequent Walleye year-classes before they reached sexual maturity, which ultimately collapsed the population.

Documented commercial Walleye harvest between 1970 and 1996 fluctuated widely. During booms, harvests frequently exceeded the 650,000 lb (294,835 kg) quota and approached 1 million lb (453,592 kg), whereas harvest during busts failed to reach 100,000 lb (45,359 kg; Figure 2). Commercial fishing effort fluctuated as well, with membership in the Fisheries Association peaking near 700 independent netters. Estimated angler harvest from creel surveys conducted on state waters also fluctuated during this period and ranged from 13,000 to 180,000 lb (5,897 to 81,647 kg). Actual harvest in both fisheries was likely considerably higher during population booms due to illegal commercial fish sales and excessive overlimits in the recreational fishery. The magnitude of illegal harvest is not known, but fishery managers deem that it was at least equal to reported harvest (4–5 lb/surface acre [4.5–5.6 kg/surface ha). Retrospectively, that level of reported harvest would not be considered excessive under our current harvest management framework, suggesting that undocumented harvest was indeed substantial.

Walleye abundance remained exceptionally low through the 1990s prior to recovery efforts (Figure 3). In response to the collapse of the Walleye population, discussions between management agencies began in 1996. The MNDNR commissioner and Band chairman authorized the formation of an interagency Red Lake Task Force to develop a recovery plan. This facilitated the sharing of biological data between agencies and inclusion of federal agencies, academia, and the Red Lake Commercial Fisheries Association into fishery management discussions. Once details of a recovery plan were agreed upon (see below), a memorandum of understanding (MOU) was signed in 1999 among the Red Lake Band of Chippewa, state of Minnesota, and Bureau of Indian Affairs. The MOU provided a formal commitment to the recovery plan and formalized the Red Lake Fisheries Technical Committee (RLFTC) as the body to review and assess fish stock status and work to fully rehabilitate the Walleye population. The RLFTC, which remains functional at the time of this writing, includes representatives from the Red Lake Band of Chippewa (Red Lake Department of Natural Resources), state of Minnesota (MNDNR), Bureau of Indian Affairs, Department of the Interior (U.S. Fish and Wildlife Service), University of Minnesota, and Red Lake Fisheries (the commercial fishing operation).

THE RECOVERY PLAN

The RLFTC was committed to implementation of a plan that would recover the Walleye population as quickly as possible with the highest likelihood of success. An aggressive approach was favored due to the dire cultural, economic, and recreational consequences of the collapse. The Recovery Plan for Red Lakes Walleye Stocks (hereafter, "Recovery

Plan"; MOU 1999) outlined a three-pronged approach to population rehabilitation. The primary components were (1) a complete Walleye fishing moratorium with strict enforcement, (2) a rigorous, standardized population assessment program that shared data across jurisdictions to ensure science-based management of the Walleye fishery, and (3) an intense, short-term Walleye fry stocking program.

A complete Walleye fishing moratorium was a drastic measure but identified as the best course of action for rapid recovery. Further, a complete closure of all legal Walleye harvest and possession would help authorities identify illegal activity. The success of the closure was further enhanced by initiating communications between law enforcement divisions in each jurisdiction to identify and curtail illegal Walleye harvest.

Although each jurisdiction already had fish population assessment programs in place, sampling efforts were standardized between the jurisdictions for the first time. This provided a clearer picture of the status of the population as a whole and increased the power of indices such as year-class strength and spawning stock biomass (SSB). Assessment data from each jurisdiction were compiled in a shared database and used to produce the jointly prepared annual technical report for the RLFTC.

The third aspect of the Recovery Plan prescribed a short-term, intensive restoration stocking program. This aspect was not initially a preferred restoration option within the committee. Doubts existed that supplemental stocking could be effective. Conventional thought was that fry production was generally not a limiting factor in large, natural Walleye systems. For example, MNDNR's Lake Management Planning Guide (MNDNR 1982) stated that Walleye stocking "is generally unnecessary and not practical in natural walleye lakes over 15,000 acres [6,070 ha]." Concerns existed that if stocking was used, any restoration success could be falsely attributed to stocking efforts, regardless of actual contribution. Managers were also well aware of the social popularity of stocking and the difficulty in curtailing stocking programs, even when ineffective or no longer necessary. The potential for precedence setting for other large, natural Walleye lakes was also a concern. However, objective analysis of the shared population data guided the eventual decision to include stocking as part of the Recovery Plan. This decision leaned heavily on the use of a gill-net selectivity model developed for estimating total population from the experimental gill-net catch and size distribution (Anderson 1998) that was adapted to provide estimates of total female SSB for the system. Using fecundity estimates and a range of potential hatch rates from the literature, a series of wild fry densities was predicted. This exercise was informative and provided context as to how little reproductive potential remained in the population. Even using optimistic hatch rates, wild fry densities predicted were low, which suggested a high probability that stocking could be effective at enhancing Walleye recruitment and speeding Walleye recovery. Consensus opinion on the RLFTC was that the remaining spawners would be sufficient for eventual recovery, but a judicious supplemental stocking program would likely hasten the process.

Despite the conclusion that stocking would likely expedite recovery, committee consensus was only achieved by clearly defining protocols and establishing limits on the duration of the stocking. Fry stocking was guided by strict a priori criteria that were established to maximize the likelihood of success while avoiding long-term maintenance stocking. The target fry density (1,000 fry/littoral acre [2,471 fry/littoral ha]) was aggressive, but it was imperative to stock a quantity capable of making a difference rather than proceed with a superficial stocking effort. First, managers had to identify a Walleye egg source that could supply enough eggs and that was a close genetic match. Second, the Recovery Plan dictated that fry would only be stocked in years when natural fry production was projected to be less than 500 fry/littoral acre (1,236 fry/littoral ha) to avoid density-dependent effects. In addition, if the previous year's stocking was determined to have produced a strong year-class, subsequent stocking events were delayed 1 year due to anticipated suppression effects from that cohort (Li et al. 1996). The Recovery Plan also defined criteria that would result in the permanent cessation of stocking: wild fry production in excess of 1,000 fry/littoral acre (2,471 fry/littoral ha) or five stocking attempts that failed to produce a strong year-class.

Most importantly, the Recovery Plan called for an evaluation program to assess the effects of stocking on the recovery process. All stocked fry were marked with oxytetracycline to facilitate the identification of stocked fish (see Logsdon 2006; Logsdon et al. 2016). The original intent of fry marking was to determine if, and to what extent, stocked fry were contributing to recovery efforts. Although useful to illustrate the role of stocked fry in recovery, using mark–recapture methods to estimate wild fry production and relate wild fry densities to various levels of SSB turned out to be even more valuable. Quantifying how many wild fry were produced by a given level of SSB allowed the committee to use SSB, a routinely measured parameter, to define population recovery because the committee was able to determine the range of fry densities known to produce strong year-classes. After recovery, this relationship was used to develop a stock–recruitment relationship that identified optimal levels of SSB, which became the basis for the initial "Harvest Plan for Red Lakes Walleye Stocks" (Red Lakes Fisheries Technical Committee, 2006 "Harvest Plan for Red Lakes Walleye Stocks," unpublished; hereafter "Harvest Plan") and subsequent revision (Red Lakes Fisheries Technical Committee, 2015 "Harvest Plan for Red Lakes Walleye Stocks," unpublished).

IMPLEMENTING THE RECOVERY PLAN

Although the signing of an MOU in 1999 formalized the implementation of the Recovery Plan, some recovery steps were initiated prior to signing. Commercial fishing was closed in 1997, limiting harvest on Band waters to subsistence only (i.e., for personal use). In 1998, Minnesota reduced the sportfishing bag limit from six to two fish and the Band placed a complete harvest moratorium on Walleyes from reservation waters. Commercial fishing had already reached a point where the processing plant could no longer operate and angling pressure on state waters was nearly nonexistent. But these initial steps were important in setting a tone of commitment and cooperation, and a gradual tightening of harvest regulations early may have helped soften the social and cultural impact of the eventual total Walleye fishing moratorium implemented on state waters in 1999.

Concurrent with implementation of steps necessary for biological recovery, the technical committee was well aware of the need for public support for recovery efforts to be successful. Considerable effort was directed at public involvement early in the planning stages and frequently throughout the implementation process. Without public support, managers knew the implementation of restrictive fishing regulations, particularly the complete Walleye fishing moratorium, would be difficult.

The Band had numerous public meetings to educate the fishermen and the public on why the Walleye population was collapsing. Many Band members denied this assertion early on, claiming that the lakes were too large to be fished out and the fishery had always bounced back in the past. However, when they could no longer catch Yellow Perch, the species that commercial fishers had been targeting since the Walleye population had crashed, the realization hit that the fishery was in trouble. The membership of the Red Lake Fisheries Association voted to suspend commercial fishing in April 1997. The commercial fishers voluntarily gave up their livelihood without receiving any financial assistance, an indication of how serious they were about recovering the Walleye population.

The state of Minnesota formed a Citizen Advisory Committee (CAC) made up of numerous local organizations and business interests but also including broader regional and statewide fishing and tourism interests. The focus of those early CAC meetings was to provide information to stakeholders, receive their feedback, and gain support for all aspects of the Recovery Plan. For example, recovery stocking could not proceed without anglers and business interests accepting a harvest closure and increased enforcement efforts. Public support was vital to attain these objectives. Although CAC members did not have official capacity on the RLFTC, they were invited to attend and participate in meetings. This helped assure transparency and build trust in the RLFTC process.

Once the decision to stock Walleye fry was made, the CAC reviewed several potential egg sources for the recovery stockings. The MNDNR Walleye egg take operation on the Pike River, a tributary to Lake Vermilion (Figure 1),was selected as the desired source because of the capability of providing a large-enough quantity of eggs, as well as the genetic similarity of the two populations. Representatives from Lake Vermilion area associations were included on the CAC, recognizing the importance of this contribution to the recovery effort. They needed to be assured that being a donor lake for this effort would not affect the Lake Vermilion Walleye population. Concerns about this were addressed through MNDNR's policy of returning 10% of the number of eggs taken from Walleye spawning sites to the source lake as newly hatched fry. Thus, stocking densities for Lake Vermilion increased in years when the Red Lakes were stocked because the number of eggs needed for the recovery stockings were greater than the typical quota from this egg source.

Each fry stocking event was intended to stock at a rate so that total fry density (projected wild + stocked fry) was near 1,000 fry/littoral acre (2,471 fry/littoral ha), a standard fry stocking density prescribed by MNDNR (1995). However, the total Walleye fry density achieved from the initial 1999 stocking was only 608 fry/littoral acre (1,502 fry/littoral ha). This level was partially due to fewer eggs from the donor system than proposed, but more so due to the considerably lower wild fry production than predicted (Table 1). One of the most striking realizations in 1999 was the low hatch rate (0.54%). The preliminary estimate of SSB density identified in the recovery plan (1.5 lb/surface acre [1.68 lb/surface ha]) was based on the amount of egg production (25,000 eggs/lb [55,125 eggs/kg]) using a 2% hatch rate to achieve 1,000 fry/littoral acre (2,471 fry/littoral ha). Two percent was thought to be a conservative value from the range of estimates (0.5–13%) available in the literature. Admittedly, the measure of hatch rate (fry produced/SSB egg potential) is a measure of natality and not the percent of deposited eggs that hatch, but regardless, wild fry production was orders of magnitude lower than expected. Despite not achieving the target fry density, shoreline seining catch rates in July 1999 were the highest observed since standardized sampling began in 1989 (Figure 4). Thus, the 1999 year-class was deemed to be strong and no stocking was programmed for the following spring.

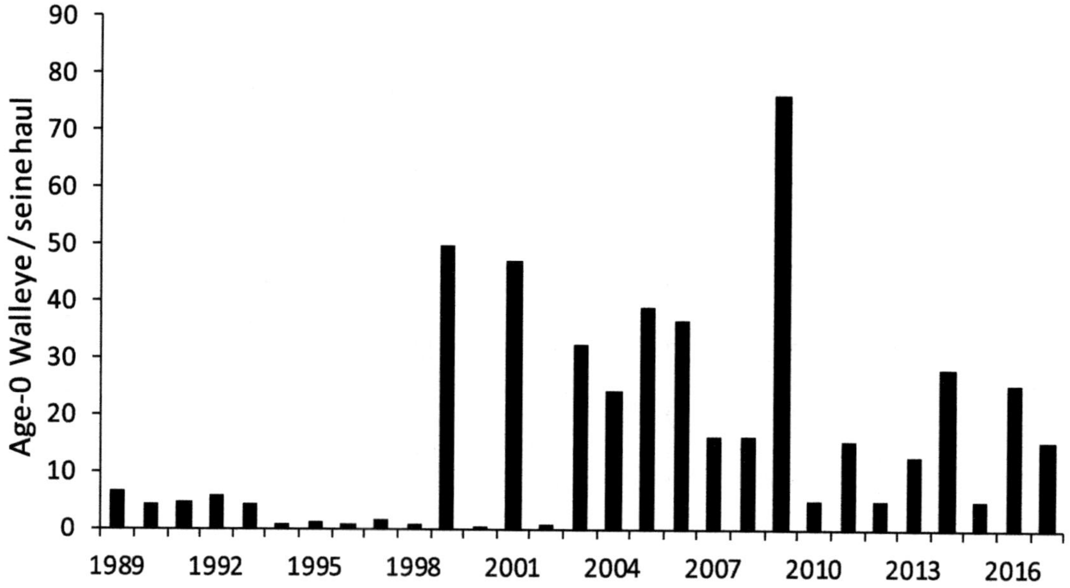

FIGURE 4. Catch per unit effort of age-0 Walleyes in shoreline seine hauls in the Red Lakes, 1989–2017.

Early indications were that the 2000 year-class, which was entirely naturally produced, was very weak (seine CPUE = 0.2 age-0 Walleyes/haul). This observation was not unexpected as SSB, and thus wild fry projections using revised hatch rates, remained low (0.33 lb SSB/surface acre = 62 fry/littoral acre [0.37 kg/surface ha = 153 fry/littoral ha]). The direct observation of a poor year-class produced at low SSB in 2000, in contrast to the exceptional year-class produced in 1999 through fry stocking with even lower SSB (0.17 lb/surface acre [0.19 kg/surface ha]), further confirmed that the Red Lakes were indeed spawner-limited and rehabilitation stockings were contributing to recovery efforts.

After a second stocking event in 2001, total Walleye fry density (575 fry/littoral acre [1,421 fry/littoral ha]) was similar to 1999 (Table 1), and seining results (Figure 4) suggested that the RLFTC should expect another strong year-class. Thus, the third stocking event was delayed a year and the 2003 stocking (425 fry/littoral acre [1.050 fry/littoral ha]) produced yet another strong year-class. It should be noted that the eventual year-class strength values for these cohorts were not known at the time, but rather projected based on shoreline seining CPUE. The RLFTC was confident that these catch rates were indicative of strong year-classes because shoreline seining catch rates for all three stocked years were at least five times higher than the previous record.

By 2004, a large proportion of female Walleye from the 1999 year-class had reached sexual maturity at age 4, resulting in a surge in SSB. Using the revised mean hatch rate from three previous wild fry estimates, predicted wild fry for 2004 exceeded the Recovery Plan threshold (1,000 fry/littoral acre [2.471 fry/littoral ha]) where stocking would be discontinued (i.e., "restoration has sufficiently progressed and stocking was no longer needed" [MOU 1999]). While this prediction marked a significant milestone in the recovery effort, the RLFTC recognized a unique opportunity to collect an additional mark–recapture fry estimate, this time when wild fry densities were projected to be high

TABLE 1. Walleye fry abundance and density in the Red Lakes and corresponding year-class strength rank during the period immediately before recovery efforts began until immediately after fry stocking experiments were completed. Percentile ranks are for the period 1987 to 2015.

Year	Stocked fry (millions)	Age-0 *N*	Marked %	Total fry (millions)	Wild fry (millions)	Total fry density (fry/littoral acre)	Year-class strength percentile
1998	0			0.7	0.7	9	29
1999	41.1	360	86	47.9	6.8	608	96
2000	0			4.9	4.9	62	36
2001	31.5	730	70	45.3	13.8	575	93
2002	0			10.8	10.8	137	43
2003	32.6	304	97	33.5	0.9	425	68
2004	10.0	766	9	114.5	104.5	1,452	50
2005	3.8	460	13	26.7	22.9	339	86
2006	0			54.0	54.0	685	79

(1,325 fry/littoral acre [3,274 fry/littoral ha]). Additional estimates would improve our understanding of the SSB/wild fry relationship and could confirm if our preliminary recovery target of SSB exceeding 1.5 lb/surface acre (1.68 ka/surface ha) was, in fact, adequate for sustainable levels of fry production.

Despite initial concerns during the planning stage about using stocking as a recovery tool, RLFTC members agreed to deviate from the recovery plan guidelines. Recovery-level stocking events would cease as prescribed, but two additional, low-density evaluation stockings to facilitate wild fry estimates would proceed. Thus, a relatively small number of marked fry were stocked in 2004 (10 million fry; 127 fry/littoral acre [314 fry/littoral ha]) and in 2005 (3.8 million fry; 48 fry/littoral acre [119 fry/littoral ha]). These evaluation stockings proved to be invaluable to developing a better understanding of wild fry production and hatch rates across a range of SSB densities. This relationship between wild fry production based on SSB and subsequent year-class strength was used to identify the optimal amount of spawners to manage for, and guide, overall population management after recovery. This history highlights the need for guiding documents to be well conceived but flexible, to allow for mangers to adapt as new information emerges.

The Recovery Plan specified that harvest should not resume until SSB exceeded 1.5 lb/surface acre (1.68 kg/surface ha) for three consecutive years. Spawning stock biomass had remained low during the first few years of recovery efforts, but with the onset of sexual maturation of the 1999 year-class, it jumped from 0.74 lb/surface acre (0.83 kg/surface ha) in 2002 to 8.29 lb/surface acre (9.29 kg/surface ha) in 2003. Although the Recovery Plan never assigned a timeline, the RLFTC had anticipated that it would take at least 10 years to recover SSB to levels that would allow Walleye harvest to resume. However, the 2003 SSB estimate put the technical committee on alert that Walleye harvest may return sooner than expected. The RLFTC began to develop the Harvest Plan, which outlined the annual harvest management framework that would be adjusted based on key indicators of population sustainability.

THE 2006 HARVEST PLAN

The initial "Harvest Plan for Red Lakes Walleye Stocks" was drafted by a subcommittee of the RLFTC prior to the reopening of the Walleye fishery in May 2006. Given that the collapse of the Red Lakes' Walleye population was due to overharvest, particularly overharvest of females prior to their reaching sexual maturity, it was logical that the initial Harvest Plan would focus on maintaining adequate levels of SSB to ensure population sustainability. Spawning stock biomass estimates that would provide fry densities of 400–600 fry/littoral acre (988–1,483 fry/littoral ha) were identified as the optimal SSB condition (2–3 lb/surface acre [2.24–3.36 kg/surface ha]) because this range of fry densities produced strong year-classes during recovery efforts (Logsdon 2006). Optimal SSB differed from the original Recovery Plan goal (1.5 lb/acre) because although empirical hatch rates were lower than expected, so were optimal fry densities. In addition to ensuring maintenance of enough spawning stock, the plan also advised that SSB be age-diverse to reduce the dependence on a single cohort providing a preponderance of the reproductive potential, as was the case during the boom-and-bust cycles prior to the collapse. This objective would be met when the Shannon's diversity index value for the cohorts in the spawning stock was greater than 0.6 (Figure 5; Gangl and Pereira 2003).

The RLFTC knew that careful harvest management would be critical to avoid mistakes from the past but wanted to avoid managing with hard quotas. Quota management can result in closing a thriving fishery when fishing is good or "chasing quota" when fishing is slow. The preferred approach was to define a range of harvests thought to be sustainable and to set fishing regulations to target the center of that range. Target harvest ranges were determined using 18 empirical predictors of fish yield similar to methods described in MNDNR Special Publication 151 (MNDNR 1997). Predicted yields varied considerably between models, so the interquartile range was used to define a target harvest range for each basin. Predicted yields were about 25% lower for the deeper, less-productive lower lake than for the upper lake, but the RLFTC agreed that a common target harvest for both basins combined would be the best approach for managing a shared resource so that allocations to each jurisdiction were equal on a per-acre basis. The values for the Upper Red Lake and Lower Red Lake were averaged and a sustainable harvest range of 1.75–3.5 lb/surface acre (1.96–3.92 kg/surface ha) was established.

Given the recent history of overharvest and variability in model predictions, the initial Harvest Plan tended toward conservative harvest recommendations. In general, target harvest ranges for each SSB condition increased with increasing mature female biomass, except that harvest was the same for optimal and surplus conditions due to the uncertainty of managing the Walleye population in this condition. When the plan was developed, SSB was anticipated to possibly reach surplus levels during the recovery due to artificial spikes in recruitment because of stocking efforts. This condition also could occur once the population stabilized if regulations were overly successful at protecting SSB or when new, strong year-classes first reach sexual maturity. However, these events were believed likely to be ephemeral in nature and last only a year or two before SSB returned to the optimal condition.

The Harvest Plan created a tiered harvest management framework where target harvest ranges were adjusted for different SSB conditions: closed, marginal, optimal, and surplus (Table 2). The SSB condition was determined based on the fall gill-net assessment conducted immediately prior to the start of the harvest year (December 1 to November 30). The closed SSB condition occurred when SSB was low (<1 lb/surface acre [2.47 kg/surface ha]) and allowed no harvest. The marginal SSB condition (1–2 lb SSB/surface

(A)

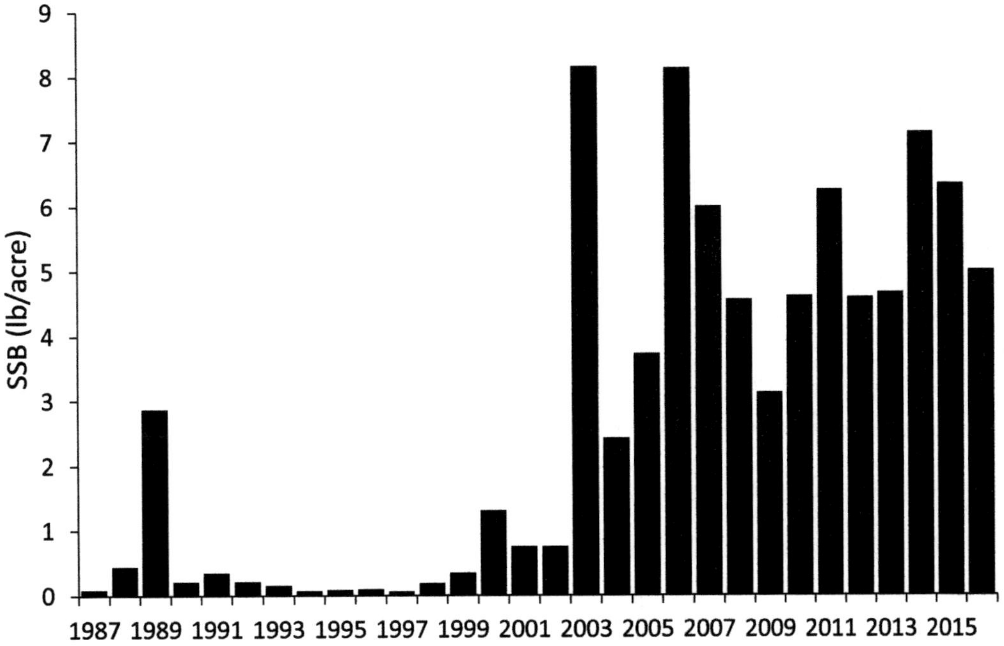

(B)

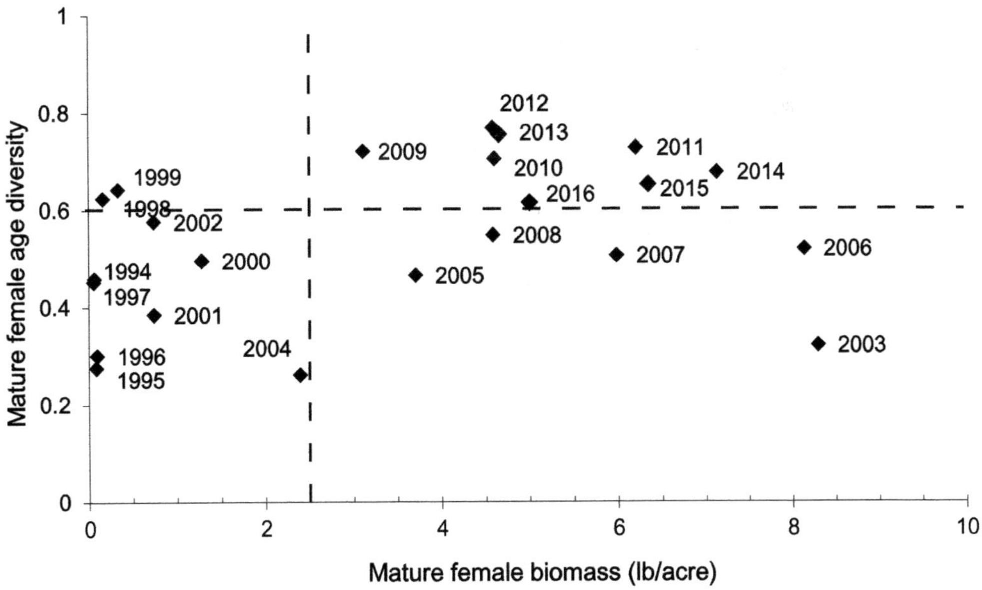

FIGURE 5. **(A)** Spawning stock biomass (SSB) for the Red Lakes, 1987–2017. **(B)** Spawning stock biomass plotted against spawner age diversity (Shannon's diversity index of the number of sexually mature cohorts). Data points in the upper right quadrant fulfill both recovery objectives, thus indicating full recovery.

TABLE 2. Definitions of spawning stock biomass (SSB) condition and associated harvest ranges outlined in the original (2006) and revised (2015) harvest plans for Red Lakes Walleye Stocks documents.

SSB lb/acre	SSB condition	Harvest zones (lb/acre)			
		Opportunity	Target	Caution	Cap
		Original harvest plan (2006)			
>3	Surplus	0–1.75	1.75–3.5	3.5–5.0	5.0
2–3	Optimal	0–1.75	1.75–3.5	3.5–4.5	4.5
1–2	Marginal	0	0–2.0	2.0–2.5	2.5
<1	Closed	0	0	0	0
		Revised harvest plan (2015)			
>4.5	Surplus	0–5.0	5.0–7.0	7.0–8.5	8.5
2.5–4.5	Optimal	0–2.5	2.5–5.0	5.0–6.0	6.0
1–2.5	Marginal	0	0–2.5	2.5–3.0	3.0
<1	Closed	0	0	0	0

acre [1.12–2.24 kg/surface ha]) allowed a modest level of harvest but less than levels allowed when SSB was in the optimal condition. The target harvest range for the optimal SSB condition (2–3 lb SSB/surface acre [2.24–3.36 kg/surface ha]) was set equal to the sustainable harvest range identified by the modeling exercise. Target harvest ranges for the surplus SSB condition (>3 lb SSB/surface acre [3.36 kg/surface ha]) were set equal to the optimal condition. This exemplifies the conservative approach taken by the initial Harvest Plan due to uncertainty about how the population would respond to, or how long it would remain in, surplus condition.

In addition to the target harvest ranges for each SSB condition, the plan included ranges to direct management actions for actual harvest outside the target range. Annual harvest was categorized into one of the of the following harvest ranges: opportunity, target, caution, or cap (Table 2). Ideally, Walleye harvest would fall within the target range annually. However, the RLFTC understood the inherent variability in Walleye populations, fishing pressure, vulnerability to angling, and other factors that determine annual harvest, and thus, the Harvest Plan required staying within the target harvest range on a 3-year running average.

Annual harvest within the opportunity range would indicate that an opportunity may exist to increase harvest (i.e., increase effort, relax regulations) the following year. Harvest within the caution range was permissible within a given year without closing the fishery but required adjustments to reduce harvest the following year. However, if Walleye harvest reached the harvest cap at any point during the harvest year, the Walleye fishery would be closed immediately. Reaching the harvest cap had similar consequences to reaching a quota (i.e., closure of the fishery), though there was no incentive to try to achieve it because harvest in subsequent years would have to be reduced to maintain the 3-year running average within the target harvest range.

STATUS OF THE WALLEYE POPULATION AND FISHERIES, 2006–2015

When Walleye harvest resumed in 2006, total Walleye biomass from the previous fall was nearly 22 lb/surface acre (24.7 kg/surface ha; Figure 3). Spawning stock biomass during the 2005 gill-net assessment was 3.7 lb/surface acre (4.15 kg/surface ha), which put

SSB in the surplus condition setting the target harvest range at 1.75–3.5 lb/surface acre (1.96–3.92 kg/surface ha) for the 2006 harvest year (Figure 5; Table 2). While harvest levels were dictated by the SSB condition, the Harvest Plan was clear that harvest methods and regulations needed to manage harvest within the target range were jurisdictional decisions. The fishery on state waters would continue to be a sport fishery, and the Band had always intended to return to a combination of subsistence and commercial fishing.

The Walleye population during the mid-2000s was still relatively young, with few fish older than age 7, but growth rates were fast, and fish from 14 to 20 in (356 to 508 mm) dominated the population. These fish were almost entirely recruits from the three recovery stockings, and angler catch rates were exceptional, exceeding 2 fish/angler-hour during peak time periods. In addition to the stocked cohorts, relative abundance in shoreline seine hauls and mark–recapture fry estimates indicated that natural reproduction in 2005 and 2006 was excellent and these fish would begin recruiting to the fishery in another 2 to 3 years (Figure 4; Table 1). The strong 2005 year-class, which was 86% wild origin, and the 2006 year-class, which was entirely naturally produced, confirmed that the Red Lakes Walleye population had recovered such that reproductive potential was high. The fishing moratorium was lifted prior to achieving sufficient age diversity in the spawning stock (Figure 5B) due to the presence of a number of strong, young (i.e., immature) year-classes in the population and a short-term (i.e., 3 year) spawning stock forecast. The presence of these immature year-classes that were poised to recruit to the SSB in the next few years gave the RLFTC confidence that the age-diversity objective would soon be reached and extending the closure any longer would result in continued, local economic recession for little conservation gain.

Fishing regulations set by each jurisdiction for the first year of fishing were conservative. State anglers were restricted by a two-fish bag limit with a 17- to 26-in (432 to 660 mm) protected slot limit (PSL). The conservative bag limit was necessary due to the uncertainty of how much fishing pressure would be generated by the highly anticipated return of a popular Walleye fishery and the expectation of high catch rates, as was observed during peaks in the boom-and-bust cycles. The PSL was designed to protect SSB since the Walleye population was now maintained entirely through natural reproduction.

The Band restricted Walleye harvest to hook and line only with modest bag limits for subsistence and commercial fishing. Subsistence anglers were allowed a 10-fish-daily bag limit with 30 fish allowed in possession, whereas commercial anglers could harvest 50 fish daily for sale to the Band's fish processing plant. Regulations for both commercial and subsistence anglers were regulated by an 18- to 28-in (457 to 711 mm) PSL. The commercial fishery was restructured after holding numerous public meetings and sending out a public survey with several options. The meetings and survey were an effort to gain public support for the new set of regulations. The majority of people wanted regulations that would allow everyone to prosper from the lake but, at the same time make sure that adequate protection was in place so that history would not repeat itself. A small, very vocal minority of ex-commercial fishermen wanted to go back to the old methods that allowed independent commercial gill netting. The tribal council knew that this approach would not be sustainable and enacted regulations to protect the Walleye population but also allow all Band members to benefit from the restored fishery.

With the return of an open Walleye season, each jurisdiction was obligated under the MOU to monitor Walleye harvest. An angler creel survey was conducted on state waters during the open Walleye season with harvest estimates produced biweekly to ensure that harvest did not exceed the harvest cap. The Band monitored Walleye harvest through creel surveys in the summer and winter and direct measurement of harvest at the fish processing plant. It was quickly realized that more than 85% of the harvest was being turned into the processing plant, which greatly assisted in the accuracy of the Band's harvest estimates.

Sportfishing regulations on state waters were incrementally relaxed during the first few years after Walleye harvest resumed. These regulations were adjusted to provide additional harvest opportunities and utilized a greater portion of the target harvest range. The 17- to 26-in (432 to 660 mm) PSL on state waters remained in use during this time, but the bag limit was adjusted several times (range, 2–4 fish), at times using different bag limits for the open-water and winter portions of the harvest year to manage harvest within the target harvest range. Managers met with the CAC semiannually to revisit fishery objectives and review fish population and creel survey information, as well as to provide regulation options for the upcoming season.

The state waters of Upper Red Lake were becoming well known for exceptionally high Walleye catch rates, particularly in May–June and December. Anglers quickly became acquainted with this pattern, and open-water fishing pressure and harvest were heavily concentrated in May and June. In response, an annual, mid-season PSL adjustment to 20–26 in (506–660 mm) was employed beginning June 15, 2009 to provide additional harvest opportunity and encourage fishing pressure during late season periods when angler effort was low. Though biologically insignificant, anglers also liked the idea that larger fish could be kept during summer periods when water temperatures were high and release mortality was highest (Reeves and Brusewitz 2007).

The harvest on Band waters ramped up slowly between 2006 and 2008, with a mean annual Walleye harvest of 179,500 lb (81,420 kg). To increase harvest, the Band relaxed regulations by adjusting the PSL to 20–28 in (508–711 mm) in 2007 and increasing daily bag limits for commercial anglers from 50 to 100 fish in 2009. By 2009, managers realized that supplemental net crews were going to be needed to further boost harvest levels. In addition, net crews were able to supply enough fish to keep the processing plant operating through the summer months as hook-and-line fishing declined. These changes proved to be successful, and mean Band harvest between 2009 and 2016 increased to 687,000 lb (311,618 kg) or 83% of the targeted harvest.

Though fishing resumed in 2006, the Walleye population was not declared fully recovered until 2009. Multiple year-classes established during recovery had now matured and the second recovery threshold, age diversity of the spawning stock, was achieved (Figure 5B). The RLFTC continued to transition their focus from population recovery to harvest management. Biologists communicated frequently to coordinate sampling between jurisdictions and share assessment and harvest data. Formal committee meetings were held twice each year to review annual harvest data, population status, and future harvest plans.

By 2011, Walleye population density was greater than 35 lb/surface acre (86.5 kg/surface ha; Figure 3) and comprised of fish from ages 0–12. Unexpectedly, SSB had remained in the surplus condition since 2005. This status was likely due to conservative regulations that overprotected mature females. The 2009 year-class was strong and entirely naturally produced, as were all cohorts since 2006. Fish greater than 20 in (229 mm) remained uncommon and growth rates had declined to a record low in 2009, likely due to density-dependent effects. Gill-net CPUE for Yellow Perch, a primary forage fish in the Red Lakes, also reached a record low as a result of intense predation by the thriving Walleye population. Further, the Yellow Perch population was composed almost entirely of fish that were larger than 9 in (229 mm) and likely too large to provide forage for most Walleye in the population. While these adult Yellow Perch were still producing substantial numbers of age-0 fish each year as evidenced by relative abundance in seining efforts, few survived the heavy predation pressure to recruit to the adult population (Figure 6).

Over the next few years, an increasing weight of evidence indicated that the Walleye population could not only sustain higher harvest than in the recent past, but that increased harvest may be beneficial. Managers reasoned that increased harvest would alleviate the negative effects of high population density on Walleye growth and provide predation relief for the highly pressured Yellow Perch population. Further evidence that harvest could be increased without risk of overharvest was provided by total annual morality estimates. Since reopening the fishery, total annual mortality for fish age 2 and older remained similar to levels observed during the harvest closure and considerably lower than during the collapse. While being overexploited, total annual mortality was 75–86%. In contrast, total annual mortality during the Walleye fishing moratorium ranged from 3% to 55% (mean = 39%) and from 3% to 52% after reopening the fishery (mean = 35%; Figure 7). The similarity in total annual mortality during the harvest closure and reopened fishery suggests that harvest mortality during the first several years after reopening the fishery was likely compensatory. In essence, harvesting fish was not additive, increasing total mortality, but rather removing fish from the population that would have otherwise experienced natural mortality.

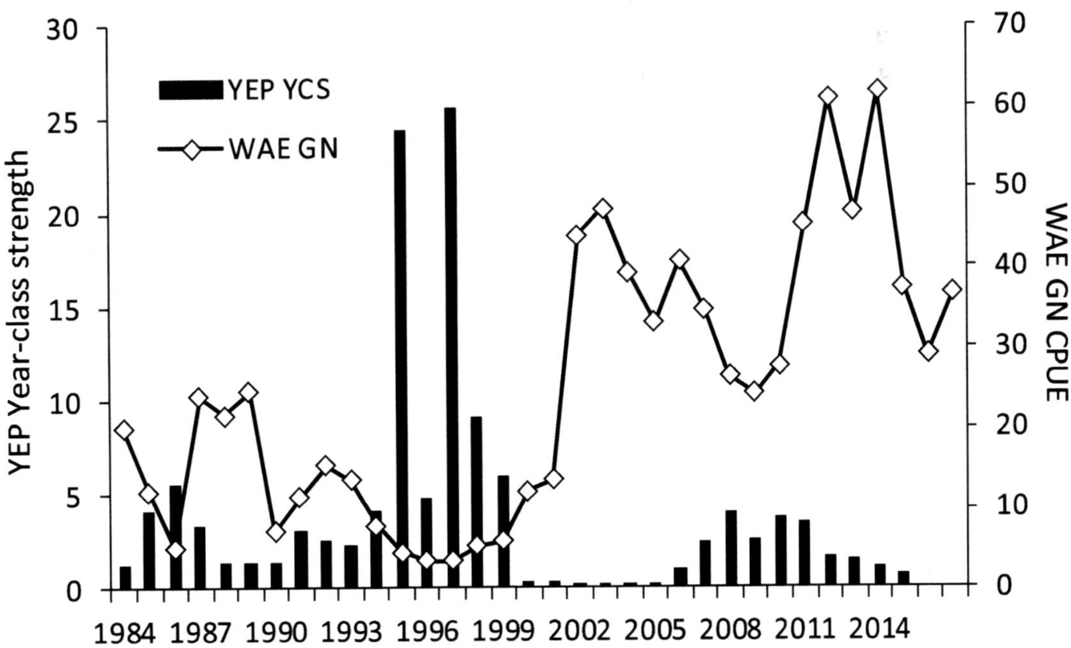

FIGURE 6. Yellow Perch year-class strength (YEP YCS) and Walleye relative abundance (WAE GN; number of fish per gill net) in state waters of Upper Red Lake.

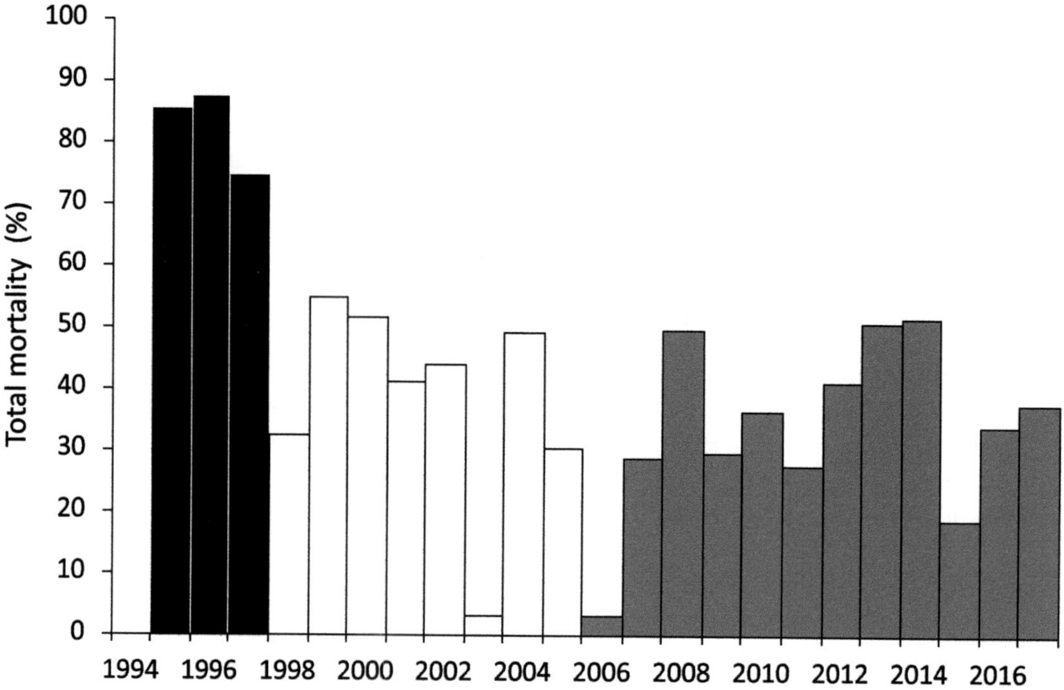

FIGURE 7. Total annual mortality of Walleyes age 2 and older in the Red Lakes, 1995 to 2017. Bars represent years during which the fishery was overexploited (black), closed (white), and reopened (gray) under the joint harvest plan.

Perhaps the most influential data that convinced the RLFTC that increasing harvest may be beneficial was examination of the stock–recruitment relationship. The stock–recruitment relationship was constructed using fry density in place of SSB to account for artificially high fry densities unrelated to SSB levels in stocked years. Fry densities for unstocked years were projected using the relationship established between SSB and wild fry production in stocked years. Year-class strength was based on gill-net catches of each cohort at ages 2–5. High fry densities (>850 fry/littoral acre [2,100 fry/littoral ha]) produced stronger year-classes than those observed at low fry densities but were markedly weaker than the extraordinary year classes produced at moderate densities (339–850 fry/littoral acre [838–2,100 fry/littoral ha]; Figure 8), likely due to density dependence.

Recognizing that reducing overall Walleye abundance and SSB could be useful, both jurisdictions attempted to increase harvest near the top of the target harvest range to observe the effects on the population (Figure 9). In 2013 and 2014, state managers employed a 20- to 26-in (508 to 660 mm) PSL, except during May and early June (i.e., postspawn) when large fish are concentrated and extremely vulnerable to angling. Despite the desire to remove some SSB, a more restrictive 17- to 26-in (432 to 660 mm) PSL was used from the beginning of the Walleye season in mid-May until June 14 to prevent yield from accumulating too rapidly and reaching the harvest cap prior to the end of the fishing season, which would close the Walleye fishery. Annual harvests in 2013 and 2014 would have certainly exceeded the target harvest range, and likely the harvest cap, if fish from 17 to 20 in (432 to 508 mm) were allowed in

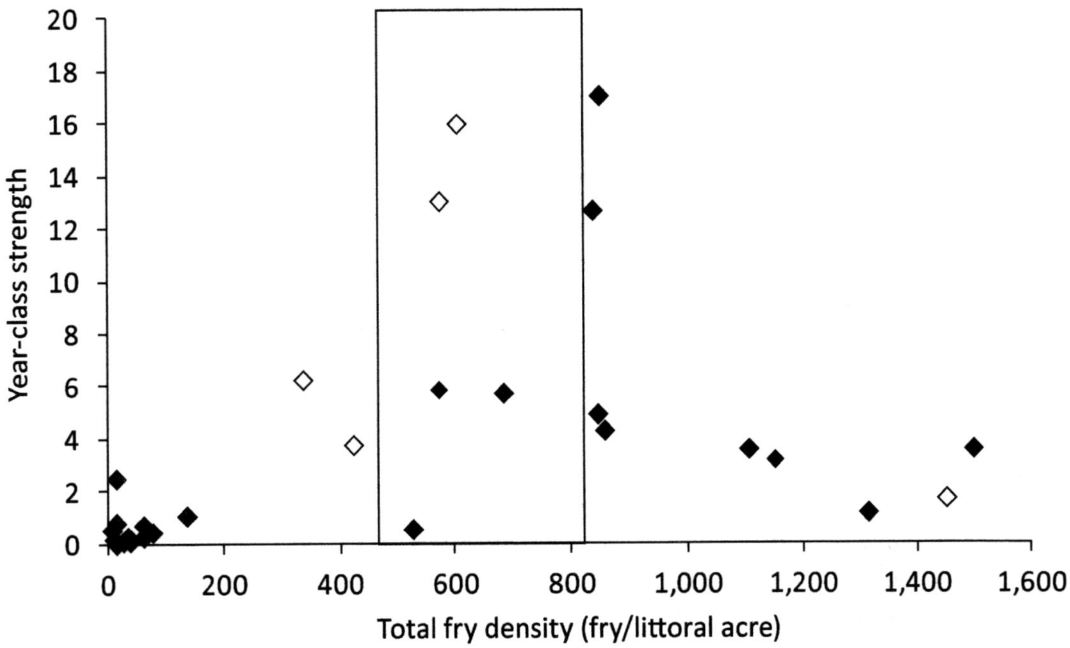

FIGURE 8. Stock–recruitment relationship for the Red Lakes using fry density in place of mature female biomass to account for fry stocking. Fry stocked years are represented by open symbols and are empirical estimates. Solid black symbols were projected based on the relationship between mature female biomass and wild fry density. The box shows the optimal fry density range outlined in the 2015 "Harvest Plan for Red Lakes Walleye Stocks" (Red Lake Fisheries Technical Committee, unpublished), which corresponds to 2.5–4.5 lb of spawning stock biomass per acre.

the state harvest at that time. The early season harvest would have triggered a mid-season Walleye fishing closure, an action the CAC had identified as something to be strongly avoided. The PSL was limiting harvest more effectively than the bag limit, making it impossible to reduce SSB while harvesting within the management framework outlined in the 2006 Harvest Plan.

The Band also attempted to increase harvest at this time to provide more harvest opportunity for subsistence and commercial anglers and to reduce SSB. The PSL on Reservation waters was adjusted to 22–28 in (559–711 mm) in 2014 to increase the vulnerability of mature females. The daily processing capacity of the fish plant limits harvest capacity during certain times of the year when angler success is high and the number of participants and total catch is high. When anglers are having difficulty supplying adequate numbers of Walleye to keep the fish plant operating efficiently, net crews are used to supplement the catch. Facility improvements were made to increase processing capacity to approximately 1 million lb (453,592 kg) of Walleye per year, and net crews were mobilized earlier in the summer and increased effort in an attempt to increase their annual harvest.

The strong 2009 year-class, with support from adjacent year-classes (Figure 10), was producing record-high catch rates, and fishing pressure on state waters of Upper Red Lake had increased to unprecedented levels (Figure 11). When the 2011 year-class, the strongest ever recorded on the Red Lakes, recruited to the fishery during the 2015 har-

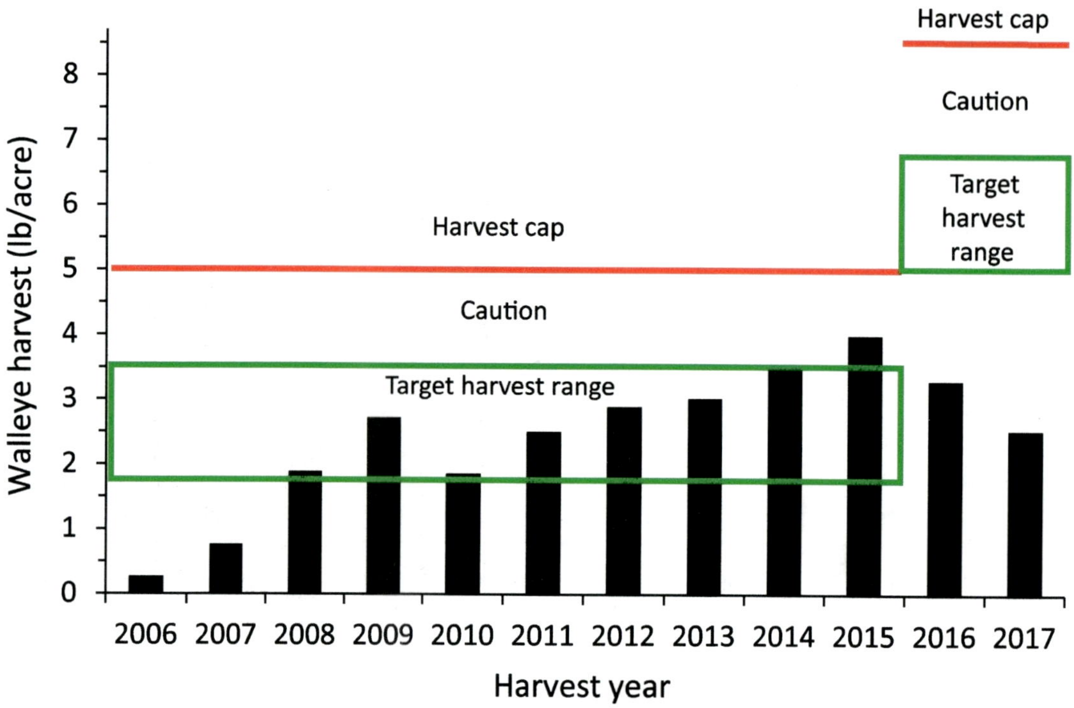

FIGURE 9. Walleye harvest for the Red Lakes in the context of the harvest management framework outlined in the 2006 and 2015 versions of the "Harvest Plan for Red Lakes Walleye Stocks" (Red Lake Fisheries Technical Committee, unpublished).

vest year, the bag limit on state waters had to be lowered twice and the PSL returned to 17–26 in (432–660 mm) to reduce yield and avoid reaching the harvest cap. Although this harvest reduction was unnecessary from a biological perspective, and perhaps even detrimental to the population given the evidence of negative density-dependent effects, managing within the Harvest Plan was imperative to preserve the integrity of the MOU. State fishery managers did not want to risk compromising the relationship between the state and Band that had taken years to build.

The RLFTC had developed and refined a forecast model to project SSB 3 years forward. Model output predicted that continued harvest at the maximum end of the target harvest range would not reduce SSB below the surplus condition. Realizing the need for increased flexibility in the Harvest Plan to allow for harvest of mature females (i.e., large fish, which disproportionately increase yield), the RLFTC tasked a subcommittee to revise the 2006 Harvest Plan. The revision incorporated the RLFTC's increased understanding of the Red Lakes' Walleye population, with particular emphasis on the stock–recruitment relationship, and the fishing pressure and harvest dynamics of the recreational, commercial, and subsistence fisheries. The RLFTC's forecast model was the primary tool used to adjust target harvest levels for the Harvest Plan revision. Retrospective analyses were conducted to compare SSB forecasts to empirical estimates. Once the model was tuned such that forecasts from previous years approximated observed values, modeled harvest levels were incrementally increased to determine the level of harvest necessary to reduce mature female biomass from the surplus to the optimal condition (Red Lakes Fisheries Techni-

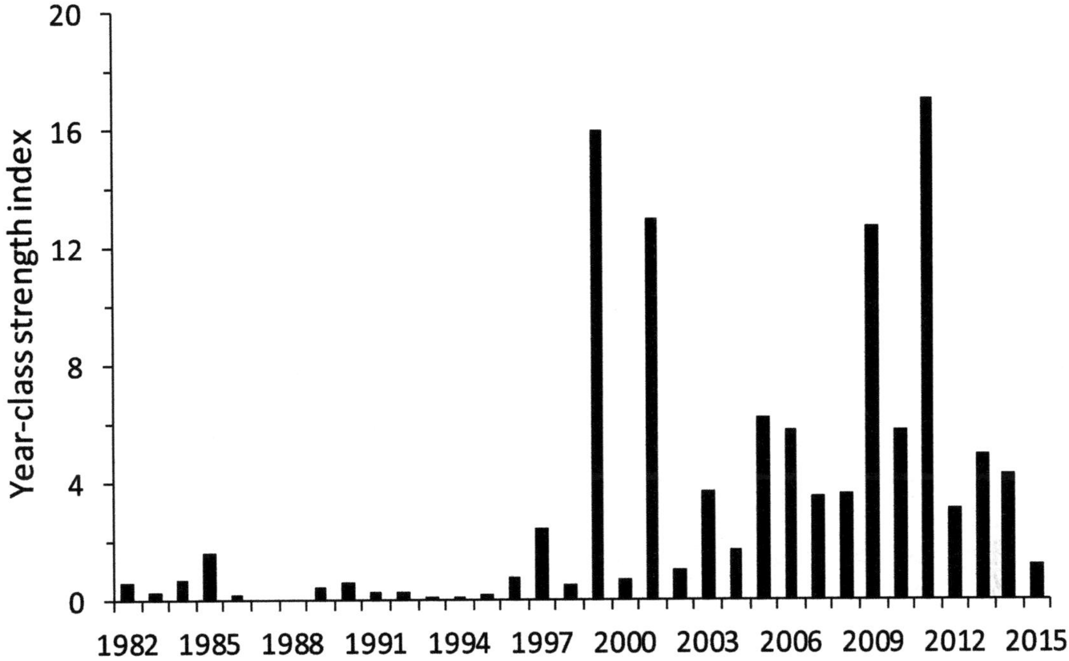

FIGURE 10. Year-class strength index values for Walleyes in the Red Lakes, 1982 to 2015 year-classes.

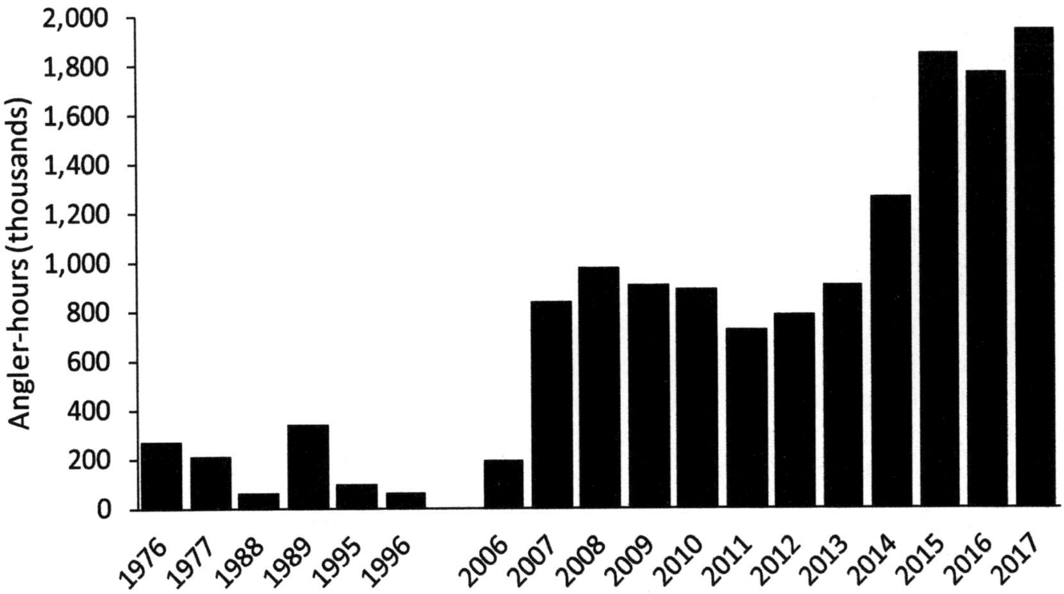

FIGURE 11. Total fishing pressure (angler-hours) expended on state waters of Upper Red Lake.

cal Committee, 2015 "Harvest Plan for Red Lakes Walleye Stocks," unpublished).

THE 2015 REVISION OF THE ORIGINAL HARVEST PLAN

The review of the 2006 Harvest Plan modified the boundaries of the optimal and surplus SSB conditions, adjusted the harvest management framework for each SSB condition, and recommended actively managing the population for the optimal SSB condition. Modifications to SSB conditions were based on the updated stock–recruitment relationship. The optimal condition in the 2006 Harvest Plan had been derived by bracketing the range of fry densities that produced strong year-classes during recovery efforts. Incorporating additional data points from 2006 to 2015 into the stock–recruitment curve greatly improved our understanding of the impact of fry densities on recruitment, particularly the negative effects at elevated (>850 fry/littoral acre [2,100 fry/littoral ha]) levels.

The addition of 10 data points since the original Harvest Plan was drafted showed that the stock–recruitment curve peaked at a broader and somewhat higher fry density range (339–850 fry/littoral acre [838–2,100 fry/littoral ha]) than those observed during the recovery stockings (425–608 fry/littoral acre [1,050–1,502 fry/littoral ha]). However, fry densities in excess of 850 fry/littoral acre (2,100 fry/littoral ha) were not merely surplus but rather negatively affected recruitment through density-dependent effects (Figure 8). Thus, the definition of the optimal SSB condition was adjusted to 2.5–4.5 lb/surface acre [2.8–5.0 kg/surface ha], corresponding to fry densities produced at the peak of the revised stock–recruitment curve (Table 2; Figure 8). A broader and higher optimal range of SSB than previously thought will make it somewhat easier to maintain SSB within this condition while reducing the risk of inadvertent declines into marginal condition.

The 2015 Harvest Plan advises that harvest should be increased when SSB is in the surplus condition to actively manage SSB towards the optimal condition to improve the likelihood of producing strong year-classes. While cohorts produced when SSB is in the surplus condition could sustain a Walleye population and modest fisheries, the probability of producing robust year-classes that provide high angler catch rates and a larger harvestable surplus is reduced. However, simply being in the optimal SSB condition does not guarantee the production of a strong year-class as multiple factors in addition to SSB and fry density contribute to year-class strength in any given year (e.g., environmental conditions during the spawning period [Smith and Krefting 1954; Johnson 1961;Busch et al. 1975], and strength of adjacent year-classes [Li et al. 1996]).

Actively managing for the optimal SSB condition rather than merely maintaining the SSB condition at optimal or surplus represents a subtle, yet significant change in management approach. Given what was known when the fishery reopened, the RLFTC was justified in being conservative when making harvest management decisions. Actively managing for the optimal condition under the revised Harvest Plan represents a more aggressive approach to harvest management but is balanced by the upward adjustment of the optimal condition to assure that SSB in surplus condition is truly in excess of what was needed to sustain recruitment.

Target harvest ranges and their associated caps were increased to provide the necessary capacity to remove large fish while avoiding the need to use a size restriction as the primary regulation to reduce yield. Model projections suggest that considerable removal of immature fish before they recruit to the spawning stock is necessary to moderate SSB from reaching surplus levels, particularly following the production of strong year-classes.

Allowing anglers to harvest large fish (i.e., fish greater than 17 in [432 mm]) was essential to enable active management of SSB in state waters. In addition to direct SSB management, harvesting fish from the entire length range of the population distributes the effects of harvest across multiple year-classes and states of maturity, rather than focusing harvest immediately below the protected slot, and is a management strategy that contributes to long-term sustainability of the fishery itself (Garcia et al. 2012). On reservation waters, lower angling pressure than on state waters and large bag limits reduce sorting of fish, which results in harvesting over a broad length range, including mature females. Size selectivity in the gill-net fishery can be managed by using different mesh sizes to target more or less SSB in the catch.

In addition to modifying SSB conditions and target harvest levels, the Harvest Plan acknowledged that the record-high overall abundance levels observed in the years immediately prior to the revision were not sustainable. These elevated densities had negative density-dependent effects on the Walleye population (e.g., slow growth, high natural mortality) and put tremendous strain on mid-sized (5–7 in [127–178 mm]) forage items, particularly Yellow Perch. As a result, the RLFTC understood that total population densities should be reduced, although the plan stopped short of identifying at what level total population densities should be maintained.

These Harvest Plan revisions represent a significant change in management philosophy that required continued public outreach through CAC meetings, newsletter articles, and news releases to maintain support. Communications focused on clearly defining population and fishery objectives and how management actions are intended to meet those objectives. A change in management direction does not mean that the previous approach was wrong, but rather that we learned along the way to restoration. Cautious conservative harvest was appropriate while the population was still recovering and gradually increasing harvest was the best way to observe how the recovered population would respond to various levels of exploitation. The inclusion of CAC and tribal council members at RLFTC meetings inspired trust and confidence in the technical committee process. Harvest Plan adjustments were approved by the RLFTC in September 2015 and represented an adaptive management approach where managers learned from what they did and implemented what they learned.

CURRENT STATUS OF THE WALLEYE POPULATION AND FISHERY

After reaching a peak of 45 lb/surface acre (111 kg/surface ha) in 2012, total Walleye biomass in 2016 was 28 lb/surface acre (69 kg/surface ha). This decline in total biomass was the result of increased harvest, high natural mortality, and modest recruitment. Spawning stock biomass peaked (7.1 lb/surface acre [8.0 kg/surface ha]) 2 years after the total biomass peak and has remained in the surplus condition (2016 SSB = 5.0 lb/surface acre [5.6 kg/surface ha]). This parameter may reach the optimal condition for the first time in fall of 2018 as a result of increased harvest and moderate recruitment since 2012.

Given their life history, characterized by surplus reproductive and recruitment potential, Walleye populations fluctuate even in a healthy, recovered state. As a result, we expect that Walleye abundance and SSB will continue to fluctuate with the natural dynamics of periodic production and maturation of exceptional year-classes. However, we also expect fishery performance even at low points in the population fluctuations to remain at acceptable, sustainable levels for each jurisdiction, rather than the historic boom-and-bust cycles prior to the collapse. These levels will be achieved through active management of SSB to ensure that the Red Lakes do not become spawner-limited again.

The age-frequency distributions from the past several annual gill-net assessments have included representatives from each cohort from ages 1 to 10, with signatures from two to three exceptional year-classes. Currently, the strong 2009 and 2011 year-classes dominate the population, and recruitment of cohorts after 2011 has been moderate. The moderate recruitment was likely the result of surplus spawning stock biomass (>4.5 lb/surface acre [5.0 kg/surface ha]) and elevated fry densities. We anticipate that once SSB moderates to optimal levels, strong year-classes will be produced when environmental conditions are favorable.

Although recovery efforts were initiated nearly 20 years ago, fish greater than 20 in (508 mm) remain uncommon and comprised only 1% of the gill-net catch in the Red Lakes in 2016. In general, female Walleyes begin to reach 20 in (508 mm) by age 5, but although the 2016 assessment sampled fish up to age 16, only one fish was larger than 24 in (610 mm). We hypothesize that limited availability of mid-sized forage has had negative effects on growth rates and increased mortality of large fish.

Immediately prior to the Harvest Plan revision, both jurisdictions were making a conscious effort to harvest near the top of the target harvest range due to concerns about high density. Now that target harvest ranges for both the optimal and surplus conditions are more aggressive, managers will again be targeting the midpoint of the target range. To increase harvest and include mature females as a portion of this harvest, different size restrictions were needed.

The 17- to 26-in (432 to 660 mm) PSL had been an effective tool for managing the sport fishery to limit harvest and protect SSB during the first several years of Walleye fishing while the population was still progressing towards full recovery. Once SSB reduction became a management objective, managers needed a regulation that would remove a portion of SSB without risking overharvest. Thus, the protected slot limit was replaced with a "one over" regulation, which allows one fish in the possession limit over 17 in (432 mm), beginning with the 2016 harvest year. This regulation allows anglers to harvest fish from the full length range of fish in the population and results in a more balanced harvest. The increased target harvest range for the surplus condition in the 2015 Harvest Plan can absorb the higher yields that result from harvesting larger fish. Managers can now regulate the magnitude of the harvest by adjusting the daily bag limit rather than focusing harvest on small fish simply to avoid accumulating yield.

The Band continues to make improvements to the fish processing plant to allow for greater processing capacity. Tribal anglers have successfully adapted from a history of gill-net use to hook-and-line fishing over the past 10 years and have become very proficient at harvesting fish in May–June and during late-ice conditions in March and April. The current PSL is between 22 and 28 in (559 and 711 mm) and the commercial angling daily bag limit has been increased to 150 in an attempt to increase hook-and-line harvest of Walleyes. Two to three commercial gill-net crews are deployed starting in early June and continue to fish into November each year. Netting effort is adjusted in response to the amount of fish being turned in by the anglers to better use the remaining daily processing capacity with the goal of harvesting approximately 1 million lb (453,592 kg) annually.

During the first 11 years of Walleye fishing after reopening in 2006, the sport fishery on state waters has generated more than 11 million angler-hours. Anglers have harvested approximately 1.4 million lb (635,029 kg) of Walleye from the eastern 48,000 acres (19,425 ha) of Upper Red Lake. The average annual fishing pressure and harvest over the past three seasons was 1.6 million angler-hours and 199,000 lb (90,265 kg), respectively. Fishing-related businesses and local economies are thriving as Upper Red Lake has regained its reputation as a premier destination Walleye fishery.

The Band's commercial fish processing plant currently employs between 35 and 50 full-time employees. This employment does not include the several hundred tribal anglers that turn in Walleye each year to help supplement their income. Commonly, the processing facility pays out more than US$1 million to tribal anglers in a single harvest season. This industry is not only important to the reservation, but also to the surrounding communities to the reservation where this money is spent for goods and services.

Band members using reservation waters and resorts and anglers on state waters have adopted a sense of pride and ownership in the joint recovery effort and restored Walleye fishery. Genuine commitment exists to protect this fishery and ensure that the mistakes made in the past are not repeated. The cooperative management effort has led to a greater understanding and acceptance of how each jurisdiction values and uses the resource.

LESSONS LEARNED

The Walleye population in the Red Lakes functions as a single population oblivious to jurisdictional boundaries. Early Red Lakes fisheries investigators had concluded this aspect by the early 1950s—decades before the formation of the RLFTC (Smith et al. 1952). However, this realization did not result in joint management of the Red Lakes' Walleye fishery until a crisis point was reached (Figure 12). Unfortunately, a lack of organized efforts to jointly manage interjurisdictional fisheries prior to population collapse is a common theme of fish management case histories (e.g., Richards and Rago 1999; Vandergoot et al. 2019, this volume). Successful management of a shared resource can only be achieved by a coordinated effort to reach common goals.

Garnering public support among the distinct user groups that share this resource was critical to successful restoration and long-term sustainability of the Red Lakes' Walleye population. Collapse of the Walleye fisheries occurred with plenty of potential flash points, including tribal netting, black markets, excessive overlimits, and boundary issues. Making progress toward recovery required that both jurisdictions moved beyond placing blame and invest considerable effort into establishing public trust in the Recovery Plan and a sense of ownership in the restored fishery.

Managers in both jurisdictions were successful in developing public trust and commitment to the Recovery and Harvest plans. However, the development of public trust in fisheries managers was a long-term endeavor and can be eroded by taking it for granted. No matter how secure the relationship between fisheries managers and stakeholders' seems, efforts to maintain and nurture these relationships must be ongoing. When the public accepts controversial management recommendations that you make as a manager, then you know that you still have the support of your stakeholders and that you have not taken their trust for granted.

Recovery and Harvest plans need to be prescriptive to clearly define future management actions based on current data, but also flexible to adapt as new information becomes available. Plans should include population and fishery objectives that are clearly defined, compatible, and developed using the best available information. However, it is imperative that ongoing population and harvest monitoring programs are in place to evaluate

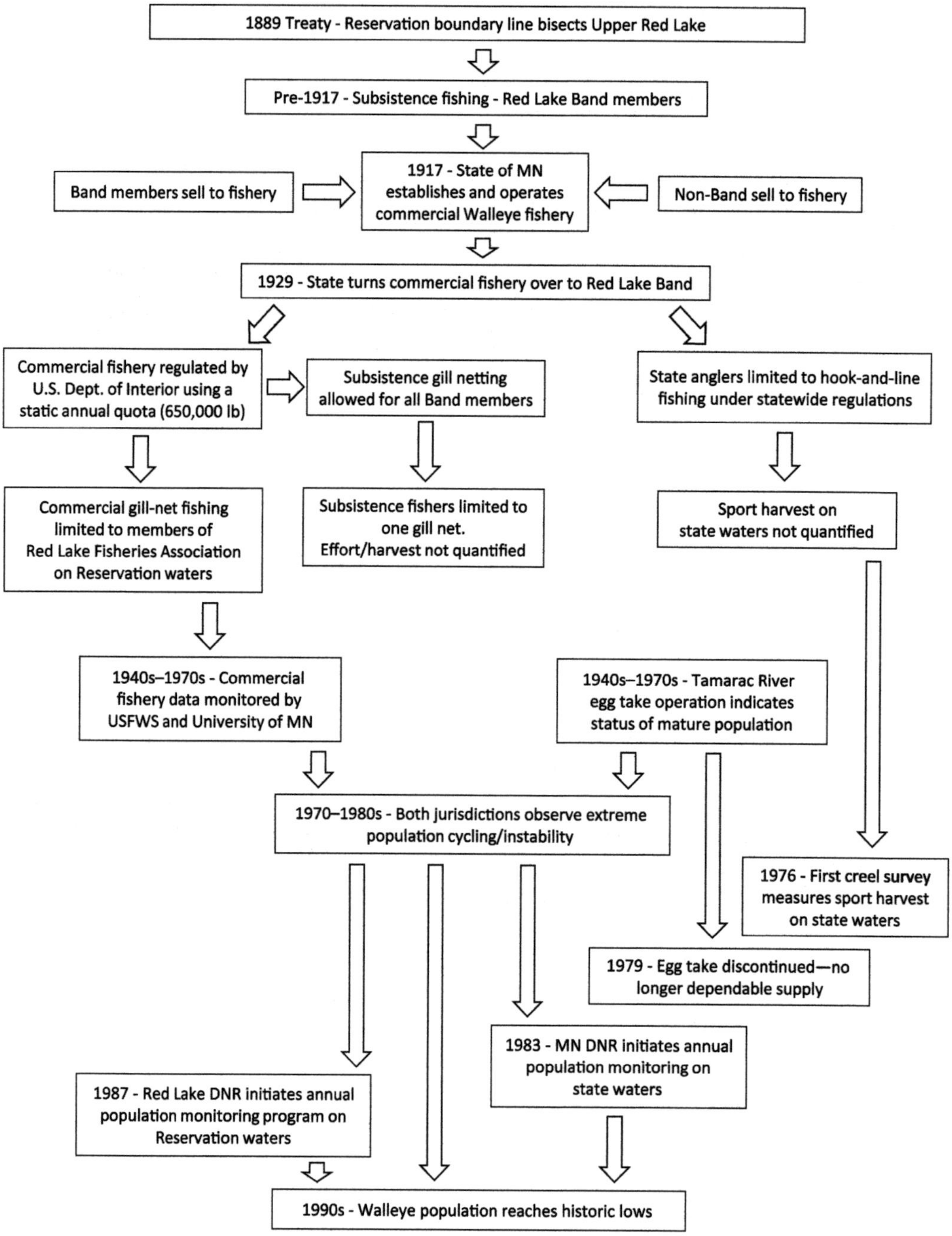

FIGURE 12. Flow chart and approximate timeline of fisheries management on the Red Lakes.

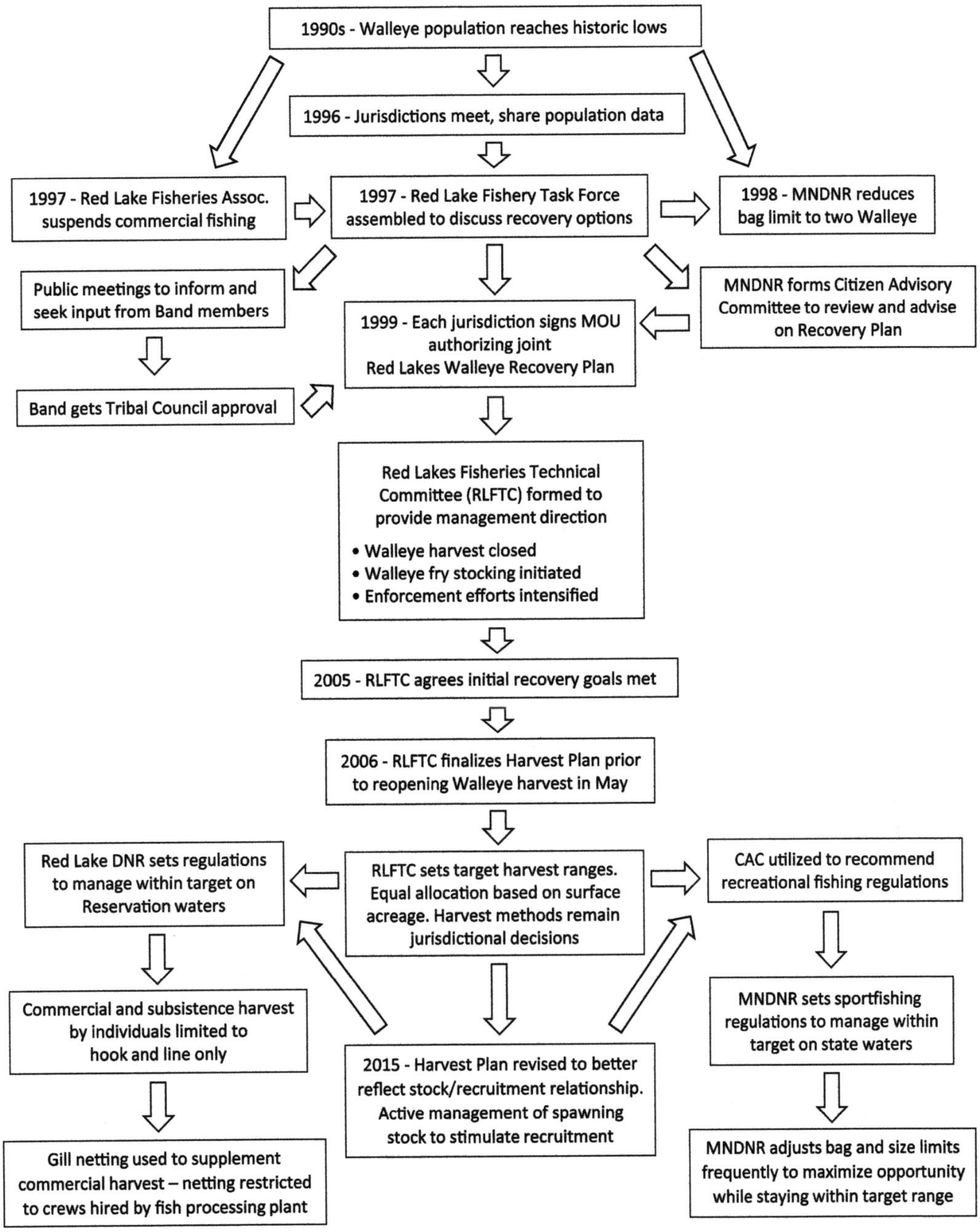

FIGURE 12. Continued.

management actions and that managing authorities are flexible and responsive to take necessary corrective measures.

Managers should not be constrained by conventional thought. Many of the assumptions regarding the effects of supplemental fry stocking of large Walleye lakes, natural hatch rates, and wild fry densities were proven wrong through a series of empirical estimates. The success of fry stocking in the Red Lakes does not suggest that stocking is the sole answer to recover Walleye populations. Recovery stocking was effective because spawning stock was limited. Similarly the idea that surplus SSB and fry production are simply extra (more than necessary but of little consequence) was challenged. Surplus SSB and fry production can have negative effects on recruitment, growth, and forage associated with excessive SSB and wild fry production; thus, supplemental stocking of systems where fry production is not the limiting factor may be counterproductive.

The collapse of the Red Lakes' Walleye fisheries was catastrophic, but it provided an incredible opportunity to increase our understanding of Walleye population dynamics and fishery management. Refining large-scale fry-marking techniques (see Logsdon et al. 2004) to estimate wild fry production was a big step forward in our understanding of the stock–recruitment relationship. Although these techniques have been used on other walleye fisheries (e.g., Zorn 2015; Fielder et al. 2019, this volume), only the Red Lakes have offered the chance to examine the effects of fry density on year-class strength from near zero to densities exceeding 1,500 fry/littoral acre (3,707 fry/littoral ha). The broad fry density range allowed us to identify the optimal fry density range to maximize the likelihood of producing strong year-classes. Fishery managers often struggle to fit stock–recruitment relationships to their data sets, but often the data are from a narrow range of SSB (i.e., fry densities), which limits the ability to observe the effects of density extremes.

Restoring a Walleye population to a naturally reproducing, self-sustaining state and managing a sustainable Walleye fishery with multiple diverse user groups are distinctly different challenges. Wild fry estimates confirmed that the collapse of the Red Lakes Walleye population was due to overharvest of spawning stock resulting in recruitment overfishing and ultimately inadequate fry production. Once spawning stock was restored beyond minimum levels needed for adequate fry production, population recovery was complete and sustainable, as long as those minimum levels could be maintained. The implementation of conservative fishing regulations that limit total yield and protect spawning stock is likely all that is needed to ensure that the population does not collapse again. Any recruitment suppression that results from excessive SSB and wild fry production would not jeopardize the sustainability of the population because less recruitment is needed to replace modest levels of mortality. However, the expectations for the Walleye fisheries on the Red Lakes are far more complex. Reservation waters have an extensive history of robust commercial and subsistence fisheries that utilize the high productivity of this system. The sport fishery on state waters had a reputation for high angler catch rates and was a harvest-oriented fishery. Upper Red Lake resorts catered to fishing, with Walleyes targeted by 97% of anglers. Low to moderate harvest levels under conservative fishing regulations are not conducive to high recruitment and strong year-classes that drive catch rates and yield potential. Managing for the optimal SSB condition and intensively harvesting expendable surplus is the best way to optimize yield and, in turn, meet both population and fishery objectives for the Red Lakes.

REFERENCES

Anderson, C. S. 1998. Partitioning of total selectivity of gill nets for Walleye into encounter, contact and retention components. Canadian Journal of Fisheries and Aquatic Sciences 55:1854–1863.

Brown, W. P., and A. L. Thompson. 2004. Red Lakes Walleye Recovery Project annual report to the Red Lakes Fisheries Technical Committee. Available from Minnesota Department of Natural Resources, Bemidji Area Fisheries Office, Bemidji.

Busch, W. D. N., R. L. Scholl, and W. L. Hartman. 1975. Environmental factors affecting the strength of Walleye year-classes in western Lake Erie, 1960–70. Journal of the Fisheries Research Board of Canada 32:1733–1743.

Fielder, D. G., and J. P. Baker. 2019. Recovery of Saginaw Bay Walleye, Lake Huron. Pages 411–430 *in* C. C. Krueger, W. W. Taylor, and S. Youn, editors. From catastrophe to recovery: stories of fish management success. American Fisheries Society, Bethesda, Maryland.

Gangl, R. S., and D. L. Pereira. 2003. Biological performance indicators for evaluating exploitation of Minnesota's large lake Walleye fisheries. North American Journal of Fisheries Management 23:1303–1311.

Garcia, S. M., J. Kolding, J. Rice, M.-J. Rochet, S. Zhou, T. Arimoto, J. E. Beyer, L. Borges, A. Bundy, D. Dunn, E. A. Fulton, M. Hall, M. Heino, R. Law, M. Makino, A. D. Rijnsdorp, F. Simard, and A. D. M. Smith. 2012. Reconsidering the consequences of selective fisheries. Science 335:1045–1047.

Johnson, F. H. 1961. Walleye egg survival during incubation of several types of bottom in Lake Winnibigoshish, Minnesota, and connecting waters. Transactions of the American Fisheries Society 90:312–322.

Li, Jingyin, Y. Cohen, D. H. Schupp, and I. R. Adelman. 1996. Effects of Walleye stocking on year-class strength. North American Journal of Fisheries Management 16:840–850.

Logsdon, D. E. 2006. Contribution of fry stocking to the recovery of the Walleye population in the Red Lakes. Minnesota Department of Natural Resources, Section of Fisheries Investigational Report 535, St. Paul.

Logsdon, D. E., C. S. Anderson, and L. M. Miller. 2016. Contribution and performance of stocked Walleyes in the recovery of the Red Lakes, Minnesota, fishery. North American Journal of Fisheries Management 36:828–843.

Logsdon, D. E., B. J. Pittman, and G. C. Barnard. 2004. Oxytetracycline marking of newly hatched Walleye fry. North American Journal of Fisheries Management 24:1071–1077.

MNDNR (Minnesota Department of Natural Resources) 1982. Lake management planning guide. MNDNR, Fisheries Division, Special Publication 132, St. Paul.

MNDNR (Minnesota Department of Natural Resources). 1995. Walleye stocking guidelines for Minnesota fisheries managers. MNDNR, Fisheries Division, Special Publication 150, St. Paul.

MNDNR (Minnesota Department of Natural Resources). 1997. Potential, target, and current yields for Minnesota's 10 large Walleye lakes. MNDNR, Fisheries Division, Special Publication 151, St. Paul.

MOU (Memorandum of Understanding Between The Red Lake Band of Chippewa Indians, The Minnesota Department of Natural Resources, And The United States Department of Interior, Bureau of Indian Affairs). 1999. Recovery Plan for Red Lakes Walleye Stocks, Attachment A. Available: www.leg.state.mn.us/docs/2015/other/150858.pdf. (September 2018).

Reeves, K. A., and R. E. Bruesewitz. 2007. Factors influencing the hooking mortality of Walleyes caught by recreational anglers on Mille Lacs, Minnesota. North American Journal of Fisheries Management 27:443–452.

Richards, R. A., and P. J. Rago. 1999. A case history of effective fishery management: Chesapeake Bay Striped Bass. North American Journal of Fisheries Management 19:356–375.

Smith, L. L., Jr. 1958. Indian fishery on the Red Lakes. Minnesota Conservation Volunteer 21:43–47.

Smith, L. L., Jr. 1977. Walleye (*Stizostidion vitreum vitreum*) and Yellow Perch (*Perca flavescens*) populations and fisheries of the Red Lakes, Minnesota, 1930–75. Journal of the Fisheries Research Board of Canada 34:1774–1783.

Smith, L. L., Jr., and L.W. Krefting. 1954. Fluctuations in production and abundance of commercial species in the Red Lakes, Minnesota, with special reference to changes in the Walleye population. Transactions of the American Fisheries Society 83:131–160.

Smith, L. L., Jr., L.W. Krefting, and R. L. Butler. 1952. Movements of marked Walleyes, *Stizostidion vitreum vitreum* (Mitchill), in the fishery of the Red Lakes, Minnesota. Transactions of the American Fisheries Society 81:179–196.

Vandergoot, C. S., M. D. Faust, J. T. Francis, D. W. Einhouse, R. Drouin, C. Murray, and R. L. Knight. 2019. Back from the brink: sustainable management of the Lake Erie Walleye fishery. Pages 431–466 *in* C. C. Krueger, W. W. Taylor, and S. Youn, editors. From catastrophe to recovery: stories of fishery management success. American Fisheries Society, Bethesda, Maryland.

VanOosten, J., and H. J. Deason. 1957. History of Red Lakes fishery, 1917–38, with observations on population status. U.S. Fish and Wildlife Service Special Scientific Report Fisheries 229.

Zorn, T. G. 2015. Contribution of hatchery-reared Walleyes to populations of in northern Green Bay, Lake Michigan. North American Journal of Aquaculture 77:409–422.

Community Management of Fish Refuges in Cambodian Rice-Field Fisheries

Miratori Kim

WorldFish Cambodia Office
House #34, Street 228, Sangkat Chaktomuk, Phnom Penh, Cambodia

Elizabeth R. Bageant

Dyson School of Applied Economics and Management, Cornell University
Warren Hall, Ithaca, New York 14850, USA

Shakuntala H. Thilsted

WorldFish
Jalan Batu Maung, Batu Maung, 11960 Bayan Lepas, Penang, Malaysia

Kathryn J. Fiorella*

Department of Population Medicine and Diagnostic Sciences, and
Atkinson Center for a Sustainable Future, Cornell University
Shurman Hall, Ithaca, New York 14850, USA

Abstract.—Home to nearly 500 aquatic species, Cambodian freshwater fisheries are among the world's most diverse (Baran 2005). These diverse fisheries support the livelihoods and the food and nutrition security of millions of Cambodian people. However, these fisheries also face a range of threats, including conflicts with dam projects that disrupt fish movement and spawning, increasing harvest pressure, and unpredictable flood seasons. Rice-field fisheries support as much as 30% of all fish catch in Cambodia and are the site of a unique management strategy that has been implemented to sustainably improve fish catch and strengthen community governance: community fish refuges (CFRs). Community fish refuges have been encouraged by the government of Cambodia and are community-managed protected areas designed to increase fish productivity and protect biodiversity. We use governance scores to examine how support for CFRs from 2012 to 2015 has improved their governance across five dimensions (structure, planning, representation, fundraising, and networking) and the relationship between governance and fish biomass within CFRs. We find some associations between the governance scores and the biomass and biodiversity of the refuges, suggesting that improving governance may affect fishery productivity in this setting. Important lessons learned include (1) that supporting community leadership and technical knowledge enables communities to better understand and justify the underlying reasons for management actions and negotiate their implementation; (2) that in changing ecosystems, community-based approaches allow communities to responsively adapt to the unique, localized environmental conditions they face; (3) that community-based approaches capitalize on a community's unique knowledge of their ecological setting and community; and (4) that engagement of community leadership with the wider fisheries context is essential to facilitating and encouraging community efforts.

* Corresponding author: kfiorella@cornell.edu

INTRODUCTION: HISTORICAL CONTEXT OF FISHERY MANAGEMENT IN CAMBODIA

Before 1880, the King of Cambodia was designated as the owner of all natural resources (Dina and Sato 2014). In practice, however, fisheries and other natural resources were only loosely managed by the monarchy (Dino and Sato 2014). The French colonized Cambodia and, in the 1880s, the colonial administrators privatized fisheries with the introduction of fishing lots (Dina and Sato 2014). Fishing lots were fishing concessions granting exclusive fishing rights and management authority to the lot owner, who, in many cases, included wealthy locals or foreign nationals. Fishing lots were geographically demarcated and located within the Tonle Sap Lake and sometimes included adjacent rivers, beaches, seasonally flooded land, and flooded forest (Degen and Thouk 2000). The fishing lot system endured into the 21st century.

As a private-property system, fishing lots initially provided a source of tax revenue for the French colonial government (Dina and Sato 2015). In the early 1900s, fishery lots were dominated by Chinese businessmen, while Vietnamese nationals and Khmer Muslims held small lots (Cooke 2011). In contrast, the Khmer Buddhist people, the majority of the Cambodian population, were relegated to the smallest and least-profitable lots (Cooke 2011). Beginning in the early 1900s, official complaints were made by fishers excluded from the lot system or relegated to unproductive areas (Dina and Sato 2014). The complaints were both regarding the limited access resulting from this large-scale privatization and concerns about the conservation of fish stocks for the future (Dina and Sato 2015). Despite these complaints, the system endured largely unchanged for decades alongside the political status quo (Dina and Sato 2014).

In 1970, a civil war broke out in Cambodia. A militarized group known as the Khmer Rouge, led by Pol Pot, and aligned with communist forces in Vietnam, rose to power (Short 2006). By 1975, the Khmer Rouge ruled Cambodia and imposed a return to an agrarian society, forcing the populace to live and work in cooperative units producing rice (Short 2006). Forced labor and abuses were ubiquitous, and mass executions, malnutrition, and disease were widespread. Ultimately, 1.5–3 million Cambodians perished (Frey 2009).

Invading Vietnamese forces entered Cambodia at the end of 1978, took Phnom Penh in 1979, and precipitated the winding down of active conflict and dismantling of work camps in Cambodia, although the political descendants of the Khmer Rouge remained active in Cambodia and continue to hold positions of leadership up to the present day (Short 2006).

The fishing lots system languished throughout the civil war and during the subsequent Khmer Rouge rule (Dina and Sato 2014). During this period, the Khmer Rouge regime took control of all fishing lots and forbade fishing, although some lots remained to supply Khmer Rouge leaders with fish (Dina and Sato 2014). By the early 1980s, the fish populations, which had been largely spared from fishing pressure over the previous decade, were flourishing (Dina and Sato 2014). The socialist government of the 1980s used the Fishery Law of 1987 to revive the fishery lot system to generate revenue and to introduce regulations of fishing gear and activities (Dina and Sato 2015; Piseth 2003).

By the early 2000s, fishing access and boundaries were a concern for many local fishers, who found traditional fishing grounds increasingly encroached upon by lot owners (Piseth 2003). At the time, researchers debated whether the traditional lot system or community fisheries was the best approach to conserving fisheries and their productivity (Dina and Sato 2015). Many asserted that fishing lots provided fish stock protection by controlling fishing pressure and incentivizing protection of the flooded forests, which are critical fish spawning grounds (Thouk and Sina 1997; Thouk and van Zalinge 1999; Degen and Thouk 2000; Virkom and Sithirith 2008).

An alternative to the fishing lots, community fisheries rely on co-management of fisheries or a partnership between local fishers and the government that "recognizes fishers' participation, local stewardship, and shared decision-making in the management of fisheries" (Pomeroy and Rivera-Guieb 2006). Others countered that community fisheries would better protect fish populations and be a more appropriate and egalitarian strategy than the fishing lot system for management of fisheries and flooded forests (Gum 1998; Somony 2002; Piseth 2003; Evans et al. 2004; Ratner et al. 2011).

As the debate over the use of community fisheries or the fishing lots system progressed, Cambodian fisheries reached the point of an institutional crisis (Kim et al. 2013). Elites held large lots and controlled them strictly, charging substantial rents while paying little in taxes and evading effective accountability by the Fisheries Administration (Jones and Say 2015). In response, a series of fishery reforms were implemented by the Cambodian government in the early 2000s to release areas held as fishing lots in order to appease fishers by improving their access (Levinson 2002). Despite the reforms, the status quo of the lot system was largely maintained, and the reforms proved insufficient to address the challenges the system faced (Dina and Sato 2015).

Tensions among local fishers regarding the lot system continued until, in March 2012, the fishing lot system was abruptly abolished (Dina and Sato 2015). Motivated by the ruling government's desire to secure fishers' votes in a close national election, lots were entirely converted to community fishing grounds (Dina and Sato 2015). Though these quickly enacted reforms presented a range of implementation challenges (Kim et al. 2013; K. Kaing, paper presented at the Voluntary Guidelines on the Responsible Governance of Tenure of Land, Fisheries and Forests in the Context of National Food Security, 2013), the demise of fishing lots precipitated the expansion of co-management of Cambodian fisheries (Chap et al. 2016). This system first began with community fisheries and soon expanded to include community fish refuges (CFRs).

In this chapter, we discuss the implementation of the CFR strategy, outline the evidence from observational research that successful governance has occurred using this strategy, and put forward its strengths and challenges. We conclude by outlining lessons learned in Cambodia and the potential applicability of this strategy to other locations.

RICE-FIELD FISHERIES

The Tonle Sap floodplain is a dynamic, complex, and multifunctional agroecosystem that provides habitat for fish and aquatic species, contributes to household food and nutrition security, and supports the livelihoods of rural Cambodians. With more than 500 fish and other aquatic species in the Tonle Sap and surrounding waters, the Mekong River basin has the second highest fish biodiversity in the world (Campbell et al. 2006; Baran et al. 2014). These include migratory fish species that travel between the lake and the floodplain, as well as unique and highly adapted fish species that survive year-round within the flooded forests and rice fields. Such unique fish species, such as the Climbing Perch *Anabas testudineus*, can traverse over land to reach the limited water available on the dry floodplain and may survive the dry season with minimal water access (Hortle et al. 2004).

The Tonle Sap floodplain is fed by the Tonle Sap Lake, which swells from approximately 2,600 km^2 during the dry season (December–April) up to 10,500 km^2 during the wet season (May–November) (Frenken

2011). The Tonle Sap Lake connects to the lower Mekong River via the Tonle Sap River. The system is highly seasonal and connects to a range of adjacent habitats (e.g., flood forest, rice fields, etc.) throughout the course of the year. During the wet season, fish migrate from the lake into a network of local tributaries, flooded forests, and rice fields. Fish and other aquatic animals rely on this habitat for food and spawning grounds. This extensive flooding is driven by the flow of the Mekong River, which annually reverses the flow of the Tonle Sap River and brings the floodwaters that swell the Tonle Sap Lake. As the floodwaters recede, the declining water levels signal the return of fish to perennial water bodies (Brooks and Sieu 2016).

Within these flooded regions lie the preponderance of Cambodian rice fields. The flooding serves the production of rice and also provides an aquatic habitat used extensively by Cambodia's incredible diversity of fish and other aquatic animals (Brooks and Sieu 2016). In addition to providing tremendous natural diversity, the rice-field ecosystem provides critical resources for rural people. Rice farming communities depend on the annual migration of fish to surrounding rice fields as a source of nutritious animal-source food and income from the sale of fish, as well as a natural source of fertilizer from fish waste for rice paddies. In 2006, the fisheries sector of the Tonle Sap Lake comprised 11.4% of Cambodia's gross domestic product and employed approximately 4.5 million Cambodians (Lymer et al. 2008). In 2013, 20% of fish production (550,000 metric tons) came from rice-field fisheries and small-scale family fisheries, and the rice-field habitat served as a particularly important source for fisheries productivity (Un et al. 2014).

Environmental changes, population growth, and development within Cambodia and elsewhere upstream in the Mekong River basin threaten the productivity and sustainability of the rice-field fisheries. Changes that impact the timing, intensity, and duration of the flood may be particularly consequential for inland fisheries (Kummu and Sarkkula 2008; Arias et al. 2014). For example, changes in flood pulse timing changes may negatively impact fish species that rely on signals from the flood timing about when to breed. In addition, increased demand for fish and freshwater resources; poor habitat management (e.g., pesticide use in rice production); and regional dams, irrigation, and roads threaten rice-field fisheries sustainability and productivity. In particular, dams and dikes impede fish movement within rivers and wetlands as well as the flow of eggs and larvae, which for many species is critical to their development. As dam projects reshape the Mekong River, they will alter the flow dynamics of water in ways that will affect patterns of flooding, fish migrations, and the flow of fish eggs within the ecosystem and thereby impact local communities (Arias et al. 2014). Importantly, potential environmental changes are likely to have interannual effects that also shape the size and composition of the fish population in subsequent flood seasons as well as the current season.

Within rice-field fisheries, rice farming, wage labor, and fishing are the dominant sources of household income, with fish being the primary animal-source food and micronutrient-rich food for households (Brooks and Sieu 2016). The rice-field fisheries production system plays a particularly important role in smoothing seasonal availability of high-quality foods as households rely predominantly on rice for food and income during some seasons and fish in others (Brooks and Sieu 2016).

COMMUNITY FISH REFUGES

To meet the demand for fish and counteract environmental changes within the system, the Cambodian government is in the process of registering CFRs, as outlined in the government's National Strategic Development Plan (2009–2013) and Strategic Planning Framework for Fisheries (2010–2019) (Fisheries Administration 2009). A CFR is a conservation measure that is intended to improve rice-field fishery productivity by using a perennial water body to support populations of fish and other aquatic species (Joffre et al. 2012). No fishing is permitted within CFRs. Fish and other aquatic animals are protected particularly when they are concentrated and vulnerable during the November–April dry season so that during the subsequent wet season, they are available to migrate out to the surrounding waterways for growth and reproduction.

Across Cambodia, 779 CFRs have been established in the Tonle Sap floodplain (Fisheries Administration 2009). Community fish refuges aim to place rice-field fisheries governance in the hands of local communities and to sustainably manage fishery resources so they may contribute to local incomes and food and nutrition security and provide micronutrient-rich small fish to fishing households (Brooks and Sieu 2016). While increasing fisheries productivity remains a key goal, in creating managed, protected areas, CFRs also potentially conserve and protect not only fish and aquatic biodiversity but also regional birds and reptiles.

Community fish refuges can be considered a type of aquatic protected area, similar to marine protected areas (e.g., Ünal and Kizilkaya 2019, this volume). The potential of freshwater protected areas such as CFRs is widely recognized (Fitzsimons and Robertson 2005; Abell et al. 2007; Suski and Cooke 2007; Hedges et al. 2010; Pittock et al. 2015). Freshwater protected areas have been established in India (Sarkar et al. 2013; Gupta et al. 2015), East Africa's Lake Victoria (Kiwango et al. 2013), Lao People's Democratic Republic (Baird 2000; Baird and Flaherty 2005), and, as noted here, Cambodia's Tonle Sap Lake (Chheng 2010), but their effectiveness has not been well measured or evaluated. Case studies on freshwater protected areas, or fish sanctuaries, show promise despite a range of management and institutional challenges (Baird and Flaherty 2005; Chheng 2010; Sarkar et al. 2013). For example, management may not be strong enough to enforce rules, and funding for protected areas may be unstable (Baird and Flaherty 2005; Chheng 2010; Sarkar et al. 2013).

To improve CFR functioning within Cambodia, with support from USAID Feed the Future, WorldFish, collaborating nongovernmental organizations (NGOs), and the Cambodia Fisheries Administration, the local CFR committees implemented a range of interventions over a period of 3 years (2012–2015) within 40 CFRs. WorldFish, the lead institution, is an international research organization that aims to reduce poverty and hunger by improving fisheries and aquaculture. The project aimed to improve physical habitat and to improve governance (e.g., reduce illegal fishing, ensure water management in dry season) to support fish production and diversity within the CFR. Consequently, interventions were tailored to the individual sites' unique biophysical and governance characteristics. For example, interventions included the development of a management plan, patrolling the sites, regular committee meetings, and training of the committee members, as well as biophysical management, such as removal of excess vegetation or deepening ponds.

The 40 CFRs (Figures 1 and 2) were drawn from those designated by the Cambodian Fisheries Administration and were selected purposively to include sites that were likely to be successfully engaged based on the judgment of WorldFish, the Cambodian Fisheries Administration, and its partner NGOs, as well as to allow for comparisons across four provinces (Siem Reap, Kampong Thom, Battambang, and Pursat) and hydrological types. Two types of CFRs exist: (1) a community pond

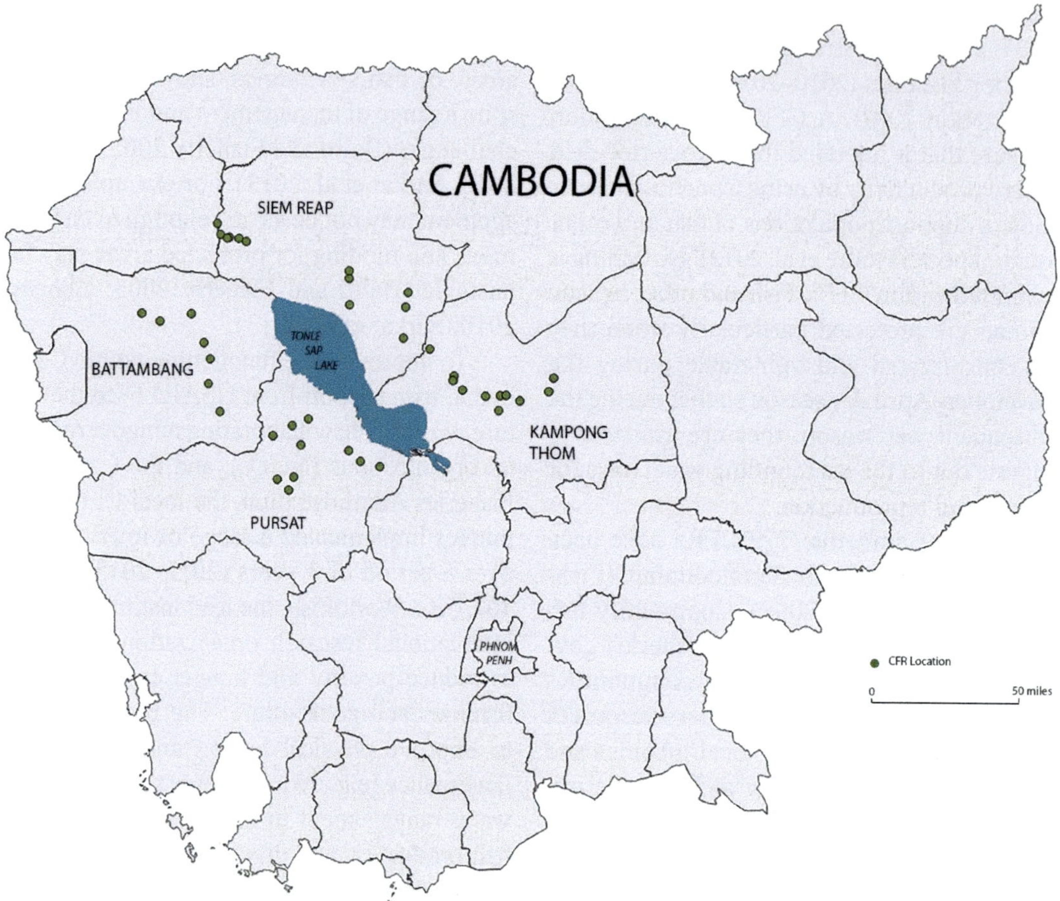

FIGURE 1. Map of community fish refuge (CFR) sites (green dots) within the Tonle Sap floodplain. Dark gray lines represent provincial boundaries.

set in agricultural land, and (2) a designated conservation area within a larger water body (reservoir or lake). To address differences in pond hydrology, these two refuge types were then further classified through community consultation and focus-group discussions with fisheries experts as four hydrologically distinct categories to aid in comparative analysis and develop beneficial management approaches suited to the features of the refuges (WorldFish 2015). The categories of CFR included (1) a irrigation reservoir, (2) a community pond without flooding, (3) a community pond with flooding, and (4) segments of a large water body (Brooks et al. 2015).

Selected CFRs differed across several important dimensions. Hydrologically, CFRs included 7 (17.5%) category 1 (i.e., irrigation reservoir), 14 (35%) category 2, 10 (25%) category 3, and 9 (22.5%) category 4 refuges. They were located in the Battambang (7, 17.5%), Kampong Thom (11, 27.5%), Pursat (10, 25%), and Siem Reap (12, 30%) provinces. Further, they were a distance from Tonle Sap Lake, ranging from 9 to 72 km, with an average distance to Tonle Sap Lake of 38 km with an SD of 16 km. During the dry season in 2012, their sizes ranged from 0.15 to 10 km^2, with a mean size of 1.9 km^2 and SD of 2.5 km^2. On av-

FIGURE 2. **(A)** A community fish refuge, where fishing is prohibited, and guard house, and **(B)** a fisher fishing in a nearby channel that connects rice fields and the community fish refuge.

FIGURE 2. Continued.

erage, they served an estimated 860 households (WorldFish 2015).

Each CFR supported by WorldFish is managed by a local CFR committee. Committee members are a group of volunteers elected by the local community and representative of the gender and ethnicity of each community. The committee's management plan is tailored to each refuge's unique characteristics and physical attributes, such as its area, the proximity of vegetable production, and the distance to the Tonle Sap Lake. Through habitat improvement and enforcement of fishing prohibitions, management is geared to increase fish populations during the wet season when fishing pressure is most intense in the rice fields. The community prohibits fishing in the refuge to protect fish and other aquatic animals. Enforcement of this prohibition is organized by the CFR committee and typically includes signage, demarcation of CFR boundaries, and, in many cases, patrolling of the region.

Differences in hydrology, fish migration patterns, and household demographics across CFRs shaped the management strategies each CFR committee adopted. Consequently, heterogeneous approaches were tailored to meet specific needs (Brooks et al. 2015). For example, within a category-1 CFR (an irrigation reservoir), the water resources have been used for irrigation purposes. In that case, effective management required negotiating use of water resources to retain water for fish habitat.

In addition to negotiating jurisdiction, CFR committees also undertook biophysical interventions. These interventions aimed to improve retention of water in ponds during the dry season, enhance water quality within ponds, and improve connectivity to rice-field fisheries. Together, these improvements to the CFR environment, waterways, and rice-field environment aimed to enhance fish productivity and diversity by improving fish habitat, particularly during the critical dry season when water levels fall and water quality diminishes after the prolonged period with no rain. Interventions were tailored to each site and included deepening of the basin, removal of excess vegetation from within refuges, creation of inlets and outlets to foster connectivity, and digging of channels.

The CFR communities were also provided support to enhance their governance capacity and general community education. Committees were supported by WorldFish and its partners while developing community management plans, electing representatives to governing committees, executing fundraising strategies, implementing fishing prohibitions, and developing water-use agreements. Other governance support included improved coordination between CFR committees and local authorities and engagement of a wide stakeholder group (Miratori and Brooks 2015). Finally, CFR committees and communities were provided with education regarding rice-field fisheries' environment and management, as well as local capacity building in the areas of bookkeeping, financial planning, fundraising, and public speaking. This capacity building was designed to meet the particular management goals of each CFR committee.

METHODS: ASSESSING THE RELATIONSHIP BETWEEN FISH BIOMASS AND GOVERNANCE

To monitor ongoing progress, WorldFish collected extensive data on fish biomass and species richness in the CFRs. In addition, in the process of supporting governance capacity building, WorldFish collected programmatic indicators that can be used to understand governance effectiveness. We focus here on two types of data: (1) CFR fish biomass, and (2) CFR committee's peer-governance scores.

At each of the 40 CFRs, fish biomass and water quality data were collected every 3 months, beginning in November 2012 and ending in November 2015, by trained local staff and members of CFR committees. Fish and other aquatic animal samples were collected using gill nets specific to the size of the CFR.[1] Fish caught were grouped by species, and the number and total weight of fish of each species were recorded. At a given time, total weight in kilograms of fish was averaged across replicates to calculate biomass and divided by the gill-net area (i.e., length by width) to adjust for the harvest effort. Species richness, or a count of the number of species, was adjusted for harvest effort by including gill-net area as covariates in regression analyses.

Community participation was integral to the planning and implementation of this project, and participatory governance activities were fully integrated into the project, beginning in 2013. Five essential characteristics of good governance of CFRs were determined using participatory methods and agreed upon by CFR committee members: (1) organizational structure and management, (2) planning and implementation, (3) representation and participation, (4) resource mobilization, and (5) linkages and networking (Miratori and Brooks 2015; described in detail in Table 1).

Organizational management described the regularity of meetings, communication with members, and establishment of roles for the CFR committee members. Planning and

[1] In ponds larger than 0.25 ha, long (18 m) gill nets were used. In small ponds, short (9 m) gill nets were used. Both long and short gill nets were 2.5 m in height with alternating mesh sizes (1.5 m/mesh size) of 1.5, 2, 4, 6, 8, and 10 cm (Minns and Hurley 1988; Jensen 1990). Gill nets were set between 1600 and 1700 hours and remained in the water 12–14 h; gill nets were set eight times per time point when ponds were larger than 0.35 ha, and five times in ponds smaller than 0.35 ha (Minns and Hurley 1988).

TABLE 1. Governance criteria of community fish refuge (CFR) committees and attributes for different scores (reproduced from Miratori and Brooks 2015).

Criteria	Cultivating (score 1)	Growing (score 2)	Mature (score 3)
Organizational management	No regular meetings. Rules and regulations were decided by a few members of CFR committee. No regular, verbal communication to members. Most members have unclear idea of the role of the CFR committee.	Irregular meetings (less than four times per year). Rules and regulations were decided by some members of CFR committee. Irregular verbal communication to members. Some members have unclear idea of the role of the CFR committee.	Regular meetings (more than four times per year). Governance procedures followed with decisions made democratically with good communication to the community. Members have a clear idea of the role of the CFR committee.
Planning and implementation	Activities not planned and implemented. Need total support from nongovernmental organizations (NGOs).	Activities planned and implemented by few committee members. Some support from NGOs.	Activities planned and implemented by all committee members. Little support from NGOs.
Resource mobilization	No fundraising mechanism. No bookkeeping system.	Regular fundraising both internally and externally. Appropriate bookkeeping system but no circulation to other members.	Regular fundraising both internally and externally. Appropriate bookkeeping system with members being kept updated and informed.
Linkages and networking	CFR committee officially recognized. No attendance at coordination meetings. Plan not integrated in commune investment plan.	CFR committee officially recognized. Coordination meetings regularly attended. Plan integrated into commune investment plan.	CFR committee officially recognized. Coordination meetings regularly attended. Plan integrated into commune investment plan. Receive extra support from NGOs, private sector, and commune councils.
Representation and participation	Less participation by women in decision making. Less representatives from different wealth groups in decision making.	Some women's participation in decision making. More representation from various wealth groups in the community.	Most women on the committee participate in decision making. Women members are kept well informed Most wealth-group representatives participate in decision making.

implementation described the planning and implementation of activities. Resource mobilization reflected the bookkeeping system and fundraising efforts. Linkages and networks referred to the official recognition of the CFR demarcation and its committee, as well as participation in the local government structure, which are called communes in Cambodia and are responsible for administration of a budget within the local area. Finally, representation and participation reflected the participation of women and people from different wealth groups.

The five essential characteristics of good governance of CFRs were outlined as a programmatic tool. Designed to enhance the learning and governance of the refuges, these criteria were engaged to provide CFRs with governance scores. These scores were not designed as evaluation data and were not collected systematically. Consequently, they comprise a source of observational data regarding the governance by CFR committees.

In 2014 and 2015, WorldFish staff facilitated reflection workshops with CFR committees. During these workshops, CFR committees from two of the four provinces met together such that provinces were paired by proximity over the course of the research. Pairwise meetings were conducted to minimize travel (e.g., only two of four provinces traveled at each meeting point) and expenses. During the meetings, committees presented their accomplishments and goals with respect to the five criteria of governance (Table 1). These workshops generated governance assessment scores from peer Committee members, ranging from 1 (low) to 3 (high) for each criterion. Scores provided were averaged to generate a governance score for each CFR on each governance criterion (Table 2). This activity was held as a means of facilitating project monitoring and reflection, but also likely inspired and influenced the activities of each CFR committee.

In a separate set of workshops facilitated by WorldFish staff and conducted within the community, governance of each CFR was assessed in 2015. Each CFR held its own workshop, and participants included local leadership, community members, and CFR committee members. Governance scores along each dimension (e.g., organization, planning, fundraising, networking, and representation; Table 1) were agreed upon through group discussion, and consensus about the score to award was reached by those who participated in the workshop. These workshops also generated retrospective CFR scores for the onset of the project in 2012.

Ultimately, three types of governance scores (peer scores in 2014 and 2015, self-evaluation post hoc 2012 scores, self-evaluation endline 2015 scores) were recorded. These scores were derived using the same rubric to assess consistent governance dimensions; however, the processes that generated the scores differed as described above, and consequently, there were limitations to their comparability.

In addition to reporting data from quantitative analyses, we draw from extensive qualitative information. Qualitative findings derive from the experience of project team that supported CFR implementation, as well as from reports made by CFR committee and community members.

FINDINGS AND INTERPRETATION

Across the subscales of organization, planning, fundraising, networking, and representation, recorded governance scores ranged from 1 to 2.83 in 2012 and were between 1.2 and 3 in 2015 (score range: 1–3; Table 2). Across the intervention period from 2012 to 2015, mean governance scores rose across all dimensions of governance, with the highest mean scores recorded in organization, planning, and networking (Figure 2).

TABLE 2. Governance scores recorded by time point. In 2014 and Oct 2015, scores were peer-governance scores generated in reflection workshops attended by community fish refuge committee members from two provinces. Endline (December 2015) scores were generated in workshops, with each community fish refuge attended by community members, community fish refuge committees, and project staff; baseline (November 2012) scores were generated retrospectively at endline workshops.

	Variable	Observations	Mean	SD	Min	Max
November 2012 (all provinces)	Structure	40	1.98	0.35	1.20	2.67
	Planning	40	1.80	0.37	1.20	2.58
	Representation	40	2.08	0.39	1.30	2.83
	Fundraising	40	1.48	0.38	1.00	2.50
	Networking	40	1.85	0.40	1.30	2.75
January 2014 (Siem Reap, Kampong Thom)	Structure	23	1.91	0.26	1.25	2.50
	Planning	23	1.82	0.34	1.18	2.92
	Representation	23	1.78	0.32	1.33	2.50
	Fundraising	23	1.74	0.52	1.00	2.92
	Networking	23	1.89	0.28	1.42	2.58
March 2014 (Battambong, Pursat)	Structure	17	2.54	0.22	2.18	3.00
	Planning	17	2.04	0.21	1.71	2.47
	Representation	17	1.99	0.28	1.65	2.65
	Fundraising	17	1.54	0.45	1.06	2.53
	Networking	17	2.13	0.20	1.82	2.59
October 2014 (Battambong, Pursat)	Structure	17	2.39	0.21	2.06	2.83
	Planning	17	1.92	0.20	1.61	2.33
	Representation	17	1.88	0.26	1.56	2.50
	Fundraising	17	1.46	0.43	1.00	2.39
	Networking	17	2.02	0.19	1.72	2.44
October 2015 (Siem Reap, Pursat)	Structure	21	2.55	0.19	2.09	2.82
	Planning	21	2.51	0.23	1.82	2.82
	Representation	21	2.51	0.40	1.73	3.00
	Fundraising	21	2.36	0.26	1.82	2.82
	Networking	21	2.24	0.28	1.64	2.73
December 2015 (all provinces)	Structure	40	2.65	0.26	1.83	3.00
	Planning	40	2.66	0.25	1.80	3.00
	Representation	40	2.54	0.28	1.92	3.00
	Fundraising	40	2.49	0.35	1.20	3.00
	Networking	40	2.67	0.24	2.00	3.00

We examined the association between governance scores and biomass and species richness within CFRs using multiple multivariate linear regression models. Our outcome was fish biomass at the endline, or after the study intervention, in November 2015. Independent variables included endline governance scores (December 2015) and the surface area of the CFR during the wet season. Surface area was included to control for the CFR size, which was likely to be associated with both biomass and biodiversity. Governance scores were normalized to Z-scores, so they were centered at 0, and 1 corresponded to 1 SD.

An association existed between some subscales of governance and endline fish biomass in CFRs. A 1 SD increase in structure score was associated with a statistically significant increase in species richness (coefficient: 2.56, SE: 1.12, $p < 0.05$). Similarly, a 1 SD increase in total score was associated with a statistically significant increase in species richness (coefficient: 2.540, SE: 1.20, $p < 0.05$). At the $p < 0.1$ significance level, increases in fundraising over time were associated with increases in fish biomass and species richness. These findings are associations based on observational, programmatic data and do not demonstrate causal relationships.

The use of CFRs has shown promise as a successful governance mechanism to support sustained fishery productivity in seasonally flooded rice fields. Using observational data, we observe an association between governance scores and fish biomass and biodiversity in 2015. Our results suggest that governance quality may improve fishery outcomes. While we did not measure the benefits of fish availability for households here, previous research supports the widespread reliance on fish as a key animal-source food and rice-field fisheries as a critical resource (Balzer et al. 2006; Hap et al. 2016). Careful attention to governance and a participatory process throughout the project may have strengthened CFR committees' ability to govern effectively across complex ecological, sociological, economic, and political dimensions and changing environmental conditions.

Over the course of the 2012–2015 study, fish biomass and biodiversity across the Tonle Sap floodplain were influenced by annual effects that drive system productivity and differed considerably each year. The timing, duration, and intensity of the seasonal flooding of the floodplain plays an established role in driving fishery productivity (Kummu and Sarkkula 2008) and biodiversity (Thomas et al. 2017). In particular, 2015 was the final year of our study but also the start of an El Nino Southern Oscillation event that substantially disrupted flood patterns (Ward et al. 2016). Further, new research has demonstrated that CFR biomass and biodiversity are strongly associated with site characteristics, annual differences, and seasonality (Fiorella and colleagues, unpublished data).

MANAGEMENT CHALLENGES AND SOLUTIONS

The initial support provided to CFR committees by WorldFish was designed to improve water quality and site connectivity, as well as to increase knowledge and governance capacity. Initial support also included identification of avenues for financial support (represented in the fundraising governance score; Figure 3) and coordination with local government (represented by networking), so that CFRs can continue to sustain fishery management into the future. Community fish refuges were designed flexibly to allow opportunities to modify habitat in response to changing conditions (e.g., rainfall quantities, refuge water levels, and nutrient run-off) and allow committees to respond to their communities' needs. In striving to establish jurisdiction of community resources and provide effective governance, CFR committees met with a range of challenges: ecological, sociological, environmental, economic, and political. Below, we outline a selection of salient challenges that CFRs faced around water quality, water use, and illegal fishing.

Maintenance of acceptable water quality within CFRs is essential to supporting fish in the system and particularly challenging during the dry season as receding floodwaters and limited rainfall and high water temperature affect fish by reducing the quantity of water and, thereby, its quality (e.g., reduced dissolved oxygen, higher volumes of nutrient

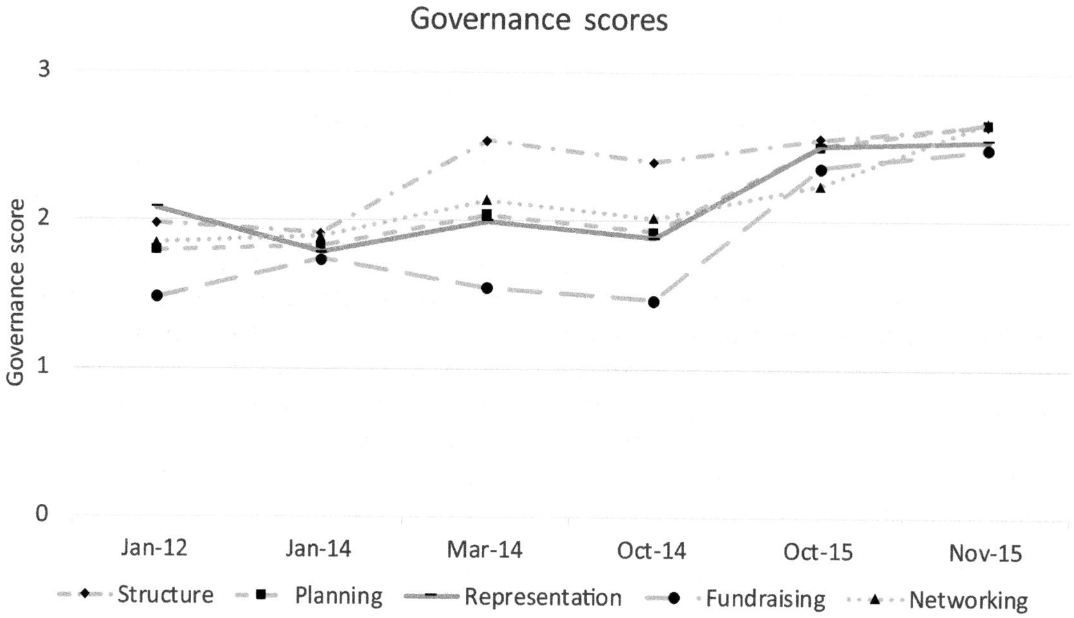

FIGURE 3. Governance scores across time for structure, planning, representation, fundraising, and networking. The governance scores at the final time point (November 2015) are used in subsequent analyses.

concentrations). During the water-limited dry season, water is needed not only to support fish within CFRs, but also to support other activities. Communities need water for crop irrigation and livestock, as well as for household uses (e.g., washing kitchen utensils and clothing, bathing) and drinking water. In perennial ponds that were converted to CFRs, often community members previously used the water to depletion. Consequently, a range of conflicts over water use arose with the establishment of CFRs. In some cases, conflicts over water access were solved through building a separate holding pond to continue to supply water to meet community irrigation and household needs. In each CFR, a color-coded measuring pole (e.g., green, yellow, and red) was installed to establish guidelines as to when water could be withdrawn, and use was continually negotiated between different users.

In some cases, CFR committee jurisdiction presented challenges that were less readily surmounted. Where CFRs were part of a larger water body, CFR committees were often in conversation with other management regimes, such as farmer water user groups, to negotiate for the enactment of policies, particularly prohibitions on fishing within the confines of the CFR. In some cases, CFR committees were never able to effectively establish jurisdiction, and governance proved far less successful.

Community fish refuges also required continued maintenance and governance to respond to changing environmental conditions and address illegal fishing. For example, some refuges required annual removal of excess vegetation to provide optimal fish habitat. In other scenarios, addressing illegal fishing required patrols to control and prevent illegal fishing by people within or outside a community. To sustain support for these labor-intensive activities, CFR committees were trained and encouraged to seek ongoing support in two ways: fundraising and engaging with local government.

First, CFR committees were trained to conduct their own fundraising operations, and committees shared ideas with each other about ways to garner donations. Fundraising efforts were diverse. Communities sought funds from wealthy community members currently living in nearby cities or through collections from nearby Buddhist temples, as animals, including fish, are prized by monks. We were not able to capture data on in-kind donations, but in many cases, CFR habitat was also buoyed by community members who gave their labor and, in rare cases, by relationships that provided access to digging equipment. In addition, at several sites, communities coordinated with construction companies to expand ponds free of charge in exchange for the soil.

Second, CFR committees also engaged local governance mechanisms to garner support for the maintenance of ponds. Committees were encouraged to connect with the local Commune Council, a local governance body that manages a small budget, to support their plans. In the majority of cases, CFRs were adopted into Commune Council plans. In some instances, members of CFR committees subsequently were elected to leadership positions within the commune council.

By generating resources to enact their initiatives (i.e., fundraising, networking), many CFR committees strengthened their governance. Committee members reported that, as a result, they were able to improve and more frequently enforce fishery prohibitions, enhance community engagement (e.g., build structures for meetings, provide signage), improve habitat (e.g., plant trees on banks), clear vegetation, and monitor water quality.

In addition to planning and implementation activities, the CFR management engaged committees in an evaluation process. The evaluation phase was unique in that it involved (1) a group setting involving local authorities, NGO partners, and all CFR committees, (2) learning exchanges to share ideas and best practices across committees and NGOs, and (3) making commitments with accountability to the wider group. This motivated all participants (committee members, NGO field staff, office staff and leaders, commune councils, and the Cambodian Fisheries Administration authorities) to invest effort and resources to ensure the achievement of their commitments. The result was a high level of engagement throughout the entire planning, implementation, and evaluation process, which, in turn, led to continued learning and improvements in CFR management.

LESSONS LEARNED

Reflecting on the challenges and successes of the project, we identify a series of lessons learned:

1. Supporting community leadership and technical knowledge enables communities to better understand and justify the underlying reasons for management actions and negotiate their implementation. Engaging communities in the management of fisheries provides not only control over resources, but also an opportunity to integrate technical and scientific information with communities' ecological knowledge. For example, the monitoring of water quality and water volume provides for an assessment of the resource that can be observed through measurement changes before negative ramifications of poor water quality have substantial effects on fish health. Similarly, technical training provides potential solutions to specific water quality problems (e.g., tree planting to provide shade in response to high temperatures). Through this process, communities are able to further understand and assess the challenges their fisheries face and adapt their management.

2. In changing ecosystems, community-based approaches allow communities to respond to unique, localized environmental conditions that communities face. By engaging communities in the monitoring of environmental conditions, such as water quality and fish biomass, communities are able to respond to the localized environmental conditions that they face. In changing environments, such as those where weather patterns may be inconsistent and local hydrology plays an important role in fish production, community understanding of environmental conditions enables quick, appropriate responses.
3. Community-based approaches allow communities to capitalize on their unique strengths, capacities, and understanding of their setting. For example, communities were able to fundraise from different sources with respect to what was available to them or to seek in-kind donations of labor or equipment usage.
4. Engagement of community leadership with the wider community leadership context (e.g., Commune Councils) and through peer networks is essential to facilitating and encouraging community efforts. By encouraging learning and public commitments by resource managers, community leaders and CFR committee members were motivated and empowered to sustain efforts in management and continue to improve their governance structure, planning, representation, fundraising, and networking.

Together, the lessons learned from this project will inform the expansion of CFRs in Cambodia and beyond.

CONCLUSION: NEXT STEPS

Community fish refuges are linked to a complex and interacting network of people, communities, rice fields, and water bodies. Aquatic resources, the rice-field fishery system, CFR committees, and other stakeholders and resource users interact in this dynamic system, creating feedbacks among environmental changes, governance actions, and users. By empowering local committees of resource users to govern their resource system, the CFR approach enables adaptive management and engagement of community knowledge and capacity to negotiate resource conservation. The participatory and community-based nature of the CFR approach and the ability to adapt to suit governance and biophysical needs of each individual CFR and community are among the strengths of this project and underlie its success.

A range of opportunities to further expand the CFR approach within Cambodia have been identified. In addition to expanding the number of supported CFRs, WorldFish has identified new areas of emphasis. In the next phase of CFR support, management of freshwater for multiple uses (e.g., managing water for drinking, household use, and vegetable irrigation) and highlighting the nutritional contributions of rice-field fisheries within communities will be priorities. By strengthening communities' use of fish resources, and particularly of small, micronutrient-rich fish in the diets of young children, expansions of this project seek to extend its impact.

A similar approach to CFRs is also relevant to many other regions of Asia as well as to parts of Africa where rice-field fisheries are common. In bringing this approach to other settings, the key lessons would emphasize creating a community-managed and adaptive approach so that communities can respond to the local environmental changes and management challenges they face now and in the future.

ACKNOWLEDGMENTS

We gratefully acknowledge the CFR committees, communities, and partner NGO staff who supported data collection on this project. We thank Yumiko Kura and Sarah Freed for their feedback on earlier drafts. We are very grateful to Charles Krueger, William Taylor, and So-Jung Youn for their feedback and editing. Financial support for this project was provided by the American people through the U.S. Agency for International Development (USAID). Cornell University's Atkinson Center for a Sustainable Future Postdoctoral Fellows Program also provided support (to KJF). The contents and opinions expressed herein are those of the authors and do not necessarily reflect the views of USAID or the U.S. Government.

REFERENCES

Abell, R., J. D. Allan, and B. Lehner. 2007. Unlocking the potential of protected areas for freshwaters. Biological Conservation 134:48–63.

Arias, M. E., T. H. Cochrane, M. Kummu, H. L. Gordon, W. Holtfrieve, J. Koponen, and T. Piman. 2014. Impacts of hydropower and climate change on drivers of ecological productivity of Southeast Asia's most important wetland. Ecological Modelling 272:252–263.

Baird, I. G. 2000. Integrating community-based fisheries co-management and protected areas management in Lao PDR: opportunities for advancement and obstacles to implementation. International Institute for Environment and Development, Evaluating Eden Series Discussion Paper 14, London.

Baird, I. G., and M. S. Flaherty. 2005. Mekong river fish conservation zones in southern Laos: Assessing effectiveness using local ecological knowledge. Environmental Management 36:439–454.

Balzer, P., T. Balzer, D. Bartley, X. Choulamany, S. Funge-Smith, H. Guttman, M. Halwart, A. Luo, J. Margraf, E. Meusch, and S. Pon. 2006. Use and availability of aquatic biodiversity in rice-based ecosystems of Southeast Asia. Journal of Food Composition and Analysis 19:765–766.

Baran, E. 2005. Cambodian inland fisheries: facts, figures and context. WorldFish, Penang, Malaysia.

Baran, E., C. Phen, L. Vuthy, J. Nasielski, S. Saray, T. Bunthang, J. Tress, K. Khim, and T. Sokhom. 2014. Fish resources in Cambodia (2001–2011). Pages 37–48 *in* S. Brod and R. Sandoval, editors. Atlas of Cambodia: maps on socio-economic development and environment. Save Cambodia's Wildlife, Phnom Penh, Cambodia.

Brooks, A., M. Kim, C. Sieu, V. Sean, and V. Try. 2015. A characterization of community fish refuge typologies in rice field fisheries ecosystems. WorldFish, Penang, Malaysia.

Brooks, A., and C. Sieu. 2016. The potential of community fish refuges (CFRs) in rice field agro-ecosystems for improving food and nutrition security in the Tonle Sap region. WorldFish, Penang, Malaysia.

Campbell, I. C., C. Poole, W. Giesen, and J. Valbo-Jorgensen. 2006. Species diversity and ecology of Tonle Sap Great Lake, Cambodia. Aquatic Sciences 68:355–373.

Chap, S., P. Touch, and J. C. Diepart. 2016. Fisheries reforms and right-based fisheries: insights from community fisheries across Cambodia. The Learning Institute, Phnom Penh, Cambodia.

Chheng, P. 2010. Fish sanctuaries in the Tonle Sap Great Lake (Cambodia). Master's thesis. Universiti Sains Malaysia, Penang.

Cooke, N. 2011. Tonle Sap processed fish: from Khmer subsistence staple to colonial export commodity. In E. Tagliacozzo, and W.-C. Chang, editors. Chinese circulations: capital, communities, and networks in Southeast Asia. Duke University Press, Durham, North Carolina.

Degen, P., and N. Thouk. 2000. Historical, legal, and cultural perspectives on the fishing lot system in Cambodia. Department of Fishery, Phnom Penh, Cambodia.

Dina, T., and J. Sato. 2014. Is greater fishery access better for the poor? Explaining deterritorialisation of the Tonle Sap, Cambodia. The Journal of Development Studies 50:962–976.

Dina, T., and J. Sato. 2015. The cost of privatizing the commons: overlapping property systems in Tonle Sap, Cambodia. International Journal of the Commons 9:261–280.

Evans, P., M. Marschke, and K. Paudyal. 2004. Flood forests, fish, and fishing villages: Tonle Sap, Cambodia. Asia Forest Network, Research Network Report No. 12. Bohol, Philippines.

Fisheries Administration. 2009. Government's national strategic development plan 2009–2013 and in the strategic planning framework (2010–2019). Ministry of Agriculture, Forestry, and Fisheries, Phnom Penh, Cambodia.

Fitzsimons, J. A., and H. A. Robertson. 2005. Freshwater reserves in Australia: directions and challenges for the development of a comprehensive, adequate and representative system of protected areas. Hydrobiologia 552:87–97.

Frenken, K. 2011. Irrigation in southern and eastern Asia in figures. AQUASTAT Survey – 2011. FAO (Food and Agriculture Organization of the United Nations), Water Report 37, Rome.

Frey, R. J. 2009. Genocide and international justice. Facts on File, Inc., New York.

Gum, W. 1998. Natural resource management in Tonle Sap Biosphere Reserve in Battambang Province. Consultancy report for the European Commission Support Program to the Environmental Sector in Cambodia, Phnom Penh.

Gupta, N., K. Sivakumar, V. B. Mathur, and M. A. Chadwick. 2015. Terrestrial protected areas and managed reaches conserve threatened freshwater fish in Uttarkhand, India. PARKS: The International Journal of Protected Areas and Conservation 21.1:89–101.

Hap N., S. Un, and J. Nasielski. 2016. A review of socioeconomic studies in the fisheries sector in Cambodia. Inland Fisheries Research and Development Institute of the Fisheries Administration and WorldFish, Pnom Penh, Cambodia.

Hedges, K. J., M. A. Koops, N. E. Mandrak, and O. E. Johannsson. 2010. Use of aquatic protected areas in the management of large lakes. Aquatic Ecosystem Health and Management 13:135–142.

Hortle, K. G., S. Lieng, and J. Valbo-Jorgensen. 2004. An introduction to Cambodia's inland fisheries. Mekong River Commission, Phnom Penh, Cambodia.

Jensen, J. W. 1990. Comparing fish catches taken with gill nets of different combinations of mesh sizes. Journal of Fish Biology 37:99–104.

Joffre, O., M. Kosal, Y. Kura, S. Pich, and T. Nao. 2012. Community fish refuges in Cambodia: lesson learned. WorldFish, Phnom Penh, Cambodia.

Jones, R. W., and S. Say. 2015. Impacts and implications of deep fisheries reforms on the governability of small-scale fisheries in Tonle Sap Lake, Cambodia. Pages 539–557 *in* S. Jentoft and R. Chuenpagdee, editors. Interactive governance for small-scale fisheries. Springer, MARE Publication Series 13, New York.

Kim, S., K. Mam, I. Oeur, S. So, and R. Blake. 2013. Fishery reforms on the Tonle Sap Lake: risks and opportunities for innovation. Cambodian Development Review 17(2):1–4.

Kiwango, Y., G. Moshi, W. Kibasa, and B. Mnaya. 2013. Papyrus wetlands creation, a solution to improve food security and save Lake Victoria. Wetlands Ecology and Management 21:147–154.

Kummu, M., and J. Sarkkula. 2008. Impact of the Mekong River flow alteration on the Tonle Sap flood pulse. AMBIO: A Journal of the Human Environment 37:185–192.

Levinson, J. 2002. An examination of the community fisheries sub-decree in Cambodia: changes and developments during the drafting process. Network of Aquaculture Centres in Asia-Pacific, STREAM Regional Office, Bangkok, Thailand.

Lymer, D., S. Funge-Smith, J. Clausen, and W. Miao. 2008. Status and potential of fisheries and aquaculture in Asia and the Pacific 2008. Food and Agriculture Organization of the United Nations, Regional Office for Asia and the Pacific, Bangkok, Thailand.

Minns, C. K., and D. A. Hurley. 1988. Effects of net length and set time on fish catches in gill nets. North American Journal of Fisheries Management 8:216–223.

Miratori, K., and A. Brooks. 2015. Good governance of rice field fishery management. WorldFish, Penang, Malaysia.

Piseth, C. 2003. Contested legitimation of access to fisheries: a case study of everyday practices among fishers in Tonle Sap, Cambodia. Master's thesis. Chiang Mai University, Chiang Mai, Thailand.

Pittock, J., M. Finlayson, A. H. Arthington, D. Roux, J. H. Matthews, H. Biggs, I. Harrison, E. Blom, R. Flitcroft, R. Froend, V. Hermoso, Dr. W. Junk, R. Kumar, S. Linke, J. Nel, C. N. da Cunha, A. Pattnaik, S. Pollard, W. Rast, M. Thieme, E. Turak, L. Turpie, L. van Niekerk, D. Willems, and J. Viers. 2015. Managing freshwater, river, wetland and estuarine protected areas. Pages 569–608 *in* G. L. Worboys, M. Lockwood, A. Kothari, S. Feary and I. Pulsford, editor. Protected area governance and management. ANU Press, Canberra, Australia.

Pomeroy, R. S., and R. Rivera-Guieb. 2006. Fisheries co-management: a practical handbook. CABI Publishing, Wallingford, UK and International Development Research Centre, Ottawa.

Ratner, B., G. Halpern, and K. Mam. 2011. Catalyzing collective action to address natural resource conflict: lesson from Cambodia's Tonle Sap. International Food Policy Research Institute, CAPRi Working Paper 103, Washington, D.C.

Sarkar, U. K., A. K. Pathak, L. K. Tyagi, S. M. Srivastava, P. Singh, and V. K. Dubey. 2013. Biodiversity of freshwater fish of a protected river in India: comparison with unprotected habitat. Revista de Biologia Tropical 61:161–172.

Short, P. 2006. Pol Pot: anatomy of a nightmare. Henry Holt, New York.

Somony, T. 2002. Concept of fisheries co-management in Cambodian context: a case study in fishing lots #3 and #6 in Siem Reap Province, Cambodia. Master's thesis. Royal College of London, London.

Suski, C. D., and S. J. Cooke. 2007. Conservation of aquatic resources through the use of freshwater protected areas: opportunities and challenges. Biodiversity and Conservation 16:2015–2029.

Thomas P., H. Gordon, E. Vittoria, M. Kevin, M. Bailey, R. Neil, S. Adrianne, P. Thach, C. Michael, C. Sydney, Pool, T., G. Holtgrieve, V. Elliott, K. McCann, B. McMeans, N. Rooney, A. Smits, T. Phanara, M. Cooperman, S. Clark, C. Phen, and S. Chhuoy. 2017. Seasonal increases in fish trophic niche plasticity within a flood-pulse river ecosystem (Tonle Sap Lake, Cambodia). Ecosphere [online serial] 8:e01881.

Thouk, N., and L. Sina. 1997. Review of the fishery sector in Cambodia. Mekong River Commission and Department of Fishery, Phnom Penh, Cambodia.

Thouk, N., and N. van Zalinge. 1999. The present status of Cambodia's fisheries and management implications. Mekong River Commission and Department of Fishery, Phnom Penh.

Un, S., P. Chheng, J. Tress, E. Baran, and V. Simpson. 2014. Fish productivity by aquatic habitat and estimated sh production in Cambodia. Inland Fisheries Research and Development Institute and WorldFish, Phnom Penh, Cambodia.

Ünal, V., and Z. Kizilkaya. 2019. A long and participatory process towards successful fishery management of Gökova Bay, Turkey. Pages 509–532 *in* C. C. Krueger, W. W. Taylor, and S. Youn. From catastrophe to recovery: stories of fishery management success. American Fisheries Society, Bethesda, Maryland.

Virkom, M., and M. Sithirith. 2008. Entitlements and the community fishery in the Tonle Sap: is the fishing lot system still an option for inland fisheries management? Fisheries Action Coalition Team, Phnom Penh, Cambodia.

Ward, P. J., M. Kummu, and U. Lall. 2016. Flood frequencies and durations and their response to El Nino Southern Oscillation: global analysis. Journal of Hydrology 539:358–378.

WorldFish. 2015. Program brief: the potential of community fish refuges (CFR) in rice-field agro-ecosystems for improving food and nutrition security and recommendations for scaling CFR lessons in the Tonle Sap region. WorldFish, Phnom Penh, Cambodia.

Species Salvation: Saving the Owens Pupfish from Extinction

Edwin Philip Pister*
Desert Fishes Council
437 South Street, Bishop, California 93514, USA

Abstract.—The Owens Pupfish *Cyprinodon radiosus* is a small fish (<6 cm [2.5 in]) in the killifish family once thought extinct but rediscovered in the early 1960s in Fish Slough in the Death Valley drainage area of eastern California, USA. At the time of discovery, the species was comprised of a single population of approximately 200 individuals. The species was listed as endangered on March 11, 1967 under the U.S. Endangered Species Preservation Act of 1966. During the summer of 1969, a spring that was feeding water to Fish Slough was discovered to have much reduced flow due to unusual precipitation patterns the previous 6 months and possibly water removals, and thus threatened complete extinction of the species. Quick actions by fish biologists prevented extinction by transporting the entire species in two buckets to nearby refuge waters similar to Fish Slough. The transplantation was successful, and six populations now exist 50 years later that each number from the hundreds to perhaps more than 10,000 fish. The species remains listed as endangered—but it did not go extinct!

EDITORS' NOTE

The following represents a very different story from all other stories in this book; however, common to all chapters is an account of the personal testimony of extraordinary individual commitment to natural resources and the courage to take action. Most of the material below was previously published in the magazine of the American Museum of Natural History titled Natural History, January 1993, which we have used here with their permission, as "Species in a Bucket" (Pister 1993). We include this story here because we believe in its value to motivate and inspire future fish biologist professionals to have courage and continue the good fight even as things look its bleakest.

The Owens Pupfish Cyprinodon radiosus is a small fish in the killifish family (Cyprinodontidae) that rarely exceeds 6 cm (2.5 in) in length (Miller and Pister 1971; U.S. Fish and Wildlife Service 1998, 2009). Females are dusky olive-green in color with several dark vertical bars aligned in a row along the sides, and males are bright blue, particularly during the spring and summer spawning season (Figure 1). The Owens Pupfish is restricted to the Owens Valley portion of the Owens River, Mono and Inyo counties, California, USA. Historically, Owens Pupfish were widespread along the Owens River, occurring in clear waters of springs, sloughs, irrigation ditches, swamps, and flooded pastures from Fish Slough in Mono County south to Lone Pine in Inyo County (Figure 2; U.S. Fish and Wildlife Service 1998, 2009). These fish once were so common as to comprise a food source for Native Americans, who seined them with sieve-like baskets and dried them for wintertime sustenance (Wilke and Lawton 1976; Pister 2001).

* Corresponding author: phildesfish123@gmail.com

FIGURE 1. Male Owens Pupfish are bright blue, particularly during the spring and summer spawning season. Photo by California Department of Fish and Wildlife.

FIGURE 2. Fish Slough in the Death Valley drainage area of eastern California, USA, with Sierra Nevada mountains in the background. Photo by Denise Waterbury.

The Owens Pupfish was thought to be extinct in 1942 as the result of the disappearance or marked reduction of marshy areas due to removal of much of the water supply through the Los Angeles aqueduct (Miller and Pister 1971). A remnant population of approximately 200 individuals were rediscovered in 1964 in Fish Slough in the Death Valley drainage area of eastern California, USA (Miller and Pister 1971). The species was listed as endangered on March 11, 1967 (U.S. Office of the Federal Register 1967) and listed as endangered by the state of California in 1971. The initial recovery plan was completed in 1984 under the U.S. Endangered Species Act (Pister 2001). The last of approximately 800 descendants of the rediscovered fish were rescued when springs providing water to Fish Slough began to fail due a wet winter, causing plant growth and excessive transpiration and possibly irrigation removals (Miller and Pister 1971). Below is a first-hand account by Phil Pister, a witness to their rescue in 1969.

INTRODUCTION

I sit here at my computer reminiscing over what has happened to me (Pister), and the results of my decisions, during a career that began nearly 70 years ago. I began work in 1950 as a fishery research biologist with the U.S. Fish and Wildlife Service while a graduate student under A. Starker Leopold at the University of California, Berkeley. Although Starker was wildlife-oriented, his great advice transcended specific career orientation and was as valid for an aspiring fishery biologist as it was for wildlife students. Starker was the eldest son of the legendary Aldo Leopold, who gave us *A Sand County Almanac* (Leopold 1949).

I officially "retired" in 1990 from service with the California Department of Fish and Wildlife as a research/management biologist. My job involved assuring the biological integrity of about a thousand lakes and streams along the crest of the Sierra Nevada mountains. I, however, remain active in research, teaching, publishing and administering the affairs of the Desert Fishes Council (www.desertfishes.org). I am writing this introduction at the age of 89 years.

Where I worked (Inyo National Forest) is likely one of the world's most valuable recreational areas. It accommodates more recreational demand than Yellowstone, Grand Canyon, and Glacier national parks combined. Most of this use comes from urban Southern California where the population exceeds more than 20 million people. As one can imagine, such a population brings with it a diversity of demands on natural resources. The trials and conflicts a fish manager face in this arena are remarkable. To illustrate the management challenges, I often used a baseball analogy of a shortstop having to handle hot grounders from a rocky infield. I never knew from day to day what I would be required to do or where the ball might bounce.

For purposes of this chapter, I have selected perhaps my most popular paper regarding the recovery of the Owen Pupfish for your examination. "Species in a Bucket" (Pister 1993) summarizes the most dramatic episode of my career. At one time, I held the entire world population of the Owens Pupfish in two buckets, one in either hand, as I walked in the dark over a field covered with barbed wire and gopher holes. My destination was my pickup truck and its salvation of fresh, cool water in buckets and battery-powered aerators that I hoped would keep these fish alive. This episode teaches many things, primarily why species salvation is important and the role of a concerned field biologist in bringing such frightening episodes to a successful conclusion. Below is my story.

SPECIES IN A BUCKET - reprinted from *Natural History*, January 1993, copyright (c) Natural History Magazine, Inc., 1993.

For a few frightening moments, there was only myself standing between life and extinction...

> [The naturalist] looks upon every species of animal and plant now living as the individual letters which go to make up one of the volumes of our earth's history; and, as a few lost letters may make a sentence unintelligible, so the extinction of the numerous forms of life which the progress of cultivation invariably entails will necessarily render obscure this invaluable record of the past. It is, therefore, an important object [to preserve them].... If this is not done, future ages will certainly look back upon us as a people so immersed in the pursuit of wealth as to be blind to higher considerations.
>
> Alfred Russel Wallace
> *Journal of the Royal Geographical Society*
> (1863)

When I retired in 1990, I built a small office in my backyard, equipped it with a phone and word processor, and began to reflect seriously upon a career that began in 1951 and continues even in retirement. I remain keenly aware of the legendary biologist Aldo Leopold's admonition that one of the penalties of an ecological education is that one lives alone in a world of wounds.

Virtually my entire career was spent as a district fishery biologist for the California Department of Fish and Game in the state's vast eastern sierra and desert regions. I worked on a great variety of management and research programs – from trying to keep millions of anglers supplied with trout to preserving the biological integrity of desert springs that support life forms totally unknown to most Americans and even to most scientists.

Having studied wildlife conservation at Berkeley in 1948 under the tutelage of Aldo Leopold's son, A. Starker Leopold, I was exposed to the Leopolds' passionately held values regarding the natural world. Impressed by their view that nonconformity is the highest evolutionary attainment of social animals, I carefully avoided the usual career track that would have landed me in one of my department's major offices in a big city. As a graduate student, I had specialized in limnology, the study of freshwater lakes, and was given the responsibility for nearly a thousand bodies of water extending from the crest of the Sierra Nevada eastward to the Nevada state line. I was especially intrigued by the diversity of the landscape in my charge. If I left the roadhead near the base of 14,494 ft Mount Whitney at 9:00 a.m., I could make a leisurely drive to the east and have my lunch 282 ft below sea level on the floor of Death Valley. This area's life forms are commensurately diverse.

Today I sit at my desk surrounded by forty little pocket diaries, each one summarizing a year of my career. So many memories and experiences are packed into these 2.5- by 4-in volumes, which, together, fill less than a shoe box. Daily entries recall a multitude of experiences: scaling through the usual routine meetings, conducting a 27-year project to restore the California golden trout within the Golden Trout Wilderness (still in progress) [Stephens et al. 2019, this volume], fighting scores of ill-considered and highly destructive entrepreneurial invasions of valuable habitats and recreation areas, managing a legendary reservoir fishery where success is measured by tons of trout harvested, then moving 180 degrees from consumption to conservation by helping save the Devil's Hole Pupfish *Cyprinodon diabolis*, a battle carried successfully to the U.S. Supreme Court.

In 1976, the Court's landmark decision protected Devil's Hole---a swimming-pool-sized window into the underground aquifer and a disjunct portion of Death Valley National Monument and its dependent life forms from the impact of a nearby ranching operation. The ranchers were consuming vast quantities of un-replenishable groundwater from an aquifer that had been undisturbed since the Pleistocene. The smallest and most highly evolved of the Death Valley system pupfishes, the Devil's Hole pupfish has been isolated from nearby pupfish populations for approximately 44,000 years. It exists in probably the most confined habitat of any vertebrate animal in the world: the ten- by fifty-foot pool in which it has evolved since its isolation.

Of more than ten thousand entries contained in my diaries, the date August 18, 1969, stands alone as the most dramatic and meaningful. Written with naive understatement: "Transplanted Cyprinodon at Fish Slough; purchased alkaline D-cells, $2.00," this cryptic entry summarized a series of events that, had they not gone right, would have accompanied the greatest tragedy of my career. As it turned out, what happened that day simply underscored the lessons I had learned earlier from the Leopolds and other ecological mentors. Perhaps such an experience was necessary for me to fully comprehend that a person's values, which serve as a compass in uncertain times, are in the long run vastly more important than the sport-fishing technologies that have often created more problems than they have solved.

During the several pluvial periods of the Pleistocene epoch, much of the Great Basin of the American West was covered by large, freshwater lakes. With the approach of the Holocene, these waters shrank and largely disappeared, and fishes were isolated within the few remaining permanent aquatic habitats. In North America, only the Cuatro Cienegas of Coahuila, Mexico, have as many well-defined local populations (species confined to the very small, isolated habitats in which they evolved). The Death Valley drainage area of eastern California and western Nevada is comparable to Charles Darwin's Galapagos Islands and their finch populations. They constitute, in effect, islands of water in a sea of sand.

One such habitat exists in eastern California's Owens Valley, where the Owens Pupfish *C. radiosus* has been evolving since the Pleistocene. Because of major habitat changes and the introduction of predacious gamefishes (a deadly combination) during the early part of the 20th century, the Owens Pupfish was gradually eliminated from a range that once covered vast marshlands. By the time it was scientifically described in 1948, the species was believed to be extinct. One of the Death Valley area pupfishes, all of which evolved in the absence of predatory fishes, the Owens Pupfish is almost totally defenseless against such introduced predators as largemouth bass, which I call "chainsaws with fins." The Owens Pupfish was among the first fishes to be designated an endangered species, a status that it unfortunately still retains.

Pupfishes (named for their frolicsome, playful behavior) are members of the killifish family, a group of fishes very popular among aquarium enthusiasts. The Owens Pupfish is the largest of the nine Death Valley pupfishes, occasionally reaching two inches in length; the Devil's Hole Pupfish rarely exceeds one inch. Habitats are varied. The Owens Pupfish thrives in the shallow, warm water that hot summer days bring to desert marshes; this same habitat may be covered with an inch or two of ice during wintertime, when air temperatures drop below zero (°F). Conversely, the Devil's Hole Pupfish lives in the upper reaches of a cavern so vast that its depth has never been determined, and in water at a constant 92°F. All pupfishes are feeding opportunists, consuming immature insects and algae. They are also highly territorial.

To survive in these rigorous habitats, pupfishes have evolved specialized adaptations. Some live in water that exceeds 100°F., and can tolerate up to 113°F for short periods; daily fluctuations may be as much as

36°F. Others live in pools with several times the salinity of seawater. The potential for research on the pupfishes is exciting. What they could tell us about kidney function, temperature tolerance, adaptation, and other areas of vertebrate physiology alone would justify our concern for preserving them. In recent years, however, it has been heartening to note a shift in emphasis from what they can do for us to what we can do for them, regardless of their potential value.

In 1964 researchers located a remnant population of Owens pupfish in a desert marshland called Fish Slough [Figure 2], a few miles from my home in Bishop, California. A recovery effort was started by gradually reintroducing them into a few apparently suitable habitats, thereby getting a jump on the more sophisticated recovery programs made possible later under the Endangered Species Act of 1973. These early preservation efforts for fishes preceded the relatively recent, and highly commendable, formalization of the science of conservation biology.

However, an unusual set of circumstances that began to coalesce in the late 1960s brought the Owens pupfish to the brink of extinction. Without constant surveillance, which even now is very difficult for harried state biologists to maintain, the Pupfish gradually disappeared from their new homes and finally were confined to a room-sized pond a short distance below Fish Slough's northwest headwater springs. The winter of 1968–69 had brought heavy rains to the Owens Valley, but by August the unusually thick vegetation was throwing off a great deal of moisture, and an unexplained reduction in spring flow contributed to the rapid depletion of the pond. It was almost completely dried up when an alert assistant came into my office and announced: "Phil, if we don't get out to Fish Slough immediately, we are going to lose the species." His pronouncement was no exaggeration. It was the hard truth!

I stopped work on a trout management program for a major reservoir (the relative importance of the two projects has long since served as a source of humor for me), shouted a few words of explanation to our receptionist, and bolted for the door. Grabbing buckets, dip nets, and aerators, we were joined by another colleague and immediately headed for Fish Slough, normally a fifteen-minute drive north of our office in Bishop (we shaved at least five minutes off the usual driving time). We hastened to the drying pond and carefully removed 800 remaining individuals, placing them in three wire mesh cages within the main northwest channel of the slough, in a diminishing flow already less than two cubic feet per second. We planned to move them later to safer locations within the same general area.

Having done all we could for the moment, we decided to take a quick dinner break before returning to move half of the fish (about 400) across the slough to a location supplied by another spring source. In endangered species preservation work, a cardinal rule is always to place your eggs in more than one basket. We had come very close to witnessing a species extinction or, nearly as bad, a population so reduced in numbers as to eventually effect the same tragic consequence.

Temporarily alone in the marsh, I decided to make one final check (sometimes it pays to be a worrier). A glance into the nearest mesh cage showed that we were not yet out of the woods. In our haste to rescue the fish, we had unwisely placed the cages in eddies away from the influence of the main current. Reduced water velocity and accompanying low dissolved oxygen were rapidly taking their toll. When taken from their natural habitat, pupfish are fragile creatures. They were overcrowded in their cages and had been stressed by unavoidably rough treatment on a hot summer afternoon.

A number of dead and dying fish were already floating belly up or swimming irregularly, and it was clear that both mesh cages and fish would have to be moved immediately upstream to more favorable conditions nearer the springheads. I ran to my pickup truck and found only two buckets (the other two were on their way back to town). However, there were two aerators available in addition to the all-important dip net.

I netted the surviving fish into the buckets, wincing as each dead one forcefully demonstrated the fragility of life. I then relocated the cages and returned to the buckets, trusting that the battery-powered aerators had not failed during my brief absence. Although the passage of time has obscured my exact words and thoughts as I lugged two heavy buckets and their precious cargo (each weighing more than thirty pounds) over the treacherous marsh terrain, I remember mumbling something like: "Please don't let me stumble. If I drop these buckets we won't have another chance!" I distinctly remember being scared to death. I had walked perhaps fifty yards when I realized that I literally held within my hands the existence of an entire vertebrate species. If I had tripped over a piece of barbed wire or stepped into a rodent burrow, the Owens Pupfish would now be extinct! But good fortune smiled upon us, and the recovery continues today.

Efforts to preserve endangered desert life forms never end, but essentially constitute only a temporary reprieve as aquatic habitats gradually decline throughout North America. Indiana University's Lynton Caldwell, speaking of our environmental crisis, observed that while endangered species are part of this lamentable phenomenon, "more importantly, the crisis is concerned with the kind of creatures we are and what we must become in order to survive."

We have received adequate warning from our prophets. Aldo Leopold's "Land Ethic," published more than forty years ago in *A Sand County Almanac*, redefined Gifford Pinchot's "resource conservation ethic" (the greatest good for the greatest number in the long run) and placed humans as simply another species within the global ecosystem. This concept has since become painfully obvious as we learn more about ourselves in relation to our environment.

Having spent much of the past two decades responding to the cynical question: "What good are they?" (in reference to my efforts on behalf of the pupfish and similar "insignificant" organisms), I have made use of an effective counter query: "What good are you?" (a very thoughtful question). I then add a Leopold corollary: "To keep every cog and wheel is the first precaution of intelligent tinkering."

Rank-and-file American citizens have been generally apathetic about the conservation of biological diversity, but one would hope not to find similar unconcern within the scientific community. Yet there is much complacency among professionals, particularly among those biologists trapped within a tenure track and faculty advancement syndrome that often ranks quantity over quality in the research endeavor. If such scientists express an interest in conservation, they usually are of the opinion (naively and incorrectly) that someone else will attend to saving species. At the 1992 annual meeting of the American Society of Ichthyologists and Herpetologists, for instance, only a small percentage of 385 research papers related to the specific area of conservation.

Workers in the pragmatic field of conservation biology, frustrated by a critical need for answers to questions posed by species recovery programs, draw analogies of mowing the lawn while the house burns down. The possibility always exists, of course, that any research, no matter how seemingly esoteric, may someday be of value in saving a species. Albert Einstein put it this way: "I have little patience with scientists who take a board of wood, look for its thinnest part, and drill a great number of holes where the drilling is easy." Unfortunately, the deadly serious matter of preserving biodiversity generally places one in the position of facing unpredictably thick boards, full of knots, and then being forced to drill holes with a bit significantly dulled by the bureaucratic process.

As I walked back to my truck following the final transplant within Fish Slough, the sun had long ago set. In my dip net remained a few dead pupfish. I glanced up at the darkening desert sky and thought of Pierre Teilhard de Chardin's concept of the infinitely large, the infinitely small, and the infinitely complex, represented here (in order) by the Milky Way, the pupfish, and the difficulty in pointing out the paramount value of such things to an increasingly materialistic society.

The day had been long. We had won an early round in a fight that will inevitably continue as long as we have a habitable planet. As a realist, I could not help but ponder the ultimate fate not only of the Owens pupfish but of all southwestern fishes and species in general. I wondered about our future. Can the values driving the industrialized nations be modified sufficiently to allow for the perpetuation of all species, including humans? Will we ever realize the potential implicit in our specific designation as *Homo sapiens*, the wise species? I hope the day will come when public policy will be guided by the wisdom of Aldo Leopold: "A thing is right when it tends to preserve the integrity, stability, and beauty of the biotic community. It is wrong when it tends otherwise." Such recognition could constitute perhaps the first major step toward creating the sustainable society upon which our long-term survival obviously depends.

That August day twenty-three years ago (from 1993 [meaning 1969]) had been a very humbling experience for me. The principles of biogeography and evolution I had learned many years before at Berkeley had taught me why the pupfish was here; it took the events of those few hours in the desert to teach me why I was. Such are the reflections of a biologist who, for a few frightening moments long ago, held an entire species in two buckets, one in either hand, with only himself standing between life and extinction. [end of **SPECIES IN A BUCKET** – reprinted from *Natural History*, January 1993, copyright (c) Natural History Magazine, Inc., 1993]

CONCLUSION

Well, there you have it—now more than 25 years since publication. This is the best I could do to point out the fact that today's students will be given many chances to make things better during the course of their careers. I cannot think of anything I might change as I fade into the sunset. I am very grateful for the circumstances I have been given, although this might be a good place to invoke the wisdom of Pogo Possum (Kelly 2011) when he stated (as you will find) that he was sometimes "faced with insurmountable opportunities." I wish you the very best as you take over. The future is bright.

EDITORS' POSTSCRIPT

One half of the 800 rescued Owens Pupfish were successfully transplanted into BLM Spring, a small, natural desert spring in Fish Slough, and from there, over subsequent years, 83 transfers involving 31,995 individuals were placed into 27 habitats (Finger et al. 2013). From these introduction efforts, six extant populations exist today, numbering from a few hundred fish to possibly as many as 10,000 pupfish, cumulatively occupying a habitat of 0.125 ha (Finger et al. 2013). The translocations and numbers of individuals used in the founding events have led to some inadvertent losses of genetic variation through founder effects and demographic fluctuation (Finger et al. 2013). The Owens Pupfish remains listed under the Endangered

Species Act but is not extinct thanks to the heroic and extraordinary efforts of Phil Pister and his team!

REFERENCES

Finger, A. J., S. Parmenter, and B. P. May. 2013. Conservation of the Owens Pupfish: genetic effects of multiple translocations and extirpations. Transactions of the American Fisheries Society 142:1430–1443.

Kelly, W. 2011. Pogo: the complete syndicated comic strips, volume 1. Through the wild blue wonder. Fantagraphics, Seattle.

Leopold, A. 1949. A sand county almanac. Oxford University Press, New York.

Miller, R. R., and E. P. Pister. 1971. Management of the Owens Pupfish, *Cyprinodon radiosus*, in Mono County, California. Transactions of the American Fisheries Society 100:502–509.

Pister, E. P. 1993. Species in a bucket. Natural History 102:14–19.

Pister, E. P. 2001. Threatened fishes of the world: *Cyprinodon radiosus* Miller, 1948 (Cyprinodontidae). Environmental Biology of Fishes 61:370.

Stephens, M. R., S. J. Stephens, and C. C. Krueger. 2019. Lessons learned from more than one-hundred years of Golden Trout management and recovery in California. Pages 19–43 *in* C. C. Krueger, W. W. Taylor, and S. Youn, editors. From catastrophe to recovery: stories of fishery management success. American Fisheries Society, Bethesda, Maryland.

Wilke, P. J., and H. W. Lawton, editors. 1976. The expedition of Capt. J. W. Davidson from Fort Tejon to the Owens Valley in 1859. Ballena Press, Socorro, New Mexico.

U.S. Fish and Wildlife Service. 1998. Owens Basin wetland and aquatic species recovery plan, Inyo and Mono Counties, California. U.S. Fish and Wildlife Service, Region 1, Portland, Oregon.

U.S. Fish and Wildlife Service. 2009. Owens Pupfish (*Cyprinodon radiosus*) 5-year review: summary and evaluation. U.S. Fish and Wildlife Service, Ventura Fish and Wildlife Office, Ventura, California.

U.S. Office of the Federal Register. 1967. Notices: Native fish and wildlife endangered species. Federal Register 32:48(11 March 1967):4001.

Section III

Laurentian Great Lakes

Restoration of Lake Trout in Lake Superior through Interagency Cooperative Management

Michael J. Hansen*
U.S. Geological Survey, Great Lakes Science Center, Hammond Bay Biological Station 11188 Ray Road, Millersburg, Michigan 49759, USA

Charles R. Bronte
U.S. Fish and Wildlife Service, Green Bay Fish and Wildlife Conservation Office 2661 Scott Tower Drive, New Franken, Wisconsin 54229, USA

Abstract.—The Lake Trout *Salvelinus namaycush* is a keystone species in the Laurentian Great Lakes that supported valuable fisheries throughout the basin until the 1950s. However, Lake Trout populations declined to near extirpation in nearly all of the lakes by the 1960s because of the combined effects of overfishing, Sea Lamprey *Petromyzon marinus* predation, and habitat degradation. To restore self-sustaining Lake Trout populations in Lake Superior, state, provincial, federal, and tribal agencies agreed to an interjurisdictional management framework that allowed them to articulate and institute (1) clear and common goals and actions for recovery, (2) early and intensive lakewide stocking of hatchery-reared Lake Trout to enhance failing stocks, (3) early and effective lakewide controls on mortality caused by Sea lampreys and fisheries, and (4) standardized lakewide evaluations of population trajectories and performance. Stocking was initiated in Lake Superior in 1950 and expanded after 1953, prior to effecting Sea Lamprey or fishery controls, thereby introducing large numbers of hatchery-origin fish that grew to maturity shortly after mortality was reduced. Abundant suitable nearshore spawning habitat was widely available for naive lean hatchery-origin Lake Trout, and native lean Lake Trout persisted in some areas. The Sea Lamprey-selective pesticide TFM (3-trifluoromethyl-4-nitrophenol) was applied first in Lake Superior in 1958 because of the presence of remnant native Lake Trout populations, which set the stage for closure of fisheries and good survival of newly stocked and remnant wild fish. As a consequence of these four factors, stocked fish exceeded historical density of wild fish by the 1980s in many areas and thereby generated enhanced reproductive potential when combined with remnant wild fish. Lake Trout recovery in Lake Superior is an extraordinary example of agency cooperation toward a common goal for managing recovery of an ecologically important shared resource.

INTRODUCTION

The Laurentian Great Lakes contain the largest volume of surface freshwater in the world and once supported valuable fisheries in Canada and the United States (Moll et al. 2013). Prior to European settlement, the lakes contributed to the sustenance, transportation, and trade of native peoples that lived in the region (Bogue 2000). Following European settlement, the lakes sustained fisheries that

* Corresponding author: michaelhansen@usgs.gov

initially supported local markets and later supported regional, national, and international markets as preservation and transportation systems were developed (Brenden et al. 2013). Fisheries initially developed unencumbered by regulations so exploitation occurred with little regard for cultural and biological impacts on the lakes or their surrounding drainage basins (Brenden et al. 2013). As a consequence of unbridled fishery development (Brenden et al. 2013), intended and accidental introductions of exotic species (Mandrak and Cudmore 2013), and alterations in aquatic habitats (Hayes 2013), most Great Lake fisheries underwent dramatic, and sometimes catastrophic, changes (Brenden et al. 2013). The Lake Trout *Salvelinus namaycush* was an important food to human settlements on each of the Great Lakes and underwent catastrophic collapses in each lake in the 19th and 20th centuries (Muir et al. 2013).

In Lake Superior, Lake Trout stocks were rebuilt through an interjurisdictional management program that we describe herein as an example of successful interagency cooperation in the restoration of an ecologically and socially important species and fishery.

HISTORICAL CONTEXT

The fish fauna of the Laurentian Great Lakes arose within the past 10,000 years after the Pleistocene glaciation (Lawrie and Rahrer 1973) and includes 28 families, 71 genera, and 174 species (Bailey and Smith 1981). The Lake Trout differentiated within each lake into different morphological forms (Khan and Qadri 1970; Goodier 1981; Krueger and Ihssen 1995). Lake Trout formed discrete spawning stocks that used tributary streams, offshore shoals, rocky shorelines, and deepwater sites throughout the Great Lakes, particularly in Lake Huron (Berst and Spangler 1973), Lake Michigan (Wells and McLain 1973), and Lake Superior (Lawrie and Rahrer 1973). Accounts of the 19th and early 20th century fisheries of the Great Lakes described many forms of Lake Trout, including a lean form that lived mostly in shallow water; a fat form, the siscowet, that lived mostly in deep water; and an intergrade form, the humper, that lived between the shallow-water and deepwater forms (Sweeny 1890; Brown et al. 1981; Goodier 1981).

Settlement of the Great Lakes basin brought about massive changes in its fisheries, through excessive fishery exploitation, introductions of deleterious nonnative species, and alteration of habitat quality and quantity (Berst and Spangler 1973; Christie 1973, Hartman 1973; Lawrie and Rahrer 1973; Wells and McLain 1973). The Lake Trout had one of the highest commercial values of all Great Lakes fish so excessive fishery exploitation has been implicated in its demise in all five of the Great Lakes. Because of its large size relative to other species, the Lake Trout was also the principal prey for the Sea Lamprey, which invaded the lakes upstream from Niagara Falls after 1900, so Sea Lamprey-induced mortality has also been implicated in the demise of Lake Trout populations in each lake (Smith and Tibbles 1980). Alteration of Lake Trout habitat has been blamed less often for the collapse of Lake Trout stocks and associated fisheries, although damming of rivers has been implicated in the early loss of river-spawning Lake Trout populations.

Lake Trout stocks collapsed after 1900 in Lake Ontario and in the other lakes throughout the century as a consequence of Sea Lamprey *Petromyzon marinus* predation, fishery exploitation, and habitat loss (Krueger and Ihssen 1995). Comprehensive programs to control the Sea Lamprey, stock hatchery-reared Lake Trout, and regulate fisheries saved a few nearshore lean Lake Trout and most offshore siscowet stocks in Lake Superior (Hansen et al. 1995a), but nearly all other populations in the Great Lakes were eliminated by these stressors. Four Lake Trout forms still survive in Lake Superior: the lean form, which inhabits most oligotrophic (coldwater) lakes in North America; the siscowet form, which also occurs in Great Slave Lake, Northwest Territories (Zimmerman et al. 2006; Hansen et al. 2016a); the humper form, which also occurs in Rush Lake, Michigan (Chavarie et al. 2017) and Lake Mistassini, Quebec (Zimmerman et al. 2007; Hansen et al. 2012); and the redfin form, which was recently redescribed in Lake Superior (Muir et al. 2014; Hansen et al. 2016b). Unless otherwise specified, most discussion, especially related to harvest and stocking, refers to the nearshore lean form that has been the focus of fishery management because of its great economic value, although it always represented a small fraction of total Lake Trout biomass in Lake Superior (Bronte et al. 2003a).

Fishery Exploitation

Native peoples were the first users of Great Lakes fish resources prior to European settlement. These peoples fished with spears, hooks, traps, and gill nets that were primitive by modern standards so their exploitation was limited to local stocks that spawned or lived in nearshore habitats or rivers (Lawrie and Rahrer 1973). Lake Trout were targeted by these early fisheries, along with Lake Sturgeon *Acipenser fulvescens*, Lake Whitefish *Coregonus clupeaformis*, suckers *Catostomus* spp., and Cisco *Coregonus artedi*. Early explorers of the 17th and 18th centuries also caught fish for their own subsistence. The size of these presettlement fisheries has not been quantified but was likely only a small fraction of the size of postsettlement fisheries (Groop 2013).

Europeans settled the Great Lakes basin from east to west, so fisheries developed earlier in Lakes Ontario and Erie than in Lakes Michigan, Huron, and Superior (Groop 2013). Fur trading outposts and forts facilitated development of fisheries by engaging in the sale of salted fish after the trade in beaver *Castor canadensis* pelts declined. In the 1830s, the American Fur Company reorganized their network of workers and outposts to harvest and pack fish for delivery to markets in eastern North America (Nute 1926). The estimated annual yield of such outpost fisheries in Lake Superior was 45,000–91,000 kg, principally of siscowet and lean Lake Trout and Lake Whitefish (Lawrie and Rahrer 1973). The fishery grew slowly to a peak of 450,000 kg in 1839, just before failure of the American Fur Company, when demand for food in eastern markets was met by fishery and agricultural output from the Ohio River valley.

Fisheries development in Lake Superior was renewed during the latter part of the 1800s when the area was settled by immigrants from Norway, Sweden, and Finland, who had strong fishing traditions and diets that preferred fish products. This influx of fishermen and new innovations in harvesting (nets and boats) and transportation (preservation and transportation) methods in the late 1800s facilitated expansion of fisheries for Lake Trout and other species in the upper Great Lakes. Steam-powered fishing tugs were introduced in 1871 to replace smaller Mackinaw sailboats, and steam-powered gill-net lifters were introduced in 1890 to replace hand lifting of nets and baited setlines (Berst and Spangler 1973; Lawrie and Rahrer 1973; Wells and McLain 1973). These were later replaced by gas and diesel marine engines and mechanical/hydraulic gill-net lifters. Fishing operations were therefore able to travel farther away from ports of call and deploy greater amounts of gear than before. Marketing of fish was improved during the period by packing fish on ice, taken from the lakes in the winter, and then rapidly transporting the fish to markets by freighter and newly built rail lines. Fisheries grew rapidly in each lake in the 1870s and 1880s as fisheries expanded from local, inshore fishing grounds to more distant, offshore fishing grounds. In Lake Superior, Lake Trout yield was sustained through 1950, but declined from 2.1 million to 0.23 million kg during 1950–1960 as stocks collapsed from excessive mortality caused by fishing and Sea Lamprey predation (Figure 1; Baldwin et al. 1979). Lean Lake Trout stocks in Lake Superior persisted only at offshore islands and reefs (Lawrie 1978). Siscowet Lake Trout also declined, even under relatively light exploitation but, unlike lean Lake Trout, persisted throughout the lake (Bronte and Sitar 2008).

Excessive fishery exploitation was primarily responsible for the collapse of Lake Trout stocks in Lake Superior in the 1950s (Hile et al. 1951a), where yield was unrelated to abundance because abundance was being suppressed by increasing fishing intensity:

> The upward trend of fishing pressure and the high negative correlation between abundance and fishing intensity...invite the belief that the two may be related causally—that rising fishing pressure has brought about a decrease of abundance and that the fishermen in the face of this reduced availability have intensified their efforts in order to keep production at a good level. If so, a continuation of present trends until fishing becomes unprofitable is to be anticipated. On the other hand, it is possible that the rising fishing pressure has been merely the result of a natural decline in abundance but has not contributed materially to it. In this event, we can hope for a natural upturn of abundance which would relieve the precarious status of the fishery. Regardless of the causes that led to the present situation, the fact remains that the Lake Trout stocks ... of Lake Superior are fast nearing a dangerously low level and are in poor condition to withstand the impending ravages of a growing population of Sea Lampreys. [From Hile et al. 1951a.]

Needless to say, a natural upturn of Lake Trout abundance did not occur in Lake Superior, as Hile et al. (1951a) hoped for, and stocks continued to decline through the early 1960s as Sea Lamprey predation increased.

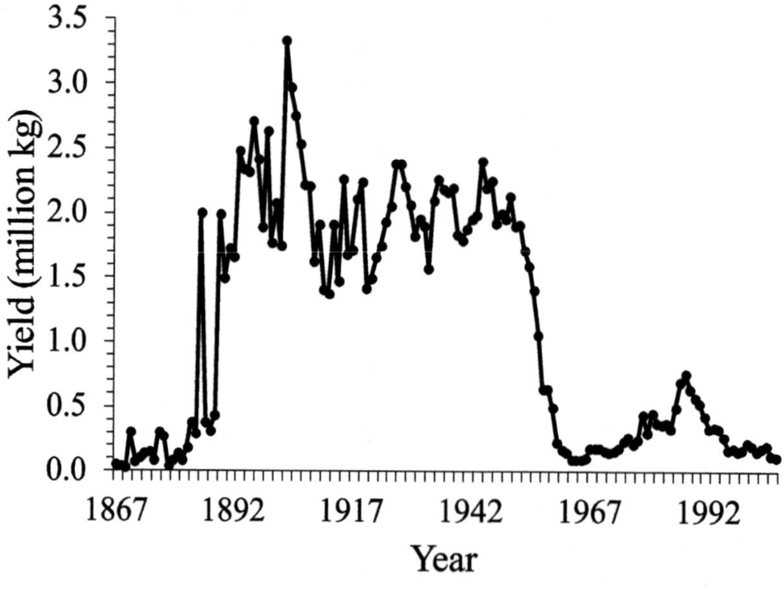

FIGURE 1. Yield of all Lake Trout forms to commercial fisheries in Lake Superior during 1867–2006 (data from Baldwin et al. 1979).

Pycha and King (1975) expanded the analysis of commercial fishery data through 1970 and into Wisconsin's waters of Lake Superior and concluded that "intensive fishing, aided by the introduction of nylon gillnets (from cotton which increased efficiency), was the principal factor involved in the early years of the post-World War II decline of Lake Trout in Lake Superior."

Sea Lamprey Predation

The Sea Lamprey was thought to have invaded Lake Ontario through the Oswego Bypass of the Erie Canal by 1835, but some have argued that they were native to that lake (Waldman et al. 2004, 2009). Completion of the Welland Canal in 1829 permitted Sea Lampreys to bypass Niagara Falls and reach Lake Erie from Lake Ontario, but the first Sea Lamprey was not found in Lake Erie until 1921 and the first spawning run was not observed until 1932 (Smith 1972). Colonization of the upper Great Lakes was more rapid, as Sea Lampreys were first found in Lake St. Clair in 1934, Lake Huron in 1937, Lake Michigan in 1936, and Lake Superior in 1946. The date of first encounter in Lake Michigan, however, suggests that Sea Lampreys made their way into Lake Huron before they were first found in Lake Michigan, perhaps as early as 1932. Spawning runs of Sea Lampreys were monitored only in Lake Superior, but long-term monitoring of runs in the Ocqueoc River in Lake Huron and Hibbards Creek in Lake Michigan suggest that numbers of adult Sea Lampreys were too few to have caused initial declines of Lake Trout populations in any of the upper Great Lakes (Smith 1968).

The relative impacts of fisheries and Sea Lampreys on Lake Trout stocks in the upper Great Lakes is unclear because of the relative timing of arrival of Sea Lampreys in each lake in relation to the onset of stock decline. Unlike in the other Great Lakes, Lake Trout abundance in Lake Superior declined continuously after 1929 (Hansen 1999), before Sea Lampreys were first observed (1946) or were numerous enough (mid 1950s and later) to have caused the collapse of Lake Trout stocks after 1950 (Hile et al. 1951a; Pycha and King 1975; Wilberg et al. 2003; Bronte and Sitar 2008; Figure 2). Jensen (1978) used a surplus production model based on commercial fishery data from central Michigan waters of Lake Superior to infer that overfishing of Lake Trout stocks began in 1945, whereas Sea Lampreys were first reported in 1946. Coble et al. (1990) used regression analysis of trends in commercial fishery data to determine that abundance of Lake Trout (of all forms) began to decline lakewide as early as 1939 in the Michigan gill-net fishery, well before the Sea Lamprey became abundant. Although not the primary cause of Lake Trout stock collapse in Lake Superior, Sea Lamprey predation contributed toward the end of the collapse that was driven by overfishing (Hansen 1999).

Other Exotic Species

Colonization of the Great Lakes basin by exotic species has altered fish communities in each lake. The Alewife *Alosa pseudoharengus* and Sea Lamprey colonized the Great Lakes basin using the same pathway, so the Alewife was first found in each lake within a decade after the Sea Lamprey (Smith 1972). The Alewife became abundant in Lake Huron, Lake Michigan, and Lake Ontario during the 1950s, where water temperatures were suitable and predators were virtually nonexistent, but not in Lake Superior, where temperatures were too cold (Bronte et al. 1991), or in Lake Erie, where abundant predators existed (Smith 1972). The Rainbow Smelt *Osmerus mordax*, in contrast to the Alewife, colonized the upper Great Lakes basin from Crystal Lake, Michigan, where they were stocked and escaped into Lake Michigan in 1912, and colonized Lake Ontario from multiple sources (Smith 1972; Bergstedt 1983). Therefore, the Rainbow Smelt was found be-

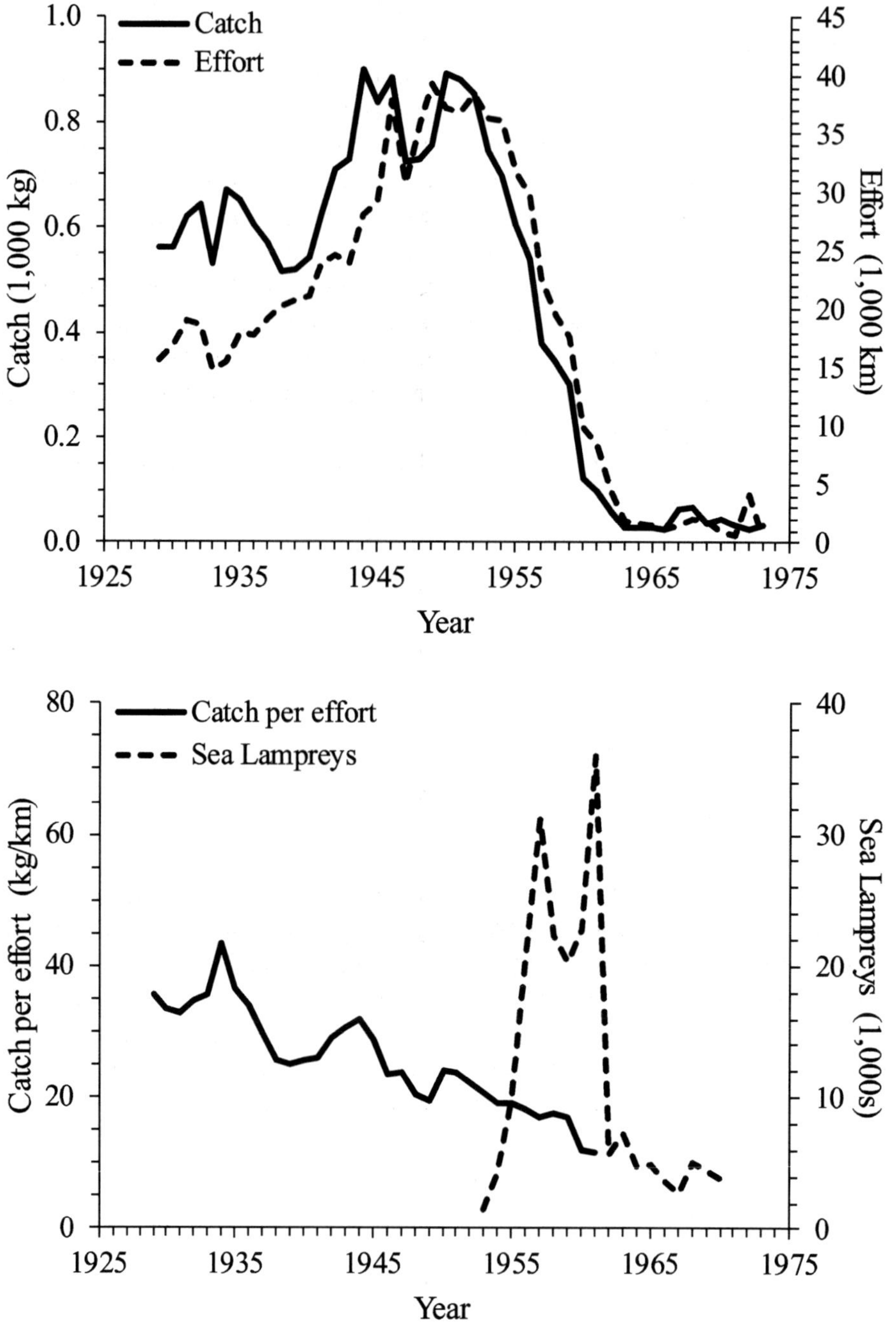

FIGURE 2. Upper panel: catch of all Lake Trout forms and large-mesh gill-net fishing effort (data from Jensen and Buettner 1976). Lower panel: Lake Trout catch per effort in Michigan large-mesh gill-net fisheries (adjusted for conversion from cotton to nylon twine by Pycha and King 1975) and Sea Lamprey abundance (data from Smith 1968) in Michigan waters of Lake Superior in 1929–1961.

fore either the Alewife or the Sea Lamprey in Lake Michigan, Lake Huron, and Lake Superior, but after both species arrived in Lakes Erie and Ontario (Smith 1972). The Rainbow Smelt became predominant only in Lakes Erie and Superior, compared to the other lakes, where the Alewife became the predominant nonnative forage (Smith 1972). The Rainbow Smelt has been the predominant prey for Lake Trout in Lake Superior since the mid-1950s (Dryer et al. 1965; Conner et al. 1993; Ray et al. 2007). Ciscoes were the most common prey for Lake Trout in 1950–1953 (52.2–87.5% by volume), whereas Rainbow Smelt increased from 1.7% of food volume in 1950 to 42.1% in 1952, 27.3% in 1953, and 65.6% in 1963 (Dryer et al. 1965).

Habitat Degradation

Alteration of Lake Trout habitat has not often been blamed for the collapse of Lake Trout stocks and associated fisheries in Lake Superior, although damming of rivers has been implicated in the early loss of river-spawning Lake Trout stocks that supported small, local populations (Loftus 1958; Jones et al. 2018). Most such river-spawning populations were lost before the period of record, so their presence in specific streams is not known with certainty. In the other Great Lakes, habitat loss or degradation has occurred to a certain extent, including destruction of stream and inshore spawning habitat, eutrophication and consequent oxygen depletion of open waters, loading by organic contaminants and consequent disruption of spawning viability, and colonization by exotic species and consequent alteration of fish communities (Berst and Spangler 1973; Christie 1973; Hartman 1973; Lawrie and Rahrer 1973; Wells and McLain 1973). However, effects of habitat degradation on Lake Trout stocks have been difficult to impute in Lake Superior, where water quality and inshore and offshore spawning habitat has not been degraded enough to impair Lake Trout populations. For example, organic contaminants were suggested to cause reproductive problems in Lake Trout because of coherence of trends in ambient contaminant levels, egg hatching, and fry survival, but definitive proof of effects on Lake Trout reproduction in the wild has been lacking (Mac and Edsall 1991). Concentrations of contaminants were never high enough to impair early survival of Lake Trout, even in Lake Michigan and Lake Ontario (Fitzsimons 1995), where contaminant loadings exceeded that of Lake Superior.

RECOVERY COORDINATION

The decline of Lake Trout and Sea Lamprey invasion of the upper Great Lakes led to interagency cooperation in fishery management and research, which, in conjunction with relatively pristine habitat and presence of remnant native Lake Trout of multiple morphs, enabled Lake Trout stocks to recover throughout Lake Superior. Interagency cooperation enabled agencies to articulate and institute (1) clear and common goals and actions for recovery, (2) early and intensive lakewide stocking of hatchery-reared Lake Trout to enhance failing stocks, (3) early and effective lakewide controls on mortality caused by Sea Lampreys and fisheries, and (4) standardized lakewide evaluations of population trajectories and performance. The first such efforts were formalized as the Great Lakes Lake Trout and Sea Lamprey committees in 1946 (Smith 1972), which were merged in 1952, renamed the Great Lakes Fishery Committee in 1953, and then replaced by the Great Lakes Fishery Commission in 1955, under the Convention on Great Lakes Fisheries between the United States and Canada, to undertake Sea Lamprey control and coordinate fishery research and management (Gaden et al. 2013). The U.S. Fish and Wildlife Service and the Fisheries Research Board of Canada were contracted to perform Sea Lamprey control. In 1966, the Sea Lamprey control program in

Canada was transferred to the Department of Fisheries and later the Department of Fisheries and Oceans. During the late 1940s, the U.S. Congress directed the Fish and Wildlife Service to locate sites to build Lake Trout hatcheries. Shortly thereafter, federal restocking began in 1953 in Lake Superior to join state and provincial stocking that was already underway (Figure 3).

Under the auspices of the Great Lakes Fishery Commission, Lake Committees for each of the Great Lakes made up of fishery managers from each management agency were formed to coordinate interagency fishery management actions (Gaden et al. 2008, 2013). Each Lake Committee appointed a Lake Technical Committee made up of fishery biologists and researchers from each agency to serve as technical advisors on matters of fishery science and management advice. By the mid-1980s, Lake Technical Committees drafted Lake Trout restoration or rehabilitation plans for each lake that included lakewide targets for Lake Trout stock restoration or rehabilitation, management actions to enhance recruitment and survival, pertinent fishery research topics, and standards for coordinating stock assessment programs. The importance of these two committees in advancing Lake Trout recovery in Lake Superior cannot be overemphasized because fishery biologists and managers from all jurisdictions were forced to meet face to face several times each year to discuss Lake Trout recovery across the entire lake. This process fostered trust among agencies that led to cooperation and coordination and, more importantly, implementation of steps to achieve lakewide recovery of Lake Trout stocks.

Lake Trout restoration or rehabilitation targets were developed for each lake to mimic historically sustainable yield, a need to enhance recruitment through stocking, and a need to limit mortality through Sea Lamprey control and fishery regulation. Plans for each lake had targets for sustainable yield that anticipated (rightly or wrongly) a future state that would be at least reasonably simi-

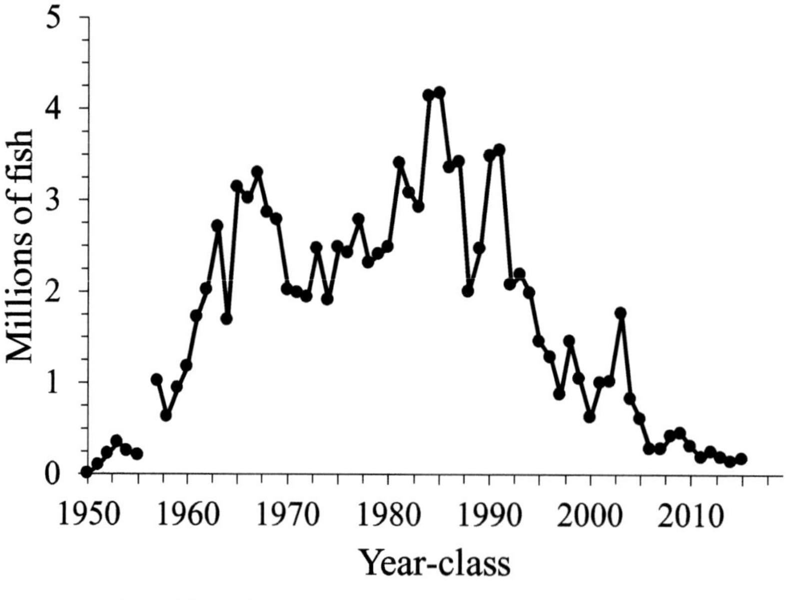

FIGURE 3. Annual stocking of lean Lake Trout (yearling equivalents = yearlings + fingerlings converted into yearling equivalents using average relative survival) by all agencies in all jurisdictions of Lake Superior for the 1951–2014 year-classes.

lar to historic Lake Trout production. Targets for stocking were virtually all driven by early evaluations in Lake Superior (see below) that were adapted to each lake based on the distribution and amount of Lake Trout spawning habitat. Targets for mortality were based on a review by Healey (1978), which suggested that self-sustaining Lake Trout populations persisted only if total annual mortality was at or below 50% but were adapted to each lake based on growth and maturity rates for Lake Trout (e.g., Hansen 1996).

In Lake Superior, the goal of Lake Trout restoration was set in 1986 and reiterated in 1996 to achieve a sustained annual yield of 2 million kg from naturally reproducing stocks (LSLTTC 1986; Hansen 1996). Because contemporary commercial fisheries were restrained below historical yield levels, to allow stocks to rebuild, an intermediate objective was set to restore wild recruitment by stocking and reducing total mortality to levels consistent with rehabilitation. The plan therefore focused on strategies for ensuring that this intermediate objective was achieved: (1) stocking yearling Lake Trout at an average size of 40–55 fish/kg, derived from wild strains that spawned in stocking areas, at a rate of 232–347 fish/km^2 of Lake Trout habitat; and (2) limiting total annual mortality to 42% (indexed as 50% on age 7–11 Lake Trout caught in 114-mm, 4.5-in stretch-mesh gill nets). In 1996, the restoration plan was refocused on criteria to eliminate stocking while sustainably managing self-sustaining populations (Hansen 1996) because increased natural recruitment suggested that stocking rates could be reduced in Michigan (Hansen 1988), Wisconsin (Hansen 1989), and Ontario (Powell and Atkinson 1991) waters.

CHALLENGES, SOLUTIONS, AND ACTIONS

Fishery managers had a relatively short time to react to the collapse of Lake Trout stocks in the upper Great Lakes, and most were convinced that Sea Lamprey predation was the primary cause, rather than overharvest (Hile 1949; Hile et al. 1951b). Before a selective Sea Lamprey toxicant was discovered in 1957 (Smith 1971; Smith et al. 1974), lean Lake Trout stocks were mostly driven to extirpation. As described above, Lake Trout yield collapsed in one decade in Lake Superior (1950–1960). State and provincial fishery management agencies in Lake Superior initiated stocking in 1950, and the U.S. Fish and Wildlife Service began stocking in 1953. Sea Lamprey control was initiated in 1950 by installing barriers in streams to block adult Sea Lampreys from migrating up streams to spawn and in 1958 by applying a selective toxicant (3-trifluoromethyl-4-nitrophenol [TFM]) to kill larval Sea Lampreys living in streams (Smith and Tibbles 1980; Heinrich et al. 2003). Lake Trout fisheries were closed to protect remaining stocks in 1962, when Sea Lamprey control was observed to be effective (Pycha and King 1975; Hansen et al. 1995a). Subsequently, management of Lake Trout in Lake Superior focused on restoration of self-sustaining stocks, especially where native stocks persisted in the face of overfishing and Sea Lamprey predation.

Stocking

Stocking of hatchery-reared juvenile Lake Trout was initiated in the early 1950s in Lake Superior by the states of Michigan, Minnesota, and Wisconsin, the province of Ontario, and, under Congressional mandate, by the U.S. Fish and Wildlife Service to stave off the complete demise of failing stocks (Lawrie and Rahrer 1973; Lawrie 1978). Stocking increased from less than 500,000 yearlings in the 1950s to a peak of nearly 5 million yearlings in 1985 (Figure 3). Targets for Lake Trout stocking in the Great Lakes were largely based on analyses of survival of early plantings in Lake Superior (Pycha and King 1967).

In Keweenaw Bay, Lake Trout reproduced when 347 yearlings/km^2 of habitat area ≤73 m (bottom depth) were stocked for 5 years (1965–1969), followed by 232 yearlings/km^2 (1.0 per acre) in the 1970s, as stocking expanded to more areas (Hansen et al. 1995a). Based on these observations, agencies agreed to targets for stocking in the lakewide Lake Trout recovery plan for each management area across all jurisdictions in the lake, which agencies fulfilled from the network of hatchery systems (Figure 3). Lake Trout stocking rates in Lake Superior were generally lower than targets because of limited hatchery production capacity, which slowed the rate of stock rebuilding: 105 yearlings/km^2 (330,000/year) since 1951 in Wisconsin, 124 yearlings/km^2 (796,000/year) since 1952 in Michigan, 153 yearlings/km^2 (949,000/year) since 1957 in Ontario, and 543 yearlings/km^2 (333,000/year) since 1962 in Minnesota (Hansen et al. 1995a). Yearlings survived 4–10 times better than fingerlings (Buettner 1961; Pycha and King 1967), so yearlings predominated over fingerlings in each jurisdiction: 81% in Wisconsin, 91% in Michigan, 89% in Ontario, and 88% in Minnesota (Hansen et al. 1995a). Stocking of fertilized eggs in artificial turf bundles was also attempted to imprint Lake Trout to historically important offshore spawning reefs (in contrast to inshore areas, where inshore stocking was successful) and was successful in recolonizing a spawning population at Devils Island Shoal in Wisconsin (Bronte et al. 2002).

Sea Lamprey Control

The Great Lakes Fishery Commission implemented coordinated basinwide Sea Lamprey control on behalf of the U.S. and Canadian governments by contracting Sea Lamprey control to the U.S. Fish and Wildlife Service and Department of Fisheries and Oceans Canada (Gaden et al. 2013). The earliest attempts to control Sea Lampreys used mechanical and electrical weirs and traps in spawning streams (Smith and Tibbles 1980; Heinrich et al. 2003). In Lake Superior, mechanical weirs were installed in two streams in 1950 and 1951, and electrical weirs were tested in 1952 and installed in 162 streams by 1960. Electrical weirs were used to monitor changes in adult Sea Lamprey populations in varying numbers of streams after 1958 but did not clearly reduce adult populations. Permanent barriers have also been built on several streams. Other types of barriers were tested for integration into the control program, including velocity barriers to block slow-swimming Sea Lampreys without blocking faster swimming fish, barriers that inflate only during periods of Sea Lamprey spawning migrations, and portable traps and fish wheels to intercept portions of particular spawning runs of Sea Lamprey.

In the search for a chemical means of controlling Sea Lamprey, ~6,000 candidate chemicals were screened for their selective toxicity to the Sea Lamprey using standard bioassay methods before TFM was selected to control larval Sea Lampreys in streams (Smith and Tibbles 1980; Heinrich et al. 2003). The use of TFM began in Lake Superior in 1958. Over the years, all streams that produced Sea Lampreys were treated at least once, and many have been treated numerous times, mostly on a 3-year cycle that mimics the time spent by larvae in streams before out-migrating to the lakes to feed and mature for a year or two before returning to streams to reproduce.

Releases of sterile-male Sea Lampreys, which can theoretically reduce recruitment through reduction in spawning success, were tested in Lake Superior during 1987–1996 and then integrated into the overall control program on the St. Marys River, the outlet to Lake Superior that flows into northern Lake Huron, beginning in 1997 (Heinrich et al. 2003). The sterile-male program was discontinued in 2013 in favor of targeted toxicant applications that were judged to be more effective for suppressing Sea Lamprey abundance than releases of sterile-male Sea Lampreys.

Fishery Regulation

The Great Lakes Fishery Commission coordinates cooperative lakewide fishery research and management among state, provincial, tribal, and federal agencies through annual meetings of lake technical committees and lake committees (Gaden et al. 2013). Agencies cooperated in closing Lake Trout fisheries throughout the upper Great Lakes in 1962, when use of TFM had effectively reduced Sea Lamprey abundance in Lake Superior and Lake Trout could therefore be saved from predation (Pycha and King 1975). Since then, agencies focused on protecting recovering Lake Trout populations from commercial harvest while allowing limited harvest by recreational and treaty-reaffirmed fisheries (Cornelius et al. 1995; Elrod et al. 1995; Eshenroder et al. 1995; Hansen et al. 1995a; Holey et al. 1995). Various strategies have been used to limit harvest in commercial fisheries within sustainable levels: prohibition of lethal gears such as large-mesh gill nets, mandatory release of all Lake Trout caught while fishing for other species, use of catch quotas in conjunction with mandatory tagging of all fish caught legally under quota prior to sale, prohibition of fishing during the Lake Trout spawning season, establishment of primary restoration areas where commercial fishing is not allowed but recreational fishing is permitted, and establishment of refuge areas where all fishing is prohibited. Nonetheless, fishery agencies gradually allowed commercial fisheries for Lake Whitefish to expand in Wisconsin after 1970 and in Michigan after 1980, with catch quotas and tagging systems on Lake Trout that limited the number of fish reaching markets but not the number of Lake Trout that died in fishing operations (Hansen et al. 1995a). Consequently, stock recovery and natural reproduction were slower than if large-scale fisheries were not allowed to redevelop.

Commercial fisheries in the upper Great Lakes expanded in the 1970s when state and federal courts reaffirmed treaty-based fishing rights by various Native American tribes in Michigan and Wisconsin (Eshenroder et al. 1995; Hansen et al. 1995a; Holey et al. 1995). Tribal fishing, under the Treaty of 1836, was reaffirmed in 1971 for all Michigan waters of eastern Lake Superior (Western District Court of Michigan 2000). Tribal fishing under the Treaties of 1842 and 1854 was reinstated in 1971 for western Michigan waters (Lac Courte Oreilles v. Lester P. Voight 1983) and in 1972 for eastern Wisconsin waters (State of Wisconsin v. Gurnoe 1972) of Lake Superior. Tribal waters in U.S. waters of Lake Superior were reserved for maximizing Lake Whitefish yield, with limited Lake Trout bycatch, and some areas, like Whitefish Bay, were designated as deferred Lake Trout rehabilitation areas that were not stocked because Lake Trout bycatch mortality in Lake Whitefish fisheries would have been too high to allow populations to recover. Tribal fishing effort, mostly with large-mesh gill nets (≥108 mm stretch mesh), increased rapidly in the ensuing decade in eastern Wisconsin and in the 1980s in Michigan waters, and tribes attempted to control nontarget bycatch mortality using quotas. Quotas were initially set by modeling year-classes of stocked and wild fish exposed to fixed natural mortality and age-specific Sea Lamprey and fishing mortality (Ebener et al. 1985). Statistical catch-at-age models are now used to estimate stock size and other population metrics while accounting for uncertainty in input data and parameter estimates (Lenart and Caroffino 2015).

Regulation of recreational fisheries has been through seasonal and area restrictions, daily bag limits, and size limits but with little coordination among agencies. In Lake Superior, angling is prohibited in October and November when Lake Trout spawn (except in Michigan), the creel limit has been 3 fish/d in all jurisdictions, and there is a minimum size limit of 152 mm in Minnesota, 254 mm in Michigan, and 381 mm in Wisconsin (only one can exceed 635 mm) and no length limit in Ontario. Subsistence fisheries have been allowed by Native American fishery management agencies that ensure catches are low by limiting effort and prohibiting sale of the catch (e.g., Western District Court of Michigan 2000).

ASSESSMENT AND EVALUATION

Lake Trout stocks increased dramatically in each of the Great Lakes within a decade of the institution of stocking, chemical suppression of Sea Lampreys, and regulation of fisheries, but natural reproduction was restored only in Lake Superior, the lake where habitat quality and quantity was least perturbed; residual nearshore populations were not completely eliminated; and most offshore stocks persisted (Eshenroder et al. 1984; Hansen 1999; Bronte and Sitar 2008). Widespread reproductive failure of hatchery-origin Lake Trout led the Great Lakes Fishery Commission to convene the Conference on Lake Trout Research in 1983 that was intended to identify possible impediments to wild recruitment, recommend priorities for research, identify hypotheses to be tested, outline experimental designs, and encourage sharing of tasks among agencies and institutions (Eshenroder et al. 1984). In 1994, a decade after the Conference on Lake Trout Research, another conference was convened to critically evaluate Lake Trout restoration in the Great Lakes basin (Selgeby 1995). Impediments were revisited again in the 2000s for Lakes Michigan (Bronte et al. 2003b, 2008), Erie (Markham et al. 2008), and Ontario (Lantry et al. 2014).

Abundance

Since the 1960s, Lake Trout stocks in the Great Lakes have been monitored by standardized spring or summer fishery-independent surveys. In Lake Superior, the first such survey was instituted in Michigan waters by the U.S. Fish and Wildlife Service, where commercial fishermen were contracted to fish for Lake Trout using standardized gear beginning in the late 1950s. Similar programs were instituted in Wisconsin and Minnesota waters immediately after fisheries were closed in 1962 (Pycha and King 1975). Lake Trout density was indexed annually at semi-fixed locations using standardized gill nets in Lake Trout management areas (Hansen et al. 1995a; Figure 4). Density of stocked Lake Trout increased quickly in nearly all jurisdictions during the 1960s and 1970s, but later in Minnesota waters, where stocking began 10 years later than elsewhere. Unlike other Great Lakes where Lake Trout were mostly extirpated, wild fish persisted in some nearshore (<80 m) and in all offshore waters (>80 m; mostly siscowet), were detected from the start in all jurisdictions, and increased in abundance in the 1970s and 1980s. By the 1990s, wild fish densities equaled or exceeded historical densities during the late 1920s and 1930s in most areas of Michigan waters, which suggested that stock density was restored (Wilberg et al. 2001; Figure 5). Consequently, Lake Trout stocking was discontinued in most areas in 1996 due to this predominance of wild fish and declining survival of hatchery-reared fish (Hansen et al. 1994a, 1996), so fishery managers declared that Lake Trout restoration was successful in 1996.

Mortality

Simulations of Lake Trout population dynamics suggested that Lake Trout stocks could

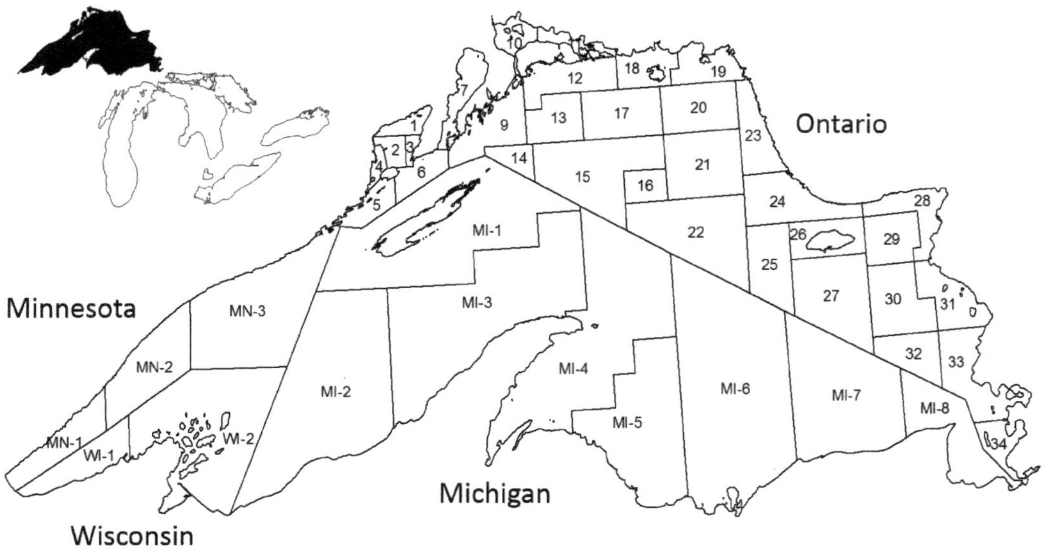

FIGURE 4. Lake Trout management areas in Michigan (MI), Minnesota (MN), Wisconsin (WI), and Ontario (numbered) waters of Lake Superior (reproduced from Hansen et al. 1995a). Inset shows Lake Superior within the Laurentian Great Lakes basin.

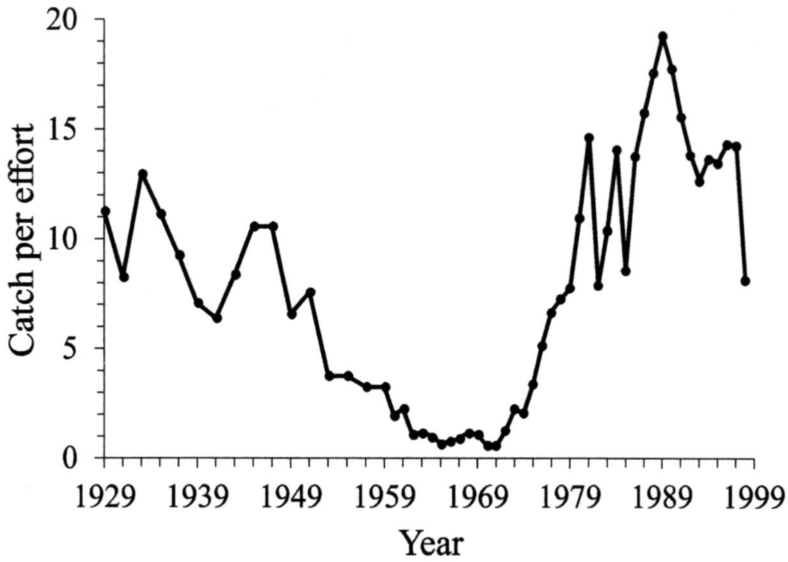

FIGURE 5. Catch per effort (number per km of gill net set for one night) of wild lean Lake Trout in Michigan waters of Lake Superior during 1929–1998 (modified from Wilberg et al. 2003).

withstand mortality from Sea Lampreys and fisheries if both were controlled (Lett et al. 1975; Walters et al. 1980). In Lake Superior, Lake Trout stocks declined in response to Sea Lamprey predation but recovered in the absence of fishing when Sea Lampreys were suppressed. Similarly, stocks declined in response to fishing but recovered rapidly in the absence of Sea Lampreys when stocking was instituted. Clearly, the focus by all fishery management agencies was to reduce total mortality initially, although stock recovery and natural reproduction would have likely occurred sooner had large-scale fisheries not been allowed to redevelop. Having residual populations of wild fish was also advantageous in further promoting wild recruitment, an advantage lacking in the lower lakes where progress toward restoration has been slower (Muir et al. 2013).

Sea Lamprey Effects

Use of TFM to control larval Sea Lampreys began in 1958 and reduced spawning runs by 80–90% within 4–6 years after the first round of treatments throughout Lake Superior (Smith and Tibbles 1980; Heinrich et al. 2003). Barriers that blocked upstream spawning migrations and sterile-male releases were also used later to reduce Sea Lamprey recruitment (Heinrich et al. 2003). Sea Lamprey control was effective enough to allow Lake Trout stocks to recover (Hansen et al. 1995a; Heinrich et al. 2003), but Sea Lampreys still killed large numbers of Lake Trout, particularly in inshore areas of western Lake Superior (Hansen et al. 1994b). In 1968–1978, the rate of annual Sea Lamprey mortality on Lake Trout in inshore Michigan waters was higher (16–66%) than annual fishing mortality (13–16%; Pycha 1980). To allow further advances in Lake Trout restoration, a goal for Sea Lamprey control in Lake Superior was set in 1990 to reduce Sea Lampreys 50% from current levels by the year 2000 and 90% by the year 2010 (Busiahn 1990). Goals were refined after 2010 to include a lakewide adult abundance index of 9,700 adult Sea Lampreys and a Sea Lamprey wounding rate of 5 wounds per 100 Lake Trout greater than 532 mm total length. Since 1999, indices of Sea Lamprey abundance and wounding rates exceeded these goals but declined to meet both goals in 2015 in response to increased control effort in 2000 and 2006 (Figure 6).

Fishery Exploitation

After 1970, effectiveness of commercial and recreational fishery regulations varied among jurisdictions in Lake Superior, and ineffective regulation contributed to excessive total annual mortality in some jurisdictions (Hansen et al. 1995a, 1996). For example, state and tribal fishery managers allowed commercial large-mesh gill-net fisheries for Lake Whitefish to expand in Michigan and Wisconsin so Lake Trout restoration stalled (as a consequence of high bycatch mortality on Lake Trout in Lake Whitefish fisheries) in the 1980s in eastern Wisconsin, near the Keweenaw Peninsula in western Michigan, and around Grand Marais and Whitefish Bay in eastern Michigan (Hansen et al. 1995a). More recently, large-mesh gill-net fishing effort was constrained in eastern Wisconsin but has not yet been effectively regulated in Whitefish Bay, where agreements among the state of Michigan, five tribes of Chippewa and Ottawa (Odawa) Native Americans, and the Department of Interior forgo Lake Trout stocking and restoration in favor of large-mesh gill-net fisheries for Lake Whitefish (Western District Court of Michigan 2000).

Reproduction

Stocked Lake Trout aggregate and spawn in many areas of the Great Lakes but have only rebuilt self-sustaining populations in Lake Superior, although significant progress has

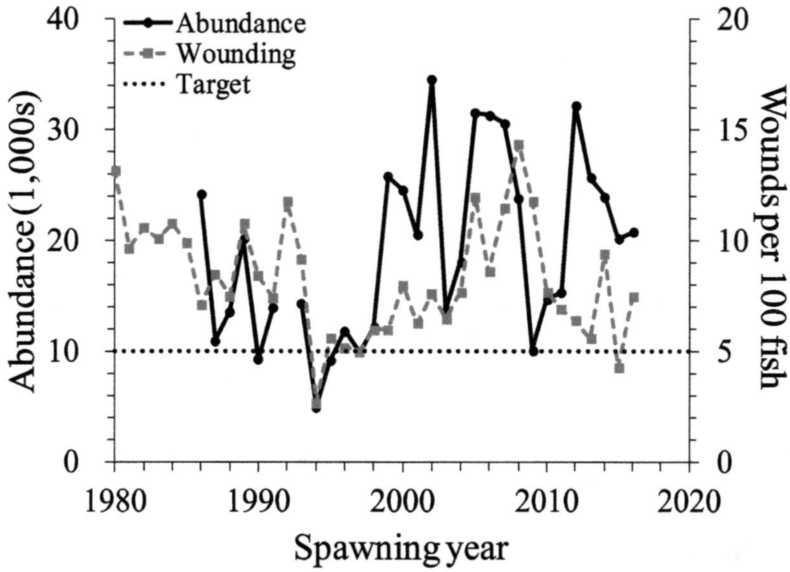

FIGURE 6. Index of adult Sea Lamprey abundance, number of fresh wounds per 100 Lake Trout >532 mm caught during standardized assessments, and targets for abundance and wounding in Lake Superior versus Sea Lamprey spawning years 1980–2016 (modified from Mullett and Sullivan 2016).

been achieved in Lake Huron (Riley et al. 2007; Johnson et al. 2015) and is beginning in southwestern Lake Michigan (Hanson et al. 2013; Madenjian et al. 2016). In Lake Superior, stocking caused a rapid increase of inshore stocks that reproduced in areas with widely distributed inshore spawning habitat (Hansen et al. 1995a, 1995b; Wilberg et al. 2003). Wild Lake Trout replaced stocked Lake Trout in most areas of Lake Superior by the mid-1980s, through reproduction by stocked Lake Trout inshore, remnant wild lean Lake Trout offshore, and siscowet and humper Lake Trout in deep areas (Hansen et al. 1995a). Survival of stocked Lake Trout declined sharply in the late 1970s and 1980s (Hansen et al. 1994a) and was attributed to increased tribal gill-net fishing in Michigan and Wisconsin and to predation by or competition with increased densities of wild Lake Trout in Minnesota (Hansen et al. 1996). Large-scale stocking was discontinued in 1996 when wild fish dominated most areas of the lake (Schreiner and Schram 1997; Muir et al. 2013). Even with redevelopment of gill-net fisheries, restoration of adequate parental stocks and wild recruitment was possible and suggested that overall mortality was below levels of reproductive output and growth (Schreiner and Schram 1997; Hansen 1999; Muir et al. 2013).

LESSONS LEARNED

Interagency cooperation in management, in conjunction with relatively pristine habitat and presence of remnant native Lake Trout of multiple morphs, enabled Lake Trout stocks to recover throughout Lake Superior. Interagency cooperation enabled agencies to articulate and institute (1) clear and common goals and actions for recovery, (2) early and intensive lakewide stocking of hatchery-reared Lake Trout to enhance failing stocks, (3) early and effective lakewide controls on mortality caused by Sea Lampreys and fisheries, and (4) standardized lakewide evaluations of population trajectories and performance. Here, we discuss successes and failures of each of these elements of the Lake Trout restoration program on Lake Superior.

Clear and Common Lakewide Goals and Actions for Restoration

Agencies were organized under the auspices of the Great Lakes Fishery Commission in 1986 because decades of earlier management actions were less effective than desired. Once organized, agencies agreed to an overall goal for Lake Trout recovery that mimicked a sustained annual lakewide yield of 2 million kg that had been observed prior to stock collapse (LSLTTC 1986; Busiahn 1990; Hansen 1996). To rebuild Lake Trout stocks throughout the lake, agencies set quantitative and measurable lakewide objectives to restore wild recruitment by stocking hatchery-reared fish and to increase survival by reducing Sea Lamprey and fishing mortality (LSLTTC 1986; Hansen 1996). Agencies also coordinated in reporting on progress toward the agreed-upon goal through annual public meetings and periodic reports on progress (Hansen 1990, 1994; Ebener 2007; Gorman et al. 2010; Pratt et al. 2016). Articulation of measurable lakewide goals and actions, in conjunction with a publicly accessible and transparent process for reporting on progress, was essential for agencies to remain committed to a process that unfolded over multiple decades, despite setbacks that protracted temporal progress. One failing was the overly optimistic goal of population recovery that would support historical yield, which, in retrospect, was not sustainable because it led to population declines. Evaluation of this goal was also unmeasurable because the nature and size of contemporary fisheries were restricted to a point where historical harvest, if possible, could not be attained. We suggest setting realistic populations goals that recognize changes in the ecosystem and fishery governance because returning to precollapse states of both harvest potential and stock size may be unlikely.

Early and Intensive Lakewide Stocking of Hatchery-Reared Lake Trout to Enhance Failing Stocks

Stocking was instituted intensively throughout Lake Superior after 1953, prior to effecting Sea Lamprey or fishery controls, thereby introducing large numbers of hatchery-origin fish that grew to maturity shortly after mortality was reduced. Despite stocking rates that were below targets because of limited hatchery capacity, by the 1980s, density of stocked fish in many areas exceeded historical density of wild fish in the 1930s and thereby generated greatly enhanced reproductive potential when combined with remnant wild fish (Richards et al. 2004). The presence of widespread good-quality nearshore spawning habitat was crucial for successful reproduction by naive hatchery-origin Lake Trout in Lake Superior (Hansen et al. 1995a). Further, persistence of remnant native Lake Trout populations in some areas undoubtedly enhanced the temporal trajectory of stock recovery throughout Lake Superior, a condition that was lacking elsewhere in the Great Lakes. Had historical precollapsed Lake Trout densities been assessed earlier in the restoration process, stocking could have been curtailed earlier and may have prevented declines in survival of hatchery fish after 1980. Earlier stocking reductions in Lake Superior could have been redirected to other Great Lakes where stocking densities were far below what was required (i.e., Lake Michigan).

Early and Effective Lakewide Controls on Mortality Caused by Sea Lampreys

Because of the presence of remnant Lake Trout populations, Lake Superior was selected first for lakewide application of the newly discovered and highly selective Sea Lamprey pesticide TFM. As a result of lakewide application of TFM in 1958–1960, in conjunction with widespread and continued use of electrical and mechanical barriers, Sea Lamprey abundance declined precipitously by autumn 1961, which set the stage for agencies to institute lakewide closures of fisheries and enabled high survival of newly stocked and remnant wild fish. Early control of Sea Lamprey populations across Lake Superior was a requisite for the early and sustained success of all other management actions.

Early and Effective Lakewide Controls on Mortality Caused by Fisheries

After fishery management agencies observed the precipitous decline in Sea Lamprey abundance in 1962, agencies instituted lakewide closures of fisheries that, combined with lower Sea Lamprey mortality, enabled high survival of newly stocked and remnant wild fish. Fishery closures were relaxed too soon to maximize the speed of Lake Trout population growth, but, nonetheless, Lake Trout stocks recovered lakewide, to the point where stocking could be ceased in much of the lake by the 1990s. Undoubtedly, Lake Trout population recovery would have been much faster if fishing mortality had been held in check earlier and longer, but the trade-off between an intermediate level of fishing mortality and a longer time to recovery eventually succeeded in restoring self-sustaining populations. To expedite restoration, yield could have been foregone until populations were self-sustaining, and fishery expansion could have been regulated through gear restrictions (entrapment gear) that directed fishing mortality toward the target species, Lake Whitefish, while minimizing mortality on recovering Lake Trout stocks.

Standardized Lakewide Evaluations of Recovery Progress

Agency cooperation in lakewide research and assessment provided information that was crucial to both setting management goals and to measuring progress toward those goals. The first such lakewide assessment was a large-mesh (114 mm stretch mesh) gill-net assessment fishery that was used initially to evaluate early progress and then, later, to monitor relative abundance (catch/effort of adult fish), mortality (age structure), and Sea Lamprey wounding rate, growth, and recruitment (catch per effort of juvenile fish; Hansen 1996). For example, an early study of catch per effort enabled development of targets for stocking rates that produced a rapid buildup of abundance in central Michigan waters. Similarly, continued use of the lakewide assessment provided baseline information for monitoring progress toward recovery (Hansen 1990, 1994; Ebener 2007; Gorman et al. 2010; Pratt et al. 2016). While the use of a single-mesh gill-net index was limited in its measurement of a wide range of sizes and ages, it allowed direct comparison to historical fishery catch per effort that extended back to 1929 and thereby allowed a more temporally comprehensive assessment of restoration progress. Standardized assessments should be established as early as possible and should be designed to be easily adopted and repeated by all jurisdictions. Assessment gears should provide metrics for monitoring biological changes of the population in support of population objectives that are consistent with population recovery, such as age structure, parental stock size, and recruitment. No single gear can accurately sample all life stages, sizes, and ages present in a population, but temporally consistent use of a single gear provides an index of underlying population parameters that would be disrupted by changes in gears or methods. Assessment data should be stored and analyzed collectively among jurisdictions while avoiding unilateral analyses that may not be comparable among jurisdictions that can lead to erroneous conclusions.

CONCLUSIONS

While restoration of Lake Trout populations in Lake Superior is generally considered to be a great success, the primary measure of that success has been population size. Only recently has the ecological and genetic diversity of these populations been studied, results of which suggest the story may be more nuanced (Muir et al. 2014; Baillie et al. 2016, Hansen et al. 2016b; Perreault-Payette et al. 2017). For example, lean and siscowet populations of Lake Trout have recovered to precollapse levels of abundance in most areas of Lake Superior, but the genetic and ecological diversity of Lake Trout is still likely lower than was historically present in Lake Superior (Guinand et al. 2003, 2012). Despite this reduced ecological and genetic diversity, Lake Trout recovery in Lake Superior is an extraordinary example of agency cooperation toward a common goal for managing recovery of an ecologically important shared resource.

Lake Superior's restoration outcome can also be credited to the focus on the Lake Trout as the primary salmonine of concern, in contrast to other Great Lakes that centered management on Pacific salmon (Claramunt et al. 2012). Outside Lake Superior, Lake Trout restoration has been viewed as less important than maintaining a nonnative predator–prey system that supports a valuable recreational fishery, which may now be in jeopardy of collapse (Dettmers et al. 2012; Clark et al. 2016). For example, the unified focus on Lake Trout restoration in Lake Superior enabled agencies to devote most hatchery production (Bronte et al. 2003b) and assessments to Lake Trout restoration and evaluation, whereas Lake Trout stocking outside Lake Superior is a much smaller fraction of total hatchery production, which generally favors stocking of nonnative trout and salmon species to support recreational fisheries. Recovery of Lake Trout in Lake Superior is viewed as a major fishery management success but has been largely unrecognized outside the Great Lakes and underutilized to inform restoration efforts in other Great Lakes. This lack of reference, combined with a limited awareness of the range of life history and ecomorphtypic diversity expressed throughout the species' range (Muir et al. 2013, 2016), further impedes informed progress elsewhere. Oligotrophication of Lakes Michigan and Huron will likely force fishery managers to favor Lake Trout, the historic keystone predator that is uniquely adapted to live in environments not unlike those that occurred historically (Smith 1972; Christie 1974).

ACKNOWLEDGMENTS

We thank the many individuals who served on the Lake Superior Technical Committee, especially Richard G. Schorfhaar, Mark P. Ebener, James W. Peck, James H. Selgeby, Donald R. Schreiner, and Stephen T. Schram. Richard L. Pycha, the first chair of that committee, provided an invaluable written legacy of Lake Trout research and an indefatigable willingness to write about and discuss his personal observations of the Lake Trout. Mark Vinson and three anonymous reviewers provided valuable comments on an earlier version of the manuscript. Any use of trade, product, or firm names is for descriptive purposes only and does not imply endorsement by the U.S. Government. The findings and conclusions in this article are those of the authors and do not necessarily represent the views of the U.S. Fish and Wildlife Service.

REFERENCES

Bailey, R. M., and G. R. Smith. 1981. Origin and geography of the fish fauna of the Laurentian Great Lakes Basin. Canadian Journal of Fisheries and Aquatic Sciences 38:1539–1561.

Baillie, S. M., A. M. Muir, M. J. Hansen, C. C. Krueger, and P. Bentzen. 2016. Genetic and phenotypic variation along an ecological gradient in Lake Trout *Salvelinus namaycush*. BMC Evolutionary Biology [online serial] 16:219.

Baldwin, N. S., R. W. Saalfeld, M. A. Ross, and H. J. Buettner. 1979. Commercial fish production in the Great Lakes 1867–1977. Great Lakes Fishery Commission, Technical Report 3, Ann Arbor, Michigan.

Bergstedt, R. A. 1983. Origins of Rainbow Smelt in Lake Ontario. Journal of Great Lakes Research 9:582–583.

Berst, A. H., and G. R. Spangler. 1973. Lake Huron: the ecology of the fish community and man's effects on it. Great Lakes Fishery Commission, Technical Report 21, Ann Arbor, Michigan.

Bogue, M. 2000. Fishing in the Great Lakes: an environmental history. University of Wisconsin Press, Madison.

Brenden, T. O., R. W. Brown, M. P. Ebener, K. Reid, and T. J. Newcomb. 2013. Great Lakes commercial fisheries: historical overview and prognoses for the future. Pages 339–397 *in* W. W. Taylor, A. J. Lynch, and N. J. Leonard, editors. Great Lakes fisheries policy and management, 2nd edition. Michigan State University Press, East Lansing.

Bronte, C. R., M. P. Ebener, D. R. Schreiner, D. S. DeVault, M. M. Petzold, D. A. Jensen, C. Richards, and S. J. Lozano. 2003a. Fish community change in Lake Superior, 1970–2000. Canadian Journal of Fisheries and Aquatic Sciences 60:1552–1574.

Bronte, C. R., J. Jonas, M. E. Holey, R. L. Eshenroder, M. L. Toneys, P. McKee, B. Breidert, R. M. Claramunt, M. P. Ebener, C. C. Krueger, and G. Wright. 2003b. Possible impediments to Lake Trout restoration in Lake Michigan. Lake Michigan Committee, Great Lakes Fishery Commission, Ann Arbor, Michigan.

Bronte, C. R., C. C. Krueger, M. E. Holey, M. L. Toneys, R. L. Eshenroder, and J. L. Jonas. 2008. A guide for the rehabilitation of Lake Trout in Lake Michigan. Great Lakes Fishery Commission, Miscellaneous Publication 2008-01, Ann Arbor, Michigan.

Bronte, C. R., S. T. Schram, J. H. Selgeby, and B. L. Swanson. 2002. Reestablishing a spawning population of Lake Trout in Lake Superior with fertilized eggs in artificial turf incubators. North American Journal of Fisheries Management 22:796–805.

Bronte, C. R., J. H. Selgeby, and G. L. Curtis. 1991. Distribution, abundance, and biology of the Alewife in US waters of Lake Superior. Journal of Great Lakes Research 17:304–313.

Bronte, C. R., and S. P. Sitar. 2008. Harvest and relative abundance of siscowet Lake Trout in Michigan waters of Lake Superior, 1929–1961. Transactions of the American Fisheries Society 137:916–926.

Brown, E. H., Jr., G. W. Eck, N. R. Foster, R. M. Horrall, and C. E. Coberly. 1981. Historical evidence for discrete stocks of Lake Trout (*Salvelinus namaycush*) in Lake Michigan. Canadian Journal of Fisheries and Aquatic Sciences 38:1747–1758.

Buettner, H. J. 1961. Recoveries of tagged, hatchery-reared Lake Trout from Lake Superior. Transactions of the American Fisheries Society 90:404–412.

Busiahn, T. R., editor. 1990. Fish community objectives for Lake Superior. Great Lakes Fishery Commission, Special Publication 90-1, Ann Arbor, Michigan.

Chavarie, L., A. M. Muir, S. M. Baillie, M. S. Zimmerman, M. J. Hansen, N. A. Nate, D. L. Yule, T. Middel, P. Bentzen, and C. C. Krueger. 2017. Challenge to the model for Lake Charr evolution: co-existence of shallow- and deep-water morphs in a small postglacial lake. Biological Journal of the Linnean Society 120:578–603.

Christie, W. J. 1973. A review of the changes in the fish composition of Lake Ontario. Great Lakes Fishery Commission, Technical Report 23, Ann Arbor, Michigan.

Christie, W. J. 1974. Changes in the fish species composition of the Great Lakes. Journal of the Fisheries Board of Canada 31:827–854.

Claramunt, R. M., C. P. Madenjian, and D. F. Clapp. 2012. Pacific salmonines in the Great Lakes. Pages 609–650 *in* N. J. Leonard, C. P. Ferreri, and W. W. Taylor, editors. Great Lakes fisheries policy and management: a binational perspective, 2nd edition. Michigan State University Press, East Lansing.

Clark, R. D., Jr., J. R. Bence, R. M. Claramunt, J. E. Johnson, D. Gonder, N. D. Legler, S. R. Robillard, and B. D. Dickinson. 2016. A spatially explicit assessment of changes in Chinook Salmon fisheries in Lakes Michigan and Huron from 1986 to 2011. North American Journal of Fisheries Management 36:1068–1083.

Coble, D. W., R. E. Bruesewitz, T. W. Fratt, and J. W. Scheirer. 1990. Lake Trout, Sea Lampreys, and overfishing in the upper Great Lakes: a review and reanalysis. Transactions of the American Fisheries Society 119:985–995.

Conner, D. J., C. R. Bronte, J. H. Selgeby, and H. L. Collins. 1993. Food of salmonine predators in Lake Superior, 1981–87. Great Lakes Fishery Commission, Technical Report 59, Ann Arbor, Michigan.

Cornelius, F. C., K. M. Muth, and R. Kenyon. 1995. Lake trout rehabilitation in Lake Erie: a case history. Journal of Great Lakes Research 21(Supplement 1):65–82.

Dettmers, J. M., C. I. Goddard, and K. D. Smith. 2012. Management of Alewife using Pacific salmon in the Great Lakes: whether to manage for economics or the ecosystem? Fisheries 37(11):495–501.

Dryer, W. R., L. F. Erkkila, and C. L. Tetzloff. 1965. Food of Lake Trout in Lake Superior. Transactions of the American Fisheries Society 94:169–176.

Ebener, M. P., editor. 2007. The state of Lake Superior in 2000. Great Lakes Fishery Commission, Special Publication 07-02, Ann Arbor, Michigan.

Ebener, M. P., C. R. Bronte, and T. R. Busiahn. 1985. Biological and commercial catch statistics from inter-tribal assessment fishing in western Michigan waters, and initial recommendations for management of the fishery. Great Lakes Indian Fish and Wildlife Commission, Odanah, Wisconsin.

Elrod, J. H., R. O'Gorman, C. P. Schneider, T. H. Eckert, T. Schaner, J. N. Bowlby, and L. P. Schleen. 1995. Lake Trout rehabilitation in Lake Ontario. Journal of Great Lakes Research 21(Supplement 1):83 107.

Eshenroder, R. L., N. R. Payne, J. E. Johnson, C. Bowen, and M. P. Ebener. 1995. Lake Trout rehabilitation in Lake Huron. Journal of Great Lakes Research 21:108–127.

Eshenroder, R. L., T. P. Poe, and C. H. Oliver. 1984. Strategies for rehabilitation of Lake Trout in the Great Lakes: proceedings of a conference on Lake Trout research, August 1983. Great Lakes Fishery Commission, Technical Report 40, Ann Arbor, Michigan.

Fitzsimons, J. D. 1995. A critical review of the effects of contaminants on early life stage (ELS) mortality of Lake Trout in the Great Lakes. Journal of Great Lakes Research 21(Supplement 1): 267–276.

Gaden, M., C. Goddard, and J. Read. 2013. Multi-jurisdictional management of the shared Great Lakes fishery: transcending conflict and diffuse political authority. Pages 305–337 *in* W. W. Taylor, A. J. Lynch, and N. J. Leonard, editors. Great Lakes fisheries policy and management, 2nd edition. Michigan State University Press, East Lansing.

Gaden, M., C. Krueger, C. Goddard, and G. Barnhart. 2008. A joint strategic plan for management of Great Lakes fisheries: a cooperative regime in a multi-jurisdictional setting. Aquatic Ecosystem Health and Management 11:50–60.

Goodier, J. L. 1981. Native Lake Trout (*Salvelinus namaycush*) stocks in the Canadian waters of Lake Superior prior to 1955. Canadian Journal of Fisheries and Aquatic Sciences 38:1724–1737.

Gorman, O. T., M. P. Ebener, and M. R. Vinson, editors. 2010. The state of Lake Superior in 2005. Great Lakes Fishery Commission, Special Publication 10-01, Ann Arbor, Michigan.

Groop, R. 2013. Demographic and economic patterns in the Great Lakes region. Pages 31–49 *in* W. W. Taylor, A. J. Lynch, and N. J.

Leonard, editors. Great Lakes fisheries policy and management, 2nd edition. Michigan State University Press, East Lansing.

Guinand, B., K. S. Page, M. K. Burnham-Curtis, and K. T. Scribner. 2012. Genetic signatures of historical bottlenecks in sympatric Lake Trout (*Salvelinus namaycush*) morphotypes in Lake Superior. Environmental Biology of Fish 95:323–334.

Guinand, B., K. T. Scribner, K. S. Page, and M. K. Burnham-Curtis. 2003. Genetic variation over space and time: analyses of extinct and remnant Lake Trout populations in the upper Great Lakes. Proceedings of the Royal Society B 270:425–433.

Hansen, M. J. 1988. Report to the Lake Superior Committee by the Lake Superior Technical Committee. Pages 9–22 *in* Lake Superior Committee 1988 annual meeting minutes. Great Lakes Fishery Commission, Ann Arbor, Michigan.

Hansen, M. J. 1989. Report to the Lake Superior Committee by the Lake Superior Technical Committee. Pages 43–64 *in* Lake Superior Committee 1989 annual meeting minutes. Great Lakes Fishery Commission, Ann Arbor, Michigan.

Hansen, M. J., editor. 1990. Lake Superior: the state of the lake in 1989. Great Lakes Fishery Commission, Special Publication 90-3, Ann Arbor, Michigan.

Hansen, M. J., editor. 1994. The state of Lake Superior in 1992. Great Lakes Fishery Commission, Special Publication 94-1, Ann Arbor, Michigan.

Hansen, M. J., editor. 1996. A Lake Trout restoration plan for Lake Superior. Great Lakes Fishery Commission, Ann Arbor, Michigan.

Hansen, M. J. 1999. Lake Trout in the Great Lakes: basin wide stock collapse and binational restoration. Pages 417–453 *in* W. W. Taylor and C. P. Ferreri, editors. Great Lakes fishery policy and management: a binational perspective. Michigan State University Press, East Lansing.

Hansen, M. J., M. P. Ebener, R. G. Schorfhaar, S. T. Schram, D. R. Schreiner, and J. H. Selgeby. 1994a. Declining survival of Lake Trout stocked in U. S. waters of Lake Superior during 1963–1986. North American Journal of Fisheries Management 14:395–402.

Hansen, M. J., M. P. Ebener, R. G. Schorfhaar, S. T. Schram, D. R. Schreiner, J. H. Selgeby, and W. W. Taylor. 1996. Causes of declining survival of Lake Trout stocked in U.S. waters of Lake Superior in 1963–1986. Transactions of the American Fisheries Society 125:831–843.

Hansen, M. J., M. P. Ebener, J. D. Shively, and B. L. Swanson. 1994b. Lake trout. Pages 13–34 *in* M. J. Hansen, editor. The state of Lake Superior in 1992. Great Lakes Fishery Commission, Special Publication 94-1, Ann Arbor, Michigan.

Hansen, M. J., N. A. Nate, L. Chavarie, A. M. Muir, M. S. Zimmerman, and C. C. Krueger. 2016a. Life history differences between fat and lean Lake Trout morphs in Great Slave Lake, Northwest Territories, Canada. Hydrobiologia 783:21–35.

Hansen, M. J., N. A. Nate, C. C. Krueger, M. S. Zimmerman, H. G. Kruckman, and W. W. Taylor. 2012. Age, growth, survival, and maturity of Lake Trout morphotypes in Lake Mistassini, Quebec. Transactions of the American Fisheries Society 141:1492–1503.

Hansen, M. J., N. A. Nate, A. M. Muir, C. R. Bronte, M. S. Zimmerman, and C. C. Krueger. 2016b. Life history variation among four Lake Trout morphs at Isle Royale, Lake Superior. Journal of Great Lakes Research 42:421–432.

Hansen, M. J., J. W. Peck, R. G. Schorfhaar, J. H. Selgeby, D. R. Schreiner, S. T. Schram, B. L. Swanson, W. R. MacCallum, M. K. Burnham-Curtis, G. L. Curtis, J. W. Heinrich, and R. J. Young. 1995a. Lake Trout (*Salvelinus namaycush*) populations in Lake Superior and their restoration during 1959–1993. Journal of Great Lakes Research 21(Supplement 1):152–175.

Hansen, M. J., R. G. Schorfhaar, J. W. Peck, J. H. Selgeby, and W. W. Taylor. 1995b. Abundance indices for determining the status of Lake Trout restoration in Michigan waters of Lake Superior. North American Journal of Fisheries Management 15:830–837.

Hanson, S. D., M. E. Holey, T. J. Treska, C. R. Bronte, and T. H. Eggebraaten. 2013. Evidence of wild juvenile Lake Trout recruitment in western Lake Michigan. North American Journal of Fisheries Management 33:186–191.

Hartman, W. L. 1973. Effects of exploitation, environmental changes, and new species on the fish habitats and resources of Lake Erie. Great Lakes Fishery Commission, Technical Report 22, Ann Arbor, Michigan.

Hayes, D. B. 2013. Issues affecting fish habitat in the Great Lakes basin. Pages 51–80 *in* W. W. Taylor, A. J. Lynch, and N. J. Leonard, editors. Great Lakes fisheries policy and management, 2nd edition. Michigan State University Press, East Lansing.

Healey, M. C. 1978. The dynamics of exploited Lake Trout populations and implications for management. Journal of Wildlife Management 42:307–328.

Heinrich, J. W., K. M. Mullett, M. J. Hansen, J. V. Adams, G. T. Klar, D. A. Johnson, G. C. Christie, and R. J. Young. 2003. Sea Lamprey abundance and management in Lake Superior, 1957 to 1999. Journal of Great Lakes Research 29:566–583.

Hile, R. 1949. Trends in the Lake Trout fishery of Lake Huron through 1946. Transactions of the American Fisheries Society 76:121–147.

Hile, R., P. H. Eschmeyer, and G. F. Lunger. 1951a. Status of the Lake Trout fishery in Lake Superior. Transactions of the American Fisheries Society 80:278–312.

Hile, R. O., G. F. Lunger, and P. H. Eschmeyer. 1951b. Decline of the Lake Trout fishery in Lake Michigan. Government Printing Office, Washington, D.C.

Holey, M. E., R. R. Rybicki, G. W. Eck, E. H. Brown, Jr., J. E. Marsden, D. S. Lavis, M. L. Toneys, T. N. Trudeau, and R. M. Horrall. 1995. Progress toward Lake Trout restoration in Lake Michigan. Journal of Great Lakes Research 21(Supplement 1):128–151.

Jensen, A. L. 1978. Assessment of the Lake Trout fishery in Lake Superior, 1929–1950. Transactions of the American Fisheries Society 107:543–549.

Jensen, A. L., and H. J. Buettner. 1976. Lake Trout, whitefish, chubs, and Lake Herring yield and effort data for state of Michigan waters of the upper Great Lakes: 1929–1973. Michigan Sea Grant, Technical Report 52, Ann Arbor.

Johnson, J. E., J. X. He, and D. G. Fielder. 2015. Rehabilitation stocking of Walleyes and Lake Trout: restoration of reproducing stocks in Michigan waters of Lake Huron. North American Journal of Aquaculture 77:396–408.

Jones, N. E., M. Parna, S. Parna, and S. Chong. 2018. Evidence of Lake Trout (*Salvelinus namaycush*) spawning and spawning habitat use in the Dog River, Lake Superior. Journal of Great Lakes Research 44:1117–1122.

Khan, N. Y., and S. U. Qadri. 1970. Morphological differences in Lake Superior Lake Char. Journal of the Fisheries Research Board of Canada 27:161–167.

Krueger, C. C., and P. E. Ihssen. 1995. Review of genetics of Lake Trout: molecular genetics, physiology, and strain comparisons of Great Lakes stocks. Journal of Great Lakes Research 21(Supplement 1):348–363.

Lac Courte Oreilles v. Lester P. Voight et al. 1983. 700 F.2d 341.

Lantry, J., T. Shaner, and T. Copeland. 2014. A management strategy for the restoration of Lake Trout in Lake Ontario, 2014 update. Great Lakes Fishery Commission, Ann Arbor, Michigan.

Lawrie, A. H. 1978. The fish community of Lake Superior. Journal of Great Lakes Research 4:513–549.

Lawrie, A. H., and J. F. Rahrer. 1973. Lake Superior: a case history of the lake and its fisheries. Great Lakes Fishery Commission, Technical Report 19, Ann Arbor, Michigan.

Lenart, S. J., and D. C. Caroffino, editors. 2015. Technical Fisheries Committee administrative report 2015: status of Lake Trout and Lake Whitefish populations in the 1836 treaty-ceded waters of Lakes Superior, Huron and Michigan, with recommended yield and effort levels for 2015. Available: www.michigan.gov/documents/dnr/2015StatusStocksReport_498685_7.pdf. (July 2018).

Lett, P. F., F. W. H. Beamish, and G. J. Farmer. 1975. System simulation of the predatory activities of Sea Lampreys (*Petromyzon marinus*) on Lake Trout (*Salvelinus namaycush*). Journal of the Fisheries Research Board of Canada 32:623–631.

Loftus, K. H. 1958. Studies on river-spawning populations of Lake Trout in eastern Lake Superior. Transactions of the American Fisheries Society 87:259–277.

LSLTTC (Lake Superior Lake Trout Technical Committee). 1986. A Lake Trout rehabilitation plan for Lake Superior. Pages 47–62 *in* S. D. Morse, editor. Lake Superior Committee 1986 annual meeting minutes. Great Lakes Fishery Commission, Ann Arbor, Michigan.

Mac, M. J., and C. C. Edsall. 1991. Environmental contaminants and the reproductive success of Lake Trout in the Great Lakes: an epidemiological approach. Journal of Toxicology and Environmental Health 33:375–395.

Mandenjian, C. P., B. Breidert, D. Boyarski, C. Bronte, B. Dickinson, K. Donner, M. Ebener, R. Gordon, D. Hanson, M. Holey, J. Janssen, J. Jonas, M. Kornis, E. Olsen, S. Robillard, T. Treska, B. Weldon, and G. Wright. 2016. Lake Michigan Lake Trout Working Group report. Great Lakes Fishery Commission, Ann Arbor, Michigan.

Mandrak, N. E., and V. Cudmore. 2013. Fish species at risk and non-native fishes in the Great Lakes basin: past, present, and future. Pages 167–202 *in* W. W. Taylor, A. J. Lynch, and N. J. Leonard, editors. Great Lakes fisheries policy and management, 2nd edition. Michigan State University Press, East Lansing.

Markham, J. L., A. Cook, T. MacDougall, L. Witzel, K. Kayle, M. Murray, M. Fodale, E. Trometer, F. Neave, J. Fitzsimons, J. Francis, and M. Stapanian. 2008. A strategic plan for the rehabilitation of Lake Trout in Lake Erie, 2008–2020. Great Lakes Fishery Commission, Miscellaneous Publication 2008-02, Ann Arbor, Michigan.

Moll, R. A., C. Sellinger, E. S. Rutherford, J. L. Johnson, M. R. Fainter, and J. E. Gannon. 2013. The Great Lakes: an overview of their formation, geology, physics, and chemistry. Pages 3–29 *in* W. W. Taylor, A. J. Lynch, and N. J. Leonard, editors. Great Lakes fisheries policy and management, 2nd edition. Michigan State University Press, East Lansing.

Muir, A. M., C. R. Bronte, M. S. Zimmerman, H. R. Quinlan, J. D. Glase, and C. C. Krueger. 2014. Ecomorphological diversity of Lake Trout at Isle Royale, Lake Superior. Transactions of the American Fisheries Society 143:972–987.

Muir, A. M., M. J. Hansen, C. R. Bronte, and C. C. Krueger. 2016. If Arctic Charr *Salvelinus alpinus* is "the most diverse vertebrate," what is the Lake Charr *Salvelinus namaycush*? Fish and Fisheries 17:1194–1207.

Muir, A. M., C. C. Krueger, and M. J. Hansen. 2013. Re-establishing Lake Trout in the Laurentian Great Lakes: the past, present, and future. Pages 533–588 *in* W. W. Taylor, A. J. Lynch, and N. J. Leonard, editors. Great Lakes fishery policy and management: a binational perspective, 2nd edition. Michigan State University Press, East Lansing.

Mullett, K., and P. Sullivan. 2016. Sea Lamprey control in the Great Lakes 2016: Annual Report to the Great Lakes Fishery Commission. Available: www.glfc.org/pubs/slcp/annual_reports/ANNUAL_REPORT_2016.pdf. (July 2018).

Nute, G. L. 1926. The American Fur Company's fshing enterprises on Lake Superior. The Mississippi Valley Historical Review 12:483–503.

Perreault-Payette, A., A. M. Muir, F. Goetz, C.

Perrier, E. Normandeau, P. Sirois, and L. Bernatchez. 2017. Investigating the extent of parallelism in morphological and genomic divergence among Lake Trout ecotypes in Lake Superior. Molecular Ecology 26:1477–1497.

Powell, M. J., and J. Atkinson. 1991. Lake Superior Lake Trout stocking review. Ontario Ministry of Natural Resources, Toronto.

Pratt, T. C., O. T. Gorman, W. P. Mattes, J. T. Myers, H. R. Quinlan, D. R. Schreiner, M. J. Seider, S. P. Sitar, D. L. Yule, and P. M. Yurista. 2016. The state of Lake Superior in 2011. Great Lakes Fishery Commission, Special Publication 16-01, Ann Arbor, Michigan.

Pycha, R. L. 1980. Changes in mortality of Lake Trout (*Salvelinus namaycush*) in Michigan waters of Lake Superior in relation to Sea Lamprey (*Petromyzon marinus*) predation, 1968–78. Canadian Journal of Fisheries and Aquatic Sciences 37:2063–2073.

Pycha, R. L., and G. R. King. 1967. Returns of hatchery-reared Lake Trout in southern Lake Superior, 1955–62. Journal of the Fisheries Research Board of Canada 24:281–298.

Pycha, R. L., and G. R. King. 1975. Changes in the Lake Trout population of southern Lake Superior in relation to the fishery, the Sea Lamprey, and stocking, 1950–70. Great Lakes Fishery Commission, Technical Report 28, Ann Arbor, Michigan.

Ray, B. A., T. R. Hrabik, M. P. Ebener, O. T. Gorman, D. R. Schreiner, S. T. Schram, S. P. Sitar, W. P. Mattes, and C. R. Bronte. 2007. Diet and prey selection by Lake Superior Lake Trout during spring, 1986–2001. Journal of Great Lakes Research 33:104–113.

Richards, J. M., M. J. Hansen, C. R. Bronte, and S. P. Sitar. 2004. Recruitment dynamics of the 1971–1991 year-classes of Lake Trout in Michigan waters of Lake Superior. North American Journal of Fisheries Management 24:475–489.

Riley, S. C., J. X. He, J. E. Johnson, T. P. O'Brien, and J. S. Schaeffer. 2007. Evidence of widespread natural reproduction by Lake Trout *Salvelinus namaycush* in the Michigan waters of Lake Huron. Journal of Great Lakes Research 33:917–921.

Schreiner, D. R., and S. T. Schram. 1997. Lake Trout rehabilitation in Lake Superior—maintaining our progress. Fisheries 22(7):12–14.

Selgeby, J. H. 1995. Introduction to the proceedings of the 1994 International conference on restoration of Lake Trout in the Laurentian Great Lakes. Journal of Great Lakes Research 21(Supplement 1):1–2.

Smith, B. R. 1971. Sea Lampreys in the Great Lakes of North America. Pages 207–247 *in* M. W. Hardisty and I. C. Potter, editors. The biology of lampreys, volume 1. Academic Press, New York.

Smith, B. R., and J. J. Tibbles. 1980. Sea Lamprey (*Petromyzon marinus*) in Lakes Huron, Michigan, and Superior: history of invasion and control, 1936–78. Canadian Journal of Fisheries and Aquatic Sciences 37:1780–1801.

Smith, B. R., J. J. Tibbles, and B. G. H. Johnson. 1974. Control of the Sea Lamprey (*Petromyzon marinus*) in Lake Superior. Great Lakes Fishery Commission, Technical Report 26, Ann Arbor, Michigan.

Smith, S. H. 1968. Species succession and fishery exploitation in the Great Lakes. Journal of the Fisheries Research Board of Canada 25:667–693.

Smith, S. H. 1972. Factors of ecologic succession in oligotrophic fish communities of the Laurentian Great Lakes. Journal of the Fisheries Research Board of Canada 29:717–730.

State of Wisconsin v. Gurnoe. 1972. 53 Wis. 2d 390.

Sweeny, R. O. 1890. The siskiwit. Transactions of the American Fisheries Society 19:84–85.

Waldman, J., R. Daniels, M. Hickerson, and I. Wirgin. 2009. Mitochondrial DNA analysis indicates Sea Lampreys are indigenous to Lake Ontario: response to comment. Transactions of the American Fisheries Society 138:1190–1197.

Waldman, J. R., C. Grunwald, N. K. Roy, and I. I. Wirgin. 2004. Mitochondrial DNA analysis indicates Sea Lampreys are indigenous to Lake Ontario. Transactions of the American Fisheries Society 133:95–960.

Walters, C. J., G. Steer, and G. Spangler. 1980. Responses of Lake Trout (*Salvelinus namaycush*) to harvesting, stocking, and lamprey reduction. Canadian Journal of Fisheries and Aquatic Sciences 37:2133–2145.

Wells, L., and A. L. McLain. 1973. Lake Michigan: man's effects on native fish stocks and other biota. Great Lakes Fishery Commission, Technical Report 20, Ann Arbor, Michigan.

Western District Court of Michigan. 2000. United States v. State of Michigan (Consent Decree). Case No. 2:73-CV-26 (M26 73).

Wilberg, M. J., M. J. Hansen, and C. R. Bronte. 2003. Historic and modern abundance of wild lean Lake Trout in Michigan waters of Lake Superior: implications for restoration goals. North American Journal of Fisheries Management 23:100–108.

Zimmerman, M. S., C. C. Krueger, and R. L. Eshenroder. 2006. Phenotypic diversity of Lake Trout in Great Slave Lake: differences in morphology, buoyancy, and habitat depth. Transactions of the American Fisheries Society 135:1056–1067.

Zimmerman, M. S., C. C. Krueger, and R. L. Eshenroder. 2007. Morphological and ecological differences between shallow-and deep-water Lake Trout in Lake Mistassini, Quebec. Journal of Great Lakes Research 33:156–169.

Recovery of Saginaw Bay Walleye, Lake Huron

David G. Fielder*

Michigan Department of Natural Resources, Alpena Fisheries Research Station
160 East Fletcher Street, Alpena, Michigan 49707, USA

James P. Baker

Michigan Department of Natural Resources, Bay City Customer Service Center
3580 State Park Drive, Bay City, Michigan 48706, USA

Abstract.—Saginaw Bay is a large coolwater region of Lake Huron and Walleye *Sander vitreus* is the apex predator. From the time of first settlement to the mid-1940s, the bay's Walleye population was the target of a loosely regulated commercial fishery characterized by periods of overharvest and recovery but was sustained for more than half a century at an average annual yield of about 495 metric tons. The fishery collapsed due to a series of year-class failures attributed to declining water quality, habitat degradation, and effects of invasive species. The degraded and collapsed condition lasted until the early 1980s. With improving water quality stemming from clean water legislation and the closure of the commercial fishery, a new period of improvement was achieved. Walleye fingerling stocking was implemented and a recreational fishery soon emerged. Research and assessment sought to monitor stock mortality, growth, and exploitation rates as well as contribution of stocked fish to the fishery. Recovery plans were drafted that sought to improve spawning habitat and improve survival of Walleye fry by creating a predation barrier to the predatory effects of the invasive Alewife *Alosa pseudoharengus* through increased Walleye stocking. A series of cascading food-web changes took place in Lake Huron, resulting in the sudden collapse of Alewives, and Walleye natural reproduction surged beginning in 2003. Walleye stocking was discontinued in 2006 and recovery targets were first achieved in 2009. Management and research shifted from recovery efforts to enhanced stock assessment efforts and modeling, a clear sign of success! Key lessons learned include (1) eliminating or at least reducing obstacles to reproduction (such as habitat and water quality) are essential first steps to laying the foundation for recovery, (2) maintaining populations (via of stocking in this instance) will help ensure that broodfish are available for spawning when conditions improve, (3) ecosystems are resilient and when released from stressors (Alewives in this instance) natural processes can resume, (4) great value exists in survey/assessment investment and long-term data sets for guiding restoration, and (5) resolve and commitment by natural resource professionals, administrators, and stakeholders is critical for sustaining restoration efforts and the investment they require.

* Corresponding author: fielderd@michigan.gov

INTRODUCTION—THE EARLY HISTORY

Saginaw Bay is a large (2,770 km^2), shallow embayment of Lake Huron in the Laurentian Great Lakes, North America (Figure 1). The inner half of Saginaw Bay (defined by a line between Point Lookout and Sand Point) averages 5.1 m deep and 13.7 m in the outer bay (Johengen et al. 2000). As a result, the bay has a gradient from the eutrophic inner bay to the mesotrophic outer bay and is warmer and more productive than the oligotrophic main basin of Lake Huron. The bay includes seven tributaries, the largest of which is the Saginaw River system that includes the Shiawassee, Tittabawassee, Cass, and Flint rivers. The landscape of the modern-day Saginaw Bay watershed is a combination of agriculture and urban/industry. The bay includes a substantial costal wetland complex (Albert 2003).

The fish community of Saginaw Bay is a percid-dominated fish community with Walleye *Sander vitreus* as the apex predator (Schneider and Leach 1977; Keller et al. 1987; Fielder and Thomas 2014). Historically, the bay was used also as a spawning and nursery ground for species that otherwise were regarded as more characteristic of the main basin community, such as Cisco *Coregonus artedi*, Lake Whitefish *C. clupeaformis*, and Lake Trout *Salvelinus namaycush* (Organ et al. 1979; Goodyear et al. 1982). Much of what is known about the early fish community comes from commercial fishery landing records, which date back to 1885 (Baldwin et al. 2009). The Walleye commercial fishery of that era began as early as the 1830s (Lanman 1839; Bogue 2000) and was only second in size in the Great Lakes behind that of Lake Erie, averaging more than 495 metric tons in yield per year (Schneider and Leach 1977; Baldwin et al. 2009). This fishery collapsed in the mid-1940s (Figure 2).

Walleyes spawn in the spring when water temperatures reach the 5–7°C range (Bozek et al. 2011), which occurs between late March and mid-April in the Saginaw Bay area. Walleyes will spawn in both rivers and in open water, and site choice is believed to be a heritable trait (Jennings et al. 1996). In Saginaw Bay, spawning historically took place in both tributaries and offshore rock reefs, mostly in the inner bay (Schneider and Leach 1977; 1978). Habitat degradation affected both habitats, but research indicates that spawning has only returned to tributaries, although trace offshore egg deposition has been documented (Kalejs 2017). Several tributaries likely host spawning Walleyes in Saginaw Bay, but the tributaries to the Saginaw River system (Figure 1) are believed to be the principle sources, and of those, the Tittabawassee River is likely the largest producer. Once hatched, fry disperse by river or lake currents and immediately start feeding on zooplankton. Fry must find food within the first few days and avoid predators. Surviving fry grow over the spring and summer and begin feeding on small forage fish. Young-of-the-year Walleyes can reach 152 mm in total length by autumn. Age at first maturity varies with growth and density, but typically, males mature by age 2 and fully by age 3 and females beginning at age 3 and fully by age 4 (Fielder and Thomas 2014).

The commercial fishery in Saginaw Bay increased simultaneously with the burgeoning logging industry of Michigan from 1860 to 1880. The fishery expansion was part of a boom that broadly harvested the state's natural resources (Tody 2003) and was partly attributed to the adoption of pound nets (entrapment gear set with stakes) and the use of steam power for lake-going fishing boats. The first declines in harvest were evident in 1862 based on anecdotal evidence, just before reliable landing records were kept. Thus, yields of Walleye prior to this time

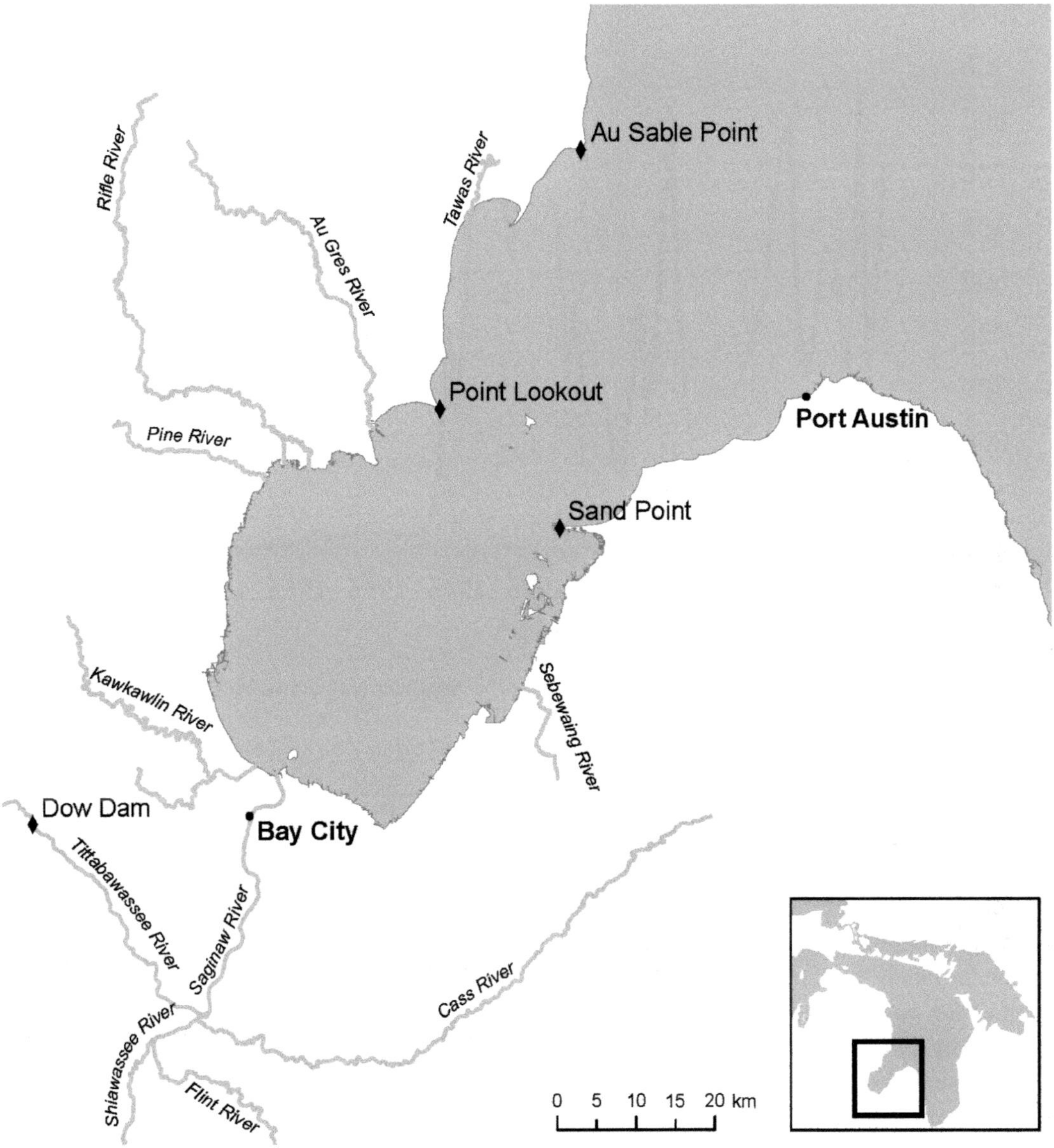

FIGURE 1. Saginaw Bay, Lake Huron.

were likely even greater than historically recorded (Figure 2).

First attempts at fishery management came in 1865. The Michigan legislature implemented a minimum mesh size on pound nets to address complaints of excessive mortality of fish too small to market (Bogue 2000; Tody 2003). Continuing concerns over the sustainability of the commercial fishery in Lake Huron and Saginaw Bay led to the creation of the State Fisheries Commission in 1873, but the emphasis was on the culture and stocking of Lake Whitefish and not on Walleye (Tody 2003). Michigan increasingly began to rely on the federal government's newly formed Commission of Fish and Fisheries for management of Great Lakes resources (Bogue 2000). In the 1930s, a minimum size limit of 680 g (about 432 mm in length) was established for the commercial Walleye

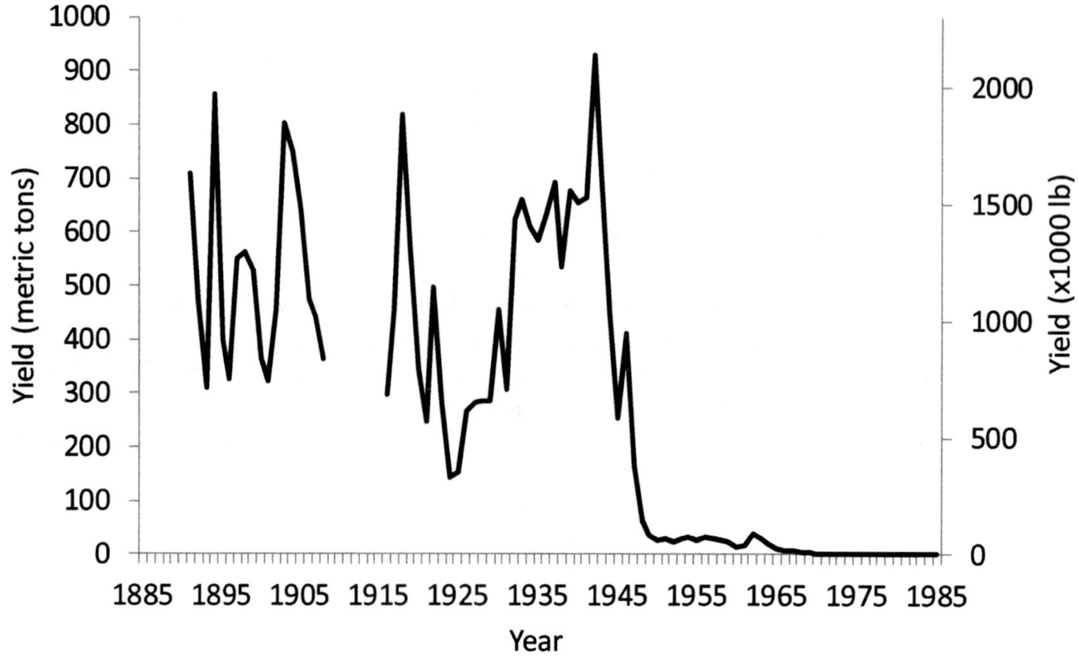

FIGURE 2. Yield (round weight) of Walleye in the Saginaw Bay commercial fishery from 1885 to 1985 (from Baldwin et al. 2009).

fishery (Schneider 1977). During this time, a municipal fish hatchery in Bay City, Michigan adjacent to Saginaw Bay propagated Walleyes using gametes collected by commercial fishermen in the spring that were stocked into the bay as newly hatched fry. In spite of these efforts, the commercial Walleye fishery remained largely unregulated and the population showed periods of overharvest (recruitment overfishing) and then recovery as evidenced in the peaks and valleys of yield (Figure 2). Schneider and Leach (1978) postulated that the commercial fishery increasingly harvested immature Walleyes prior to the eventual collapse, but they did not believe that overfishing was a primary contributor to the collapse.

This chapter reviews the history of the collapse and recovery of the Saginaw Bay Walleye population of Lake Huron, identifying the causes, management actions, and lessons learned that have contributed to the successful restoration of the population and its fishery.

COLLAPSE OF SAGINAW BAY WALLEYE

The intensive exploitation of the era likely hastened the decline of the population; however, environmental degradation of the bay and its tributaries are believed to have been the main culprit in the collapse of Saginaw Bay Walleye during the mid-1940s. The collapse was attributed to a series of year-class failures associated with a string of anthropogenic perturbations. The reproductive failure was tied to spawning habitat degradation in the tributaries (U.S. Fish and Wildlife Service 1969, cited by Schneider and Leach 1977). Rivers were subject to increasing sediment loading, first from deforestation during the logging era and then later due to increasing agricultural activity. Spawning substrate became smothered with silt, especially in the Saginaw River system of tributaries, which lead to increased turbidity in the bay. A rise in abundance of Sauger *Sander canadensis*, another percid (which prefer greater turbidity than Walleyes), became evident during this time and was seen as further evidence of effects of increased turbidity, but that species eventually disappeared too (Schneider 1977; Schneider and Leach 1977).

In addition to sediment loading, man-made barriers were another important factor that contributed to the Walleye population collapse. Dams and spillways aggressively constructed in the early 20th century impeded access to spawning habitat in the bay's tributaries (Mrozinski et al. 1991; Haas and Schaeffer 1992). Of the barriers erected during this period, almost 300 dams or spillways persist in the Saginaw Bay watershed today (Fielder and Baker 2004), and when combined with the effects of sedimentation, tributary Walleye recruitment was lost. Recruitment from the offshore rock reefs within the bay sustained the Walleye population for a time until the 1940s (Schneider 1977; Keller et al. 1987). Recruitment from this source finally succumbed as well to the effects of sedimentation and poor water quality. The Bay City Fish Hatchery closed due to the lack of Walleye egg availability in 1945, and a cessation of Walleye fry stocking ensued (Schneider 1977). Although Walleye fry stocking had taken place from 1908 to 1944, it was generally regarded as ineffectual (Hile and Buettner 1959) and the population collapse occurred before the cessation of stocking.

Once the sources of Walleye recruitment were gone, additional forces emerged to further suppress population recovery. Water quality continued to decline due to cultural eutrophication and pollution (Freedman 1974; Keller et al. 1987). Schneider (1977) noted that Bay City and Saginaw, Michigan did not install primary sewage treatment plants until the early 1950s. Water quality particularly suffered during both world wars when heavy manufacturing in the area was greatest and the few regulations in place for environmental protection were suspended. Oil and phenol discharges into the Saginaw River system were common. Dissolved oxygen depletion was severe enough in the bay to kill fish caught in commercial trap nets (Schneider 1977).

Last, the establishment of invasive species contributed to the decline and suppression of Walleye stocks in Lake Huron. Two species that had important effects on Walleye were the predacious and competitive effects of Rainbow Smelt *Osmerus mordax* and Alewife *Alosa pseudoharengus*. Both of these species occupied Saginaw Bay for spawning and nursery grounds, entering the bay about the same time when Walleye fry were emerging in the spring. Research has documented that these species are formidable predators on newly hatched Walleye fry (Crowder 1980; Kohler and Ney 1980; Brandt et al. 1987; Evans and Loftus 1987; Brooking et al. 1998) and likely had a deleterious effect on Walleye reproduction. Environmental degradation and invasive species became the legacy of the bay through the 1950s and 1960s. In those decades, the Great Lakes ecosystem, including Saginaw Bay, was in ruins, severely degraded, polluted, dominated by invasive species, and managed only for remnants of commercial fisheries, principally by the U.S. Bureau of Commercial Fisheries (Dempsey 2001).

THE 1960S—A PARADIGM SHIFT IN MANAGEMENT JURISDICTION AND PRIORITY

Spurred on by the budding environmental movement in the United States and the Great Lakes, the deplorable environmental condition of Saginaw Bay improved during the late 1960s. The Michigan Department of Conservation, predecessor of the modern Department of Natural Resources (DNR), established a renewed interest in the Great Lakes. The decision to manage Great Lakes fisheries rather than continue to default to federal management stemmed from a reorganization of fishery management by Michigan DNR Director Ralph MacMullan who acted on the findings and recommendations of the "McFadden Report" (McFadden et al. 1964; Tody 2003). The state invoked the public trust doctrine privileges and responsibilities of states' rights to assume management for the commercial fisheries (Keller et al. 1987; Dochoda 1999; Dempsey 2001). By then, the commercial fishery had shifted, becoming almost entirely a gill-net fishery, a nonselective gear that kills fish regardless of species (Dempsey 2001). The state moved quickly in 1966 to close most of the commercial fisheries and convert remaining fisheries to live-capture trap nets, allowing sorting by species and size. Director MacMullan hired Dr. Howard Tanner to lead the newly reorganized Fisheries Division of the DNR, who, along with Dr. Wayne Tody, made bold and drastic decisions for the Great Lakes, culminating with the introduction of Coho Salmon *Oncorhynchus kisutch* and later Chinook Salmon *O. tshawytscha*. The result was a boom of recreational fishing on the Great Lakes. Importantly, this paradigm shift from commercial to recreational fisheries emphasis (Dempsey 2001; Tody 2003) set the stage for a later focused management approach to Saginaw Bay Walleye. The Saginaw Bay Walleye population remained in a collapsed state during these years (1960s through the 1970s). The commercial harvest was formally closed in 1970 to protect any remaining stock (Keller et al. 1987) and over human-health concerns of mercury contamination.

RECOVERY OF SAGINAW BAY WALLEYE

While the elimination of gill nets as a state-licensed commercial gear and the closure of commercial harvest of Walleyes was a prerequisite to Walleye restoration, the foundation for population recovery was the passage of the Clean Water Act of 1972 and the Great Lakes Water Quality Agreement of 1978 (Haas and Schaeffer 1992; Fielder and Baker 2004). Saginaw Bay's water quality proved responsive to pollution controls over industry and municipalities as well as improving agriculture practices and sediment abatement efforts (Dunbar and May 1995). Attempts at Walleye restoration were reinitiated during the 1970s when the Michigan DNR began fry stocking (Keller et al. 1987; Mrozinski et al. 1991; Haas and Schaeffer 1992). By the early 1980s, improvements in pond-rearing techniques of spring-fingerling Walleyes (40–60 mm in size) that were larger than fry resulted in increasingly dependable quantities for stocking purposes (Haas and Schaeffer 1992). Fry stocking was discontinued in 1984 and only fingerlings were stocked thereafter because they likely had a greater potential for survival (Table 1).

Adult Walleye abundance grew quickly in the early 1980s, as evidenced by the rise of recreational fishing in the bay and its tributaries (Ryckman 1986; Fielder et al. 2014). The Michigan DNR was now firmly established as the management authority on the Michigan waters of the Great Lakes through creation of fisheries research stations, including one on each of the four Great Lakes within its boundaries. The Michigan DNR began a series of survey efforts; foremost were fishery harvest (creel) surveys that provided estimates of angler harvest, effort, and catch rate with the first conducted in 1983. The survey was conducted annually from 1986 onward (Figure 3). At this time, the burgeoning recreational fishery was managed with statewide Walleye harvest regulations that included a five fish daily possession limit, a 381 mm (15 in) minimum length limit and a closed season in the rivers from March 15 through the last Saturday in April each year to protect spawning fish. The re-emergence of the recreational

TABLE 1. Walleye stocking in Saginaw Bay and its tributaries by life stage since 1972 and whether a mark was applied at the hatchery. OTC denotes immersion marking with buffered oxytetracycline hydrochloride. Fry are larvae <1 week old. Fingerlings are spring fingerlings that have been reared for approximately 2 months.

Year	Fry	Fingerlings	Mark
1972	50,000,000		
1973	50,000,000		
1974		5,500	
1975	300,000		
1976	300,000		
1977	4,575,000		
1978	300,000	25,000	
1979	1,200,000	334,400	
1980	600,000	10,000	
1981	1,300,000	294,700	
1982		269,500	
1983	1,745,735	869,000	
1984	2,000,000	947,800	
1985		954,200	
1986		871,300	
1987		632,200	
1988		345,500	
1989		834,400	
1990		850,100	
1991		622,700	
1992		787,700	
1993			
1994		1,283,000	
1995		717,500	
1996			
1997		1,006,300	OTC
1998		1,106,000	OTC
1999		640,000	OTC
2000		675,000	OTC
2001		951,000	OTC
2002		1,052,000	OTC
2003		1,829,000	OTC
2004		2,140,614	OTC
2005		398,758	OTC
2006			
2007			
2008			
2009			
2010			
2011			
2012			
2013			
2014			
2015			
2016			
2017			

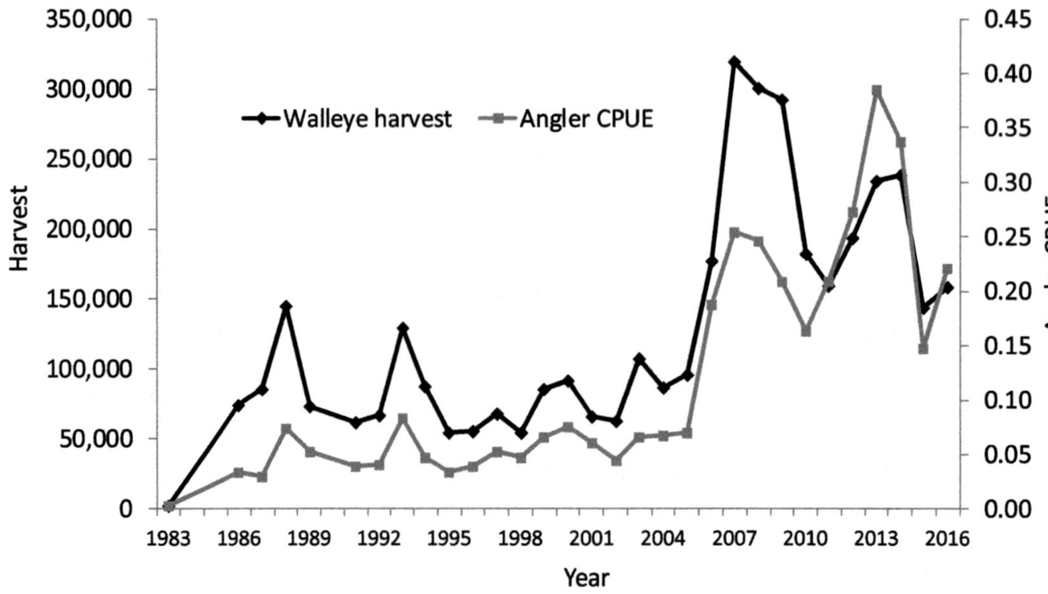

FIGURE 3. Saginaw Bay (including tributaries) Walleye recreational harvest and angler harvest rate (catch per unit effort [CPUE]) 1983–2016. Values include all modes of fishing and charter harvest for open water season and the same year's ice fishery. From Fielder et al. (2014), Ryckman (1986), and Michigan Department of Natural Resources, unpublished data.

fishery in the bay marked a new era of investigation through research, and annual population and fishery-independent surveys.

The first declaration of a recovery goal by the Michigan DNR was articulated by Keller et al. (1987) as "to reestablish a population and fishery entirely supported by natural reproduction." Later, the Great Lakes Fishery Commission also published fish community objectives (FCOs) for Lake Huron that represented a joint planning effort between the state of Michigan and the Province of Ontario. The Walleye FCO was stated as to "reestablish and/or maintain Walleye as the dominant cool-water predator over its traditional range with populations capable of sustaining a harvest of 0.7 million kg" (DesJardine et al. 1995). While this was a lakewide objective based on historical yields, most recognized that its success hinged on successful recovery of the Saginaw Bay Walleye stock.

The initial question confronting fishery managers and researchers was what is the origin of Walleyes (i.e., stocked or naturally reproduced) constituting the new Saginaw Bay Walleye population and fishery? Fingerling stocking was ongoing, but a convenient and reliable means to differentiate hatchery fish from wild fish did not exist until 1997 (Table 1) when stocked fingerlings were marked with oxytetracycline hydrochloride (OTC). At this time, initial analysis suggested that the population and its fishery were too large to be explained by stocking alone (unless survival of stocked fingerlings was extraordinarily high). The conclusion was that some natural reproduction must be taking place or the fishery was being supplemented by immigrants from outside the bay, mostly likely from Lake Erie, which had achieved Walleye recovery in the mid-1980s (Muth and Wolfert 1986; Hatch et al. 1987; Keller et al. 1987; Vandergoot et al. 2019, this volume).

The efficacy of the ongoing Walleye fingerling stocking was part of the emerging debate within the Michigan DNR and the public on how best to further recover the bay's population. The Lake Erie Walleye population had recovered without stocking playing a central role, and some believed that Saginaw Bay would do the same and that stocking was either unnecessary or even detrimental. Genetic analysis of Walleyes from spawning rivers and mixed stocks confirmed that most Saginaw Bay Walleyes were genetically distinct from Walleyes from Lake Erie and the connecting Huron–Erie corridor (Todd and Haas 1993; McParland et al. 1999). Some of the Walleyes inhabiting the bay appeared to have come from Lake Erie but were not deemed the main source. These studies also indicated that most Walleyes resident in the bay were likely native Walleyes from the Tittabawassee River, the primary spawning river in the Saginaw Bay watershed.

Oxytetracycline hydrochloride marking revealed that an average of 80% of age-0 (young of the year) Walleyes collected in the fall were of hatchery origin between 1997 and 2003 (Fielder 2002; Fielder and Thomas 2006). Given that stocking was the majority of each year-class in these studies, early fingerling stocking efforts probably were critical to restarting the Saginaw Bay Walleye population (Johnson et al. 2015). Subsequently, other successful Walleye restorations elsewhere, such as in Red Lake, Minnesota, were also built on the contributions of stocked fish (Barnard et al. 2019, this volume). Considering the environmental improvements in the 1980s, likely some natural reproduction also occurred at this time in Saginaw Bay.

The success of fingerling stocking and the apparent lack of success of early fry stocking gave managers and researchers clues as to remaining bottlenecks still limiting natural reproduction, namely the competitive and predatory effects of invasive species. The success of stocking Walleyes at more advanced life stages (i.e., spring fingerlings) than fry was consistent with the idea that fry were affected by Alewives and Rainbow Smelt, either by competition for first foods like zooplankton or from direct predation on fry. Spawning habitat also was limited by the effects of sedimentation and dams. Fielder (2002) determined that nearly all the natural recruitment originated from the tributaries, with only trace levels of reef spawning in the bay. Most of the historic reef complexes of the inner bay were still degraded from sedimentation.

Dynamics of the partially recovered Walleye population and its fisheries were closely monitored by the Michigan DNR, with an annual fall trawling survey occurring since 1970, an annual gill net survey (timed with the trawling work) since 1989 (Fielder and Thomas 2014), and a Walleye jaw tagging study to estimate mortality rates, exploitation rates, and movement since 1981 (Fielder 2014). These studies were consistently sustained over the decades and served as the foundation for gauging recovery status and management planning.

The recovery of the Walleye population was evident through the increase of the mean age of Walleyes in the annual spawning run in the Tittabawassee River. Mean age was just 2.7 years in 1981 and increased to 8.7 years in 2000 before declining again (Fielder 2014). Total annual mortality rate was high initially (55% to 70% between 1981 and 1986) but eventually declined and has averaged 37% over time (Fielder 2014; Fielder and Bence 2014). Exploitation of the stock by the recreational fishery rarely exceeded 10% annually before 2003. The adult Walleye population (age 2+) reached more than 3 million fish in 1988 but then steadily declined until coming to a new equilibrium of about 1.3 million during the late 1990s and early 2000s (Figure 4). Some believed that the first peak of Walleye abundance in 1988 might have signaled that recovery had been reached. However, no one knew then if the

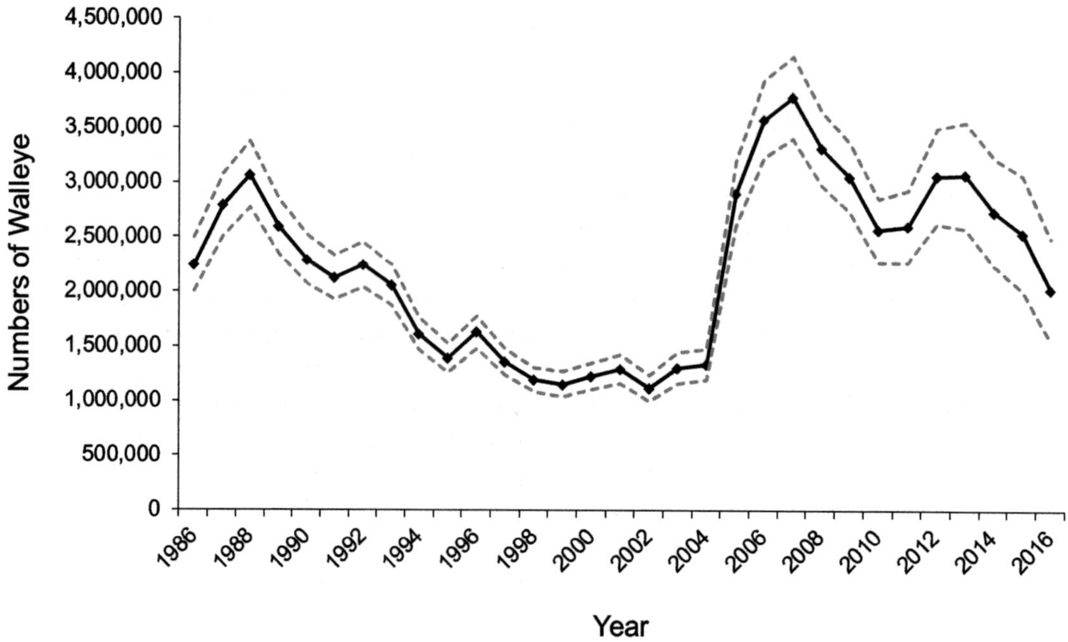

FIGURE 4. Population size of the Saginaw Bay stock of Walleye (age 2 and older) from 1986 to 2016. Dashed lines represent confidence intervals based on ±1.96 SEs of the estimate. From Fielder and Bence (2014) and Michigan Department of Natural Resources, unpublished data.

Walleyes present were stocked or naturally reproduced, and no recovery could be declared until it was certain that the population was reproductively self-sustaining.

The initial resurgence in the late 1980s of Saginaw Bay Walleyes coincided with the peak of the recovery of the Lake Erie stock of Walleyes; however, Fielder and Bence (2014) explicitly accounted for the contribution of Lake Erie Walleyes in their population model, such that Figure 4 represents only the abundance of Saginaw Bay origin of Walleyes. It appears that the initial resurgence in Saginaw Bay produced numbers that rivaled the modern day recovery but for some reason was not sustained. Survival of stocked fish may have declined or declines may have traced back to trends in Alewife. Other invasions occurred during this time, such as zebra mussels *Dreissena polymorpha* and Round Gobies *Neogobius melanostomus*, but subsequent research indicated that it was principally the effects of Alewives that suppressed Walleyes in the bay (Fielder et al. 2007).

MODERN RECOVERY MANAGEMENT OF THE 2000S

By the late 1990s, the Fisheries Division of the Michigan DNR had enough information from research and population assessment to create new recovery definitions and targets for the Saginaw Bay Walleye population. This information was used to revisit the original recovery goal articulated by Keller et al. (1987) and develop new strategies to achieve the targets. The new initiative included measurable reference points to gauge recovery, informed decision making, and proposed management actions to better move the population towards those goals. The management plan was published as "Strategy and Options for Completing the Recovery of Walleye in Saginaw Bay, Lake Huron" (Fielder and Baker 2004).

The new recovery goals built on those described by Keller et al. (1987) and were intended to support the interjurisdictional Lake Huron FCOs. These goals were to achieve predator–prey balance in the Saginaw Bay fish community, a Walleye population size that was commensurate with the carrying capacity of the prey base and adult habitat, and a self-sustaining Walleye population (not reliant on stocking). Guided by previous research findings, Fielder and Baker (2004) considered spawning habitat limitations to be the principal limiting factor to Walleye recovery. This perspective was attributed to the ongoing presence of dams in the watershed, continued sedimentation in the Saginaw River system (which was still estimated to be twice the predevelopment rate), and the degraded condition of offshore spawning reefs. Recognized in the plan as a secondary factor was the ongoing abundance of Alewives in Saginaw Bay.

The recovery plan laid out six strategies. Fish passage and dam removal were deemed essential to restore access to historic spawning grounds for migrating Walleyes. It was estimated that 72% of the river reaches in the watershed were inaccessible to spawning Walleyes. Prescribed was dam removal, where and when possible, especially if it included the first barrier to allow for fish passage. If removal was not possible, the plan suggested investing in alternate fish passage strategies for dams. The plan identified six rivers as top priorities, all but one of which were part of the Saginaw River system. Achieving fish passage at the Dow Dam on the Tittabawassee River was key to this strategy.

The second restoration strategy included spawning reef reclamation. This strategy recognized the importance reef-based spawning, which was believed to have been historically significant in the bay. Therefore, progress in restoring reef habitat would potentially contribute to Walleye recovery by increasing and diversifying recruitment sources. The plan emphasized investment in recreating natural rock reefs rather than building artificial reefs consisting of construction debris.

Control of Alewives was the next strategy outlined in the recovery plan. The plan cited Walters and Kitchell's (2001) paper advancing the concept of depensatory cultivation, that to achieve good reproduction of predators such as Walleye, high densities of adults were needed to "groom" the fish community and prey base of species predatory on their fry. The concept was that some threshold of predator abundance may exist that is needed to achieve reproductive success to become self-sustaining. Fielder and Baker's (2004) prescription to achieve Alewife control in Saginaw Bay was to create a predation barrier that would consume the majority of Alewives migrating into the bay each spring. Increased Walleye fingerling stocking could develop the predator base to accomplish this strategy. Prescribed was an annual stocking of 2.5 million Walleye fingerlings, an increase of 1.8 million over what had been occurring. This vision exceeded the fingerling production capacity of the Michigan DNR fish hatcheries, but the strategy could serve as a basis for new investment and expanded production capacity in the future.

The final strategies pertained to encouraging improved sediment abatement in the watershed by working with soil conservation organizations and environmental regulatory agencies. Also prescribed was the maintenance of existing harvest regulations in the recreational fishery and the continued closure of the commercial Walleye fishery.

The authors purposely worded the recovery goal of "abundance of Walleyes at the carrying capacity of the adult habitat and prey base" because they wanted to depart from the past practice of defining recovery goals based on historic condition. Most Great Lakes FCOs, including those for Lake Huron

(DesJardine et al. 1995), were premised on restoring a fish population to levels that could sustain historic yields, which may not be realistic given the modern food web in light of all the exotic invasions and the habitat alteration that has taken place. To define recovery of Walleye as a population at the modern-day carrying capacity, the authors conceded that (1) they would declare success if the population was at this density regardless of how that compared to the historic condition, and (2) they did not know what that capacity was.

Some of the most useful components of the Saginaw Bay Walleye recovery plan were reference and decision points to guide future management. To overcome the inability to know what the carrying capacity of Walleyes were in modern-day Saginaw Bay, Fielder and Baker (2004) drew upon the advice of Colby et al. (1994). This guidance document cited the value of using growth rates of Walleye to gauge recovery. Because Walleye growth is usually density-dependent, it will be fast when populations exist below the carrying capacity and slower or more average at capacity. Schneider (2000) had previously defined state-average Walleye growth rates in Michigan depending on the time of year collected. Walleye growth rates in Saginaw Bay before recovery averaged 127% of the state-average rate. Fielder and Baker (2004) defined a target definition of the Saginaw Bay Walleye population being at carrying capacity when age-3 growth was at or below 110% of the state-average rate for three out of five consecutive years. This approach was attractive because it was based on a metric that was easily measured during annual population assessments and was based on the intrinsic characteristics of the Walleye population.

The second key element of defining full recovery was the ability to self-sustain the population of Walleye via natural reproduction. After the implementation of OTC marking to differentiate hatchery fish, gauging the magnitude of natural reproduction became much easier. Fielder and Baker (2004) established that Walleye fingerling stocking would be suspended once three out of five consecutive year-classes were 50% wild origin or greater. Last, the authors also adopted a yield objective with a population capable of sustaining a 0.45 million kg harvest so as to tie the recovery plan back to the FCOs yield-based objectives.

UNEXPECTED LAKE HURON FOOD WEB CHANGE AND FULL RECOVERY ATTAINED

Implementation of the plan began while under development by increasing Walleye fingerling stocking pursuant to creating the prescribed predation barrier to Alewives (Table 1). Walleye fingerling stocking was increased to more than 2 million by 2004 but was preempted by an unprecedented turn of events in the food web of Lake Huron. Prior to this time, Alewives had been declining in abundance and their age structure became more and more compressed until each year-class was merely replacing itself (Riley et al. 2008; Riley and Roseman 2013). By 2003, fall forage fish surveys conducted by the U.S. Geological Survey's Great Lakes Science Center revealed a virtually complete absence of adult Alewives although an enormous year-class of age-0 Alewives occurred. By 2004, no substantial number of any age-class of Alewives was sampled in Lake Huron surveys. Much debate ensued as to the reason, including bottom-up forces (Barbiero et al. 2011) causing the loss of the macroinvertebrate *Dioperia* sp., which was attributed in turn to the cascading effects of dreissenid mussel invasion in addition to top down forces in the form of high predation from Chinook Salmon and other predators (He et al. 2016). The loss of Alewives was mirrored with substantial declines in Rainbow Smelt as well and the changes had dire consequences for the recreationally important Pacific Salmon fishery in the lake. It seemed Lake Huron had undergone an unprecedented and unexpected shift in its ecology and fish community almost overnight. The change expressed itself as a new form of ecological equilibrium.

Simultaneous to the collapse of Alewives and severe contraction of Rainbow Smelt was an explosion in reproductive success of Percids in Saginaw Bay. Both Walleye and Yellow Perch *Perca flavescens* showed record abundances, as indicated by the annual trawl series (Figure 5). Fielder et al. (2007) tested a series of possible explanatory variables, and the most predictive yet parsimonious model was Alewife trends as a predictor of Walleye year-class strength. A large year-class of Walleyes was also produced in Lake Erie the same year (2003) so likely some climatic component may have contributed as well (Vandergoot et al. 2019). However, the trend was sustained in subsequent years in Saginaw Bay but not elsewhere. The conclusion was that the abrupt explosion in reproductive success in Saginaw Bay was because of the collapse of Alewives. The proportion of stocked fish as indicated by the OTC marking confirmed that the increase in Walleye abundance came from natural reproduction. In accordance with the stocking decision criteria in the recovery plan, greater than 50% of wild age-0 fish for the third year occurred in 2005, and thus, stocking was discontinued in 2006. The Saginaw Bay Walleye population thereafter has been sustained entirely by natural reproduction.

The growth rate of age-3 Saginaw Bay Walleyes slowed precipitously as predicted by Fielder and Baker (2004) and Colby et al. (1994), reaching the 110% state average rate consistently from 2007, with the third year criteria achieved in 2009. The Michigan DNR

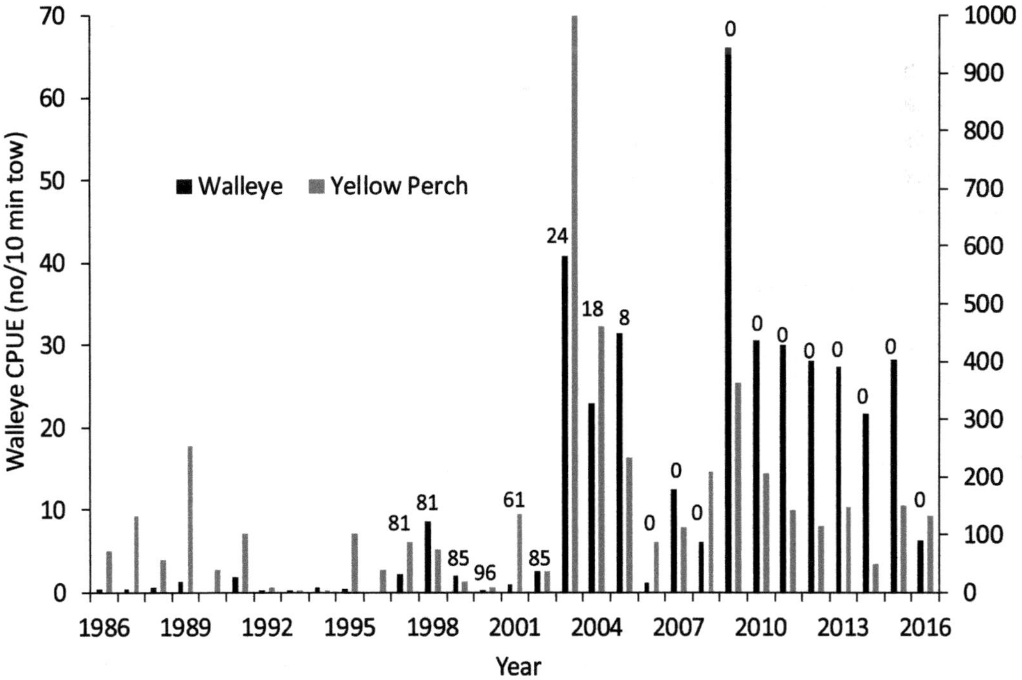

FIGURE 5. Mean catch per unit effort (CPUE) in annual Saginaw Bay trawl tows from 1986 to 2016 for age-0 Walleye and Yellow Perch. Numbers are the percentage of stocked fish to the year-class composition for Walleye. Yellow Perch CPUE for 2003 is 2,451, but the *y*-axis is limited to 1,000. From Fielder and Thomas (2014) and Michigan Department of Natural Resources, unpublished data.

declared the Saginaw Bay stock of Walleyes fully recovered that year (Fielder and Thomas 2014). It was a remarkable turn of events. Fishery experts who worked on the system and developed the management plan doubted they would ever see recovery in their lifetimes let alone within their own careers, yet in the end, it happened in just 6 years' time!

Recovery of the Saginaw Bay Walleye stock had a number of consequences. The harvest and harvest rate by anglers increased 238% and 387%, respectively (Figure 3). Harvest rate of anglers specifically targeting Walleyes was as great as 0.76 Walleyes per hour in 2013. Angler parties achieving their limit of Walleye (five fish per angler) increased nearly 10-fold after recovery. The decline in growth rate meant a lower frequency of trophy-sized fish in the recreational fishery and an overall increase of small Walleyes than previously occurred in the Saginaw Bay fishery.

An unanticipated effect was a sharp decline in Yellow Perch survival rate, specifically age-0 fish to yearling and older ages. Yellow Perch had suffered loss of the predation buffer provided by Alewives. Diet analysis of Walleyes revealed a shift from clupeids (principally Alewife) to Yellow Perch (Fielder and Thomas 2014). Other predators were also using young Yellow Perch for forage (DeBruyne et al. 2017). The overall prey index based on trawling catch rates in the fall survey indicated a steady decline of the prey base in the bay from 2003 to 2014. Clearly, the greatly increased predator population was having an effect on the fish community.

Yellow Perch had been declining since the mid-1980s and the sport and commercial fisheries were virtually collapsed since the mid-2000s. Yet the bay historically sustained robust populations of both Walleyes and Yellow Perch. The profound food web changes in Lake Huron had broken some important food web linkages. Alewives likely provided a predation buffer by making an onshore migration and using the bay as a nursery grounds for their young. Likely, Ciscos served as a predation buffer for Yellow Perch before Alewives dominated that niche in the lake. After the collapse of Alewives, Ciscos have not recovered and the broken linkage between the main basin and the nearshore likely had dire new consequences for Yellow Perch.

In spite of Walleye recovery in the bay, recreational fishing effort (angler-hours) steadily declined from a high in 1986 to only 26% of that level by 2016. This decline was partly driven by Walleye anglers who needed less time to achieve their bag limits or level of satisfaction since recovery. Primarily, however, the decline in effort paralleled the decline in the availability of Yellow Perch. Analysis indicated that half the variability in recreational fishing effort trends in the bay was explained by the angler catch rate of Yellow Perch, with higher rates predicting more participation (Fielder et al. 2014). Despite the popularity of Walleye in the region, for Saginaw Bay, much of the fishing participation hinged on the quality of the Yellow Perch fishery.

With Walleye recovery attained, assessment and management transitioned from a focus on recovery to sustainable management. To facilitate this transition, research biologists developed stock assessment models, including a statistical catch-at-age model (Fielder and Bence 2014) and a stochastic simulation model of the Saginaw Bay Walleye population and fisheries. The Michigan DNR used that tool to conduct a management strategy evaluation via a decision analysis (Fielder et al. 2016). The management question asked of the model was whether the recovered population could sustain a greater level of recreational exploitation than it was currently experiencing. The conclusion was that that the recreational fishing mortality could be increased by as much as 50% without threatening sustainability. After stakeholder consultation, a liberalized bag limit of eight Walleyes per day and a reduction of minimum length limit to 330 mm (13 in) was instituted in October 2015. The intent on the part of the fishery managers was to begin to tie annual recreational harvest regulation to current population status, as indicated by model projections and survey findings. These regulations will be revisited annually each winter and periodic adjustments made to the recreational harvest regulations by spring of the same year. This approach serves to maximize the use of the Walleye population within the limits of sustainability, make management more responsive to any population or fishery level changes, and adopt a process similar to that used on Lake Erie Walleye management employed by the Lake Erie Committee of the Great Lakes Fishery Commission (Vandergoot et al. 2019).

LESSONS LEARNED

Ecosystems can be resilient. The remarkable recovery of the Saginaw Bay Walleye population illustrates the recuperative powers of the Great Lakes when released from stressors. Clearly, the deleterious effects of invasive species, principally Alewives in this case, were likely the dominating factor affecting the Walleye population. Taken as a whole over time, prerequisites to restoration also included improvements to water quality and control of the early commercial fishery. Without basic water quality being met, combined with the opportunity for Walleye to mature and spawn, no recovery would have been possible, even in the absence of Alewives. This lesson about ecosystem resiliency is also supported by the remarkable recovery of Lake Erie (Vandergoot et al. 2019) and Red Lake (Barnard et al. 2019).

The presence of adult broodfish in the system permitted recovery to occur once ecological conditions became favorable. Managing a Walleye fishery as large as Saginaw Bay with stocking was not without controversy. Managers, however, saw that as a powerful tool to position the population for recovery as well as generate fishing opportunity and maintain public interest in the fishery. Although some local natural recruitment was ongoing, the same rapid recovery may not have occurred if recovery had been solely dependent on naturally reproduced adults.

Obstacles that stand in the way of restoration should be identified so that they can be addressed. Fishery managers and researchers had thought that spawning habitat limitations were the principal obstacle to recovery and that Alewives were a secondary factor when the opposite was true. Managers and researchers should be credited, however, for getting the suite of factors correct even if in the wrong order. The distribution and abundance of available spawning habitat remains an important variable that could affect the sustainability of the recovery. The ongoing dependence on just tributary-based spawning and the lack of reef-based spawning means that the population may be less resilient and more vulnerable to trends in watershed development and climate change. Despite the achievement of recovery targets, a team of researchers and managers, including the Michigan Department of Environmental Quality, is actively pursuing offshore reef restoration projects originally identified in the recovery plan. Fish passage in the form of rock ramps has been built at the face of two existing dams in the Saginaw River watershed with the hope of promoting Walleye spawning migration.

Long-term data series are important for the development of recovery strategies and gauging recovery progress. Survey data provided feedback information from the population in response to fishery management actions and changes in the food web. The fish community assessment work dates back to 1970 and the Walleye jaw tagging work to 1981. Creel surveys date back to 1983. These investments in information were essential to informing management decision making along the way and are as relevant as ever as management now transitions to sustaining the recovered population. State-of-the-art stock assessment models were only possible in the data-rich environment of Saginaw Bay. Credit goes to the far-sighted fishery managers and researchers of the 1970s and early 1980s for starting the studies and to the administrators who have kept them consistently funded over the decades.

Resolve to accomplish the recovery goal by fishery biologists and managers, administrators, and stakeholders is essential. This resolve, over more than half a century, when combined with the significant contributions of federal agencies and academic partners, made recovery possible. Success was achieved by defining goals, identifying obstacles, choosing and implementing actions to solve those problems, and then assessing and monitoring the response of Walleyes in the bay (Krueger and Decker 1999). This approach along with some fortuitous food-web changes in Lake Huron made large-scale recovery of fishery resources possible.

A LOOK AHEAD

The recovery of a fishery is only as good as the will to sustain it. The sustainability of Walleye recovery on Saginaw Bay is by no means guaranteed in the future. Considerable challenges and threats will continue to face the Saginaw Bay Walleye population. Still needed is a more complete management planning and visioning effort for this stock and the supporting fish community. Definitions of sustainability were only defined for interim purposes. Also needed is consideration of quality definitions for the fishery in terms of size of fish, catch, and catch per unit effort, which in turn require input from stakeholders.

Simultaneous to the recovery, new telemetry-based research revealed what had long been suspected, that substantial proportions as great as 37% of the Saginaw Bay adult Walleye population make long-distance movements into the main basin of Lake Huron (Hayden et al. 2014). This information suggests that the forces of mortality shaping the Walleye stock are not limited to the recreational fishery in the bay (Fielder and Bence 2014). While stock assessment models can account for these mortality sources, Michigan fishery managers do not have direct authority over fisheries that operate in Ontario waters of Lake Huron or the tribal-based fisheries in the northern one-third of the lake. While believed to be a relatively minor component of total mortality, these fisheries could grow, which will necessitate accommodating or working through the international Great Lakes Fishery Commission and tribal treaty provisions to prevent overharvest and threaten the recovery.

Moving from management for restoration to management of a recovered resource can also lead to new fisheries. Recently proposed is an allocation of Saginaw Bay Walleyes to commercial harvest premised on the existing bycatch and by-kill mortality already occurring in the bay's commercial fishery for Yellow Perch and other fisheries in the main basin. Whether management of this stock will come full circle and again include commercial harvest is unknown.

While the new food web of Lake Huron is regarded by most as a new paradigm, whether it has established a new stable equilibrium remains to be seen. Alewives have declined in Lake Michigan as well, but they persist. Conceivably, Alewives could make a recovery in Lake Huron if environmental conditions shifted again. If so, Walleye reproductive success would be negatively impacted.

While the identified challenges to ongoing sustainability of the recovered condition are not trivial, the greatest threats may be those that cannot be readily anticipated. The Great

Lakes have a long and sad history of destructive nonnative invasions and many of those migratory pathways still exist today. If Asian carp species (e.g., Black Carp *Mylopharyngodon piceus*, Bighead Carp *Hypophthalmichthys nobilis*, and Silver Carp *H. molitrix*) ever succeed in invading from the Illinois River drainage into the Great Lakes (Lohmeyer and Garvey 2009), it is likely productive environments like Saginaw Bay that will experience the greatest impacts. Eurasian Ruffe *Gymnocephalus cernua*, already in the Great Lakes, have made it as close as Lake Huron's Thunder Bay 90 km away, but so far they have not invaded Saginaw Bay. This warmwater percid-like fish would likely find ideal living conditions if it were ever to reach Saginaw Bay. These two examples are the invasive species that resource professionals know of; others, perhaps even more detrimental, may be on the horizon coming from unknown sources.

Lastly, climate change is projected to cause warmer Great Lakes nearshore waters than in the past, possibly making them more vulnerable to invasions, and could promote Alewife recovery. Less ice cover and the possibility of increased hypoxia events in the bay both seem likely. All of the Great Lakes are vulnerable to climate change, but very possibly nearshore fish communities may see the greatest effect if left unabated. Walleye, being a coolwater species, may find the bay only marginally or entirely unsuitable in the future. The future of the Saginaw Bay Walleye population and its recovered status will ultimately hinge on the resolve of the public to support the management necessary to safeguard the resource and to stay vigilant not only against future threats, but also to continue to improve the habitat to ensure the resilience needed to sustain the recovered population.

ACKNOWLEDGMENTS

The authors would like to thank Paige Gephardt, resource analyst of the Michigan DNR's Forest Resources Division, for assistance with Figure 1. Our appreciation is also extended to various individuals for critical reviews of early versions of this manuscript. The authors especially acknowledge and credit the resource professionals over the years who have dedicated themselves to the recovery of the Saginaw Bay stock of Walleye; notably, these include Jim Schneider, Bob Haas, Mike Thomas, Leo Mrozinski, Ray Shepherd, John Weber, Myrl Keller, Jim Johnson, Randy Eshenroder, and Kathrin Schrouder.

REFERENCES

Albert, D. A. 2003. Between land and lake: Michigan's Great Lakes coastal wetlands. Michigan State University Extension, Extension Bulletin E-2902, East Lansing.

Baldwin, N. A., R. W. Saalfeld, M. R. Dochoda, H. J. Buettner, and R. L. Eshenroder. 2009. Commercial fish production in the Great Lakes 1867–2006 [online database]. Great Lakes Fishery Commission, Ann Arbor, Michigan. Available: www.glfc.org/databases/commercial/commerc.php. (July 2018).

Barbiero, R. P., B. M Lesht., and G. J. Warren. 2011. Evidence for bottom–up control of recent shifts in the pelagic food web of Lake Huron. Journal of Great Lakes Research 37:78–85.

Barnard, G. C., A. J. Kennedy, W. P. Brown, and D. L. Pereira. 2019. Walleye management in the Red Lakes, Minnesota: collapse, recovery, and cooperative management. Pages 323–352 *in* C. C. Krueger, W. W. Taylor, and S. Youn, editors. From catastrophe to recovery: stories of fishery management success. American Fisheries Society, Bethesda, Maryland.

Bogue, M. B. 2000. Fishing the Great Lakes: an environmental history, 1783–1933. The University of Wisconsin Press, Madison.

Bozek, M. A., D. A. Baccante, and N. P. Lester. 2011. Walleye and Sauger life history. Pages 233–301 *in* B. A. Barton, editor. Biology, management, and culture of Walleye and Sauger. American Fisheries Society, Bethesda, Maryland.

Brandt, S. B., D. M. Mason, D. B. MacNeill, T. Coates, and J. E. Gannon. 1987. Predation by Alewives on larvae of Yellow Perch in Lake Ontario. Transactions of the American Fisheries Society 116:641–645.

Brooking, T. E., L. G. Rudstam, M. H. Olson, and A. J. VanDeValk. 1998. Size dependent Alewife predation on larval Walleyes in laboratory experiments. North American Journal of Fisheries Management 18:960–965.

Colby, P. J., C. A. Lewis, R. L. Eshenroder, R. C. Haas, and L. J. Hushak. 1994. Walleye-rehabilitation guidelines for the Great Lakes area. Great Lakes Fishery Commission, Ann Arbor.

Crowder, L. B. 1980. Alewife, Rainbow Smelt and native fishes in Lake Michigan: competition or predation? Environmental Biology of Fishes 5:225–233.

DeBruyne, R. L., D. G. Fielder, E. F. Roseman, and P. H. Butchko. 2017. Exploring potential effects of cormorant predation on the fish community in Saginaw Bay, Lake Huron. Journal of Great Lakes Research 43:387–393.

Dempsey, D. 2001. Ruin and recovery, Michigan's rise as a conservation leader. The University of Michigan Press, Ann Arbor.

DesJardine, R. L., T. K. Gorenflo, R. N. Payne, and J. D. Schrouder. 1995. Fish-community objectives for Lake Huron. Great Lakes Fishery Commission, Special Publication 95-1, Ann Arbor.

Dochoda, M. R. 1999. Authorities, responsibilities, and arrangements for managing fish and fisheries in the Great Lakes ecosystem. Pages 93–110 *in* W. W. Taylor and C. P. Ferreri, editors. Great Lakes fisheries policy and management. Michigan State University Press, East Lansing.

Dunbar, W. F., and G. S. May. 1995. Michigan: a history of the wolverine state. Wm. B. Eerdmans Publishing, Grand Rapids, Michigan.

Evans, D. O., and D. H. Loftus. 1987. Colonization of inland lakes in the Great Lakes region by Rainbow Smelt, *Osmerus mordax*: their freshwater niche and effects on indigenous fishes. Canadian Journal of Fisheries and Aquatic Sciences 44(Supplement 2):249–266.

Fielder, D. G. 2002. Sources of Walleye recruitment in Saginaw Bay, Lake Huron. North American Journal of Fisheries Management 22:1032–1040.

Fielder, D. G. 2014. Mortality, exploitation, movement, and stock size of Saginaw Bay Walleyes, 1981–2011: 31 years of tag return analysis. Michigan Department of Natural Resources, Fisheries Report 04, Lansing.

Fielder, D. G., and J. P. Baker. 2004. Strategies and options for completing the recovery of Walleye in Saginaw Bay, Lake Huron. Michigan Department of Natural Resources, Fisheries Special Report 29, Lansing.

Fielder, D. G., and J. R. Bence. 2014. Integration of auxiliary information in statistical catch-at-age analysis of the Saginaw Bay stock of Walleye in Lake Huron. North American Journal of Fisheries Management 34:970–987.

Fielder, D. G., M. J. Jones, and J. R. Bence. 2016. Use of a structured approach to analysis of management options and value of Information for a recreationally exploited fish population: a case study of Walleyes in Saginaw Bay. North American Journal of Fisheries Management 36:407–420.

Fielder, D. G., J. S. Schaeffer, and M. V. Thomas. 2007. Environmental and ecological conditions surrounding the production of large year classes of Walleye (*Sander vitreus*) in Saginaw Bay, Lake Huron. Journal of Great Lakes Research 33:118–132.

Fielder, D. G., and M. V. Thomas. 2006. Fish population dynamics of Saginaw Bay, Lake Huron 1998–2004. Michigan Department of Natural Resources, Fisheries Research Report No. 2083, Lansing

Fielder, D. G., T. L. Kolb, T. M. Goniea, D. L. Wesander, and K. S. Schrouder. 2014. Fisheries of Saginaw Bay, Lake Huron 1986–2010. Michigan Department of Natural Resources, Fisheries Report 02, Lansing.

Fielder, D. G., and M. V. Thomas, 2014. Status and trends of the fish community of Saginaw Bay, Lake Huron 2005–2011. Michigan Department of Natural Resources, Fisheries Report 03, Lansing.

Freedman, P. L. 1974. Saginaw Bay: an evaluation of existing and historical conditions. U.S. Environmental Protection Agency, Report #905/9-74-003, Washington, D.C.

Goodyear, C. D., T. A. Edsall, D. M. Ormsby Dempsey, G. D. Moss, and P. E. Polanski. 1982. Atlas of spawning and nursery areas of Great Lakes fishes. U.S. Fish and Wildlife Service, FWS/OBS-82/52, Washington, D.C.

Hayden, T. A., C. M. Holbrook, D. Fielder, C. S. Vandergoot, R. A. Bergstedt, J. M. Dettmers, C. C. Krueger, and S. J. Cooke. 2014. Acoustic telemetry reveals large-scale migration patterns for Walleye in Lake Huron. PLOS (Public Library of Science) ONE [online serial] 9(12):e114833.

Haas, R. C., and J. S. Schaeffer. 1992. Predator-prey and competitive interactions among Walleye, Yellow Perch, and other forage fishes in Saginaw Bay, Lake Huron. Michigan Department of Natural Resources, Research Report No. 1984, Lansing.

Hatch, R. W., S. J. Nepszy, K. M. Muth, and C. T. Baker. 1987. Dynamics of the recovery of the western Lake Erie Walleye (*Stizostedion vitreum vitreum*) stock. Canadian Journal of Fisheries and Aquatic Sciences 44(Supplement 2):15–22.

He, J. X., J. R. Bence, E. F. Roseman, D. G. Fielder, and M. P. Ebener. 2016. Using time-varying asymptotic length and body condition of top piscivores to indicate ecosystem regime shift in the main basin of Lake Huron: a Bayesian hierarchical modeling approach. Canadian Journal of Fisheries and Aquatic Sciences 73:1092–1103.

Hile, R., and H. J. Buettner. 1959. Fluctuations in the commercial fisheries of Saginaw Bay 1885–1956. U. S. Fish and Wildlife Service, Research Report No. 51, Washington, D.C.

Jennings, M. J., Claussen, J. E., and Philipp, D. P. 1996. Evidence for heritable preferences for spawning habitat between two Walleye populations. Transactions of the American Fisheries Society 125:978–982.

Johengen, T. H., T. F. Nalepa, G. A. Lang, D. L. Fanslow, H. A. Vanderploeg, and M. A. Agy. 2000. Physical and chemical variables of Saginaw Bay, Lake Huron in 1994–1996. NOAA Technical Memorandum GLERL-115.

Johnson, J. E., J. X. He, and D. G. Fielder. 2015. Rehabilitation stocking of Walleye and Lake Trout: restoration of reproducing stocks in Michigan waters of Lake Huron. North American Journal of Aquaculture 77:396–408.

Kalejs, N. I. 2017. An assessment of reef restoration potential in Saginaw Bay, Lake Huron. Master's thesis. Purdue University, West Lafayette, Indiana.

Keller, M., J. C. Schneider, L. E. Mrozinski, R. C. Haas, and J. R. Weber. 1987. History, status, and management of fishes in Saginaw Bay, Lake Huron, 1981–1986. Michigan Department of Natural Resources, Fisheries Technical Report 87-2, Lansing.

Kohler, C. C., and J. J. Ney. 1980. Piscivory in a landlocked Alewife (*Alosa pseudoharengus*) population. Canadian Journal of Fisheries and Aquatic Sciences 37:1314–1317.

Krueger, C.C., and D.J. Decker. 1999. The process of fisheries management. Pages 31–59 *in* C. C. Kohler and W. A. Hubert, editors. Inland fisheries management in North America, 2nd edition. American Fisheries Society, Bethesda, Maryland

Lanman, J. 1839. History of Michigan: civil and topographical in a compendious form with a view of surrounding lakes. E. French, New York.

Lohmeyer, A. M., and J. E. Garvey. 2009. Placing the North American invasion of Asian carp in a spatially explicit context. Biological Invasions 11:905–916.

McFadden, J. T., M. J. DeBoer, J. T. Wilkinson, W. T. Tody, and R. G. Wicklund. 1964. A management program for Michigan's fisheries. Michigan Department of Conservation, Fish Section, Lansing.

McParland, T. L., M. M. Ferguson, and A. P. Liskauskas. 1999. Genetic population structure and mixed-stock analysis of Walleyes in the Lake Erie–Lake Huron corridor using allozyme and mitochondrial DNA markers. Transactions of the American Fisheries Society 128:1055–1067.

Mrozinski, L. E., J. C. Schneider, R. C. Haas, and R. E. Shepherd. 1991. Rehabilitation of Walleye in Saginaw Bay, Lake Huron. Pages 63–84 *in* P. J. Colby, C. A. Lewis, and R. L. Eshenroder, editors. Status of Walleye in the Great Lakes: case studies prepared for the 1989 workshop. Great Lakes Fishery Commission, Special Publication 91-1, Ann Arbor, Michigan

Muth, K. M., and D. R. Wolfert. 1986. Changes in growth and maturity of Walleyes associated with stock rehabilitation in western Lake Erie, 1964–1983. North American Journal of Fisheries Management 6:168–175.

Organ, W. L., G. L. Towns, M. O. Walter, R. B. Pelletier, and D. A. Riege. 1979. Past and presently known spawning grounds of fishes in the Michigan coastal waters of the Great Lakes. Michigan Department of Natural Resources, Technical Report 79-1, Lansing.

Riley, S. C., and E. F. Roseman. 2013. Status of the offshore demersal fish community. Pages 21–35 *in* S. C. Riley, editor. The state of Lake Huron in 2010. Great Lakes Fishery Commission, Special Publication 13-01, Ann Arbor, Michigan.

Riley, S. C., E. F. Roseman, S. J. Nichols, T. P. O'Brien, C. S. Kiley, and J. S. Schaeffer. 2008. Deepwater demersal fish community collapse in Lake Huron. Transactions of the American Fisheries Society 137:1879–1890.

Ryckman, J. R. 1986. A creel survey of sport fishing in Saginaw Bay, Lake Huron, 1983–84. Michigan Department of Natural Resources, Fisheries Division, Lansing.

Schneider, J. C. 1977. History of the Walleye fisheries of Saginaw Bay, Lake Huron. Michigan Department of Natural Resources, Fisheries Research Report No. 1850, Lansing.

Schneider, J. C. 2000. Manual of fisheries survey methods II: with periodic updates. Michigan Department of Natural Resources, Fisheries Special Report No. 25, Lansing.

Schneider, J. C., and J. H. Leach. 1977. Walleye (*Stizostedion vitreum vitreum*) fluctuations in the Great Lakes and possible causes, 1800–1975. Journal of Fisheries Research Board of Canada 34:1878–1889.

Schneider, J. C., and J. H. Leach. 1978. Walleye stocks in the Great Lakes 1800–1975: fluctuations and possible causes. Great Lakes Fishery Commission, Technical Report No. 29, Ann Arbor, Michigan.

Todd, T. N., and R. C. Haas. 1993. Genetic and tagging evidence for movement of Walleyes between Lake Erie and Lake St. Clair. Journal of Great Lakes Research 19:445–452.

Tody, W. H. 2003. A history of Michigan's fisheries, a fabulous renewable resource along with our forests and wildlife. Self-published and printed by Copy Central, Traverse City, Michigan.

U.S. Fish and Wildlife Service. 1969. Fish and wildlife as related to water quality of the Lake Huron basin. A special report on fish and wildlife resources.

Vandergoot, C. S., M. D. Faust, J. T. Francis, D. W. Einhouse, R. Drouin, C. Murray, and R. L. Knight. 2019. Back from the brink: successful interagency management of the Lake Erie Walleye fishery. Pages 431–466 *in* C. C. Krueger, W. W. Taylor, and S. Youn, editors. From catastrophe to recovery: stories of fishery management success. American Fisheries Society, Bethesda, Maryland.

Walters, C., and J. F. Kitchell. 2001. Cultivation/depensation effects on juvenile survival and recruitment: implications for the theory of fishing. Canadian Journal of Fisheries and Aquatic Sciences 58:39–50.

Back from the Brink: Sustainable Management of the Lake Erie Walleye Fishery

Christopher S. Vandergoot*[1]

U.S. Geological Survey, Great Lakes Science Center, Lake Erie Biological Station
6100 Columbus Avenue, Sandusky, Ohio 44870, USA

Matthew D. Faust

Ohio Department of Natural Resources, Division of Wildlife
Sandusky Fisheries Research Station
305 East Shoreline Drive, Sandusky, Ohio 44870, USA

James T. Francis

Michigan Department of Natural Resources, Division of Fisheries
Lake Erie Management Unit
7806 Gale Road, Waterford, Michigan 48327, USA

Donald W. Einhouse

New York State Department of Environmental Conservation
Lake Erie Fisheries Research Unit
178 Point Drive North, Dunkirk, New York 14048, USA

Richard Drouin

Ontario Ministry of Natural Resources and Forestry, Lake Erie Management Unit
659 Exeter Road, London, Ontario N6E1LE, Canada

Charles Murray

Pennsylvania Fish and Boat Commission, Lake Erie Research Unit
7895 West Lake Road, Fairview, Pennsylvania 16415, USA

Roger L. Knight

Great Lakes Fishery Commission
2100 Commonwealth Boulevard, Suite 100, Ann Arbor, Michigan, 48105, USA

Abstract.—This chapter describes management actions implemented after a large-scale population decline of the Walleye *Sander vitreus* population in Lake Erie, one of the Laurentian Great Lakes in North America. Intensive fishery exploitation during the 1950s combined with declining water quality conditions collapsed the Walleye stock during the early 1960s. The fishery persisted at low levels until 1970 when the fishery was closed (1970–1972) due to elevated mercury concentrations in Walleye tissue. Lake Erie fishery managers at the time recognized the need for a coordinated, multi-agency approach to protect this ecologically, economically, and socially important resource. The harvest ban was lifted in 1973 when

* Corresponding author: vandergo@msu.edu

[1] Current address: Department of Fisheries and Wildlife, Michigan State University, 115 Manly Miles Building, East Lansing, Michigan 48823, USA

mercury levels dropped below advisory levels. In 1976, an interagency management framework was established, which relied on a coordinated, science-based management philosophy consisting of estimating safe harvest levels, performing applied research, and conducting annual population assessments. The population rebounded during the 1980s in response to improving environmental conditions, regulated harvest, and a series of strong recruitment events. Declines in harvest and population size were again observed during the late 1990s and mid-2000s, likely due to variation in natural processes controlling recruitment, and fishery managers enacted harvest practices during this period to promote long-term sustainability. Today, Lake Erie Walleyes support one of the largest self-sustaining freshwater fisheries in North America. Throughout the years, Lake Erie managers have iteratively adopted changes to their population assessment model and altered harvest policies to avoid future fishery and population collapses. More than 40 years later, Lake Erie continues to support commercial and recreational fisheries lake wide. Lessons learned from the stock recovery and subsequent coordinated management for fishery sustainability include the importance of conducting routine population assessments, using science-based research to address key uncertainties, adopting modern stock assessment approaches, incorporating stakeholder input into the quota setting process, and addressing environmental concerns collaboratively at the lake level.

INTRODUCTION

The Laurentian Great Lakes support some of the largest and most economically valuable freshwater fisheries in the world (Brenden et al. 2013; Thayer and Loftus 2013). In Lake Erie, native, self-sustaining Walleye *Sander vitreus* stocks have supported fisheries throughout the lake since the 1800s (Regier et al. 1969; Regier and Hartman 1973; Roseman et al. 2013; Kayle et al. 2015). Commercial fisheries in Canada and the United States accounted for most of the fishing pressure in Lake Erie until the 1970s, when recreational fisheries effort increased substantially in U.S. waters (Hatch et al. 1987). During 1975–2016, the combined harvest from Lake Erie fisheries averaged 4.3 million fish per year (WTG 2017), representing one of the largest freshwater fisheries in the world.

Managing this renowned resource has not been without challenges (Hatch et al. 1987). Before 1970, degraded nearshore habitats, intense fishing pressure, and a lack of interjurisdictional coordination threatened the long-term sustainability of the Lake Erie Walleye population and fishery. By the 1960s, population abundance was insufficient to support the demand from the commercial fishing industry and annual yields of Walleye fell to the lowest level since the 1930s (Regier et al. 1969). From 1970 to 1975, several events aligned to foster a dramatic recovery of the Walleye population that extended into the mid-1980s: (1) the commercial sale of Walleye fillets was banned in both countries due to mercury contamination, (2) regulations to improve water quality were enacted in both countries, and (3) coordinated, interjurisdictional fisheries management was formally established (Regier et al. 1969; Kutkuhn et al. 1976; Hatch et al. 1987). In 1976, the first interjurisdictional fishery quota management program was instituted in the Great Lakes basin, providing an annual total allowable catch (TAC) for Walleye from commercial and recreational harvests in Lake Erie (Hatch et al. 1987). By the mid-1980s, binational actions under the Great Lakes Water Quality Agreement caused major improvements in environmental conditions (Makarewicz and Bertram 1991), fishing mortality was managed at sustainable levels, and the Walleye population abundance peaked (Hatch et al. 1987; Knight 1997; Roseman et al. 2013; Kayle et al. 2015). This turnabout of events was contrary to the predictions of Regier et al. (1969): "we do not, however, expect that Walleyes will again become as abundant as they were in the early 1950s." Despite continued challenges involving environmental conditions, cooperative interjurisdictional fisheries management has persisted since 1977 with a mean annual TAC of 5.5 million fish per year.

Herein, we provide a history of how various environmental and social-economic factors affected the dynamic Lake Erie Walleye fishery, with a focus on management actions taken to facilitate recovery of a collapsed population in the 1960s to one that has sustained prominent fisheries for more than 40 years. First, we describe the geophysical and environmental conditions of Lake Erie that support Walleye production and how those conditions were affected by various anthropogenic factors over time. Next, we review the major events affecting the Walleye population and associated fisheries during three key periods: a pre-1970 era of increasing fishery pressure and deteriorating environmental conditions, a 1970–1975 period of improving ecological conditions and low fishing mortality, and a final era (1976–present) of cooperative management, where actions to manage fishery mortality under varying environmental and social-economic conditions have sustained a vibrant fishery. Finally, we highlight the key "lessons learned" that contributed to restoring Lake Erie Walleye as one of the most prolific freshwater fisheries in the world.

ENVIRONMENTAL CONDITIONS OF LAKE ERIE

Consisting of three distinct geophysical basins (Figure 1), Lake Erie is the shallowest

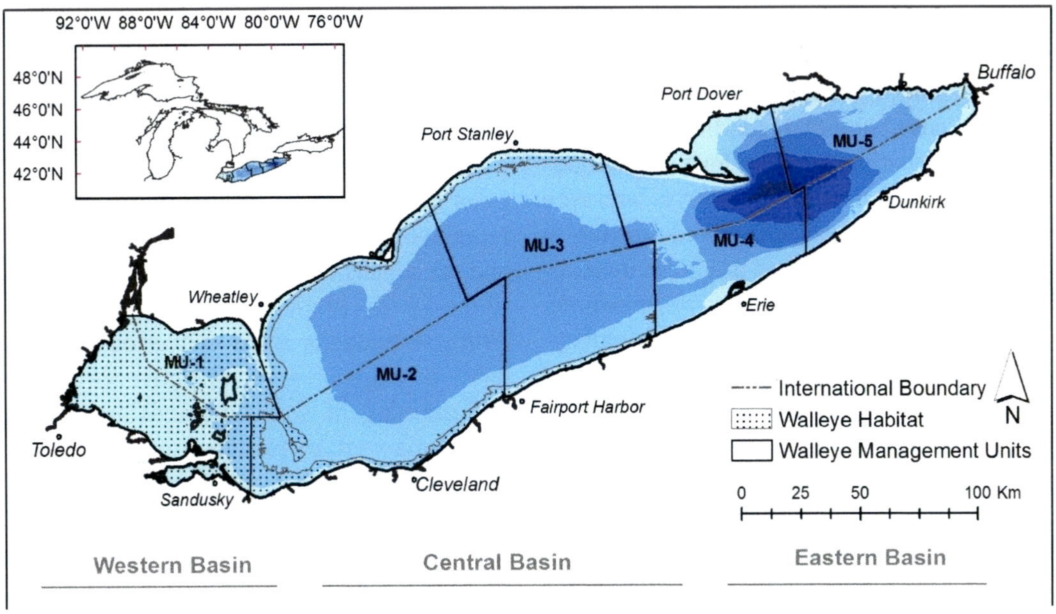

FIGURE 1. Lake Erie bathymetry (10 m, shaded from shallow [0–10 m, light blue] to deep [>80 m, dark blue] contours Lake Erie with the relative location of the western, central and eastern basins. Also shown is the extent of the western-basin Walleye stock habitat (Walleye Habitat) as defined by Kutkuhn et al. (1976) within management units (MU) 1–3. Inset shows the location of Lake Erie (highlighted area) in relation to the other Laurentian Great Lakes located in North America.

and most productive lake within the Laurentian Great Lakes, with an average depth of 19 m and a maximum depth of 64 m (Ryan et al. 2003). The western basin is the shallowest and most productive of the three basins, receiving more than 90% of the lake's annual water inputs from the Detroit, Maumee, and Sandusky rivers, and the water column remains isothermal (same temperature from surface to bottom) during most years. Historically, most of the western basin's shoreline consisted of wetlands. In contrast, the eastern basin is deep, with oligotrophic offshore areas and mesotrophic nearshore zones, and contains the lake's only outflow of water through the Niagara River to Lake Ontario. The central basin is of intermediate depth and productivity relative to the other basins, with sandy/clay bluffs along most of its shoreline. Although both central and eastern basins thermally stratify during summer (July–September), the thin (<2 m) hypolimnion of the central basin routinely becomes anoxic (dissolved oxygen <2 mg/L), whereas the eastern basin hypolimnion typically remains well oxygenated (dissolved oxygen >2 mg/L) throughout the year.

The environmental conditions of Lake Erie prominently affect Walleye populations and associated fisheries. Walleyes are well adapted to reproduce, grow, and survive in Lake Erie due to the lake's productivity (Carlander 1977) and availability of spawning, juvenile, and adult habitats in river, embayment, wetland, and nearshore areas (Colby et al. 1991). Discrete river- and reef-spawning aggregations of Walleyes are located throughout the lake (Colby et al. 1991; DuFour et al. 2015), with the largest self-sustaining stocks located in the western basin and smaller, locally important spawning aggregations in the central and eastern basins (Ryan et al. 2003; Vandergoot et al. 2010). Consequently, Walleye and Sauger *Sander canadensis* were most abundant in shallower western Lake Erie and Blue Pike *S. vitreus glaucus* (a phenotypic variant of western-basin Walleyes; Haponski and Stepien 2014) were most abundant in deeper eastern portions of Lake Erie. After spawning, adult western-basin Walleyes move throughout Lake Erie, including some that exit the lake north to Lake Huron (Wang et al. 2007; Vandergoot and Brenden 2014; Raby et al. 2018). Their propensity to move throughout the lake allows Walleyes to seek habitats with preferred water clarity, dissolved oxygen, thermal conditions, and prey fishes that vary spatially and temporally within and among the basins. Walleye movements affect their seasonal abundance and availability (and vulnerability) to commercial and recreational fisheries operating in each basin.

Cultural stresses from human activities have adversely, and in some cases permanently, affected the environmental conditions of Lake Erie, including physical alterations to water flow, chemical changes in water quality, and additions of nonindigenous species and loss of native species. Many of these alterations affected tributary, wetland, and nearshore areas that provided essential habitats for Walleyes and other fishes. European-origin settlements were well established around Lake Erie by the late 1700s, but watersheds remained largely intact with vegetated forests and savannahs and undammed tributaries until the 1800s (Regier and Hartman 1973). By 1900, watersheds were converted to farmlands through clearing of native vegetation and draining of wetlands, and hundreds of mill dams were constructed in rivers, blocking access to fish (including Walleyes) on spawning runs (Trautman 1981). Drainage of the vast (~1 million ha) Great Black Swamp was particularly damaging to spawning and nursery habitats in southwestern Lake Erie (Kaatz 1955). Increased sediment loads from the modified landscapes covered spawning reefs with silt and increased turbidity in nearshore areas, which resulted in a loss of submerged aquatic vegetation. Nutrient loads increased due to the altered landscapes, triggering major increases in primary productivity, a prevalence of blue-green algae, and increased levels of anoxia in areas of the lake. Large cities (e.g., Cleveland, Detroit, Buffalo, and Toledo) became established during the 1800s, intensifying impacts on water quality from point-source additions of toxic contaminants and untreated sewage as the cities became increasingly industrialized through the mid-1900s. Commerce from waterborne transportation of goods was an underlying basis for the establishment of these municipalities and required major alterations to the lake's hydrology, including the channelization and dredging of major tributaries and the construction of the Welland and Erie shipping canals around Niagara Falls. These canals provided direct pathways for nonindigenous species to invade from the Atlantic Ocean and colonize Lake Erie, and indirect pathways through ballast-water exchange by trans-Atlantic freight ships. Several of these invading species became established additions to the Lake Erie fish community (e.g., Sea Lamprey *Petromyzon marinus*, White Perch *Morone americanus*, Round Goby *Neogobius melanostomus*, and Alewife *Alosa pseudoharengus*), while others (e.g., dreissenid mussels) joined the lake's invertebrate community. Another exotic, Rainbow Smelt *Osmerus mordax*, intentionally introduced into the Great Lakes basin in 1912 and found in Lake Erie by the 1930s (Trautman 1981), likely contributed to the declines in the Walleye population through their predation on small fish (Regier et al. 1969).

Despite major perturbations to the environmental conditions of Lake Erie and in the presence of intensive fishery demand for them, Walleyes persisted (albeit at varying levels of abundance) in the lake when other species vanished (e.g., Cisco *Coregonus artedi*, Blue Pike, and Sauger). In the next section, the decline of Walleyes in response to these ecological and human-induced factors is discussed.

EVENTS LEADING TO POPULATION AND FISHERY COLLAPSE: PRE-1970

Percids (Walleye, Sauger, Blue Pike, and Yellow Perch *Perca flavescens*) were not primary targets of commercial or recreational fisheries in Lake Erie until the late 1800s (Bouge 2000). Earlier major declines in the abundance of preferred fishes (Lake Trout *Salvelinus namaycush*, Lake Whitefish *Coregonus clupeaformis*, and Cisco) and increased public demand for other fish promoted the efficient and intense exploitation of Walleyes and other percids. Advancements in fishing technology, overcapitalized fisheries (i.e., profit-oriented business entities), lax or unenforced fishery regulations, increased knowledge of fish behavior and improved processing, preservation, and transportation of fish products (Bouge 2000) further increased percid fishing effort. Fluctuations in abundance and harvest were observed for all four percids and were attributed to habitat loss and fishery overexploitation (for a thorough description, see Regier et al. 1969; Leach and Nepszy 1976; Schneider and Leach 1977; Hatch et al. 1987). Percids, specifically Walleyes, were intensely exploited from the 1940s to the late 1950s, when their scarcity made commercial fishing unprofitable (Parsons 1970). By 1970, Blue Pike and Sauger populations were extirpated, Walleyes were at a historically low abundance with limited commercial harvest, and Yellow Perch began to sustain commercial fisheries (Regier and Hartman 1973). After the collapse of Blue Pike, a small eastern-basin Walleye fishery developed in the 1960s and has persisted since then (Wolfert 1981). In the following sections, we describe the commercial and recreational fisheries targeting the Lake Erie Walleye population before subsequent fishery closure in 1970.

Commercial Fishery

Walleye fisheries were present in all three basins of Lake Erie (Figure 1) during the 1900s. Most of the commercial fishery effort and harvest occurred in western Lake Erie, where U.S. fisheries used primarily trap nets and the Canadian fishery primarily employed gill nets (Regier et al. 1969). Combined (Canadian and U.S.) trap-net effort ranged from 26,000 to 89,000 lifts annually in western Lake Erie during 1948–1961, a period of intensive fishing (Regier et al. 1969). Combined effort declined to 8,664 lifts in 1969 after the Walleye fishery collapsed (Kutkuhn et al. 1976). Annual gill-net effort for the Ontario commercial fishery averaged 5,166 km of large-mesh (>3″) and 9,528 km of small-mesh (<3″) gill nets fished in western Lake Erie during 1948–1961 (Regier et al. 1969) and increased nearly threefold between 1950 and 1957 (Parsons 1970). By 1969, about 1,977 km of large-mesh gill-net effort (Ontario) were directed at Walleyes in western Lake Erie (Kutkuhn et al. 1976). Due to the depth of water Walleyes inhabited, gill nets were the predominant gear used by commercial fisheries in the central and eastern basins in both countries.

Development of the fishing industry benefited from gains in gear technology during this period. Early inshore fisheries (shore seines, weirs, and dip nets in rivers/bays) were supplemented with pound nets (circa 1850), gill nets (circa 1852), and trap nets (circa 1885) that could be fished offshore in Lake Erie. Improvements to gill nets were particularly important. Early hand-sewn nets of coarse cotton were replaced with commercially produced nets of fine cotton or linen by 1870, followed by multifilament nylon web (1950) and monofilament nylon web by 1970 (Regier et al. 1969; E. T. Cox, Ontario Ministry of Natural Resources and Forestry [retired], unpublished manuscript). Compared to cotton, nylon twine was light, rot-resistant, and less prone to algal fouling, which increased the industry's capacity to fish efficiently during summer (i.e., presumably due to reduced visibility of this gear). Regier et al. (1969) estimated that gill-net catch efficiency (catch per unit effort) for Walleyes increased 25-fold between 1948 and 1961, owing to the shift from cotton-linen to nylon twine. Before being banned throughout the lake, the bull gill net was a common gear used by the commercial fishery between 1905 and 1934 (Regier and Hartman 1973). This net was particularly efficient in the shallow western basin (<12 m) because it covered such a large area of the water column (i.e., 4× deeper than a typical gill net, approximately 7 m deep; Jensen 2007).

Catch efficiency of Walleyes in commercial nets also increased as fishers learned of fish behavior. Repeated patterns of seasonal movements by Walleyes among basins informed the timing and location of fishery effort. The tendency of Walleyes to feed at intermediate depths in the water column led to the suspending (canning) of gill nets a few meters below the surface. Compared to a traditional method of setting gill nets on the lake bottom, canned nets caught mostly Walleyes with few benthic species of low market value (e.g., Common Carp *Cyprinus carpio*, Channel Catfish *Ictalurus punctatus*, and Freshwater Drum *Aplodinotus grunniens*). Canned gill nets became prevalent after the early 1950s (Regier et al. 1969). The strategic use of mesh sizes and placement of gears (i.e., canned and bottom sets) allowed fishers to target specific sizes of Walleyes in response to changes in environmental conditions, movements, market demand/dock value, and year-class abundance.

Lake Erie's commercial fishing industry also benefited from advances in vessel design, fish processing, and transportation technology (Cox, unpublished). Vessel improvements, including steam-power (1840s), mechanical net lifters (1890s), steel hulls (1920s), gas/diesel engines (1920s), and electronics (radio telephones, sonar, auto-pilot; 1940s), allowed the industry to substantially increase fishery effort in terms of quantities of nets and spatial/seasonal coverage of the lake (Figure 2). The use of stored lake ice as a food preservative began about 1830, and ice was common aboard vessels by 1845. During the 1860s, ice-making, freezing, and refrigeration technology enabled the storage of large quantities (68 metric tons [mt] or more) of fish at some processing facilities; consequently, the use of salt as a preservative diminished by the 1890s. Major improvements to transportation included the construction of navigational canals during 1832–1856 that connected Lake Erie (at Cleveland and Toledo, Ohio) to the Ohio River, the use of railroads to transport preserved fish (1870s), improved navigation after the authorization of federal dredging of U.S. harbors and channels (1871), and the use of trucks to transport preserved fish (1940s). By the 1930s, the Lake Erie U.S. commercial fishing industry employed more than 1,400 people, with at least 70 fishing vessels registered annually and 40 or more fishery-related wholesale and manufacturing establishments (Cox, unpublished).

Records of annual commercial harvests in Lake Erie extend back to 1867 but are incomplete until 1914 (Figure 3; Baldwin et al. 2009). From 1915 to 1933, annual commercial yield of Walleyes varied from 337 to 1,366 mt (792 mt, mean), and most (80%) occurred in U.S. waters. Annual yields increased from 649 mt in 1933 to a peak of 6,988 mt in 1956. Most of the harvest (80%) came from U.S. fisheries until 1954, when harvest levels became similar between countries after fishery effort increased in Ontario waters (Regier et al. 1969). Fisheries were especially intense during 1948–1961, removing an average of 3,330 mt of Walleyes annually. Yields declined sharply from 6,046 mt in 1957 to 325 mt in 1962, when the Walleye stock collapsed (i.e., fishery removals exceeded number being produced; Regier et al. 1969), and averaged 610 mt during 1963–1968. Only 215 mt were harvested commercially in 1969. After elevated levels of mercury were detected in fish tissue samples, government agencies in the United States and Canada closed commercial harvest of Walleyes in the western basin between 1970 and 1972. Limited commercial fisheries in the western basin reopened (in Ontario only) after mercury levels declined in 1973; however, the commercial fisheries in Michigan and Ohio never reopened (Hatch et al. 1987).

Walleye harvest before the 1970s was affected by various biological/environmental and socioeconomic variables. From 1915 to the 1930s, annual fishery yields of Walleye were relatively low in comparison to Cisco and Lake Whitefish. Walleye recruitment appeared to respond positively to increasing eutrophication of Lake Erie (Parsons 1970), and by the 1940s, Ciscos and Lake Whitefish were gone or declining and the fishery shifted to abundant Walleye (Regier and Hartman 1973). However, continued eutrophication of the lake increased the frequency and duration of hypolimnetic anoxia in the central basin by the 1950s (Beeton 1961), constraining the fish community during the summer to the well-oxygenated areas of the water column. As a result, the increased fish density in the upper water column may have negatively affected Walleyes through heightened interspecific competition for food and by increasing their catchability to fisheries (Regier et al. 1969). A series of especially weak year-classes occurred during the 1950s, each contributing only 1 to 3 years to the fishery (Parsons 1970). Despite failing recruitment, fishing intensity increased dramatically during the 1950s, particularly in Canadian waters (Regier et al. 1969; Parsons 1970), and compensatory responses (i.e., accelerated growth

FIGURE 2. Photographs of commercial gill-net fishing vessels used to harvest Walleyes from the western basin of Lake Erie. **(A)** The *Archie W*, a wooden hulled vessel used by commercial gill net fisherman on Lake Erie circa 1914 (courtesy of the Historical Collections of the Great Lakes, Bowling Green State University) and **(B)** the MI *Taylor Maid*, a metal hulled vessel currently in use today circa 2008.

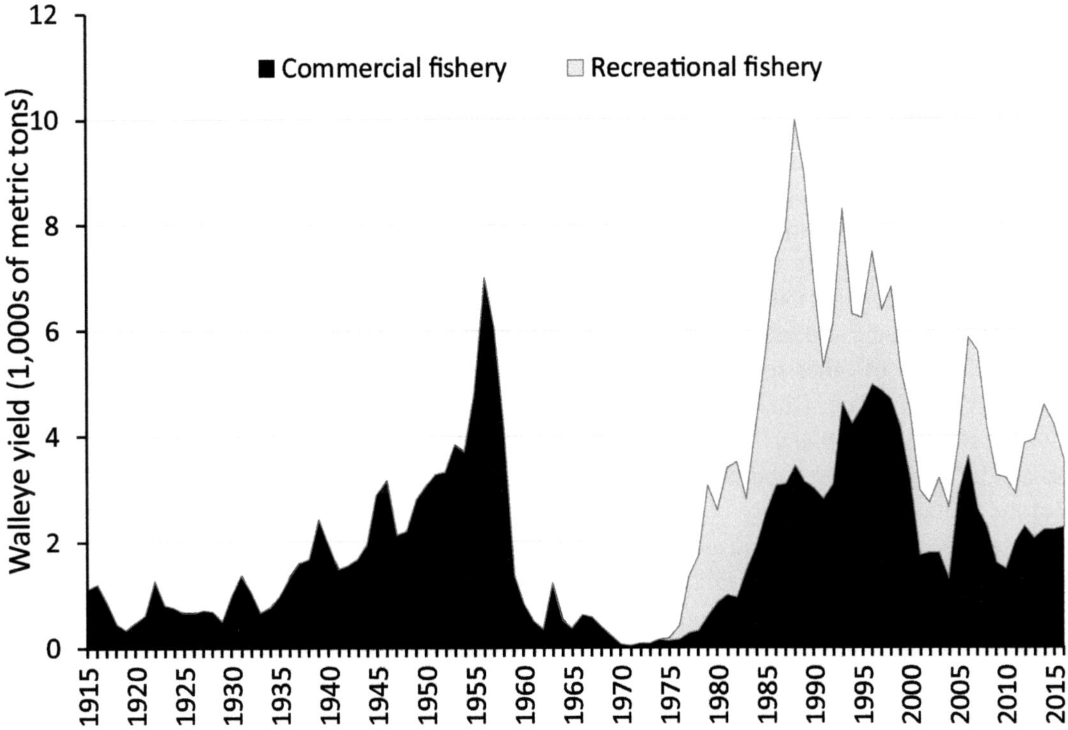

FIGURE 3. Commercial and recreational fishery Walleye yield (thousands of metric tons) harvested from Lake Erie, 1915–2016. Data sources: commercial landing data taken from Baldwin et al. (2009); recreational fishery yield provided by the individual Lake Erie Committee agencies.

and sexual maturation) were evident in the Walleye population (Wolfert 1969; Parsons 1970). However, increased growth rates also caused Walleyes to be harvested before most females could reproduce; hence, growth and, to a lesser extent, recruitment overfishing were implicated as major factors contributing to the fishery collapse by the 1960s (Regier et al. 1969). Another variable that likely affected Walleye and Blue Pike recruitment during the 1950s was predation of Walleye fry by an increasing population of nonnative Rainbow Smelt (Regier et al. 1969). Fluctuations in Walleye abundance triggered the commercial fisheries to shift effort toward abundant sizes with the highest market value (i.e., younger, small individuals).

Management actions to limit exploitation (e.g., length limits, closed areas/seasons, gear restrictions) were agency-specific with minimal interjurisdictional coordination and little overall effectiveness (Regier et al. 1969). Moreover, fishers believed that fry stockings from agency hatcheries provided a buffer against overfishing (Regier at al. 1969). Fishery regulations were met with resistance and evasive behavior by some, including misreporting/nonreporting of catch or effort, harvesting fish of sublegal lengths, fishing outside of legal areas or seasons, and using illegal gears. A history of employment, economic, and cultural benefits from the commercial fishing industries in both countries, as well as uncertainty about factors that were causing the stock to decline, made agency managers reluctant, unwilling, or unable to establish and/or enforce effective regulations, especially in the western basin where the fishing industry was largest (Bouge 2000). During 1963–1968, a period of low Walleye abundance, the fishery removed about 84% of the fishable stock annually (Kutkuhn et al. 1976).

Recreational Fishery

Compared to commercial fisheries, less is known about the dynamics of recreational Walleye fisheries before the 1970s. Angling effort was driven by weather conditions, Walleye abundance and seasonal movements, and availability of other species (e.g., Cisco, Sauger, Yellow Perch, Blue Pike, and Smallmouth Bass *Micropterus dolomieu*) also prized by anglers for food and recreation. Before 1900, angling largely occurred in inshore areas, rivers, bays, and islands of the western basin during spring, fall, and winter when Walleyes were present. Night fisheries for Walleye and Blue Pike became popular in nearshore areas and piers of the central and eastern basins by the 1890s (Cox, unpublished). Winter fisheries were insignificant until about 1890 when commercial angling fisheries developed (i.e., using ice shanties) for Cisco and later Walleye (e.g., 3 mt were caught during 1 d of 1910 in the Bass Islands area; Doan 1944). Ironclad boats were used to haul fish to island and mainland markets by 1900 (Doan 1944), and ice fishing became a major source of income to island residents through the 1950s. As populations of Walleye and Blue Pike declined, Yellow Perch became the primary target of most anglers by the 1960s (Keller 1965). Recreational fisheries expanded after the World Wars, with technological advances in vessels (outboard engines 1920s, steel hulls 1930s, fiberglass hulls 1960s), electronics (e.g., radio, sonar, navigation systems, 1950s), and gear (e.g., spin casting reels, fiberglass rods, monofilament and braided lines, and plastic lures, 1950s).

Recreational creel surveys were rarely conducted by agencies before the 1970s, and when they occurred, surveys were confined to specific regions or seasons. A few published reports provided a snapshot of the magnitude of some fisheries. For example, the recreational fishery in Ohio waters of the western basin during 1958–1960 encompassed about 47,000 vessels, expended about 135,000 angler-hours, but only harvested about 7,000 Walleyes annually during a May–November fishing season (Keller 1964). Ice fisheries in Ohio waters of the western basin during 1958–1960 annually consisted of about 12,000 anglers and 11,000 shanties, expending roughly 89,000 angler-hours per year but catching less than 800 Walleyes over both years (Keller 1964). Another creel survey of the Bass Islands ice fishery in 1963 reflected less overall effort (30,000 h) than in 1958–1960 and a catch of only 113 Walleyes (Keller 1965). In comparison, the ice fishery harvested more than 11,000 mt of Walleyes (60% of the total harvest) during a 1942 creel census of the Bass Islands winter fishery (Doan 1944), reflecting higher Walleye abundance during the 1940s than in the 1950s and 1960s.

Changes in the Walleye population from habitat loss (e.g., reduced water quality and increased shoreline development) and fishery exploitation affected the abundance and sizes of fish available to recreational anglers. Large (>4 kg) Walleyes commonly caught by western-basin ice anglers around 1900 were rare by the 1940s (Doan 1944). Regulations (sportfishing license requirement, closed seasons, bag limits, and length limits) were used to manage the sport fisheries in varying degrees by all agencies on Lake Erie, but management focus was directed mostly on commercial rather than on sport fisheries during this period. Tension between the commercial fishery and developing open-lake sport fishery was low (compared to later years), and allocation of the Walleye catch between these groups was not a major concern of managers before the 1970s.

In summary, Walleyes were increasingly targeted by commercial fisheries as more preferred species declined in abundance during the late 1800s and early 1900s. Walleyes eventually experienced major declines in abundance due to deteriorating environmental conditions and overfishing. The lack of coordinated management efforts among the jurisdictions contributed substantially to the collapse of the fishery. By 1970, the Walleye population and valued fishery were believed to be at a point of no return, but conditions began to change during the next decade.

REBUILDING AND PROTECTING THE POPULATION: 1970–1975

In the wake of the collapsed fishery, Regier et al. (1969) recognized that a more holistic approach was needed that not only involved fishery managers from all jurisdictions, but also incorporated an improved understanding of the ecology and population dynamics of the western-basin Walleye stock. Their findings, along with substantial improvements in the environmental conditions of Lake Erie, fostered an opportunity for the restoration of the Walleye population in western Lake Erie during 1970–1975. The primary drivers behind this opportunity were substantial reductions in fishing mortality and in point-source pollutants (particularly total phosphorus and mercury; Hayes 2013; Murphy et al. 2013) that supported production and expansion of the Walleye population.

Improving Environmental Conditions

The 1960s ended with a societal awakening in Canada and the United States to decades of environmental impacts from anthropogenic stresses on the Great Lakes (Regier and Hartman 1973). For instance, a highly publicized (Time magazine, June 1969) fire on the Cuyahoga River (Cleveland, Ohio, USA) prompted key legislative actions in both countries, including the creation of water quality management agencies (U.S. Environmental Protection Agency in 1970 and Environment Canada in 1971), environmental regulations (the Ontario Water Resources Act and the U.S. Federal Water Pollution Control Act, both in 1972), and the signing of the binational Great Lakes Water Quality Agreement in 1972. These actions reduced pollutants from industrial, municipal, and agricultural point sources through the regulation of discharges into the atmosphere and surface waters. Wastewater treatment facilities were required to add secondary treatment of sewage, which was particularly important to Lake Erie given the existence of large municipalities (Detroit, Toledo, Sandusky) along its major tributaries to the western basin (e.g., Detroit River, Maumee River, and Sandusky River). Reductions in the amount of nutrients (e.g., phosphorus, Bertram 1993; Dolan 1993) and contaminants (e.g., mercury, Herdendorf 1983) entering the lake greatly improved its water quality.

Another factor that may have contributed to the recovery of Walleyes during this period (1970–1975) was the low (i.e., relative to other time periods) abundance of harmful nonindigenous species. Regier et al. (1969) suggested that Rainbow Smelt negatively affected Walleye recruitment during the 1950s through predation or competition for zooplankton resources with larval Walleyes, though evidence of such impacts was inconclusive during the 1960s–1970s (Nepszy 1977). While the establishment and subsequent expansion of Alewife in Lake Michigan led to Walleye recruitment failure in Saginaw Bay through competition/predation at the larval stage (Fielder and Baker 2019, this volume), Alewife populations never became fully established in Lake Erie and were not associated with the initial population decline (Regier et al. 1969; Leach and Nepszy 1976). White Perch were first detected in Lake Erie in 1953 (Larsen 1954), but population establishment and expansion did not occur until the early 1970s. Nepszy (1977) concluded that the impacts of White Perch as predators on Walleye eggs and larvae or as competitors with Walleye larvae were low during 1970–1975. Unclear is when other invasive species (e.g., *Dreissena* sp., spiny waterflea *Bythotrephes longimanus*, fishhook waterflea *Cercopagis pengoi*, and Round Goby known to negatively affect fish recruitment and dynamics elsewhere in the Great Lakes) entered Lake Erie through the exchange of ballast water by transoceanic freighters, reproducing populations of these organisms did not become established until after the mid-1980s.

The Population and Fishery Responds

Along with improving environmental conditions, the Walleye population responded to relief from decades of intense fishing pressure as fishery managers began to establish safe harvest levels commensurate with recruitment. Mercury levels in Walleye tissue, which had exceeded the U.S. Food and Drug Administration's threshold (0.5 ppm) for safe consumption during the late 1960s, prompted fishery managers to close the fishery between 1970 and 1972 throughout the western basin. The harvest ban was lifted in 1973 when mercury levels dropped below advisory levels (Hatch et al. 1987). Mean annual fishing mortality on adult Walleyes declined from 90% during 1963–1969 to 28% during 1970–1975 (Kutkuhn et al. 1976), coinciding with the mercury-driven harvest ban (1970–1972) and a subsequent (post-1972) closure of the commercial Walleye fisheries in Michigan and Ohio waters (Hatch et al. 1987). With favorable environmental conditions and reduced mortality of adult Walleyes, recruitment increased during 1970–1975, highlighting the reproductive resiliency of the western basin stock. In contrast to other large Walleye systems where hatchery produced fish were stocked to bolster recruitment after a fishery collapse (Barnard et al. 2019, this volume; Fielder and Baker 2019), efforts to supplement natural recruitment with stocked fish in Lake Erie were deemed unsuccessful. Year-classes produced in 1970, 1972, 1974, and 1975 were stronger than in any year during 1963–1969 (Figure 4; Hatch et al. 1987), and the fishable (i.e., age-2 fish and older) stock increased from 1.2 million fish to nearly 12 million fish between 1970 and 1975 (Kutkuhn et al. 1976). Additionally, reduced fishing pressure caused the composition of the western-basin Walleye stock to transition from a population dominated by young fish (i.e., ages 2–4 before 1970; Hatch et al. 1987) to a population with a more diverse age structure after fishery closure (WTG 2017).

The 10-fold increase in Walleye abundance during 1970–1975 boosted recreational fishing in the western basin of Lake Erie, spurring the development of marinas, boat ramps, and supporting industries (bait and tackle, lodges, eateries, and entertainment) along the Michigan and Ohio shorelines. Further, a charter-boat fishery developed in Ohio and 46 licensed guides were in operation by 1975 (ODNR 2017). Creel surveys conducted by Michigan, Ohio, and Ontario in 1975 produced the first estimates of harvest (88,000 fish) and effort (562,000 h) for the recreational fishery in all jurisdictions of the western basin (WTG 2017). The population was recovering and the fishery was back!

Beginnings of Coordinated Management

Fishery management on Lake Erie and the other Great Lakes largely occurred with little or no coordination/enforcement among jurisdictions before the mid-1900s (Gaden et al. 2013). Fishery-dependent and -independent assessment data were either absent or, worse, inaccurate (Regier et al. 1969; Regier and Hartman 1973). However, management agencies across the Great Lakes realized that

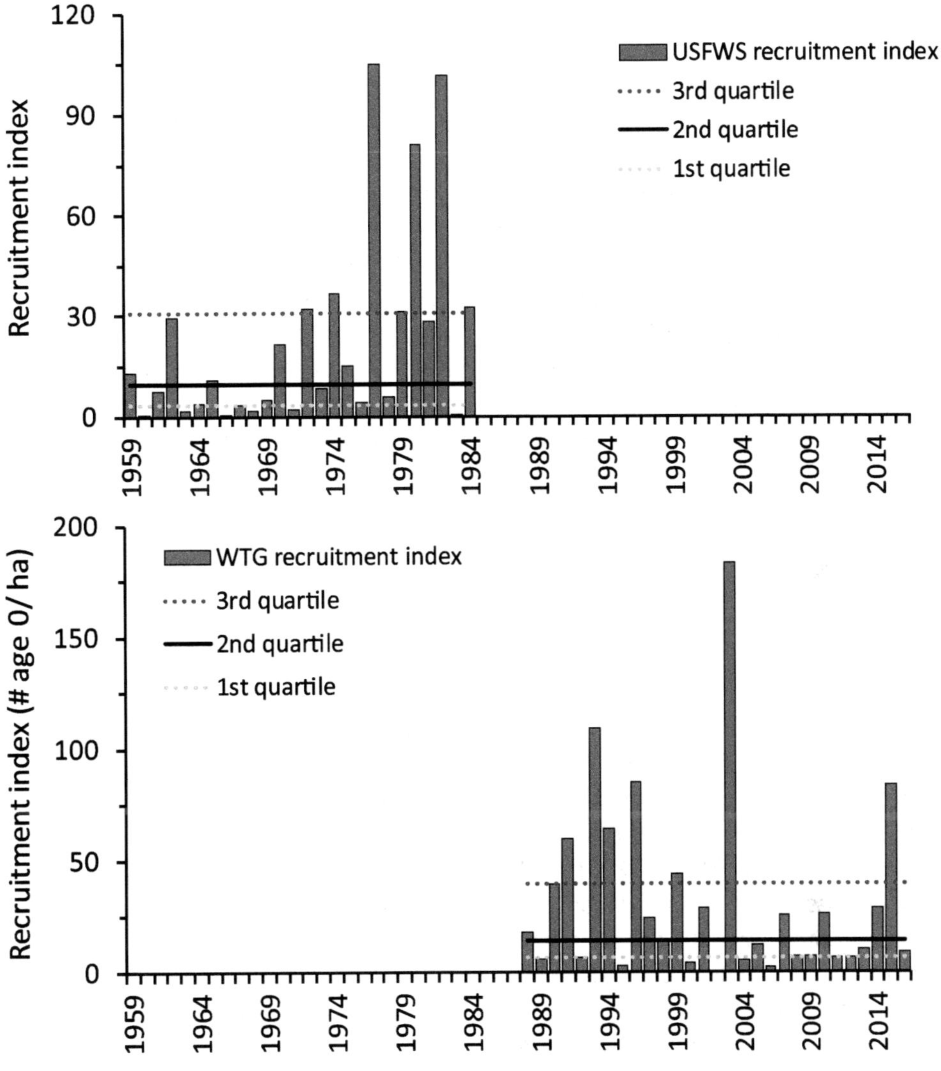

FIGURE 4. Recruitment indices for the western-basin Lake Erie Walleye stock based on August bottom trawl surveys conducted by the U.S. Fish and Wildlife Service (USFWS; top panel; 1959–1984; Hatch et al. 1987) and the Walleye Task Group (WTG; bottom panel; 1988–2016; WTG 2017) management agencies.

effective management of broadly shared fish populations required information sharing and cooperative action rather than attempting to manage fisheries independently by jurisdiction. This recognition led the Canadian and U.S. governments to ratify the Convention on Great Lakes Fisheries in 1954 and create the Great Lakes Fishery Commission (hereafter referred to as the "Commission") a year later (Gaden et al. 2013). The Commission established and maintained lake committees that were convened annually after 1964 to provide a forum to report agency activities and share information (GLFC 2007; Gaden et al. 2013). In 1973, the Lake Erie Committee (LEC), comprised of the Ontario Ministry of Natural Resources and Forestry (OMNRF), New York State Department of Environmental Conservation, Pennsylvania Fish and Boat Commission, Ohio Department of Natural Resources, and Michigan Department of Natural Resources, met to discuss coordinated management of Walleyes in western Lake Erie, building off the recommendations from Regier et al. (1969). From this meeting, the Scientific Protocol Committee (SPC) was created and populated with federal, provincial, and state fisheries scientists and managers (Kutkuhn et al. 1976; Hatch et al. 1987; Gaden et al. 2013).

In March 1976, the SPC presented the first interagency scientific assessment of the western Lake Erie Walleye stock to the LEC and proposed an interagency quota management system to control fishing mortality and equitably allocate TAC among jurisdictions (Ontario, Michigan, and Ohio) bordering the western portion of Lake Erie (Kutkuhn et al. 1976). Stock abundance estimates were to be obtained annually through sequential population analysis (a virtual population assessment technique) and a recruitment estimator (bottom trawl index of relative abundance) for the incoming year-class. Proportions of the TAC (e.g., quotas) were to be assigned to Michigan (8.8%), Ohio (52.4%), and Ontario (38.8%) based on lake surface areas within the defined range for the stock, which included the western basin and waters within the 7-fathom (i.e., 12.8 m) contour of the western half of the central basin (Figure 1; Kutkuhn et al. 1987). Each agency was responsible for allocating its annual quota to recreational or commercial fisheries (or both), assessing and reporting harvest, and regulating harvest relative to the quota. Pennsylvania, New York, and the eastern-basin waters in Ontario were not assigned quotas because this portion of Lake Erie was initially believed to be beyond the range of the western-basin Walleye population (Wolfert 1963). The LEC adopted the SPC's safe harvest recommendations and in 1976 implemented the first interagency quota management system for a shared fish stock in the Great Lakes (Hatch et al. 1987).

In summary, societal interests to improve environmental conditions in the Great Lakes fostered binational efforts to improve water quality in Lake Erie. At the same time, recognition of the depressed state of the western-basin Walleye stock and associated fisheries prompted managers to act. Together, coordinated fishery management actions and improved environmental conditions set the stage for the recovery of the Walleye population in Lake Erie.

SOLUTIONS AND ACTIONS TO FISHERY RESTORATION AND SUSTAINABILITY: 1976–PRESENT

As the western-basin Lake Erie Walleye stock recovered (strong recruitment events, reduced adult mortality), the LEC adopted the recommendations of Kutkuhn et al. (1976) for implementing a quota management framework that focused on fishery and population sustainability. As the fishery reopened and Walleye harvest and effort increased (Figure 3), the LEC committed to conducting coordinated harvest assessments and fishery enforcement, conducting science-based population assessments, establishing sustainable harvest levels, incorporating stakeholder input into the management process, conducting science-based research, and collaboratively addressing lakewide environmental concerns. This approach has formed the basis for how the Lake Erie Walleye population is currently managed (Kayle et al. 2015).

Coordinated Harvest Assessments and Enforcement

By establishing a quota management system for Lake Erie Walleyes, the SPC also recognized the need for monitoring the commercial and recreational fisheries for stock assessment purposes (i.e., magnitude and demographics of the harvest; Kutkuhn et al. 1976). Although commercial fishery harvest and effort information extended back to the early 1900s, the accuracy of these data were considered suspect by Regier et al. (1969), and similar data were scarce for the recreational fishery (Kutkuhn et al. 1976). Management agencies took actions to rectify this problem with annual harvest assessments across the lake since the SPC's initial report (Kutkuhn et al. 1976).

Lake Erie's commercial fishery underwent a contraction in size after the population collapsed. Commercial gillnetters in Michigan and Ohio were bought out by these states after closure of the fishery in 1970 (Hatch et al. 1987). Today, Ontario's commercial gillnet fishery is by far the largest commercial fishery operating on the lake, although a small commercial fishery in Pennsylvania's waters still exists (WTG 2017). Ontario's commercial license holders receive individual transferable quotas, which may be sold or transferred to other license holders (Brenden et al. 2013). Commercial fishers must submit daily catch reports that record fishing location, harvest, and effort data (Brenden et al. 2013) and are sampled dockside by monitoring agents, whose assessment must be consistent with the fisher's daily catch report. The Ontario Commercial Fisheries' Association is contracted by the OMNRF to maintain a database consisting of daily catch-and-effort reports (Brenden et al. 2013).

Ontario's commercial Walleye fishery consistently harvests millions of Walleyes each year. Harvest increased from 157,000 Walleyes in 1976 to 993,000 fish in 1980 (WTG 2017). Lakewide effort also increased during this same period from 3,700 km of gill net to 7,800 km of gill net. Since 1975, lakewide harvest and effort have averaged 2.0 million Walleyes per year and 18,700 km of gill net, respectively (Figure 5), although more recently (i.e., 2000–2016), both have been lower. Most of the commercial harvest and effort occurs in Lake Erie's western basin (Figure 6). For example, an average of 61% of the lakewide commercial harvest came from the western basin during 2000–2016 (WTG 2017). Commercial catch per unit effort (CPUE; Walleyes per kilometer of gill net) declined during the 1990s, peaked during the mid-2000s as the strong 2003 year-class recruited to the fishery, and declined again as that cohort aged (Figure 7, upper panel; WTG 2017).

Annual monitoring of Lake Erie's recreational Walleye fishery began in 1975 with Michigan and Ohio creel surveys of the western and west-central basins (WTG 2017); later, Ohio expanded its survey into the east-central basin in 1980 (WTG 2017). Annual monitoring of eastern-basin recreational fisheries began by New York in 1988, followed by Pennsylvania in 1996 (WTG 2017). Ontario initiated annual creel surveys in the western (1975–1997) and central (1989–1994) basins, and sporadically thereafter across Ontario waters of Lake Erie. General creel survey methods include counting fishing boats and direct contact interviews of recreational anglers at boat access sites across Lake Erie to estimate effort and harvest during the open-water (March–October) period. Specific survey methods vary by agency but typically consist of some form of stratified sampling design (NYSDEC 2017; ODNR 2017). Recreational fishing effort estimates are derived from aerial or marina access boat counts and completed trip interviews. Age composition of the harvest is determined via age samples (i.e., scales and otoliths) collected during LEC agency creel surveys.

Lake Erie's recreational fishery has consistently harvested 1 million fish annually since the Walleye population recovered and the fishery reopened in 1975 (Figure 8). Lakewide harvest and effort peaked in the 1980s, and although both have since de-

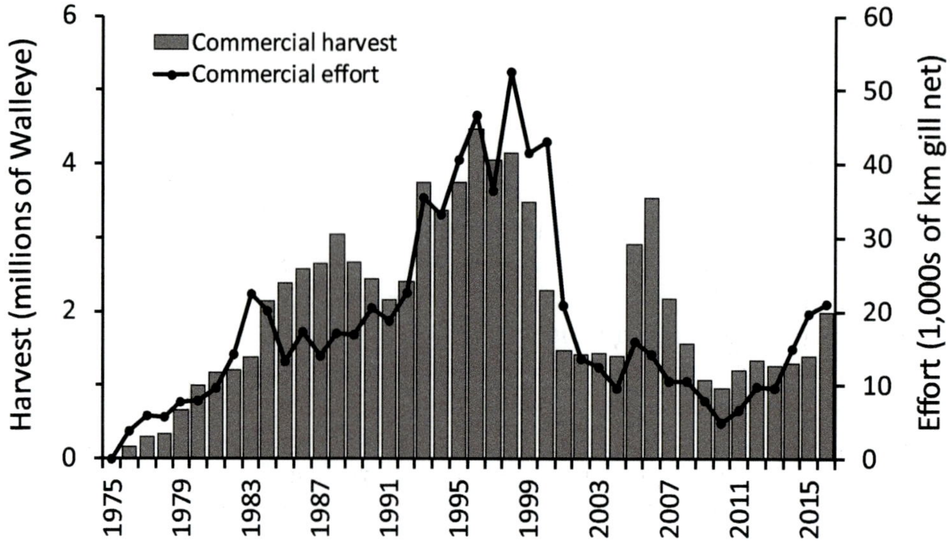

FIGURE 5. Lake Erie Walleye commercial harvest (millions of Walleyes) and effort (thousands of kilometers of gill nets fished), 1975–2016.

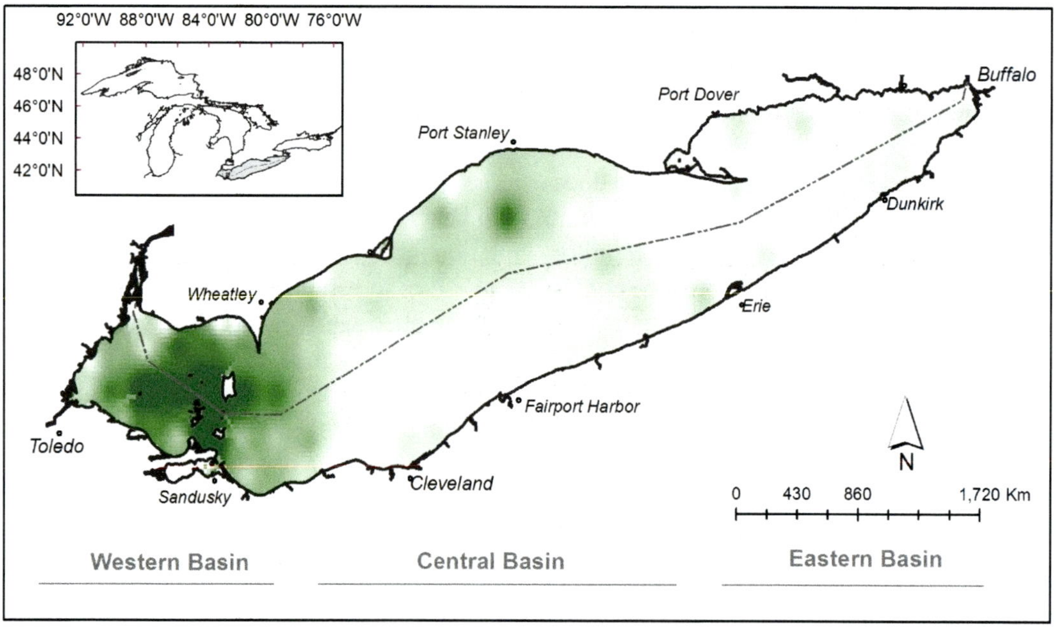

FIGURE 6. Average number of Walleyes harvested (millions) between 2011 and 2015 per 10-min grid in the Ontario (commercial fishery) and the U.S. (recreational fishery) waters of Lake Erie. Distribution of Walleye harvest throughout Lake Erie. Green shading represents areas of high harvest (approximately 200,000 Walleyes harvested/10 min grid), and white areas represent areas of low (0 Walleyes harvested/10 min grid) harvest. Average number of Walleyes harvested taken from annual estimates of harvest reported between 2011 and 2015 (WTG 2017). Inset: proximity of Lake Erie (gray shading) relative to the other Laurentian Great Lakes.

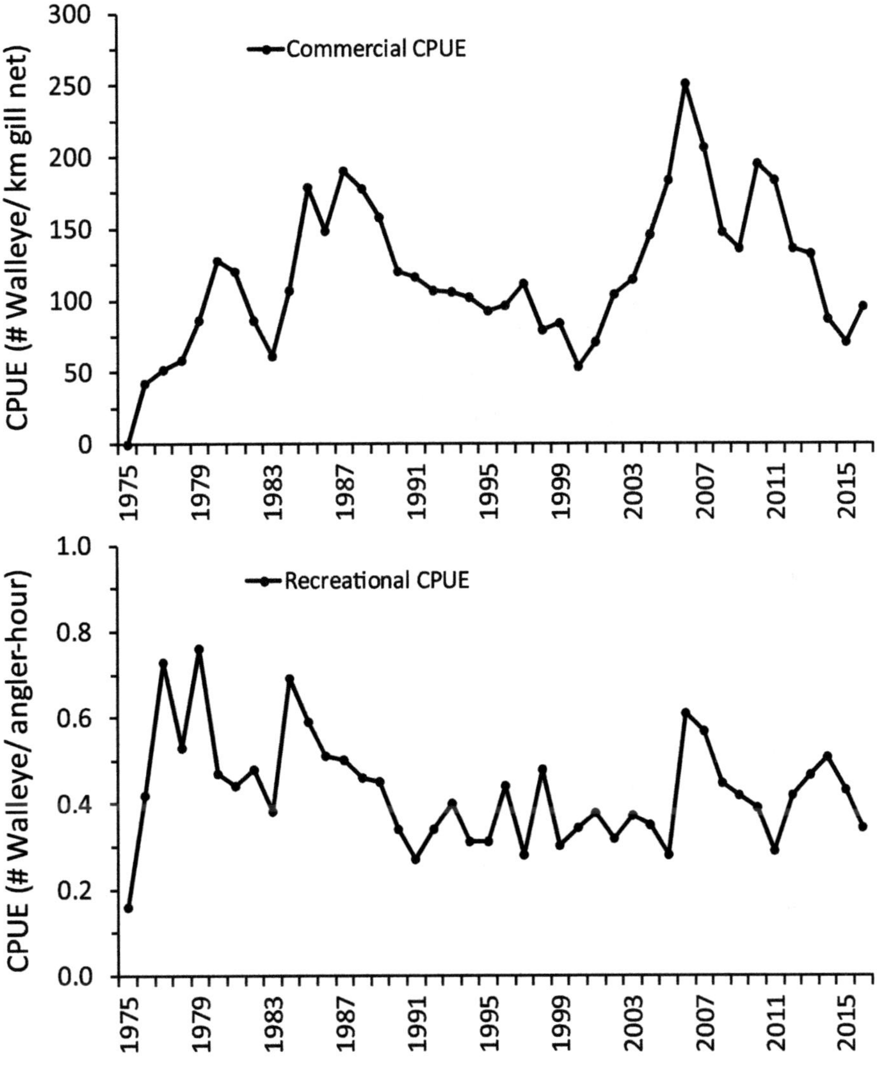

FIGURE 7. Catch per unit of effort (CPUE) for the commercial (# Walleyes/km of gill net; top panel) and recreational (# Walleyes/angler-hour; bottom panel) Walleye fisheries operating in Lake Erie, 1975–2016.

clined from those highs, fishery performance appears to have stabilized since the early 2000s. Similar to the commercial fishery harvest, recreational harvest is highest in the western basin (Figure 6). Recreational fisheries in the central and eastern basins have been much smaller in scale than those in the western basin. For instance, the recreational Walleye fishery outside of the western basin accounted for only 36% and 40% of the lakewide harvest and effort, respectively, in 2016 (WTG 2017). Recreational CPUE (Walleyes harvested per angler-hour) has remained relatively stable around a lakewide average of 0.43 Walleyes per angler-hour during 1975–2016 (Figure 7, lower panel; WTG 2017). Recreational CPUE is highest in the western basin (0.46) and declines in the central (0.34) and eastern basins (0.22). Mean age of harvested Walleyes increases from the western

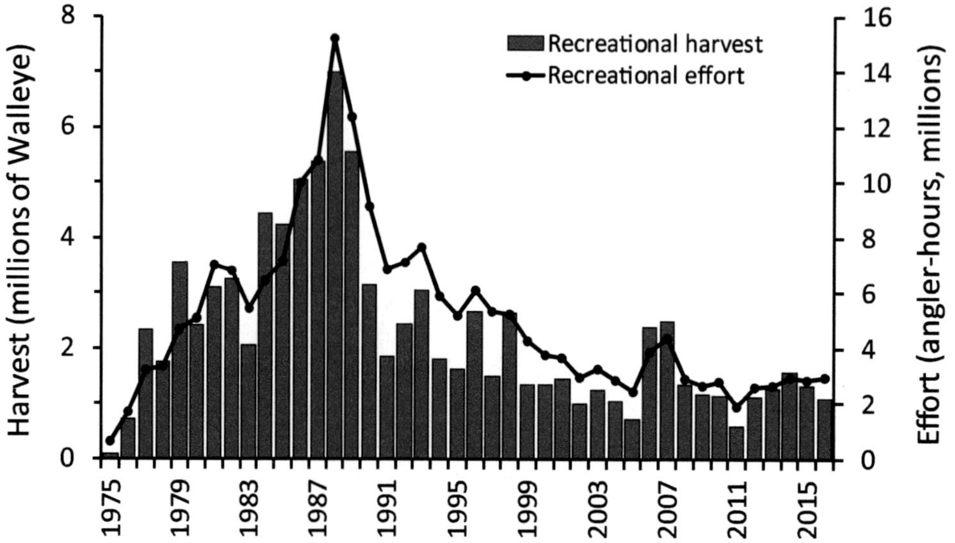

FIGURE 8. Lake Erie Walleye recreational fishery harvest (millions of Walleyes) and effort (millions of angler-hours), 1975–2016.

to eastern basin (WTG 2017), with larger individuals harvested in the central and eastern basins than in the western basin, which requires managing for a broad population age structure to maintain these fisheries across the lake (Kayle et al. 2015).

Although the LEC implemented the first interagency quota management system in the Great Lakes, the success of this system relied upon improved law enforcement. Despite harvesting commercially valuable fishes to a point of scarcity, laws designed to protect these species were slow developing, and even where they existed were rarely enforced due to inadequate resources (i.e., personnel and equipment) or political will (Bouge 2000). After the 1940s, more stringent fishery regulations were developed and began to be enforced; however, binational enforcement cooperation was lacking (Regier and Hartman 1973). Not until the western-basin Walleye population closed in 1970 and lake managers realized that overexploitation contributed to the collapse did discussions about multijurisdictional enforcement efforts began. Today, coordinated law enforcement activities occur effectively and efficiently throughout the Great Lakes basin (McGarrell et al. 2013) thanks to *A Joint Strategic Plan for the Management of Great Lakes Fisheries* (Gaden et al. 2008). On Lake Erie, enforcement agencies share enforcement information, inspect fishery facilities (e.g., fish processing plants, transportation vehicles), issue cross-border summons, participate in surveillance and intelligence gathering, share equipment, and work directly with bordering agencies through a memoranda of understanding agreement (J. Collingwood, Ohio Department of Natural Resources, Division of Wildlife, personal communication).

Science-Based Population Assessments

In response to recommendations put forward by the SPC after closure in 1970 of the western-basin fishery (Hatch et al. 1987), the LEC initiated fishery-independent assessment surveys (i.e., recruitment indices and relative abundance trends) to better understand the dynamics of the western-basin stock (Kayle et al. 2015). These data, together with fishery-dependent effort and harvest data, have been used to estimate year- and age-specific population abundances and mortalities that ultimately inform discussions to establish safe harvest levels (i.e., TACs; Kayle et al. 2015; Jones et al. 2016). Initially conducted in the western basin, these surveys now extend to the central and eastern basins as research showed that western-basin Walleye spawning populations contributed to lakewide fisheries.

Walleye recruitment (i.e., relative abundance of age-0 fish) has been assessed via bottom trawl surveys throughout the western basin of Lake Erie (Tyson et al. 2006). Bottom trawl surveys conducted by the U.S. Fish and Wildlife Service in the late 1950s provided the earliest insight into the dynamics of western-basin Walleye recruitment patterns (Parsons 1972). In 1969, Ohio began a bottom trawl survey at six fixed sites throughout the western basin (Hatch et al. 1987), which served as a precursor to the current western-basin interagency trawl survey initiated in 1987 (Tyson et al. 2006). The current survey, consisting of trawl sites in Ohio and Ontario waters of the western basin, uses standardized gear and sampling protocols to generate annual recruitment indices for the western-basin Walleye stock (Figure 4; Kayle et al. 2015; WTG 2017). Though not formally part of the interagency trawling survey, additional bottom trawl surveys conducted in the central and eastern basins of Lake Erie provide insight into Walleye recruitment dynamics in these areas of the lake (Figure 9).

To monitor trends in relative abundance, growth rates, and population demographics

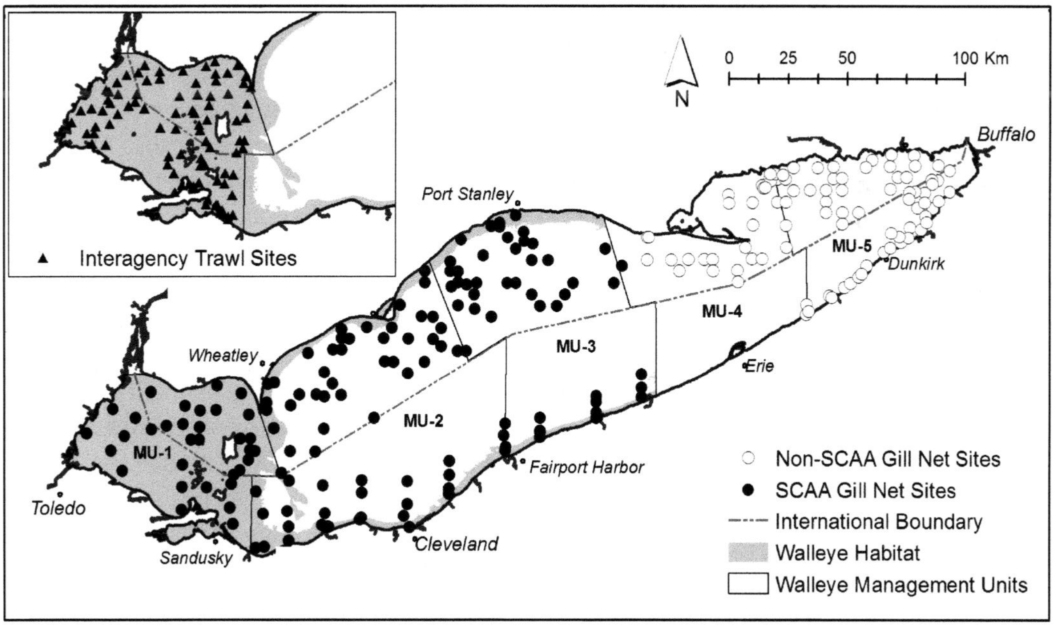

FIGURE 9. Distribution of interagency bottom trawl (inset) and gill-net (bottom panel) survey sites used to assess the Lake Erie Walleye population. Within the gill-net survey figure (bottom panel) the gill-net sites used to estimate population abundance via statistical catch at age analysis (SCAA) and gill-net sites used to better understand the spatial distribution of Lake Erie Walleye but not used to estimate population abundance (non-SCAA) are shown. Also shown is the extent of the western-basin Walleye stock habitat (Walleye Habitat) as defined by Kutkuhn et al. (1976) within management units (MU) 1–3.

(i.e., sex and age composition), annual autumn (September–November) gill-net surveys occur across the lake to assess the status of the adult Walleye population (Figure 9). In contrast to the interagency western-basin bottom trawl survey (using similar gear and sampling protocols), substantial differences exist among the gill-net surveys conducted by the LEC agencies. For instance, gill-net configurations vary by mesh type (multifilament and monofilament), mesh size, and fishing depth (i.e., set near the surface and bottom sets). Consequently, gear differences persist out of a concern to not compromise the long-term monitoring data sets maintained by each agency (Kraus et al. 2017). Michigan and Ohio initiated similar western-basin gill-net surveys (i.e., with respect to gear type and sampling methodology) in 1978. As resources became available and a better understanding of the movement dynamics of the western-basin stock became known, Ohio expanded their survey into the central basin in 1983. Ontario's partnership gill-net survey, a cooperative assessment program between OMNRF and the commercial fishing industry conducted throughout Ontario waters, began in 1989. The gill-net survey administered in New York waters of Lake Erie by the New York State Department of Environmental Conservation was implemented in 1981 but was modified after 1992 to better align with methods of the Ontario partnership survey.

Estimating Walleye abundance was integral to restoring the western-basin stock after collapse, maintaining the stock after recovery and continues to be a crucial component of ensuring the sustainability of the population and fishery. Since the late 1970s, numerous population modeling changes have occurred to ensure that the most scientifically defensible analytic methods are used (Kayle et al. 2015). The initial method used to estimate population size after the fishery collapse in 1970 was a sequential projection model (Kutkuhn et al. 1976; Hatch et al. 1987). However, after the Walleye Task Group (WTG; a panel of fishery biologists from the LEC agencies) expressed concerns about the ability of this method to accurately describe observed recruitment patterns and the age structure of the fishable stock, this model was abandoned in 1990 in favor of a more complex, age-structured model (CAGEAN, Deriso et al. 1985) that better tracked patterns observed in fishery-independent assessments and fisheries performance (Kayle et al. 2015). Age-structured or statistical catch-at-age (SCAA) models provide detailed insight into interannual stock dynamics and the ability to track individual cohorts through time, but also require more detailed fishery-dependent and -independent information (Hilborn and Walters 1992). Consequently, the LEC agencies focused on expanding and refining existing assessment surveys initiated during the current quota management era to address these increased data requirements. Another population modeling change occurred when the WTG again grew concerned that CAGEAN was overestimating Walleye abundance; specifically, individual cohorts appeared to increase in abundance with successive model runs (Kayle et al. 2015). A new SCAA model (i.e., using Automated Differentiation Model Builder [ADMB] software) was proposed by the WTG and eventually adopted in 2005 after external review (Myers and Bence 2001). The new SCAA-ADMB model allowed for greater programming flexibility and incorporated multiple data sources into the estimation process (Quinn and Deriso 1999). With some exception to changes incorporated during the Lake Erie Percid Management Advisory Group (LEPMAG) process to stay consistent with contemporary stock assessment approaches (Jones et al. 2016), Walleye abundance and stock dynamics in the binational quota management area continue to be assessed using a similar approach.

Establishing Sustainable Harvest Levels

Developing and implementing harvest policies that support commercial and recreational fisheries throughout the lake have been a critical component of the western-basin Walleye stock recovery. Early harvest policies (1977–1989) were based on determining the stock's maximum sustainable yield (MSY; Gulland 1971) and selecting a fishing mortality rate (*F*) thought to promote sustainability. Along with favorable recruitment events, establishing safe harvest limits (i.e., TACs) for the first time in the fishery's history assisted with stock recovery efforts after closure (Hatch et al. 1987). When the WTG transitioned to estimating population size with CAGEAN in 1990 (see Science-Based Population Assessments section), the LEC adopted a new harvest policy designed to optimize yield to the fishery using a Beverton–Holt yield-per-recruit model (Hilborn and Walters 1992). This approach generated annual optimal fishing mortality rates (F_{opt}) for the western-basin Walleye stock during 1990–2000. The F_{opt} harvest strategy was a major departure from how the fishery had been exploited under the MSY approach, as it allowed different segments of the population to be fished more aggressively than others (e.g., older versus younger fish). However, under this policy, actual fishing mortality rates often exceeded target fishing mortality rates when implemented (Kayle et al. 2015).

After a series of below-average recruitment events during the late 1990s (Figure 4), downward trends in fishery performance (e.g., commercial fishery catch rates; Figure 7, upper panel), and concerns surrounding CAGEAN model estimates, the LEC implemented a fixed harvest level from 2001 to 2003. Referred to as the coordinated percid management strategy (CPMS; Locke et al. 2005), LEC managers used this strategy to return the western-basin Walleye stock to a favorable condition (Kayle et al. 2015). Objectives of the CPMS period were to (1) reverse declines and rebuild percid stocks to achieve a broad distribution of benefits throughout the lake, and (2) improve approaches used to estimate percid abundance and determine sustainable harvest levels (Locke et al. 2005). Several noteworthy changes arose during this period regarding how the western-basin Walleye stock was managed. In addition to adopting a new SCAA population model in response to the tendency of CAGEAN to inflate population estimates with successive model runs (see Science-Based Population Assessments section), the LEC adopted a harvest policy known as a "sliding *F* policy" (Locke et al. 2005). The feedback loop built into this policy reduced *F* as the population decreased in abundance (and vice versa). The sliding *F* policy also recognized broad fishery management objectives (Wright et al. 2005), including promoting an age structure necessary for supporting fisheries in the central and eastern basins consistent with the fish community goals and objectives developed for Lake Erie (Ryan et al. 2003).

The current harvest policy, in place since 2014 and used to promote western-basin Walleye stock stability, addressed stakeholder concerns after implementation of the sliding *F* policy (Kayle et al. 2015). Specifically, stakeholders expressed a desire to be formally involved in defining fishery objectives. Two significant stakeholder concerns were that (1) the sliding *F* policy was too risk-averse, and (2) this policy resulted in highly variable annual TACs that were incompatible with maintaining stable, long-term yields of percids from Lake Erie. To address these concerns, the LEC sought to formally engage stakeholders during fishing policy development through the Lake Erie Percid Management Advisory Group (LEPMAG) process (Jones et al. 2016).

Facilitated by the Quantitative Fisheries Center at Michigan State University and the Commission, representatives from the commercial and recreational fisheries across the lake participated in a series of meetings where (1) fishery objectives were developed, (2) analytical modifications to the population model were made, and (3) fishery performance under different harvest policy scenarios were evaluated. Although this effort proved time-consuming and sometimes arduous, the adopted policy addressed two significant stakeholder concerns by (1) developing a method to quantify the biological risk or risk of fishery collapse associated with selecting any level of harvest, and (2) placing a ±20% constraint on how much TACs could vary from year to year. In addition to the population estimate and harvest policy, social and economic considerations are included during the annual TAC setting process (Kayle et al. 2015). Age-specific abundance estimates (WTG 2017) based on model projections and year-specific TACs were developed as a result (Figure 10). A thorough description of how the LEPMAG process shaped policy development has been provided by Kayle et al. (2015) and Jones et al. (2016).

Incorporating Stakeholder Input into Management Process

Incorporating input from diverse stakeholder groups into the fishery management process is complex, particularly when fishery performance expectations differ between user groups (Krueger and Decker 1999). Managing these expectations on Lake Erie is compounded by the spatial variation in the expectations among and within commercial and recreational fishery components (Figure 11); consequently, reaching consensus regarding how the fishery should be managed sustainably is challenging (Gaden et al. 2013; Roseman et al. 2013). Commercial fishers often cite maximizing yield or profit as management objectives, whereas recreational anglers value maximizing catch rates or attaining

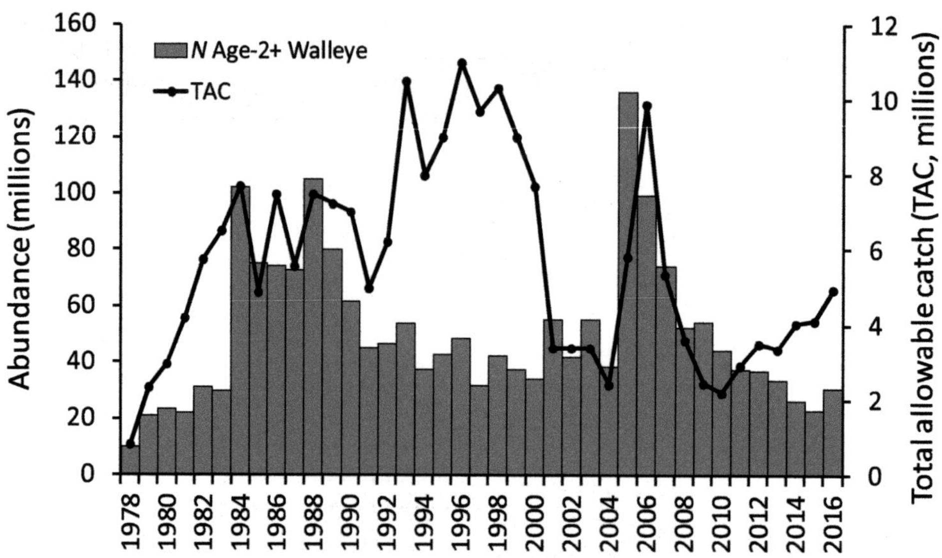

FIGURE 10. Annual population estimates (*N* age-2+ Walleyes) for the Lake Erie western-basin Walleye stock (WTG 2017). Also shown are the annual total allowable catch (TAC) levels set by the Lake Erie Committee using year-specific harvest polices described by Locke et al. (2005) and Kayle et al. (2017).

FIGURE 11. The Lake Erie Walleye population supports a diversity of fishery components throughout the lake. Walleyes are prized by both trophy and chartered anglers (top and middle, respectively) and commercial fishers (bottom).

daily bag limits and often have preferences as to the size of the catch (e.g., trophy versus meat fisheries; Larkin 1977; Pereira and Hansen 2003; Ihde et al. 2011). During the LEPMAG process, commercial fishery stakeholders identified 1,812 mt as a minimum annual yield for maintaining economic viability, and recreational stakeholders felt that catch rates below 0.4 Walleyes/h would negatively effort fishing effort and economic value (Jones et al. 2016). Additionally, over time, an increased understanding evolved that western-basin spawning stocks contribute to fisheries throughout the lake (Stepien et al. 2010; Zhao et al. 2011), with old, large fish migrating toward the eastern basin (Wang et al. 2007; Vandergoot and Brenden 2014; Raby et al. 2018). As such, Lake Erie managers worked to develop harvest strategies that promote sustainable commercial and recreational fisheries over the entire lake (Kayle et al. 2015).

When interagency management began in the 1970s and 1980s, LEC managers relied on informally soliciting stakeholder input from their respective constituents when developing harvest regulations and strategies. Although this approach sufficed when the fishery was performing well (i.e., average to above average harvest rates; Figure 7), the decline in Walleye abundance during the late 1990s and early 2000s (Figure 10) precipitated a contentious period of management on Lake Erie (Gaden 2007), and the need to formally include stakeholders' input in the decision-making process became evident. Under the auspices of the Commission, LEPMAG was formed in 2010 with the express purpose of establishing a formal dialogue between Lake Erie's diverse Walleye and Yellow Perch fishery stakeholders and LEC lake managers (Jones et al. 2016). Comprised of commercial and recreational fishery representatives and lake managers from each LEC jurisdiction, this forum provided a transparent, collaborative, and structured mechanism for incorporating stakeholder input and providing feedback regarding percid management issues. Management decisions based on sound science and developing partnerships to achieve stable and sustainable harvests for all jurisdictions is the overarching goal of this endeavor. As part of the collaborative process, LEPMAG members participate in defining fishery objectives, evaluating different harvest policies, and providing LEC managers with first-hand perspectives on the needs of the lake's different fisheries. These interactions provide stakeholders and managers the opportunity to better understand how management decisions affect fisheries throughout Lake Erie. Evidence of the success of this endeavor is that the LEPMAG forum continues to provide management advice to the LEC 7 years after its inception (Jones et al. 2016).

Commitment to Science-Based Research

Because a comprehensive understanding of the ecological and demographic processes underlying an exploited population is essential for sustainable management, the LEC agencies have invested considerable resources into furthering their understanding in these areas. Perhaps the best example of this investment is a long-term mark–recapture study initiated in 1990 to better understand the seasonal distribution of Walleye stocks within Lake Erie and to better discern contributions of fishing and natural mortality to total mortality (Einhouse and Haas 1995). Spawning Walleyes were collected annually during 1990–2007 in the western and eastern basins, marked with conventional jaw tags, and recovered by fishers and management agencies (Vandergoot et al. 2012). Analysis of these data suggested that western-basin stocks have a greater tendency to move into the central and eastern basins or out of Lake Erie (via the Detroit River) than eastern stocks that typically remain within the eastern basin (Wang et al. 2007; Vandergoot and Brenden 2014). Also, older Walleyes were more likely to move into the central and eastern basins than younger fish (Wang et al. 2007; Vandergoot and Brenden 2014). Lake-wide movements of western-basin stocks may also contribute significantly to harvest in the eastern basin (Zhao et al. 2011). These mark–recapture data have also been useful in estimating F and natural mortality (Zhao et al. 2011; Vandergoot and Brenden 2014). For instance, historic and contemporary stock assessment models rely on a fixed value for natural mortality across all ages, but age-specific estimates suggested that natural mortality rates decline with age (Vandergoot and Brenden 2014). Further, eastern-basin stocks experienced lower F rates and had accompanying higher survival rates than western-basin stocks (Zhao et al. 2011). More recently, numerous interagency studies have taken advantage of a broad-scale acoustic telemetry network in the Great Lakes region (Krueger et al. 2018) to gain a better understanding of the spatial ecology, spawning behavior, bioenergetics, and survival rates of Lake Erie Walleyes (Pritt et al. 2013; Peat et al. 2015; Hayden et al. 2017; WTG 2017; Madenjian et al. 2018; Raby et al. 2018).

Lake Erie's management agencies have long recognized the value of maintaining long-term, standardized fishery-independent data sets, but also understand the need to critically examine these data sources and seize opportunities to improve existing surveys through time. For example, agency research vessels have a finite life span and are occasionally replaced, which may alter gear catchability despite standardized survey gear and protocols. Prompted by Ohio replacing their research vessel in the early 2000s, an interagency trawl comparison study among vessels occurred in western Lake Erie. This study relied on a series of 35 side-by-side trawls designed to minimize environmental variability (e.g., fish distribution, lake conditions) and sought to develop fishing power correction factors for Walleyes and other fishes collected by five research vessels historically used to conduct trawl surveys in the western and central basins (Knight and Vondracek 1993; Tyson et al. 2006). More recently, an evaluation of fishery-independent gill nets occurred because of concerns that survey gear used by Michigan and Ohio since 1978 did not adequately sample large, old Walleyes, which could bias age composition estimates of the adult population (Vandergoot et al. 2011). To address these concerns, a new experimental gill net was designed that used monofilament, rather than multifilament, and included a greater range of mesh sizes and varied filament diameters (Vandergoot et al. 2011). Both old and new gear configurations were fished side by side at sites across Lake Erie during 2007–2008, and again during 2010–2013 (Vandergoot et al. 2011; Kraus et al. 2017). Although the study results demonstrated that the net configurations compared produced similar mean catches and length distributions of fish caught (Kraus et al. 2017), the historical multifilament gear did not sample the available size distribution of Walleyes as well as the new monofilament configuration (Vandergoot et al. 2011), leading Ohio to change survey gear in 2016.

Addressing Environmental Concerns

While fishing mortality directly affects adult (age 2+) Walleye abundance, changes in environmental conditions indirectly affect reproduction success (i.e., recruitment). Several environmental variables have been linked to year-class strength of Lake Erie Walleyes, including spring warming rates (Busch et al. 1975; Koonce et al. 1977), winter severity (Hokanson 1977), storm events (Roseman et al. 2001), lake currents (Zhao et al. 2009), and river discharge (Mion et al. 1998). Other environmental conditions that affect Walleye reproductive success are prey fish abundance (Madenjian et al. 1996), egg predation (Wolfert et al. 1975; Roseman et al. 2006), and production from lower-trophic-level organisms (Gopalan et al. 1998). Degradation of spawning and nursery habitats from cultural stresses on hydrology (dams, channelization, dredging, and tiling), water quality (sediments, nutrients, and contaminants), and shoreline modifications (development, armoring, and wetland removal/diking) also affect Walleye recruitment in Lake Erie (Regier et al. 1969; Regier and Hartman 1973).

After implementation of the Great Lakes Water Quality Agreement, Walleye abundance increased as water quality steadily improved. Targeted annual loads for total phosphorous (11,000 mt) into Lake Erie were attained by 1981, providing desired mesotrophic conditions (Ryan et al. 2003) throughout most of Lake Erie during the 1980s. Walleyes annually produced about 26 million age-2 fish during the 1980s, with at least 7 million age-2 fish originating from all year-classes during this period (WTG 2017). However, the improved water quality in major shipping ports perhaps also created favorable conditions for invasive organisms that were introduced via ballast water from transoceanic freighters. The most prominent of these invaders, zebra mussel *Dreissena polymorpha* and quagga mussel *D. bugensis*, became abundant throughout Lake Erie during the 1990s, densely colonizing bottom substrates, reducing pelagic productivity, and creating oligotrophic-like conditions (Leach 1993; Hecky et al. 2004). Walleye recruitment decreased to 16.8 million age-2 fish annually during the 1990s and dropped below 4 million fish in 1992 and 1995, which was the lowest level of recruitment observed since 1976 (WTG 2017). In the late 1990s, phosphorus concentrations began to increase in western Lake Erie, due mostly to increased levels of agricultural runoff, and hypoxic conditions returned in the central basin by the mid-2000s (Scavia et al. 2014). Walleye recruitment was highly variable during the 2000s; the highest level of recruitment was observed in 2003 (109.4 million age-2 fish), with recruitment failure occurring the year before in 2002 (0.4 million age-2 fish; WTG 2017). Walleye recruitment patterns improved slightly during 2010–2016, with average recruitment during this period of 12.4 million age-2 fish, but only one cohort (2015 year-class) exceeded the long-term (1976–2016) average (i.e., 17.9 million fish; WTG 2017).

Historically, the management of environmental conditions for desired levels of fish production in Lake Erie has been challenging for fishery managers. The linkage between environmental conditions and fish production is generally indirect, is often poorly understood, and few if any options for management may exist (e.g., climatic factors like spring storms and winter severity). Consequently, managers have few potential actions to implement at effective temporal and spatial scales. Moreover, fishery managers lack the regulatory authority necessary to develop and implement remedial actions and are dependent on land and watershed managers to appropriately address fish habitat priorities. For example, the western-basin Walleye stock likely would not have recovered as quickly or as fully in the absence of improved water quality from pollution abatement efforts during the 1970s due to the scope and scale of the perturbations affecting the lake at the time. However, those efforts were driven solely by the intent of the 1972 Great Lakes Water Quality Agreement (Knight 1997), which did not include linkage to fish production. To combat this challenge, Great Lakes fishery managers, including the LEC, recognize this dependency and each lake committee is responsible for developing and communicating to the appropriate environmental agencies and also implementing fishery priorities (Gaden et al. 2008). The LEC followed through with these commitments by identifying lakewide and basin-specific priorities for desired fish community structure in the Lake Erie fish community goals and objectives (Ryan et al. 2003) and associated environmental objectives (LEC 2005). These priorities include recognizing Walleye as an apex predator throughout the lake and protecting and/or restoring coastal nearshore and tributary habitats important for sustaining natural recruitment.

In summary, a commitment to interagency management by LEC member agencies was instrumental to restoring the collapsed western-basin Walleye stock. Additionally, using science-based research to address key uncertainties, using modern stock assessment approaches, incorporating stakeholder input into the quota setting process, and addressing environmental concerns collaboratively at the lake level have been key to the recovery and subsequent sustainability of the Lake Erie Walleye population and fishery for more than four decades.

LESSONS LEARNED: KEYS TO LONG-TERM SUSTAINABILITY

1. Interagency Cooperation Is Vital to Long-Term Fishery Sustainability

The need for a coordinated, multi-agency approach (i.e., population assessment and enforcement) to sustainably manage the broadly shared Lake Erie fishery resource was one of the most prominent lessons learned by fishery managers after the collapse of the western-basin Walleye stock (Hatch et al. 1987). While interagency collaboration of a shared resource seems intuitive, multiple attempts over decades failed to implement interagency management, primarily because the sovereign authority of each agency to manage its fishery was jealously guarded and not properly recognized within proposed early agreements (Gaden et al. 2008). The tide turned when the Commission formed interagency lake committees in 1964, which fostered development of working relationships and the eventually signing of *A Joint Strategic Plan for Management of Great Lakes Fisheries* in 1981 (Gaden et al. 2008), a nonbinding agreement rooted in developing consensus on fishery management practices, agency accountability, information sharing, and ecosystem management. The events surrounding closure of the Lake Erie Walleye fishery highlighted the need for interagency collaboration for monitoring the dynamics of this economically and ecologically important fish stock. Specifically, it highlighted the need for a comprehensive and coordinated approach to measure harvest and monitor population status, develop and evaluate sustainable harvest policies, and enforce effective regulations. Today, these principles not only guide current management efforts on Lake Erie and other Great Lakes fisheries, but also serve as a model for similar multi-jurisdictional fisheries across the globe.

2. Management Decisions Need to Be Based on Best Available Science

Scientific research has played a central role in monitoring stock status and developing sustainable harvest polices for Lake Erie fishery management. For example, the SPC relied on an early mark–recapture study conducted by Wolfert (1963) to define the spatial range of the western Walleye stock for a quota management program. Subsequent tagging studies (e.g., Wang et al. 2007; Zhao et al. 2011; Raby et al. 2018) helped refine the current understanding of Lake Erie Walleye movement patterns, which was vital for developing harvest strategies that provided fishery benefits to stakeholders across the basin. Additionally, the use of fishery-independent surveys (e.g., Tyson et al. 2006; Vandergoot et al. 2011; Kraus et al. 2017) to provide accurate assessments of the western Walleye stock are yet another example of making management decisions based on research findings. Through the LEPMAG process, stakeholders gained awareness of how scientific findings (e.g., performance of alternative harvest policies) inform decisions regarding the management of this complex fishery resource. Furthermore, assessing interlake movement and survival rates through the Great Lakes Acoustic Telemetry Observation System network (Krueger et al. 2018; Raby et al. 2018) have contributed to developing a comprehensive understanding of the spatial and temporal distribution of the western-basin Lake Erie Walleye stock, which previously would have been impossible due to the extensive movements shown by these fish (WTG 2017).

3. External Reviews Benefit Stock Assessment Practices

As within any discipline, the tendency to become complacent or too narrowly focused can plague the fishery management process. To combat this tendency, the LEC relies on external stock assessment scientists to conduct formal program reviews as circumstances dictate. External reviews have been used to evaluate proposed changes to harvest policies, how stock assessment surveys are conducted, how harvest is estimated, and population abundance modeling methods (Myers and Bence 2001; Lester et al. 2005; Wright et al. 2005; Jones et al. 2016). The external review process provides an objective way to ensure current management practices are appropriate and remain consistent with up-to-date science-based approaches for managing this ecologically and economically important fishery.

4. All Models Are Wrong, but Some Are Useful

Generally credited to Box (1976), this adage is worth remembering in the context of using statistical models to estimate the abundance of fish stocks. Although decision makers trust that the model being used is right, generally, most agree that it is likely not the case. For instance, the WTG (Locke et al. 2005; Kayle et al. 2015) believed that estimates of annual abundance from ADMB were perhaps more scientifically and mathematically defensible than those from CAGEAN (Myers and Bence 2001), which produced estimates that were 29% higher (on average) but followed similar trends during 1999–2009 (r = 0.98; Figure 12). While both models suggested that abundance was high during the 1980s and declining during the 1990s, it was impossible to know for certain which (if either) of the models accurately predicted stock size in any year. Another caveat associated with estimating population size is the tendency for model estimates to be least precise when a cohort first enters the fishery and the SCAA model attempts to determine abundance. This early problem also presents fishery managers with uncertainty as quotas are set (i.e., based on population estimates) before the model produces a more stable estimate of how many recruits originated from a particular year-class after subsequent years of data are collected (i.e., a cohort is surveyed in the fishery-dependent and -independent gears). To avoid these problems, managers and biologists can work to routinely evaluate model results and performance via internal and external reviews (see Lesson Learned #3), be cautious about relying solely on abundance estimates when making management decisions, and provide explicit rationale (e.g., ecological or economic) for deviating from the harvest policy.

5. The Benefits of Engaging Stakeholders in Management Process Outweighs the Challenges

Incorporating stakeholder input into the management process can be complex, more so when the stock is highly migratory and sup-

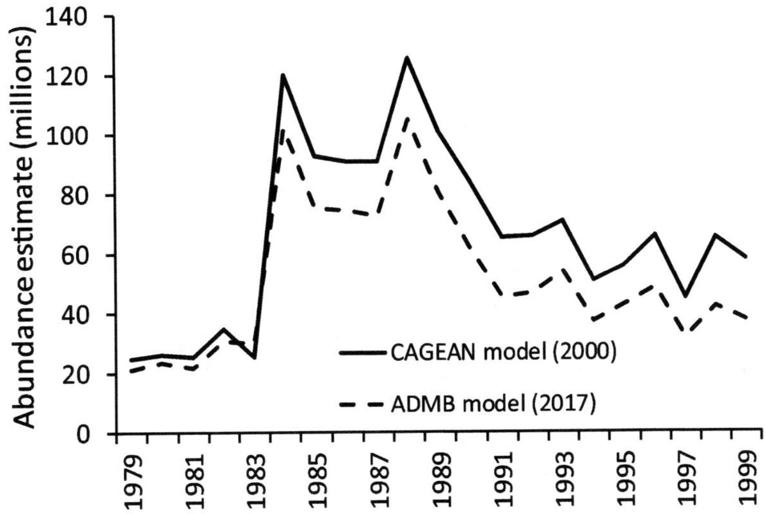

FIGURE 12. Comparison of the Lake Erie Walleye abundance estimates from CAGEAN and the most recent ADMB model projection.

ports multiple fishery components. The spatial ecology of Lake Erie Walleye is complex; consequently, management decisions (e.g., setting safe harvest levels) for this stock affects commercial and recreational fishery stakeholders throughout the lake. Through the LEPMAG process, fishery objectives were identified and discussed collectively for the first time. These interactions provided fishery and LEC representatives an opportunity to hear what was important to each group, develop a harvest strategy that promoted population and fishery resiliency, and promote communication between managers and stakeholders. Despite competing fishery objectives and perspectives, a mechanism for transparency and open dialogue among stakeholders was beneficial to the process by helping fishers move toward common/shared fishery objectives (Jones et al. 2016).

6. Engaging Land, Shoreline, and Water Quality Managers Is Imperative to Future Sustainability

Cultural stressors (e.g., point and nonpoint source pollution, land development) within the Lake Erie watershed affect Walleye habitat, creating challenges for fishery managers trying to meet fish community goals and objectives (Ryan et al. 2003). To address regional watershed stressors, LEC members work collaboratively with water quality managers in Canada and the United States through the Great Lakes Water Quality Agreement. Within jurisdictions, fishery managers' work with regulators and managers of municipalities, watershed groups, and landowners to soften shorelines, protect or improve riparian corridors and coastal wetlands, reduce runoff, remove barriers to fish access, or address other priorities (Markham and Knight 2017).

CONCLUSION

Despite periodic declines in abundance and fishery performance, population resiliency (i.e., favorable recruitment patterns) and the collaborative actions implemented by fishery managers have contributed to Lake Erie's Walleye being considered one of the most productive (with respect to fishery yields) freshwater fisheries in North America. As mentioned throughout the chapter, the keys to ensuring sustainable management once the Walleye population recovered were conducting coordinated harvest assessments and fishery enforcement, science-based population assessments; establishing sustainable harvest levels; incorporating stakeholder input into the management process; conducting science-based research; and collaboratively addressing lakewide environmental concerns. Although these efforts led to the recovery of the western-basin Walleye population (with expansion into the eastern basin), restoration came too late for the Blue Pike population, as that species is considered extinct. In the future, increased attention may need to focus on management actions associated with environmental issues (e.g., invasive species, water quality, habitat loss and degradation, and climate change) and continued refinement of stock assessment practices and sustainable harvest policies.

ACKNOWLEDGMENTS

The authors thank past and present the members of the Lake Erie Committee and Walleye Task Group for maintaining and updating the annual harvest and population assessment information used to assemble this chapter. We would also thank the Great Lakes Fishery Commission (www.glfc.org) for the role they play coordinating management activities since inception. All Great Lakes Science Center sampling and handling of fish during research are carried out in accordance with guidelines for the care and use of fishes by the American Fisheries Society (http://fisheries.org/docs/wp/Guidelines-for-Use-of-Fishes.pdf). Any use of trade, firm, or product names is for descriptive purposes only and does not imply endorsement by the U.S. Government.

REFERENCES

Baldwin, N. A., R. W. Saalfeld, M. R. Dochoda, H. J. Buettner, and R.L. Eshenroder. 2009. Commercial fish production in the Great Lakes 1867–2006. Great Lakes Fishery Commission, Ann Arbor, Michigan. Available: www.glfc.org/great-lakes-databases.php. (August 2017)

Barnard, G. C., A. J. Kennedy, W. P. Brown, and D. L. Pereira. 2019. Walleye management in the Red Lakes, Minnesota: collapse, recovery, and cooperative management. Pages 323–352 *in* C. C. Krueger, W. W. Taylor, and S. Youn, editors. From catastrophe to recovery: stories of fish management success. American Fisheries Society, Bethesda, Maryland.

Beeton, A. M. 1961. Environmental changes in Lake Erie. Transactions of the American Fisheries Society 90:153–159.

Bertram, P. E. 1993. Total phosphorus and dissolved oxygen trends in the central basin of Lake Erie, 1970–1991. Journal of Great Lakes Research 19:224–236.

Bouge, M. B. 2000. Fishing the Great Lakes, an environmental history, 1783–1933. University of Wisconsin Press, Madison

Box, G. E. P. 1976. Science and statistics. Journal of the American Statistical Association 71:791–799.

Brenden, T. O., R. W. Brown, M. P. Ebener, K. Reid, and T. J. Newcomb. 2013. Great Lakes commercial fisheries: historical overview and prognosis for the future. Pages 339–398 *in* W. W. Taylor, A. J. Lynch, and N. J. Leonard, editors. Great Lakes fisheries policy and management: a binational perspective. Michigan State University Press, East Lansing.

Busch, W. D.-N., R. L. Scholl, and W. L. Hartman. 1975. Environmental factors affecting the strength of Walleye (*Stizostedion vitreum vitreum*) year-classes in western Lake Erie, 1960–1970. Journal of the Fisheries Research Board of Canada 32:1733–1743.

Carlander, K. D. 1977. Biomass, production, and yields of Walleye (*Stizostedion vitreum vitreum*) and Yellow Perch (*Perca flavescens*) in North American lakes. Journal of the Fisheries Research Board of Canada 34:1602–1612.

Colby, P. J., C. A. Lewis, R. L. Eshenroder. 1991. Status of Walleye in the Great Lakes: case studies prepared for the 1989 workshop. Great Lakes Fishery Commission, Special Publication 91-1, Ann Arbor, Michigan. Available: http://www.glfc.org/pubs/SpecialPubs/Sp91_1.pdf. (August 2017).

Deriso, R. B., T. J. Quinn, III, and P. R. Neal. 1985. Catch-age analysis with auxiliary information. Canadian Journal of Fisheries and Aquatic Sciences 42:815–824.

Doan, K. H. 1944. The winter fishery in western Lake Erie, with a census of the 1942 catch. The Ohio Journal of Science 44:69–74.

Dolan, D. M.. 1993. Point source loadings of phosphorus to Lake Erie: 1986–1990. Journal of Great Lakes Research 19:212–223.

DuFour, M. R., C. May, E. Roseman, C. M. Mayer, S. A. Ludsin, C. S. Vandergoot, J. J. Pritt, M. E. Fraker, J. T. Tyson, J. Miner, and E. A. Marschall. 2015. Portfolio theory as a management tool to guide conservation and restoration of multi-stock fish populations. Ecosphere [online serial] 6(12):296.

Einhouse, D. W., and R. Haas. 1995. A preliminary examination of walleye distribution and exploitation in the eastern basin of Lake Erie using tag-recovery data. New York Department of Environmental Conservation, Division of Fish and Wildlife, Dunkirk.

Fielder, D. G., and J. P. Baker. 2019. Recovery of Saginaw Bay Walleye, Lake Huron. Pages 411–430 *in* C. C. Krueger, W. W. Taylor, and S. Youn, editors. From catastrophe to recovery: stories of fish management success. American Fisheries Society, Bethesda, Maryland.

Gaden, M. E. 2007. Bridging jurisdictional divides: collective action through a joint strategic plan for management of Great Lakes fisheries. Doctoral dissertation. University of Michigan, Ann Arbor.

Gaden, M., C. Krueger, C. Goddard, and G. Barnhart. 2008. A joint strategic plan for management of the Great Lakes fisheries: a cooperative regime in a multi-jurisdictional setting. Aquatic Ecosystem Health and Management 11:50–60.

Gaden, M. E., C. Goddard, and J. Read. 2013. Multi-jurisdictional management of the shared Great Lakes fishery: transcending conflict and diffuse political authority. Pages 305–337 *in* W. W. Taylor, A. J. Lynch, and N. J. Leonard, editors. Great Lakes fisheries policy and management: a binational perspective. Michigan State University Press, East Lansing.

GLFC (Great Lakes Fishery Commission, editors). 2007. A joint strategic plan for management of Great Lakes fisheries (adopted in 1997 and supersedes 1981 original). Great Lakes Fishery Commission, Miscellaneous Publication 2007-01, Ann Arbor, Michigan. Available: www.glfc.org/fishmgmt/jsp97.pdf. (August 2017).

Gopalan, G., D. A. Culver, L. Wu, and B. K. Trauben. 1998. Effects of recent ecosystem changes on the recruitment of young-of-the-year fish in western Lake Erie. Canadian Journal of Fisheries and Aquatic Sciences 55:2572–2579.

Gulland, J. A. 1971. Science and fishery management. ICES Journal of Marine Science 33:471–477.

Haponski, A. E., and C. A. Stepien. 2014. A population genetic window into the past and future of the Walleye *Sander vitreus*: relation to historic Walleye and the extinct "Blue Pike" *S. v.* "*glaucus*." BMC Evolutionary Biology 14:133–154.

Hatch, R. W., S. J. Nepsy, K. M. Muth, and C. T. Baker, 1987. Dynamics of the stock recovery of the western Lake Erie Walleye (*Stizostedion vitreum vitreum*) stock. Canadian Journal of Fisheries and Aquatic Sciences 44 (Supplement 2):15–22.

Hayden, T. A., T. R. Binder, C. M. Holbrook, C. S. Vandergoot, D. G. Fielder, S. J. Cooke, J. M. Dettmers, and C. C. Krueger. 2017. Spawning site fidelity and apparent survival of Walleye (*Sander vitreus*) differs between a Lake Huron and Lake Erie tributary. Ecology of Freshwater Fish 27:339–349.

Hayes, D. B. 2013. Issues affecting fish habitat in the Great Lakes basin. Pages 51–80 *in* W. W. Taylor, A. J. Lynch, and N. J. Leonard, editors. Great Lakes fisheries policy and management: a binational perspective. Michigan State University Press, East Lansing.

Hecky, R. E., R. E. H. Smith, D. R. Barton, S. J. Guildford, W. D. Taylor, M. N. Charlton, and T. Howell. 2004. The near shore phosphorus shunt: a consequence of ecosystem engineering by dreissenids in the Laurentian Great Lakes. Canadian Journal of Fisheries and Aquatic Sciences 61:1285–1293.

Herdendorf, E. 1983. Lake Erie water quality 1970–1982: a management assessment. Center for Lake Erie Area Research, CLEAR Technical Report No. 279, Columbus, Ohio.

Hilborn, R., and C. J. Walters. 1992. Quantitative fisheries stock assessment: choice, dynamics and uncertainty. Kluwer Academic Publishers, Dordrecht, Netherlands.

Hokanson, K. E. 1977. Temperature requirements of some percids and adaptations to the seasonal temperature cycle. Journal of the Fisheries Research Board of Canada 34:1524–1556.

Ihde, T. F., M. J. Wilberg, D. H. Secor, and T. J. Miller. 2011. FishSmart: harnessing the knowledge of stakeholders to enhance U.S. marine recreational fisheries with application to the Atlantic King Mackerel fishery. Pages 75–93 *in* A. M. Orlov and T. Iwanoto, editors. Grenadiers of the world oceans: biology, stock assessment, and fisheries. American Fisheries Society, Symposium 73, Bethesda, Maryland.

Jensen, T. 2007. Wooden boats and iron men: history of commercial fishing in northern Lake Michigan & Door County 1850–2005. Paisa Publishing Company, De Pere, Wisconsin.

Jones, M. L., M. J. Catalano, L. K. Peterson, and A. M. Berger. 2016. Stakeholder-centered development of a harvest control rule for Lake Erie Walleye *Sander vitreus*. Pages 163–183 *in* C. T. T. Edwards and D. J. Dankel, editors. Management science in fisheries: an introduction to simulation-based methods. Routledge, New York.

Kaatz, M. R. 1955. The black swamp: a study in historical geography. Annals of the Association of American Geographers 45:1–35.

Kayle, K., K. Oldenburg, C. Murray, J. Francis, and J. Markham. 2015. Lake Erie Walleye management plan 2015–2019. Great Lakes Fishery Commission, Ann Arbor, Michigan. Available: www.glfc.org/pubs/lake_committees/erie/LEC_docs/position_statements/walleye_managment_plan.pdf. (August 2017).

Keller, M. 1964. Lake Erie sport fishing survey. Ohio Department of Natural Resources, Division of Wildlife, Publication W-316, Columbus.

Keller, M. 1965. The winter fishery of South Bass Island with a census of the 1963 catch. The Ohio Journal of Science 65:327–334.

Koonce, J. F., T. B. Bagenal, R. F. Carline, K. E. F. Hokanson, and M. Nagiec. 1977. Factors influencing year-class strength of percids: a summary and a model of temperature effects. Journal of the Fisheries Research Board of Canada 24:1900–1909.

Knight, R. L. 1997. Successful interagency rehabilitation of Lake Erie Walleye. Fisheries 22:16–17.

Knight, R. L., and B. Vondracek. 1993. Changes in prey fish populations in western Lake Erie, 1969–88, as related to Walleye, *Stizostedion vitreum*, predation. Canadian Journal of Fisheries and Aquatic Sciences 50:1289–1298.

Kraus, R. T., C. S. Vandergoot, P. M. Kocovsky, M. W. Rogers, H. A. Cook, and T. O. Brenden. 2017. Reconciling catch differences from multiple fishery independent gill net surveys. Fisheries Research 188:17–22.

Krueger, C. C., and D. J. Decker. 1999. The process of fisheries management. Pages 31–59 *in* C. C. Kohler and W. A. Hubert, editors. Inland fisheries management in

North America, 2nd edition. American Fisheries Society, Bethesda, Maryland.

Krueger, C. C., C. M. Holbrook, T. R. Binder, C. S. Vandergoot, T. A. Hayden, D. W. Hondorp, N. Nate, K. Paige, S. C. Riley, A. Fisk, and S. J. Cooke. 2018. Acoustic telemetry observation systems: problems, solutions, and lessons learned in the Laurentian Great Lakes. Canadian Journal of Fisheries and Aquatic Sciences 75:1755–1763.

Kutkuhn, J., W. Hartman, A. Holder, R. Kenyon, S. Kerr, A. Lamsa, S. Nepszy, M. Patriarche, R. Scholl, W. Shepherd, and G. Spangler. 1976. First technical report of the Great Lakes Fishery Commission Scientific Protocol Committee on interagency management of the Walleye resource of western Lake Erie. Great Lakes Fishery Commission, Ann Arbor, Michigan.

Larkin, P. A. 1977. An epitaph for the concept of maximum sustained yield. Transactions of the American Fisheries Society 106:1–11.

Larsen, A. 1954. First record of White Perch (*Morone americana*) in Lake Erie. Copeia 1954:134.

Leach, J. H., and S. J. Nepszy. 1976. The fish community in Lake Erie. Journal of the Fisheries Research Board of Canada 33:622–638.

Leach, J. H. 1993. Impacts of the zebra mussel (*Dreissena polymorpha*) on water quality and fish spawning reefs in western Lake Erie. Pages 381–397 *in* T. F. Nalepa and D. W. Schloesser, editors. Zebra mussels: biology, impacts, and control. Lewis Publishers, Chelsea, Michigan.

LEC (Lake Erie Committee). 2005. Lake Erie environmental objectives: report of the Environmental Objectives Sub-Committee of the Lake Erie Committee. Great Lakes Fishery Commission, Ann Arbor, Michigan.

Lester, N., A. Bingham, B. Clark, K. Pollock, and P. Sullivan. 2005. Report of the Blue Ribbon Panel for review of procedures used to estimate percid harvest in Lake Erie. Report to the Great Lakes Fishery Commission, Ann Arbor, Michigan. Available: www.glfc.org/pubs/lake_committees/erie/STC_docs/other_reports_and_docs/LE_percid_harv_rev_2005.pdf. (August 2017).

Locke, B., M. Belore, A. Cook, D. Einhouse, R. Kenyon, R. Knight, K. Newman, P. Ryan, and E. Wright. 2005. Lake Erie Walleye management plan. Great Lakes Fishery Commission, Ann Arbor, Michigan. Available: www.glfc.org/pubs/lake_committees/erie/WTG_docs/other_reports_and_docs/wmp20051207.pdf. (August 2017).

Madenjian, C. P., T. A. Hayden, T. B. Peat, C. S. Vandergoot, D. G. Fielder, A. M. Gorman, S. A. Pothoven, J. M. Dettmers, S. J. Cooke, Y. Zhao, and C. C. Krueger. 2018. Temperature regimes, growth, and food consumption for adult Walleye in Lake Huron and Lake Erie: a bioenergetics analysis. Canadian Journal of Fisheries and Aquatic Sciences 75:1573–1586.

Madenjian, C. P., J. T. Tyson, R. L. Knight, M. W. Kershner, and M. J. Hansen. 1996. First-year growth, recruitment, and maturity of Walleyes in western Lake Erie. Transactions of the American Fisheries Society 125:821–830.

Makarewicz, J. C., and P. Bertram. 1991. Restoration of the Lake Erie ecosystem. Bioscience 41:216–223.

Markham, J. L., and R. L. Knight. 2017. The state of Lake Erie in 2009. Great Lakes Fishery Commission, Ann Arbor, Michigan. Available: www.glfc.org/pubs/SpecialPubs/Sp17_01.pdf. (July 2017).

McGarrell, E. F., M. Suttmoeller, and C. Gibbs. 2013. Great Lakes fisheries law enforcement. Pages 455–472 *in* W. W. Taylor, A. J. Lynch, and N. J. Leonard, editors. Great Lakes fisheries policy and management: a binational perspective. Michigan State University Press, Ann Arbor.

Mion, J. B., R. A. Stein, and E. A. Marschall. 1998. River discharge drives survival of larval Walleye. Ecological Applications 8:88–103.

Murphy, C. A., S. P. Bhavsar, and N. Gandhi. 2013. Contaminants in Great Lakes fish: historic, current, and emerging concerns. Pages 203–258 *in* W. W. Taylor, A. J. Lynch, and N. J. Leonard, editors. Great Lakes fisheries policy and management: a binational perspective. Michigan State University Press, East Lansing.

Myers, R. A., and J. R. Bence. 2001. The Walleye of western and central Lake Erie. Unpublished technical review for the Lake Erie Committee of the Great Lakes Fishery Commission, Ann Arbor, Michigan.

Nepszy, S. J. 1977. Changes in percid populations and species interactions in Lake Erie. Journal of the Fisheries Research Board of Canada 34:1861–1868.

NYSDEC (New York State Department of Environmental Conservation). 2017. Lake Erie 2016 annual report. NYSDEC, Albany. Available: www.dec.ny.gov/docs/fish_marine_pdf/2016lerpt.pdf. (August 2017).

ODNR (Ohio Department of Natural Resources). 2017. Ohio's Lake Erie fisheries, 2016. ODNR, Division of Wildlife, Columbus. Available: http://wildlife.ohiodnr.gov/portals/wildlife/pdfs/fishing/LakeErieStatus.pdf. (August 2017).

Parsons, J. W. 1970. Walleye fishery of Lake Erie in 1943–62 with emphasis on contributions of the 1942–61 year-classes. Journal of the Fisheries Research Board of Canada 27:1475–1489.

Parsons, J. W. 1972. Life history and production of Walleyes of the 1959 year-class in western Lake Erie, 1959–62. Transactions of the American Fisheries Society 101:655–661.

Peat, T. B., T. A. Hayden, L. F. G. Gutowsky, C. S. Vandergoot, D. G. Fielder, C. P. Madenjian, K. J. Murchie, J. M. Dettmers, C. C. Krueger, and S. J. Cooke. 2015. Seasonal thermal ecology of adult Walleye (*Sander vitreus*) in Lake Huron and Lake Erie. Journal of Thermal Biology 53:98–106.

Pereira, D. L., and M. J. Hansen. 2003. A perspective on challenges to recreational fisheries management: summary of the symposium on active management of recreational fisheries. North American Journal of Fisheries Management 23:1276–1282.

Pritt, J. J., M. R. DuFour, C. M. Mayer, P. M. Kocovsky, J. T. Tyson, E. J. Weimer, and C. S. Vandergoot. 2013. Including independent estimates and uncertainty to quantify total abundance of fish migrating in a large river system: Walleye (*Sander vitreus*) in the Maumee River, OH. Canadian Journal of Fisheries and Aquatic Sciences 70:803–814.

Quinn, T. J., and R. B. Deriso. 1999. Quantitative fishery dynamics. Oxford University Press, New York.

Raby, G. D., C. S. Vandergoot, T. A. Hayden, M. D. Faust, R. T. Kraus, J. M. Dettmers, S. J. Cooke, Y. Zhao, A. T. Fisk, and C. C. Krueger. 2018. Does behavioural thermoregulation underlie seasonal movements in Lake Erie Walleye? Canadian Journal of Fisheries and Aquatic Sciences 75:488–496.

Regier, H. A., V. C. Applegate, and R. A. Ryder. 1969. The ecology and management of the Walleye in western Lake Erie. Great Lakes Fishery Commission, Technical Report 15, Ann Arbor, Michigan. Available: www.glfc.org/pubs/TechReports/Tr15.pdf. (August 2017).

Regier, H. A., and W. L. Hartman. 1973. Lake Erie's fish community: 150 years of cultural stresses. Science 180:1248–1255.

Roseman, E. F., R. Drouin, M. Gaden, R. L. Knight, J. Tyson, and Y. Zhao. 2013. Managing inherent complexity for sustainable Walleye fisheries in Lake Erie. Pages 475–493 *in* W. W. Taylor, A. J. Lynch, and N. J. Leonard, editors. Great Lakes fisheries policy and management: a binational perspective. Michigan State University Press, East Lansing.

Roseman, E. F., W. W. Taylor, D. B. Hayes, A. L. Jones, and J. T. Francis. 2006. Predation on Walleye eggs by fish on reefs in western Lake Erie. Journal of Great Lakes Research 32:415–423.

Roseman, E. F., W. W. Taylor, D. B. Hayes, R. L. Knight, and R. C. Haas. 2001. Removal of Walleye eggs from reefs in western Lake Erie by a catastrophic storm. Transactions of the American Fisheries Society 130:341–346.

Ryan, P.A., R. Knight, R. MacGregor, G. Towns, R. Hoopes, and W. Culligan. 2003. Fish-community goals and objectives for Lake Erie. Great Lakes Fishery Commission, Special Publication 03-02, Ann Arbor, Michigan. Available: www.glfc.org/pubs/SpecialPubs/Sp03_2.pdf. (August 2017).

Schneider, J. C., and J. H. Leach. 1977. Walleye (*Stizostedion vitreum*) fluctuations in the Great Lakes and possible causes, 1800–1975. Journal of the Fisheries Research Board of Canada 34:878–889.

Scavia, D., J. D. Allan, K. K. Arend, S. Bartell, D. Beletsky, N. S. Bosch, S. B. Brand, R. D. Briland, I. Daloğlu, J. V. DePinto, D. M. Dolan. M. A. Evans, T. M. Farmer, D. Goto, H. Han, T. O. Höök, R. Knight, S. A. Ludsin, D. Mason, A. M. Michalak, R. P. Richards, J. J. Roberts, D. K. Rucinski, E. Rutherford, D. J. Schwab, T. M. Sesterhenn, H. Zhang, and Y. Zhou. 2014. Assessing and addressing the re-eutrophication of Lake Erie: central basin hypoxia. Journal of Great Lakes Research 40:226–246.

Stepien, C. A., D. J. Murphy, R. N. Lohner, A. E. Haponski, and O. J. Sepulveda-Villet. 2010. Status and delineation of Walleye (*Sander vitreus*) genetic stock structure across the Great Lakes. Pages 189–223 *in* E. F. Roseman, P. Kocovsky, and C. Vandergoot, editors. Status of Walleye in the Great Lakes: proceedings of the 2006 symposium. Great Lakes Fishery Commission, Technical Report 69, Ann Arbor, Michigan.

Thayer, S. A., and A. J. Loftus. 2013. Great Lakes recreational fisheries and their role in fisheries management and policy. Pages 399–440 *in* W. W. Taylor, A. J. Lynch, and N. J. Leonard, editors. Great Lakes fisheries policy and management: a binational perspective. Michigan State University Press, East Lansing.

Trautman, M. B. 1981. The fishes of Ohio. Ohio State University Press, Columbus.

Tyson, J. T., T. B. Johnson, C. T. Knight, and M. T. Bur. 2006. Intercalibration of research survey vessels on Lake Erie. North American Journal of Fisheries Management 26:559–570.

Vandergoot, C. S., and T. O. Brenden. 2014. Spatially varying population demographics and fishery characteristics of Lake Erie Walleyes inferred from a long-term tag recovery study. Transactions of the American Fisheries Society 143:188–204.

Vandergoot, C. S., and T. O. Brenden. 2015. Accuracy and precision of fishery and demographic estimates from a spatial tag-recovery model when inter-regional movements are treated as fixed. Fisheries Research 164:8–25.

Vandergoot, C. S., T. O. Brenden, M. V. Thomas, D. W. Einhouse, H. A. Cook, and M. W. Turner. 2012. Estimation of tag shedding and reporting rates for Lake Erie jaw-tagged Walleyes. North American Journal of Fisheries Management 32:221–223.

Vandergoot, C. S., H. A. Cook, M. V. Thomas, D. E. Einhouse, and C. Murray. 2010. Status of Walleye in western Lake Erie, 1985–2006. Pages 123–150 *in* E. R. Roseman, P. Kocovsky, and C. S. Vandergoot, editors. Status of Walleye in the Great Lakes: proceedings of the 2006 symposium. Great Lakes Fishery Commission, Technical Report 69, Ann Arbor, Michigan. www.glfc.org/pubs/TechReports/Tr69.pdf. (August 2017).

Vandergoot, C. S., P. M. Kocovsky, T. O. Brenden, and W. Liu. 2011. Selectivity evaluation of two experimental gill net configurations used to sample Lake Erie Walleyes. North American Journal of Fisheries Management 31:832–842.

Wang, H., E. S. Rutherford, H. A. Cook, D. W. Einhouse, R. C. Haas, T. B. Johnson,

R. Kenyon, B. Locke, and M. W. Turner. 2007. Movement of Walleyes in Lakes Erie and St. Clair inferred from tag returns and fisheries data. Transactions of the American Fisheries Society 136:539–551.

Wolfert, D. R. 1963. The movement of Walleyes tagged as yearlings in Lake Erie. Transactions of the American Fisheries Society 92:539–551.

Wolfert, D. R. 1969. Maturity and fecundity of Walleyes from the eastern and western basins of Lake Erie. Journal of the Fisheries Research Board of Canada 26:1877–1888.

Wolfert, D. R. 1981. The commercial fishery for walleyes in New York waters of Lake Erie, 1959–1978. North American Journal of Fisheries Management 1:112–126.

Wolfert, D. R., W. Busch, and C. T. Baker. 1975. Predation by fish on Walleye eggs on a spawning reef in western Lake Erie, 1969. The Ohio Journal of Science 75:118–125.

Wright, E., M. Belore, A. Cook, B. Culligan, D. Einhouse, T. Johnson, K. Kayle, R. Kenyon, R. Knight, and K. Newman. 2005. Decision analysis application for Lake Erie Walleye management: final report to the Lake Erie Committee. Great Lakes Fishery Commission, Ann Arbor, Michigan. Available: www.glfc.org/pubs/lake_committees/erie/WTG_docs/other_reports_and_docs/daalwm.pdf. (August 2017).

WTG (Walleye Task Group). 2017. Report for 2016 by the Lake Erie Walleye Task Group. Great Lakes Fishery Commission, Ann Arbor, Michigan. Available: www.glfc.org/pubs/lake_committees/erie/WTG_docs/annual_reports/WTG_report_2017.pdf. (August 2017).

Zhao, Y., D. W. Einhouse, and T. MacDougall. 2011. Resolving some of the complexity of a mixed-origin Walleye population in the east basin of Lake Erie using a mark–recapture study. North American Journal of Fisheries Management 31:379–389.

Zhao, Y., M. L. Jones, B. J. Shuter, and E. F. Roseman. 2009. A biophysical model of Lake Erie Walleye (*Sander vitreus*) explain interannual variations in recruitment. Canadian Journal of Fisheries and Aquatic Sciences 66:114–125.

Lake Ontario Deepwater Sculpin Recovery: An Unexpected Outcome of Ecosystem Change

Brian C. Weidel*

U.S. Geological Survey, Great Lakes Science Center
17 Lake Street, Oswego, New York 13126, USA

Michael J. Connerton

New York State Department of Environmental Conservation
Lake Ontario Fisheries Unit
541 East Broadway, Cape Vincent, New York 13618, USA

Maureen G. Walsh

U.S. Fish and Wildlife Service
1601 Balboa Avenue, Panama City, Florida 32405, USA

Jeremy P. Holden

Ontario Ministry of Natural Resources and Forestry, Lake Ontario Management Unit
41 Hatchery Lane, Picton, Ontario K0K 2T0, Canada

Kristen T. Holeck

Cornell Biological Field Station, Cornell University
900 Shackelton Point Road, Bridgeport, New York 13030, USA

Brian F. Lantry

U.S. Geological Survey, Great Lakes Science Center
17 Lake Street, Oswego, New York 13126, USA

Abstract.—Fish population recoveries can result from ecosystem change in the absence of targeted restoration actions. In Lake Ontario, native Deepwater Sculpin *Myoxocephalus thompsonii* were common in the late 1800s, but by the mid-1900s the species was possibly extirpated. During this period, mineral nutrient inputs increased and piscivore abundance declined, which increased the abundance of the nonnative planktivores Alewife *Alosa pseudoharengus* and Rainbow Smelt *Osmerus mordax*. Deepwater Sculpin larvae are pelagic and vulnerable to predation by planktivores. Annual bottom trawl surveys did not capture Deepwater Sculpin from 1978 to 1995 (n = 6,666 tows) despite sampling appropriate habitat (trawl depths: 7–170 m). The absence of observations during this time resulted in an elevated conservation status for the species, but no restoration actions were initiated. In 1996, three individuals were caught in bottom trawls, the first observed since 1972. Since then, their abundance has increased, and in 2017, they were the second most abundant Lake Ontario prey fish. The food-web changes that occurred from 1970 through the 1990s contributed to this recovery. Alewife and Rainbow Smelt abundance declined during this period due to predation by stocked salmonids and legislation that reduced nutrient

* Corresponding author: bweidel@usgs.gov

inputs and food web productivity. In the 1990s, proliferation of nonnative, filter-feeding dreissenid mussels dramatically increased water clarity. As light penetration increased, the early-spring depth distribution of Alewife and Rainbow Smelt shifted deeper, away from larval Deepwater Sculpin habitat. The intentional and unintentional changes that occurred in Lake Ontario were not targeted at Deepwater Sculpin restoration but resulted in conditions that favored the species' recovery. While standard surveys documented the recovery, more diverse information (e.g., observations in deep habitats and early-life stages) would have improved our understanding of why the species recovered when it did. Annual Lake Ontario trawl surveys have collaboratively expanded their spatial extent and diversified habitat sampled, based on lessons learned from the Deepwater Sculpin recovery.

INTRODUCTION

Deepwater Sculpin *Myoxocephalus thompsonii* are a relatively uncommon species, found only in the Laurentian Great Lakes and a few oligotrophic Canadian lakes (Sheldon et al. 2008). They inhabit the deepest portions of lakes and have been captured at depths greater than 360 m (Scott and Crossman 1973). They often co-occur with the glacial-relict invertebrates *Mysis* and *Diporeia* spp., both of which are common prey items for sculpin (Kraft and Kitchell 1986; O'Brien et al. 2009; Londer 2011). As a prey fish, Deepwater Sculpin are often found in Lake Trout *Salvelinus namaycush* (Dymond 1928; Madenjian et al. 1998; Gamble et al. 2011) and Burbot *Lota lota* stomachs (Fratt et al. 1997). Deepwater Sculpin's limited biogeographic range, role in deep lake food webs, and sensitivity to anthropogenic change make them a species of conservation interest.

Primary threats to Deepwater Sculpin populations include eutrophication and nonnative species (COSEWIC 2006). Deepwater Sculpin populations in inland lakes are sensitive to increased nutrient inputs, which can reduce dissolved oxygen concentration in deep habitats (COSEWIC 2006; Sheldon et al. 2008). In the Laurentian Great Lakes, Deepwater Sculpin were historically found in profundal habitats of all five lakes, but declines and disappearances have been attributed to nonnative planktivores. Deepwater Sculpin are demersal; however, they have a pelagic larval life stage (Mansfield et al. 1983). This pelagic behavior makes them vulnerable to nonnative planktivores such as Alewife *Alosa pseudoharengus* and Rainbow Smelt *Osmerus mordax* (Crowder 1980; Geffen and Nash 1992; Madenjian et al. 2005a). Larval Deepwater Sculpin have been most commonly observed in relatively shallow waters (<50 m) and are believed to transition from pelagic to demersal habitats in water less than 75 m (Mansfield et al. 1983; Geffen and Nash 1992).

In the Great Lakes, proliferation of nonnative planktivores has been implicated in Deepwater Sculpin population declines in Lake Michigan (Wells and McLain 1973; Crowder 1980; Madenjian et al. 2008) and Lake Ontario (Geffen and Nash 1992; Lantry et al. 2007). Deepwater Sculpin have been commonly caught in trawl surveys in Lake Superior, Lake Michigan, and Lake Huron since the 1970s; however, abundance indices suggest that those populations have generally declined over the past 30 years (Bronte et al. 2003; Madenjian et al. 2005a; Riley and Adams 2010). In Lake Erie, recent observations have been limited to larval Deepwater Sculpin caught immediately downstream of Lake Huron (Roseman et al. 1998). In Lake Ontario, Deepwater Sculpin abundance has fluctuated over the past 70 years. They were historically described as common in the early 1900s (Stone 1947; Wells 1969), were possibly extirpated in the 1980s (Christie 1973; Brandt 1986), and in 2017 were one of the most abundant benthic prey fishes in the lake (Weidel et al. 2018).

The decline and subsequent recovery of Lake Ontario Deepwater Sculpin over the past century provides an example where a species' recovery resulted from changing ecological conditions rather than management actions specifically directed at their recovery. The mechanisms that contributed to their recovery are unknown. We offer evidence that nonnative planktivore declines (Alewife and Rainbow Smelt) and habitat shifts in nonnative planktivores contributed to Deepwater Sculpin recovery. We identify knowledge gaps in our understanding of the recovery to illustrate how long-term surveys might adapt to provide the necessary observations for quantifying the next 50 years of change in the Laurentian Great Lakes.

HISTORIC DECLINE OF LAKE ONTARIO DEEPWATER SCULPIN

The Lake Ontario Deepwater Sculpin population decline was poorly chronicled. No fishery-independent sampling occurred in the 1800s and early 1900s, and historic commercial fisheries did not consistently report incidental catches (Lantry et al. 2007). The earliest fishery-independent records come from Dymond et al. (1929), who reported Deepwater Sculpin as abundant based on Lake Ontario gill-net catches fished at depths from 91 to 124 m. Commercial gill-net fishers mentioned incidental catches of Deepwater Sculpin as late as 1950 (Christie 1973), but by the 1960s, accounts of the species in gill nets and bottom trawls were rare. A whole-lake, cross-season sampling effort in 1972 yielded three Deepwater Sculpin, which were the last known collections until the mid-1990s (Christie and Thomas 1981; Lantry et al. 2007).

During the Deepwater Sculpin's population decline in the 1900s, Lake Ontario's watershed land use, food web, and fish community changed substantially. During this time, sewage inputs increased and phosphorus-laden fertilizers and detergents were adopted, elevating nutrient loads and ultimately changing the trophic state of Lake Ontario from oligotrophic to mesotrophic (Schelske 1991; Sly 1991; Estepp and Reavie 2015). Also during the early 1900s, commercial fishing and Sea Lamprey *Petromyzon marinus* predation reduced the abundant native Lake Trout population, effectively eliminating piscivores by the mid-1900s (Christie 1973; Elrod et al. 1995). Increased food web productivity and low piscivore abundance provided ideal conditions for nonnative prey fish, such as Alewife and Rainbow Smelt, to proliferate in Lake Ontario. Alewife density increased to a level where their population often suffered mass mortality events due to poor fish condition and harsh winters (O'Gorman and Schneider 1986). These mass mortality events were chronicled as early as the 1890s (Crossman and Van Meter 1979) and continued into the 1970s (O'Gorman and Schneider 1986). By 1972, the Lake Ontario fish community was dominated by Alewife, Rainbow Smelt, and Slimy Sculpin *Cottus cognatus*, and Deepwater Sculpin were rare (Christie and Thomas 1981).

Numerous studies have provided compelling arguments that introduced planktivores, such as Alewife and Rainbow Smelt, have caused declines of native Great Lake fishes, including Deepwater Sculpin (Smith 1970; Wells and McLain 1973; Crowder 1980; Eck and Wells 1987). Crowder (1980) noted that declining native species all shared a common life-history trait, pelagophile eggs or larvae, which could be consumed by nonnative planktivores. A review of Lake

Michigan sculpin dynamics concluded that Alewife had the strongest negative effect on Deepwater Sculpin abundance (Madenjian et al. 2005a) and that the negative relationship was also hypothesized to have affected Deepwater Sculpin in Lake Huron and Lake Ontario (Madenjian et al. 2008). Negative interactions between Deepwater Sculpin and Rainbow Smelt have not been evaluated. However, Rainbow Smelt larval predation has negatively impacted other native Great Lakes' fishes that have spring larval stages, including Cisco *Coregonus artedi* and Lake Whitefish *C. clupeaformis* (Loftus and Hulsman 1986; Myers et al. 2009). Brandt (1986) proposed that Slimy Sculpin could have had negative effects on Deepwater Sculpin, but this mechanism was not supported by Lake Michigan observations (Madenjian et al. 2005a). Multiple lines of evidence suggest a negative interaction between Deepwater Sculpin and nonnative planktivores in the Laurentian Great Lakes.

The lack of Deepwater Sculpin in Lake Ontario in the 1980s and early 1990s caused resource management agencies to elevate the species' conservation status. No Deepwater Sculpin were caught in Lake Ontario annual bottom trawl surveys from 1978 to 1995 (total trawl tows = 6,666, mean = 373 trawls/year). These surveys employed methods to detect rare species in trawl catches and sampled at depths (7–170 m) where Deepwater Sculpin are frequently captured elsewhere (Madenjian and Bunnell 2008). The lack of Deepwater Sculpin in catches caused the Committee on the Status of Endangered Wildlife in Canada (COSEWIC) to designate the Great Lakes–Western St. Lawrence Deepwater Sculpin populations as special concern in 1987 (COSEWIC 2006). The state of New York classified the species as endangered in the mid-1980s. Further, Therres (1999) noted that the species was of conservation concern throughout the northeastern United States. These designations did not mandate any actions be taken to recover the species. In the context of Great Lakes native species restoration, Zimmerman and Krueger (2009) asked whether live transfer of Deepwater Sculpin was feasible for reintroduction into Lake Ontario; however, this action was never implemented. Binational management plans listed the species as a restoration candidate (Stewart et al. 2017), but no management actions were directed at the species (e.g., stocking, habitat restoration).

RECOVERY

Despite a lack of directed management actions to restore Lake Ontario Deepwater Sculpin, one individual was caught in 1996 during routine bottom trawling, southeast of Prince Edward County, Ontario, Canada. Subsequent targeted trawling found two other specimens in Canadian waters in 1996 and a single individual in U.S. waters in 1998 (Lantry et al. 2007). Since 2004, the species has been consistently caught in annual bottom trawl surveys (Weidel et al. 2018). Density estimates generally increased from 1996 to 2017 (Figure 1), and an exponential model estimated the rate of population increase at ~59% per year (Weidel et al. 2018) Bottom-trawl catches suggested that by 2016, Lake Ontario Deepwater Sculpin density and biomass were similar to Lake Superior (Weidel et al. 2017b), which is often described as the least anthropogenically affected Great Lake.

Throughout the recovery, bottom-trawl catches have illustrated that depth distribution of Deepwater Sculpin increased with ontogeny, and this pattern provided insights into their recovery (Figure 2). The smallest, postlarval Deepwater Sculpin (total length = 23–30 mm) caught in bottom trawls were at depths of 30–40 m. This relatively shallow depth range, where early demersal stage

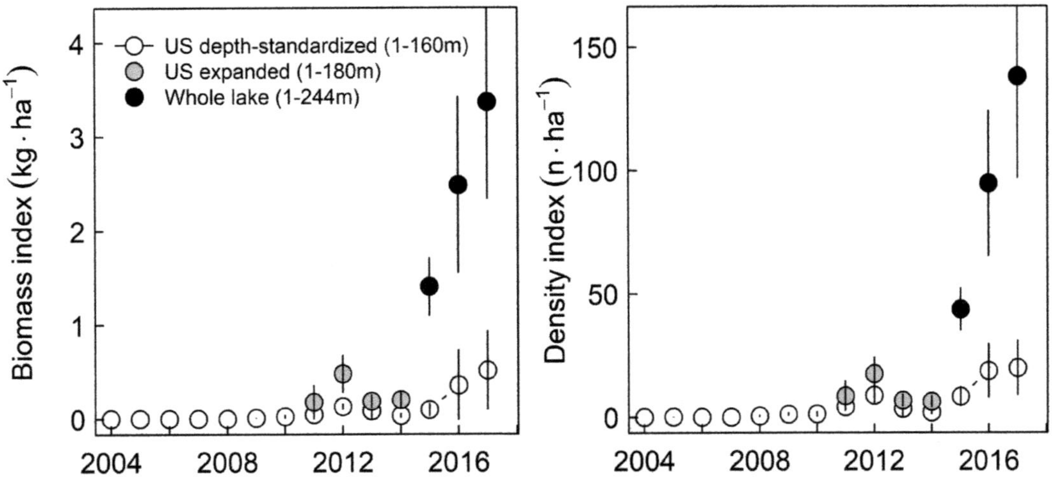

FIGURE 1. Lake Ontario Deepwater Sculpin biomass and density from bottom trawls, 2004–2017. Point estimates represent annual means and standard errors, weighted by the proportion of the lake area in each 20-m strata (Cochrane 1977). Open circles represent values from the original survey design that sampled U.S. waters within depth strata from 0 to 160 m. Gray circles denote years when the survey depths in U.S. waters were expanded to 180 m. Black circles represent estimates from whole-lake surveys (U.S. and Canadian waters) and included depths from 0 to 240 m. Bottom trawl area swept is based on trawl wing width and estimated distance towed; however, values are noted as indices because estimates are not adjusted for trawl catchability.

Deepwater Sculpin were found, is consistent with Lake Michigan observations (Geffen and Nash 1992). The consistent ontogenic change in depth distribution that we observed (Figure 2) suggests that the 6,666 bottom trawl surveys conducted 1978–1995 would likely have captured Deepwater Sculpin if they were present and reproducing in Lake Ontario over this period.

Whether the current Deepwater Sculpin population came from a remnant Lake Ontario population, from upstream Great Lakes' populations, or from a combination of both is unknown. Microsatellite loci data suggested that contemporary Lake Ontario Deepwater Sculpin were more similar to Deepwater Sculpin from the upper Great Lakes (Superior, Michigan, and Huron) than to Lake Ontario museum specimens from the 1930s and 1940s (Welsh et al. 2017). Recolonization from upstream sources could have resulted from passive drift (Roseman et al. 1998). Alternatively, pelagic larval Deepwater Sculpin could have been transported within ship ballast water from Lakes Huron, Superior, or Michigan and discharged into Lake Ontario (Roseman et al. 1998). Briski et al. (2012) found the ballast of vessels operating within the Great Lakes contained more pelagic nonindigenous species than transoceanic vessels entering the Great Lakes. Within-lake transport by vessels has been proposed as a mechanism for spreading other benthic fishes with pelagic larvae in the Great Lakes, such as Round Goby Neogobius melanostomus (Hensler and Jude 2007). Regardless of the origin of Lake Ontario Deepwater Sculpin, changes in environmental conditions that occurred in the 1970s to 1990s favored natural recovery of the species.

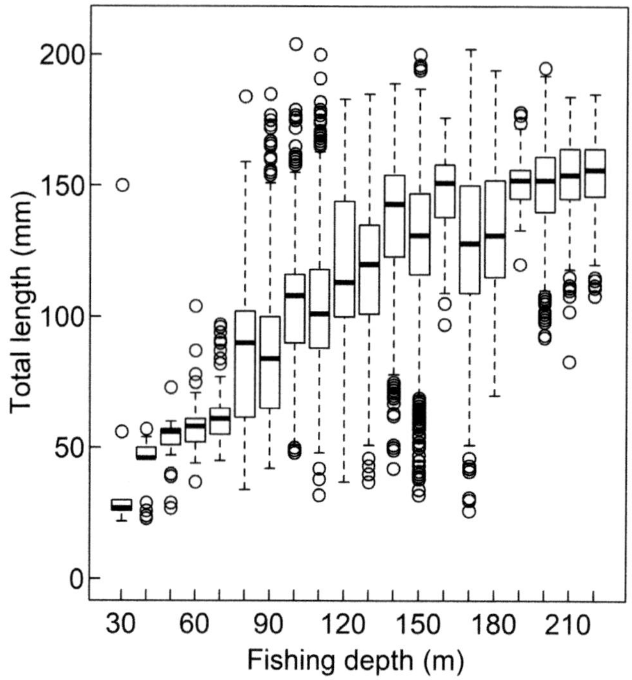

FIGURE 2. Deepwater Sculpin total lengths captured at different depths in Lake Ontario bottom trawls, 1998–2017. Lengths have been combined into 10-mm size-classes.

SYSTEM CHANGES ASSOCIATED WITH RECOVERY

Deepwater Sculpin recovery followed a series of intentional and unintentional changes in the Lake Ontario ecosystem; however, none of the intended actions were specifically directed at recovering this species. Some of the more notable changes that occurred during the 1970s through the 1990s included nutrient load reductions, piscivore stocking, nonnative planktivore declines, water clarity increases, and nonnative planktivore habitat use. Which events, or series of conditions, specifically aided Deepwater Sculpin recovery are difficult to determine because many of the changes occurred simultaneously and could have acted in concert. For instance, decreasing the lake's productivity and stocking predators both reduced nonnative planktivore abundance. Discriminating between contributing variables is also hampered by the information available from long-term monitoring efforts. Surveys that began in the 1970s were not designed to evaluate the presence or absence of Deepwater Sculpin and, therefore, did not sample the deepest lake habitats, or across all Deepwater Sculpin life stages (larval, juvenile, and adult).

Ecologically, nutrient reductions and piscivore increases both act to reduce planktivore abundance (Carpenter et al. 1985; Downing and Plante 1993). The U.S. Clean Water Act and the Great Lakes Water Quality Agreement (1972, revised 1978) changed land use and water discharge practices. These changes caused Lake Ontario phosphorus concentration to declines from ~25 to ~7 µg/L (Figure 3) from the late 1970s to the 1990s (Dove and Chapra 2015; Holeck et al. 2015). Over this time, Lake Ontario shifted from mesotrophic to oligotrophic, which was a trophic state more similar to lakes where

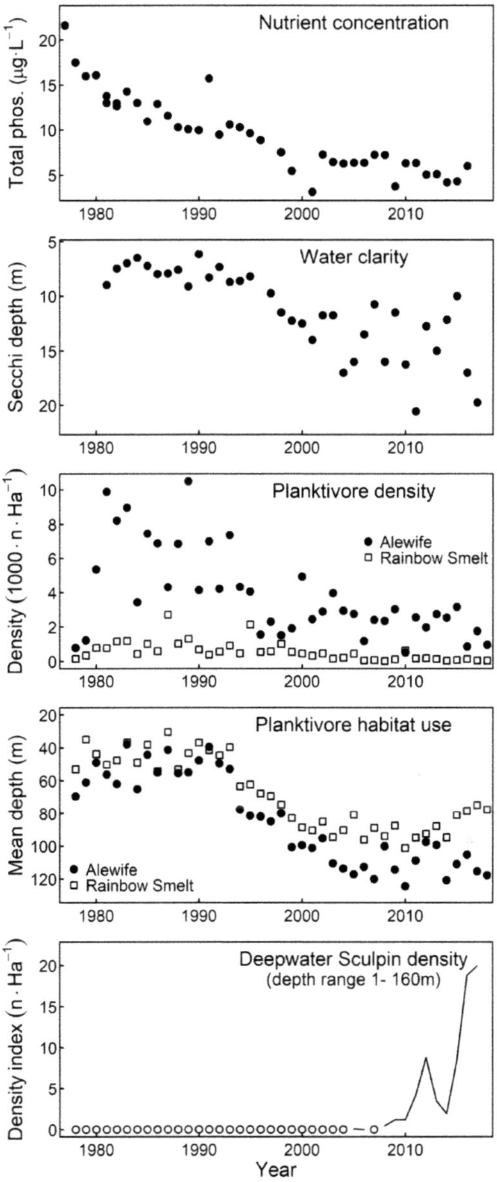

FIGURE 3. Lake Ontario total phosphorus concentration, Secchi disk depth, nonnative planktivore density and depth distributions and Deepwater Sculpin density. Total phosphorus concentrations represent mean, epilimnetic April values at sites greater than 100 m from multiple sources (Stevens and Neilson 1987; Johannsson et al. 1998; Holeck et al. 2016). Secchi disk depth data were collected in April (Weidel 2016). Planktivore density is based on April–May bottom trawl surveys in U.S. waters (Weidel et al. 2017a). Alewife density estimates from 1978 to 1996 were corrected for differences in area swept between current and historic trawls (multiplied by 1.82 based on comparison towing). No correction was used for Rainbow Smelt density. Mean depth of capture for Alewife and Rainbow Smelt are from surveys in April and May. Deepwater Sculpin density is based on U.S. waters within depth strata from 0 to 160 m. Open points in the final figure illustrate years when no Deepwater Sculpin were caught.

Deepwater Sculpin were found throughout their range (Wells and McLain 1973; Sheldon et al. 2008). Human-accelerated eutrophication threatens inland Deepwater Sculpin population by reducing dissolved oxygen in deep habitats; however, that mechanism did not likely influence Lake Ontario Deepwater Sculpin. Dissolved oxygen near the lake bottom was close to saturation, even during peak nutrient loading of the 1960s (Allen 1969). Rather, when nutrient inputs declined, food web productivity reduced and thus contributed to declines in abundance of Alewife and Rainbow Smelt (Walsh et al. 2016; Weidel et al. 2017b, Figure 3). Concurrent with declining nutrient levels, piscivore abundance increased from the late 1960s to the mid-1980s when stocking of salmon and trout peaked, with approximately 8 million salmonids annually added to the lake (Mills et al. 2003). Increased salmonid abundance in the Great Lakes has caused declines in abundance of prey species such as Alewife (Madenjian et al. 2002, 2005b) and Rainbow Smelt (Owens et al. 2003; Gorman 2007). Because nutrient declines coincided with piscivore stocking, the relative role of each variable in changing nonnative planktivore density cannot be determined. Regardless, the observed declines in Alewife and Rainbow Smelt probably created an environment more suitable for Deepwater Sculpin recovery.

Changes in nonnative planktivore habitat use also likely contributed to Deepwater Sculpin recovery. In late winter and early spring (e.g., April), Alewife and Rainbow Smelt are typically observed near bottom in deep water because inverse thermal stratification established during the winter causes the warmest water to sink to the lake bottom (Wells 1968; Bergstedt and O'Gorman 1989). In the 1970s through the early 1990s, Alewife and Smelt were most abundant in trawls from 25 to 75 m, but since 1994, the depth distributions of Rainbow Smelt and Alewife have shifted deeper (Figure 4). This shift has been attributed to water clarity increase that has recently occurred (O'Gorman et al. 2000). Water clarity increases were partly due to declining mineral nutrients but more importantly to filtering by the recently established nonnative driessenid mussels (Mills et al. 2003; Holeck et al. 2015). These habitat shifts acted to reduce planktivore biomass by more than 90% in shallow (20–50 m) habitats (Figure 4). Pelagic larval Deepwater Sculpin are believed to transition to their demersal stage in habitats <75 m (Eck and Wells 1987; Geffen and Nash 1992). The small sizes of Deepwater Sculpin caught in bottom trawls at depths of 30–40 m illustrated a similar pattern in Lake Ontario (Figure 2). Declines in planktivore populations observed in the 1980s and 1990s would have likely benefited Deepwater Sculpin recovery; however, shifts in planktivore depth distribution probably caused a substantial decline in predation by planktivores within critical larval Deepwater Sculpin habitats. Planktivore depth distribution shifts, caused by dreissenid mussels and increased water clarity, likely improved larval Deepwater Sculpin survival and also contributed to the species' recovery (Geffen and Nash 1992; Owens et al. 2003; Weidel et al. 2017b).

SURVEY ADAPTATION, COLLABORATION, AND COMPARISONS

Long-term surveys are critical for illustrating human impact on fish communities, but information needs from long-term surveys can change over time. The original benthic fish trawl survey design provided limited observations for understanding Lake Ontario Deepwater Sculpin dynamics. Initiated in 1978, surveys were primarily designed to assess the most abundant benthic prey fish of the time, Slimy Sculpin, which was critical prey for native Lake Trout (Elrod and

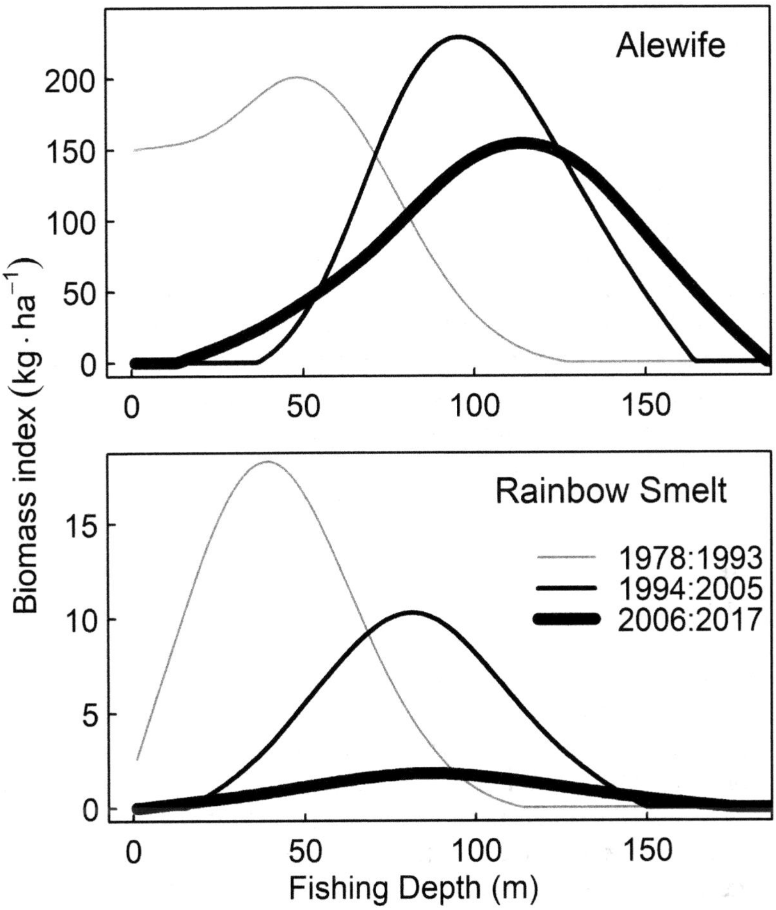

FIGURE 4. Lake Ontario Alewife and Rainbow Smelt April habitat use has shifted deeper over time. Depth distribution are illustrated using density predictions from generalized additive models with fishing depth as a smoothed predictor. Bottom trawl data collected in April represents late-winter and early-spring distributions (Wells 1968). Number of bottom trawls for each period are 1,772, 1,135, and 1,632, respectively. The habitat shift has been attributed to increased water clarity from dreissenid mussel introduction and reduced nutrient inputs (O'Gorman et al. 2000). The timing of the habitat use change coincides with the reappearance of Deepwater Sculpin.

O'Gorman 1991; Elrod et al. 1995). That survey produced precise relative abundance estimates but sampled only in the U.S. waters of the main lake at depths from 8 to 150 m. From 1990 to 2010, the Lake Ontario benthic community changed substantially due to colonization by dreissenids mussels and Round Goby and population increases of Deepwater Sculpin. These changes resulted in decision makers of fishery management needing more spatially distributed information on the entire benthic fish community than the historic survey could provide. The survey was redesigned to meet current information needs using a collaborative, adaptive approach with improvements guided by independent trawl survey critiques (MacNeill et al. 2005). By coordinating vessels and staff from the New

York Department of Environmental Conservation and the Ontario Ministry of Natural Resources and Forestry, the survey was expanded to Canadian waters and to all lake depths. Changes were not readily accepted by all agency personnel. Those who had operated the surveys in the same manner for decades were reluctant to alter the long-term procedures. To make the necessary changes, we found that it was critical to carefully illustrate and communicate the need for changes to the information users (managers, policymakers), as well as to vessel staff and others who supported the data gathering process. Our current understanding of Lake Ontario Deepwater Sculpin lakewide density and biomass would differ greatly if the depth-limited information (0–160 m) were used instead of the current survey that samples all depth strata within the lake (Figure 1). As resource management adopts more ecosystem-based perspectives, long-term surveys focused on single-species or on a limited habitat range may need to be adapted and broadened.

Comparing Lake Ontario Deepwater Sculpin abundance metrics to other Great Lakes ecosystems was critical to providing context for the extent of the sculpin recovery. Historic reference points for Lake Ontario Deepwater Sculpin density were lacking because the species disappeared before annual fishery-independent surveys began. Recent Lake Ontario biomass estimates ranged between 2 and 2.5 kg/ha while Lake Superior have been 1–3 kg/ha (Table 1). Lake Huron and Lake Michigan reported densities generally lower (<1. 0–1.0 kg/ha) than reported for Lake Superior and Lake Ontario; however, sampling in those lakes generally does not occur deeper than 110 m (Table 1). In Lake Ontario, Deepwater Sculpin densities were highest at depths greater than 110 m, suggesting that Lake Huron and Lake Michigan estimates may potentially be biased low. Despite survey differences, similarly observed biomasses in Lake Ontario and the other Great Lakes suggests that the Lake Ontario Deepwater Sculpin has recovered (Weidel et al. 2017b).

The impact of Deepwater Sculpin recovery on the Lake Ontario food web may depend on the extent to which they are consumed by piscivores. Observations from Lake Superior and Lake Michigan suggest that Lake Trout and Burbot in Lake Ontario may eventually consume Deepwater Sculpin (Madenjian et al. 2005a; Gamble et al. 2011). Lake Ontario Lake Trout primarily consume Alewife, with Round Goby and Rainbow Smelt comprising smaller proportions of their diet (Brandt 1986; Dietrich et al. 2006; Happel et al. 2017). High thiaminase levels in these species, such as Alewife and Rainbow Smelt (Honeyfield et al. 2012), have been suggested to lower Lake Trout reproductive success and inhibit that species' restoration (Fitzsimons et

TABLE 1. Deepwater Sculpin biomass from bottom trawl surveys in the Laurentian Great Lakes.

Lake	Biomass (kg/ha)	Year range	Depths sampled (m)	Maximium lake depth (m)	Source
Superior	1–3	2011–2016	90–315	406	(Vinson et al. 2017)
Michigan	<1–1	2010–2016	9–110	281	(Bunnell et al. 2017)
Huron	<1–1	2010–2016	9–110	229	(Riley et al. 2016)
Erie	0		Not reported	64	(Roseman et al. 1998)
Ontario	2.0–2.5	2015–2017	6–225	244	(Weidel et al. 2017b)

al. 2007). If Deepwater Sculpin become more common in Lake Trout diets, this dietary change could potentially improve reproduction of this native Lake Ontario piscivore.

LESSONS LEARNED

Recovery Is Possible without Targeted Actions

In Lake Ontario, native Deepwater Sculpin were common in the late-1800s, but by the mid-1900s, the species was possibly extirpated. No Deepwater Sculpin were captured from 1978 to 1995 and the species conservation status was elevated; however, no restoration actions were initiated. During that period, piscivore stocking and the Clean Water Act both served to reduce populations of nonnative Alewife and Rainbow Smelt, which prey on larval fishes including Deepwater Sculpin. Additionally, nonnative, filter-feeding, dreissenid mussels substantially increased Lake Ontario water clarity in the 1990s. Increased water clarity and light penetration caused Alewife and Rainbow Smelt habitat use to shift deeper, away from larval Deepwater Sculpin habitats. In 1996, trawls caught three Deepwater Sculpin, and since then, their abundance has increased to where they are one of the most abundant Lake Ontario prey fishes captured. The Deepwater Sculpin in Lake Ontario is an example of a population recovering after environmental and food web conditions changed but where no management actions were targeted at restoring the species.

Understanding Recovery Mechanisms Requires More Diverse Data

Long-term trawl surveys tracked the Lake Ontario Deepwater Sculpin recovery but were inadequate for understanding why and how the population recovered. It was difficult to know whether the species had been extirpated because the lake's deepest habitats, where the species should have been most common, were not historically sampled. Similarly, we had no information on potential sources of larval Deepwater Sculpin. Quantifying larval drift from upstream sources, as done in Lake Erie (Pritt et al. 2014), or from ship's ballast waters (Briski et al. 2012) would have provided evidence for the source of the recovering population. Long-term survey data are critical for measuring fish community change, but their value may be best realized when complimented with studies that identify the mechanisms contributing to those changes.

Survey Critique and Adaptation Increase Learning

Information from long-term surveys can change over time, but the fear of data incomparability often results in reluctance to alter survey designs. In Lake Ontario, a bottom trawl survey in a portion of U.S. waters was not providing the lakewide information on a variety of benthic prey fishes needed by management decision makers. The survey was redesigned to provide data that were comparable to historic values but also provide new observations. To achieve those survey changes, we found it was critical to use independent survey critique recommendations (MacNeill et al. 2005) and communicate the need for changes to the information users (managers, policymakers), as well as to vessel staff and others who supported the data gathering process.

Collaboration Works

Having new agencies and scientists contribute to the Lake Ontario benthic prey fish survey allowed the survey to be expanded lakewide but, more importantly, diversified ideas and promoted information sharing. With multiple agencies contributing, survey results were not viewed as being proprietary information of an individual scientist or agency. The effective changes realized provided an example of the value of a collaborative, adaptive, and data-sharing approach to long-term fishery research. Students specifically trained in using a collaborative and diverse research approach are most likely to further our understanding of Great Lakes fish, fisheries, and ecosystem ecology.

Cross-Lake Comparisons Provide Context to Gauge Recovery

To quantify the extent of the Lake Ontario recovery, biomass was compared to populations in less-impacted lakes (e.g., Lake Superior) because we lacked historic reference points. Reporting Lake Ontario Deepwater Sculpin catches in area-standardized units (density, biomass) quantified at a lake scale facilitated comparisons with Lake Superior. Similar observations of Deepwater Sculpin biomass in the deepest habitats of Lakes Huron, Michigan, and Erie would further improve our understanding of this unique species and its habitat across the Great Lakes basin.

ACKNOWLEDGMENTS

Data from this chapter are publicly available using this digital object identifier: 10.5066/F74M92QQ or at www.sciencebase.gov/catalog/item/57ed1b5de4b0908250 11d3db. We thank the vessel crews, support staff, and biological staff of the New York State Department of Environmental Conservation, the Ontario Ministry of Natural Resources and Forestry (OMNRF), and the U.S. Geological Survey (USGS) research vessels. Special thanks to T. Bower, S. Furgal, T. Strang, T. Lewchanin, E. Taber, I. Harding, M. Cooper, P. Schulze, R. Haehn, J. Palumbo, A. Gatch, N. Balk, and S. Minihkeim for their tireless effort supporting all aspects of this research. Funding for this research was from the USGS Ecosystems, Fisheries and Status and Trends programs (ND00GAP), the Federal Aid in Sportfish Restoration (FA-5-R, F-48-R, F-55-R, F-61-R, F-62-R, F-62-R-1), as well as Ontario provincial funding to implement OMNRF priorities under the Canada-Ontario Agreement on Great Lakes Water Quality and Ecosystem Health and Ontario's Great Lakes Strategy. Any use of trade, firm, or product names is for descriptive purposes only and does not imply endorsement by the U.S. Government. All Great Lakes Science Center sampling and handling of fish during research are carried out in accordance with guidelines for the care and use of fishes by the American Fisheries Society (http://fisheries.org/docs/wp/Guidelines-for-Use-of-Fishes.pdf).

REFERENCES

Allen, H. F. 1969. Chemical characteristics of Lake Ontario. Pages 1–18 *in* Limnological survey of Lake Ontario, 1964. Great Lakes Fishery Commission, Ann Arbor, Michigan.

Bergstedt, R. A., and R. O'Gorman. 1989. Distribution of alewives in southeastern Lake Ontario in autumn and winter: a clue to winter mortalities. Transactions of the American Fisheries Society 118:687–692.

Brandt, S. B. 1986. Disappearance of the Deepwater Sculpin (*Myoxocephalus thompsoni*) from Lake Ontario: the keystone predator hypothesis. Journal of Great Lakes Research 12:18–24.

Briski, E., C. J. Wiley, and S. A. Bailey. 2012. Role of domestic shipping in the introduction or secondary spread of nonindigenous species: biological invasions within the Laurentian Great Lakes. Journal of Applied Ecology 49:1124–1130.

Bronte, C. R., M. P. Ebener, D. R. Schreiner, D. S. DeVault, M. M. Petzold, D. A. Jensen, C. Richards, and S. J. Lozano. 2003. Fish community change in Lake Superior, 1970–2000. Canadian Journal of Fisheries and Aquatic Sciences 60:1552–1574.

Carpenter, S. R., J. F. Kitchell, and J. R. Hodgson. 1985. Cascading trophic interactions and lake productivity. BioScience 35:634–639.

Christie, W. J. 1973. A review of the changes in the fish species composition of Lake Ontario. Great Lakes Fishery Commission, Technical Report 23, Ann Arbor, Michigan.

Christie, W. J., and N. A. Thomas. 1981. Biology and chemistry. Pages 325–340 *in* E. J. Aubert and T. L. Richards, editors. IFYGL—the international field year for the Great Lakes. National Oceanic and Atmospheric Administration, Great Lakes Environmental Research Laboratory, Ann Arbor, Michigan.

Cochrane, W. G. 1977. Sampling techniques, 3rd edition. Wiley, New York.

COSEWIC (Committee on the Status of Endangered Wildlife in Canada. 2006. COSEWIC assessment and update status report on the deepwater sculpin *Myoxocephalus thompsonii* (Western and Great Lakes-Western St. Lawrence populations) in Canada. COSEWIC, Ottawa.

Crossman, E. J., and H. D. Van Meter. 1979. Annotated list of the fishes of the Lake Ontario watershed. Great Lakes Fishery Commission, Ann Arbor, Michigan.

Crowder, L. B. 1980. Alewife, Rainbow Smelt and native fishes in Lake Michigan: competition or predation? Environmental Biology of Fishes 5:225–233.

Dietrich, J. P., B. J. Morrison, and J. A. Hoyle. 2006. Alternative ecological pathways in the eastern Lake Ontario food web: Round Goby in the diet of Lake Trout. Journal of Great Lakes Research 32:395–400.

Dove, A., and S. C. Chapra. 2015. Long-term trends of nutrients and trophic response variables for the Great Lakes: Great Lakes nutrient trends. Limnology and Oceanography 60:696–721.

Downing, J. A., and C. Plante. 1993. Production of fish populations in lakes. Canadian Journal of Fisheries and Aquatic Sciences 50:110–120.

Dymond, J. R. 1928. Some factors effecting the production of Lake Trout (*Cristivomer namaycush*) in Lake Ontario. Publications of the Ontario Fisheries Research Laboratory 33, Toronto.

Dymond, J. R., J. L. Hart, and A. L. Pritchard. 1929. The fishes of the Canadian waters of Lake Ontario. Publications of the Ontario Fisheries Research Laboratory 37, Toronto.

Eck, G. W., and L. Wells. 1987. Recent changes in Lake Michigan's fish community and their probable causes, with emphasis on the role of the Alewife (*Alosa pseudoharengus*). Canadian Journal of Fisheries and Aquatic Sciences 44:53–60.

Elrod, J. H., and R. O'Gorman. 1991. Diet of juvenile Lake Trout in southern Lake Ontario in relation to abundance and size of prey fishes, 1979–1987. Transactions of the American Fisheries Society 120:290–302.

Elrod, J. H., R. O'Gorman, C. P. Schneider, T. H. Eckert, T. Schaner, J. N. Bowlby, and L. P. Schleen. 1995. Lake Trout rehabilitation in Lake Ontario. Journal of Great Lakes Research 21:83–107.

Estepp, L. R., and E. D. Reavie. 2015. The ecological history of Lake Ontario according to phytoplankton. Journal of Great Lakes Research 41:669–687.

Fitzsimons, J. D., B. Williston, G. Williston, L. Brown, A. El-Shaarawi, L. Vandenbyllaardt, D. Honeyfeld, D. Tillitt, M. Wolgamood, and S. B. Brown. 2007. Egg thiamine status of Lake Ontario salmonines 1995–2004 with emphasis on Lake Trout. Journal of Great Lakes Research 33:93–103.

Fratt, T. W., D. W. Coble, F. Copes, and R. E. Brusewitz. 1997. Diet of Burbot in Green Bay and western Lake Michigan with comparison to other waters. Journal of Great Lakes Research 23:1–10.

Gamble, A. E., T. R. Hrabik, J. D. Stockwell, and D. L. Yule. 2011. Trophic connections in Lake Superior part I: the offshore fish community. Journal of Great Lakes Research 37:541–549.

Geffen, A. J., and R. D. M. Nash. 1992. The life-history strategy of Deepwater Sculpin, *Myoxocephalus thompsoni* (Girard), in Lake Michigan: dispersal and settlement patterns during the first year of life. Journal of Fish Biology 41:101–110.

Gorman, O. T. 2007. Changes in a population of exotic Rainbow Smelt in Lake Superior:

boom to bust, 1974–2005. Journal of Great Lakes Research 33:75–90.

Happel, A., R. Pattridge, M. Walsh, and J. Rinchard. 2017. Assessing diet compositions of Lake Ontario predators using fatty acid profiles of prey fishes. Journal of Great Lakes Research 43:838–845.

Hensler, S. R., and D. J. Jude. 2007. Diel vertical migration of Round Goby larvae in the Great Lakes. Journal of Great Lakes Research 33:295–302.

Holeck, K. T., L. G. Rudstam, J. M. Watkins, C. Hotaling, R. D. McCullough, D. Lemon, W. Pearsall, J. Lantry, M. J. Connerton, S. LaPan, Zy Biesinger, B. F. Lantry, M. Walsh, and B. C. Weidel. 2016. 2015 status of the Lake Ontario lower trophic levels. New York State Department of Environmental Conservation, Annual Report of Lake Ontario, Albany.

Holeck, K. T., L. G. Rudstam, J. M. Watkins, F. J. Luckey, J. R. Lantry, B. F. Lantry, E. S. Trometer, M. A. Koops, and T. B. Johnson. 2015. Lake Ontario water quality during the 2003 and 2008 intensive field years and comparison with long-term trends. Aquatic Ecosystem Health and Management 18:7–17.

Honeyfield, D. C., M. E. Daniels, L. R. Brown, M. T. Arts, M. G. Walsh, and S. B. Brown. 2012. Survey of four essential nutrients and thiaminase activity in five Lake Ontario prey fish species. Journal of Great Lakes Research 38:11–17.

Johannsson, O. E., E. S. Millard, K. Ralph, D. D. Myles, D. M. Graham, W. D. Taylor, B. Giles, and R. E. Allen. 1998. The changing pelagia of Lake Ontario (1981 to 1995) a report of the DFO long-term biomonitoring (Bioindex) program. Canadian Technical Report of Fisheries and Aquatic Sciences 2243.

Kraft, C. E., and J. F. Kitchell. 1986. Partitioning of food resources by sculpins in Lake Michigan. Environmental Biology of Fishes 16:309–316.

Lantry, B. F., R. O'Gorman, M. G. Walsh, J. M. Casselman, J. A. Hoyle, M. J. Keir, and J. R. Lantry. 2007. Reappearance of Deepwater Sculpin in Lake Ontario: resurgence or last gasp of a doomed population? Journal of Great Lakes Research 33(Supplement 1):34–45.

Loftus, D. H., and P. F. Hulsman. 1986. Predation on larval Lake Whitefish (*Coregonus clupeaformis*) and Lake Herring (*C. artedii*) by adult Rainbow Smelt (*Osmerus mordax*). Canadian Journal of Fisheries and Aquatic Sciences 43:812–818.

Londer, J. G. 2011. Diet overlap and food habits of Slimy Sculpin, Deepwater Sculpin, and Round Goby during winter and spring in offshore Lake Michigan. Master's thesis. University of Michigan, Ann Arbor.

MacNeill, D. B., J. S. Ault, S. Smith, and S. Murawski. 2005. A technical review of the Lake Ontario forage base assessment program. New York Sea Grant, YSGI-T-05-001, Ithaca.

Madenjian, C. P., and D. B. Bunnell. 2008. Depth distribution dynamics of the sculpin community in Lake Michigan. Transactions of the American Fisheries Society 137:1346–1357.

Madenjian, C. P., T. J. Desorcie, and R. M. Stedman. 1998. Ontogenic and spatial patterns in diet and growth of Lake Trout in Lake Michigan. Transactions of the American Fisheries Society 127:236–252.

Madenjian, C. P., G. L. Fahnenstiel, T. H. Johengen, T. F. Nalepa, H. A. Vanderploeg, G. W. Fleischer, P. J. Schneeberger, D. M. Benjamin, E. B. Smith, J. R. Bence, E. S. Rutherford, D. S. Lavis, D. M. Robertson, D. J. Jude, and M. P. Ebener. 2002. Dynamics of the Lake Michigan food web, 1970–2000. Canadian Journal of Fisheries and Aquatic Sciences 59:736–753.

Madenjian, C. P., D. W. Hondorp, T. J. Desorcie, and J. D. Holuszko. 2005a. Sculpin community dynamics in Lake Michigan. Journal of Great Lakes Research 31:267–276.

Madenjian, C. P., T. O. Höök, E. S. Rutherford, D. M. Mason, T. E. Croley, E. B. Szalai, and J. R. Bence. 2005b. Recruitment variability of alewives in Lake Michigan. Transactions of the American Fisheries Society 134:218–230.

Madenjian, C. P., R. O'Gorman, D. B. Bunnell, R. L. Argyle, E. F. Roseman, D. M. Warner, J. D. Stockwell, and M. A. Stapanian. 2008. Adverse effects of Alewives on Laurentian Great Lakes fish communities. North American Journal of Fisheries Management 28:263–282.

Mansfield, P. J., D. J. Jude, D. T. Michaud, D. C. Brazo, and J. Gulvas. 1983. Distribution and abundance of larval Burbot and Deepwater Sculpin in Lake Michigan. Transactions of the American Fisheries Society 112:162–172.

Mills, E. L., J. M. Casselman, R. Dermott, J. D. Fitzsimons, G. Gal, K. T. Holeck, J. A. Hoyle, O. E. Johannsson, B. F. Lantry, J. C. Makarewicz, E. S. Millard, I. F. Munawar, M. Munawar, R. O'Gorman, R. W. Owens, L. G. Rudstam, T. Schaner, and T. J. Stewart. 2003. Lake Ontario: food web dynamics in a changing ecosystem (1970–2000). Canadian Journal of Fisheries and Aquatic Sciences 60:471–490.

Myers, J. T., M. L. Jones, J. D. Stockwell, and D. L. Yule. 2009. Reassessment of the predatory effects of Rainbow Smelt on Ciscoes in Lake Superior. Transactions of the American Fisheries Society 138:1352–1368.

O'Brien, T. P., E. F. Roseman, C. S. Kiley, and J. S. Schaeffer. 2009. Fall diet and bathymetric distribution of Deepwater Sculpin (*Myoxocephalus thompsonii*) in Lake Huron. Journal of Great Lakes Research 35:464–472.

O'Gorman, R., J. H. Elrod, R. W. Owens, C. P. Schneider, T. H. Eckert, and B. F. Lantry. 2000. Shifts in depth distributions of Alewives, Rainbow Smelt, and age-2 Lake Trout in southern Lake Ontario following establishment of dreissenids. Transactions of the American Fisheries Society 129:1096–1106.

O'Gorman, R., and C. P. Schneider. 1986. Dynamics of Alewives in Lake Ontario following a mass mortality. Transactions of the American Fisheries Society 115:1–14.

Owens, R. W., R. O'Gorman, T. H. Eckert, and B. F. Lantry. 2003. The offshore fish community in southern Lake Ontario 1972–1998. Pages 407–441 *in* M. Munawar, editor. State of Lake Ontario: past, present, and future. Backhuys Publishers, Leiden, Netherlands.

Pritt, J. J., M. R. DuFour, C. M. Mayer, E. F. Roseman, and R. L. DeBruyne. 2014. Sampling little fish in big rivers: larval fish detection probabilities in two Lake Erie tributaries and implications for sampling effort and abundance indices. Transactions of the American Fisheries Society 143:1011–1027.

Riley, S. C., and J. V. Adams. 2010. Long-term trends in habitat use of offshore demersal fishes in western Lake Huron suggest large-scale ecosystem change. Transactions of the American Fisheries Society 139:1322–1334.

Riley, S. C., E. F. Roseman, D. W. Hondorp, M. A. Chriscinske, and M. Smith. 2016. Status and trends of the Lake Huron offshore demersal fish community, 1976–2015. Pages 51–63 *in* Compiled reports to the Great Lakes Fishery Commission of the annual USGS bottom trawl and acoustics surveys. U.S. Geological Survey, Great Lakes Science Center, Ann Arbor, Michigan.

Roseman, E. F., D. J. Jude, M. K. Raths, T. G. Coon, and W. W. Taylor. 1998. Occurrence of the Deepwater Sculpin (*Myoxocephalus thompsoni*) in western Lake Erie. Journal of Great Lakes Research 24:479–483.

Schelske, C. L. 1991. Historical nutrient enrichment of Lake Ontario: paleolimnological evidence. Canadian Journal of Fisheries and Aquatic Sciences 48:1529–1538.

Scott, W. B., and E. J. Crossman. 1973. Deepwater Sculpin. Pages 842–848 *in* Freshwater fishes of Canada. Fisheries Research Board of Canada Bulletin 184.

Sheldon, T. A., N. E. Mandrak, and N. R. Lovejoy. 2008. Biogeography of the Deepwater Sculpin (*Myoxocephalus thompsonii*) a neararctic glacial relict. Canadian Journal of Zoology 86:108–115.

Sly, P. G. 1991. The effects of land use and cultural development on the Lake Ontario ecosystem since 1750. Hydrobiologia 213:1–75.

Smith, S. H. 1970. Species Interactions of the Alewife in the Great Lakes. Transactions of the American Fisheries Society 99:754–765.

Stevens, R. J. J., and M. A. Neilson. 1987. Response of Lake Ontario to reductions in phosphorus load, 1967–82. Canadian Journal of Fisheries and Aquatic Sciences 44:2059–2068.

Stewart, T. J., A. Todd, and S. Lapan. 2017. Fish community objectives for Lake Ontario. Great Lakes Fisheries Commission Special Publication, Ann Arbor, Michigan.

Stone, U. B. 1947. A study of the Deep-water Cisco fishery of Lake Ontario with particular reference to the Bloater *Leucichthys hoyi* (Gill). Transactions of the American Fisheries Society 74:230–249.

Therres, G. D. 1999. Wildlife species of regional conservation concern in the northeastern United States. Northeast Wildlife 54:93–100.

Vinson, M. R., L. M. Evrard, O. T. Gorman, and D. L. Yule. 2017. Status and trends in the Lake Superior Fish Community, 2016. Pages 13–24 *in* Compiled reports to the Great Lakes Fishery Commission of the annual USGS bottom trawl and acoustics surveys. U.S. Geological Survey, Great Lakes Fishery Commission, Ann Arbor, Michigan.

Walsh, M. G., B. C. Weidel, M. J. Connerton, and J. P. Holden. 2016. Status of Alewife and Rainbow Smelt in the U.S. waters of Lake Ontario, 2015. New York State Department of Environmental Conservation, Lake Ontario Annual Report, Albany.

Weidel, B. C. 2016. Lake Ontario Deepwater Sculpin bottom trawl catch and biological data, 1996–2016. U.S. Geological Survey data release, doi:10.5066/F74M92QQ.

Weidel, B. C., M. J. Connerton, and J. P. Holden. 2018. Bottom trawl assessment of Lake Ontario prey fishes. Section 12 *in* New York State Department of Conservation Lake Ontario Annual Report 2017. New York State Department of Conservation, Albany.

Weidel, B. C., M. G. Walsh, M. J. Connerton, and J. P. Holden. 2017a. Trawl-based assessment of Lake Ontario pelagic prey fishes including Alewife and Rainbow Smelt. 2016 Annual Report, Bureau of Fisheries Lake Ontario Unit and St. Lawrence River Unit to the Great Lakes Fishery Commission's Lake Ontario Committee, Ann Arbor, MI.

Weidel, B. C., M. G. Walsh, M. J. Connerton, B. F. Lantry, J. R. Lantry, J. P. Holden, M. J. Yuille, and J. A. Hoyle. 2017b. Deepwater Sculpin status and recovery in Lake Ontario. Journal of Great Lakes Research 43:854–862.

Wells, L. 1968. Seasonal depth distribution of fish in southeastern Lake Michigan. Fishery Bulletin 67:1–15.

Wells, L. 1969. Fishery survey of U. S. waters of Lake Ontario. Pages 51–59 *in* Limnological survey of Lake Ontario, 1964. Great Lakes Fishery Commission, Ann Arbor, Michigan.

Wells, L., and A. L. McLain. 1973. Lake Michigan: man's effects on native fish stocks and other biota. Great Lakes Fishery Commission, Technical Report 20, Ann Arbor, Michigan.

Welsh, A. B., K. Scribner, W. Stott, and M. G. Walsh. 2017. A population on the rise: the origin of deepwater sculpin in Lake Ontario. Journal of Great Lakes Research 43:863–870.

Zimmerman, M. S., and C. C. Krueger. 2009. An ecosystem perspective on re-establishing native deepwater fishes in the Laurentian Great Lakes. North American Journal of Fisheries Management 29:1352–1371.

Section IV

Ocean and Sea

Caspian Kutum *Rutilus kutum*: A Story of Exploitation, Survival, and Revival

Nukhkadi I. Rabazanov
Dagestan State University
43a, Gadzhiyev Street, Makhachkala 367000, Russia
and
Caspian Institute of Biological Resources
Dagestan Scientific Center of the Russian Academy of Sciences
45, Gadzhiyev Street, Makhachkala 367000, Russia

Alexei M. Orlov*
Russian Federal Research Institute of Fisheries and Oceanography
17, Verkhnyaya Krasnoselskaya Street, Moscow 107140, Russia
and
A.N. Severtsov Institute of Ecology and Evolution, Russian Academy of Sciences
33, Leninsky Prospekt, Moscow 119071, Russia
and
Dagestan State University
43a, Gadzhiyev Street, Makhachkala 367000, Russia
and
Caspian Institute of Biological Resources
Dagestan Scientific Center of the Russian Academy of Sciences
45, Gadzhiyev Street, Makhachkala 367000, Russia
and
Tomsk State University
36, Lenina Street, Tomsk, 634050, Russia

Akhma S. Abdusamadov
Caspian Fisheries Research Institute, Dagestan Branch
104, Abubakarov Street, Makhachkala 367022, Russia

Ruslan M. Barkhalov
Dagestan State University
43a, Gadzhiyev Street, Makhachkala, 367000, Russia
and
Caspian Institute of Biological Resource
Dagestan Scientific Center of the Russian Academy of Sciences 45
Gadzhiyev Street, Makhachkala 367000, Russia
and
Dagestan State Nature Reserve
120, Gagarina Street, Makhachkala, 367010, Russia

* Corresponding author: orlov@vniro.ru

Kurban M. Akhemedkhanov
Dagestan State University
43a, Gadzhiyev Street, Makhachkala, 367000, Russia

Abstract.—This chapter provides the history of the Caspian Kutum *Rutilus kutum* (Kamensky 1901) fishery in the Caspian Sea, analyzes long-term changes of stock condition and the main causes of fluctuations in abundance, and describes conservation measures that allowed resumption of fishing. Caspian Kutum (Cyprinidae family) is an endemic, semi-anadromous, medium-sized fish, reaching 53–67 cm in total length (rarely 71 cm) and weighing up to 4.0 kg (rarely 5.0 kg). Commercially important fisheries occur in Russia, Azerbaijan, Iran, and Turkmenistan. Flesh and roe are enjoyed as food and have a high price in markets. Variability in sea level, construction of hydroelectric power plants on rivers, water irrigation withdrawals, industrial and domestic pollution, overfishing, and illegal fishing resulted in a sharp decline of Caspian Kutum abundance and resulted in a total ban on harvest in Russia between 1995 and 2004. In Iran, fishing for Caspian Kutum continued due to their stocking program. Conservation measures for Caspian Kutum stocks (e.g., listing in federal and local Red Data books, fishing ban, fight against illegal fishing), as well as an increase of artificial propagation in Iran, Azerbaijan, and Dagestan (Russia) during subsequent years, have allowed the recovery of stocks in Russian waters to 1990s levels as well as the resumption of fishing. The follow lessons may be applicable to fishery management programs elsewhere:

- Control of fishing, legal and illegal, is required if sustained harvests are to be realized.
- Fishing should be carried out within the limits of scientifically based total allowable catches and in accordance with fishing regulations—for example, about specific fishing gear, mesh/hook size, prohibited periods, and minimum fishing length of caught fish.
- Artificial propagation of fish and their release can have a significant impact on fisheries throughout species' range and should be considered as a possible tool for restoration if appropriate management measures, conservation of habitat, and spawning areas are in force.
- Continuous monitoring of stock status is needed that allows an annual assessment of total allowable catches and monitors changes in exploited populations.
- Success in the conservation of transboundary fish stocks can be achieved through coordinated actions to protect stocks by all countries that exploit them.

INTRODUCTION

The Caspian Sea is located just west of the Black Sea and the Mediterranean Sea between Europe and Asia (Figure 1). The Caspian Sea is bordered by Kazakhstan in the northeast, Russia in the northwest, Azerbaijan in the west, Iran in the south, and Tukemenistan in the southeast. The Caspian Sea basin has no outflows and is the largest enclosed inland body of water on Earth. The primary tributaries to the Caspian Sea are the Volga River,

FIGURE 1. Caspian Sea, tributaries, and surrounding countries (http://eurodialogue.org/Caspian-Sea-Map).

Ural River, Kura River, and Terek River. Its average salinity is about 12.9‰, being about one-third that of ocean water, and has an average depth of 211 m with a maximum depth of 1,025 m.

The Caspian Kutum *Rutilus kutum* (Kamensky 1901) is one of the most economically important commercial fish species of the southern Caspian Sea and is an important source of food for the local population (Figure 2; Jafari et al. 2011; Abdolhay et al. 2012; Karimpour et al. 2013; Khodorevskaya et al. 2014; Safari 2016). A high demand exists among the local residents of the region due to the high nutritional value and delicious taste of Caspian Kutum (Shikhshabekov et al. 2006; Kavan et al. 2009). As a result, Caspian Kutum has supported commercial fisheries in the Caspian Sea waters of Russia, Azerbaijan, Iran, and Turkmenistan since the early 20th century.

Caspian Kutum is a species endemic to the Caspian Sea. They can reach a length of 67 cm (rarely 71 cm) and a weight of 4 kg (rarely 5 kg), although the average lengths in commercial catches are typically 40–49 cm and average weights are 1.0–1.7 kg (Trushinskaya 1975; Belyaeva et al. 1989; Vasil'eva and Luzhnyak 2013). Adult fish feed on benthic organisms, mainly mollusks and other aquatic invertebrates (insect larvae, crustaceans). Maturation typically occurs after 4 years and fish can live as long as 12 years. Migration from sea to freshwater (rivers) for spawning begins when water temperatures reach 9–10°C, around mid- to late February. The peak of the spawning run is observed from mid-March to the beginning of April, when water temperatures reach 12.3–12.6°C. Iteroparity is characteristic for this species and spawning occurs in reeds and water lilies in the river outflows at a temperature of 13–15°C from late March to mid-April. Caspian Kutum fecundity depends on size and varies from 30,000 to 300,000 eggs per female. Beginning in mid-May, fry (larvae) begin their gradual migration to the sea, where they will grow to maturity in 3 to 4 years and subsequently spawn.

FIGURE 2. Catch of Caspian Kutum in Dagestan (Russia).

The Caspian Kutum fishery on the coast of Dagestan in Russia has traditionally been conducted as coastal fishing in the spring season, from March to the end of April, and in the autumn season, from September 1 to December 31. In general, commercial fishing is carried out using stationary nets (stretched mesh size from 55 to 70 mm) off the Sulak River and the coast of Derbent. Commercial fishing occurs mainly in the sea (Voskoboinikova 2006) using motor boats. A large number of fish are caught by illegal fishers and amateur fishermen. Caspian Kutum is caught illegally in closed seasons using illegal gear, in prohibited areas, undersized, or in numbers that exceed allowable amount. The fishing for Caspian Kutum by recreational or sport (amateur) fishermen is carried out almost throughout the year (except for short periods of prohibition) with the coverage of migration routes, wintering, spawning and feeding grounds in freshwater bodies and marine coastal waters. In Iran, Caspian Kutum, among the bony fishes, is the most important commercial target of coastal fishing in the Caspian Sea (Karimpour et al. 2013). Fishing is carried out mainly by beach seines with a stretched mesh of 35 mm (Strukova et al. 2016).

Caspian Kutum have a high nutritional value with a caloric content of 80.5 kcal per 100 g of meat, a fat content of 1.6–5.8%, and 18% protein. Half of the fish can be used for consumption (edible parts are 47–54% of the fish; Shikhshabekov et al. 2006). Preparation of Caspian Kutum (meat and roe) for food is varied and they are used fresh, salted, or smoked and dried (Figure 3). The roe is also highly valued as a food item (Voskoboinikova 2006; Abdolmalaki and Ghaninejad 2008). The price of dried Caspian Kutum in Dagestan (Russia) reaches 1,000–2,000 rubles (ca. US$15–30) per fish depending on the weight. In Iran, from 2004 to 2008, the value of the annual catch of Caspian Kutum varied from $34 million to $50 million (Strukova et al. 2016) and comprised most of the total annual export value ($49–88 million) of fishery products from Iran between 1997 and 2008 (Karimpour et al. 2013).

FIGURE 3. Dried Caspian Kutum at food market in Makhachlaka, Dagestan (Russia)..

In this chapter, we describe the distribution of Caspian Kutum, their fishery harvest and stock status, causes for depletion, and measures that have been used to restore their fishery. We conclude with the actions used to solve problems facing the species as lessons for restoration.

DISTRIBUTION

Caspian Kutum is a Ponto-Caspian semi-anadromous species inhabiting the Caspian Sea in the main east-central and southwestern areas (Figure 4). Caspian Kutum are distributed in the Caspian Sea waters of Azerbaijan, Russia, Iran, Kazakhstan, and Turkmenistan (Kavan et al. 2009; Khodorevskaya et al. 2014). Their range extends from the mouth of the Terek River in the North Caucasus (Russia) to the Pahlavi Bay (small bay west of the Safid River, Iran) in the south and further east to Turkmenbashi (formerly Krasnovodsk) Bay and mouth of Atrek River on the southern coast of Turkmenistan (Figure 1; Kazancheev 1981; Belyaeva et al. 1989; Voskoboinikova 2006; Safari 2016). The main populations are concentrated in the southwestern part of the Caspian Sea between the Kura River in Azerbaijan and the Safid (Sefidrud) River in Iran (Voskoboinikova 2006; Borujeni et al. 2015). Caspian Kutum are not found in the northeastern or eastern coasts and only sporadically occur in the mouths of the Volga and Ural rivers of the northern region (Figure 1; Ustarbekov et al. 2009; 2014). A 1953 report exists of a recorded catch of 1.1 metric tons (mt) from the Volga River delta (Berlyand 1954). During periods of high abundance, Caspian Kutum notably extend their range and appear in the northern Caspian Sea entering the Volga and Ural rivers. No records of this fish in the northern Caspian Sea and Volga River basin have occurred since the late 1930s (Zavialov et al. 2006); however, recently (early 2000s), they appeared in the Volga and Ural rivers (Sultangaliev 2009; Bokova et al. 2010; A. I. Kushnarenko, Caspian Fisheries Research Institute, unpublished data), which also testifies to the current stock recovery.

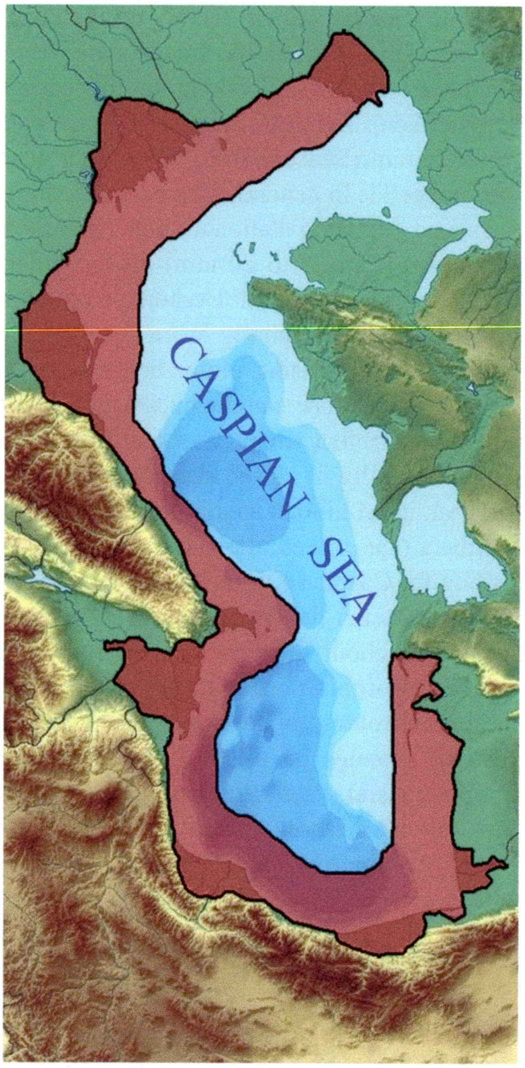

FIGURE 4. Natural range of Caspian Kutum in the Caspian Sea (shaded in red).

OVERVIEW OF FISHERY CATCH AND TRENDS IN STOCK ABUNDANCE

Catch data of the Caspian Kutum fishery from Russia and Azerbaijan are not continuous over time because Azerbaijan was a part of the Soviet Union prior to 1991, and then, after the break-up of the USSR, data were not collected separately until 2001. However, in the published literature, some data exist that characterizes the historical catch in Azerbaijan during the Soviet period.

The commercial stocks of Caspian Kutum in the Caspian Sea are subject to sharp fluctuations in abundance, which is reflected in the variation of catches. In the period from 1900 to 1964, catches were rarely below 500 mt, with most catches exceeding 1,000 mt (Figure 5). Since the mid-1950s, catches of Caspian Kutum declined everywhere through the end of the 1960s, when they reached their minimum levels (Trushinskaya 1975), but now have substantially recovered.

Russia

Caspian Kutum had commercial importance in Russia/the Soviet Union during the first half of the 20th century (Figure 5). Currently, this species has little commercial value and is caught mainly for local consumption by individual fishermen or small fishing cooperatives. This small-scale commercial fishery occurs mainly in coastal areas and uses motorboats with seines and various nets. A recreational (amateur) fishery occurs both in the sea and in rivers. Illegal fishing occurs with the use electrofishing and nets in rivers and various nets in the sea.

From 1900 to 1936 in Russia/the Soviet Union, an increase occurred in Caspian Kutum catches from 790 to 1,980 mt (Figure 5). Between 1940 and 1945, catches were below 1,000 mt as a result of reduced fishing activity during the World War II. The maximum catch (2,200 mt) was recorded in 1949, after which a precipitous decrease in catch occurred, dropping to 9 mt in 1968. In the period between 1969 and 1977, catch of Caspian Kutum ranged from 15 to 94 mt, much lower than historically observed during the early 1900s. The period from 1980 to 1985 was characterized by increased catches; in some years, this exceeded 180 mt. Since 1985, a substantial decrease in catch occurred, decreasing from 67 (1985)

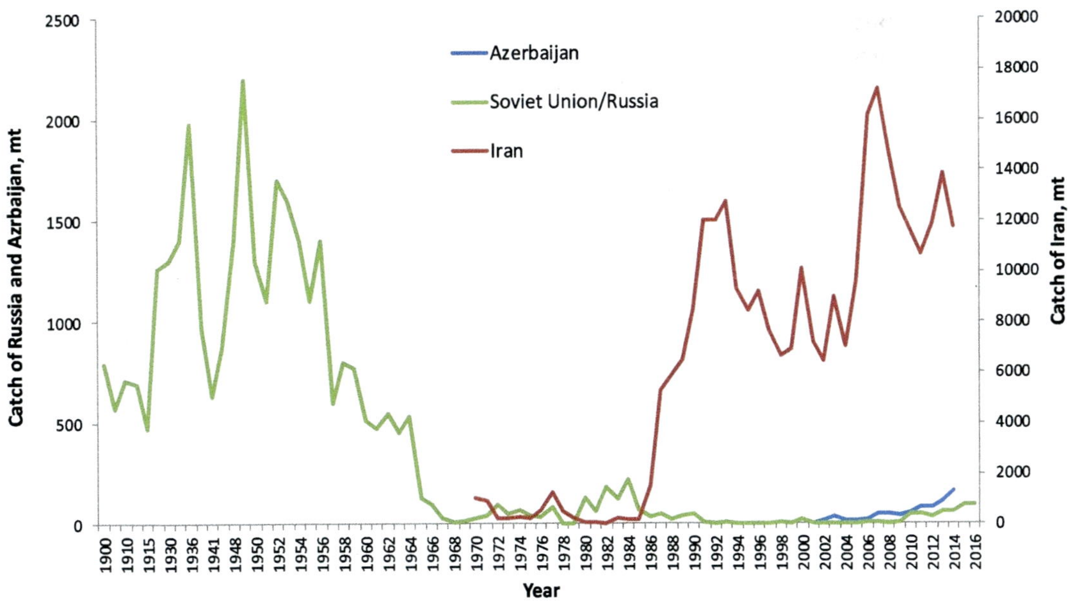

FIGURE 5. Annual Caspian Kutum catches of Russia, Azerbaijan, and Iran, 1900–2016 (according to Food and Agriculture Organization of the United Nations and VNIRO [Russian Federal Research Institute of Fisheries and Oceanography] statistics).

to 2 mt (1995). This decline in catch resulted in Caspian Kutum being listed as species of concern in the *Red Data Book of the Russian Federation* from 1995 to 2004 (Vasil'eva 2000). As a listed species, no commercial fishing was allowed for Caspian Kutum along the entire Russian coastline of the Caspian Sea. Fishing resumed in 2006 based on scientific advice, with a total allowable catch set at 9 mt. Catch rates (total allowable catch) have since been constantly increasing, exceeding 90 mt in 2015 and 2016. This change over the past decade provides clear evidence of the recovery of the species in this region.

The sharp decline in catches of Caspian Kutum in Dagestan in the early 1960s was attributed to hydroprojects for irrigation, which, due to water withdrawals, deteriorated Caspian Kutum spawning grounds in the wide and complex Terek River delta (discussed below; Belyaeva et al. 1989). Beginning in 1998, increased numbers of Caspian Kutum were observed off the Dagestan coast of the Caspian Sea (Figure 6), which resulted in an expansion of feeding and spawning areas (Ustarbekov and Guseynov 2009). The abundance and biomass of Caspian Kutum after 2010 demonstrated another upward trend. The size of commercial stocks off the Dagestan coast, for a number of years, was estimated at about 800 mt (Khodorevskaya et al. 2014). Recent data have indicated a slight decrease in the commercial biomass of Caspian Kutum off the Dagestan coast of the Caspian Sea—from 725 mt in 2016 to 530 mt in 2017. The decrease in Caspian Kutum stocks in the region during this recent time period is mainly due to overfishing, especially during spawning migrations in the rivers. A huge impact on the population is from the use of electrofishing in rivers by illegal fishers during spawning migrations of Caspian Kutum.

Azerbaijan

Caspian Kutum is an important commercial species of an artisanal fishery in some coastal regions of Azerbaijan, where it is caught using motorboats with seines (Strukova et al. 2016). The development of the Azerbaijan Caspian Kutum fishery dates back to the 1890s. Especially high catches were recorded in 1903, 1911, and 1916, with a maximum catch of

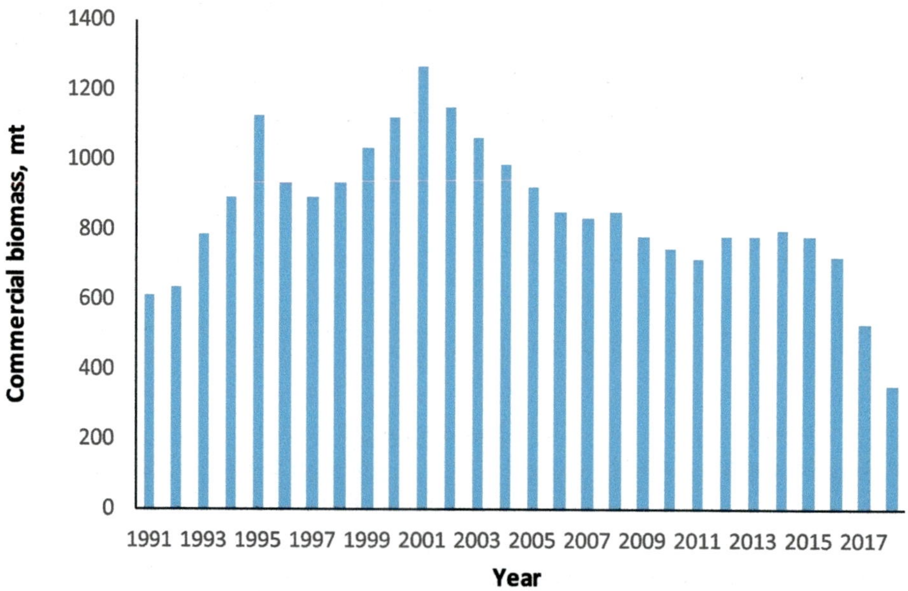

FIGURE 6. Dynamics of Caspian Kutum biomass off Dagestan coast, 1991–2017.

4,010 mt reached in 1932. Up to 1956, the catches remained strong and did not show any systematic decline (Trushinskaya 1975). As a result of Caspian Kutum's habitat degradation in the rivers, however, catches began to decline from 1957 to 1963, with average catches ranging from 400 to 800 mt (Voskoboinikova 2006). Since 2002, when the catch was only 16 mt, Azerbaijan has showed an upward trend in Caspian Kutum harvest, providing further evidence of successful recovery due to improvement of habitat, reduced fishing pressure, and increased reproduction and stocking (Figure 5). In 2010, catch exceeded 50 mt (Khodorevskaya et al. 2014), and in 2014, harvest reached 162 mt.

Official catch statistics of Caspian Kutum in Dagestan (Russia) and Azerbaijan are not reported as total catches; thus, catch does not include those from the recreational fishery or from illegal fishers. As a result, probably only 5–10% of total Caspian Kutum catches are reported (Ustarbekov et al. 2009), and the catch statistics shown in Figure 5 for these areas are likely underestimated.

No data exist to assess the Caspian Kutum stocks in the waters of Azerbaijan.

Kazakhstan

In the late 1990s, Caspian Kutum was recognized as a species with reduced abundance everywhere, and therefore was included in the *Red Data Book of Kazakhstan* (Dukravets and Mitrofanov 1996). In recent years, the number of Caspian Kutum in Kazakhstan waters has increased (Bokova et al. 2010). In 2009, more than 40 individual Caspian Kutum, weighing 4–6 kg with spawning coloration, were caught in the Ural River, which may indicate that successful reproduction of this species has begun in this region (Sultangaliev 2009).

Turkmenistan

In Turkmenistan, Caspian Kutum is a commercial fish species fished in the Caspian Sea and in the Atrek River (Rustamov 2011; Strukova et al. 2016). However, no catch statistics are available for the fisheries to assess the dynamics of Caspian Kutum populations.

Iran

Caspian Kutum is an important target of artisanal fishing (local fishing in small fishing cooperatives) in Iran, which uses mainly beach seining. In Iran, the minimum catch (98 mt) was recorded in 1981 (Karimpour et al. 2013). During the period from 1927 to 2011 (Figures 5 and 7), three periods of recorded high catches of Caspian Kutum occurred in Iran's waters. The first peak came in the 1930s and 1940s when as much as 4,000–6,000 mt were caught. After these years, a sharp decrease in catches occurred with a collapse of Caspian Kutum populations, resulting in the need for artificial propagation and release of Caspian Kutum fingerlings (Figure 7; Abdolmalaki and Ghaninejad 2008).

Caspian Kutum hatchery programs in Iran began in 1939 and showed positive results with increased catches (Figure 7). In 1990–1994, catches reached 11,000 mt and significantly exceeded levels observed in 1939 (6,000 mt), when all fish were of natural origin (Abdolmalaki and Ghaninejad 2008). However, in 1995, a decrease in the catch of Caspian Kutum was associated with the closure of fishing grounds for sturgeons, which likely affected Caspian Kutum catch because similar locations and fishing gears were used (Figure 7). From 1999 to 2004, the average catch of Caspian Kutum was 7,500 mt (Abdolmalaki and Ghaninejad 2008), after which catch substantially increased. Until 2007, the catches stabilized at about 17,000 mt per year, but then, during the next 3 years, gradually decreased to 14,800, 12,500, and 11,600 mt, respectively (Abdolhay et al. 2012). In recent years, Caspian Kutum harvest has remained stable and has varied between 12,000 and 14,000 mt per year, much higher than observed historically.

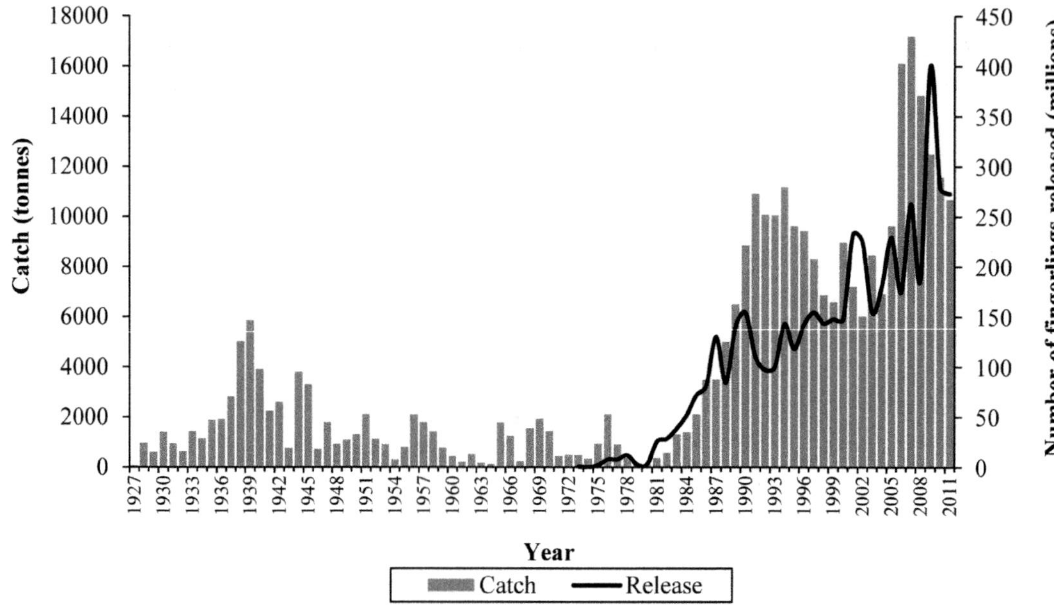

FIGURE 7. Iranian catch of Caspian Kutum (metric tons, left axis) and release of fry (millions, right axis) in the Caspian Sea (after Borujeni et al. 2015).

Recovery and success of the present-day fishery is, in part, due to the substantial propagation and release of Caspian Kutum fry by Iran.

A reduction in the stocks of Caspian Kutum in the waters of Iran has been observed since 1975 to the late 1980s (Abdolhay et al. 2012; Safari 2016). An increase in the recruitment in numbers to the fishery has occurred, from 5 million in 1989 to 40 million fish in 2007. Maximum spawning biomass (27,000–30,000 mt) was recorded in 2000–2002 (Borujeni et al. 2015). In recent years, the stocks of Caspian Kutum in the southern part of the Caspian Sea have been stable, associated with the successful release of propagated fingerlings from Iranian hatcheries (Khodorevskaya et al. 2014).

CAUSES FOR PAST STOCK DECLINES

Reasons for Caspian Kutum stock declines in the Caspian Sea include variability of sea levels; reduction of river flows; irrigation; increased pollution by agricultural, municipal and industrial wastes; excavation of sand and gravel from spawning rivers; habitat degradation; construction of bridges and dams that alter or block access to natural spawning grounds; overfishing; and illegal fishing (Kavan et al. 2009; Abdolhay et al. 2011; Jafari et al. 2011; Borujeni et al. 2015). Of these reasons, the reduction of river flows, overfishing, and illegal fishing were among the most important causes for stock declines.

Variability of Sea Level

Periodic sea level fluctuations are typical for the Caspian Sea, associated mainly with climatic phenomena affecting rainfall and evaporation. The last period of sea level decline of the Caspian Sea ended in 1977, when sea level reached its lowest point in the past 400 years at −29.0 m (below ocean sea level; Figure 8).

The level of the Caspian Sea began to rise from 1978 to the next maximum level in 1996 (–26.5 m), after which a new stage of water level decline began, which continues to this day (ca. –28 m).

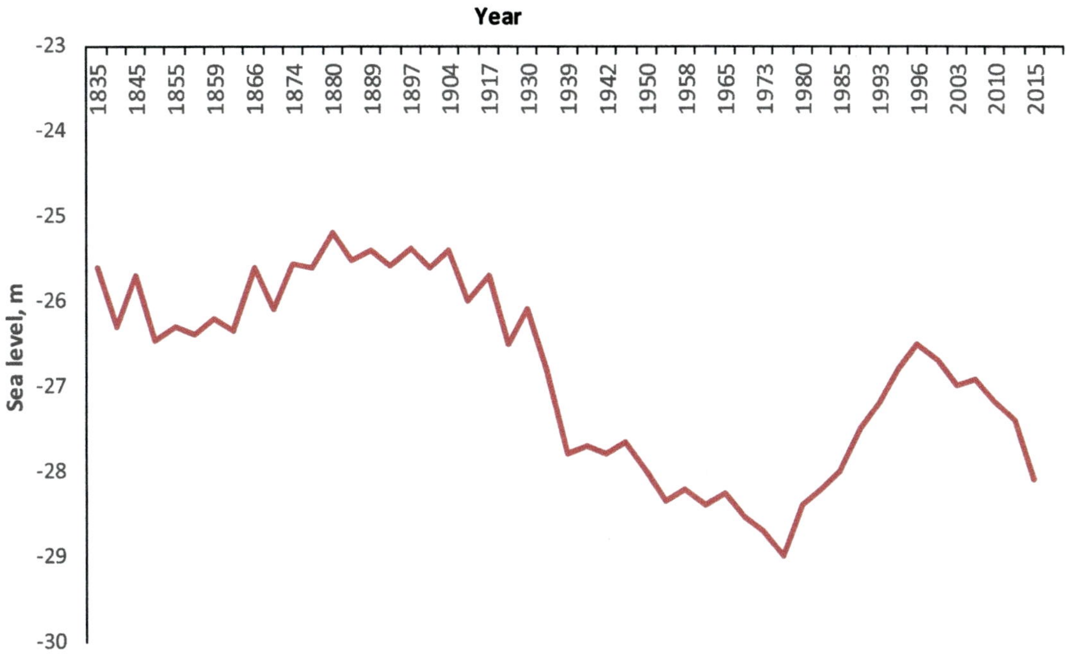

FIGURE 8. Variability of the Caspian Sea level, 1835–2015.

The drop in sea level has been suggested as one of the reasons for the sharp decline in catches of semi-anadromous fish, including Caspian Kutum, which practically disappeared from catches in the 1960s (Gimbatov et al. 1999). However, the conditions for reproduction and feeding of Caspian Kutum significantly differ from those of other Caspian Sea fish. Its spawning occurs in the rivers flowing into the Caspian Sea, in the deltas and not far upstream in tributaries; therefore, fluctuations in sea level do not exert a strong influence on spawning grounds. Unlike most Caspian Sea fish, Caspian Kutum feed in the deep southern and middle Caspian Sea, where fluctuations in sea level may not be a significant variable.

On the other hand, the rise of sea level resulted in the erosion of the dam at the Samursky fish-breeding station, which has reduced the area of the reservoir used for rearing Caspian Kutum from 100 to 16 ha and, thus, decreased Caspian Kutum fry stocking from 15 million to 3.0–3.5 million (Gimbatov et al. 1999).

Regulation and Transformation of Runoff of Spawning Rivers

Russia.—Construction of hydropower electric stations (HPESs) is among the top reasons for the decline in numbers and the cause for Caspian Kutum to be a listed species in the *Red Data Book of the Russian Federation* (Vasil'eva 2000).

In the 1930s in Dagestan, the construction of a cascade of HPESs on the Sulak River created reservoirs for energy, irrigation, and water supply. Currently, eight large and seven small HPESs operate on the Sulak and other rivers in Dagestan. The large hydroelectric power stations have created an area of about 7,000 ha of reservoir lakes, changing the geomorphology of rivers as well as their thermal and hydrochemical composition. When water is collected, especially in the spring–summer period, a large number of juveniles of various fish species enter spillways leading to generators and die. Low-water discharge in the spring during the spawning period has led to insufficient water provided to the downstream spawning grounds.

The construction of HPESs does not have a significant negative impact on blocking access by the stocks of Caspian Kutum to spawning grounds because their spawning grounds are located below the dams. In rare cases, HPESs can even have had a positive effect on the natural reproduction of Caspian Kutum. Thus, the creation of a reservoir on Sulak River and the construction of the Chirkeiskaya HPES changed the hydrological pattern of the river by reducing flow fluctuations, reducing silt loads, and improving water clarity. As a result, the Sulak River now functions as an important Caspian Kutum spawning ground, whereas before, it was not a spawning ground (Belyaeva et al. 1989).

Azerbaijan.—In the 1960s, the Kumbashinka River was severely impacted as a result of the construction of hydroelectric dams and lost its value as a fishery. The reclamation of the Malyi Kizil-Agachsky Bay in 1956 by increasing its area led to abundant growth of vegetation in the reservoir and a significant decrease in overall water exchange between the bay and main body of the Caspian Sea. The reduced water exchange made the area no longer attractive to Caspian Kutum for spawning (Trushinskaya 1975). The actual impact of the construction of hydraulic structures and regulation of river flow on the stocks of Caspian Kutum in Azerbaijan is unknown, but in the 1960s, in response to the construction of HPESs on Azerbaijan rivers, the artificial propagation and stocking of sturgeon and some cyprinid species, including Caspian Kutum, were implemented over concerns about the negative ecological effects of these projects (Rzaev 1966).

Water Withdrawals for Irrigation

Russia and Azerbaijan.—The withdrawals of water was another major cause for the decline of Caspian Kutum populations and for the species' inclusion in the *Red Data Book of the Russian Federation* (Vasil'eva 2000).

Within Dagestan, the Terek River has been highly impacted since 1932, when the need for irrigation required construction of three large dams in the delta: Karagalinsky, Pavlodol'sky, and Malo-Kabardinsky. The impetus for this construction was the intensive development of irrigated agriculture, livestock, and rice growing. The withdrawals of water into the irrigation canals reduced flooding of spawning grounds and flows into the lower reaches of the river. This reduced flow lowered the water levels of the lakes in the lower reaches of the Terek River and its delta, which are connected to the Agrakhansky Gulf. The construction of these irrigation hydrosystems blocked the free access of Caspian Kutum adults to spawning grounds. At the same time, more than 100,000 ha of flooded water bodies (Nizhne-Terskie and Arakumskie lakes and Agrakhansky Gulf) lost much of their importance as reproduction and rearing areas in the 1950s (Trushinskaya 1975), which was reflected in the subsequent sharp decrease in catches in the 1960s (Figure 5). In response, Caspian Kutum natural reproduction declined (Magomedov et al. 1987; Gimbatov et al. 1999).

The high water-flow velocities (4.8–5.0 m/s) in front of the Karagalinsky dam also have had a negative effect on fish movement and spawning migrations, and these high velocities occur even during low flows. The dam is opened once a week for 20 h, allowing for fish passage, which is clearly not enough to ensure a normal migration to spawning grounds. Pumping stations that draw water for irrigation often do not have devices to prevent fish from entering, which leads to the mass death of larvae and juveniles.

The increase in the scale of water withdrawals from the Terek River for irrigation is illustrated by the following: in 1930, water consumption was only 8%; in 1970, it had increased to 40%; in 1977, it reached 50–60%; and in the 1980s, it increased to 70% of the total runoff. In contrast, the allotment of water supplied to hatchery propagation (filling the Arakumskoe, Nizhne-Terskoe and Karakol'skoe spawning-and-growing farms [SGFs]), has been three to four times less than actually needed.

The Samur River, located in the Dagestan territory of Russia and Azerbaijan, is one of the most important spawning rivers of Caspian Kutum in this region. Nevertheless, about 90% of the river is used for irrigation of agricultural lands in Azerbaijan and Dagestan. Because only 10% of water is discharged, most of the discharge occurs in the autumn–winter period, leaving water levels insufficient for natural reproduction during the springtime.

Iran.—The decrease in catches from the Iranian coast during the 1960s–1970s was associated with a significant reduction in spawning habitat due to the withdrawal for irrigation of water from the main spawning rivers at the height of spawning period in March–April (Trushinskaya 1975).

Pollution, Oil Production, and Seismic Surveys

Russia.—No negative effects on Caspian Kutum stocks appear to have occurred from pollution in the rivers and sea waters of the Dagestan coast based on data available.

Azerbaijan.—Progressive pollution caused by the rapid development of offshore oil production in the Caspian Sea led to the closure of the Karadagsky SGF in 1963. In the spring of 1962, the seismic surveys and the use of underwater explosions, which precedes offshore oil production, caused a large fish kill of Caspian Kutum. Fish were thrown ashore by the explosions, especially in the area of Nordostovy Kultuk Bay (Trushinskaya 1975).

Overfishing Caused by Illegal Fishing

Russia.—Overfishing of Caspian Kutum is another key reason for the decline in Caspian Kutum abundance (Shikhshabekov et al. 2006) and their entry into the *Red Data Book of the Russian Federation* (Vasil'eva 2000).

Illegal fishing of Caspian Kutum—fishing without a license, in prohibited periods and locations, and by prohibited fishing gear—has contributed to overfishing and was especially damaging when conducted on their spawning grounds. Ubiquitously, Caspian Kutum are illegally caught by anglers during their spawning run from the sea to the rivers (Figure 9). Illegal fishing of all forms is carried out from spring to autumn in the fish channels and directly in the reservoirs. In addition, illegal electrofishing of Caspian Kutum is carried out on spawning grounds, which causes great damage to both eggs and juveniles. These illegal catches are suspected to be many times higher than official reported statistics and may reach total catch amounts of 200–300 mt annually (Ustarbekov and Guseynov 2009; Khodorevskaya et al. 2014). According to our assessment, unreported and illegal catches of Caspian Kutum averaged 286 mt (Figure 10) during all phases of life—marine, spawning migrations, and on spawning grounds (Rabazanov et al. 2017). At the same time, in the period from 2006 to 2016, a decrease in illegal catches of Caspian Kutum from 412 to 176 mt occurred (Figure 10), probably due to the strengthening fight against illegal fishing.

Before Caspian Kutum was entered in the *Red Data Book of the Russian Federation*, immature, undersized individuals were caught in large quantities as bycatch via folding seines in the herring fishery. At the same time, bycatch of Caspian Kutum often exceeded the catch of the herring itself, which probably adversely affected Caspian Kutum stocks.

Despite the intensified fight against illegal fishing by Russian authorities (Federal Agency for Fisheries and Border Guard

FIGURE 9. Sport-fishing capture of Caspian Kutum off the Dagestan coast of the Caspian Sea.

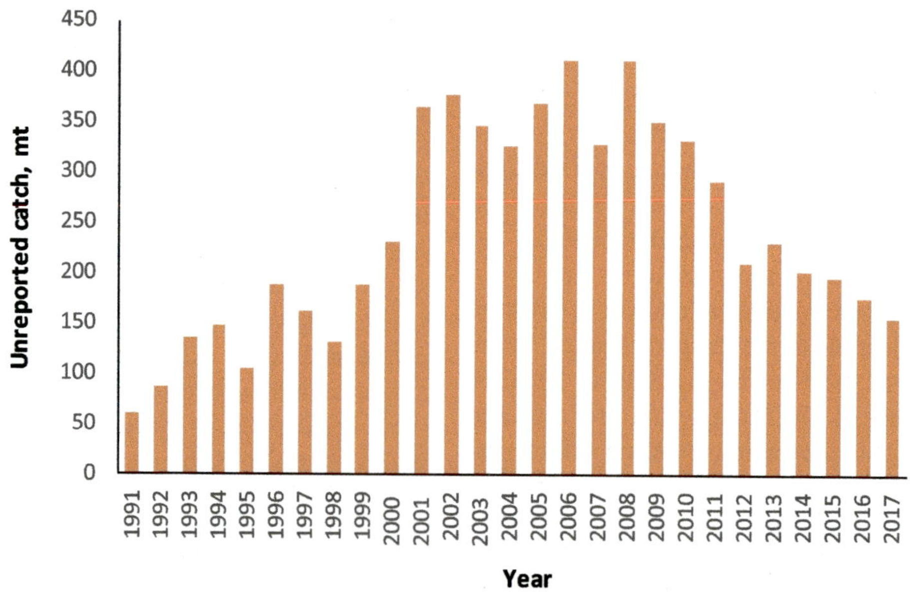

FIGURE 10. Unreported and illegal Caspian Kutum catches in Dagestan, 1991–2017.

Service of the Federal Security Service), the problem of illegal fishing of Caspian Kutum in Dagestan has not yet been solved. Excessive illegal harvesting of Caspian Kutum occurs because of their high monetary value and the high demand for Caspian Kutum. For example, in the city of Makhachkala, the cost of dried Caspian Kutum can reach 1000–2000 rubles (ca. US$15–30) per fish.

Azerbaijan.—No official fisheries data exist on the illegal catch of Caspian Kutum in the waters of Azerbaijan. However, the Azerbaijanian press (Ismailov 2014) reported that about 98% of the catch in this country is from uncontrolled fishing.

Iran.—According to Iranian scientists (Abdolmalaki and Ghaninejad 2008), overfishing of Caspian Kutum in Iran began in the 1930s–1940s, when the annual catch reached 4,000–6,000 mt. Currently, 35% of all catches of Caspian Kutum in Iran is estimated to be taken illegally (Abdolmalaki and Ghaninejad 2008).

Kazakhstan.—Unregulated fishing was the main reason for the decline of Caspian Kutum stocks in the waters of Kazakhstan (Dukravets and Mitrofanov 1996; Bokova et al. 2010).

MANAGEMENT ACTIONS TO RECOVER THE STOCKS

Introduction of Caspian Kutum to New Waters

Russia and Ukraine.—In the early 1950s, experiments were conducted on the cultivation of Caspian Kutum in fish farms along the Volga River to increase, through stocking, its abundance in the northern Caspian Sea (Berlyand 1954; Zhukinsky and Balan 1959). These experiments were not successful and the number of Caspian Kutum in the northern Caspian Sea and in the Volga and Ural rivers remained low (Ustarbekov et al. 2009).

In 1957, the introduction of Caspian Kutum outside of its native range began into the basin of the Sea of Azov (Figure 11). In the same year, Caspian Kutum were introduced into the basin of the Kuban River in the Black Sea basin, Russia (Figure 11). In 1958, work began on the introduction of Caspian Kutum into the basin of the lower Don River (basin of the Azov Sea, Russia) and the lower Dnieper River in the Black Sea basin, Ukraine (Doroshin 1959; Zhukinsky and Balan 1959; Trushinskaya 1962; Voskoboinikova 2006). In the early 1990s, experiments were carried out to introduce Caspian Kutum in several Ukrainian estuaries on the northwestern coast of the Black Sea (Tolokonnikov 1994).

Success of these Caspian Kutum introductions was mixed. In 1961, in the Azov Sea, Caspian Kutum reached maturity and successfully spawned (Doroshin 1961), yet despite these large-scale introductions, the number of Caspian Kutum in the Azov Sea did not reach levels to support commercial fisheries, the species did not naturalize, and, after stocking stopped, the Sea of Azov population of Caspian Kutum declined greatly in abundance to barely detectable levels (Vasil'eva and Luzhnyak 2013). Caspian Kutum still occur in very low numbers in the Azov Sea estuaries and the lower reaches of the Kuban River (Pashkov et al. 2004).

Harvest Regulation

Russia.—Caspian Kutum was included in the *Red Data Book of the Russian Federation* in 1997, and for 7 years, all fishing was banned, reducing overfishing and significantly contributing to the restoration of its stocks. Caspian Kutum were removed from the *Red Data Book of the Russian Federation* in 2004 (Khodorevskaya et al. 2014). However, the main reason for the restoration of stocks in the waters of Dagestan has been due to a sharp increase in the scale of natural reproduction and stocking in Iranian, Azerbaijanian, and

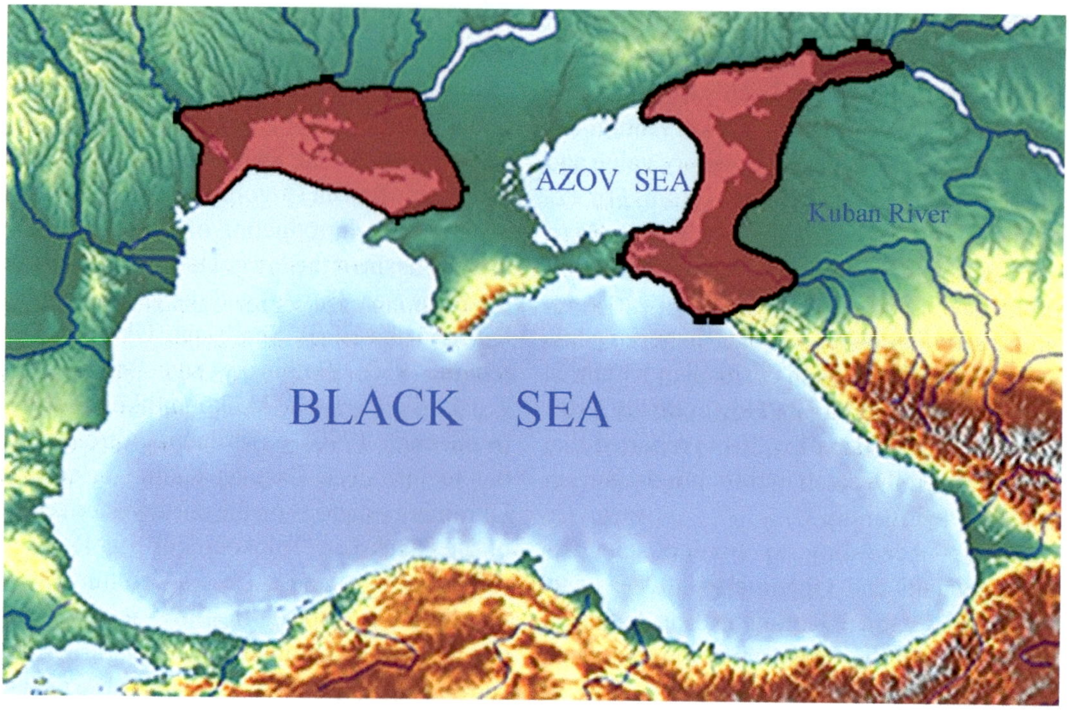

FIGURE 11. Areas of introductions of Caspian Kutum into the Black and Azov seas (shaded in red).

Dagestanian waters. The other reason might be an increase in water temperature, leading to an improvement in reproduction conditions due to global warming.

Iran.—At the beginning of the new millennium, Caspian Kutum was listed as one of four most threatened freshwater species in Iran (Coad 2000). Nevertheless, in 2000, the International Union for Conservation of Nature evaluated the conservation status of the species as "data deficient" (Kavan et al. 2009). Currently, Caspian Kutum are considered to be "not evaluated" (Esmaeili et al. 2017).

Kazakhstan.—Caspian Kutum is currently listed in the *Red Data Book of Kazakhstan* (Mitrofanov and Mamilov 2015). As a precautionary measure, catches of Caspian Kutum in Kazakhstan are prohibited, and artificial propagation and stocking are recommended for the restoration of stocks. Specification of the state of abundance and distribution boundaries in the waters of Kazakhstan is lacking (Dukravets and Mitrofanov 1996).

Artificial Propagation and Stocking

Caspian Kutum is one of the most important targets of artificial propagation and stocking in Dagestan (Russia), Azerbajan, and Iran. Some differences occur in technologies of Caspian Kutum artificial propagation in these countries. In Dagestan, fish-breeding stations release both larvae (weight 0.002 g) and fingerlings (weight 0.66–1.22 g). It should be noted that in the past (1960–1970s), the weight of Caspian Kutum fingerlings released in Dagestan and Azerbajan was higher than it is currently and was 1.0–2.6 g per fish (Trushinskaya 1975). According to published literature (Belyaeva et al. 1989; Abdullaev et al. 2014), Caspian Kutum in Azerbajan were released both as larvae and fingerlings (1 month old). Currently, most fish released are larvae (Abdullaev et al. 2014). In Iran, both larvae and fingerlings are produced and released, but most of the Caspian Kutum progeny released (68.1% by weight and 81.0% by number) are 1.0–1.5 g fingerlings, with length approximately 5 cm (Abdolhay et al. 2011; Jafari et al. 2011).

Russia.—Experiments on artificial reproduction at the Samursky fish-breeding station were started in 1924, and at the Sulaksky SGF since 1995. Technologies for artificial propagation of Caspian Kutum in Russia were developed by the mid-1950s (Berlyand 1954). In the years 1948–1965, at the Samursky fish-breeding station, 22.3–136.7 million larvae were produced, with an average of 34.0 million stocked (Trushinskaya 1975; Ustarbekov et al. 2009). At present, the propagation and stocking of Caspian Kutum in Dagestan is carried out at the Sulaksky sturgeon SFG (Sulak River, annual production of 1.2–2.2 million fingerlings of Caspian Kutum), the Primorskiy fish hatchery (Samur River, annual production of 2.2–2.3 million larvae and 5.5–6.6 million fingerlings), the Biryuzyaksky fish hatchery (Kuma River, annual production of 2.1–2.2 million of larvae and 1.0–9.5 million of fingerlings), and the Mekhtebsky SGF (Sulak River, annual production of 2.4–4.4 million of fingerlings). In the early 2000s, these facilities produced up to 13 million of juvenile Caspian Kutum for release (Gimbatov et al. 2001; Ustarbekov and Guseynov 2009). In the past 3 years, the release of Caspian Kutum in Dagestan has grown to 18.5–25.5 million larvae and fingerlings (Figure 12).

Since 1960, in Dagestan, improvement of spawning habitats for Caspian Kutum was started. In the lower reaches of the Terek River, the Arakumsky, the Nizhnetersky, and the Karakol'sky spawning-and-growing reservoirs (SGR) more than 40,000 ha of rearing habitat were created, which has provided an increase of juvenile production in the amount of 3.5–4.0 million fish (Gimbatov et al. 1999).

To improve the reproduction of Caspian Kutum, the Samursky SGR was built. The area of Samursky Lake was increased from 8 to 180 ha. The SGR is connected from the lake to the Caspian Sea by a fish channel with a length of 1.5 km. During the spawning run, mature Caspian Kutum are allowed to pass through the sluice and natural spawning takes place in the fish channel. Once they have

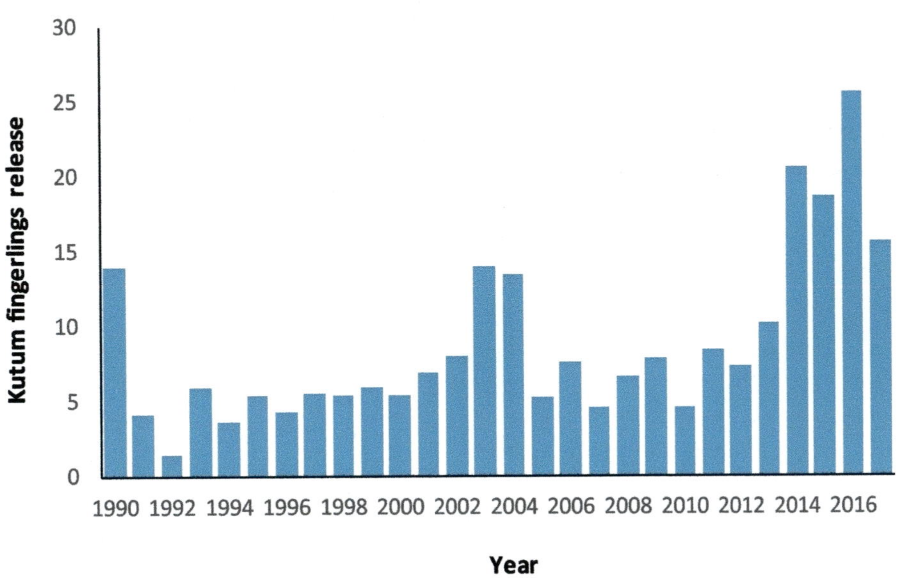

FIGURE 12. Release of Caspian Kutum fingerlings (millions) in the Caspian Sea off the Dagestan coast, 1990–2017.

reached the appropriate size, fry migrate into the Caspian Sea. In 1971, 103,400 Caspian Kutum spawned in the Samurskoye Lake, producing 7.6 million fry that eventually returned to the sea. Based on average long-term data, 100,000–200,000 mature Caspian Kutum entered the spawning grounds annually, producing an estimated 10–15 million juveniles (Trushinskaya 1975; Gimbatov et al. 1999). After the destruction of the dam, as a result of the rise in sea level, the area of the lake was reduced to 16 ha and the annual number of juveniles entering the sea decreased to 3.0–3.5 million fish (Gimbatov et al. 1999).

The dynamics of natural reproduction of Caspian Kutum in Dagestan show that from 1996 to 2008 their levels increased and 84.6 million juveniles were produced, evidence of successful recovery (Figure 13). The average annual recruitment yield of Caspian Kutum in this region is 44.5 million fish. Taking into account the average annual output of larvae and fingerlings from industrial breeding in the west-Caspian region, the total recruitment of Caspian Kutum fry is estimated to have reached 53 million annually.

In addition, in the Krivaya Balka River (35 km north of Makhachkala), Dagestan, more than 1,000 mature Caspian Kutum have migrated to the spawning grounds into the lake located in the mouth of this river (about 3,000 m^2), and up to 2.5 million juveniles returned back to the sea (Trushinskaya 1975; Magomedov et al. 1987).

Azerbaijan.—Experiments on artificial breeding began in 1925 on the Kumbashinka River, and techniques for artificial propagation of Caspian Kutum in Azerbaijan were fully developed by the 1950–1960s (Abbasov and Agalarov 1962; Rzaev 1966). In 1948–1955, the annual output of propagated Caspian Kutum larvae was 137–544 million, an average of 377.3 million per year (Trushinskaya 1975).

Caspian Kutum are also cultivated in the Maly Kizil-Agachsky SGF (since 1956) and the Kirovsky SGF (since 1966), but the production is much less than the design capacity of these facilities, in part due to insufficient operational funds. For example, the capacity of the Maly Kizil-Agachsky SGF is 112.5 million fingerlings. In 1968–1970, both

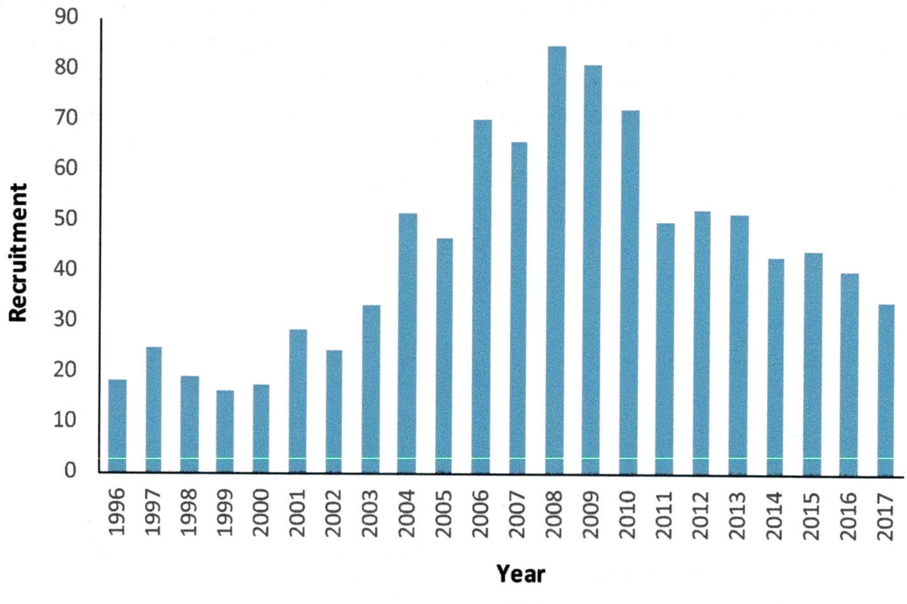

FIGURE 13. Dynamics of Caspian Kutum natural recruitment (millions) in Dagestan, 1996–2017.

SGFs only produced a total of 0.2–41.5 million fingerlings (Trushinskaya 1975). However, in the early 1980s, the output of Caspian Kutum fingerlings from these facilities significantly increased because of improved funding for the facilities. In 1980, 32-million 1-month-old fingerlings were produced, and since 1982, they have produced and released 50 million Caspian Kutum fingerlings annually (Belyaeva et al. 1989).

One of the most important reservoirs for the natural reproduction of Caspian Kutum is the Divichinsky estuary (Lake Agzybir). Spawning takes place in both the reservoir and the exit channel leading to the Caspian Sea and provides suitable feeding habitat for juveniles and adult fish. Up to 2006, the number of Caspian Kutum larvae produced from Divichinsky estuary was more than 30% of the total for this species in Azerbaijan. In 2007–2011, minimum and maximum production declined to 12.3% and 27.4%, respectively (Abdullaev et al. 2014).

The numbers of adults using the Divichinsky estuary for spawning has varied over the years. In the 1960s, 7,000–12,000 adult Caspian Kutum spawned and an average of 88 million larvae were produced, with annual fluctuations from 0.034 to 184 million larvae. Between 1995 and 1996, when 65,600 and 20,300 spawners, respectively, used the Divichinsky estuary, spawning was much greater than in the past. From 2007 to 2011, the number of mature Caspian Kutum that entered the estuary for spawning declined, and 1,130 adults (range of 880–1,360 fish) and 10.7 million larvae, on average, were produced, with annual variability from 9.5 to 12.1 million (Abdullaev et al. 2014). Recently, an increase of spawners has been observed over the past 3 years.

Iran.—Artificial propagation of Caspian Kutum in Iran is considered the main reason for maintaining stability and improving its stocks (Borujeni et al. 2015). The construction of four large propagation complexes in the southern provinces on the Caspian coast of Iran for the artificial reproduction of Caspian Kutum has significantly increased the abundance and catches of this species (Jafari et al. 2011; Karimpour et al. 2013).

Artificial reproduction and release of Caspian Kutum fingerlings in Iran began in 1939 and, since 1973, has become a routine practice (see Figure 7). In 1969, 71.35 million larvae and fingerlings were produced and released. In 1974, a small reproductive center was built, which produced 2.1 million fingerlings. In 1977, this facility produced 13 million fingerlings and 50 million larvae. Between 1974 and 1977, the production of larvae and fingerlings was respectively 250 and 28.5 million. After the Islamic revolution in 1979, artificial reproduction of Caspian Kutum was suspended. In 1983, artificial reproduction of Caspian Kutum resumed, and by 1990, the number of fry produced and released reached 100 million fish. Since 2000, the release of Caspian Kutum has exceeded 200 million fish annually (Abdolhay et al. 2011; Jafari et al. 2011; Karimpour et al. 2013; Khodorevskaya et al. 2014; Safari 2016). Three release strategies are used in Iran: (1) spot planting with releasing of larvae and fingerlings into the receiving waters at the same time, (2) scatter planting with release of fry at several sites in the same region, and (3) trickle planting with release of fingerlings in the same region over a period of time (Abdolhay et al. 2011).

CONCLUSION

Conservation and management measures for Caspian Kutum stocks, notably to enhance natural reproduction, as well as an increase of artificial propagation in Iran, Azerbaijan, and Dagestan (Russia) have allowed for recovery of their stocks to a stable level and have allowed for the resumption of fishing. This review of the Caspian Kutum indicates that at present in Dagestan, the problem of water pollution and its negative impact on the condition of the Caspian Kutum population are not a priority to solve. Also, the effect of the construction of hydroelectric power stations on rivers is insignificant, as the main spawning grounds of Caspian Kutum are historically located below the dams in operation. Variability in the level of the Caspian Sea affects the fish species living in the northern shallow-water zone of the Caspian Sea differently; however, its impact on the population of Caspian Kutum is likely minimal. The main negative variables affecting Caspian Kutum abundance have been excessive uncontrolled illegal catch, irrigation, construction of hydroelectric power stations, and development of irrigated agriculture in the region, which have led to the reduction in the quantity and quality of Caspian Kutum spawning grounds.

On August 12, 2006, the Framework Convention for the Protection of the Marine Environment of the Caspian Sea, a regional international convention signed in Tehran, Iran, on November 4, 2003 by representatives of all five Caspian Sea countries (Russia, Kazakhstan, Turkmenistan, Iran, and Azerbaijan) entered into force. The convention provides for a number of measures for the protection, conservation, restoration, and rational use of the biological resources of the Caspian Sea, namely the restoration of the populations of marine commercial species to levels that allow a sustainable harvest; protection of endemic, rare, and endangered species; and the sustainable conservation of biodiversity and habitat of aquatic organisms of the Caspian Sea. Despite the recognition by all the Caspian Sea countries of the need to take immediate steps to preserve the aquatic biological resources and solve the environmental problems of the Caspian Sea, all attempts to fully establish international cooperation in this area have not been particularly successful because of disagreements on the international status of the Caspian Sea. Regardless of the solution to this issue, Caspian Kutum remains a common living resource of the five countries in the region. In this regard, the solution to problems related to the research and use of the Caspian Sea bioresources, including Caspian Kutum, likely will become an integral part of the future legal regime of the Caspian Sea.

Due to actions recently performed for improvement of Caspian Kutum stocks, they are now at a good level. This improvement is evidenced by the growth of Caspian Kutum catch in Dagestan and Azerbaijan in recent years (Figure 5) and by the growth of natural reproduction in Dagestan in the 2000s as compared to the 1990s (Figure 13), largely due to an increase in the artificial reproduction and release of Caspian Kutum (Figure 12) and a decrease in illegal catches (Figure 10). Despite some decrease in catch of Caspian Kutum in Iran (Figure 5), their levels remain high, and some decrease is probably due to a decrease in artificial reproduction (Figure 7), which after 2005 still remains at a consistently high level. Recovery of Caspian Kutum stocks might also be confirmed by their recent appearance in the Volga and Ural rivers (Sultangaliev 2009; Bokova et al. 2010; Kushnarenko, unpublished data), where no previous records of this fish has occurred for decades.

Further enhancement of Caspian Kutum stocks in Dagestan would be possible by taking the management actions listed below to promote the conservation and reproduction of Caspian Kutum:

- Declare a 5-year moratorium on the commercial fishing of Caspian Kutum in Russia to reduce fishing mortality and allow more fish to reach maturity and reproduce. Further restrictions on fishing everywhere would further rehabilitate Caspian Kutum stocks.
- Implement special enforcement measures

to protect Caspian Kutum during its marine life stage and on its migration routes to reduce illegal catch. Further efforts to control illegal fishing must be expanded to reduce overharvest.

- Carry out large-scale improvements of migration routes (fish channels) and spawning grounds of Caspian Kutum.
- Raise the level of artificial reproduction of Caspian Kutum at fish-breeding facilities, bringing the total output to more than 100 million fish released annually.

Based on our experiences of managing Caspian Kutum in the Caspian Sea, the lessons that follow may have application to fishery management programs elsewhere.

- Control of fishing, both legal and illegal, is required if sustained harvests are to be realized.
- Fishing should be carried out within the limits of scientifically based estimates of total allowable catches and in accordance with the fishing regulations that regulate, for example, the use of specific fishing gear, mesh/hook size, prohibited periods, and the minimum fishing length of caught fish.
- Artificial propagation of fish and their release can have a significant impact on fisheries in marine settings and should be considered as a possible tool for restoration, provided habitat and spawning areas are suitable.
- Continuous monitoring of stock status is needed to provide an annual assessment of total allowable catches and to monitor changes in the state of exploited populations.
- Success in the conservation and enhancement of straddling and/or transboundary fish stocks can largely be achieved through coordinated actions to protect and reproduce these stocks by all countries that exploit them.

ACKNOWLEDGMENTS

The authors thank their colleagues Nina Yanovskaya and Lyubov' Kononova (VNIRO, Moscow) for long-term data on Caspian Kutum catch statistics and Svetlana Orlova (VNIRO, Moscow) for drawing the maps of distribution. We also thank the International Fisheries Section of the American Fisheries Society for their editorial assistance with the English language.

REFERENCES

Abbasov, G. S., and G. D. Agalarov. 1962. Experiments on the artificial reproduction of Kutum in Azerbaijan. Rybnoye Khoziaistvo 10:11–12. (In Russian.)

Abdolhay, H. A., S. K. Daud, S. Rezvani Ghilkolahi, M. Pourkazemi, S. S. Siraj, and M. K. Abdul Satar. 2011. Fingerling production and stock enhancement of Mahisefid (*Rutilus frisii kutum*) lessons for others in the south of Caspian Sea. Reviews in Fish Biology and Fisheries 21:247–257.

Abdolhay, H. A., S. K. Daud, S. Rezvani, M. Pourkazemi, S. S. Siraj, F. Laloei, A. Javanmard, and M. Hassanzadeh Saber. 2012. Population genetic structure of Mahi Sefid (*Rutilus frisii kutum*) in the South Caspian Sea: implications for fishery management. Iranian Journal of Animal Biosystematics 8:15–26.

Abdolmalaki, S., and D. Ghaninejad. 2008. Dynamics of Kutum (*Rutilus frisii kutum*) stocks and their regulation in Iranian coastal waters of the Caspian Sea. Pages 18–21 *in* Comprehensive approach to the problem of conservation and restoration of bioresources of the Caspian basin. CaspNIRKh Press, Astrakhan, Russia. (In Russian.)

Abdullaev, A. I., S. N. Nadirov, and M. M. Akhundov. 2014. Evaluation of the role of Divichinsky Liman in reproduction of commercial stocks of semi-anadromous cyprinid fishes. Pages 11–16 *in* Current state of inland waters biological resources. Proceedings of the second all-Russian conference with foreign partners. November 6–9, 2014, Borok, Russia. POLIGRAF-PLUS, Moscow. (In Russian.)

Belyaeva, V. N., E. N. Kazancheev, and V. M. Paspopov. 1989. The Caspian Sea: ichthyofauna and commercial resources. Nauka, Moscow. (In Russian.)

Berlyand, T. B. 1954. On the directed formation of stocks of cyprinid fish in the southern seas of the USSR and industrial cultivation of Kutum. Trudy Ikhtiologicheskoi Komissii AN SSSR 7:259–267. (In Russian.)

Bokova, E. B., G. M. Dukravets, and S. K. Kaldybaev. 2010. To the evaluation of population condition of Kutum in Kazakhstan sector of Caspian Sea. Selevinia 2010:178–179.

Borujeni, P. A., S. I. Steinshamn, S. A. Mortazavi, and H. Salehi. 2015. Enhance and advance: the benefits of recruitment enhancement in the case of the Iranian Kutum fishery. Marine Policy 61:23–32.

Coad, B. W. 2000. Criteria for assessing the conservation status of taxa (as applied to Iranian freshwater fishes). Biologia, Bratislava 55:539–557.

Doroshin, G. Y. 1959. Caspian Kutum in the Azov Sea. Rybnoye Khoziaistvo 6:31–32. (In Russian.)

Doroshin, G. Y. 1961. First spawning migration of Azov Sea Kutum. Rybnoye Khoziaistvo 9:33–34. (In Russian.)

Dukravets, G. M., and V. P. Mitrofanov. 1996. Rutilus frisii kutum - Kutum. Red Data Book of the Kazakhstan [online database]. Available: www.redbookkz.info/species.php?lang=ru&num=61. (In Russian.)

Esmaeili, R., H. Mehraban, K. Abbasi, Y. Keivany, and B. Coad. 2017. Review and updated checklist of freshwater fishes of Iran: taxonomy, distribution and conservation status. Iranian Journal of Ichthyology 4(Supplement 1):1–114.

Gimbatov, I. M, Z. M. Magomedrasulov, and F. M. Magomaev. 1999. Kutum as an object of artificial reproduction in Dagestan water bodies. Pages 29–30 *in* Resource-saving technologies in aquaculture. Publishing House "Zdravsvuite," Krasnodor, Russia. (In Russian.)

Gimbatov, I. M., M. B. Magomedov, N. K. Kuzina, and F. N. Magomedov. 2001. Artificial propagation of Kutum in Dagestan. Pages 28–29 *in* Problems and prospects of aquaculture development in Russia. Publishing House "Zdravsvuite," Krasnodor, Russia. (In Russian.)

Ismailov, A. 2014. This fish inaccessible to Azerbaijan (again about corruption). Haqqin.az. November 4, 2014. Available: https://haqqin.az/news/33153. (December 2018). (In Russian.)

Jafari, M., M. S. Kamarudin, C. R. Saad, A. Arshad, S. Oryan, and M. H. T. Guilani. 2011. Effects of different diets on growth: survival and body composition of *Rutilus frisii kutum* larvae. Journal of Fisheries and Aquatic Sciences 6:662–668.

Karimpour, M., M. M. Harlioglu, A. A. Khanipour, S. Abdolmalaki, and Ö. Aksu. 2013. Present status of fisheries in Iran. Journal of FisheriesSciences.com [online serial] 7:161–177.

Kavan, L. S., S. R. Gilkolaei, G. Vossoughi, S. M. R. Fatemi, R., Safari, and S. Jamili. 2009. Population genetic studies of *Rutilus frisii kutum* (Kamensky 1901) from the Caspian Sea, Iran and Azerbaijan regions, using microsatellite markers. Journal of Fisheries and Aquatic Sciences 4:316–322.

Kazancheev, E. N. 1981. Fishes of the Caspian Sea (field guide). Lyogkaya I Pishchevaya Promyshlennost', Moscow. (In Russian.)

Khodorevskaya, R., Y. Kim, R. Shahifar, E. Mammadov, D. Katunin, B. Morozov, M. Akhundov, O. Muradov, and V. Velikhova. 2014. State and dynamics of the bioresources in the Caspian Sea. Pages 1–84 *in* The handbook of environmental chemistry. Springer-Verlag, Berlin, Germany.

Magomedov, G. M., D. A. Aliev, and V. O. Proskurina. 1987. Reproduction of Kutum off Dagestan coast of the Caspian Sea. Rybnoye Khoziaistvo 1:37–39. (In Russian.)

Mitrofanov, I. V., and N. S. Mamilov. 2015. Fish

diversity and fisheries in the Caspian Sea and Aral–Syr Darya basin in the Republic of Kazakhstan at the beginning of the 21st century. Aquatic Ecosystems Health and Management 18:160–170.

Pashkov, A. N., G. K. Plotnikov, and I. V. Shutov. 2004. New data on composition and distribution of acclimated species in ichthyocoenes of continental water bodies of the northwestern Caucasus. Pages 68–70 *in* Problems of lithodynamics and ecosystems of the Azov Sea and Kerch Strait. CVVR Ltd., Rostov-na-Donu, Russia. (In Russian.)

Rabazanov, N. I., A. M. Orlov, A. S. Abdusamadov, R. M. Barkhalov, K. M. Akhmedkhanov, and K. G. Buzulutskaya. 2017. Condition of stocks, fishery, and artificial reproduction of the Caspian Kutum. Trudy VNIRO 166:55–71. (In Russian with English summary.)

Rustamov, A. K. 2011. Wildlife of Turkmenistan and its conservation (on the example of vertebrates). Ylym, Ashgabat, Turmenistan. (In Russian.)

Rzaev, Z. A. 1966. Caspian Kutum (fishery, biology, systematics, physiology, and reproduction of stocks). Doctoral dissertation. S.M. Kirov Azerbaijan State University, Baku, Azerbaijan. (In Russian.)

Safari, R. 2016. Population structure of *Rutilus frisii kutum* in Iranian coastline of the Caspian Sea using microsatellite markers. Environmental Resources Research 4:65–74.

Shikhshabekov, M. M., M. I. Karpyuk, G. M. Abdurakhmanov, and N. I. Rabazanov. 2006. Biological resources of Dagestan part of the central Caspian Sea. KaspNIRH, Astrakhan, Russia. (In Russian.)

Strukova, E., O. Guchgeldiyev, E. Evans, D. Katunin, R. Khodorevskaya, Y. Kim, M. Akhundov, T. Mammadli, R. Shahivar, O. Muradov, Mammadov, and V. Velikhova. 2016. Exploitation of the Caspian Sea bioresources (with focus on economics and bioresources utilization). Pages 1–44 *in* The handbook of environmental chemistry. Springer, Berlin, Germany.

Sultangaliev, M. 2009. Kutum felt cramped in the "Red Data Book." Ak Zhaiyk, December 10, 2009. Available: http://azh.kz/ru/news/view/3670. (December 2018). (In Russian.)

Tolokonnikov, G. Y. 1994. Acclimation of Kutum off Black Sea coast. Rybnoye Khoziaistvo 3:42–44. (In Russian.)

Trushinskaya, M. B. 1962. Settling of Kutum into the Azov Sea. Rybnoye Khoziaistvo 11:32–36. (In Russian.)

Trushinskaya, M. B. 1975. The status of the stock and reproduction of Kutum under condition of regulated rivers. Trudy VNIRO 107:57–64. (In Russian with English summary.)

Ustarbekov, A. K., and A. D. Guseynov. 2009. Morpho-ecological variability of Kutum *Rutilus frisii kutum* (Kamensky, 1901) and Danube Bleak *Chalcalburnus chalcoides* (Güldenstädt, 1772) in rivers of the western part of middle Caspian Sea. RGZhT, Makhachkala, Russia. (In Russian)

Ustarbekov, A. K., A. D. Guseynov, and Z. M. Kurbanov. 2014. Variation of morphological characteristics of *Rutilus frisii kutum* (Kamensky, 1901) in the Caspian Sea. Vestnik of Astrakhan State Technical University: Fisheries 1:46–54. (In Russian with English summary.)

Ustarbekov, A. K., Z. M. Kurbanov, D. A. Ustarbekova, and A. D. Guseinov. 2009. The peculiarities of Kutum (*Rutilis frisii kutum*) biology in the rivers of the western part of the middle of the Caspian Sea basin. Rybnoye Khoziaistvo 3:60–64. (In Russian with English summary.)

Vasil'eva, E. D. 2000. Kutum - *Rutilus frisii kutum. In* Red data book of the Russian Federation [online database]. Available: http://biodat.ru/db/rb/rb.php?src=1&vid=192. (In Russian.)

Vasil'eva, E. D., and V.A. Luzhnyak. 2013. Fishes of the Azov Sea basin. Southern Scientific Center of the Russian Academy of Sciences, Rostov-na-Donu, Russia. (In Russian.)

Voskoboinikova, O. S. 2006. Kutum - *Rutilus frisii kutum* (Kamensky, 1901). Pages 210–

212 *in* O. F. Gritsenko, A. N. Kotlyar, and B. N. Kotenev, editors. Commercial fishes of Russia, volume 1. VNIRO Publishing, Moscow. (In Russian.)

Zavialov, E. V., G. V. Shlyakhtin, A. B. Ruchin, and V. A. Shashulovsky. 2006. Ichthyofauna of the northern lower-Volga River region: modern trends in distribution and abundance of rare and endangered species. Povolzhskiy Journal of Ecology Special Issue: 57–77. (In Russian with English summary.)

Zhukinsky, V. N., and A. I. Balan. 1959. Acclimation of Kutum in waters of Ukraine. Rybnoye Khoziaistvo 4:23–26. (In Russian.)

A Long and Participatory Process towards Successful Fishery Management of Gökova Bay, Turkey

Vahdet Ünal*
Faculty of Fisheries, Ege University
Bornova, İzmir 35119, Turkey
and
Mediterranean Conservation Society
Doğa Park Villaları, 3360 Sokak, No 7/16, Kalabak, Urla, İzmir 35430, Turkey

Zafer Kizilkaya
Mediterranean Conservation Society
Doğa Park Villaları, 3360 Sokak, No 7/16, Kalabak, Urla, İzmir 35430, Turkey

Abstract.—In Gökova Bay (Turkey), in the Mediterranean Sea, a number of problems were encountered within the small-scale fishing sector, including illegal fishing activities and declines in landings of valuable species, especially groupers (Serranidae) and shrimps (Decapoda crustaceans). Fishing income was not sufficient to sustain the livelihood of fishers who depended solely on fishing. Thanks to a collaboration with academics, government administrators, nongovernmental organizations (NGOs), and local fishery cooperatives, six no-fishing zones (NFZ) were officially declared in July 2010 to support small-scale fisheries in Gökova Bay. Establishing NFZs, however, increased illegal fishing due to the lack of enforcement and made problems worse rather than providing a solution. In 2012, the Mediterranean Conservation Society (an NGO) established a marine ranger system by training and employing local fishermen as marine rangers, equipping them with fast boats, and working in close cooperation with the Coast Guard. These actions and many others (e.g., supporting marketing of invasive species) have led to conservation of fish stocks as well as an increase in fishers' incomes. Dramatic increases in abundance for some species have occurred in comparison to previous years. Biomass of predator fishes, such as groupers, increased significantly in NFZs. Mean fishing income per vessel reached the highest level ever recorded. Today, we are near the end of a long process to reach a solution for all stakeholders (fishery cooperatives, academics, NGOs, and government officials) in Gökova Bay. Lessons learned in the process were the importance of working together, building trust and cooperation among stakeholders, implementing a marine ranger system for enforcement, creating marketing opportunities for invasive species, and developing a data collection system for improved monitoring. These lessons may be replicated in other areas of Turkey and elsewhere in support of sustainable fisheries.

* Corresponding author: vahdetunal@gmail.com

INTRODUCTION—GÖKOVA BAY

Gökova Bay is located on the southern Mediterranean coast of Turkey within the Mediterranean Basin global biodiversity hotspot and World Wildlife Fund Global 200 ecoregion (Olson and Dinerstein 2002). Part of the Gökova Bay, including marine and land areas, was designated a special environmental protection area (SEPA) by Decree No. 88/13019 of the Cabinet of Ministers of Turkey (June 12, 1988). Two years later, a border change of the Gökova SEPA was approved to make it larger by Decree No. 90/1117 of the Cabinet of Ministers (October 22, 1990) (Aktaş et al. 2011). Considering only the marine area, Gökova Bay is the largest marine protected area (MPA), including no-fishing zones (NFZs), in Turkey (Figures 1 and 2).

The bay is an 82,700-ha MPA and one of Turkey's 305 listed Natura 2000 sites. Kıraç and Veryeri (2010) reported that 73% of all fish species recorded in Turkish waters can be found here (indicating that this marine environment is rich in terms of marine biodiversity. The economy of the region depends on tourism, agriculture, and fishing. Small-scale fisheries have developed in the region due to the conditions of the coastal line, the geography, and the productivity of the bay, as well as the status of the area pertaining to conservation.

The Gökova Bay MPA has high species diversity, with 723 macroscopic species (both from fauna and flora) present belonging in 19 systematic groups. Thirty-four of these species are protected under national and international treaties. Twenty-six nonnative species have invaded the Mediterranean through various routes, and some have even become dominant over local species (Okuş et al. 2006).

This chapter chronicles the fisheries of Gökova Bay, the problems it faces, the information gained from research, management actions taken to solve problems, interactions with stakeholders, and the response of the fisheries and fish populations to these actions.

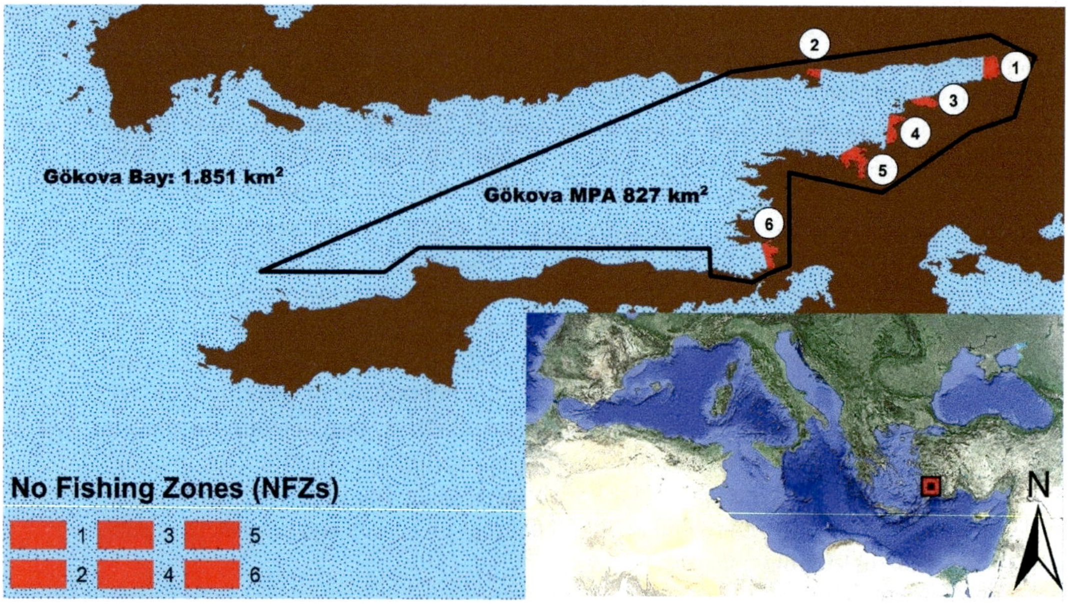

FIGURE 1. Map of Gökova Bay, marine protected area (MPA), and no-fishing zones (southern Aegean Sea), Turkey.

FIGURE 2. Gökova Bay, Turkey marine protected area (Photo credit: Z. Kızılkaya).

GÖKOVA BAY FISHERIES

The fisheries in Gökova Bay are similar in targeted species and types of gears used in fisheries elsewhere in the Mediterranean Sea. Fishing is one of the major livelihoods in Gökova Bay; three main settlements in the study area (Akçapınar, Akyaka, and Sarnıç-Akbük) have their own fishery cooperatives. The Akcapinar Fishery Cooperative of Gökova was established in 1973. Fishers in Akyaka established the Akyaka Fishery Cooperative in 1992. Last, in 1999, a limited number of fishers from Sarnıç Village came together to establish the Sarnic-Akbük Fishery Cooperative. All cooperatives had the aim of marketing members' products at better prices than in the past, to sell fish directly to markets, and to eliminate fishmongers, sellers (middlemen) between the members and the consumers. Cooperatives helped to stop dynamite fishing through better enforcement and are now fighting against illegal spearfishing practiced with scuba diving equipment (especially during nighttime, using lights). Also, many of the fishery cooperatives have helped their members to market their own products and to solve problems related to legislative procedures (e.g., renewal of fishing licenses of fishers when they expire). Today, the fishery cooperatives represent 66 vessel owner members who fish in Gökova Bay. In all three fishery cooperatives, groups are small enough to retain the interest of fishers; additionally, no evidence exists of corruption, larceny, or other dishonest activities in any of the cooperatives. All had been formed on the basis of local initiatives, in response to the fishers themselves. All three cooperatives have managers. Two of these (Akyaka and Akçapınar), which also offer marketing services, each have a full-time employee. Salaries of the managers are symbolic as they are below minimum wage. None of the cooperatives in the region has a professional manager. Almost 99% of the cooperatives' revenues come from marketing fish. Species that belong to Sparidae and Epinephelinae are the focus for the fisheries in this region. Nevertheless, Randall's Threadfin Bream *Nemipterus randalli*, an invasive alien species first recorded in the bay in 2013 (Gülşahin and Kara 2013), constitutes the maximum percentage (18%) in total catch (Figure 3).

More than 100 vessels fish in the Gökova MPA, with approximately 15 additional vessels occasionally arriving from neighboring areas. Gill-net and longline fisheries dominate the small-scale fishery in the bay. In addition to the approximately 115 small-scale fishing boats presently operating, two local purse seiners also fish in the bay. Small-scale fishing boats are 6–12 m in length and locally built of wood. Usually, each boat is operated by one or two fishermen. They use gill nets, trammel nets, and longlines. Additionally, the small-scale fishery provides considerable employment (approximately 200 fishers directly work on the boats).

PROBLEMS FACING GÖKOVA BAY FISHERIES

Illegal fishing mostly performed by spearfishers from outside the area is considered to be the most important problem among local small-scale fishers. It has disastrous environmental and socioeconomic impacts on Gökova Bay by causing depletion of local and high-value fish stocks that play a vital role in traditional fishing. Unregulated fishing contributes to the decreasing income of local fishermen and yields unfair competition because of price differences of legal and illegal fishers' products. Fishery cooperatives and the Coast Guard successfully stopped fishing by dynamite at the beginning of 2000s, but illegal spearfishing has replaced dynamite fishing (Ünal and Erdem 2009). Illegal spearfishing is practiced by professional divers (usually by people outside the region) with a light source and scuba equipment during night in many different locations of the bay. Cooperative authorities report that 2.5 metric tons of groupers, 1 ton

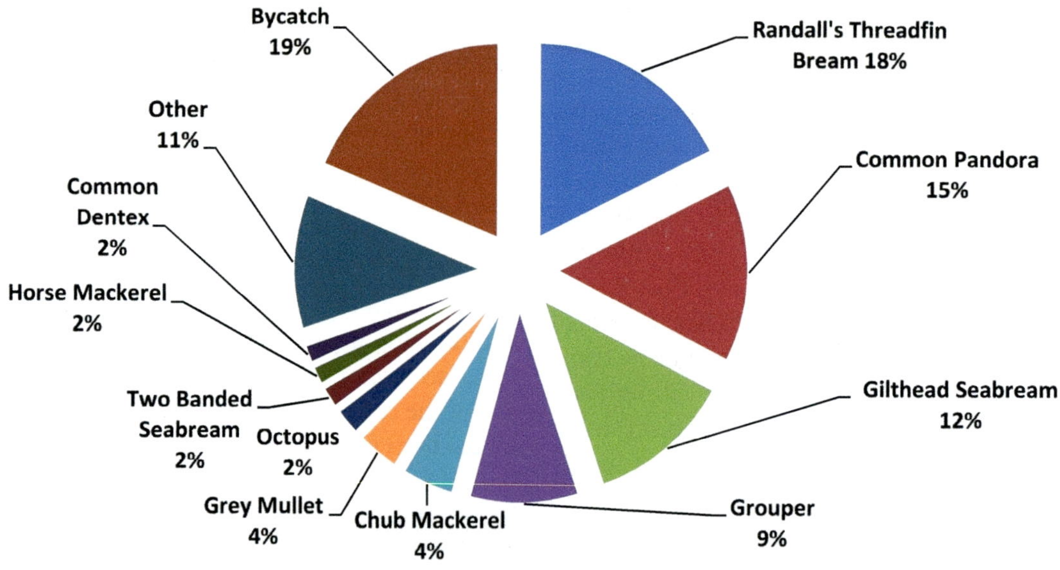

FIGURE 3. Main target species and their percentage of total catch of the Akyaka Fishery Cooperative in 2016.

of Dentex *Dentex dentex*, and 1.5–2 metric tons of Sea Bass (also known as European Bass) *Dicentrarchus labrax* (also *Morone labrax*) are caught this way each season. Based on the reports of cooperative authorities, the amount of illegally caught groupers (White Grouper *Epinephelus aeneus*, Golden Grouper *E. alexandrinus*, and Dusky Grouper *E. guaza*), Dentex, and Sea Bass is larger than the cooperative's annual catch of these species (C. Gorgun, Akyaka Su Ürünleri Koooperatifi, personal communication).

Another problem is an invasive pufferfish, the Silver-cheeked Toadfish *Lagocephalus sceleratus*, first observed and recorded in 2003 in Gökova Bay in the Mediterranean by Akyol et al. (2005), and that poses a serious problem to the gears and catches of fishermen (Ünal et al. 2015a; Ünal and Göncüoğlu Bodur 2017). Silver-cheeked Toadfish, an aggressive predator, is among the most devastating invasive species to both fisheries and habitats. This species, after a very short period from its first appearance, established a population that has been colonizing new territories of the eastern Mediterranean Sea at a rapid rate. The Silver-cheeked Toadfish is a carnivorous species that feeds on crustaceans, including prawns, during their juvenile phase and then shifts to a mollusk diet as it matures and increases in size. It is one of the largest members of its family, reaching 110 cm and 7 kg. This species threatens local crustacean and mollusk populations, and it can destroy fishing nets by attacking entangled fishes, creating large-scale impacts on local fisheries. A few years after the Silver-cheeked Toadfish was first reported in Gökova Bay, the prawn fishery, which had previously been the backbone of the Gökova Bay fishing economy, collapsed completely. Similarly, in 2008, landings of octopus dropped to negligible amounts. As a result, overfishing, illegal fishing, and invasive species have negatively impacted the fish stocks in Gökova Bay.

According to Ünal et al. (2015a), pufferfish create great problems for fishers by damaging fishing nets and eating fish caught in fishing gears, requiring extra labor and gear modification costs. The economic impact of pufferfish populations in Gökova Bay on incomes for fisherman using longlines was estimated at US$1,302 annually, and for those using trammel nets, $1,616 per year. In Turkey, the overall annual loss for small-scale fishers caused by pufferfish was estimated at approximately $2,4 million for 2011. The loss caused by pufferfish in 2013 reached $6 million, more than double when compared to 2011 (Ünal and Göncüoğlu Bodur 2017).

An international research team from Centre D'estudis Avançats De Blanes Spain (CSIC), with participation from other Mediterranean countries, conducted research on a rocky reef in the northern Mediterranean Sea in 2008 for the assessment of reef health, fish stocks, and other benthic life. Fish census studies showed that Turkish coasts of the eastern Mediterranean Sea had the lowest fish biomass while Gökova Bay had the lowest biomass (less than 4 g/m^2) among all research sites evaluated (Sala et al. 2012). The research also revealed that apex predators were rarely seen on the Gökova Bay reefs; on the other hand, the majority of the fish biomass were invasive species of rabbitfishes (Dusky Spinefoot *Siganus luridus* and Marbled Spinefoot *S. rivulatus*) due to the changes of species composition as a result of climate change impacts (Cochrane et al. 2009).

A research project conducted in 2009 revealed the massive impacts of invasive rabbitfish species (Dusky Spinefoot and Marbled Spinefoot), which are herbivores. Sala et al. (2011) reported that invasive rabbitfish species targeted macroalgae and the epilithic algal matrix, which contains early life history stages of algae, therefore preventing macroalgae from becoming established. Indeed, adult algae in regions dominated by rabbitfish were only present below a depth of 25 m hidden within crevices or in wave-washed regions where low water temperature restricts rabbitfish distribution. This massive scale of deforestation of aquatic vegetation caused by tropical rabbitfishes poses a major threat to Mediterranean biodiversity and ecosystem services. As the algal forest disappeared in the region, main nursery and feeding grounds of many juvenile fish and invertebrates disappeared unnoticed by fishers and management organizations. Chronic overfishing in Gökova Bay and other eastern Mediterranean Sea areas may have released predation pressure on rabbitfish and other herbivores that then facilitated their survival in the eastern part of the basin. Additionally, apex species such as large groupers and Dentex were missing in regions (Sala et al. 2012), mostly because of overfishing and habitat loss. Skliris et al. (2012) stated that the whole Mediterranean Sea displayed a significant warming trend of 0.037°C per year from 1985 to 2008, especially in the eastern Mediterranean Sea. As the sea-surface temperatures have risen over the past decades, the eastern Mediterranean Sea has become increasingly hospitable to the growth, reproduction, and survival of tropical species. This change in temperature has led to an increasing migration of numerous invasive marine species to the area, with profound effects on the local ecosystem and the fisheries sector.

FISHERIES MANAGEMENT OF GÖKOVA BAY

Currently, no local-level fishery management programs exist as all management is vested in the central government of Turkey. Territorial-use rights fishery management, spatial-based fishery management, or ecosystem-based fishery management has not been practiced anywhere in Turkey. Fishing activities are governed centrally in accordance with the first comprehensive fishery law (no. 1380) passed in 1971 and amended in 1986 (no. 3288) and 2003 (no. 4950). Recently, a new fishery law, developed with consultation of all stakeholders, is waiting for official approval. The office responsible for the management of Turkey's fisheries is the Ministry of Food, Agriculture, and Livestock (MFAL), which is now the Ministry of Agriculture and Forestry as of September 7, 2018. Fisheries management operates through legal restrictions and prohibitions on certain fishing gears, areas, and fishing times based on biological concerns rather than socioeconomic concerns, or a combination of both. Although a formal system has not yet been established where all concerned stakeholders could participate in decision making, MFAL does call a consultative meeting before the release of a fishing notification and considers the proceedings of the meeting when revising the notification.

In the entire Gökova Bay, fishing activities are regulated by central regulations (fishing notifications), but not based on any local fisheries management plan. Until 2010, conventional management measures such as mesh size regulation, hook-size regulation, minimum landing size for some species, and temporal and spatial restrictions for large-scale fisheries were in practice in Gökova Bay fishery. However, these management measures have not been sufficient to ensure sustainable fishing. Critically, only spatial and temporal restrictions were used to manage large-scale fisheries in the bay. Open access to the bay's fisheries created a free-for-all that damaged the livelihoods of local fishermen and, in combination with invasive species and climate change, eroded ecosystem function. As one example, the two staples of the local fishing economy, White

Grouper and caramote prawn *Penaeus kerathurus*, accounted for 31% and 10% respectively, of total fishing incomes of fishery cooperative in 2006 (Ünal et al. 2009). In the 2009–2010 fishing season, Ünal (2010) reported that those numbers dropped to 10% and 0.02%, respectively, with the caramote prawn almost disappearing completely from the bay's ecosystem. Since 2010, NFZs were introduced to the Gökova Bay and fisheries have rebounded. Detailed current regulations related to Gökova Bay fishery are given below.

CURRENT MANAGEMENT MEASURES AND REGULATIONS IN GÖKOVA BAY FISHERY

The fishing notification contains all the relevant regulations and restrictions related to fishing in all over the seas, bays (including the Gökova Bay) in Turkey. Those relevant to Gökova Bay, regardless of whether they are common in Turkey or specific to for Gökova, are summarized below.

Bycatch

A total of 44 species have a minimum size limit based on biological studies. Undersized catches are permitted for certain species such as European Anchovy *Engraulis encrasicolus*, Sardine (also known as European Pilchard) *Sardina Pilchardus*, and Horse Mackerel (also known as Mediterranean Scad) *Trachurus mediterraneus* up to 15% of total catch, and in other fish species and fisheries up to 5%. Five percent of the catch of Atlantic Bluefin Tuna *Thunnus thynnus* between 8 and 30 kg or 75–115 cm also is permitted (Anonymous 2016).

Restrictions on Species and Hook Sizes

- Because they are toxic, fishing of Silver-cheeked Toadfish, Half-smooth Golden Pufferfish *Lagocephalus spadiceus*, and Ocean Sunfish *Mola mola* is strictly prohibited along the Turkish coastline.
- Fishing of Atlantic Bonito *Sarda sarda* by any fishing gear is prohibited between April 1 and August 31. However, Atlantic Bonito fishing is allowed by trolling between August 14 and 31.
- Atlantic Bonito nets may not be set between 0700 and 1900 hours in the Gökova MPA.
- Fishing of Turbot *Psetta maxima* (also known as *Scophthalmus maximus*) is prohibited between April 15 and June 15.
- Fishing of Common Sole *Solea solea* and *Pleuronectes* spp. is prohibited between December 15 and January 15. Mesh size of the nets must not be smaller than 80 mm.
- Fishing Atlantic Bluefin Tuna is prohibited during the 11 months between June 25 and May 25.
- Fishing Leerfish *Lichia amia* is prohibited between April 15 and May 15.
- Leerfish nets may not be set between 0700 and 1900 hours in the Gökova MPA.
- Fishing Dolphinfish *Coryphaena hippurus* is prohibited between January 1 and August 14.
- Fishing Swordfish *Xiphias gladius* is prohibited between February 15 and March 15 and between October 1 and November 30. Minimum gape of hooks is 2.8 cm and minimum catch size is 125 cm.
- Fishing of shrimp by all gears (except by gill net) is prohibited between April 15 and September 15 in the Mediterranean Sea and between April 15 and August 31 in the Aegean Sea.
- Fishing Dusky Grouper and White Grouper is totally prohibited due to the sharp decrease in their catch amounts according to fishery statistics by the Turkish Statistical Institute (TUIK).
- Fishing of the common octopus *Octopus vulgaris* is prohibited between April 15 and October 31.

Temporal Restrictions

- Fishing with longlines and gill nets is allowed throughout the year. However, fishing with these gears is prohibited within 500 m of harbor and creek mouths. No fishing may occur inside harbors and creek mouths.
- Fishing with bottom trawls is prohibited between April 15 and September 15.
- Fishing with purse seiners is prohibited between April 15 and August 31.

Regulations on Fishing Gears

- Fyke nets, beach seines, boat seines, and monofilament nets are prohibited.
- All trawling and purse seining activity is prohibited in the areas that are in the east of the line connecting two points, Ören Cape (37°01.055′ N–27°56.751′ E) and Teke Cape (36°54.410′ N–28°00.921′ E).

Depth Regulations

- Trawling is prohibited at depths greater than 1,000 m.
- Use of purse seines is prohibited in waters shallower than 24 m depth.

Regulation on Closed Areas

- All fishing activities have been prohibited in six selected areas within the Gökova SEPA since July 10, 2010. These areas are designated as NFZs, and all fishing activities, including recreational fishing, within these areas are banned. These six NFZs are shown and numbered in Figure 1.

According to the fishing notification published in 2016, MFAL permitted shore-based fishing activities in three of the six NFZs (Akyaka, Akbük, and Çamlı) due to the low impact of these fisheries on fish stocks and small-scale fishing. The size of NFZs 4–6 was further extended to include more coralline habitats (Figure 1; Anonymous 2016):

Recently, Ünal et al. (in press) indicated that the existing management regulations related to Gökova Bay are having a positive effect, with some rules and regulations being more effective than others; however, further measures are still needed to support sustainable fisheries (Table 1). Effectiveness of management measures was determined by local fisheries stakeholders during the second stakeholder meeting of the Food and Agricultural Organization of the United Nation's East Mediterranean Project (FAO-EastMed Project). Meanwhile, buyback of certain large-scale fishing gears could be a good option for the Gökova Bay. Ünal et al. (2016) analyzed fisheries buyback programs in Turkey and suggested that a buyback program could be used for eliminating trawlers and purse seiners that operate in the Gökova Bay MPA. Once a buyback program is completed, this type of fishing activity could be prohibited within the Gökova Bay MPA.

AWAKENING OF THE STAKEHOLDERS

Fisheries stakeholders have learned to confront the challenges and the problems of fisheries in the bay and to participate directly in the management of its fisheries. They participate in meetings, raise their voices, and develop joint actions and help decision making (Figure 4).

The stakeholders realized that they had to do something for the sustainability of the fishery in the bay; otherwise, they would lose their jobs. The disappearance of shrimp and the large decrease in the abundance of groupers have threatened their livelihoods. More importantly, the decrease in fishing incomes has made it almost impossible for fishing activity to be viable. For the first time, the prohibition of fishing in spawning and nursery areas of Mediterranean seagrass *Posidonia oceanica* and

TABLE 1. Present management measures in practice in the Gökova marine protected area (MPA; Ünal et al., in press). Evaluation of effectiveness was carried out at the second stakeholder meeting to develop a management plan for the small-scale fishery (SSF) in Gökova Bay in accordance with the ecosystem approach to fisheries. Effectiveness of management measures was considered as follows: 1 = weak, 2 = fairly good, and 3 = very good.

Management measures	Description	Effectiveness
Spatial restrictions (e.g., closed areas, MPAs, etc.)	The MPA of Gökova is 827 km^2. Within the MPA, there are six no-fishing zones officially designated in July 2010. Trawling and purse seining are prohibited within the MPA.	2
Temporal restrictions (e.g., closed seasons)	There is no closed season fro SSF in general, but temporal restrictions for certain species are in practice.	2
Gear restrictions (e.g., forbidden gears, limits to mesh size, etc.)	Beach seines and boat seines are prohibited in the Gökova MPA as well as all Turkish seas. It is also prohibited to use monofilament nets. Minimum mesh size for Common Sole is 80 mm. The minimum mesh size of gill nets is 32 mm. Maximum length of fishing nets (gill nets, trammel nets) is 6 km per fishing vessel.	3
Minimum size	There is minimum landing-size regulation for many species. It is shown in the fishing notification, which covers all relevant regulations in fisheries.	3
Participatory restrictions (e.g., licensing, territorial use rights fisheries, etc.)	Licensing system is in practice.	3
Limits to fishing capacity (e.g., maximum number of vessels, fleet reduction, etc.)	No. Three buyback programs were applied to the fisheries at the national level.	2
The Fishing Vessels Monitoring System (VMS)	The VMS is obliged to be used for fishing vessels that are longer than 12 m.	2
Others; restrictions on depth for purse seiners	There is a 24-m-depth regulation for purse seiners.	3

reproduction areas for endangered species such as the Sandbar Shark *Carcharhinus plumbeus* was discussed among fisheries stakeholders. This prohibition was the first effective step towards declaring and designating NFZs. Fishers and their leaders have organized small, but effective group meetings for discussing the challenges facing the sustainability of their fisheries (Figure 5).

FIGURE 4. Local fishers and fisheries stakeholders in the meeting on the proposal of no-fishing zones, 2009 (Photo credit: V. Ünal).

FIGURE 5. Small-group meeting on determination of potential no-fishing zones in Gökova Bay, 2009 (Photo credit: [left] V. Ünal, [right] H. Göncüoğlu).

PARTICIPATORY APPROACH, ESTABLISHMENT OF NO-FISHING ZONES, AND MARINE RANGER SYSTEM

Meetings with stakeholders were held to facilitate collaboration with academics, government administrators, and local fishery cooperatives to agree on overall goals. As a result, six NFZs were officially declared in July 2010 to support biodiversity and small-scale fisheries in Gökova Bay. Ünal et al. (2015b) provides detail information on the establishment of the NFZs and convincing fishers to do so. Unfortunately, establishing NFZs increased illegal fishing and made the problems worse rather than better in the short term due to the lack of enforcement by official responsible bodies. This situation created unfair competition because while responsible fishers were following the rules, others were not and continued fishing in these areas. This problem spurred the foundation of the Mediterranean Conservation Society (AKD) in 2012 as a locally based organization engaging with both local knowledge and scientific data to develop management programs for conservation and poverty reduction in Gökova Bay. Establishment of the AKD may be considered a milestone for the restoration of the fishery in the bay. The founders of the AKD were academics, fishers, businesspeople, and environmentalists. The primary objective of the organization's work has been to institute and enforce a series of interconnected NFZs in which fishing of any type is prohibited. Although well-enforced MPAs offer opportunities for ecosystem protection (Sala et al. 2012), NFZs are critical to fully restoring marine ecosystem function after years of overfishing. No-go, no-fishing zones in the entire Mediterranean Sea (2,516,900 km^2), however, currently account for less than 1% (0.04%) of the total marine area of the Mediterranean Sea (MedPAN and UNEP-MAP-SPA/RAC 2017). The AKD envisions raising that number to 10% through a combination of local empowerment and community capacity building, as well as organizing meetings with stakeholders similar to methods used in the past on the way to establishing NFZs. The organization has a high level of respect for small-scale fishermen, who have accumulated generations of knowledge about local fishing conditions in Gökova Bay. The integration of traditional knowledge with multi-stakeholder partnerships and enforcement mechanisms enables the AKD to serve as a model for other marine areas around the world.

In 2012, the AKD established a marine ranger system in the bay that puts small-scale fishermen at the forefront of monitoring compliance. At the core of this approach is a participatory marine resource management model that engages a wide range of stakeholders (e.g., the Coast Guard, fishery cooperatives, the General Directorate for Protection of Natural Assets, and the Directorate-General for Fisheries and Aquaculture [DG-Fisheries]) in monitoring, evaluation, and program design. This approach also ensures that local fishermen have a high level of ownership and are drivers of the management process. To build the capacity of local communities to enforce and monitor designated NFZs, the organization provides marine guard training that integrates a combination of theoretical knowledge and practical training. Training topics range from protected area patrolling to boat safety and from marine biodiversity assessments to public awareness raising strategies. Local fishing cooperatives choose trainees for the program, selecting experienced fishermen with the most knowledge of fishing grounds and also the types of illegal activities that need to be addressed. The AKD provides tools and resources to ensure that each community marine guard unit is fully equipped to effectively enforce and monitor NFZs. The AKD trained and employed four local fishermen as marine rangers and equipped them with fast boats. Since January 2013, they have been working in close cooperation with Turkey's Coast Guard in Gökova Bay.

The organization has worked with the government to legitimize and amplify the reach of community-based surveillance and patrol activities through support from the national authorities in Turkey. Although community rangers cannot prosecute and fine illegal fisherman, they can record incidents with cameras and report illegal activities to the Coast Guard and submit evidence. Equipped with this information—photo and video evidence—the Coast Guard can then take legal action against violators. In some cases, local officers from the DG-Fisheries have joined community rangers during their patrols, bolstering the alliance between government and community members. This cooperation with government authorities has dramatically improved communication, monitoring, and, in turn, enforcement of fisheries regulations. A central part of this enforcement campaign has been raising local awareness about the location and importance of NFZs. In collaboration with marine guard units, the organization has set up 20 prominent signboards in the NFZs with detailed information about NFZ boundaries. All entry points to NFZs have likewise been marked with signposts to raise awareness among community members and recreational fishermen (e.g., anglers, spearfishers). The organization also produced and disseminated a "Codes of Good Practice" document to all local resource users, detailing proper conduct in and around NFZs. Raising awareness among stakeholder has been complemented by rigorous enforcement with marine guard units actively surveying NFZs three times daily to monitor activity and apprehend any violators. The AKD has provided a framework and methodology for establishing community-based enforcement strategies and empowering local fishermen to lead cooperative efforts with the support of government. This approach provides a high level of community ownership of the NFZ concept, the benefits of NFZs, and their need to be properly enforced.

RESULTS OF ACTIONS AND THEIR EFFECTS

On Fishers and Fishery Cooperatives

The AKD has supported community-based enforcement strategies as a means of protecting and restoring marine ecosystems. The initiatives of AKD have led to positive improvements in local fishery management, which has meant increases in catches (Figure 6) and incomes of local fishermen (Figure 7). The increase in income is due to stricter enforcement of NFZs, which has led to both a reduction in illegal fishing, demonstrated by the decrease in an average number of fishing boats and an increase in fish stocks. The Akyaka Fishery Cooperative reports a more than fourfold increase in the average income of small-scale fishers 4 years after implementing NFZs. This increase is indicative of improvements within the other Gökova Bay fishing cooperatives.

One of the notable returning species is White Grouper, which, by itself, accounted for 31% of fishing incomes in 2006. In the 2009–2010 fishing season, this income contribution dropped to 10%, but by 2015, 5 years after implementing NFZs, the species had increased to 25%. In 2016, the DG-Fisheries prohibited any kind of grouper fishing in all Turkish waters without exception. They made this decision based on statistics from the TUIK, showing decreasing catch statistics of groupers overall, but they did not check the statistics of Gökova Bay, which were higher. The ban was put into practice everywhere in Turkey (Figure 8).

Akyaka Fishery Cooperative members employed as rangers and data collection managers now have new skills in communi-

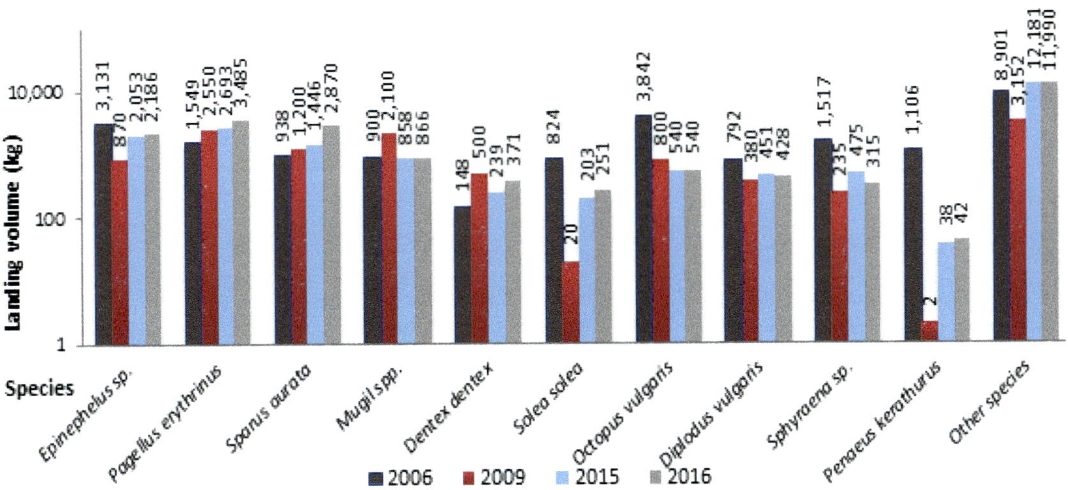

FIGURE 6. Comparison of catch data in Gökova marine protected area by year (2006, 2009, 2015, and 2016).

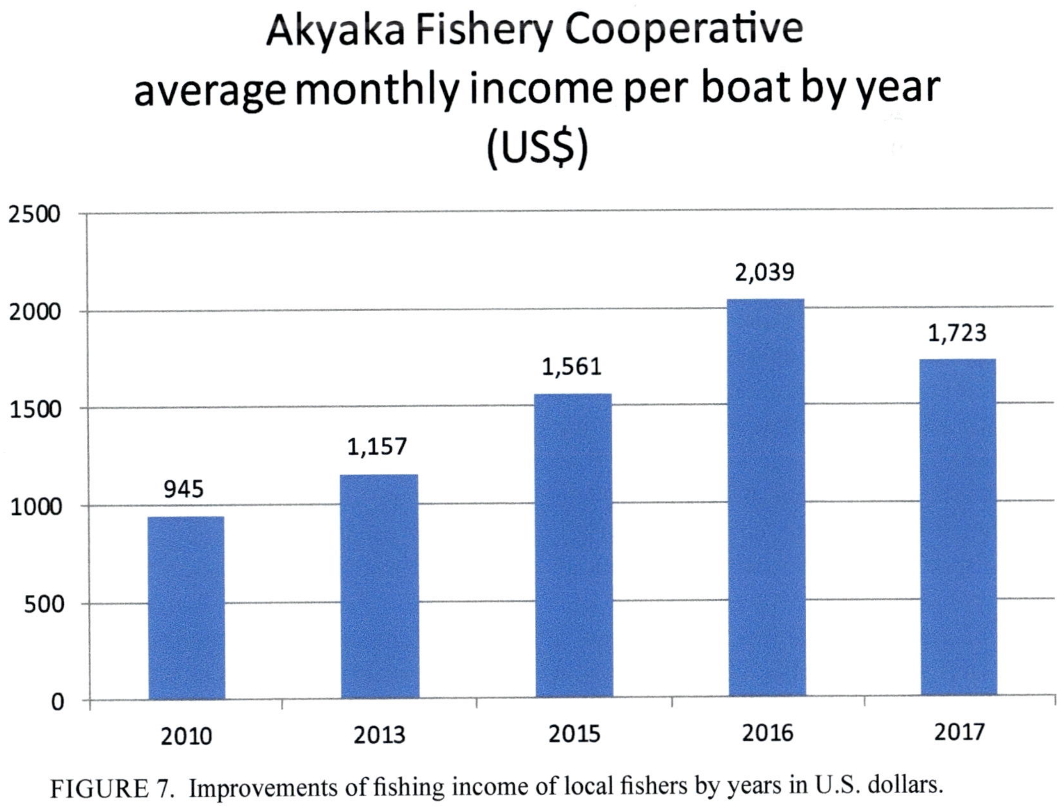

FIGURE 7. Improvements of fishing income of local fishers by years in U.S. dollars.

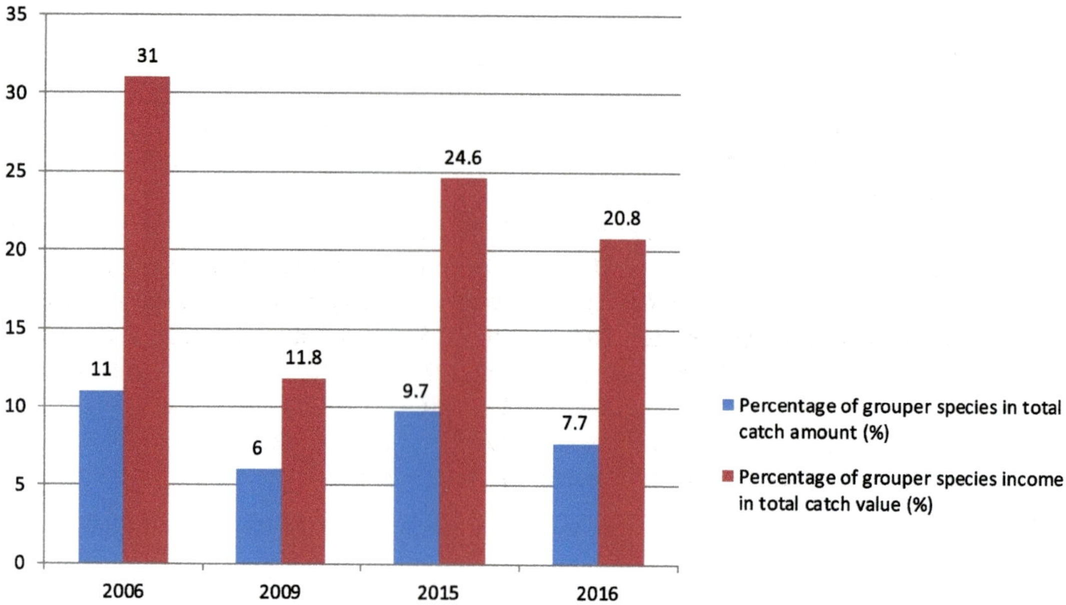

FIGURE 8. Improvements on landing and value of groupers in the Akyaka Fishery Cooperative.

cations, management, and technology. The Coast Guard and DG-Fisheries staff are now their co-management partners and are working together in close cooperation.

To document the effect of invasive species on local fisheries and monitor fish landings in the bay, AKD draws on the reporting system used by local fishing cooperatives. These data have established that invasive species between 2013 and 2016 constituted more than 25% of the total catch of Gökova Bay fisheries. Stopping the spread of invasive species is nearly impossible in a marine environment; however, many of these species are edible but unknown in local markets. The AKD has made significant efforts to promote consumption of invasive species and transform these fish into a commercially viable species for the local fisheries industry. Increasing public awareness and market price of these invasive species has been so successful that Randall's Threadfin Bream, alone, constituted 31% of the total income in 2015 in the Akyaka Fishery Cooperative. This example shows that the success in terms of increasing catch quantity and fishing income is also highly linked to the adaptation of the fisheries to invasive species in the bay.

In addition, improvements in market acceptance are sought for the invasive Brushtooth Lizardfish *Saurida undosquamis*, Marbled Spinefoot, and Goldband Goatfish *Upeneus moluccensis*. However, impact of commercial fishing on the overall abundance of these invasive species in Gökova Bay is not known.

Fishery stakeholder interest in data collection and the implementation of the new data analytic software SMART (Spatial Monitoring and Reporting Tool) has been strong and collaborative. Currently, the Akyaka Fishery Cooperative data set is one of the most accurate and sophisticated fishery data sets available in the country. This reliable 2-year data set now enables us to conduct preliminary analyses and consider variation in species abundance to better advise on fisheries management needs. A key example of this capability relates to White Grouper, which was the subject of a nationwide fishing ban declared on September 1, 2016. At that point, White Grouper accounted for approximately 30% of the cooperative income and, thus, was very important to the community. Armed with a data set showing the recovery of the White Grouper in Gökova Bay, AKD presented these results to the government at an international grouper workshop held in Bodrum, Turkey, in October 2016. On reviewing the evidence, the government representative agreed to re-evaluate the ban in the near future.

The first marine software program in the world, SMART was developed for Gökova Bay rangers by the Zoological Society of London and has been in use since December 2016. Using this program, field data from wildlife observations to illegal activities and patrolling routes are provided in real time and such information is easily shared with rule makers and government officials. SMART also enables the exploration of relationships between fishing data and other variables such as tracking fuel efficiency and identification of hotspots for illegal fishing activities with precise times and locations. SMART allows rangers to directly record all activities effortlessly and enables managers to download them instantly. This capability also helps managers to monitor how often rangers implement regular patrolling times and routes. Now a new cooperative phase has begun, with Gökova Bay rangers as an extension of the Coast Guard. However, one of the most important issues facing the Gökova Bay fishing community is creating a sustainable finance mechanism to pay for marine rangers and their activities in the near future.

Fisherwomen have traditionally been marginalized and excluded from marine management decisions and fishery cooperative membership in Turkey. However, in Gökova Bay and the Datça-Bozburun Peninsula, families have been fishing for generations and fishing responsibilities are shared by men and women (Göncüoğlu et al. 2015). The Underwater Research Society conducted a project in 2012 to (1) document that fishing, generally known in Turkey as a job for men, is also carried out by women; (2) assess the needs of the active fisherwomen of Gökova Bay; and (3) help them continue fishing. The project reached a total of 21 fisherwomen actively working on fishing vessels. In 2013, AKD initiated a project to support more than 70 fisherwomen in the Bozburun-Datça area of Gökova Bay. The AKD has linked, via a cooperative, fisherwomen in this area who work in predominantly poor and challenging conditions without many forms of security, and has also provided the opportunity to discuss challenges such as overfishing and the resulting declines in income. This network supports members' use of sustainable, environmentally friendly fishing gear and ability to actively participate in the cooperative governance of fisheries management. The women are given training in sustainable fishing, labor rights, women's leadership, and microfinance. The cooperative is notable for its activities and achievements, but perhaps even more noteworthy is its successes with women's groups in a male-dominated profession. The cooperative has proven to be a unique and empowering platform for local women to express themselves, to engage in dialogue and debate about resource management, and to lead resource governance.

On Fish Stocks

The AKD is involved in the regular evaluation of protected areas using standard protocols, contributing to a Mediterranean-wide monitoring effort initiated by their partner organization, Mediterranean Protected Areas Network (Med-PAN). Environmental monitoring includes fish populations, algae, and macrobenthic species. Fish biomass, however, is the single most important indicator of the health of MPAs that have previously suffered overexploitation.

A study conducted by the Ecological Research Society in 2013, 1 year after the inception of the NFZs, demonstrated that total fish abundance had risen 27% and that populations of the endangered Dusky Grouper had increased 19% in the English Bay and Akyaka NFZs (EKAD 2013). Furthermore, Kızılkaya et al. (2015) reported that at the end of the third year after NTZ implementation, monitoring of NTZs and surrounding areas showed a substantial increase in fish biomass, and the local fishery cooperative records showed increases in revenues. Revenues of cooperative members over this time period increased 53% according to the Akyaka Fishery Cooperative's accounting statistics.

An in-depth study by AKD throughout Gökova Bay demonstrated that between 2013 and 2016, the biomass of apex predator fishes increased significantly in several NFZs, with the English Bay NFZ demonstrating an average biomass nearly 25 times that observed in unprotected sites (Figure 9). This same study additionally showed that the biomass of apex predators increased most dramatically of all species studied and led to a reduction of invasive herbivorous species (Figure 10). As Sala et al. (2011) have shown, invasive species have had a negative impact on overall ecosystem health in the Mediterranean Sea, so this increase in apex predator biomass and the decrease in invasive herbivorous species biomass are a promising signal for ecosystem conservation in the region.

Although changes in fish biomass and certain sponge species have been documented within a few years of the implementation of NFZs, some of the invertebrate and algal communities typically take longer to recover because invasive herbivore species, such as Dusky Spinefoot and Marbled Spinefoot, have had a great impact on algal canopies, creating barrens (Sala et al. 2011). These species are not able to be caught by fishers in considerable amounts as they graze on shallow bottoms. Furthermore, invasive pufferfish have affected mollusk and crustaceans while increased water temperature has created thermal stress for some species

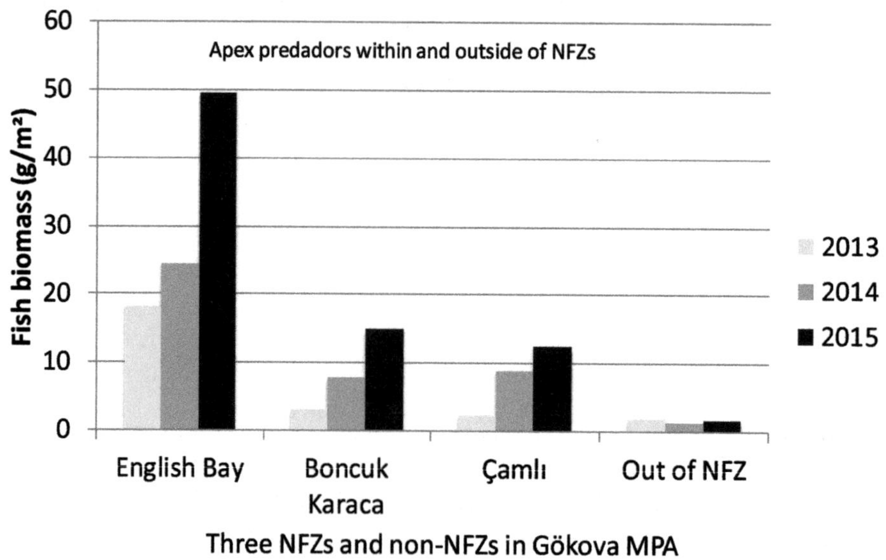

FIGURE 9. Biomass of apex predators inside the no-fishing zones (NFZs) and outside the NFZs in Gökova Bay (2013–2015).

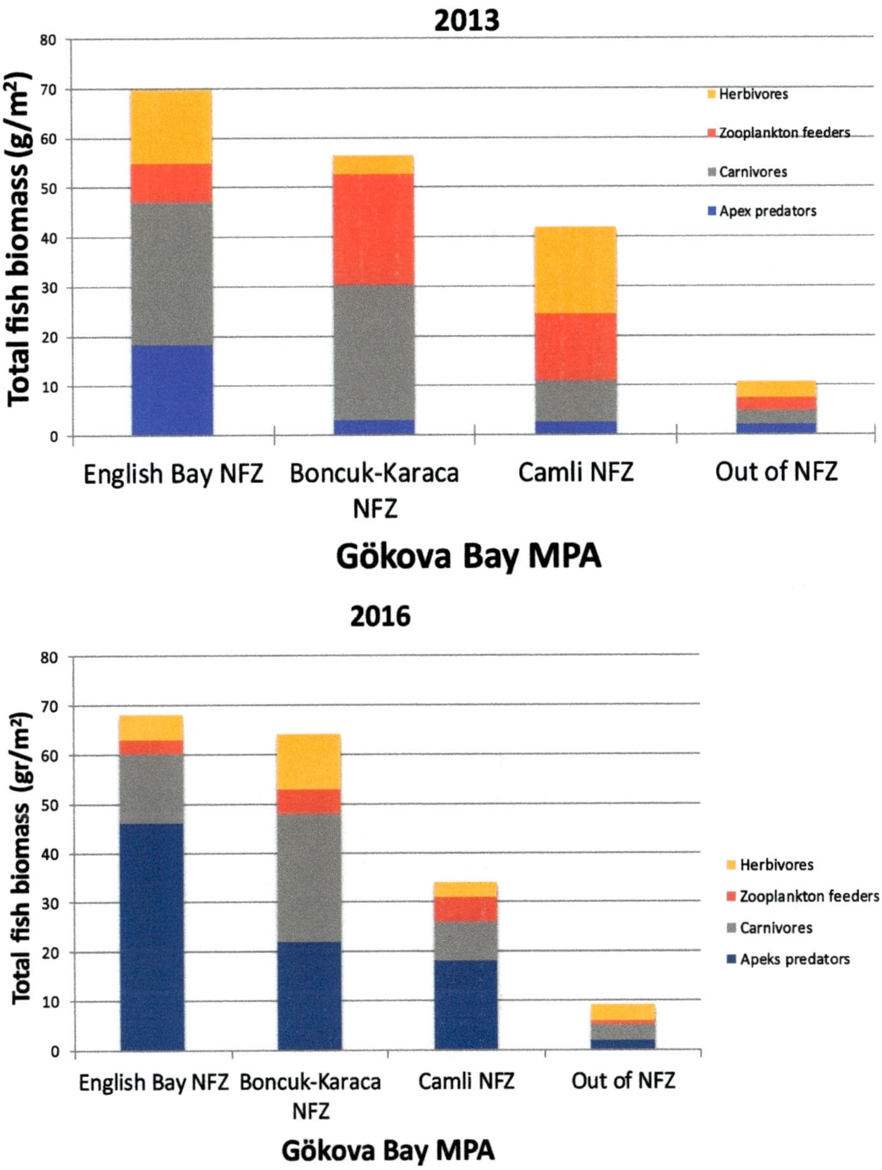

FIGURE 10. Comparision of the biomass status inside the no-fishing zones (NFZs) and outside of the NFZs.

(Figure 11). The increase in water temperature in 2015 and 2016 in the Gökova MPA caused the thermocline to be deeper in 2016. Thus, monitoring efforts need to document changes in the species composition over the next 5 to 10 years. The AKD also started a project in 2015 that uses camera traps in monk seal *Monachus monachus* breeding caves and in Sandbar Shark nursing groups within the MPA to monitor populations of these charismatic endangered species.

The fishery cooperative landing data over 2 years also show restoration in commercial stocks. White Grouper, Pandora *Pagellus erythrinus*, and Gilthead Bream *Sparus auratus* stocks have shown significant improvement. The new data collection system established by AKD in the Akyaka

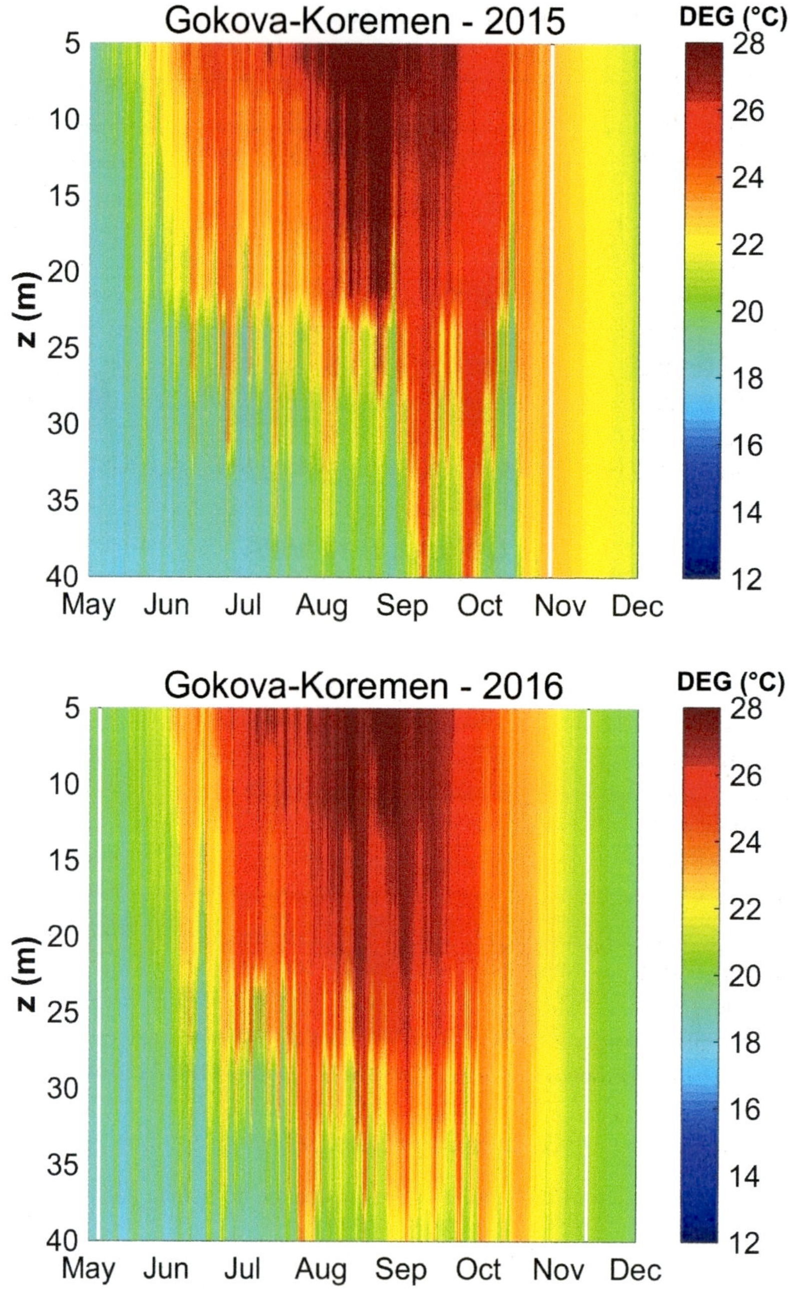

FIGURE 11. Water column temperature in 2015 and 2016 in Gökova marine protected area.

Fishery Cooperative permits comparison of present data to data from previous years. In recent years (2015–2016), the catches of 7 out of 10 species in the Gökova Bay MPA (groupers, Pandora, Common Sole, Twoband Bream *Diplodus vulgaris*, Yellowmouth Barracuda *Sphyraena viridensis*, and caramote prawn), as well as total catch of other species (Randall's Threadfin Bream, Pacific Chub Mackerel *Scomber japonicas*, Horse Mackerel, Red Mullet *Mullus surmuletus*, Dusky Spinefoot, Marbled Spinefoot,

Leerfish, Bluefish *Pomatamus saltatrix*, Shi Drum [also known as Corb] *Umbrina cirrhosa*, John Dory *Zeus faber*, Striped Sea Bream [also known as Sand Steenbras] *Lithognathus mormyrus*, Saddled Seabream [also known as Saddle Bream] *Oblada melanura*, Sea Bass, and Atlantic Bonito) has been greater than the catch estimated in 2009 (Figure 6).

The year 2006 was one of the golden years for fishers in terms of fishing, but data from 2015 to 2016 indicated that fishing in the Gökova Bay has started to return to these former levels. It should be noted that data from 2015 to 2016 come only from the Akyaka Fishery Cooperative while the data for 2006–2009 comes from two local fishery cooperatives (Akyaka and Akçapınar Fishery Cooperatives).

Monitoring

The location of Gökova Bay's NFZs within the Mediterranean Basin global biodiversity hotspot provides an excellent opportunity to protect many threatened and keystone species of the region. These include Mediterranean seagrass, Mediterranean monk seal, Dusky Grouper, Sandbar Shark, loggerhead sea turtle *Caretta caretta*, and green turtle *Chelonia mydas*. Within the NFZs, as fishing pressure diminished, many fish species like the Two-banded Seabream (also known as Twoband Bream) *Diplodus vulgaris* have started to appear. In 2016, two of the NFZs expanded to include another 600 ha where important rocky, coralligenous, and sea grass habitats exist. However, extensive ghost nets and fishing lines had deteriorated the habitats over years. A cleanup process is still ongoing. An interesting recovery of a sponge species, yellow tube sponge *Aplysina aerophoba*, occurred in areas where longlines had inflicted extensive damage and reduced the population. The sponge is now flourishing in the NFZs, where some populations have reached 50 cm. Similarly, an important traditional species sought by sponge divers, *Calyx nicaeensis*, is slowly recovering below 20 m depth. The species is protected under Turkish legislation since it was almost wiped out by a fungal disease and overharvesting over the past decades. However, no fungal disease has been reported among the propagating population so far. The disease was believed to be related to the increase in temperature and is being monitored.

The harmful invasive rabbitfish species, which graze on the macro algae habitats, have been recorded to be on the decline in terms of biomass within the well-enforced NFZs, proving that the resilience of the system has been increasing with the return of apex predators (Figure 12). Those areas are the only ones where the presence of macro algae has been noted during monitoring; none have been recorded in unprotected habitats.

One of the extended NFZs, Bordubet, has the richest seasgrass beds in the Gökova Bay, and now more habitat is under protection. However, in other areas, anchor damage from boats is still a major threat for sea grasses and the pina shell *Pina nobilis*, another protected species.

Trap cameras in four monk seal caves show seal presence in each of them, and sightings by the fishing community have never been this frequent in the recent past. According to the seasonal monitoring records of AKD (unpublished data), underwater surveillance cameras in the Boncuk NFZ, where sandbar sharks come for breeding, have noted the presence of the sharks and have recorded natural behavior of the species and evidence of their abundance.

Mentality of Fishers

The fishers of Gökova Bay are now acting as the greatest supporters of scientific research, fisheries monitoring, control, surveillance, and projects directly or indirectly related to fishing activities and fishery cooperatives. They have been eager to work together with NGOs, the

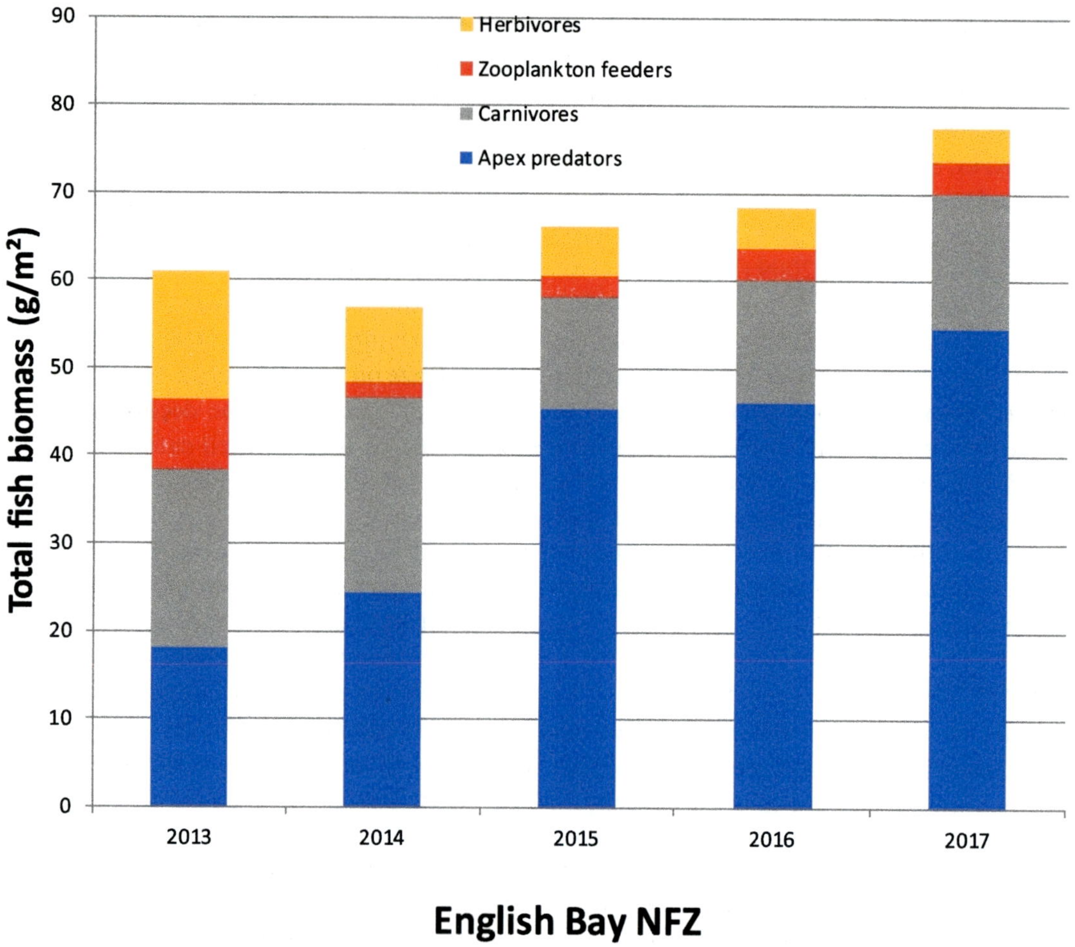

FIGURE 12. Fish biomass increase over the years in English Bay no-fishing zones (NFZ).

academic community, and official institutions. They have a keen interest and support for the FAO-EastMed Project and have contributed to the preparation of the fisheries management plan within this project.

LESSONS LEARNED AND RECOMMENDATIONS

A fairly straightforward process was followed to rebuild fish stocks and improve the economic conditions of fishers in the Gökova Bay fishery by identifying threats to the fisheries and gaps in existing management programs and actions. We tried to understand why the system did not work properly and found that weak enforcement (inspection, control, and surveillance), which allowed illegal fishing to be conducted, was a main cause. As a result, the marine ranger system was developed and implemented to overcome these problems and succeeded in solving the problem of illegal fishing. Other important problems were habitat degradation, overfishing, invasive species, and lack of monitoring. It is accepted that the problem of invasive species could not easily be solved in the short term and that it is impossible to eliminate invasive species like the Silver-cheeked Toadfish from the ecosystem. However, establishing NFZs with better enforcement also helped mitigate the impact of these species by supporting the rebuilding of the ecosystem. For other invasive species, we increased the awareness among the consumers, created market, and increased the demand and their price by organizing festivals. First, after review of available information, many meetings were organized to develop communication with stakeholders. We developed trust among fishers and other stakeholders towards building a participatory process of co-management. Central and local fisheries management authorities were invited to the meetings. The main problems of fishers were discussed on the basis of scientific studies and local traditional knowledge from the most respected fishers of the bay. As a result, six NFZs were established in 2010. Finally, the marine ranger system and fishery cooperative-based data collection system were developed by AKD. The provision of patrolling boats, training of the guards, and associated costs were important problems to solve. These obstacles were overcome with intensive work thanks to several projects. During this time, confidence increased with all stakeholders, especially fishers and heads of local fishery cooperatives. In addition, during this process, meetings were organized to educate and inform stakeholders properly, to enlighten them about good practices and the results of scientific work in other parts of the world.

The creation of NFZs that are monitored and enforced by local communities in cooperation with government institutions has answered the urgent need to protect marine resources in the Mediterranean Basin. This achievement has been amplified by AKD's successful advocacy campaign to restrict access of large-scale fishing vessels, such as purse seiners and trawlers, to the inner Gökova Bay. Today, more than 260,000 ha of the Gökova Bay MPA is closed to trawlers and purse seiners. The previous scientific studies, project carried out in the area, and traditional ecological knowledge helped in reaching success in terms of sustainable fisheries in Gökova Bay.

Despite these unprecedented achievements and successes in the region, an urgent need exists for a fisheries management plan in the bay that will address overfishing and competing fishing interests and that balances conservation needs with local incomes over the long term. Fishery cooperatives of small-scale fishers have begun to delineate the tenets of such a plan. In particular, given the success of the community guard program and the dividends seen thus far from the NFZs, local fishery cooperatives have advocated for co-management as a central aspect of any management plan. In answer to this call, the FAO-EastMed Project has chosen Gökova Bay as the site for an ecosystem-based fisheries management template. Clearly, the local communities are keen on stepping forward to steward and manage Gökova Bay, which illustrates the efficacy, impact, and results of community-based management. The work carried out in the bay, particularly those by the AKD during the past 5 years, have built a bridge between community-based conservation and government policy development that previously did not exist, providing the foundation to address the yet unmet need for a collaborative fisheries management plan in Gökova Bay.

CONCLUSION

The management of fisheries in Gökova Bay is providing a good practice example for Turkish fisheries and even the Mediterranean region. Fisheries co-management, which started with the closure of six areas to any kind of fishing activities in 2009–2010; the foundation of AKD in 2012; and the launching of effective joint venture projects played a big part in contributing to this success. The 2017 Whitley Gold Award for nature conservation was given to the president of AKD, who is also co-author of this chapter

(Z. Kizilkaya), in recognition for the success in Gökova Bay. The AKD has addressed environmental, social, economic, and political challenges in an integrated fashion. In terms of environmental impacts, a substantial improvement in marine biodiversity and ecosystem health has occurred, as indicated by increases in fish biomass and species diversity in and around the NFZs. Increases in fish stocks (whether due to the creation of NFZs and spillover effect or stopping illegal fishing or creating a market for invasive species or the overall effects of all these acts) have translated into economic benefits for local fishers, who have greater food security and a greater abundance of fish to take to market. The fact that these benefits have come about is due to community-based action and a marine guard system run by local fishers, which have led to social cohesion and a sense of local empowerment. Great progress has been achieved in combatting illegal fishing. Government policymakers have become more aware of community capacity and more willing to invest in community-based solutions. The mutually reinforcing benefits begin with proper enforcement and responsible management of marine resources by the people who rely on them most.

The case of small-scale fishery management in Gökova Bay could be replicated in other marine areas in Turkey and throughout the Mediterranean region. Gökova Bay representatives were invited to the meetings in Datça-Bozburun Peninsula and Hisarönü Bay to share the successes and challenges of the community guard program. These meetings resulted in the creation of four new NFZs in the Datça-Bozburun Peninsula and Hisarönü Bay. Although the NFZs in these areas are small compared to Gökova Bay, their inception provides evidence that the model can be successfully initiated in other areas. Although no scientific evidence exists, the local fishers believe that the model has not worked well in these areas due to lack of enforcement (Ecrüment Altınsoy, Karaköy Köyü, personal communication).

Here, we want to recall the fish biomass survey of the coast of the northern Mediterranean Sea in 2008 that showed that Gökova Bay had the least fish (4 g/m^2) of all of the Mediterranean MPAs (Sala et al. 2012). Thanks to many efforts, fish biomass, abundance, fishing income, and habitat protection have increased in the Gökova MPA. Most importantly, awareness has increased in all stakeholders, particularly fishers, of the importance of management for fisheries sustainability.

Today, we are now addressing and implementing an ecosystem approach to fisheries management in the Gökova Bay MPA. The draft management plan in accordance with the ecosystem approach to fisheries was discussed in DG-Fisheries on December 2017 and presented to the stakeholders on March 2018 in the fifth stakeholder workshop of the FAO-EastMed Project. The workshops have resulted in unanimous acceptance of the "Management Plan for the Small-Scale Fisheries in Gökova MPA" (Ministry of Agriculture and Forestry, unpublished report), but it has not yet been officially published by the Ministry of Agriculture and Forestry.

In conclusion, successful fisheries management is possible anywhere, as long as official bodies (management authorities), fishers, and other stakeholders really want it and are willing to work together closely to achieve their objectives. In some cases, such as the Gökova Bay, an NGO may be more willing than other actors (e.g., fishers or even official bodies at the beginning) and even be pioneering in its efforts. No doubt, the contribution of all parties is important, but it is clear that some are more important than others. In this regard, we observed in the Gökova Bay case that success in fisheries management could not be achieved without the willingness and support of AKD, the Akyaka Fishery Cooperative, academics, and governmental bodies. In addition, the many researchers and NGOs who have worked in the region in the past have also been important contributors to the success in Gökova Bay.

REFERENCES

Aktaş N., M. Menengiç, İ. Ünal, E. Akgül, M. O. Yapıcıoğlu, and B. Taşeli. 2011. Reflections from the special environmental protection areas of Turkey. Turkey Ministry of Environment and Forestry, Environmental Protection Agency for Special Areas, Ankara.

Akyol, O., V. Ünal, T. Ceyhan, and M. Bilecenoğlu. 2005. First confirmed record of *Lagocephalus sceleratus* (Gmelin, 1789) in the Mediterranean Sea. Journal of Fish Biology 66:1183–1186.

Anonymous. 2016. Fishing notification to regulate fishing for commercial purpose for 2016–2020 fishing period: No. 4/1. SÜR-KOOP, Official Gazette No. 29800, Ankara, Turkey.

Cochrane, K., C. De Young, D. Soto, and T. Bahri, editors. 2009. Climate change implications for fisheries and aquaculture. FAO (Food and Agriculture Organization of the United Nations) Fisheries and Aquaculture Technical Paper 530.

EKAD (Ecological Research Society). 2013. Monitoring of the fish population in fisheries restricted areas of Gökova SEPA and Selimiye & Orhaniye bays of Datça–Bozburun SEPA (June 2011–December 2013). EKAD, Ankara, Turkey.

Göncüoğlu, H., V. Ünal, and Z. Kızılkaya. 2015. Supporting fisherwomen in small-scale fisheries in Turkey. Pages 447–451 *in* A. Srour, N. Ferri, D. Bourdenet, D. Fezzardi, and A. Nastasi, editors. First regional symposium on sustainable small-scale fisheries in the Mediterranean and Black Sea. FAO (Food and Agriculture Organization of the United Nations) Fisheries and Aquaculture Proceedings 39.

Gülşahin, A., and A. Kara. 2013. Record of *Nemipterus randalli* Russell, 1986 from the southern Aegean Sea (Gökova Bay, Turkey). Journal of Applied Ichthyology 29:933–934.

Kıraç, C. and N. Veryeri. 2010. Putting PEEN to practice in marine and coastal areas: a demonstration project ensuring the ecological resilience, coherence and sustainable future of Gökova Bay SEPA in Turkey. SAD-AFAG, Ankara, Turkey.

Kızılkaya, Z., V. Ünal, and D. Yıldırım. 2015. Three years' experience with small-scale fishers and no-take-zones in Gökova Bay (eastern Mediterranean), Turkey. Pages 297–305 *in* A. Srour, N. Ferri, D. Bourdenet, D. Fezzardi, and A. Nastasi, editors. First regional symposium on sustainable small-scale fisheries in the Mediterranean and Black Sea. FAO (Food and Agriculture Organization of the United Nations) Fisheries and Aquaculture Proceedings 39.

MedPAN (Mediterranean Protected Areas Network) and UNEP-MAP-RAC/SPA (United Nations Environment Programme-Mediterranean Action Plan-Regional Activity Center/Specially Protected Areas). 2017. The 2016 status of marine protected areas in the Mediterranean. MedPAN and SPA/RAC. Available: http://d2ouvy59p0dg6k.cloudfront.net/downloads/medpan_forum_mpa_2016___brochure_a4_en_web_1_.pdf. (August 2018).

Okus, E., A. Yuksek, B. Yokes, N. Yilmaz, A. Aslan-Yilmaz, U. Karhan, N. Demirel, V. Demir, S. Zeki, S. Tas, H.I. Sur, H. Altiok, A.E. Muftuoglu, N. Balkis, A. Aksu, E. Dogan, and C. Gazioglu. 2006. Marine biological diversity assessment of Gokova special environmental protection area. Turkish Ministry of Environment and Forestry, Authority for Specially Protected Areas, Ankara.

Olson, D. M., and E. Dinerstein. 2002. The Global 200: priority ecoregions for global conservation. Annals of the Missouri Botanical Garden 89:199–224.

Sala, E., E. Ballesteros, P. Dendrinos, A. Di Franco, and F. Ferretti, et al. 2012. The structure of Mediterranean rocky reef ecosystems across environmental and human gradients, and conservation implications. PLOS (Public Library of Science) ONE [online serial] 7(2):e32742.

Sala, E., Z. Kizilkaya, D. Yildirim, and E. Ballesteros. 2011. Alien marine fishes deplete algal biomass in the eastern Mediterranean. PLOS (Public Library of Science) ONE [online serial] 6(2):e17356.

Skliris, N., S. Sofianos, A. Gkanasos, A. Mantziafou, V. Vervatis, P. Axaopoulos, and A. Lascaratos. 2012. Decadal scale variability of sea surface temperature in the Mediterranean Sea in relation to atmospheric variability. Ocean Dynamics 62:13–30.

Ünal, V. 2010. Fishery management in Gökova special environment protection area. Pages 1– 72 *in* C. O. Kıraç and N. O. Veryeri, editors. Putting PEEN to practice in marine and coastal areas: a demonstration project ensuring the ecological resilience, coherence and sustainable future of Gökova Bay SEPA in Turkey. SAD-AFAG, Ankara, Turkey.

Ünal, V., B. Ekmekci, S. Göktay, and H. Göncüoğlu. 2016. Analysis of fisheries buyback programs in Turkey: preliminary results. Page 302 *in* Türkiye deniz bilimleri konferansi. [Marine science conference of Turkey.] Middle East Technical University, Ankara, Turkey. (English abstract.)

Ünal, V., and V. Erdem. 2009. Combating illegal fishing in Gökova Bay (Aegean Sea), Turkey. Pages 80–86 *in* B. A. Çiçek and H. Öniz, editors. Proceedings of the 3rd international symposium on underwater research. Eastern Mediterranean University, Famagusta, Cyprus.

Ünal, V., M. Erdem, H. Göncüoğlu, H. Güçlüsoy, and Z. Tosunoğlu. 2009. Management paradox of groupers (Epinephelinae) fishing in the Gökova Bay, (eastern Mediterranean), Turkey. International Journal of Food, Agriculture and Environment 7:914–917.

Ünal, V., and H. Göncüoğlu Bodur. 2017. The socio-economic impacts of the silver-cheeked toadfish on small-scale fishers: a comparative study from the Turkish coast. Ege Journal of Fisheries and Aquatic Sciences 34:119–127.

Ünal, V., H. Göncüoğlu, D. Durgun, Z. Tosunoğlu, C. Deval, and C. Turan. 2015a. Silver-cheeked Toadfish, *Lagocephalus sceleratus* (Actinopterygii: Tetraodontiformes: Tetraodontidae), causes a substantial economic losses in the Turkish Mediterranean coast: a call for decision makers. Acta Ichthyologica et Piscatoria 45:231–237.

Ünal, V., Z. Kızılkaya, and D. Yıldırım. 2015b. How could we convince fisheries stakeholders to establish no-take-zones? Lessons from small-scale fishery in Gökova Bay (eastern Mediterranean), Turkey. Pages 201–208 *in* A. Srour, N. Ferri, D. Bourdenet, D. Fezzardi, and A. Nastasi, editors. First regional symposium on sustainable small-scale fisheries in the Mediterranean and Black Sea. FAO (Food and Agriculture Organization of the United Nations) Fisheries and Aquaculture Proceedings 39.

Ünal, V., Z. D. Yıldırım, and E. M. Tıraşın. In press. Implementation of the ecosystem approach to fisheries in Gokova Bay, Turkey: baseline report. FAO (Food and Agriculture Organization of the United Nations)Fisheries and Aquaculture Technical Paper.

Atlantic Striped Bass: An Interjurisdictional Fisheries Management Success Story

Ronald J. Essig*[1]
U.S. Fish and Wildlife Service (retired), Wildlife and Sport Fish Restoration Division
300 Westgate Center Drive, Hadley, Massachusetts 01035, USA

R. Wilson Laney
U.S. Fish and Wildlife Service, North Carolina State University
127 David Clark Labs, Campus Box 7617, Raleigh, North Carolina 27695, USA

Max H. Appelman
Atlantic States Marine Fisheries Commission
1050 North Highland Street, Suite A-N, Arlington, Virginia 22201, USA

Fred A. Harris[2]
North Carolina Wildlife Resources Commission (retired)
1701 Mail Service Center, Raleigh, North Carolina 27699, USA

Roger A. Rulifson
Department of Biology, East Carolina University
Greenville, North Carolina 27858, USA

Kent L. Nelson[3]
North Carolina Wildlife Resources Commission (retired)
1701 Mail Service Center, Raleigh, North Carolina 27699, USA

Abstract.—The Striped Bass *Morone saxatilis* is an extremely important commercial and recreational species with a coastal migratory stock in the United States referred to as "Atlantic Striped Bass" managed by the Atlantic States Marine Fisheries Commission (ASMFC). Atlantic Striped Bass has four major contributing stocks, including the Chesapeake Bay, which comprises 70–90%, and the Hudson River, the Delaware River, and the Albemarle Sound/Roanoke River (A/R). The collapse of Atlantic Striped Bass in the late 1970s precipitated federal funding and legislation like the Emergency Striped Bass Study for research on causative factors of the decline and potential management recommendations. The 1981 ASMFC Interstate Fishery Management Plan (ISFMP) for Atlantic Striped Bass was nonmandatory and mostly ineffective until the 1984 Atlantic Striped Bass Conservation Act provided regulatory authorities to the ASMFC and the federal government to close fisheries in states out of compliance with ISFMPs. Restrictions and moratoria on harvest imposed in several states reduced mortality, and under favorable environmental conditions and given Striped Bass

* Corresponding author: rjessig@gmail.com
[1] Current address: 681 Wendell Road, Shutesbury, Massachusetts 01072, USA
[2] Current address: 2413 Mary Marvin Trail, Fuquay Varina, North Carolina 27526, USA
[3] Current address: 2353 Kay Road, Greenville, North Carolina 27858, USA

life history, multiple years of good recruitment occurred. This allowed target thresholds for female spawning stock biomass to be achieved and the ASMFC to declare recoveries of Atlantic Striped Bass stocks from 1995 to 1998. Regulation of river flows was particularly important for the A/R stock recovery, and this stock is presented as a case study. During the 20+ years following recovery, long-term monitoring by states in support of adaptive management was primarily supported by the stable, nonappropriated funding of the Sport Fish Restoration Act. Monitoring includes spawning stock characterization and biomass estimation, juvenile abundance surveys, cooperative coastwide tagging, and harvest data collection. Future issues facing the recovered Atlantic Striped Bass include interspecies effects of relatively high abundance, management of stocks separately instead of as a single coastal stock, and ecosystem-based fisheries management. Key lessons learned in the Atlantic Striped Bass recovery are that high societal value of the species provided the political impetus to create and fund the recovery program, coordination of management and enforcement efforts among all jurisdictions was essential for this migratory species, and fully funded long-term monitoring programs are critical to adaptive population management.

INTRODUCTION

The Striped Bass *Morone saxatilis*, also commonly referred to as rockfish or striper, is an extremely important commercial and recreational species that ranges from the Canadian Maritimes to Florida and through the Gulf of Mexico to Texas (Figure 1). The Atlantic States Marine Fisheries Commission (ASMFC) has jurisdiction for the Atlantic coastal migratory stock in the United States, herein referred to as "Atlantic Striped Bass," and includes the spawning populations that contribute to this migratory stock. Atlantic Striped Bass has four major contributing stocks, including the Chesapeake Bay, which comprises 70–90% of the coastal migratory stock, and the Hudson River, the Delaware River, and the Albemarle Sound/Roanoke River (A/R) stocks (Figure 2; Berggren and Lieberman 1978). This species has been called the premier game fish on the Atlantic coast (Russell 2005) and has supported commercial fisheries for centuries. In 1669, the commercial sales of Atlantic Striped Bass helped the Massachusetts Bay Colony fund the first public school in North America (Pearson 1938). Since 1992, recreational removals (harvest mortality plus catch-and-release mortality) have exceeded commercial harvest (Figure 3). Estimates of 2015 removals indicate that 70% came from the recreational sector and 30% came from the commercial sector (ASMFC 2016).

The Atlantic Striped Bass can live more than 30 years (Secor 2000a) and spends most of its adult life in coastal estuaries and within state territorial ocean waters (i.e., 0–3 mi [3–4.8 km] from shore), although large adults are known to congregate offshore beyond state waters during the winter months (ASMFC, Maryland Department of Natural Resources [MDDNR], North Carolina Division of Marine Fisheries [NCDMF], and U.S. Fish and Wildlife Service [USFWS], unpublished data). Mature individuals (beginning about age 6 for females and age 2 for males) migrate into freshwater rivers and tributaries in the spring to spawn and then return to the sounds and the ocean. Nursery areas for larvae and juveniles are typically in the lower portions of these rivers and upper estuaries and include the zone of estuarine turbidity maxima, which serves to concentrate larvae and their prey (North and Houde 2003; Martino and Houde 2010).

FIGURE 1. Recreational anglers with large Striped Bass. Photo by Captain John McMurray.

Beginning as early as age 2, they migrate out of the estuaries to join the coastal migratory stock that moves north along the Atlantic coast in the early summer and south in the late fall to primary wintering grounds offshore from New Jersey to North Carolina (Secor and Piccoli 2007).

These life history characteristics of living various distances from shore and undergoing seasonal north–south migrations necessitate interjurisdictional fisheries management. Accordingly, the ASMFC manages Atlantic Striped Bass from Maine to North Carolina within state waters (0–3 mi) while management authority in federal waters (3–200 mi [4.8–321.9 km]) lies with the National Marine Fisheries Service (NMFS). All ASMFC member states and jurisdictions from Maine through North Carolina, including Pennsylvania, are participants in the Atlantic Striped Bass Interstate Fishery Management Plan (ISFMP). Management issues are addressed by the ASMFC Atlantic Striped Bass Management Board, which includes three commissioners from each state: an administrative commissioner, a legislative commissioner, and a governor's appointee. Other nonstate members include a representative from the District of Columbia; the Potomac River Fisheries Commission (PRFC), which is the body responsible for management of Potomac River fisheries; the USFWS; and NMFS. The board frequently uses its Advisory Panel, Technical Committee, Stock Assessment Subcommittee, Tagging Subcommittee, Plan Review Team, and Plan Development Team to collect and review all available information necessary to make informed management decisions.

This chapter describes the population collapse, recovery, and two decades of management of Striped Bass along the Atlantic coast of the United States. The A/R stock is

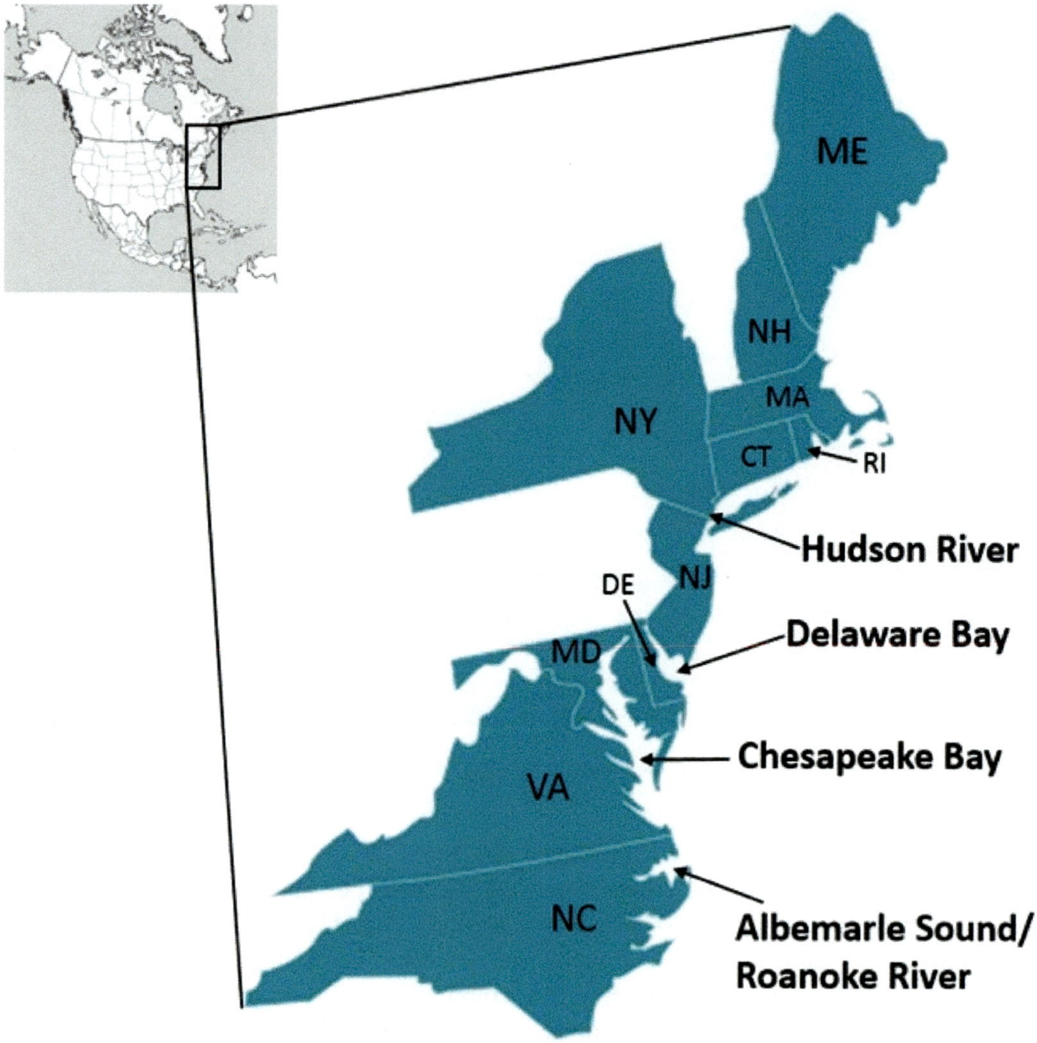

FIGURE 2. Atlantic States Marine Fisheries Commission's Atlantic Striped Bass Management Unit (Maine to North Carolina) highlighting the four primary producer areas. Source of base map: Shepherd et al. (2016).

presented as a case study due to its separate management by North Carolina via stock-specific biological reference points (BRPs) subject to ASMFC approval and its complexity resulting from the construction and operation of hydropower dams adjacent to historic spawning grounds. Interjurisdictional fisheries management of the coastal migratory population under favorable environmental conditions for recruitment, and with broadened age structure, provides a success story and lessons learned that might be applicable to other recovery programs.

HISTORY OF POPULATION COLLAPSE AND CAUSATIVE FACTORS

Commercial landings of Atlantic Striped Bass generally increased since 1950 and peaked in 1973 at approximately 14.7 million lb (6.7

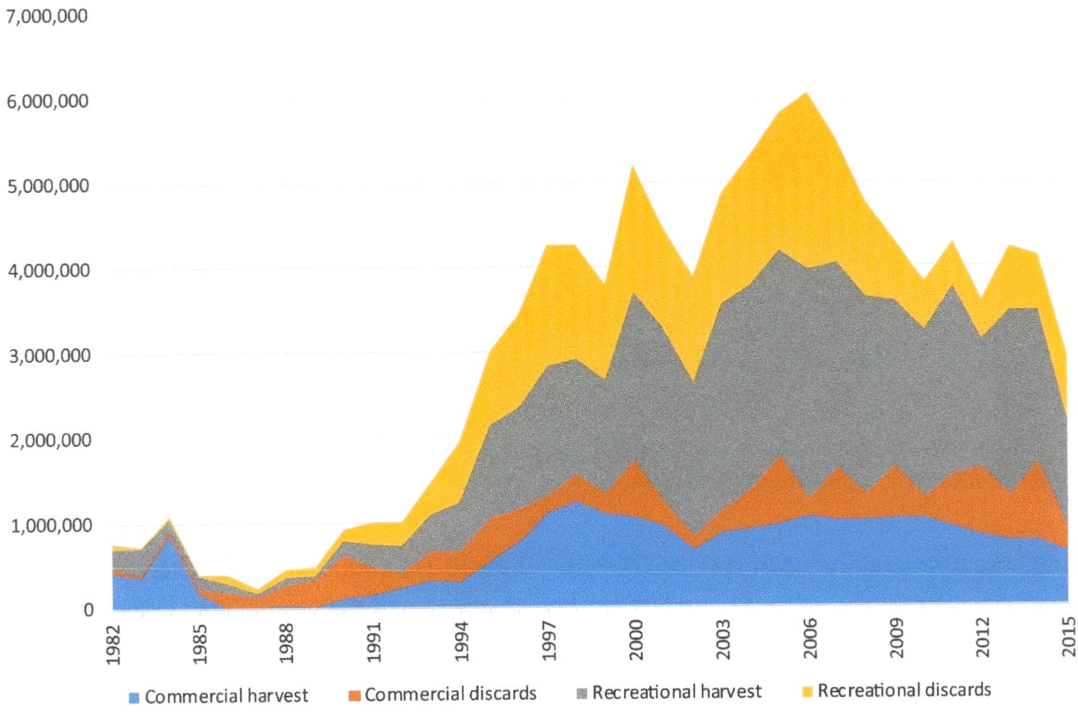

FIGURE 3. Atlantic Striped Bass harvest by sector in number of fish, 1982–2015. Discards are fish either released dead from commercial fishing operations or estimated to die from recreational catch and release. Source: ASMFC (2016).

million kg; Figure 4; ASMFC 1981). Coastwide recreational catch statistics were not collected annually prior to 1979; however, the recreational harvest was likely similar in magnitude during that time (ASMFC 2003).[4] By 1983, commercial landings fell to less than 1.7 million lb [0.8 million kg], a 90% decline from the peak a decade earlier (Figure 4).

The three Striped Bass stocks north of North Carolina differed in their stock status relative to the coastal migratory stock collapse. The Hudson River stock, which can contribute significantly to the coastal migratory stock in some years (Van Winkle et al. 1988), declined but did not collapse in part because the commercial fishery in the river was closed in the 1970s due to polychlorinated biphenyls contamination (ASMFC Striped Bass Technical, Stock Assessment, and Tagging Committees 2013). The Delaware River stock was low for much of the 20th century (Chittenden 1971) but has contributed more to the coastal migratory stock as riverine water quality improved during the 1980s (DOI and DOC 1994). As the dominant contributor to the coastal migratory stock, the Chesapeake Bay stock drove the coastal collapse. This decline was observed during the early 1970s to the early 1980s in the Maryland juvenile abundance index (JAI), which was the primary recruitment index for monitoring

[4] All of the recreational data in this chapter were queried in 2016 and therefore do not reflect recent changes to the Marine Recreational Information Program (MRIP) survey methodologies. The newly calibrated MRIP catch and effort estimates were released in July 2018. Although the magnitude of the new MRIP estimates is significantly larger, trends between the original and revised data series are similar.

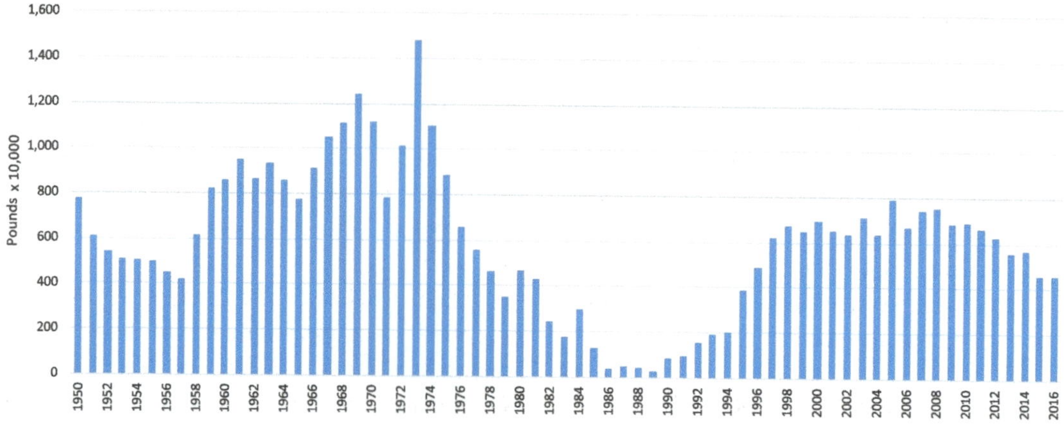

FIGURE 4. Coastwide commercial landings of Atlantic Striped Bass in pounds, 1950–2016. Source: ASMFC, Atlantic Coastal Cooperative Statistics Program, May 19, 2017.

the coastal migratory stock (Figure 5). This fishery-independent index of abundance is based on the mean number of juvenile Atlantic Striped Bass from beach seine hauls in Chesapeake Bay tributary nursery areas conducted by the MDDNR.

In response to the Atlantic Striped Bass stock collapse, Congress amended the Anadromous Fish Conservation Act in 1979 to create the Emergency Striped Bass Study (ESBS) program (U.S. Code, volume 16, section 757a–757g). The ESBS provided funding to state and federal agencies to monitor the status of Atlantic Striped Bass, to determine the cause of the decline, and to evaluate the economic importance of fisheries. Emergency Striped Bass Study results were summarized in a series of annual reports (1980–1994) to Congress, jointly authored by the U.S. Department of the Interior and the U.S. Department of Commerce. Studies examined contaminant effects on survival of early life stages, larval feeding and predation, climatic events, water use practices, disease, and instantaneous fishing mortality (*F*). For example, Goodyear (1985) modeled decreased survival of larval Atlantic Striped Bass from contaminants and decreased survival of adults from fishing. His results indicated that even with contaminant toxicity to larvae, the decline in the Chesapeake Bay stock could be reversed by a reduction in *F*. Richards and Rago (1999) synthe-

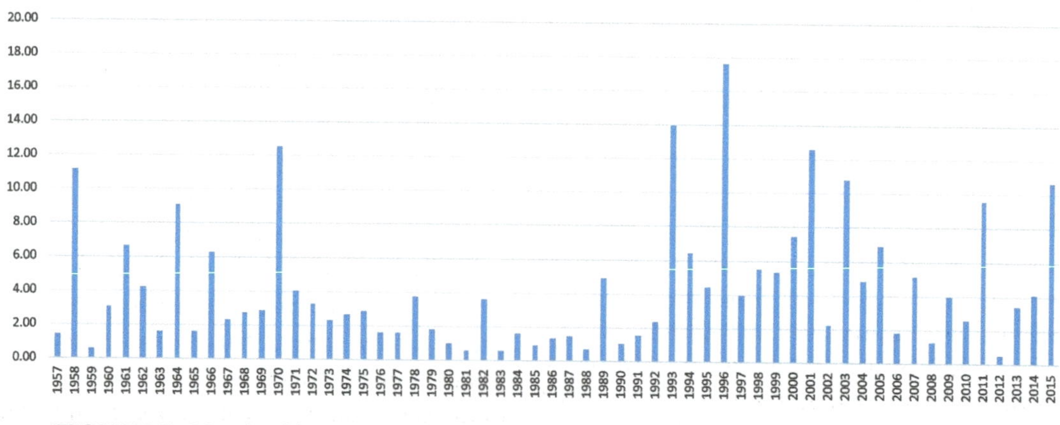

FIGURE 5. Maryland juvenile Atlantic Striped Bass abundance index. Source: ASMFC (2016).

sized many ESBS results and noted that water quality problems contributed to the decline by reducing survival of early life stages, but they also determined that high F was the primary cause. The primary conclusion of the ESBS was that overfishing was the major factor for the fishery decline, harvesting more fish than the Atlantic Striped Bass stock could produce, and that large reductions in F were required if the fishery was to recover.

LEGISLATION, MANAGEMENT, ENVIRONMENT, AND LIFE HISTORY SET A COURSE FOR RECOVERY

The collapse of Atlantic Striped Bass required major legislation and management working in concert to make up for past actions that were inadequate. Regulations for the harvest of Atlantic Striped Bass were relatively minimal prior to the 1970s. Most states north of Delaware had a 16-in (406 mm) minimum size, and some states, like Maryland and North Carolina, had 10-in or 11-in (254 or 279 mm) minimum sizes to allow liberal harvest. As Atlantic Striped Bass stocks further declined in the 1970s, more stringent management measures were implemented. In 1978, Maryland prohibited the capture of Striped Bass on the spawning grounds within state watersheds tributary to the Chesapeake Bay (Speir et al. 1999). However, that prohibition was lifted the following year with the establishment of a 14-in (356 mm) minimum size in the upper bay in summer and fall, and a closure of gill netting in the middle bay in the summer to reduce mortality on the stock.

In 1981, the ASMFC approved the first ISFMP for Atlantic Striped Bass (ASMFC 1981) with key recommendations of different minimum size limits for producer (i.e., spawning and/or nursery) areas and coastal areas and closure of major spawning areas to Atlantic Striped Bass harvest (Table 1). Amendments 1 and 2, approved in 1984, set targets to reduce F and allowed states flexibility in their approaches to do so (ASMFC 1984a, 1984b). Although states generally enacted regulations to implement the size limit provisions of this ISFMP, not all states implemented spawning area closures as recommended, primarily due to political agendas and the ASMFC's lack of enforcement authority.

The 1984 Atlantic Striped Bass Conservation Act gave the ASMFC the much needed authority to inform the Secretaries of Commerce and Interior that states were out of compliance when they failed to implement management measures consistent with the ISFMP, with a recommendation to close Striped Bass fisheries in those states (U.S. Code, volume 16, section 1851). The 1984 act ultimately served as the inspiration for passage of the Atlantic Coastal Fisheries Cooperative Management Act in 1993 (U.S. Code, volume 16, sections 5101–5108), which extended the same interstate regulatory authority approach to all ASMFC-managed species (the ASMFC currently manages or co-manages 27 species and species groups). It is important to note that a number of authors have expressed critical opinions on legal aspects of the interjurisdictional management of Atlantic Striped Bass (Bubier and Rieser 1984; Ballou 1987; Rapone 2011).

The first enforceable plan under the Atlantic Striped Bass Conservation Act was Amendment 3 to the ISFMP (ASMFC 1985). The Atlantic Striped Bass still met the generally accepted definition of a collapsed stock with its commercial landings below 10% of historic amounts, and so the objective of Amendment 3 was to protect at least 95% of the 1982 year-class females of the Chesapeake Bay stock (the first modest-sized recruitment event since the previous decade) and all subsequent year-classes until those females were able to spawn at least once. This amendment basically required states to change their size limits each year to account for growth of the

TABLE 1. History of Atlantic Striped Bass Management Actions (www.asmfc.org/species/atlantic-striped-bass#management).

Interstate Fishery Management Plan (ISFMP) for Striped Bass (1981)

- Fourteen-inch minimum size limit in producing areas (Albemarle Sound/Roanoke River (A/R), Chesapeake Bay, Delaware Bay and Hudson River, as well as their tributaries). Twenty-four-inch minimum size limit in coastal areas. Closure of major spawning areas or rivers to fishing during spawning seasons.

Amendment 1 (1984)

- Overall strategy required a reduction in fishing mortality (*F*). States were given a list of recommended management measures but had the ability to use additional measures or to use different measures in place of the recommended measures as long as the same reduction in *F* was achieved. The states could not modify the minimum size limits.

Amendment 2 (1984)

- All the states needed to achieve a 55% reduction in *F*, beyond the reduction implemented with the ISFMP.

Amendment 3 (1985)

- States prevent directed *F* on at least 95% of the 1982 year-class females, and females of all subsequent year-classes of Chesapeake Bay stock until 95% of the females of these year-classes have an opportunity to reproduce at least once. The amendment basically required states to change their size limits each year to account for growth of the 1982 year-class.

States imposed complete or partial moratorium for all or part of the period between 1985 and 1989

- CT, NY, NJ, DE, MD, Potomac River Fisheries Commission (PRFC), DC, and VA

Amendment 4 (1989): reopening during initial stages of recovery

- Initiated because the MD young-of-year index exceeded the 3-year running average of 8.0 and Amendment 3 measures lapsed once this was achieved.
- Proposed regulations: commercial season that reduces the harvest rate to 80% of historical rate; one fish creel limit for recreational fishery; minimum size limit of 28 in along the coast and 18 in in the Hudson River and Chesapeake Bay.
- Combination of size limits, seasons, and creel limits used to attain target *F*. Allowed the state to select the actual management program as long as *F* remained below the target.
- Beginning in 1991, five addenda to Amendment 4 were implemented to maintain protection of the 1982 year-class.

Striped Bass prohibition in the exclusive economic zone (EEZ) (1990)

- To provide additional protection to the Striped Bass and ensure the effectiveness of state regulations, the National Marine Fisheries Service passed a final rule (55 Federal Register 40181-02) prohibiting the possession, fishing (i.e., catch and release fishing), harvest and retention of Atlantic Striped Bass in the EEZ, with the exception of a defined transit zone within Block Island Sound. Atlantic Striped Bass may be possessed and transported through this defined area, provided that the vessel is not used to fish while in the EEZ and the vessel remains in continuous transit.
- The ban on possession, fishing, harvest and retention of Striped Bass in the EEZ remains in effect.

Atlantic States Marine Fisheries Commission declared the Chesapeake Bay stocks of Atlantic Striped Bass recovered as of January 1, 1995

- Chesapeake Bay stock contributes the greatest proportion of Striped Bass to the coastal migratory stock.

TABLE 1. Continued.

Amendment 5 (1995)
- Adopts a target F = 0.33 for the first 2 years; then, F = 0.40. Allows the states to establish regulatory programs for the recreational and commercial fisheries that will meet the target (conservation equivalency).
- Recreational fisheries: 20-in minimum size limit, one fish creel limit (except for two fish creel limit for VA recreational fisheries, and PRFC and MD charterboat fisheries), variable season lengths in the producer areas (Chesapeake Bay, Hudson River,) and 28-in minimum size limit, two-fish creel limit along the coast.
- Commercial fisheries: flexible quota, same size limits as the recreational fishery.
- **Addendum I (1997):** Maintained the interim target F = 0.33 and froze all 1997 commercial and recreational Striped Bass regulations at 1996 levels.
- **Addendum II (1997):** Lowered the target F to 0.31 and maintained minimum sizes for the 1998 fishery.
- **Addendum III (1998):** maintain status quo for 1999 and 2000 fishery; A/R and Delaware River stocks declared recovered to historical levels.
- **Addendum IV (1999):** Fiscal years 2000 and 2001:1998 F exceeded target F; Reduce F on age-8 and older Striped Bass by 14%.
- **Addendum V (2000):** Fiscal years 2001 and 2002: states can implement regulations that are equivalent to 1998–1999 or 2000. F did not exceed the target in 1998 and 1999. No additional reductions necessary.

Amendment 6 (2003)
- Adopts new reference points for F and implements for the first time a target and threshold biomass; conservation equivalency continued. Commercial fisheries restored to 100% of the base period. Recreational fisheries: two fish creel limit and 28-in minimum size limit.
- Chesapeake Bay and A/R stocks are managed by maintaining an F = 0.27; the commercial and recreational fisheries in these areas can use a 20-in minimum size limit (through conservation equivalency can go no lower than 18 in).
- **Addendum I (2007):** Established bycatch monitoring and research program; recommended development of Web-based angler education program.
- **Addendum II (2010):** Established new definition of recruitment failure to be "a value that is below 75% of all values in a fixed time series appropriate to each juvenile abundance index."
- **Addendum III (2012):** Required states to implement a commercial harvest tagging program to stem illegal harvest and sale of Striped Bass.
- **Addendum IV (2014):** Established a single set of F reference points for the entire coastal migratory complex, and required fisheries to implement harvest reductions to reduce F to a level at or below the new target.

1982 year-class. Spurred by this action, several states, beginning with Maryland and Delaware in 1985, opted for a more conservative approach and imposed a total moratorium on the harvest of Striped Bass. Connecticut, New York, New Jersey, Virginia, and the PRFC imposed complete moratoria between 1986 and 1989. North Carolina closed their ocean fishery in 1984 but still permitted the harvest of Striped Bass in inland waters with more stringent measures of an 18-in (203 mm) minimum size limit, total allowable landings (TAL), and regulation of the spring season in the Roanoke River so that a majority of the harvest was male fish. Amendment 3 contained a mechanism to trigger a decision to reopen all Atlantic

Striped Bass fisheries when the 3-year moving average of the Maryland Chesapeake Bay JAI exceeded its long-term average. The trigger was reached in 1989, and states began to reopen fisheries in 1990 under the management provisions of Amendment 4 (ASMFC 1989).

Amendment 4 to the Atlantic Striped Bass ISFMP was designed to rebuild the resource rather than maximize harvest. It implemented commercial and recreational regulations aimed to achieve a target F equal to about half of F corresponding to maximum sustainable yield. Specifically, all Atlantic Striped Bass commercial fisheries were restricted to 20% of the amount landed during the base period of 1972–1979, and recreational fisheries were constrained by a one fish creel limit and 28-in (711 mm) minimum size limit along the coast, and an 18-in minimum size limit in the Chesapeake Bay and other producer areas. Fishing mortality was only allowed to increase after the coastwide spawning stock biomass (SSB) was restored to those levels estimated for the 1960s and 1970s.

In 1990, as the states reopened their fisheries in the state waters, NMFS prohibited the harvest and possession of Atlantic Striped Bass in federal waters (i.e., the exclusive economic zone (EEZ) from 3 to 200 mi from shore). These offshore waters, with the exception of a defined transit zone within Block Island Sound, have remained closed ever since. Although a relatively low harvest occurred offshore of northern states, this EEZ closure was a major conservation measure to protect large overwintering aggregations of Atlantic Striped Bass in waters offshore of southern states including North Carolina, Virginia, and Maryland.

A number of studies have also focused on environmental and biological factors that may relate to the inconsistent recruitment of Striped Bass. Since the relationship between SSB and recruitment has been found to be weak and highly variable, year-class strength is thought to be linked to environmental conditions during the early life history stages (Richards and Rago 1999; North and Houde 2003). Secor and Houde (1995) observed a positive relationship between spring temperatures and early survival of eggs and larvae in the Patuxent River tributary to the Chesapeake Bay. Spring flows were also correlated to recruitment in the upper Chesapeake Bay and attributed to physical changes to the nursery area in relation to the transport and retention of larvae and their potential prey in a 2-year study by North and Houde (2003). Martino and Houde (2010) further found that strong recruitment of young-of-year (age 0) Striped Bass occurred when peak abundances of larvae and zooplankton prey were synchronous. This 10-year analysis of data (1993–2002) also observed the strongest age-0 Striped Bass recruitment in years of higher freshwater inflows from the Susquehanna River. The presence of older female Striped Bass has also been suggested as important to successful recruitment, with Secor (2000b) reporting a positive correlation between age diversity to the egg presence ratio and juvenile abundance (1985–1995). A robust age structure may also enhance recruitment by increasing the seasonal duration of the spawning period and contribute to a "storage effect," which Secor (2000a and 2007) proposed to explain the rapid population growth that occurs when environmental conditions favor early survival. It is clear that environmental and biological factors, along with management measures, played key roles in the recovery of Atlantic Striped Bass.

More than a decade of controls on fishing effort prescribed by science resulted in declarations by the ASMFC that Atlantic Striped Bass stocks were fully recovered when SSB achieved historical levels observed in the 1960s. These declarations were issued in 1995 for the Chesapeake Bay stock, in 1997 for the A/R stock, and in 1998 for the Delaware River stock.

ALBEMARLE SOUND-ROANOKE RIVER STOCK COLLAPSE AND RECOVERY

Numerous haul seine and pound net operations in Albemarle Sound and a wooden dam constructed in the Roanoke River between the towns of Weldon and Roanoke Rapids were thought to be responsible for causing the spawning run decline of A/R Striped Bass in the late 1800s. In response to plummeting commercial catches, A/R Striped Bass and American Shad *Alosa sapidissima* were cultured first on steamboats converted to hatchery operations in Albemarle Sound and then at the Weldon Fish Hatchery on the spawning grounds of the Roanoke River starting in 1873 (Baird 1875; Stickney 1996). During World War II, the increased demand for electrical generation focused attention on the Roanoke basin as a possible site for hydropower development. Also, periodic devastating floods clearly suggested the need for flood control. Subsequently, three dams were constructed on the upper Roanoke River, creating John H. Kerr Reservoir (1952), Roanoke Rapids Reservoir (1955) and Lake Gaston (1963).

The closure of Roanoke Rapids Dam and decay of pulp mill effluent and submerged vegetation caused severe hypoxia, resulting in an extensive fish kill downstream. Alarmed at this devastation, staff from state and federal agencies and academia formed the ad hoc Roanoke River Studies Committee in 1955 to document the fish kills and to develop minimum flow and dissolved oxygen standards for downstream reaches of the river. As a member of the committee, Dr. William W. Hassler of North Carolina State University began investigating the spawning of A/R Striped Bass downstream of Roanoke Rapids Dam in 1955. He sampled spawned Striped Bass eggs to assess viability and estimated annual spawning run sizes by mark and recapture of prespawning fish. Because the majority of the fish captured were relatively small compared to earlier records (the largest recorded from the A/R system was 125 lb in 1891; Smith 1907) and recaptured within the system, the population was thought to be primarily resident and not anadromous (Hassler et al. 1981). This paradigm held until the 2000s, when a re-examination of Hassler's data combined with continuing tagging efforts confirmed that the large fish do indeed go to the ocean and contribute to the Atlantic migratory population (Callihan et al. 2015).

The egg studies conducted by Hassler showed a decline in viability in the mid-1970s. Egg viability ranged from 80% to 96% from 1960 through 1974, but declined to 56% in 1975 and ranged from 23% to 74% in the succeeding years through 1987 (Manooch and Rulifson 1989). In 1988, egg studies showed a viability of 89%, a value coinciding with a revised discharge schedule to moderate Roanoke River flows during the spawning season. Egg viability values ranged from 42% to 58% from 1989 to 1992, and 36% in 1993 (Rulifson and Isely 1995). A long-standing criticism of the egg production and spawning population estimates was that the two indices did not appear to be correlated (i.e., large egg production years did not necessarily coincide with large year-classes found during the spawning run when the year-class reached maturity). A subsequent evaluation of egg production in relation to daily ages of age-0 Striped Bass caught in Albemarle Sound found that a significant portion of the recruitment from 1990 through 1993 occurred from eggs spawned at the end of the spawning season and after major spawning events (Rulifson et al. 1993; Rulifson and Isely 1995). A 10-year database of Striped Bass larval fish sampling combined with zooplankton, phytoplankton, and chlorophyll-*a* sampling from 1983 to 1992 indicated that zooplankton densities were orders of magnitude less than that of Chesapeake Bay, suggesting that food may be limiting for larval Striped Bass (Rulifson et al. 1992).

Hassler also developed an annual JAI for A/R Striped Bass using wing trawl sampling in Batchelor Bay and western Albemarle Sound from 1956 to 1988. The NCDMF has conducted JAI sampling annually since then (Figure 6; NCDMF and NCWRC 2013). The JAI is an integral component of population modeling and has been used to document the effects of pre- and postimpoundment flows on Striped Bass recruitment (Rulifson and Manooch 1990a; Zincone and Rulifson 1991).

The A/R Striped Bass stock declined significantly in the late 1970s, with only three successful year-classes produced from 1978 through 1992. Juvenile abundance index values in the 1980s were less than one age-0 Striped Bass per trawl, the lowest values since the survey was established. At the time, a common conclusion was that the population was diminished beyond recovery. In response, an ad hoc Roanoke River Water Flow Committee was formed in 1988, consisting of state and federal agency professionals and scientists. The committee's purpose was to understand the hydropower production strategy and reservoir management on the Roanoke River, compared to run-of-river records prior to impoundment, and to develop a strategy for water management to benefit all fish and wildlife resources. The committee combined analysis of river flows, pre- and postimpoundment, with ongoing research involving water quality, spawning, and watershed events (Manooch and Rulifson 1989; Rulifson and Manooch 1990b, 1991, 1993). At the recommendation of the committee, Kerr Reservoir discharge was modified in 1988 to reduce the daily hydropower-influenced flow fluctuations in the Roanoke River to more resemble natural flow conditions. Eventually the committee developed and recommended, and North Carolina negotiated, a more natural (preimpoundment) spawning season flow regime annually from April 15 to June 15. Prior to 1988, flows on the spawning grounds from hydropower operations could result in daily changes from 1,000 to 20,000 ft^3/s (28.3–566.3 m^3/s). In 1993, following the 14-year period dominated by recruitment failure, year-classes above the long-term average began to be produced more frequently, including several that exceeded any within the historical postimpoundment

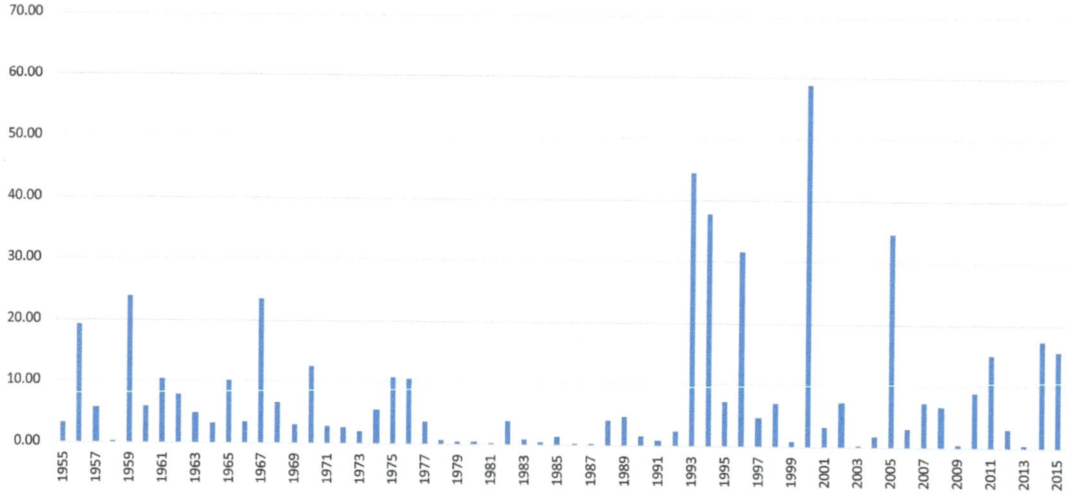

FIGURE 6. North Carolina juvenile Atlantic Striped Bass abundance index. Source: 2016 North Carolina State Compliance Report for Atlantic Striped Bass.

record. While difficult to quantify in light of increasing spawning stock, resource managers and scientists believe that decreasing the frequency of extreme flow events has contributed to enhanced reproductive success because female Striped Bass orient to flows and velocities during spawning. Rapidly fluctuating discharges alter flow characteristics and could disrupt spawning as females seek optimum conditions for releasing eggs. On higher releases, Striped Bass migrate into the upper section of the Roanoke River until blocked by the dam at Roanoke Rapids at river mile 137 (river kilometer 220.5). On lower flows, semibuoyant Striped Bass eggs could settle to the riverbed and die.

Before 1980, minimal restrictions existed on commercial and recreational harvest of A/R Striped Bass. During the 1970s, a 12-in minimum size limit and 25 fish/d creel limit statewide were in place to control the recreational fishery. Commercial harvest by gill net and large bow (landing) nets were allowed on the Roanoke River, and there were no seasonal restrictions on commercial harvest in Albemarle Sound.

As part of the reauthorization of the Atlantic Striped Bass Conservation Act (Public Law 100-589), in 1988 Congress authorized a study on the A/R Striped Bass stock, which was overseen by the North Carolina Striped Bass Study Management Board. Congress mandated that the study be undertaken by the USFWS in consultation with NMFS and others. The board consisted of representatives from the USFWS, NMFS, the North Carolina Wildlife Resources Commission (NCWRC), the NCDMF, the Virginia Department of Natural Resources, and the U.S. Army Corps of Engineers. Congress appropriated funds (1989–1991) for 11 studies to assess the abundance and condition of the A/R spawning stock, including evaluations of egg and larval dynamics, flow patterns, and land use changes (North Carolina Striped Bass Study Management Board 1991; USFWS and NOAA 1992).

To facilitate stock recovery, the TAL for recreational and commercially harvested A/R Striped Bass was capped at 117,600 lb from 1991 to 1997 based on ASMFC guidelines recommending that harvest not exceed 20% of average landings from 1972 to 1979 (ASMFC 1989). The NCMFC and NCWRC negotiated a harvest allocation between commercial and recreational fisheries and established management areas to facilitate interagency coordination. Over several years, a number of regulatory and monitoring changes were initiated, including reductions in recreational creel limits, increases in minimum size limits, and restrictions on commercial fishing gear and season so as not to exceed TAL. Since 1991, continuing surveys, primarily federally funded by the Sport Fish Restoration (SFR) Act, have monitored harvest and evaluated stock abundance and composition, including creel surveys on Roanoke River and Albemarle Sound, spring electrofishing on Roanoke River, monitoring and sampling of commercial harvest, a fishery independent gill-net survey in Albemarle Sound, and JAI sampling in Albemarle Sound. Throughout the recovery process, public information meetings were held and proposed regulatory changes were addressed annually through public hearings as required in North Carolina.

The A/R Striped Bass stock, which includes any contribution from adjacent watersheds, was declared recovered by the ASMFC in 1997 based on achieving target SSB levels. The SSB threshold was set at the recovery value of 772,588 lb with a target value of 965,735 lb and threshold F of 0.41 and target F of 0.33 (Flowers et al. 2016).

Currently, the A/R Striped Bass stock is managed under Amendment 6 to the ISFMP for Atlantic Striped Bass and its addenda (ASMFC 2003). Striped Bass stocks in other North Carolina waters to the south (Tar, Neuse, and Cape Fear rivers) are considered essentially nonmigratory but are included under the auspices of the ASMFC. The NCDMF is the state's representative agency to ASMFC and manages anadromous Striped Bass in coordination with the NCWRC based on the North Carolina Estuarine Striped Bass Fishery Management Plan (NCDMF and NCWRC 2013). Fisheries in state waters (0–3 mi from shore) operate under the same ASMFC management measures as in coastal waters of other states. The state agencies monitor harvest and conduct surveys annually as a basis for estimating population abundance and mortality rates (ASMFC 2003).

Under 2017 management regulations, the minimum size limit for Striped Bass is 18 in for both recreational and commercial fisheries. Additional restrictions on Roanoke River include a 22–27-in (559–686 mm) protective slot limit and a two-fish-per-day creel limit that may include one fish greater than 27 in/d. Creel survey data are analyzed frequently and the harvest season in the Roanoke River may be closed by proclamation to prevent exceeding the TAL. Several models have been used to estimate *F* and SSB of A/R Striped Bass: catch at age from 1988 to 1998 (Dorazio 1995), virtual population analysis from 1999 to 2006 (Carmichael 2000, 2001, 2002, 2003; Grist 2004, 2005), and statistical catch-at-age models from 2007 to present (Takade-Heumacher 2010; Mroch and Godwin 2014). Modeling indicated that A/R Striped Bass abundance peaked at 2,052,510 fish in 2007 (Mroch and Godwin 2014). The 2014 statistical catch-at-age model included updated stock-specific maturation and mortality estimates representing a new benchmark for A/R stock assessment. Hydroacoustic and tagging-based models have provided an additional measure of spawning stock abundance and estimated that between 499,000 and 715,000 Striped Bass migrated in the Roanoke River between 2011 and 2013 (Harris and Hightower 2017).

With recovery status and increasing trends in SSB based on four major Striped Bass year-classes between 1993 and 2000, TAL was increased in four steps from 117,600 in 1997 to 550,000 lb in 2004. The estimated total annual harvest averaged 431,390 lb from 2000 to 2005, peaking in 2004 at 492,767 lb (Flowers et al. 2016). A period of low recruitment occurred during this decade with estimated SSB (3.9 million lb in 2014) having declined to near recovery levels. As a result, total harvest from 2006 to 2014 declined to an average of 215,549 lb despite the increased TAL. Mroch and Godwin (2014) determined that the new benchmark *F* and SSB biological reference points were between target and threshold values. Fishing mortality for age-4 Striped Bass, the age at full selection, was estimated at $F = 0.06$ in 2014 (Flowers et al. 2016). With declining abundance, however, the TAL was not sustainable and was reduced to 275,000 lb in 2015. Recent downward trends in SSB of A/R Striped Bass indicate that even with recovery, population levels will fluctuate, thus necessitating the continuation of monitoring programs and adaptive management.

In summary, A/R Striped Bass stocks declined significantly in the late 1970s as a result of recruitment overfishing, exacerbated by the construction and operation of three dams on the upper Roanoke River. A concerted effort by federal and state agencies evaluated stock abundance and implemented significant harvest restrictions in 1991. In concert with flow modifications from upstream reservoirs, these actions promoted the growth of the A/R stock, which was declared recovered by the ASMFC in 1997.

CRITICAL ROLE OF STABLE FUNDING TO EFFECTIVELY MONITOR STRIPED BASS POPULATIONS

The ESBS mentioned earlier provided valuable funding leading up to the recovery of

Atlantic Striped Bass starting in 1995 but did not provide a long-term solution to the sustainability of this species. In 1984, Congress passed the Wallop-Breaux (WB) Amendment (U.S. Code, volume 26, section 9504) to the Federal Aid in Sport Fish Restoration (SFR) Act (U.S. Code, volume 16, section 777). The original SFR Act provided funding from excise taxes on sport fishing equipment to state fisheries agencies for work benefiting recreational finfish and providing angling opportunities. Although projects could be conducted in both freshwater and salt water, many northeast states with separate marine agencies only used SFR funds in their freshwaters. The WB Amendment mandated that a portion of the funding be spent in salt water, so this opened up a new revenue source for work on Atlantic Striped Bass. An example is the Virginia JAI for Chesapeake Bay Striped Bass, which was funded by NMFS commercial fisheries research grants from 1967 to 1972, dropped due to lack of funding, reinstituted in 1980 with ESBS funding, and then funded through SFR since 1989.

Even more important than requiring states to spend SFR funds on saltwater species was the significant increase in funding provided by the WB Amendment. It added the portion of the already existing federal tax on gasoline that was attributable to motorboat usage to the SFR trust fund. This tripled the level of funding available to states starting in 1986 and allowed revenues to increase to the current level of approximately US$350 million per year from the federal government (Figure 7). States from Maine to North Carolina used SFR funds for projects directly related to Atlantic Striped Bass (Essig and Beal 2000). A 26-year time series (1990–2015) of federal

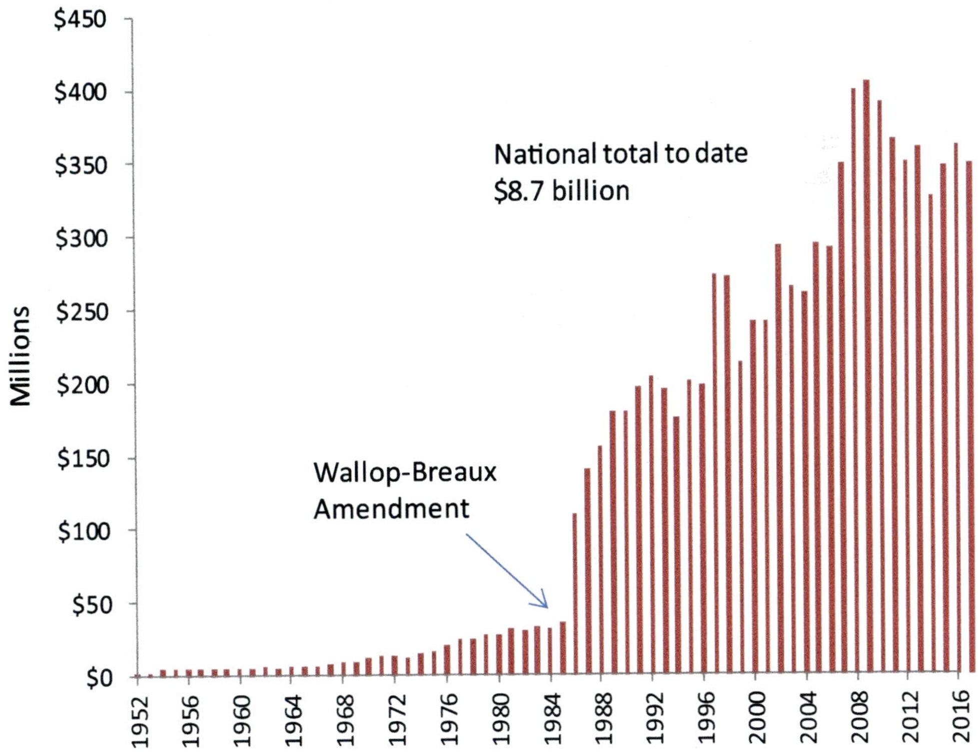

FIGURE 7. National Sport Fish Restoration funding to states federal fiscal years, 1952–2017. Source: https://wsfrprograms.fws.gov/Subpages/GrantPrograms/SFR/SFR_Funding.htm, USFWS.

plus state SFR expenditures has shown an increasing trend, but when adjusted for inflation, is relatively constant at about \$4 million in 2015 dollars per year (Figure 8). The total SFR expenditures on Atlantic Striped Bass over these 26 years, mostly postrecovery, were \$83.1 million (\$57.6 million federal). It is difficult to imagine the Atlantic Striped Bass management success and the estimated $\$6.6 \times 10^9$ and \$250 million of economic impact from recreational and commercial fisheries, respectively, in 2003 (Southwick Associates 2005), without SFR funding.

MANAGEMENT MEASURES POSTRECOVERY

Just after the recovery declaration in 1995, Amendment 5 to the Atlantic Striped Bass ISFMP (ASMFC 1995) was approved, which increased the F target to 0.33, with an increase to 0.40 after 2 years. To achieve this, recreational fisheries were managed by a 28-in minimum size limit and two fish daily creel limit along the coast, with a 20-in minimum size limit and one fish daily creel limit in the producer areas (except for two fish per day in Virginia recreational fisheries and PRFC and Maryland charter boat fisheries). Commercial fisheries were managed via a flexible quota system with the same size limits as the recreational fisheries. As before, states could implement alternative management programs deemed by the ASMFC to be a conservation equivalent to those contained in the ISFMP to meet the needs of their fisheries. For example, since Connecticut and New Jersey instituted bans on commercial harvest and sale, these states were able to transfer their respective commercial quotas to the recreational sector.

From 1997 to 2000, a series of five addenda to Amendment 5 were implemented to adjust target F in response to the latest stock status information coastwide and to implement measures to achieve each change in target F (Table 1). Several concerns were expressed about Atlantic Striped Bass management under Amendment 5 during its 8-year course, including the (1) potential inability to prevent exceedance of the target

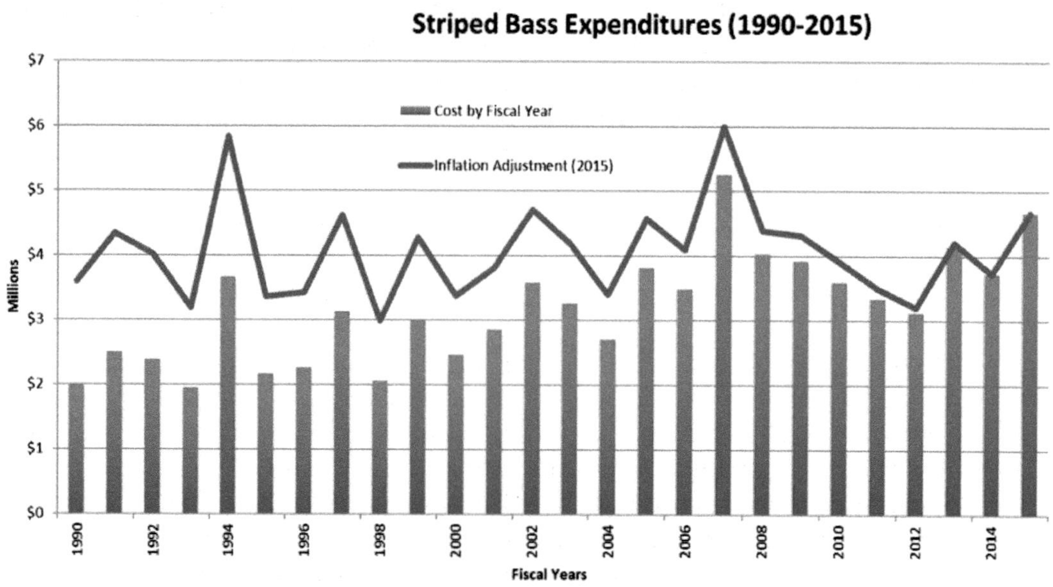

FIGURE 8. Maine to North Carolina Sport Fish Restoration expenditures (federal plus state) on Atlantic Striped Bass federal fiscal years 1990–2015. Source: Wildlife TRACS database, USFWS.

F, (2) perceived decrease in the availability of large-sized fish in the coastal migratory stock, (3) nonexistence of target and threshold biomass levels, and (4) inequitable effect of regulations on various sectors of the fishery. To address these limitations, Amendment 6 (ASMFC 2003) modified the *F* targets and thresholds and introduced a new set of BRPs (i.e., metrics used to measure the status of a fish population or stock relative to specific management objectives). The BRPs were based on the point estimate of female SSB in 1995, the year the Chesapeake Bay stock was declared restored, as well as a list of management triggers for decision making based on the BRPs, (i.e., SSB and *F* targets and *F* thresholds). These BRPs and management triggers have enabled the ASMFC Atlantic Striped Bass Management Board to be more responsive to changes in the stock.

Amendment 6 also phased in new regulations for both the commercial and recreational fisheries. Along the coast, recreational fisheries were managed by a two-fish daily creel limit and a 28-in minimum size limit. Coastal commercial fisheries were managed by a 28-in minimum size limit and state-specific quota allocations, which were restored to 100% of the states' average landings during the 1972–1979 base period. Regulatory programs for the Chesapeake Bay and other producer areas differed from the coastal migratory stock because these programs were predicated on a more conservative *F* target ($F = 0.27$). Accordingly, these jurisdictions were able to implement separate seasons, TAL, and size and creel limits as long as they remained under the *F* target for those areas. Additionally, states could implement alternative regulations through conservation equivalency enacted in Amendment 4. For example, Albemarle Sound and Chesapeake Bay fisheries were granted the ability to implement a lower minimum size limit if these jurisdictions also implemented a lower target *F* as a penalty. Amendment 6 also outlined a process for developing a bycatch reduction program; it mandated that several states conduct JAI surveys on an annual basis, and it required annual spawning stock surveys in the Hudson River, Delaware River, the Chesapeake Bay, and the A/R. To implement adaptive management, a series of four addenda to Amendment 6 (Table 1) were adopted to fine-tune the regulatory program. These addenda added a bycatch monitoring program, added a commercial harvest tagging program, and adjusted target *F* to reduce harvest.

Also crucial to the recovery and management of Atlantic Striped Bass has been the on-the-water enforcement by state and federal officers of size and creel regulations, seasonal and area closures, and fish sales. While the conservation ethic of fishers has been strong since the recoveries, particularly in the recreational sector with the prevalence of catch-and-release fishing (Russell 2005), a diligent law enforcement presence has been necessary, particularly when this migratory species is concentrated in certain locations at certain times of the year. For example, a major undercover sting operation conducted by NMFS in 2011 was successful in stopping illegal winter harvest of Striped Bass in the EEZ operating out of Rudee Inlet, Virginia, and similar law enforcement actions in the EEZ have continued since.

The Atlantic Striped Bass ISFMP continues to evolve based on the latest stock assessment and fishery performance information to maintain the coastwide Atlantic Striped Bass population above established BRPs (Table 1). In 2013, a benchmark assessment indicated that *F* was above the target for two consecutive years and female SSB had declined below the target and was approaching the threshold level, triggering management action. Accordingly, prior to the 2015 fishing year, Addendum IV implemented harvest reductions in Striped Bass commercial and recreational fisheries coastwide (i.e., fisheries operating along the coast and in the Chesapeake Bay and other producer areas) to reduce coastwide *F* to a level at or below the target, to protect the strong 2011 year-class, and to stabilize SSB. In 2016, a stock assessment update indicated that Addendum IV successfully reduced *F* below the target, but female SSB continued to be on a declining trajectory towards the threshold (Figures 9 and 10; ASMFC 2016). Nonetheless, in 2017, the ASMFC Atlantic Striped Bass Management Board considered relaxing coastwide commercial and recreational regulations after hearing repeated concerns from the Chesapeake Bay for-hire sector regarding economic hardship endured since implementation of Addendum IV. However, the board did not move forward with the addendum due to concerns that relaxing regulations would result in *F* being above the target and SSB dropping below the threshold. Instead, the board decided to wait until after review of the results of the 2018 benchmark stock assessment before considering changes to the management program.

MONITORING AND RESEARCH IN SUPPORT OF MANAGEMENT

The continued interjurisdictional fisheries management throughout the 20+ year postrecovery of Atlantic Striped Bass has required long-term monitoring programs that are mandatory for states to be in compliance with the ISFMP. These state fishery agency efforts were broadly characterized by Essig and Beal (2000) as population surveys, utilization surveys, and other activities as described below.

Population surveys are primarily focused on major spawning areas in the Chesapeake Bay, the Delaware River, the Hudson River, and the A/R. For example, the MDDNR assesses spawning Striped Bass in its portion of the Chesapeake Bay through fishery-independent gill net sampling and through boat electrofishing surveys on the spawning grounds in spring. Later in the season, the relative abundance of juveniles near spawning areas is monitored with JAI beach seine surveys in several states.

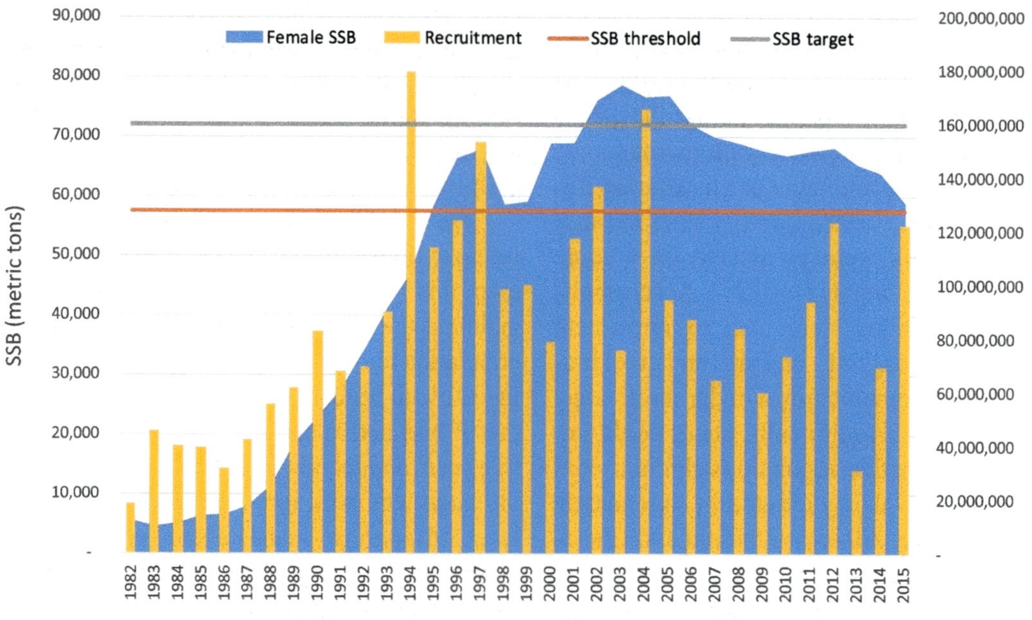

FIGURE 9. Atlantic Striped Bass female spawning stock biomass (SSB) and recruitment, 1982–2015. Source: ASMFC (2016).

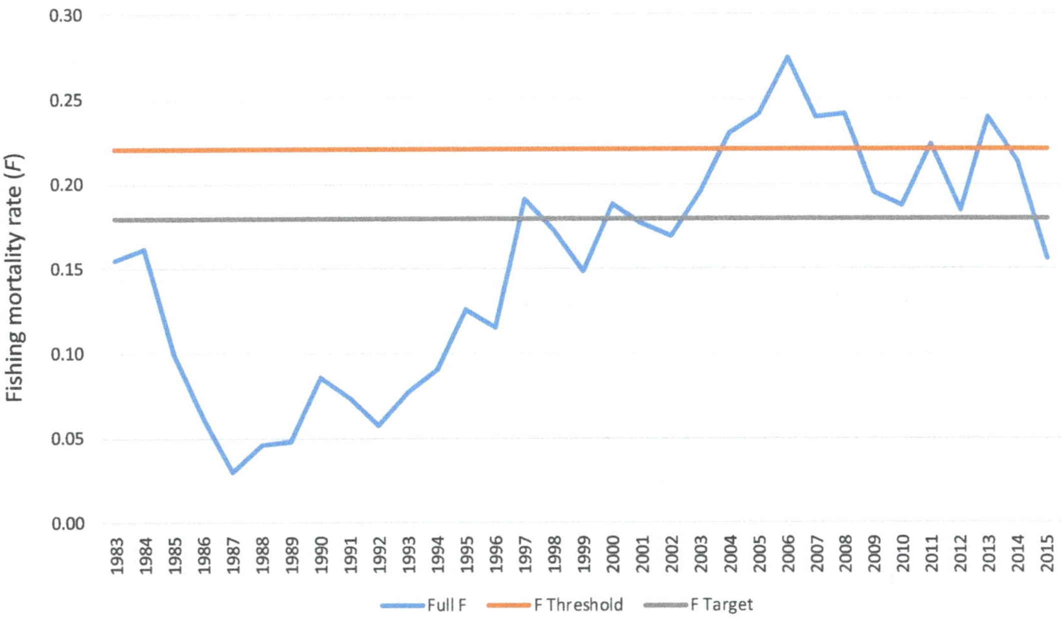

FIGURE 10. Atlantic Striped Bass fishing mortality rate (F), 1983–2015. Source: ASMFC (2016).

Utilization surveys are conducted to estimate recreational catch and harvest while census methods are used for commercial removals. The NMFS MRIP currently estimates recreational catch per unit effort through an angler intercept survey and effort through a telephone survey. North Carolina conducts its own survey of sport harvest in the A/R. Many states conduct volunteer angler logbook programs with the intent to collect size composition of Atlantic Striped Bass released alive. States also conduct biological sampling for age, size, and sex composition of commercially and recreationally harvested fish.

Other long-term Atlantic Striped Bass activities have included stocking, tagging, and interstate fishery management planning. The USFWS coordinated a stocking program within Chesapeake Bay with more than 8 million Striped Bass marked with coded wire tags from 1985 to 1995 (Upton 1993; Upton and Mangold 1996). As part of their ongoing monitoring programs, states scanned Atlantic Striped Bass for the presence of these tags. Hatchery fish survived and contributed to the spawning population (Upton and Mangold 1996); however, "the recovery of the Chesapeake Striped Bass wild population was nothing less than dramatic with two extraordinary year classes in 1989 and 1993." This result was likely due to favorable environmental conditions that enabled high survival and subsequent recruitment. The recovery of the wild component of the Chesapeake Bay stock made assessment of the stocking program benefits difficult (Upton and Mangold 1996). Richards and Rago (1999) evaluated the efficacy of the Striped Bass stocking program and concluded that the potential for gains in population recovery were far greater from reducing F than from stocking.

The ASMFC partners also have implemented and maintained a Cooperative Coastwide Striped Bass Tagging Program using conventional internal anchor tags (Rago 1987; Wooley et al. 1990; USFWS 2000). The objectives of the program are to determine the temporal and spatial distribution of Striped Bass, determine migration rates, assist in stock identification, determine instantaneous natural mortality (M) and F rates, facilitate estimation of stock size, examine effectiveness of stocking, and provide estimates of growth rate (where the length and weight of recaptured fish are provided). To date, the partners have tagged 533,812 Striped Bass, of which 93,780 have been reported as recaptures (90,354 unique captures, since some fish have been captured multiple times; Joshua Newhard, USFWS, Maryland Fish and Wildlife Conservation Office, unpublished data). The data from the program are analyzed annually by the ASMFC Striped Bass Technical Committee's Tagging Subcommittee to estimate M and F as an independent check on population assessment models and to evaluate effectiveness of management.

One component of the Cooperative Coastwide Striped Bass Tagging Program focuses on the mixed stocks on wintering grounds off the mid- and south Atlantic coast. Cooperative Winter Tagging Cruises were conducted during January–February, from 1988 through 2016, using multiple federal research vessels (Welsh et al. 2007) equipped with trawls to capture Striped Bass and other species for tagging (Figure 11). The USFWS initiated a hook-and-line component during winter using charter boats out of Rudee Inlet, Virginia in 2011. From both the trawl and hook-and-line components of this winter tagging program, 50,887 Striped Bass were tagged and released from 1988 through 2017 (USFWS and NCDMF, unpublished data). Although funding for the trawling component ended in

FIGURE 11. Cooperative Winter Tagging Cruise for Striped Bass off North Carolina (RV *Cape Hatteras*, 2004; USFWS photo).

2016, the NCDMF has continued the hook-and-line winter tagging to contribute to the annual harvest mortality estimates required by Congress. Geographic and temporal recoveries of tagged Striped Bass from the offshore mixed stock were consistent with previous studies, and estimated *F* rates fell below the management triggers for 1988–2004 (Welsh et al. 2007), suggesting that management measures designed to reduce *F* were working.

Another significant long-term activity of state and federal agency partners is the personnel time investment in interjurisdictional fishery management planning. Continued active participation on stock assessment teams, technical committees, advisory panels, and management boards has been essential to the sustainability of Atlantic Striped Bass.

A significant continuing emphasis on Striped Bass research has occurred during the postrecovery period. The ASMFC periodically polls members of its Management and Science Committee and Assessment Science Committee to determine prioritized research needs for Striped Bass (Kipp and Paine 2013). Categories emphasized include needs for fishery-dependent and fishery-independent data acquisition; modeling and quantitative needs; additional life history, biological and habitat information; and management, law enforcement, and socioeconomic needs. The ASMFC also provides updates on completed research (Kipp and Paine 2013).

One significant development during the postrecovery period that has greatly increased researchers' ability to document the behavior of individual Striped Bass has been the advent of acoustic transmitters and receivers and the deployment of a large number of acoustic receivers coastwide, along with data sharing by the researchers. The availability of this technology has resulted in a number of studies documenting Striped Bass spawning behavior, annual and seasonal migrations, and habitat use (e.g., Wingate et al. 2011; Callihan et al. 2015; Harris and Hightower 2017). For example, this information has been used to corroborate that Striped Bass return to natal rivers and spawn annually.

INTERSPECIES EFFECTS

While Atlantic Striped Bass are managed as a single species, it is recognized that stock recovery has effects on other species within the food web. Field (1997) reported that within the Chesapeake Bay some fishers claimed, "the species has become a ravenous pest" because they consumed so many juveniles of other important fishery species like blue crabs. Hartman and Margraf (2003) noted that following recovery, concern was expressed over the prevalence of disease in the Chesapeake Bay stock, and coincident with declines in fish health were changes in diets, declines of preferred prey, and reduced growth and condition. Managing the species to high levels of abundance, given its status as a top predator, can result in impacts to prey species. Grout (2006) noted that the 10-fold increase in abundance of Striped Bass from cooperative interjurisdictional management resulted in increased predation on a variety of other anadromous species, including federally listed Atlantic Salmon *Salmo salar*. For perspective, Atlantic Striped Bass SSB during the early 2000s was above target levels (Figure 9).

Laney (2009) provided a detailed historical review of Atlantic Striped Bass species interactions as both prey and predator by life stage. Multiple authors have examined Atlantic Striped Bass postrecovery species interactions through diet studies of various life stages, in both inshore and offshore habitats (Hartman 2000; Nelson et al. 2003; Uphoff 2003; Walter and Austin 2003; Rudershausen et al. 2005; Overton et al. 2008; Smith and Link 2010). Atlantic Striped Bass undergo ontogenetic dietary shifts, feeding primarily on zooplankton, insects, invertebrates, and macrobenthos when young (Hartman and Brandt 1995; Walter et al. 1995) and becoming more piscivorous with age and migration offshore (Overton et al. 2008; Laney 2009; Smith and Link 2010).

Site-specific interactions between Striped Bass and Bluefish *Pomatomus saltatrix* have been observed, with Buckel et al. (1999) reporting that Bluefish accounted for 50–100% of the total estimated loss of age-0 Striped Bass in the Hudson River estuary and selected for them at higher age-0 Striped Bass densities. In contrast, Buckel and McKown (2002) found low habitat and diet overlap during the early summer and fall in three New York Bight embayments. Similarly, Buckel et al. (2009) found low overlap in diet and habitat between these species in trawl data from the continental shelf.

Evidence exists that the abundance of forage species like Atlantic Menhaden *Brevoortia tyrannus* may, at times, be insufficient to support current Chesapeake Bay stocks (Zhang et al. 2011; Houde et al. 2016). Uphoff (2003) documented that Striped Bass in the upper Chesapeake Bay were in poor nutritional condition, suggesting that alternate prey were not offsetting decreased Atlantic Menhaden availability. This condition, combined with their high density in a habitat degraded by hypoxia, contributed to an outbreak of skin lesions in the late 1990s. Uphoff concluded that Striped Bass exceeded their carrying capacity in Chesapeake Bay in the late 1990s since more abundant and larger resident Striped Bass caused a decline in forage-sized Atlantic Menhaden. Hypoxia in certain areas of the Chesapeake Bay has affected the distribution of forage species and has resulted in improved Striped Bass foraging efficiency (Costantini et al. 2008). When combined with high Striped Bass abundance, improved foraging efficiency due to hypoxia contributed to shortages of prey fishes, resulting in poor growth and health of Striped Bass (Costantini et al. 2008). Wood and Austin (2009) investigating long-term (1968–2004) recruitment patterns of multiple Chesapeake Bay species found that a significant regime shift occurred in 1992 in favor of anadromous species at the expense of Atlantic Continental Shelf-spawning species (i.e., Atlantic Menhaden, Spot *Leiostomus xanthurus*, and Summer Flounder *Paralichthys dentatus*). The authors concluded that while the causative factors were not identified, the study suggested that stock recovery of shelf species may be complicated by the increased recruitment of anadromous species.

The prevalence of mycobacteriosis disease in Chesapeake Bay Striped Bass has generally increased since 1998 (Matsche 2016). Mycobacteriosis is an ongoing epizootic that has been reported to affect up to 75% of adults in the bay (Overton et al. 2003) and 65% of age 1–5 fish in the Maryland portion of the bay (Matsche 2016). Infected fish have a high proportion of ulcerative skin lesions that make them rather unappealing for recreational and commercial purposes. Hoenig et al. (2017) estimated that the disease-adjusted $M = 0.29$ for Striped Bass in Chesapeake Bay was double that used in population modeling by the ASMFC, suggesting that mycobacteriosis was having an impact on this stock. However, M in the assessment is for all stocks combined and coastal migratory Striped Bass have low levels of mycobacteriosis reported.

Interspecies effects involving Atlantic Striped Bass have been documented in areas other than Chesapeake Bay. Some researchers attribute the current low abundance of Weakfish *Cynoscion regalis* within Delaware Bay to an overabundance of Atlantic Striped Bass coastwide. Two different models of forage and predation predict that the M of Weakfish has increased with increased Atlantic Striped Bass abundance (ASMFC Weakfish Technical Committee 2009). In the Connecticut River, Atlantic Striped Bass may have negatively impacted the population of American Shad (Savoy and Crecco 2004) and Blueback Herring *Alosa aestivalis* (Savoy and Crecco 2004; Davis et al. 2012). In New England, numerically abundant Atlantic Striped Bass consumed, seasonally, more than three times the number of American lobster *Homarus americanus*, 1,940 times the number of rock crab *Cancer irroratus*, and 965 times the number of Atlantic Menhaden, respectively, harvested by regional and statewide fisheries, suggesting that Atlantic Striped Bass may exert considerable predation pressure on these prey populations (Nelson et al. 2003).

FUTURE ISSUES

The popularity of Atlantic Striped Bass to anglers and consumers, coupled with the complex nature of its seasonal distribution and highly variable recruitment, make sustainable management of this species challenging and politically charged. Stakeholders regularly call for the ASMFC to implement biologically, economically, and socially sound regulations within each state and sector. This presents a management conundrum as levels of abundance above SSB targets favor recreational catches and commercial harvest, yet may have the negative biological effects described above.

Now, and as it was prior to Amendment 5, Atlantic Striped Bass (excluding the A/R stock) are managed under one set of SSB and *F* threshold and target levels for the coastal migratory stock. In reality, the ASMFC-managed Atlantic Striped Bass comprises at least three biologically unique stocks (i.e., the Chesapeake Bay, Hudson River, and Delaware Bay stocks), each with varying levels of contribution to the coastal migratory stock. Although the stock assessment model and new *F* reference points do include the effects of regional-specific harvest on the coastwide stock, modeling and data limitations preclude the ability to estimate separate reference points for each unique stock, along with the inability to account for the sex ratio that exists in different areas where target fisheries occur.

Another future issue is the inclusion of environmental and ecosystem parameters within the stock assessment models used for Atlantic Striped Bass that could lead to ecosystem-based fisheries management (EBFM). Advances in ecosystem modeling, coupled with improved understanding of predator–prey interactions, will help inform managers of the appropriate level of Striped Bass harvest depending on the management objective (e.g., maintaining predator or prey populations at a desired level). Ecosystem-based fisheries management has not been a major part of ASMFC management of Atlantic Striped Bass to date, but initial efforts have occurred for the Chesapeake Bay stock where the local scientific community has drafted EBFM background papers and issue briefs (EBFM Striped Bass Species Team 2009). Martino and Houde's (2010) model accurately predicted recruitment of Chesapeake Bay Striped Bass during 2 out of 3 years from 2007 to 2009. Rerunning their model with more years of JAI data may provide more insights on model performance and applicability to other major Striped Bass stocks. Wood and Austin (2009) determined the existence of a multispecies recruitment pattern in Chesapeake Bay, which also plays a role in anadromous species recruitment success. As they noted, "to meet society's expectations of its coastal fisheries, especially considering the mounting pressures on coastal ecosystems that are likely to result from continued coastal development (Crossett et al. 2004) and potential future climate changes (Wood et al. 2002; Pyke et al. 2008), it is critically important that fisheries management continue its evolution towards ecosystem-based fisheries planning by accounting for important multispecies relationships."

Overall, the Atlantic Striped Bass stocks appear to be abundant in number and capable of producing strong year-classes in any given year, given favorable environmental conditions. However, until stock assessment data and modeling approaches produce statistically reliable stock-specific sex-based BRPs for management use, the ASMFC is limited in its ability to maximize Striped Bass fishing opportunities across all regions and sectors. As a result, sustainable Atlantic Striped Bass fishery management will likely continue to evolve and remain politically charged for years to come.

LESSONS LEARNED

The high value that society placed on Atlantic Striped Bass provided the political impetus to create and fund the recovery management program. Field (1997) observed that extreme interest in the fisheries for Atlantic Striped Bass played a pivotal role in initiation of management efforts and the acceptance of management actions by participants in these fisheries. It is hard to imagine how Congress would have passed the necessary legislation and appropriated sufficient funding if this species was not so prized. This was a key factor in the recovery and continued prevention of overfished status.

Coordination of management efforts among all jurisdictions combined with the cooperation of stakeholders was essential to the recovery of a highly mobile species such as Striped Bass. The interjurisdictional fisheries management that is needed for migratory anadromous species is not easy. In the case of Atlantic Striped Bass, it involved the 12 states and their freshwater and saltwater agencies from Maine to North Carolina, and the District of Columbia, the PRFC, NMFS, and the USFWS. Field (1997) noted that it was fortunate that the ASMFC was a well-established, functional body long before the collapse of the Chesapeake Bay stock. As members of ASMFC boards and technical committees, state and federal fisheries managers were able to communicate effectively in evaluating the Atlantic Striped Bass stocks and in establishing management measures needed for recovery and the postrecovery period. Cooperation and coordination of these fisheries managers together with other stakeholders like watershed managers for water quantity and quality were critical to reach stock recovery and then to sustain that recovery.

Federal legislation passed when the Atlantic Striped Bass had collapsed was absolutely critical in the recovery from two aspects: funding the recovery programs and enforcement. First, from a federal funding perspective, the ESBS, Atlantic Striped Bass Conservation Act and North Carolina Striped Bass Study supported necessary scientific studies of the declines, causative factors, and recommended management measures. The WB expansion of the SFR Act provided stable, long-term revenue to support the management needed for recoveries and their sustenance 20+ years since. Collection of monitoring data by so many states is expensive, and those sources of funding were key. Second, from a regulatory perspective, the Atlantic Striped Bass Conservation Act provided the threat of closed fisheries to strongly encourage states to comply with management measures with the ISFMP and its amendments. Until passage of the act, several states failed to voluntarily comply with the ISFMP (Field 1997), but the legislation changed that, with the Secretary of Commerce having the authority to close fisheries in noncompliant states. In fact, the act was so successful that it served as the model for the 1993 Atlantic Coastal Fisheries Cooperative Management Act, which had similar mandatory compliance for all ASMFC-managed species. Also critical to the recovery and management of Atlantic Striped Bass has been the state and federal enforcement of size and creel regulations, seasonal and area closures, and sale provisions.

Long-lived species require decades before population recovery can occur. Not surprisingly, 15 years elapsed since ESBS funding was first available in 1980 until the ASMFC's designation of full recovery of the Chesapeake Bay stock in 1995, the A/R stock in 1997, and the Delaware River stock in 1998. A successful year-class takes at least 6 years for females (females typically mature at 6 years) to contribute to the spawning stock. Striped Bass JAIs indicate that years of subaverage production will occur on a regular basis despite an abundant SSB, revealing the additional importance of environmental or other biotic influences on recruitment.

Monitoring programs are essential to provide feedback on effects of management actions and provide indicators when new actions need to be taken. Even with recovery, stock levels of Atlantic Striped Bass have fluctuated, thus requiring continued monitoring programs and adaptive management. These long-term efforts are also important in detecting and reacting to interspecies interactions that arise from relatively high abundance levels of recovered stocks. The Atlantic Striped Bass represents a powerful story of effective interjurisdictional fisheries management where the intensive work needed to bring the stocks to recovery was followed by decades of assessment work needed to sustain the recovered populations.

ACKNOWLEDGMENTS

The successful recovery of the Atlantic Striped Bass would not have been possible without the considerable contributions of a large fishery management "family" of federal and state agency staff; nongovernmental organizations, including fishing and conservation groups; and academic institutions, including many graduate students. The sacrifices and support of commercial and recreational fishers during the rebuilding period is especially recognized. We also thank the many ASMFC Striped Bass Management Board and Technical Committee chairs. Appreciation also is expressed to those members in Congress who had the foresight to pass the Atlantic Striped Bass Conservation Act (1984), which provided the ASMFC with needed authority to enforce Striped Bass management provisions. The authors wish to express their appreciation to the editors and two anonymous reviewers for their help improving this chapter. The opinions and ideas expressed in this chapter are those of the authors and do not necessarily reflect the positions of the Atlantic States Marine Fisheries Commission, East Carolina University, or the U.S. Fish and Wildlife Service.

REFERENCES

ASMFC (Atlantic States Marine Fisheries Commission). 1981. Interstate fisheries management plan for the Striped Bass of the Atlantic coast from Maine to North Carolina. ASMFC, Arlington, Virginia.

ASMFC (Atlantic States Marine Fisheries Commission). 1984a. Amendment 1 to the Interstate Fisheries Management Plan for Atlantic Striped Bass. ASMFC, Arlington, Virginia.

ASMFC (Atlantic States Marine Fisheries Commission). 1984b. Amendment 2 to the Interstate Fisheries Management Plan for Atlantic Striped Bass. ASMFC, Arlington, Virginia.

ASMFC (Atlantic States Marine Fisheries Commission). 1985. Amendment 3 to the Interstate Fisheries Management Plan for Atlantic Striped Bass. ASMFC, Arlington, Virginia.

ASMFC (Atlantic States Marine Fisheries Commission). 1989. Supplement to the Striped Bass Interstate Fishery Management Plan–Amendment #4. ASMFC, Fisheries Management Report No. 15, Arlington, Virginia.

ASMFC (Atlantic States Marine Fisheries Commission). 1995. Amendment #5 to the Interstate Fishery Management Plan for Atlantic Striped Bass. ASMFC, Fisheries Management Report No. 24, Arlington, Virginia.

ASMFC (Atlantic States Marine Fisheries Commission). 2003. Amendment 6 to the Interstate Fishery Management Plan for Atlantic Striped Bass. ASMFC, Fisheries Management Report No. 41, Arlington, Virginia.

ASMFC (Atlantic States Marine Fisheries Commission) Weakfish Technical Committee. 2009. Weakfish stock assessment report. ASMFC, Arlington, Virginia.

ASMFC (Atlantic States Marine Fisheries Commission) Striped Bass Technical, Stock Assessment, and Tagging Committees. 2013. Atlantic Striped Bass benchmark stock assessment. ASMFC, Arlington, Virginia.

ASMFC (Atlantic States Marine Fisheries Commission) Atlantic Striped Bass Technical Committee. 2016. Atlantic Striped Bass stock assessment update. ASMFC, Arlington, Virginia.

Baird, S. F. 1875. Report of Commissioner of Fish and Fisheries for the year ending June 30, 1874. Government Printing Office, Washington, D.C.

Ballou, R. 1987. The Atlantic Striped Bass Conservation Act: reauthorized and put to the test. Territorial Sea: Legal Developments in the Management of Interjurisdictional Resources 7 (3–4):1–12.

Berggren, T. J., and J. T. Lieberman. 1978. Relative contribution of Hudson, Chesapeake and Roanoke Striped Bass, *Morone saxatilis*, stocks to the Atlantic coast fishery. U.S. National Marine Fisheries Service Fishery Bulletin 76:335–345.

Bubier, J., and A. Rieser. 1984. Atlantic Striped Bass Conservation Act: unconstitutional amendment of an interstate compact? Territorial Sea: Legal Developments in the Management of Interjurisdictional Resources 4(4):1–11.

Buckel, J. A., D. O. Conover, N. D. Steinberg, and K. A. McKown. 1999. Impact of age-0 Bluefish (*Pomatomus saltatrix*) predation on age-0 fishes in the Hudson River estuary: evidence for density-dependent loss of juvenile Striped Bass (*Morone saxatilis*). Canadian Journal of Fisheries and Aquatic Sciences 56:275–287.

Buckel, J. A., and K. A. McKown. 2002. Competition between juvenile Striped Bass and Bluefish: resources partitioning and growth rate. Marine Ecology Progress Series 234:191–204.

Buckel, J. A., J. P. Pessutti, J. E. Rosedale, and J. S. Link. 2009. Interactions between Bluefish and Striped Bass: behavior of Bluefish under size- and number-impaired conditions and overlap in resource use. Journal of Experimental Marine Biology and Ecology 368:129–137.

Callihan, J. L., J. E. Harris, and J. E. Hightower. 2015. Coastal migration and homing of Roanoke River Striped Bass. Marine and Coastal Fisheries: Dynamics, Management, and Ecosystem Science 7:301–315.

Carmichael, J. T. 2000. Status of the Albemarle Sound–Roanoke River stock of Striped Bass. North Carolina Division of Marine Fisheries, Morehead City.

Carmichael, J. T. 2001. Status of the Albemarle Sound–Roanoke River stock of Striped Bass. North Carolina Division of Marine Fisheries, Morehead City.

Carmichael, J. T. 2002. Status of the Albemarle Sound–Roanoke River stock of Striped Bass. North Carolina Division of Marine Fisheries, Morehead City.

Carmichael, J.T. 2003. Status of the Albemarle Sound–Roanoke River stock of Striped Bass. North Carolina Division of Marine Fisheries, Morehead City.

Chittenden, M. E., Jr. 1971. Status of the Striped Bass, *Roccus saxatilis*, in the Delaware River. Chesapeake Science 12:131–136.

Costantini, M., S. A. Ludsin, D. M. Mason, X. Zhang, W. C. Boicourt, and S. B. Brandt. 2008. Effect of hypoxia on habitat quality of Striped Bass (*Morone saxatilis*) in Chesapeake Bay. Canadian Journal of Fisheries and Aquatic Sciences 65:989–1002.

Crossett, K., T. J. Culliton, P. Wiley and T. R. Goodspeed. 2004. Population trends along the coastal United States, 1980–2008. National Oceanic and Atmospheric Administration, Silver Spring, Maryland.

Davis, J. P., E. T. Schultz, and J. C. Vokoun. 2012. Striped Bass consumption of Blueback Herring during vernal riverine migrations: does relaxing harvest restrictions on a predator conserve a prey species of concern? Marine and Coastal Fisheries: Dynamics, Management, and Ecosystem Science 4:239–251.

DOI (U.S. Department of the Interior) and DOC (U.S. Department of Commerce). 1994. Striped Bass research study: report for 1992. National Marine Fisheries Service, Silver Spring, Maryland.

Dorazio, R. M. 1995. Mortality estimates of Striped Bass caught in Albemarle Sound and Roanoke River, North Carolina. North American Journal of Fisheries Management 15:290–299.

EBFM (Ecosystem-Based Fisheries Management) Striped Bass Species Team. 2009. Ecosystem-based fisheries management for Chesapeake Bay: Striped Bass background and issue briefs. Maryland Sea Grant, Publication No. UM-SG-TS-2009-07, College Park.

Essig, R. J., and R. E. Beal. 2000. Atlantic Coast Striped Bass recovery: federal aid in sport fish restoration in support of interstate fishery management. Fisheries 25(7):S17–S18.

Field, J. D. 1997. Atlantic Striped Bass management: where did we go right? Fisheries 22(7):6–8.

Flowers, J., S. Darsee, L. Lee, and C. Godwin. 2016. Stock status of Albemarle Sound–Roanoke River Striped Bass: update 1982–2014. North Carolina Division of Marine Fisheries, NCDMF SAP-SAR-2016-01, Morehead City.

Goodyear, C. P. 1985. Toxic materials, fishing, and environmental variation: simulated effects on Striped Bass population trends. Transactions of the American Fisheries Society 114:107–113.

Grist, J. 2004. Stock status of Albemarle Sound–Roanoke River Striped Bass. North Carolina Division of Marine Fisheries, Morehead City.

Grist, J. 2005. Stock status of Albemarle Sound–Roanoke River Striped Bass. North Carolina Division of Marine Fisheries, Morehead City.

Grout, D. E. 2006. Interactions between Striped Bass (*Morone saxatilis*) rebuilding programs and the conservation of Atlantic Salmon (*Salmo salar*) and other anadromous fish species in the USA. ICES Journal of Marine Science 63:1346–1352.

Harris, J. E., and J. E. Hightower. 2017. An integrated tagging model to estimate mortality rates of Albemarle Sound–Roanoke River Striped Bass. Canadian Journal of Fisheries and Aquatic Sciences 74:1061–1076.

Hartman, K. J. 2000. The influence of size on Striped Bass foraging. Marine Ecology Progress Series 194:263–268.

Hartman, K. J., and S. B. Brandt. 1995. Trophic resource partitioning, diets and growth of sympatric estuarine predators. Transactions of the American Fisheries Society 124:520–537.

Hartman, K. J., and F. J. Margraf. 2003. US Atlantic coast Striped Bass: issues with a recovered population. Fisheries Management and Ecology 10:309–312.

Hassler, W. W., N. L. Hill, and J. T. Brown. 1981. The status and abundance of Striped Bass, *Morone saxatilis*, in the Roanoke River and Albemarle Sound, North Carolina 1958–1980. North Carolina Department of Natural Resources and Community Development, Division of Marine Fisheries, Special Report 38, Morehead City.

Hoenig, J. M., M. L. Groner, M. W. Smith, W. K. Vogelbein, D. M. Taylor, F. F. Landers, Jr., J. T. Swenarton, D. T. Gauthier, P. Sadler, M. A. Matsche, A. N. Haines, H. J. Small, R. Pradel, R. Choquet, and J. D. Shields. 2017.

Impact of disease on the survival of three commercially fished species. Ecological Applications 27:2116–2127.

Houde, E. D., E. R. Annis, L.W. Harding, Jr., M. E. Mallonee, and M. J. Wilberg. 2016. Factors affecting the abundance of age-0 Atlantic Menhaden (*Brevoortia tyrannus*) in Chesapeake Bay. ICES Journal of Marine Science 73:2238–2251.

Kipp, J., and M. Paine. 2013. Research priorities and recommendations to support interjurisdictional fisheries management. Atlantic States Marine Fisheries Commission, Special Report No. 89, Arlington, Virginia.

Laney, R. W. 2009. Chapter 9: Striped Bass (*Morone saxatilis*). Pages 255–326 *in* K. E. Greene, J. L. Zimmerman, R. W. Laney, and J. C. Thomas-Blate, editors. Atlantic coast diadromous fish habitat: a review of utilization, threats, and recommendations for conservation and research needs. Atlantic States Marine Fisheries Commission, Habitat Management Series No. 9, Arlington, Virginia.

Manooch, C. S., III, and R. A. Rulifson, editors. 1989. Roanoke River Water Flow Committee report: a recommended water flow regime for the Roanoke River, North Carolina, to benefit anadromous Striped Bass and other below-dam resources and users. NOAA Technical Memorandum NMFS-SEFC-216.

Martino, E. J., and E. D. Houde. 2010. Recruitment of Striped Bass in Chesapeake Bay: spatial and temporal environmental variability and availability of zooplankton prey. Marine Ecology Progress Series 409:213–228.

Matsche, M. 2016. Health investigations of Striped Bass and other fishes in Maryland waters. Maryland Department of Natural Resources, Annapolis.

Mroch, R., and C. H. Godwin. 2014. Stock status of Albemarle Sound–Roanoke River Striped Bass. North Carolina Division of Marine Fisheries, Morehead City.

NCDMF (North Carolina Division of Marine Fisheries) and NCWRC (North Carolina Wildlife Resources Commission). 2013. Amendment 1 to the North Carolina Estuarine Striped Bass Fishery Management Plan. NCDMF, Morehead City.

Nelson, G. A., B. C. Chase, and J. Stockwell. 2003. Food habits of Striped Bass (*Morone saxatilis*) in coastal waters of Massachusetts. Journal of Northwest Atlantic Fisheries Science 32:1–25.

North, E. W., and E. D. Houde. 2003. Linking ETM physics, zooplankton prey, and fish early-life histories to Striped Bass *Morone saxatilis* and White Perch *M. americana* recruitment. Marine Ecology Progress Series 260:219–236.

North Carolina Striped Bass Study Management Board. 1991. Report on the Albemarle Sound-Roanoke River stock of Striped Bass. U.S. Fish and Wildlife Service, South Atlantic Fisheries Coordination Office, Morehead City, North Carolina.

Overton, A. S., F. J. Margraf, C. A. Weedon, L. H. Pieper, and E. B. May. 2003. The prevalence of mycobacterial infections in Striped Bass in Chesapeake Bay. Fisheries Management and Ecology 10:301–308.

Overton, A. S., C. S. Manooch, III, J. W. Smith, and K. Brennan. 2008. Interactions between adult migratory Striped Bass (*Morone saxatilis*) and their prey during winter off the Virginia and North Carolina Atlantic coast from 1994 through 2007. U.S. National Marine Fisheries Service Fishery Bulletin 106:174–182.

Pearson, J. C. 1938. The life history of the Striped Bass, or Rockfish, *Roccus saxatilis* (Walbaum). Bulletin No. 28 of the Bureau of Fisheries 49:825–851.

Public Law 100-589. 100th Congress (3 November 1988).

Pyke, C. R., R. G. Najjar, M. B. Adams, D. Brietburg, C. Hershner, M. Kemp, R. Howarth, M. Mulholland, M. Paolisso, D. Secor, K. Sellner, D. Wardrop, and R. Wood. 2008. Climate change and the Chesapeake Bay: state-of-the-science review and recommendations. Science and Technical Advisory Committee, Chesapeake Bay Program, Edgewater, Maryland.

Rago, P. J. 1987. Development, management and analysis of a Striped Bass tagging database—working paper. U.S. Fish and Wildlife Service, National Fisheries Center, Kearneysville, West Virginia.

Rapone, T. 2011. The EEZ solution to striper management: why the federal government should ban the commercial harvest of Striped Bass once and for all. Suffolk University Law Review XLIV:567–591.

Richards, R. A., and P. J. Rago. 1999. A case history of effective fisheries management: Chesapeake Bay Striped Bass. North American Journal of Fisheries Management 19:356–375.

Rudershausen, P. J., J. E. Tuomikoski, J. A. Buckel and J. E. Hightower. 2005. Prey selectivity and diet of Striped Bass in western Albemarle Sound, North Carolina. Transactions of the American Fisheries Society 134:1059–1074.

Rulifson, R. A., J. E. Cooper, D. W. Stanley, M. E. Shepherd, S. F. Wood, and D. D. Daniel. 1992. Food and feeding of young Striped Bass in the lower Roanoke River and western Albemarle Sound, North Carolina, 1984–1991. North Carolina Wildlife Resources Commission, Completion Report for Project F-27, Greenville.

Rulifson, R. A., and J. J. Isely. 1995. Striped Bass egg abundance and viability in the Roanoke River, North Carolina, and young-of-year survivorship, for 1993. Completion Report to Virginia Power, Innsbrook Technical Center, Glenn Allen, Virginia.

Rulifson, R. A., J. J. Isely and C. S. Manooch, III. 1993. Age, growth and survival of juvenile Striped Bass determined by counting daily growth rings on otoliths, 1990–1992. Pages 226–238 *in* R. A. Rulifson and C. S Manooch, III, editors. Roanoke River Water Flow Committee Report for 1991–1993. Albemarle-Pamlico Estuarine Study, Project No. APES 93-18, Raleigh, North Carolina.

Rulifson, R. A., and C. S. Manooch, III. 1990a. Recruitment of juvenile Striped Bass in the Roanoke River, North Carolina, as related to reservoir discharge. North American Journal of Fisheries Management 10:397–407.

Rulifson, R. A., and C. S. Manooch, III, editors. 1990b. Roanoke River Water Flow Committee report for 1988 and 1989. NOAA Technical Memorandum NMFS-SEFC-256.

Rulifson, R. A., and C. S. Manooch, III, editors. 1991. Roanoke River Water Flow Committee report for 1990. NOAA Technical Memorandum NMFS-SEFC-291.

Rulifson, R. A., and C.S. Manooch, III editors. 1993. Roanoke River Water Flow Committee report for 1991–1993. Albemarle-Pamlico Estuarine Study, Project No. APES 93-18, Raleigh, North Carolina.

Russell, D. 2005. Striper wars: an American fish story. Island Press, Washington, D.C.

Savoy, T. F., and V. A. Crecco. 2004. Factors affecting the recent decline of Blueback Herring and American Shad in the Connecticut River. Pages 361–377 *in* P. M. Jacobson, D. A. Dixon, W. C. Leggett, B. C. Marcy, Jr., and R. R. Massengill, editors. The Connecticut River ecological study (1965–1973) revisited: ecology of the lower Connecticut River 1973–2003. American Fisheries Society, Monograph 9, Bethesda, Maryland.

Secor, D. H. 2000a. Longevity and resilience of Chesapeake Bay Striped Bass. ICES Journal of Marine Science 57:808–815.

Secor, D. H. 2000b. Spawning in the nick of time? Effect of adult demographics on spawning behavior and recruitment in Chesapeake Bay Striped Bass. ICES Journal of Marine Science 57:403–411.

Secor, D. H. 2007. The year-class phenomenon and the storage effect in marine fishes. Journal of Sea Research 57:91–103.

Secor, D. H., and E. D. Houde. 1995. Temperature effects on the timing of Striped Bass egg production, larval variability, and recruitment potential in the Patuxent River (Chesapeake Bay). Estuaries 18:527–544.

Secor, D. H., and P. M. Piccoli. 2007. Oceanic migration rates of upper Chesapeake Bay Striped Bass (*Morone saxatilis*), determined by otolith microchemical analysis. U.S. National Marine Fisheries Service Fishery Bulletin 105:62–73.

Shepherd, G. R., R. W. Laney, M. Appelman, D. Honabarger, and C. L. Wright. 2016. Biennial report to Congress on the findings of studies of Striped Bass populations, 2015. National Marine Fisheries Service, Silver Spring, Maryland.

Smith, B. E., and J. S. Link. 2010. The trophic dynamics of 50 finfish and 2 squid species on the northeast US Continental shelf. NOAA Technical Memorandum NMFS-NE-216.

Smith, H. M. 1907. North Carolina geological and economic survey, volume II. The fishes of North Carolina. J. M. Uzzle and Co., State Printers and Binders, Raleigh, North Carolina.

Southwick Associates. 2005. The economics of recreational and commercial Striped Bass fishing. Report prepared for Stripers Forever, South Portland, Maine.

Speir, H., J. H. Uphoff, Jr., and E. Durell. 1999. A review of management of large Striped Bass and Striped Bass spawning grounds in Maryland. Maryland Department of Natural Resources, Fisheries Service, Technical Memorandum No. 15, Annapolis.

Stickney, R. R. 1996. Aquaculture in the United States: a historical survey. Wiley, Hoboken, New Jersey.

Takade-Heumacher, H. 2010. Stock status of Albemarle Sound–Roanoke River Striped Bass. North Carolina Division of Marine Fisheries, Morehead City.

Uphoff, J. H., Jr. 2003. Predator–prey analysis of Striped Bass and Atlantic Menhaden in upper Chesapeake Bay. Fisheries Management and Ecology 10:313–322.

Upton, H. F. 1993. Striped Bass (Morone saxatilis) stocking summary: 1985–1992. Atlantic States Marine Fisheries Commission, Special Report No. 28, Washington, D.C.

Upton, H. F., and M. Mangold. 1996. Evaluation of anadromous Striped Bass enhancement efforts with emphasis on Chesapeake Bay stocking activities. Atlantic States Marine Fisheries Commission, Special Report No. 53, Washington, D.C.

USFWS (U.S. Fish and Wildlife Service). 2000. Cooperative tagging program. Maryland Fisheries Resources Office, Annapolis, Maryland.

USFWS (U.S. Fish and Wildlife Service) and NOAA (National Oceanic and Atmospheric Administration). 1992. Albemarle Sound and Roanoke River basin North Carolina Striped Bass study. Report to Congress. USFWS, Washington, D.C.

Van Winkle, W., K. D. Kumar, and D. S. Vaughan. 1988. Relative contributions of the Hudson River and Chesapeake Bay Striped Bass stocks to the Atlantic coastal population. Pages 255–266 *in* L. W. Barnthouse, R. J. Klauda, D. S. Vaughan, and R. L. Kendall, editors. Science, law and Hudson River power plants. American Fisheries Society, Monograph 4, Bethesda, Maryland.

Walter, J. F., III, and H. M. Austin. 2003. Diet composition of large Striped Bass (*Morone saxatilis*) in Chesapeake Bay. U.S. National Marine Fisheries Service Fishery Bulletin 101:414–423.

Walter, J. F., III, A. S. Overton, K. H. Ferry, and M. F. Mather. 1995. Atlantic coast feeding habits of Striped Bass: a synthesis supporting a coast-wide understanding of trophic biology. Fisheries Management and Ecology 10:349–360.

Welsh, S. A., D. R. Smith, R. W. Laney, and R. C. Tipton. 2007. Tag-based estimates of annual fishing mortality of a mixed Atlantic coastal stock of Striped Bass. Transactions of the American Fisheries Society 136:34–42.

Wingate, R. L., D. H. Secor, and R. T. Kraus. 2011. Seasonal patterns of movement and residency by Striped Bass within a subestuary of the Chesapeake Bay. Transactions of the American Fisheries Society 140:1441–1450.

Wood, R. J., and H. M. Austin. 2009. Synchronous multidecadal fish recruitment patterns in Chesapeake Bay, USA. Canadian Journal of Fisheries and Aquatic Sciences 66:496–508.

Wood, R. J., D. F. Boesch, and V. S. Kennedy. 2002. Future consequences of climate change for the Chesapeake Bay ecosystem and its fisheries. Pages 171–184 *in* N. A. McGinn, editor. Fisheries in a changing climate. American Fisheries Society, Symposium 32, Bethesda, Maryland.

Wooley, C. M., N. C. Parker, B.M. Florence, and R. M. Miller. 1990. Striped Bass restoration along the Atlantic coast: a multistate and federal cooperative hatchery and tagging program. Pages 775–781 *in* N. C. Parker, A. E. Giorgi, R. C. Heidinger, D. B. Jester, Jr., E. D. Prince, and G. A. Winans, editors. American Fisheries Society, Symposium 7, Bethesda, Maryland.

Zhang, X., R. J. Wood, E. J. Martino, E. D. Houde, H. Townsend, and C. Wicks. 2011. Modeling Atlantic Menhaden recruitment in the Chesapeake Bay: is the Striped Bass recovery a problem? Pages 120–125 *in* A. L. Mason, D. Apeti, and D. Whitall, editors. National Centers for Coastal Ocean Science (NCCOS) research highlights in the Chesapeake Bay. NOAA Technical Memorandum NOS NCCOS 128.

Zincone, L. H., Jr., and R. A. Rulifson. 1991. Instream flow and Striped Bass recruitment in the lower Roanoke River, North Carolina. Rivers 2(2):125–137.

Section V

Epilogue

From Sustainability to Catastrophe to Recovery: The Need for Using an Ecosystem Approach for Responsible Fishery Management

Devin M. Bartley*

Department of Fisheries and Wildlife, Michigan State University East Lansing, Michigan 48823, USA

The title for this epilogue was chosen to highlight why many of the world's fisheries are in trouble and why an ecosystem approach to fishery management is needed. Although natural events such as tornados, floods, and droughts certainly impact fisheries, for the most part it is humans that are directly or indirectly responsible for the poor state of many of the world's fisheries today. The problem facing these fisheries may seem obvious, but the often unsuccessful efforts at developing and implementing effective fishery management programs have not adequately addressed the human component of the fishery ecosystem. Not addressed are especially those humans who may not be fishers or, at the other extreme, those humans who have fishing as their only livelihood option. Just as most of the contributions to this book addressed inland fisheries, so too will this epilogue. Inland, or freshwater, fisheries present special challenges to fishers and to fishery managers.

Evidence of early humans fishing and eating fish dates back millennia. Bones of tilapia and catfish were found with settlements of *Homo erectus* and *Homo habilis* in the Olduvai Gorge of Tanzania (Gartside and Kirkegaard 2009). The Aurignac people of the late Paleolithic period were considered the first "modern" *Homo sapiens* and began making pointed bits of bone and horn to help capture river fish (Sahrhage and Lundbeck 1992). In ancient times, hunting, gathering, and trapping on the worlds aquatic ecosystems was a low-tech activity, and as such, "fairly sustainable yields were obtained" (Sahrhage and Lundbeck 1992:61).

Although technological progress and increased efficiency have been made in fishing since ancient times, such as the use of motorized boats, synthetic mono and multifilament nets, and global positioning systems, "methods of fishing have not, in principle, changed much over thousands of years" (Sahrhage and Lundbeck 1992:2). This observation is especially true for the world's inland fisheries, where the majority of production comes from small-scale fisheries in developing countries (Bartley et al. 2016).

What has changed most over the millennia is human population growth and its impact on the environment. Around the start of the second millennium (~1,000 AD) a shift in the English diet occurred with a move from a diet high in freshwater species to more marine species, providing evidence that urbanization and associated impacts of farming had degraded inland waters and their fishes, forcing people to fish the sea for their needed food (Barrett et al. 2004). This shift to marine species was thought not to be the result of abundant marine fisheries because it corresponded with climatic conditions that did not favor marine fishes. As Barrett et al. (2004) suggests, "This counterintuitive discovery (i.e., of more fishing in the sea when

* Corresponding author: devinmichaelbartley@gmail.com

conditions didn't favor it) can be explained by the concurrent rise of urbanism and human impacts on freshwater ecosystems." The degradation of inland waters and their fisheries continues today with water removals from the world's rivers, draining of wetlands, habitat fragmentation, invasive species, pollution, and hydroelectric development combining to make freshwater fishes the most endangered group of vertebrates used by humans (Ricciardi and Rasmussen 1999; IUCN 2010).

The success stories contained in this volume demonstrate that in spite of the challenges facing inland waters, fishes and fisheries can recover. Hope for the future is possible and realistic. The Food and Agriculture Organization of the United Nations (FAO) recently reported that the proportion of marine fish stocks that are fished beyond biologically sustainable limits has increased over the past several decades (FAO 2018b). This global picture of marine stocks has been picked up and reported on extensively by the media (Neslen 2016). However, numerous examples exist where well-designed and implemented fishery management actions have improved the status of marine fisheries (e.g., Worm et al. 2009; Essig et al. 2019, this volume; Rabazanov et al. 2019, this volume; Unal and Kizilkaya 2019, this volume). The authors in this volume are to be commended for demonstrating that fishery management using an ecosystem and inclusive approach can rebuild both inland and coastal marine fisheries.

An ecosystem approach to inland fisheries is essential due to strong competition for the use of land and freshwater from stakeholders outside the fishing sector. Freshwater is an essential, but scarce, human requirement for survival. The lack of access to stable supplies of freshwater is becoming critical, with the problem expected to worsen due to rapid urbanization worldwide and the roughly 1.2×10^9 persons to be added over the next 20 years (U.S. National Intelligence Council 2008). Humans extract about 10% of the world's freshwaters, with 70% of that used for agriculture, 20% for industry, and 10% for municipalities (Molden 2007). These statistics are global averages, and extractions from some rivers for use by an ever-growing human population in such activities as agriculture, hydropower, and municipal use cause them to not even reach the sea; clearly, inland fishers and resource managers need to engage these other stakeholders or find new jobs if we want to have sustainable fisheries in the future!

Again, the authors of the case studies presented in this volume demonstrated that engagement can happen. The FAO and Michigan State University convened a global conference at FAO headquarters in Rome (Taylor et al. 2016) that highlighted challenges facing inland fisheries. The conference discussions led to the *Rome Declaration: Ten Steps to Responsible Inland Fisheries* (FAO 2016). Many of these steps are directly addressed by the case studies, but two seem especially important:

- Improve communication among freshwater users, and
- Develop collaborative approaches to cross-sectoral integration in development agendas.

For sustainable fisheries to occur, it is imperative that the various sectors using water, such as the farming, irrigation, power generation, and land development sectors, work together (across sectors) for mutually beneficial development outcomes.

The case studies depicted in this book provide several excellent examples of engagement and communication with managing agencies and stakeholders (e.g., Moore and Kulp 2019, this volume). But more is needed. As a start, we do not know much about the status of the world's inland fisheries. The FAO reported that more than half of the global inland fishery catch is not reported to species (FAO 2018b); traditional stock assessment is very difficult on most of the world's inland fisheries and efforts are underway to develop innovative approaches to inland fishery stock assessment (FAO 2018a; see also Gurdak et al. 2019, this volume).

The FAO is pleased to be a part of this work. Research papers are good at advancing knowledge, but fishers, resource managers, and countries that want to continue to provide a reliable source of freshwater fishes to their people want and need more. They need proven methods of improving fisheries, as presented in this book. The Lessons Learned sections in the chapters should be especially helpful in providing the needed information that we can use to improve the status of the world's fisheries resources!

These success stories demonstrate that it is possible to recover fisheries, despite the apparent dismal status of some fisheries. Success will depend on the objectives of fishery management. For example, if the objective was to support recreational fisheries, active management may be economically viable and result in success, whereas supporting commercial fisheries in the same area might not be sustainable; saving an endangered species (e.g., Pister 2019; Stephens et al. 2019; both this volume) may be judged as success and not include development of a fishery. As such, "success" depends on the goal and objectives of the management programs employed.

What were the causes for successes presented here? No one silver bullet was used. In some cases, stocking worked (e.g., Kline et al. 2019, this volume); in others, it did not (Kraft 2019, this volume). But clearly and perhaps most importantly, there seemed to be a common thread to having and maintaining sustainable fisheries, that being that one has to meaningfully engage as many stakeholders as possible, building trust among them, taking care of the environment and its ecosystem services, and monitoring the effects of the interventions. This compilation of case studies helps to implement *The Rome Declaration* (FAO 2016) and further the development of responsible fisheries. The authors are to be congratulated for sharing their important findings on restoring fish populations and their fisheries, which should help others to learn how to be more successful in protecting, enhancing, and restoring the fisheries of today and tomorrow!

REFERENCES

Barrett, J. H., A. M. Locker, and C. M. Roberts. 2004. The origins of intensive marine fishing in medieval Europe: the English evidence. Proceedings of the Royal Society B 271:2417–2421.

Bartley, D. M., G. de Graaf, and J. Valbo-Jorgensen. 2016. Commercial inland capture fisheries. Pages 438–448 *in* J. F. Craig, editor. Freshwater fisheries ecology. Wiley-Blackwell Scientific Publications, Oxford, UK.

Essig, R. J., R. W. Laney, M. H. Appelman, F. A. Harris, R. A. Rulifson, and K. L. Nelson. 2019. Atlantic Striped Bass: an interjurisdictional fisheries management success story. Pages 533–563 *in* C. C. Krueger, W. W. Taylor, and S. Youn, editors. From catastrophe to recovery: stories of fishery management success. American Fisheries Society, Bethesda, Maryland.

FAO (Food and Agriculture Organization). 2016. The Rome declaration: ten steps to responsible inland fisheries. FAO, Rome.

FAO (Food and Agriculture Organization) 2018a. Report of the Advisory Roundtable on the Assessment of Inland Fisheries, Rome, Italy, 8–10 May 2018. FAO, Report No. R1240, Rome.

FAO (Food and Agriculture Organization). 2018b. State of world fisheries and aquaculture 2018. FAO, Rome.

Gartside, D. F., and I. R. Kirkegaard. 2009. A history of fishing. Pages 105–139 *in* V. R.

Squires editor. The role of food, agriculture, forestry and fisheries in human nutrition, volume II. Eolss Publishers Co Ltd, Oxford, UK.

Gurdak, D. J., C. C. Arantes, L. Castello, D. J. Stewart, and L. C. Watson. 2019. Evidence of recoveries from tropical floodplain fisheries: three examples of management gains for South American Giant Arapaima. Pages 267–295 *in* C. C. Krueger, W. W. Taylor, and S. Youn, editors. From catastrophe to recovery: stories of fishery management success. American Fisheries Society, Bethesda, Maryland.

IUCN (International Union for Conservation of Nature) 2019. The IUCN red list of threatened species [online database]. Available: www.iucnredlist.org.

Kline, P. A., T. A. Flagg, C. C. Kozfkay, D. J. Baker, D. J. Maynard, and W. C. McAuley. 2019. Snake River Sockeye Salmon—a perspective on twenty-five years of effort to recover a population in central Idaho, USA. Pages 101–123 *in* C. C. Krueger, W. W. Taylor, and S. Youn, editors. From catastrophe to recovery: stories of fishery management success. American Fisheries Society, Bethesda, Maryland.

Kraft, C. 2019. Adirondack Brook Trout and acid rain: environmental legislation fosters successful restoration. Pages 299–322 *in* C. C. Krueger, W. W. Taylor, and S. Youn, editors. From catastrophe to recovery: stories of fishery management success. American Fisheries Society, Bethesda, Maryland.

Molden, D., editor. 2007. Water for food, water for life: a comprehensive assessment of water management in agriculture. International Water Management Institute, Colombo, Sri Lanka and Earthscan, London.

Moore, S. E., and M. A. Kulp. 2019. Overcoming challenges to native fish restoration in a national park. Pages 45–64 *in* C. C. Krueger, W. W. Taylor, and S. Youn, editors. From catastrophe to recovery: stories of fishery management success. American Fisheries Society, Bethesda, Maryland.

Neslen, A. 2016. Global fish production approaching sustainable limit, UN warns. The Guardian. Available: www.theguardian.com/environment/2016/jul/07/global-fish-production-approaching-sustainable-limit-un-warns. (April 2019).

Pister, E. P. 2019. Species salvation: saving the Owens Pupfish from extinction. Pages 373–381 *in* C. C. Krueger, W. W. Taylor, and S. Youn, editors. From catastrophe to recovery: stories of fishery management success. American Fisheries Society, Bethesda, Maryland.

Rabazanov, N. I., A. M. Orlov, A. S. Abdusamadov, R. M. Barkhalov, and K. M. Akhemedkhanov. 2019. Caspian Kutum *Rutilus kutum*: a long of exploitation, survival, and revival. Pages 485–508 *in* C. C. Krueger, W. W. Taylor, and S. Youn, editors. From catastrophe to recovery: stories of fishery management success. American Fisheries Society, Bethesda, Maryland.

Ricciardi, A., and J. B. Rasmussen. 1999. Extinction rates of North American freshwater fauna. Conservation Biology 13:1220–1222.

Sahrhage, D., and J. Lundbeck. 1992. A history of fishing. Springer-Verlag, Berlin.

Stephens, M. R., S. J. Stephens, and C. C. Krueger. 2019. Lessons learned from more than one-hundred years of golden trout management and recovery in California. Pages 19–43 *in* C. C. Krueger, W. W. Taylor, and S. Youn, editors. From catastrophe to recovery: stories of fishery management success. American Fisheries Society, Bethesda, Maryland.

Taylor, W. W., D. M. Bartley, C. I. Goddard, N. J. Leonard, and R. Welcomme, editors. 2016. Freshwater, fish and the future: proceedings of the global cross-sectoral conference. Food and Agriculture Organization of the United Nations, Rome; Michigan State University, East Lansing; American Fisheries Society, Bethesda, Maryland.

U.S. National Intelligence Council. 2008. Global trends 2025: a transformed world.

U.S. Government Printing Office, Washington, D.C. Available: www.files.ethz.ch/isn/94769/2008_11_global_trends_2025.pdf. (January 2019).

Ünal, V., and Z. Kizilkaya. 2019. A long and participatory process towards successful fishery management of Gökova Bay, Turkey. Pages 509–532 *in* C. C. Krueger, W. W. Taylor, and S. Youn, editors. From catastrophe to recovery: stories of fishery management success. American Fisheries Society, Bethesda, Maryland.

Worm, B., R. Hilborn, J. K. Baum, T. A. Branch, J. S. Collie, C. Costello, M. J. Fogarty, E. A. Fulton, J. A. Hutchings, S. Jennings, O. P. Jensen, H. K. Lotze, P. M. Mace, T. R. McClanahan, C. Minto, S. R. Palumbi, A. M. Parma, D. Ricard, A. A. Rosenberg, R. Watson, and D. Zeller. 2009. Rebuilding global fisheries. Science 325:578–585.

Index

Italicized page numbers indicate figures, tables, and boxes.

B

C

D

I

J

K

L

M

N

O

P

Q

R

S

T

U

V

W

X

Y

Z